The Standardized Normal Distribution

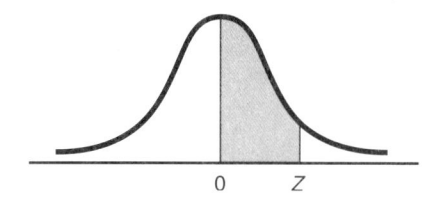

Entry represents area under the standardized normal distribution from the mean to Z

Z	.00	.01	.02	.03	.04	.05	.06	.07	.08	.09
0.0	.0000	.0040	.0080	.0120	.0160	.0199	.0239	.0279	.0319	.0359
0.1	.0398	.0438	.0478	.0517	.0557	.0596	.0636	.0675	.0714	.0753
0.2	.0793	.0832	.0871	.0910	.0948	.0987	.1026	.1064	.1103	.1141
0.3	.1179	.1217	.1255	.1293	.1331	.1368	.1406	.1443	.1480	.1517
0.4	.1554	.1591	.1628	.1664	.1700	.1736	.1772	.1808	.1844	.1879
0.5	.1915	.1950	.1985	.2019	.2054	.2088	.2123	.2157	.2190	.2224
0.6	.2257	.2291	.2324	.2357	.2389	.2422	.2454	.2486	.2518	.2549
0.7	.2580	.2612	.2642	.2673	.2704	.2734	.2764	.2794	.2823	.2852
0.8	.2881	.2910	.2939	.2967	.2995	.3023	.3051	.3078	.3106	.3133
0.9	.3159	.3186	.3212	.3238	.3264	.3289	.3315	.3340	.3365	.3389
1.0	.3413	.3438	.3461	.3485	.3508	.3531	.3554	.3577	.3599	.3621
1.1	.3643	.3665	.3686	.3708	.3729	.3749	.3770	.3790	.3810	.3830
1.2	.3849	.3869	.3888	.3907	.3925	.3944	.3962	.3980	.3997	.4015
1.3	.4032	.4049	.4066	.4082	.4099	.4115	.4131	.4147	.4162	.4177
1.4	.4192	.4207	.4222	.4236	.4251	.4265	.4279	.4292	.4306	.4319
1.5	.4332	.4345	.4357	.4370	.4382	.4394	.4406	.4418	.4429	.4441
1.6	.4452	.4463	.4474	.4484	.4495	.4505	.4515	.4525	.4535	.4545
1.7	.4554	.4564	.4573	.4582	.4591	.4599	.4608	.4616	.4625	.4633
1.8	.4641	.4649	.4656	.4664	.4671	.4678	.4686	.4693	.4699	.4706
1.9	.4713	.4719	.4726	.4732	.4738	.4744	.4750	.4756	.4761	.4767
2.0	.4772	.4778	.4783	.4788	.4793	.4798	.4803	.4808	.4812	.4817
2.1	.4821	.4826	.4830	.4834	.4838	.4842	.4846	.4850	.4854	.4857
2.2	.4861	.4864	.4868	.4871	.4875	.4878	.4881	.4884	.4887	.4890
2.3	.4893	.4896	.4898	.4901	.4904	.4906	.4909	.4911	.4913	.4916
2.4	.4918	.4920	.4922	.4925	.4927	.4929	.4931	.4932	.4934	.4936
2.5	.4938	.4940	.4941	.4943	.4945	.4946	.4948	.4949	.4951	.4952
2.6	.4953	.4955	.4956	.4957	.4959	.4960	.4961	.4962	.4963	.4964
2.7	.4965	.4966	.4967	.4968	.4969	.4970	.4971	.4972	.4973	.4974
2.8	.4974	.4975	.4976	.4977	.4977	.4978	.4979	.4979	.4980	.4981
2.9	.4981	.4982	.4982	.4983	.4984	.4984	.4985	.4985	.4986	.4986
3.0	.49865	.49869	.49874	.49878	.49882	.49886	.49889	.49893	.49897	.49900
3.1	.49903	.49906	.49910	.49913	.49916	.49918	.49921	.49924	.49926	.49929
3.2	.49931	.49934	.49936	.49938	.49940	.49942	.49944	.49946	.49948	.49950
3.3	.49952	.49953	.49955	.49957	.49958	.49960	.49961	.49962	.49964	.49965
3.4	.49966	.49968	.49969	.49970	.49971	.49972	.49973	.49974	.49975	.49976
3.5	.49977	.49978	.49978	.49979	.49980	.49981	.49981	.49982	.49983	.49983
3.6	.49984	.49985	.49985	.49986	.49986	.49987	.49987	.49988	.49988	.49989
3.7	.49989	.49990	.49990	.49990	.49991	.49991	.49992	.49992	.49992	.49992
3.8	.49993	.49993	.49993	.49994	.49994	.49994	.49994	.49995	.49995	.49995
3.9	.49995	.49995	.49996	.49996	.49996	.49996	.49996	.49996	.49997	.49997

Statistics for Managers Using Microsoft® Excel

Statistics for Managers Using Microsoft® Excel

DAVID M. LEVINE
Bernard M. Baruch College
City University of New York

MARK L. BERENSON
Bernard M. Baruch College
City University of New York

DAVID STEPHAN
Bernard M. Baruch College
City University of New York

PRENTICE-HALL, International, Inc.

This edition may be sold only in those countries to which it is consigned by Prentice-Hall International. It is not to be reexported and it is not for sale in the U.S.A., Mexico, or Canada.

Microsoft and Windows are registered trademarks of the Microsoft Corporation in the U.S.A. and other countries. Screen shots and icons reprinted with permission from the Microsoft Corporation. This book is not sponsored or endorsed by or affiliated with the Microsoft Corporation.

Acquisitions Editor: Tom Tucker
Assistant Editor: Diane Peirano
Assistant Editor: Audrey Regan
Marketing Manager: Patrick Lynch
Production Editor: Carol Lavis
Managing Editor: Katherine Evancie
Senior Manufacturing Supervisor: Paul Smolenski
Manufacturing Manager: Vincent Scelta
Senior Manager of Production and Technology: Lorraine Patsco
Electronic Page Make-up Artist: Christy Mahon
Electronic Art Supervisor: Warren Fischbach
Computer Artist: Steven Frim
Senior Designer: Suzanne Behnke
Design Director: Patricia Wosczyk
Interior Design: Edward Smith
Cover Art and Design: Kenny Beck

Copyright ©1997 by Prentice-Hall, Inc.
A Simon & Schuster Company
Upper Saddle River, New Jersey 07458

All rights reserved. No part of this book may be reproduced, in any form or by any means, without written permission from the Publisher.

Printed in the United States of America
10   9   8   7   6   5   4   3   2   1

ISBN   0-13-614587-6

Prentice-Hall International (UK) Limited, *London*
Prentice-Hall of Australia Pty. Limited, *Sydney*
Prentice-Hall Canada, Inc., *Toronto*
Prentice-Hall Hispanoamericana, S.A., *Mexico*
Prentice-Hall of India Private Limited, *New Delhi*
Prentice-Hall of Japan, Inc., *Tokyo*
Simon & Schuster of Asia Pte. Ltd., *Singapore*
Editora Prentice-Hall do Brasil, Ltda., *Rio de Janeiro*
Prentice-Hall, Inc., Upper Saddle River, *New Jersey*

To our wives,
Marilyn L., Rhoda B., and Mary N.
and to our children,
Sharyn, Kathy, Lori, and Mark

Brief Contents

Chapter 1 Introduction and Data Collection 1
Supplement Introduction to Using Microsoft Excel 27
Chapter 2 Presenting Data in Tables and Charts 51
Chapter 3 Summarizing and Describing Numerical Data 119
Chapter 4 Basic Probability and Discrete Probability Distributions 171
Chapter 5 The Normal Distribution and Sampling Distributions 225
Chapter 6 Estimation 291
Chapter 7 Fundamentals of Hypothesis Testing 339
Chapter 8 Two-Sample and c-Sample Tests with Numerical Data 379
Chapter 9 Two-Sample and c-Sample Tests with Categorical Data 449
Chapter 10 Statistical Applications in Quality and Productivity Management 485
Chapter 11 Simple Linear Regression and Correlation 533
Chapter 12 Multiple Regression Models 599
Chapter 13 Time-Series Forecasting for Annual Data 651
Appendix A Review of Arithmetic and Algebra A-1
Appendix B Summation Notation B-1
Appendix C Statistical Symbols and Greek Alphabet C-1
Appendix D Special Data Sets D-1
Appendix E Tables E-1
Appendix F Documentation for Diskette Files F-1
Index I-1
Excel Index EI-1

Contents

Preface xxi

Chapter 1 Introduction and Data Collection 1

    1.1 Why a Manager Needs to Know About Statistics 2

    1.2 The Growth and Development of Modern Statistics 3

    1.3 Statistical Thinking and Modern Management 4

    1.4 Descriptive versus Inferential Statistics 5

    1.5 Enumerative versus Analytical Studies 6

    1.6 The Need for Data 7

    1.7 Data Sources 8

    1.8 Measurement Scales 9
        1.8.1 Types of Data 9   *1.8.2 Types of Measurement Scales* 10

    1.9 Types of Samples 14

    1.10 Drawing the Simple Random Sample 15

    1.11 Ethical and Other Issues in Survey Research 18
        1.11.1 The Sample Survey 18   *1.11.2 Coverage Error* 19
        1.11.3 Nonresponse Error 19   *1.11.4 Sampling Error* 19
        1.11.5 Measurement Error 20   *1.11.6 Ethical Issues* 20

    1.12 Introduction and Data Collection: A Review and Preview 21

    Key Terms 23
    Chapter Review Problems 23
    Team Projects 25

Chapter 1 Supplement: Introduction to Using Microsoft Excel

Part 1 Microsoft Excel Orientation

    1S.1 What are Spreadsheet Application Programs? 28
        1S.1.1 Why Use Microsoft Excel? 28   *1S.1.2 Making the Most Effective Use of the Excel Sections* 28   *1S.1.3 Organization of This Supplement* 29

    1S.2 Using Windowing Environments 29

    1S.3 Using Dialog Boxes in Microsoft Excel 30

    1S.4 Using the Microsoft Excel Application Window 32

    1S.5 Organizing Excel Workbooks 34

Part 2 Getting Started with Microsoft Excel

1S.6 Designing Sheets for a Workbook 35

1S.7 Designing the Data Sheet 36

1S.8 Entering Values into a Data Sheet 37

1S.9 Using Formulas in a Data Sheet 39

1S.10 Using Copy to Facilitate the Entry of Formulas 40

1S.11 Enhancing the Appearance of a Data Sheet 41

1S.12 Using Functions in Formulas 42

1S.13 Implementing a Results Sheet 43

1S.14 Introduction to What If Analysis 45

1S.15 Using the Scenario Manager 45

1S.16 Importing Data into a Worksheet 47

1S.17 Summary 49

Key Terms 50

Chapter 2 Presenting Data in Tables and Charts 51

2.1 Introduction 52

2.2 Organizing Numerical Data: The Ordered Array and Stem-and-Leaf Display 52
  2.2.1 The Ordered Array 52   2.2.2 The Stem-and-Leaf Display 53
  2.2.3 Using Microsoft Excel to Sort Data into an Ordered Array 55

2.3 Tabulating Numerical Data: The Frequency Distribution 60
  2.3.1 Selecting the Number of Classes 61   2.3.2 Obtaining the Class Intervals 61   2.3.3 Establishing the Boundaries of the Classes 62   2.3.4 Subjectivity in Selecting Class Boundaries 63

2.4 Tabulating Numerical Data: The Relative Frequency Distribution and Percentage Distribution 65

2.5 Graphing Numerical Data: The Histogram and Polygon 67
  2.5.1 Histograms 67   2.5.2 Polygons 68

2.6 Cumulative Distributions and Cumulative Polygons 70
  2.6.1 The Cumulative Percentage Distribution 70
  2.6.2 Cumulative Percentage Polygon 71

2.7 Using Microsoft Excel to Obtain Tables and Charts for Numerical Variables 74
  2.7.1 Introduction 74   2.7.2 Using the FREQUENCY Function to Obtain a Frequency Distribution 74   2.7.3 Using the Data Analysis Tool to Obtain Frequency and Cumulative Frequency Distributions and Histograms 77   2.7.4 Using the Chart Wizard to Obtain Frequency Polygons and Histograms 80

2.8 Organizing and Tabulating Categorical Data: The Summary Table 84

2.9 Graphing Categorical Data: Bar and Pie Charts 85

2.10 Graphing Categorical Data: The Pareto Diagram 88

2.11 Tabularizing Categorical Data Using Contingency Tables   92

2.12 Using Microsoft Excel to Obtain Tables and Charts for Categorical Variables   95
    2.12.1 Introduction   95   2.12.2 Using PivotTable Wizard to Obtain One-Way Summary Tables and Contingency Tables   95
    2.12.3 Using the Chart Wizard to Obtain Bar Charts, Pie Charts, and Pareto Diagrams   100

2.13 Proper Tabular and Chart Presentation and Ethical Issues   104
    2.13.1 Failing to Compare Data Sets on a Relative Basis   105
    2.13.2 Failing to Indicate the Zero Point on the Vertical Axis   106
    2.13.3 Ethical Issues   107

2.14 Data Presentation: A Review and a Preview   107

Key Terms   109
Chapter Review Problems   109
Team Projects   116

Chapter 3 Summarizing and Describing Numerical Data   119

3.1 Introduction: What's Ahead   120

3.2 Exploring the Data   120

3.3 Properties of Numerical Data   121

3.4 Measures of Central Tendency   121
    3.4.1 The Arithmetic Mean   121   3.4.2 The Median   124
    3.4.3 The Mode   125   3.4.4 The Midrange   126
    3.4.5 The Midhinge   127
    3.4.6 Using Microsoft Excel Functions to Obtain Measures of Location   128

3.5 Measures of Variation   135
    3.5.1 The Range   135   3.5.2 The Interquartile Range   136
    3.5.3 The Variance and the Standard Deviation   137
    3.5.4 The Coefficient of Variation   141
    3.5.5 Using Microsoft Excel Functions to Obtain Measures of Variation   142

3.6 Shape   144

3.7 Using the Data Analysis Tool to Obtain Descriptive Statistics   146

3.8 Five-Number Summary and the Box-and-Whisker Plot   148
    3.8.1 Five-Number Summary   148   3.8.2 Box-and-Whisker Plot   149
    3.8.3 Simulating a Box-and-Whisker Plot Using Microsoft Excel   150

3.9 Calculating Descriptive Summary Measures from a Population   152
    3.9.1 Population Measures of Central Tendency   153
    3.9.2 Population Measures of Variation   153   3.9.3 Results   154
    3.9.4 Shape   156   3.9.5 Summarizing the Findings from the Sample and Population   157   3.9.6 Using the Standard Deviation: The Empirical Rule   158
    3.9.7 Obtaining the Population Standard Deviation and Variance from Microsoft Excel   158

3.10 Recognizing and Practicing Proper Descriptive Summarization and Exploring Ethical Issues   160

3.10.1 *Avoiding Errors in Analysis and Interpretation* 160
3.10.2 *Ethical Issues* 161

3.11 Summarizing and Describing Numerical Data: A Review 161

Key Terms 163
Chapter Review Problems 163
Team Projects 165
Case A—Kalosha Industries Employment Satisfaction Survey 166
Case B—Campus Cafeteria Nutrition Study 167

Chapter 4 Basic Probability and Discrete Probability Distributions 171

4.1 Introduction 172

4.2 Objective and Subjective Probability 172

4.3 Basic Probability Concepts 173
 4.3.1 Sample Spaces and Events 173   *4.3.2 Contingency Tables* 174

4.4 Simple (Marginal) Probability 176

4.5 Joint Probability 177

4.6 Addition Rule 178
 4.6.1 Mutually Exclusive Events 179   *4.6.2 Collectively Exhaustive Events* 180

4.7 Conditional Probability 181

4.8 Multiplication Rule 183

4.9 Using Microsoft Excel to Obtain Probabilities 186

4.10 The Probability Distribution for a Discrete Random Variable 187

4.11 Mathematical Expectation and Expected Monetary Value 188
 4.11.1 Expected Value of a Discrete Random Variable 188
 4.11.2 Variance and Standard Deviation of a Discrete Random Variable 189   *4.11.3 Expected Monetary Value* 190
 4.11.4 Using Microsoft Excel to Obtain Expected Values and Variances 191

4.12 Discrete Probability Distribution Functions 196

4.13 Binomial Distribution 197
 4.13.1 Development of the Model 199   *4.13.2 Characteristics of the Binomial Distribution* 201
 4.13.3 Using Microsoft Excel to Obtain Binomial Probabilities 203

4.14 Other Discrete Probability Distributions 207
 4.14.1 Hypergeometric Distribution 207   *4.14.2 Negative Binomial Distribution* 208   *4.14.3 Poisson Distribution* 209
 4.14.4 Using Microsoft Excel to Obtain Hypergeometric, Negative Binomial, and Poisson Probabilities 211

4.15 Ethical Issues and Probability 219

4.16 Basic Probability: A Review and a Preview 219
Key Terms 220
Chapter Review Problems 220

Chapter 5 The Normal Distribution and Sampling Distributions 225

5.1 Introduction 226

5.2 Mathematical Models of Continuous Random Variables 226

5.3 The Normal Distribution 227
5.3.1 Importance of the Normal Distribution 227   5.3.2 Properties of the Normal Distribution 227   5.3.3 The Mathematical Model 228   5.3.4 Standardizing the Normal Distribution 229   5.3.5 Using the Normal Probability Tables 231

5.4 Applications 233
5.4.1 Finding the Probabilities Corresponding to Known Values 234   5.4.2 Finding the Values Corresponding to Known Probabilities 238

5.5 Using Microsoft Excel to Obtain Normal Probabilities 244

5.6 Assessing the Normality Assumption: Evaluating Properties and Constructing Probability Plots 247
5.6.1 Exploring the Data—The Art of Data Analysis 247   5.6.2 Evaluating the Properties 247   5.6.3 Constructing the Normal Probability Plot 248   5.6.4 Interpreting the Results 252
5.6.5 Using Microsoft Excel to Obtain a Normal Probability Plot 254

5.7 Other Continuous Probability Distributions 258
5.7.1 The Exponential Distribution 258
5.7.2 Using Microsoft Excel to Obtain Exponential Probabilities 259

5.8 Introduction to Sampling Distributions 260

5.9 Sampling Distribution of the Mean 261
5.9.1 Properties of the Arithmetic Mean 261   5.9.2 Standard Error of the Mean 263   5.9.3 Sampling from Normal Populations 264   5.9.4 Sampling from Nonnormal Populations 268

5.10 Sampling Distribution of the Proportion 273

5.11 Using Microsoft Excel to Select Random Samples and Simulate Sampling Distributions 276
5.11.1 Using Microsoft Excel Functions to Select a Simple Random Sample 276   5.11.2 Using the Data Analysis Tool to Select a Random Sample from a Population 277   5.11.3 Using the Data Analysis Tool to Generate Distributions of Random Numbers 278

5.12 Sampling from Finite Populations 283

5.13 Continuous Distributions and Sampling Distributions: A Review 285

Key Terms 287
Chapter Review Problems 287
Team Projects 290

Chapter 6 Estimation 291

6.1 Introduction 292

6.2 Confidence Interval Estimation of the Mean (σ Known) 292

6.3 Confidence Interval Estimation of the Mean (σ Unknown) 297
  6.3.1 Student's t Distribution 297   6.3.2 Properties of the t Distribution 297   6.3.3 The Concept of Degrees of Freedom 299   6.3.4 The Confidence Interval Statement 299

6.4 Estimation Through Bootstrapping 301

6.5 Confidence Interval Estimation for the Proportion 304

6.6 Sample Size Determination for the Mean 306

6.7 Sample Size Determination for a Proportion 308

6.8 Estimation and Sample Size Determination for Finite Populations 311
  6.8.1 Estimating the Mean 311   6.8.2 Estimating the Proportion 311   6.8.3 Determining the Sample Size 312

6.9 Applications of Estimation in Auditing 313
  6.9.1 Introduction to Auditing 313   6.9.2 Estimating the Population Total 314   6.9.3 Difference Estimation 315

6.10 Using Microsoft Excel for Confidence Intervals and Sample Size Determination 317
  6.10.1 Using Microsoft Excel for the Confidence Interval Estimate for the Mean (σ Known) 318   6.10.2 Using Microsoft Excel for the Confidence Interval Estimate for the Mean (σ Unknown) 320   6.10.3 Using Microsoft Excel for the Confidence Interval Estimate for the Proportion 321   6.10.4 Using Microsoft Excel for Sample Size Determination 322   6.10.5 Using Microsoft Excel for Bootstrapping Estimation 325   6.10.6 Using Microsoft Excel with the Finite Population Correction Factor 325   6.10.7 Using Microsoft Excel for Applications of Estimation in Auditing 327

6.11 Estimation, Sample Size Determination, and Ethical Issues 331

6.12 Estimation and Statistical Inference: Review and a Preview 333
Key Terms 333
Chapter Review Problems 333
Team Projects 338

Chapter 7 Fundamentals of Hypothesis Testing 339

7.1 Introduction 340

7.2 Hypothesis-Testing Methodology 340
  7.2.1 The Null and Alternative Hypotheses 340   7.2.2 The Critical Value of the Test Statistic 341   7.2.3 Regions of Rejection and Nonrejection 342   7.2.4 Risks in Decision Making Using Hypothesis-Testing Methodology 343

7.3 Z Test of Hypothesis for the Mean (σ Known) 345

7.4 Summarizing the Steps of Hypothesis Testing 347

7.5 The *p*-Value Approach to Hypothesis Testing: Two-Tailed Tests 348

7.6 A Connection Between Confidence Interval Estimation and Hypothesis Testing 350

7.7 One-Tailed Tests 351

7.8 The *p*-Value Approach to Hypothesis Testing: One-Tailed Tests 354

7.9 t Test of Hypothesis for the Mean (σ Unknown)   354
    7.9.1 Introduction   354   7.9.2 Application   355
    7.9.3 Assumptions of the One-Sample t Test   357

7.10 One-Sample Z Test for the Proportion   360

7.11 Using Microsoft Excel for One-Sample Tests   364
    7.11.1 Using Microsoft Excel for the One-Sample Z test for the Mean (σ Known)   364   7.11.2 Using Microsoft Excel for the One-Sample t Test for the Mean (σ Unknown)   366   7.11.3 Using Microsoft Excel for the One-Sample Z test for the Proportion   370

7.12 Potential Hypothesis-Testing Pitfalls and Ethical Issues   371

7.13 Hypothesis-Testing Methodology: A Review and a Preview   374

Key Terms   376

Chapter Review Problems   376

Chapter 8   Two-Sample and c-Sample Tests with Numerical Data   379

8.1 Introduction   380

8.2 Choosing the Appropriate Test Procedure   380
    8.2.1 Parametric Procedures   380   8.2.2 Nonparametric Procedures   380   8.2.3 Importance of Assumptions in Test Selection   381   8.2.4 Choosing the Appropriate Test Procedure When Comparing Two Independent Samples   381

8.3 Pooled-Variance t Test for Differences in Two Means   382
    8.3.1 Introduction   382   8.3.2 Developing the Pooled-Variance t Test   383   8.3.3 Application   384   8.3.4 Summary   387
    8.3.5 Using Microsoft Excel for the Pooled-Variance t Test   387

8.4 Wilcoxon Rank Sum Test for Differences in Two Medians   395
    8.4.1 Introduction   395   8.4.2 Procedure   395
    8.4.3 Application   397
    8.4.4 Using Microsoft Excel for the Wilcoxon Rank Sum Test   400

8.5 F Test for Differences in Two Variances   405
    8.5.1 Introduction   405   8.5.2 Development   405
    8.5.3 Application   406   8.5.4 Caution   408
    8.5.5 Using Microsoft Excel for the F test for Differences in Two Variances   409

8.6 One-Way ANOVA F test for Differences in c Means   414
    8.6.1 Introduction   415   8.6.2 Development   415
    8.6.3 Application   419   8.6.4 Reflection   423
    8.6.5 Multiple Comparisons: The Tukey-Kramer Procedure   424
    8.6.6 ANOVA Assumptions   425
    8.6.7 Using Microsoft Excel for the One-Way ANOVA F test   426
    8.6.8 Using Microsoft Excel for the Tukey-Kramer Multiple Comparisons Procedure   428

8.7 Kruskal-Wallis Rank Test for Differences in c Medians   433
    8.7.1 Introduction   433   8.7.2 Development   433
    8.7.3 Application   434
    8.7.4 Using Microsoft Excel for the Kruskal-Wallis Rank Test for the Difference in c Medians   437

8.8 Hypothesis Testing Based on Two and c Samples of Numerical Data: A Review 441
Key Terms 442
Chapter Review Problems 443
Team Projects 447
Case C—Test-Marketing and Promoting a Ball-Point Pen 445

Chapter 9 Two-Sample and c-Sample Tests with Categorical Data 449

9.1 Introduction 450

9.2 Z test for Differences in Two Proportions (Independent Samples) 450
9.2.1 Introduction 450   9.2.2 Development 450
9.2.3 Application 451
9.2.4 Using Microsoft for the Z test for Differences in Two Proportions 453

9.3 χ^2 Test for Differences in Two Proportions (Independent Samples) 456
9.3.1 Introduction 456   9.3.2 Development 457
9.3.3 Application 459   9.3.4 Testing for the Equality of Two Proportions by Z and by χ^2: A Comparison of Results 462

9.4 χ^2 Test for Differences in c Proportions (Independent Samples) 463

9.5 χ^2 Test of Independence 469

9.6 Using Microsoft Excel for χ^2 Tests 475

9.7 Hypothesis Testing Based on Categorical Data: A Review 478
Key Terms 479
Chapter Review Problems 480
Case D—Airline Satisfaction Survey 481

Chapter 10 Statistical Applications in Quality and Productivity Management 485

10.1 Introduction 486

10.2 Quality and Productivity: A Historical Perspective 486

10.3 The Theory of Control Charts 487

10.4 Some Tools for Studying a Process: Fishbone (Ishikawa) and Process Flow Diagrams 489
10.4.1 Introduction 489   10.4.2 The Fishbone (or Ishikawa) Diagram 489   10.4.3 Process Flow Diagrams 491

10.5 Deming's 14 Points: A Theory of Management by Process 495

10.6 Control Charts for the Proportion and Number of Nonconforming Items—The p and np Charts 498
10.6.1 Introduction 498   10.6.2 The p Chart 498
10.6.3 The np Chart 502
10.6.4 Using Microsoft Excel for the p and np Charts 504

10.7 The Red Bead Experiment: Understanding Process Variability 512

10.8 Control Charts for the Range (R) and the Mean ($\bar{X}$) 515
10.8.1 Introduction 515   10.8.2 The R Chart: A Control Chart for Dispersion 515   10.8.3 The $\bar{X}$ Chart 517

10.8.4 Using Microsoft Excel for the R and $\bar{X}$ Charts   519

10.9   Summary and Overview   527
Key Terms   528
Chapter Review Problems   529

Chapter 11   Simple Linear Regression and Correlation   533

11.1   Introduction   534

11.2   The Scatter Diagram   534

11.3   Types of Regression Models   538

11.4   Determining the Simple Linear Regression Equation   540
11.4.1 The Least Squares Method   540   11.4.2 Predictions in Regression Analysis: Interpolation Versus Extrapolation   543

11.5   Standard Error of the Estimate   545

11.6   Measures of Variation in Regression and Correlation   547
11.6.1 Obtaining the Sum of Squares   547   11.6.2 The Coefficient of Determination   549
11.6.3 Using the Microsoft Excel Chart Wizard and TREND Function for Regression Analysis   550

11.7 Using the Data Analysis Tool for Regression   554

11.8   Correlation—Measuring the Strength of the Association   557
11.8.1 The Correlation Coefficient   557
11.8.2 Using the Microsoft Excel CORREL Function for Correlation Analysis   560

11.9   Assumptions of Regression and Correlation   561

11.10   Residual Analysis   562
11.10.1 Introduction   562   11.10.2 Evaluating the Aptness of the Fitted Model   563   11.10.3 Evaluating the Assumptions   564
11.10.4 Using Microsoft Excel for Residual Analysis   567

11.11   Measuring Autocorrelation: The Durbin-Watson Statistic   567
11.11.1 Introduction   567   11.11.2 Residual Plots to Detect Autocorrelation   567   11.11.3 The Durbin-Watson Procedure   570
11.11.4 Using Microsoft Excel to Study Autocorrelation   572

11.12   Confidence Interval Estimate for Predicting μ_{YX} 574
11.12.1 Obtaining the Confidence Interval Estimate   574
11.12.2 Using Microsoft Excel to Obtain a Confidence Interval Estimate for μ_{YX} 576

11.13   Prediction Interval Estimate for an Individual Response Y_I 579
11.13.1 Obtaining the Prediction Interval Estimate   579
11.13.2 Using Microsoft Excel to Obtain a Prediction Interval Estimate for Y_I 579

11.14   Inferences about the Population Parameters in Regression and Correlation   580

11.15   Pitfalls in Regression and Ethical Issues   584
11.15.1 Introduction   584   11.15.2 The Pitfalls of Regression   584   11.15.3 Ethical Considerations   587

11.16 Summary and Overview   587
Key Terms   589
Chapter Review Problems   589
Case E—Predicting Sunday Newspaper Circulation   596

Chapter 12 Multiple Regression Models   599

12.1 Introduction   600

12.2 Developing the Multiple Regression Model   600

12.3 Prediction of the Dependent Variable Y for Given Values of the Explanatory Variables   606

12.4 Measuring Association in the Multiple Regression Model   607

12.5 Residual Analysis in Multiple Regression   608

12.6 Testing for the Significance of the Relationship Between the Dependent Variable and the Explanatory Variables   610

12.7 Testing Portions of a Multiple Regression Model   613

12.8 Inferences Concerning the Population Values of the Regression Coefficients   617
   12.8.1 Tests of Hypothesis   617   *12.8.2 Confidence Interval Estimation*   619

12.9 Coefficient of Partial Determination   620

12.10 The Curvilinear Regression Model   621
   12.10.1 Finding the Regression Coefficients and Predicting Y 622
   12.10.2 Testing for the Significance of the Curvilinear Model   625
   12.10.3 Testing the Curvilinear Effect   626

12.11 Dummy-Variable Models   629

12.12 Other Types of Regression Models   633
   12.12.1 Interaction Terms in Regression Models   633   *12.12.2 Using Transformations in Regression Models*   634

12.13 Multicollinearity   636

12.14 Using Microsoft Excel for Multiple Regression   637
   12.14.1 Using the Regression option of the Data Analysis Tool for Multiple Regression   638   *12.14.2 Using the Correlation Option of the Data Analysis Tool for Multiple Regression*   638   *12.14.3 Using Microsoft Excel to Obtain the Coefficients of Partial Determination*   638
   12.14.4 Using Microsoft Excel for Curvilinear Regression   640
   12.14.5 Using Microsoft Excel for Dummy-Variable and Other Types of Regression Models   640   *12.14.6 Using Microsoft Excel to Obtain the Variance Inflationary Factor*   640

12.15 Pitfalls in Multiple Regression and Ethical Issues   641
   12.15.1 Pitfalls in Multiple Regression   641   *12.15.2 Ethical Considerations*   641

12.16 Summary and Overview   641
Key Terms   643
Chapter Review Problems   643
Case F—The Mountain States Potato Company   647

Chapter 13 Time-Series Forecasting for Annual Data 651

13.1 Introduction 652

13.2 The Importance of Business Forecasting 652
    13.2.1 Introduction to Forecasting 652   *13.2.2 Types of Forecasting Methods 652*   *13.2.3 Introduction to Time-Series Analysis 652*   *13.2.4 Objectives of Time-Series Analysis 653*

13.3 Component Factors of the Classical Multiplicative Time-Series Model 653
    13.3.1 Introduction 653   *13.3.2 The Classical Multiplicative Time-Series Model 654*

13.4 Smoothing the Annual Time Series: Moving Averages and Exponential Smoothing 655
    13.4.1 Moving Averages 655   *13.4.2 Exponential Smoothing 659*
    13.4.3 Using Microsoft Excel for Moving Averages and Exponential Smoothing 662

13.5 Time-Series Analysis of Annual Data: Least Squares Trend Fitting and Forecasting 668
    13.5.1 The Linear Model 668   *13.5.2 The Quadratic Model 671*
    13.5.3 The Exponential Model 672
    13.5.4 Using Microsoft Excel for Least Squares Trend Fitting 674

13.6 Autoregressive Modeling for Trend Fitting and Forecasting 680
    13.6.1 Model Development 680
    13.6.2 Using Microsoft Excel for Autoregressive Modeling 686

13.7 Choosing an Appropriate Forecasting Model 688
    13.7.1 Residual Analysis 689   *13.7.2 Measuring the Magnitude of the Residual Error 690*   *13.7.3 Principle of Parsimony 690*
    13.7.4 A Comparison of Four Forecasting Methods 690
    13.7.5 Model Selection: A Warning 693
    13.7.6 Using Microsoft Excel to Obtain the Mean Absolute Deviation 693

13.8 Pitfalls Concerning Time-Series Analysis 695

13.9 Summary and Overview 695

Key Terms 696
Chapter Review Problems 696
Case G—Currency Trading 700

Answers to Selected Problems 703

Appendix A Review of Arithmetic and Algebra A-1

A.1 Rules for Arithmetic Operations A-1

A.2 Rules for Algebra: Exponents and Square Roots A-2

Appendix B Summation Notation B-1

Appendix C Statistical Symbols and Greek Alphabet C-1

C.1 Statistical Symbols   C–1
C.2 Greek Alphabet   C–1

Appendix D   Special Data Sets   D–1
D.1 Special Data Set 1   D–1
D.1 Special Data Set 2   D–6
D.1 Special Data Set 3   D–8

Appendix E   Tables   E–1
E.1 Table of Random Numbers   E–2
E.2 The Standardized Normal Distribution   E–4
E.3 Critical Values of t E–5
E.4 Critical Values of χ^2 E–7
E.5 Critical Values of F E–8
E.6 Lower and Upper Critical Values of T_1 of Wilcoxon Rank Sum Test   E–12
E.7 Critical Values of Studentized Range Q E–13
E.8 Control Chart Factors   E–15
E.9 Critical Values of d_L and d_U of the Durbin-Watson Statistic D E–16

Appendix F   Documentation for Diskette Files   F–1
F.1 Diskette Overview   F–1
F.2 File Contents   F–1
F.3 Microsoft Excel Workbook Files   F–6
F.4 Diskette Installation Instructions   F–8

Index   I–1

Excel Index   EI–1

Preface

When planning this textbook, the authors focused on how desktop productivity tools, such as spreadsheet applications, have altered managers' decision-making processes. Whereas they once had to turn to a Management Information Systems Department or an Information Center to obtain customized summaries of corporate data, today an increasing number of managers use spreadsheet applications as the means to retrieve and analyze directly the data they need. In this context, employers now are beginning to desire, if not demand, that their college-educated, entry-level employees have more than just a cursory awareness of such tools as spreadsheet applications.

These changes, along with the realization that current spreadsheet applications can perform the type of analyses once done only by specialized statistical packages, have led us to develop *Statistics for Managers Using Microsoft Excel*. Our text contains the following features that distinguish it from the many other statistics texts available for business students (several of which have been written by two of us):

- Use of Microsoft Excel as a tool for statistical analysis throughout the text
- A streamlined version of topical coverage with sufficient breadth of coverage
- An enhanced managerial focus for statistical methods.

MAIN FEATURE: USE OF MICROSOFT EXCEL FOR STATISTICAL ANALYSIS THROUGHOUT THE TEXT

Statistics for Managers Using Microsoft Excel integrates the spreadsheet application Microsoft Excel throughout the entire text. This approach is fundamentally different from that of the many texts published and revised in the past twenty years. Since the advent of the computer revolution, statistics texts have struggled with the appropriate way to incorporate the use of statistical software packages. Most typically such packages as SAS, SPSS, and Minitab have been illustrated. A dilemma for faculty teaching this course has been how students could obtain access to (often through site licenses and student versions) the statistical software selected and how these packages could be used in the course. Often, students are not familiar with these packages prior to the statistics course, and only a limited number may use them in subsequent courses. Thus, students may view them as but one more hurdle to overcome in getting through the statistics course.

However, in the last several years, with the increasing functionality and power of spreadsheet applications, virtually all the kinds of statistical analysis taught in an introductory course are directly supported by the Microsoft Excel program (Version 5.0 or later), available for a variety of different systems including Windows 3.1, Windows 95, and Macintosh. In addition to its possible use in a statistics course, students typically learn the fundamentals of a spreadsheet application—either Microsoft Excel itself or a similar program—in an information systems course, and then use Excel in courses in accounting, finance, and other functional areas of business. Even if they are not familiar with Excel, they undoubtedly have heard of this software; and its use in the statistics course will give added relevancy to the course.

Because entering students' spreadsheet applications skills do vary, and because school and home computer facilities are sometimes limited, the demonstration of how Microsoft Excel can be incorporated into a statistics course must also vary. This text has been written for a number of situations, including the following:

1. **Statistical concepts with hands-on Excel development.** Instructors who wish to teach statistical concepts and the development of Excel-based solutions will find that the text includes detailed instructions for the design and implementation of Excel workbooks for each statistical topic discussed. These instructions are presented using a consistent developmental methodology that assists the student in developing their own solutions to statistical problems. (The methodology is summarized in Chapter 1S and is reflected in all of the Excel workbooks that are included on the diskette that accompanies this text.)

2. **Statistical concepts with Excel usage.** Instructors who wish to include Microsoft Excel in their courses but do not have the time or inclination to discuss specific steps of workbook development, can simply use the implemented examples on the diskette that accompanies this text, along with the portions of the Excel sections that analyze workbook results. The design of many of the diskette workbooks allows them to function as generalized templates into which different sets of data values can be entered and analyzed. (Each Excel section includes at least one workbook, and finished versions of all workbooks whose implementation is discussed in the text are included on the diskette that accompanies this text.)

3. **Statistical concepts with Excel exposure.** Where the instructor wishes to make students aware of the statistical applications of spreadsheets, but using Microsoft Excel directly is impractical, the results obtained from Excel for each statistical topic can be illustrated. (In these cases, the contents of the diskette can serve as the basis for optional assignments or student enrichment.)

Special Excel-related Features of the Text

1. **Excel orientation.** Because of the diverse computer backgrounds of incoming students, this text includes a comprehensive tutorial chapter, Introduction to Using Microsoft Excel (Chapter 1 Supplement), that assumes no previous experience using Excel or the windowing environments in which the program runs.

2. **Use of the workbook structure.** All the examples discussed in this text make full use of the Excel workbook feature to organize logically the data, calculations, and the results of a statistical analysis into different worksheets. In addition, all the workbooks included on the diskette that accompanies the text contain overview sheets that summarize the contents of the workbook and the relevant statistical concepts. The overview sheet that relates to the example discussed in Chapter 11 is illustrated in Figure P.1.Excel.

3. **Generalized instructions.** Instructions in the text for using Microsoft Excel will work equally well with the current versions of the program for the Macintosh, Windows 3.1, and Windows 95. (Details that are specific to particular windowing environments, such as the use of accelerator keys, have been avoided.)

4. **Alternative Excel approaches are contrasted.** Where appropriate, the text explores the differences between the use of Data Analysis tools and formula-based worksheets for statistical analysis.

5. **Coverage of an extensive set of Excel features.** The text includes a comprehensive discussion of the specialized functions, the PivotTable and Chart Wizards, the Text Import Wizard for importing text files into Microsoft Excel, the Data Analysis tools for statistical analysis, and the Scenario Manager as an aid in "What if" analyses.

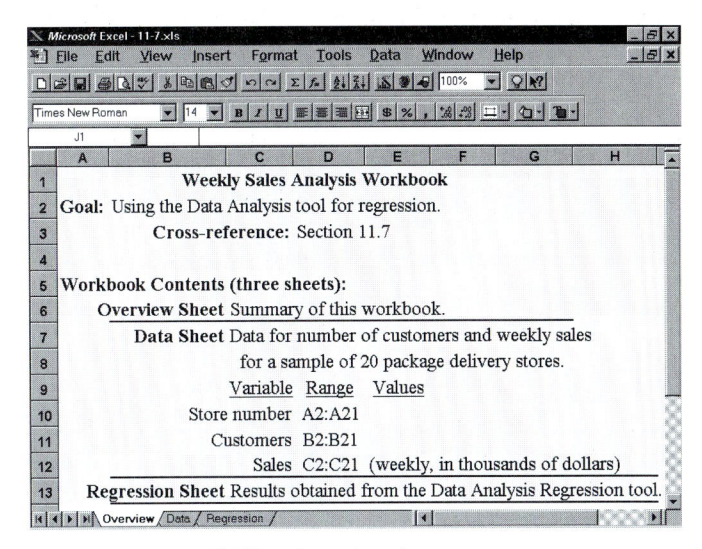

FIGURE P.1.EXCEL Overview sheet.

Excel Integration

In each chapter subsequent to Chapter 1, after a statistical topic has been covered, the use of Microsoft Excel as applied to the statistical topic is discussed in step-by-step detail. Each presentation of Excel material has the goal of ensuring that students will be able to use the standard features of Microsoft Excel to do what was just covered in the text. Thus, by the time the text is completed, students will have gained the necessary foundation in Excel to create their own workbooks to perform statistical and other types of analyses. An example of this detail can be seen in the discussion of the use of Excel to obtain descriptive statistics. In Sections 3.4.6 and 3.5.5 (pages 128–132 and 142–143), Excel functions are discussed; while in Section 3.7, the use of the Data Analysis tool is explained (pages 146–148). The output obtained from Excel is illustrated in Figure 3.7.Excel, which is duplicated below in Figure P.2.Excel.

Numerous screen shots such as this are utilized throughout the text, and any statistical output obtained is explained in detail.

FIGURE P.2.EXCEL Measures of central tendency obtained from Excel for out-of-state tuition rates for the six-school sample from Pennsylvania.

Preface   **xxiii**

FIGURE P.3.EXCEL Design for computing normal probabilities illustrated for the individually trained factory worker (with $\mu = 75$ and $\sigma = 6$).

	A	B
1	Calculating Normal Probabilities	
2		
3	Arithmetic Mean	75
4	Standard Deviation	6
5	Left Tail Probability	
6	First X Value	69
7	Z Value	-1
8	P(X<=69)	0.15865526
9	Right Tail Probability	
10	P(X>=69)	0.84134474
11	Interval Probability	
12	Second X value	81
13	P(X<=81)	0.84134474
14	P(69<X<81)	0.68268948
15	Finding a X Value	
16	Cumulative Percent	0.1
17	Z Value	-1.281550794
18	X Value	67.31069523

What If Examples

One of the advantages of using a spreadsheet application like Microsoft Excel is that it facilitates the use of "What if" analyses, which enable the student to explore the effect of changing data values. An example of such an analysis occurs in the computation of probabilities under the normal curve. Figure P.3.Excel, illustrated above, shows the use of Excel to find a probability under the normal curve.

For this situation, if we change the standard deviation to 10 we obtain the results shown in Figure P.4.Excel.

FIGURE P.4.EXCEL Design for computing normal probabilities illustrated for the individually trained factory worker (with $\mu = 75$ and $\sigma = 10$).

	A	B
1	Calculating Normal Probabilities	
2		
3	Arithmetic Mean	75
4	Standard Deviation	10
5	Left Tail Probability	
6	First X Value	69
7	Z Value	-0.6
8	P(X<=69)	0.274253065
9	Right Tail Probability	
10	P(X>=69)	0.725746935
11	Interval Probability	
12	Second X value	81
13	P(X<=81)	0.725746935
14	P(69<X<81)	0.45149387
15	Finding a X Value	
16	Cumulative Percent	0.1
17	Z Value	-1.281550794
18	X Value	62.18449206

Numerous problems throughout the text include parts that ask the student to do this type of "What if" analysis. These problems are marked with an Excel icon next to the part that requires a "What if" analysis. This type of sensitivity analysis enhances understanding of the particular topic studied. To facilitate the use of numerous "What if" analyses, the Excel Scenario Manager feature is covered in the supplement to Chapter 1.

About the Diskette that Accompanies This Text

The diskette that accompanies this text includes Excel workbooks for all examples discussed in the text and all problems denoted with an Excel icon. The diskette also contains two visual basic for application modules. The first one (MOUSING.XLS) allows novices (and all others) to practice the mousing skills needed when using Microsoft Excel. The illustration below represents one portion of this workbook.

The second module enables the user to generate a stem-and-leaf display from a set of values on a worksheet. All of these workbooks have been designed for and will load properly in Versions 5.0 or later of Microsoft Excel for the Macintosh, Windows 3.1, or Windows 95.

The diskette also includes files containing the data for the problems and examples in the text marked with a data disk icon in the margin. These files can be imported into Microsoft Excel using the Text Import Wizard as discussed in Chapter 1S. (Technical details about the diskette, including a complete list of the contents of the diskette, are given in Appendix F.)

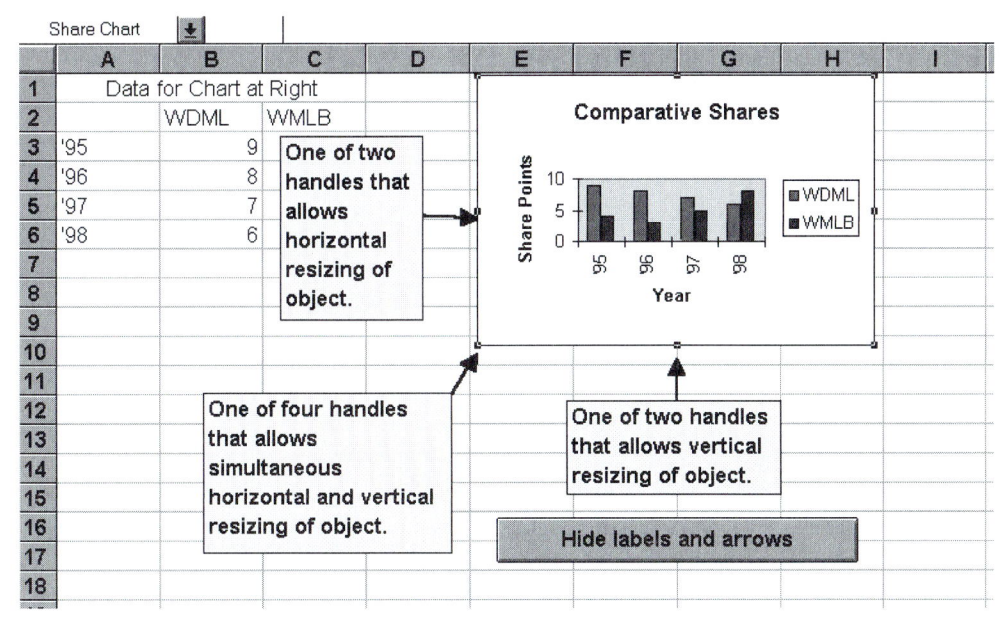

FIGURE P.5.EXCEL Level 4 of mousing program.

Using Microsoft Excel

This text and the Excel workbook files on the diskette that accompanies this text have been designed for use with any of the following three versions of Microsoft Excel:

Version 5 (or 5.0c) for Windows 3.1

Version 7 for Windows 95

Version 5 for the Macintosh

These versions share the same command set and nearly identical dialog boxes for the operations related to the statistical analyses discussed in this text. (Section 1S.3, Using Microsoft Excel Dialog Boxes, details the differences between the three versions that occur when using dialog boxes for common computing tasks such as opening, saving, and printing files.)

As noted elsewhere, the instructions in the Excel sections of the text have been written to work equally well with any of the three versions. Although many of the instructions would also apply to Version 4 of Excel, this version should *not* be used for two reasons: Many problems are associated with the statistical routines in that earlier version and because Version 4 will not open any of the Excel workbooks included on the diskette that accompanies the text.

MAIN FEATURE: A STREAMLINED VERSION OF TOPICAL COVERAGE WITH SUFFICIENT BREADTH OF COVERAGE

The statistical coverage, as can be seen from the Table of Contents, provides more than sufficient breadth of coverage, although not the depth of coverage of other, more comprehensive texts such as Berenson and Levine's *Basic Business Statistics, 6th ed.* The first three chapters provide an introduction and cover tables, charts, and descriptive statistics. Chapters 4 and 5 discuss probability and probability distributions. Chapters 6–9 examine inference and hypothesis testing. Chapter 10 describes quality management. Chapters 11 and 12 talk about regression and multiple regression, and Chapter 13 ends the text with time series forecasting.

MAIN FEATURE: AN ENHANCED MANAGERIAL FOCUS FOR STATISTICAL METHODS

Although the statistical coverage in the text represents a more concise version of Berenson and Levine's *Basic Business Statistics, 6th ed.*, extensive rewriting provides for a more managerial focus. In numerous instances, notation has been simplified, and formulas such as Equation (3.7) below are written in words as well as statistical symbols.

> The standard deviation is the square root of the sum of the squared differences around the arithmetic mean divided by the sample size minus 1.
>
> $$S = \sqrt{\frac{\sum_{i=1}^{n}(X_i - \bar{X})^2}{n-1}} \quad (3.7)$$

Many of the important pedagogical features of Berenson and Levine, *Basic Business Statistics, 6th ed.*, have been retained, including:

- Case Studies
- Thought-provoking action ACTION▶ and "light bulb"   problems

- Chapter-ending summary flow charts
- Team projects
- Discussion of ethical issues
- Extensive end-of-section and end-of-chapter problems with answers to selected problems indicated by the • symbol. (Answers begin on page 703.)
- Extensive use of real data throughout the text
- Listing of Key Terms

About the World Wide Web Icon

This text has a home page on the World Wide Web **WWW** with an address of http://www.prenhall.com/phbusiness. This home page includes a variety of information, including:

- Alternative course outlines
- Teaching tips
- Diskette additions and updates
- Links to other sites containing data appropriate for statistics courses
- Microsoft Excel–related sites

ACKNOWLEDGMENTS

We are extremely grateful to the many organizations and companies that allowed us to use their actual data for developing problems and examples throughout the text. We would like to thank *The New York Times,* Consumers Union (publisher of *Consumer Reports*), *U. S. News and World Report*, Moody's Investor Service (publishers of *Moody's Handbook of Common Stocks*), CEEPress Books, and Gale Research.

In addition, we would like to thank the Biometrika Trustees, American Cyanamid Company, the Rand Corporation, the American Society for Testing and Materials for their kind permission to publish various tables in Appendix E, and the American Statistical Association for its permission to publish diagrams from the *American Statistician*. Finally we are grateful to Professors George A. Johnson and Joanne Tokle of Idaho State University and Ed Conn, Mountain States Potato Company, for their kind permission to incorporate parts of their work as our Case Study F, "The Mountain States Potato Company."

A Note of Thanks

We would like to thank Kent S. Borowick, Baylor University; Ann Brandwein, Baruch College; Philip C. Fry, Boise State University; J. Morgan Jones, University of North Carolina; Stephen Reid, British Columbia Institute of Technology; Albert H. Segars, Clemson University; Richard Spinetto, University of Colorado; William E. Stein, Texas A & M University; Stanley D. Stephenson, Southwest Texas State University; Nancy C. Weida, Bucknell University; Peter H. Westfall, Texas Tech University; and Wayne Winston, Indiana University for their constructive comments during the writing of this textbook.

We would like to thank especially the editorial and production team at Prentice Hall for their assistance. In particular, we would like to thank our editor, Tom Tucker, without whose vision this text would not have become a reality. In addition, we would like to thank Richard Wohl, Katherine Evancie, Joanne Jay, Diane Peirano, Audrey Regan, Carol Lavis, Christy Mahon, Warren Fischbach, Steven Frim, Lorraine Patsco, Paul Smolenski, Sue Behnke, and Patricia Wosczyk. Finally, we would like to thank our spouses and children for their patience, understanding, love, and assistance in making this book a reality. It is to them that we dedicate this book.

David M. Levine
Mark L. Berenson
David Stephan

Statistics for Managers Using Microsoft® Excel

chapter 1

Introduction and Data Collection

CHAPTER OBJECTIVE To present a broad overview of the subject of statistics and its applications, and to describe the data collection process.

1.1 WHY A MANAGER NEEDS TO KNOW ABOUT STATISTICS

A century ago, H. G. Wells commented that "statistical thinking will one day be as necessary as the ability to read and write." As we approach the next millennium, the issue facing managers is not a shortage of information, but how to use the available information to make better decisions. It is from this perspective that we should consider why a manager needs to know about statistics. Among the reasons for learning statistics are the following:

1. Managers need to know how to properly present and describe information.
2. Managers need to know how to draw conclusions about large populations based only on information obtained from samples.
3. Managers need to know how to improve processes.
4. Managers need to know how to obtain reliable forecasts of variables of interest.

A road map of this text from the perspective of these four reasons for learning statistics is displayed below.

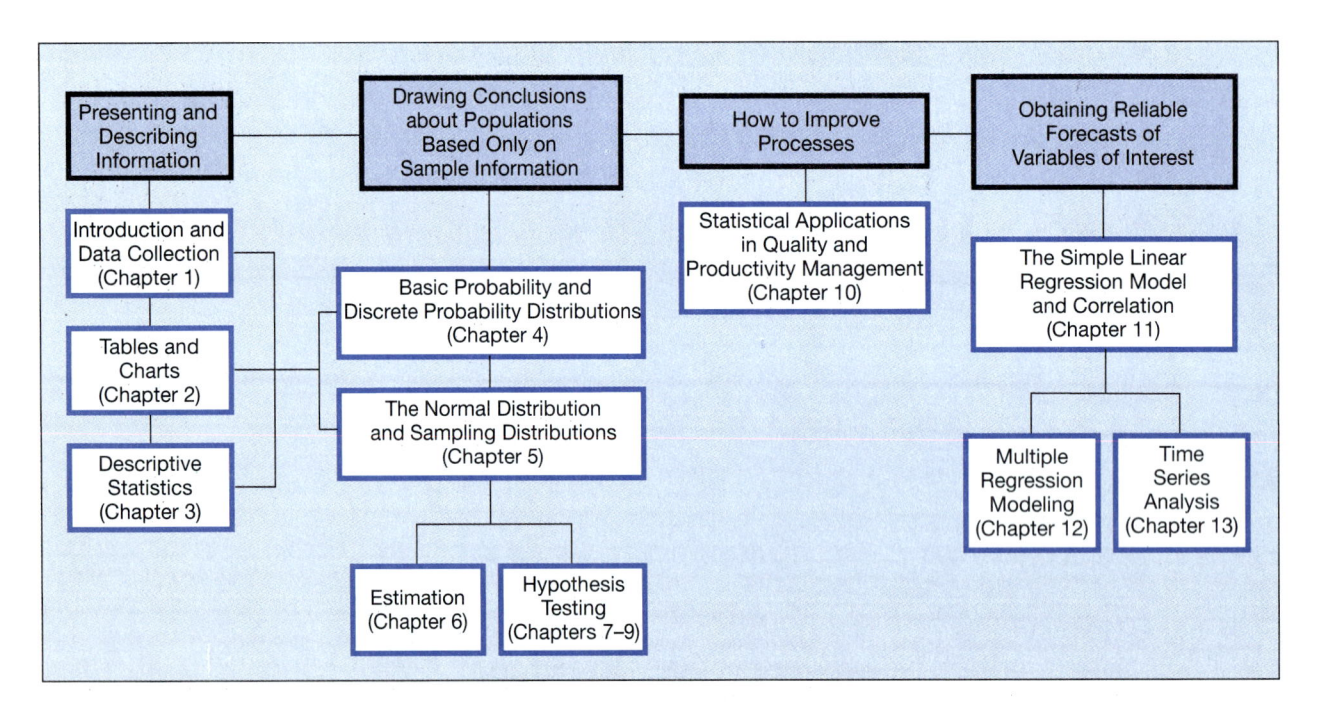

```
Presenting and          Drawing Conclusions         How to Improve        Obtaining Reliable
Describing              about Populations           Processes             Forecasts of
Information             Based Only on                                     Variables of Interest
                        Sample Information

Introduction and                                    Statistical Applications
Data Collection                                     in Quality and         The Simple Linear
(Chapter 1)             Basic Probability and       Productivity Management Regression Model
                        Discrete Probability Distributions (Chapter 10)   and Correlation
Tables and              (Chapter 4)                                        (Chapter 11)
Charts
(Chapter 2)             The Normal Distribution
                        and Sampling Distributions                          Multiple      Time
Descriptive             (Chapter 5)                                         Regression    Series
Statistics                                                                  Modeling      Analysis
(Chapter 3)                                                                 (Chapter 12)  (Chapter 13)

                        Estimation     Hypothesis
                        (Chapter 6)    Testing
                                       (Chapters 7–9)
```

From this road map we observe that the first three chapters include coverage of methods involved in the collection, presentation, and description of information. Chapters 4 and 5 provide sufficient coverage of the basic concepts of probability, the binomial, normal, and other distributions, and sampling distributions so that in Chapters 6–9 the reader will learn how to draw conclusions about large populations based only on information obtained from samples. Chapter 10 contains coverage of statistical applications in quality and productivity management that is essential for process improvement. The text concludes with chapters that focus on regression, multiple regression, and time-series analysis that provide methods for obtaining forecasts.

1.2 THE GROWTH AND DEVELOPMENT OF MODERN STATISTICS

Historically, the growth and development of modern statistics can be traced to three separate phenomena—the needs of government to collect data on its citizenry (see References 13, 14, 19, 20, and 26), the development of the mathematics of probability theory, and the advent of the computer.

Data have been collected throughout recorded history. During the Egyptian, Greek, and Roman civilizations, data were obtained primarily for the purposes of taxation and military conscription. In the Middle Ages, church institutions often kept records concerning births, deaths, and marriages. In America, various records were kept during colonial times (see Reference 26), and beginning in 1790, the federal Constitution required the taking of a census every 10 years. In fact, the expanding needs of the census helped spark the development of tabulating machines at the beginning of the twentieth century. This led to the development of large-scale mainframe computers and eventually to the personal computer revolution.

These developments have profoundly changed the field of statistics in the last 30 years. Mainframe packages such as SAS and SPSS became popular during the 1960s and 1970s. During the 1980s, statistical software experienced a vast technological revolution. Besides the usual improvements manifested in periodic updates, the availability of personal computers led to the development of new packages. In addition, personal computer versions of existing packages such as SAS, SPSS, and MINITAB (see References 18, 23, and 24) quickly became available, and, the increasing use of popular spreadsheet packages such as Lotus 1–2–3 and Microsoft Excel (see References 15 and 16) led to the incorporation of statistical features in these packages.

Although this text is appropriate for those who use statistical software packages, one of its distinctive features is the incorporation of many of the statistical and graphic aspects of the widely used Microsoft Excel spreadsheet package. Certainly, spreadsheet applications have become an important productivity tool in business. In fact, it can be argued that it was the availability of spreadsheet packages, particularly VISICALC (first introduced in 1979) and LOTUS 1–2–3 (first introduced in 1983), that fueled the personal computer revolution in the workplace by providing a rationale for the acquisition of necessary computer technology. This use of spreadsheet applications in business has accelerated with the development of integrated products such as Microsoft Office, which allow one to integrate both the textual and graphical output of the spreadsheet applications into word-processed documents generated by packages such as Microsoft Word. These developments have given managers the ability to analyze data right at their desktop. Recent versions of Microsoft Excel (beginning with version 5.0) provide extensive capability in statistics, sufficient for all statistical topics that are included in this text. Not only will output obtained from Excel be illustrated throughout, but extensive tutorials will be provided that take the reader through the process of how such output may be developed. An introduction to the Microsoft Excel spreadsheet program will be provided in the Supplement to Chapter 1.

Although statistical and spreadsheet software has made even the most sophisticated analyses feasible, problems arise when statistically unsophisticated users who do not understand the assumptions behind procedures or the limitations of the results obtained are misled by statistical results obtained from computer software. Therefore, we believe that, for pedagogical reasons, it is important that the applications of the methods covered in the text be illustrated through the use of worked-out examples.

1.3 STATISTICAL THINKING AND MODERN MANAGEMENT

In the past decade the emergence of a global economy has led to an increasing focus on the quality of products manufactured and services delivered. In fact, more than that of any other individual, it has been the work of a statistician, W. Edwards Deming, that has led to this changed business environment. An integral part of the managerial approach that contains this increased focus on quality (often referred to as **Total Quality Management**) is the application of certain statistical methods and the use of statistical thinking on the part of managers throughout a company.

> **Statistical thinking** can be defined as thought processes that focus on ways to understand, manage, and reduce variation.

Statistical thinking includes the recognition that data are inherently variable (no two things or people will be exactly alike in all ways) and that the identification, measurement, control, and reduction of variation provide opportunities for quality improvement. Statistical methods can provide the vehicle for taking advantage of these opportunities. The role of statistical methods in the context of quality improvement can be better understood if we refer to a model of quality improvement as presented in Figure 1.1.

We may observe from Figure 1.1 that the triangle consists of three portions; at the top we have management philosophy, and at the two lower corners we have statistical methods and behavioral tools. Each of these three aspects is indispensable for long-term quality improvement of either the manufactured goods or the services provided by an organization. A management philosophy provides a constant foundation for quality improvement efforts. Among the approaches available are those advocated by W. Edwards Deming (see References 5 and 6 and Section 10.5) and Joseph Juran (see References 11 and 12).

In order to implement a quality improvement approach in an organization, one needs to use both behavioral tools and statistical methods. Each of these aids in understanding and improving processes. Among the useful behavioral tools are process flow and fishbone diagrams (see Section 10.4), brainstorming, nominal group decision making, and team building. (For further discussion, see Reference 22.) Among the most useful statistical methods for

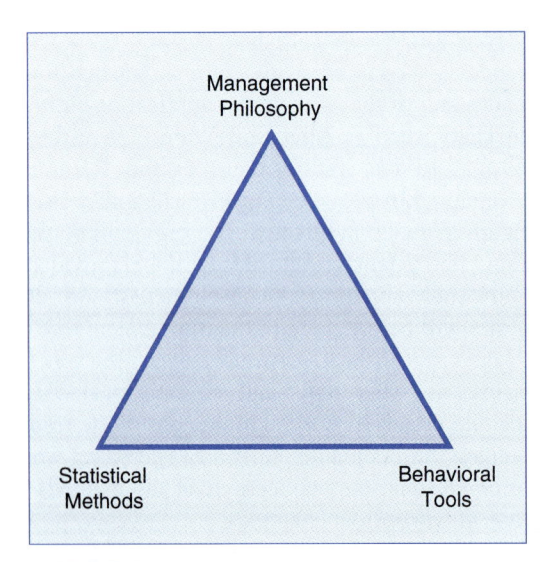

FIGURE 1.1 A model of the quality improvement process.

quality improvement are the numerous tables and charts and descriptive statistics discussed in Chapters 2 and 3 and the control charts developed in Chapter 10.

1.4 DESCRIPTIVE VERSUS INFERENTIAL STATISTICS

The need for data on a nationwide basis was closely intertwined with the development of descriptive statistics.

> **Descriptive statistics** can be defined as those methods involving the collection, presentation, and characterization of a set of data in order to describe the various features of that set of data properly.

Although descriptive statistical methods are important for presenting and characterizing data (see Chapters 2 and 3), it has been the development of inferential statistical methods as an outgrowth of probability theory that has led to the wide application of statistics in all fields of research today.

The initial impetus for formulation of the mathematics of probability theory came from the investigation of games of chance during the Renaissance. The foundations of the subject of probability can be traced back to the middle of the seventeenth century in the correspondence between the mathematician Pascal and the gambler Chevalier de Mere (see References 19 and 20). These and other developments by such mathematicians as Bernoulli, DeMoivre, and Gauss were the forerunners of the subject of inferential statistics. However, it has only been since the turn of this century that statisticians such as Pearson, Fisher, Gosset, Neyman, Wald, and Tukey pioneered in the development of the methods of inferential statistics that are widely applied in so many fields today.

> **Inferential statistics** can be defined as those methods that make possible the estimation of a characteristic of a population or the making of a decision concerning a population based only on sample results.

To clarify this, a few more definitions are necessary.

> A **population** (or **universe**) is the totality of items or things under consideration.

> A **sample** is the portion of the population that is selected for analysis.

> A **parameter** is a summary measure that is computed to describe a characteristic of an entire population.

> A **statistic** is a summary measure that is computed to describe a characteristic from only a sample of the population.

To relate these definitions to an example, suppose that the president of your college wanted to conduct a survey to learn about student perceptions concerning the quality of life on campus. The population or universe in this instance would be all currently enrolled students, while the sample would consist only of those students who had been selected to participate in the survey. The goal of the survey would be to describe various attitudes or characteristics of the entire population (the parameters). This would be achieved by using the statistics obtained from the sample of students to estimate various attitudes or characteristics of interest in the population. Thus, one major aspect of inferential statistics is the process of using sample statistics to draw conclusions about the population parameters.

The need for inferential statistical methods derives from the need for sampling. As a population becomes large, it is usually too costly, too time-consuming, and too cumbersome to obtain our information from the entire population. Decisions pertaining to the population's characteristics have to be based on the information contained in a sample of that population.

Probability theory provides the link by ascertaining the likelihood that the results from the sample reflect the results from the population.

1.5 ENUMERATIVE VERSUS ANALYTICAL STUDIES

Our discussion of the role of statistical methods for quality improvement in Section 1.3 and of statistical inference in Section 1.4 allows us to make an important distinction between two types of statistical studies that are undertaken: enumerative studies and analytical studies.

Enumerative studies involve decision making regarding a population and/or its characteristics.

The political poll is an example of an enumerative study, since its objectives are to provide estimates of population characteristics and to take some action on that population. The listing of all the units (such as the registered voters) that belong to the population is called the *frame* (see Section 1.9) and provides the basis for the selection of the sample. Thus, the focus of the enumerative study is on the counting (or measuring) of outcomes obtained from the frame.

Analytical studies involve taking some action on a process to improve performance in the future.

The investigation of the outcomes of a manufacturing or service process taken over time is an example of an analytical study. The focus of an analytical study is on the prediction of future process behavior and on understanding and improving a process. In an analytical study there is no identifiable universe, as in an enumerative study, and therefore there is also no frame. Perhaps we can highlight the distinction between enumerative and analytical studies by referring to Figures 1.2 and 1.3.

In the enumerative study, the bowl represents the population. The questions of interest revolve around the issue of "What's in the bowl?" An example of this might be wanting to know how many of the balls in the bowl are dark or what proportion of the balls is dark.

In the analytical study, there are several stages that make up a process. These stages typically include inputs that might involve some combination of people, equipment, material, and information; outputs that are in the form of a product manufactured or a service provided; and the intermediate transformation step that turns the inputs into the desired outputs. A key question of interest revolves around how any data that might be collected as part of the process (often over a period of time) can be used to improve the process in the future. This is indicated in Figure 1.3 by the presence of a feedback loop.

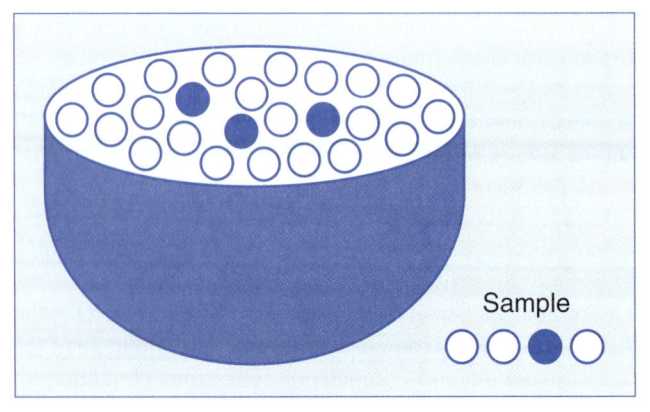

FIGURE 1.2 An enumerative study.

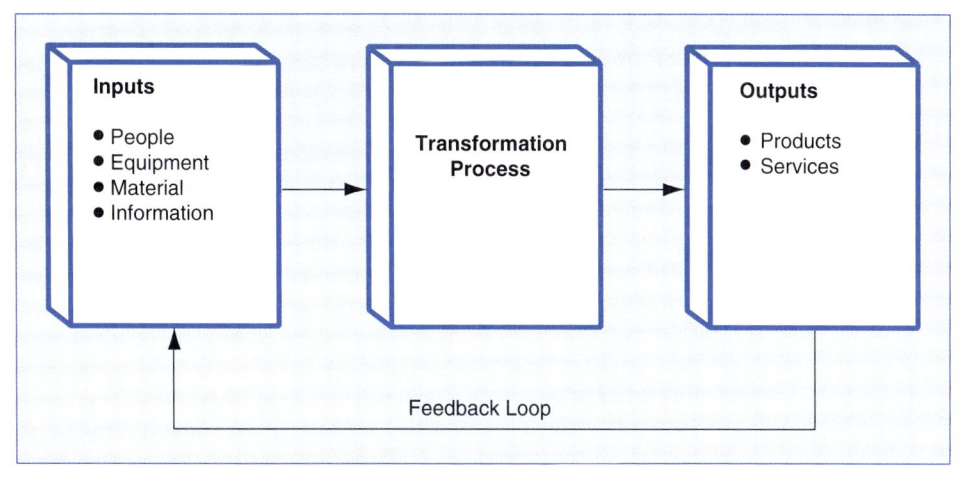

FIGURE 1.3 An analytical study.

The distinction between enumerative and analytical studies is an important one, since the methods that have been developed primarily for enumerative studies may be misleading or incorrect for analytical studies (see References 4–6). In this textbook we shall develop methods that are appropriate for these two different types of studies. Some of the methods are appropriate for either type of study. Other methods are appropriate primarily for enumerative studies or primarily for analytical studies.

Problems for Section 1.5

1.1 For each of the following, indicate whether the study is enumerative or analytical. Discuss.
   (a) A college decides to count the total number of students registering for classes that begin before 9 A.M.
   (b) A college wishes to determine whether the total number of students registering for classes that begin before 9 A.M. has increased or decreased over different semesters.
   (c) A college wishes to determine the reasons for a decrease in the number of applications for undergraduate admission.

1.2 For each of the following, indicate whether the study is enumerative or analytical. Discuss.
   (a) A magazine wishes to determine the proportion of its readership that is above 50 years of age.
   (b) A magazine would like to determine whether the availability of a discounted price on a 5-year subscription renewal will affect the number of subscriptions.
   (c) A magazine would like to determine the income level of its readership.
   (d) A magazine would like to determine how to reduce the number of errors made in billing subscribers.

1.6 THE NEED FOR DATA

Why do we need to collect data? Four main reasons could be given. Data are needed to

1. Provide the necessary input to a survey
2. Measure performance in an ongoing service or production process
3. Assist in formulating alternative courses of action in a decision-making process
4. Satisfy our curiosity

As examples:

- The manager wants to monitor a process on a regular basis to find out whether the quality of service being provided or products being manufactured are conforming to company standards.
- The market researcher looks for the characteristics that distinguish a product from its competitors.
- The potential investor wants to determine which firms within which industries are likely to have accelerated growth in a period of economic recovery.
- The pharmaceutical manufacturer needs to determine whether a new drug is more effective than those currently in use.

What exactly do we mean by data?

Data can be thought of as the numerical information needed to help us make a more informed decision in a particular situation.

For a statistical analysis to be useful in the decision-making process, the input data must be appropriate. Thus, proper data collection is extremely important. If the data are flawed by biases, ambiguities, or other types of errors, even the fanciest and most sophisticated statistical methodologies would not likely be enough to compensate for such deficiencies.

1.7 DATA SOURCES

There are many methods by which we may obtain needed data. First, we may seek data already published by governmental, industrial, or individual sources. Second, we may design an experiment to obtain the necessary data. Third, we may conduct a survey.

Owing to some of the most exciting scientific developments in this final decade of the century, we have truly reached an "age of information technology." Bar codes automatically record inventory information as products are purchased from supermarkets, department stores, and other outlets. ATMs enable the occurrence of banking transactions in which information is immediately recorded on account balances. Airline ticketing offices and travel agents have up-to-the-minute information regarding space availability on flights and at hotels. Transactions that took hours, or even days, a decade ago are now accomplished in a matter of seconds. Never before have so much up-to-date data and information been so readily available—and from so many sources.

Use of the library for research has literally taken on a new meaning. No longer does one have to visit the library to access material in printed form—from books, journals, magazines, pamphlets, and newspapers. Although we can still visit the library for these purposes, we can also use its modern multimedia facilities and obtain data electronically through information retrieval systems and on-line databases. In fact, our library "visit" can occur electronically through the use of a personal computer with a modem in our home or in our office. CD-ROM has revolutionized information access. As examples, ABI/INFORM Ondisc indexes and abstracts articles from more than 800 journals in the business fields, COMPACT D/SEC has information taken from annual and periodic company reports filed with the Securities and Exchange Commission, and NATIONAL TRADE DATA BANK contains recent trade and international data as well as foreign exchange data.

Regardless of the source used, a distinction is made between the original collector of the data and the organization or individuals who compile the data into tables and charts. The data collector is the **primary source;** the data compiler is the **secondary source.**

The federal government is a major collector and compiler of data for both public and private purposes. The Bureau of Labor Statistics is responsible for collecting data on employment as well as for establishing the well-known monthly *Consumer Price Index.* In addition to its constitutional requirement for conducting a decennial census, the Bureau of the Census is concerned with a variety of ongoing surveys regarding population, housing, and manufacturing and from time to time undertakes special studies on topics such as crime, travel, and health care.

In addition to the federal government, various trade publications present data pertaining to specific industrial groups. Investment services such as Moody's display financial data on a company basis. Syndicated services such as A. C. Nielsen provide clients with information enabling the comparison of client products with their competitors. And, of course, daily newspapers are filled with numerical information regarding stock prices, weather conditions, and sports statistics. Throughout this text, various applications will utilize data obtained from such published sources.

A second method for obtaining needed data is through experimentation. In an experiment, strict control is exercised over the treatments given to participants. For example, in a study testing the effectiveness of toothpaste, one would determine which participants in the study would use the new brand and which would not, instead of leaving the choice to the subjects. Proper experimental designs are usually the subject matter of more advanced texts, since they often involve sophisticated statistical procedures. However, in order to develop a feeling for testing and experimentation, the fundamental experimental design concepts will be considered in Chapters 7–9.

A third method for obtaining data is to conduct a survey. Here no control is exercised over the behavior of the people being surveyed. They are merely asked questions about their beliefs, attitudes, behaviors, and other characteristics. Their responses are then edited, coded, and tabulated for analysis.

Remember that there are four main reasons for collecting data: (1) to provide input to a research study, (2) to measure performance, (3) to enhance decision making, or (4) to satisfy our curiosity. To emphasize the importance of obtaining good data, researchers have adopted the term **GIGO**—garbage in, garbage out. Regardless of the method utilized for obtaining the data, if a study is to be useful, if performance is to be appropriately monitored, or if the decision-making process is to be enhanced, the data gathered must be *valid;* that is, the "right" responses must be assessed, and in a manner that will elicit meaningful measurements.

In order to design an experiment or conduct a survey, one must understand the different types of data and measurement levels.

1.8 MEASUREMENT SCALES

A survey statistician will most likely want to develop an instrument that asks several questions and deals with a variety of phenomena or characteristics. These phenomena or characteristics are called *random variables.* The data, which are the observed outcomes of these random variables, may differ from response to response.

1.8.1 Types of Data

As outlined in Figure 1.4, there are basically two types of characteristics of **random variables** that can be studied that yield the observed outcomes or data: categorical and numerical. **Categorical random variables** yield categorical responses, while **numerical random vari-**

Data Type	Question Types	Responses
Categorical →	Do you currently own U.S Government Savings Bonds?	Yes ☐ No ☐
Numerical ↗ Discrete →	To how many magazines do you currently subscribe?	_____ Number
↘ Continuous →	How tall are you?	_____ Inches

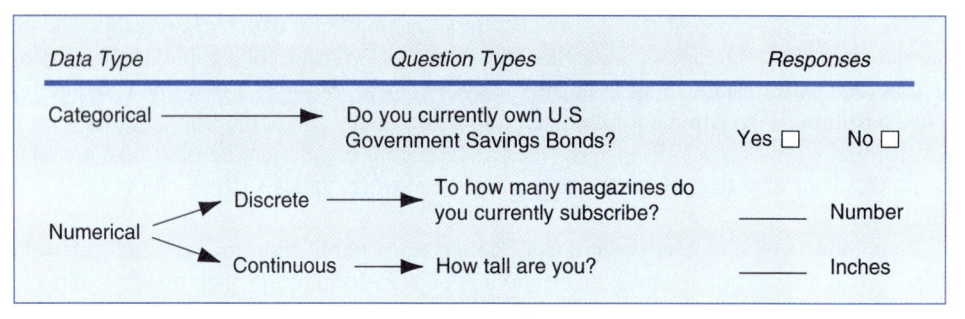

FIGURE 1.4 Types of data.

ables yield numerical responses. For example, the response to the question "Do you currently own U. S. Government Savings Bonds?" is categorical. The choices are limited to "yes" or "no." On the other hand, responses to questions such as "To how many magazines do you currently subscribe?" or "How tall are you?" are clearly numerical. In the first case, the numerical random variable may be considered as *discrete,* while in the second case, it can be thought of as *continuous.*

Discrete data are numerical responses that arise from a counting process, while **continuous data** are numerical responses that arise from a measuring process.

"The number of magazines subscribed to" is an example of a discrete numerical variable, since the response takes on one of a (finite) number of integers. The individual currently subscribes to no magazine, one magazine, two magazines, and so on. On the other hand, "the height of an individual" is an example of a continuous numerical variable, since the response can take on any value within a continuum or interval, depending on the precision of the measuring instrument. For example, a person whose height is reported as 67 inches may be measured as $67\frac{1}{4}$ inches, $67\frac{7}{32}$ inches, or $67\frac{58}{250}$ inches if more precise instrumentation is available. Therefore, we can see that height is a continuous phenomenon that can take on any value within an interval.

It is interesting to note that theoretically no two persons could have exactly the same height, since the finer the measuring device used, the greater the likelihood of detecting differences between them. However, most measuring devices are not sophisticated enough to detect small differences, and hence *tied observations* are often found in experimental or survey data even though the random variable is truly continuous.

1.8.2 Types of Measurement Scales

We could also describe our resulting data in accordance with the level of measurement attained. In the broadest sense, all collected data are "measured" in some form. For example, discrete numerical data can be thought of as arising by a process of *measurement through counting.* The four widely recognized levels of measurement are—from weakest to strongest level of measurement—the nominal, ordinal, interval, and ratio scales.

● **Nominal and Ordinal Scales**   Data obtained from a categorical variable are said to have been measured on a **nominal scale** or on an **ordinal scale.** If the observed data are merely classified into various distinct categories in which no ordering is implied, a nominal level of measurement is achieved. On the other hand, if the observed data are classified into distinct categories in which ordering is implied, an ordinal level of measurement is attained. These distinctions are depicted in Figures 1.5 and 1.6, respectively.

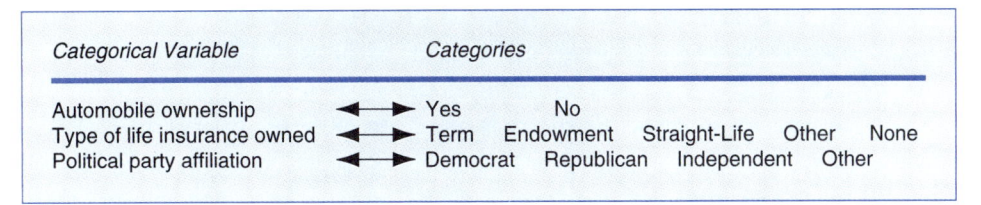

Categorical Variable	Categories
Automobile ownership	⟷ Yes No
Type of life insurance owned	⟷ Term Endowment Straight-Life Other None
Political party affiliation	⟷ Democrat Republican Independent Other

FIGURE 1.5 Examples of nominal scaling.

Categorical Variable	Ordered Categories
	(Lowest–Highest)
Student class designation	⟷ Freshman Sophomore Junior Senior
Product satisfaction	⟷ Very Unsatisfied Fairly Unsatisfied Neutral Fairly Satisfied Very Satisfied
Movie classification	⟷ G PG PG-13 R NC-17
	(Highest–Lowest)
Faculty rank	⟷ Professor Associate Professor Assistant Professor Instructor
Standard & Poor's bond ratings	⟷ AAA AA A BBB BB B CCC CC C DDD DD D
Restaurant ratings	⟷ ***** **** *** ** *
Student grades	⟷ A B C D E F

FIGURE 1.6 Examples of ordinal scaling.

Nominal scaling is the weakest form of measurement because no attempt can be made to account for differences within a particular category or to specify any ordering or direction across the various categories. Ordinal scaling is a somewhat stronger form of measurement, because an observed value classified into one category is said to possess more of a property being scaled than does an observed value classified into another category. Nevertheless, within a particular category no attempt is made to account for differences between the classified values. However, ordinal scaling is still a weak form of measurement, because no meaningful numerical statements can be made about differences between the categories. The ordering implies only *which* category is "greater," "better," or "more preferred"—not *how much* "greater," "better," or "more preferred." For instance, college basketball and college football rankings are other applications of ordinal scaling. The differences in ability between the teams ranked first and second might not be the same as the differences in ability between the teams ranked second and third, or those ranked sixth and seventh, and so on.

● **Interval and Ratio Scales**   An **interval scale** is an ordered scale in which the difference between the measurements is a meaningful quantity. For example, a noontime temperature reading of 67 degrees Fahrenheit is 2 degrees Fahrenheit warmer than a noontime temperature reading of 65 degrees Fahrenheit. In addition, the 2 degrees Fahrenheit difference in the noontime temperature readings is the same quantity that would be obtained if the two noontime temperature readings were 76 and 74 degrees Fahrenheit, so the difference has the same meaning anywhere on the scale.

If, in addition to differences being meaningful and equal at all points on a scale, there is a true zero point so that ratios of measurements are sensible to consider, then the scale is a **ratio scale.** A person who is 76 inches tall is twice as tall as someone who is 38 inches tall; in general, then, measurements of length are ratio scales. Temperature is a trickier case: Fahrenheit and Celsius (centigrade) scales are interval but not ratio scales; the "0" demarcation is arbitrary, not real. No one should say that a noontime temperature reading of 76 degrees Fahrenheit is twice as hot as a noontime temperature reading of 38 degrees

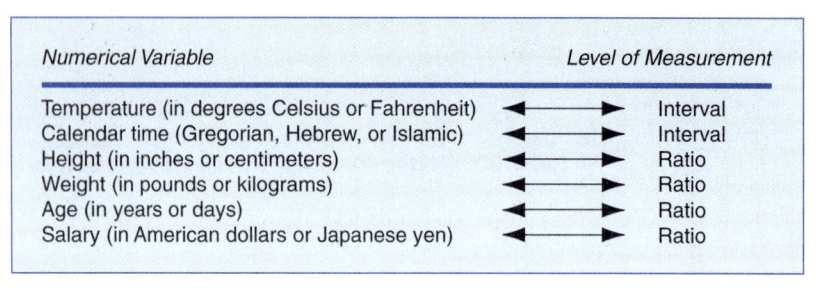

Numerical Variable	Level of Measurement
Temperature (in degrees Celsius or Fahrenheit)	Interval
Calendar time (Gregorian, Hebrew, or Islamic)	Interval
Height (in inches or centimeters)	Ratio
Weight (in pounds or kilograms)	Ratio
Age (in years or days)	Ratio
Salary (in American dollars or Japanese yen)	Ratio

FIGURE 1.7 Examples of interval and ratio scaling.

Fahrenheit. But when measured from absolute zero, as in the Kelvin scale, temperature is on a ratio scale, because a doubling of temperature is really a doubling of the average speed of the molecules making up the substance. Figure 1.7 gives examples of interval- and ratio-scaled variables.

Data obtained from a numerical variable are usually assumed to have been measured either on an interval scale or on a ratio scale. These scales constitute the highest levels of measurement. They are stronger forms of measurement than an ordinal scale because we are able to discern not only which observed value is the largest, but also by how much.

● **Caution: The Need for Operational Definitions**   Regardless of the level of measurement for our variables, *operational definitions* (see Reference 5) are needed to elicit the appropriate response or attain the appropriate outcome.

> An **operational definition** provides a meaning to a concept or variable that can be communicated to other individuals. It is something that has the same meaning yesterday, today, and tomorrow to all individuals.

As an example, take the word *round*. Although the dictionary provides a literal meaning, what is necessary is a meaning that can be used in practice. Thus, the real issue is not what is round, but how far something departs from "roundness" before we say that it is not round. This needs to be defined in a way that can be applied consistently from day to day (or, for a production worker, e.g., from product to product).

In the context of a survey, consider the question "What is your age?" To avoid problems of ambiguity, we must develop an operational definition for the responses to the question. For example, we must clarify whether age should be reported to the *nearest* birthday or as of the *last* birthday, because if your birthday is next month, you would probably choose the nearest birthday if you were turning 20; but you would be likely to report your present age if you were turning 50!

As a further example of operational definitions, consider the following headline that appeared in a suburban New York newspaper a few years ago: "Off With a Head Count: Is Suffolk more populous than Nassau? LILCO and the Census Bureau disagree."[1] The article included quotes from the Suffolk county executive ("…we are confident that Suffolk is No. 1") and the Nassau county executive ("We'll declare it a tie in the spirit of regional cooperation"). Of course, the differences in the two estimates come from the fact that the Census Bureau and the Long Island Lighting Company (LILCO) have different operational definitions used to estimate population in the two counties. The Census Bureau uses birth and death rates, migration patterns as shown on income tax returns, and a demographic formula that estimates that the average number of people per household has been shrinking in the past several years. On the other hand, for its definition, LILCO uses the number of year-round electric and gas meters, building permits, and a factor for the number of people in each house.

Problems for Section 1.8

1.3 Explain the difference between a categorical and a numerical random variable and give an example of each.

1.4 Explain the difference between a discrete and a continuous random variable and give an example of each.

1.5 If two students both score a 90 on the same examination, what arguments could be used to show that the underlying random variable (phenomenon of interest)—test score—is continuous?

• 1.6 Determine whether each of the following random variables is categorical or numerical. If numerical, determine whether the phenomenon of interest is discrete or continuous. In addition, provide the level of measurement and an operational definition for each of the variables.
   (a) Number of telephones per household
   (b) Type of telephone primarily used
   (c) Number of long-distance calls made per month
   (d) Length (in minutes) of longest long-distance call made per month
   (e) Color of telephone primarily used
   (f) Monthly charge (in dollars and cents) for long-distance calls made
   (g) Ownership of a cellular phone
   (h) Number of local calls made per month
   (i) Length (in minutes) of longest local call per month
   (j) Whether there is a telephone line connected to a computer modem in the household
   (k) Whether there is a FAX machine in the household

1.7 Suppose that the following information is obtained from students upon exiting from the campus bookstore during the first week of classes:
   (a) Amount of money spent on books
   (b) Number of textbooks purchased
   (c) Amount of time spent shopping in the bookstore
   (d) Academic major
   (e) Gender
   (f) Ownership of a personal computer
   (g) Ownership of a videocassette recorder
   (h) Number of credits registered for in the current semester
   (i) Whether or not any clothing items were currently purchased at the bookstore
   (j) Method of payment

   Classify each of these variables as categorical or numerical. If it is numerical, determine whether the variable is discrete or continuous. In addition, provide the level of measurement and an operational definition for each of the variables.

1.8 Give an example of a numerical variable that is actually discrete but might be considered continuous.

1.9 One of the variables most often included in surveys is income. Sometimes the question is phrased "What is your income (in thousands of dollars)?" In other surveys, the respondent is asked to "Place an X in the box corresponding to your income level."
   ❑ Under $20,000?   ❑ $20,000–$39,999   ❑ $40,000 or more
   (a) For each of these formats, state whether the measurement level for the variable is nominal, ordinal, interval, or ratio.
   (b) In the first format, explain why income might be considered either discrete or continuous.
   (c) Which of these two formats would you prefer to use if you were conducting a survey? Why?
   (d) Which of these two formats would likely bring you a greater rate of response? Why?

1.10 Provide an operational definition for each of the following:
   (a) An outstanding teacher
   (b) A hard worker
   (c) A nice day
   (d) Fast service

(e) A leader
(f) Commuting time to school or work
(g) A fine quarterback

1.11 Provide an operational definition for each of the following:
(a) A dynamic individual
(b) A boring class
(c) An interesting book
(d) An outstanding performance
(e) A manager
(f) An on-time plane arrival
(g) Study time

1.9 TYPES OF SAMPLES

As depicted in Figure 1.8, there are basically two kinds of samples: the **nonprobability sample** and the **probability sample.** For many studies only a nonprobability sample such as a judgment sample is available. In these instances, the opinion of an expert in the subject matter of a study is crucial to being able to use the results obtained to make changes in a process. Some other typical procedures of nonprobability sampling are quota sampling and chunk sampling; these are discussed in detail in specialized books on sampling methods (see References 1, 3, and 10).

In an enumerative study, the only way for us to make correct statistical inferences from a sample to a population is through the use of a probability sample.

A **probability sample** is one in which the subjects of the sample are chosen on the basis of known probabilities.

The four types of probability samples most commonly used are the simple random sample, the systematic sample, the stratified sample, and the cluster sample.

In a **simple random sample,** every individual or item has the same chance of selection as every other, and the selection of a particular individual or item does not affect the chances that any other is chosen. Moreover, a simple random sample may also be construed as one in which each possible sample that is drawn has the same chance of selection as any other sample that could be drawn.

A detailed discussion of systematic sampling, stratified sampling, and cluster sampling procedures can be found in References 1, 3, and 10.

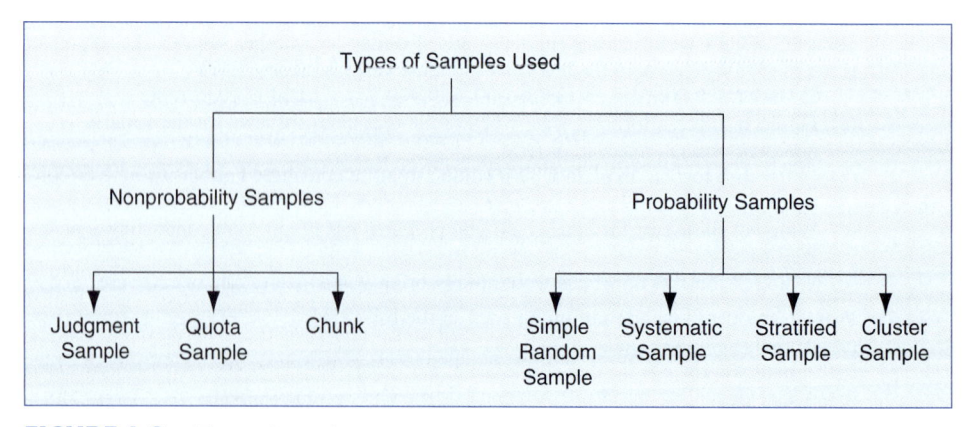

FIGURE 1.8 Types of samples.

1.10 DRAWING THE SIMPLE RANDOM SAMPLE

In this section we will be concerned with the process of selecting a simple random sample. Although it is not necessarily the most economical or efficient of the probability sampling procedures, it provides the base from which the more sophisticated procedures have evolved.

The key to proper sample selection is obtaining and maintaining an up-to-date list of all the individuals or items from which the sample will be drawn. Such a list is known as the **population frame** or simply **frame**. This population listing will serve as the **target population**, so that if many different probability samples were to be drawn from such a list, hopefully each sample would be a miniature representation of the population and yield reasonable estimates of its characteristics. If the listing is inadequate because certain groups of individuals or items in the population were not properly included, the random probability samples will only provide estimates of the characteristics of the *target* population—not the *actual* population—and biases in the results will occur.

Two basic methods could be used for selecting the sample: The sample could be obtained **with replacement** or **without replacement** from the finite population. The method employed must be clearly stated, since various formulas subsequently used for purposes of statistical inference are dependent upon the selection method.[2]

Let N represent the population size and n represent the sample size. To draw a simple random sample of size n, one could conceivably record the names of the N individual members of the population on separate index cards of the same size, place these index cards in a large fish bowl, thoroughly mix the cards, and then randomly select the n sample subjects from the fish bowl.

When sampling with replacement, the chance that any particular member in the population, say Judy Craven, is selected on the first draw from the fish bowl is $1/N$. Regardless of whoever is actually selected on the first draw, pertinent information is recorded on a master file, and then the particular index card is replaced in the bowl (sampling *with replacement*). The N cards in the bowl are then well shuffled and the second card is drawn. Since the first card had been replaced, the chance for selection of any particular member, including Judy Craven, on the second draw—regardless of whether or not that individual had been previously selected—is still $1/N$. Again, the pertinent information is recorded on a master file, and the index card is replaced in order to prepare for the third draw. Such a process is repeated until n, the desired sample size, is obtained. Thus, when sampling with replacement, every individual or item on every draw will always have the same 1 out of N chances of being selected.

But should we want to have the same individual or item possibly selected more than once? When sampling from human populations, it is generally more appropriate to have a sample of different people than to permit repeated measurements of the same person. Thus, we would employ the method of sampling without replacement so that once a particular individual is drawn, the same person cannot be selected again. As before, when sampling without replacement, the chance that any particular member in the population, say Judy Craven, is selected on the first draw from the fish bowl is $1/N$. Whoever is selected, the pertinent information is recorded on a master file, and then the particular index card is set aside rather than replaced in the bowl (sampling *without replacement*). The remaining $N - 1$ cards in the bowl are then well shuffled, and the second card is drawn. The chance that any individual not previously selected will be selected on the second draw is now 1 out of $N - 1$. This process of selecting a card, recording the information on a master file, shuffling the remaining cards, and then drawing again continues until the desired sample of size n is obtained.

Regardless of whether we sample with or without replacement, such "fish bowl" methods for sample selection have a major drawback—our ability to thoroughly mix the cards and randomly pull the sample. This became a major issue of contention in 1969 when the

Selective Service Commission developed a lottery system for choosing males to be drafted into the military because of the Vietnam War.[3]

"Fish bowl" methods of sampling as just described, although easily understandable, are not very useful. Less cumbersome and more scientific methods of selection are desirable to ensure randomization in the selection process. One such method utilizes a **table of random numbers** (see Table E.1 of Appendix E) for obtaining the sample.

A table of random numbers consists of a series of digits randomly generated and listed in the sequence in which the digits were generated (see Reference 21). Since our numeric system uses 10 digits (0, 1, 2, …, 9), the chance of randomly generating any particular digit is equal to the probability of generating any other digit. This probability is 1 out of 10. Hence if a sequence of 500 digits were generated, we would expect about 50 of them to be the digit 0, 50 to be the digit 1, and so on. In fact, researchers who use tables of random numbers usually test out such generated digits for randomness before employing them. Table E.1 has met all such criteria for randomness. Since every digit or sequence thereof in the table is random, we may use the table by reading either horizontally or vertically. The margins of the table designate row numbers and column numbers. The digits themselves are grouped into sequences of five for the sole purpose of facilitating the viewing of the table.

To use such a table in lieu of a fish bowl for selecting the sample, it is first necessary to assign code numbers to the individual members of the population. We then obtain our random sample by reading the table of random numbers and selecting those individuals from the population frame whose assigned code numbers match the digits found in the table. To better understand the process of sample selection from its inception, let us consider the following example.

Suppose that a company wanted to select a sample size of 20 full-time workers out of a population of 500 full-time employees. However, because not everyone will be willing to respond to the survey, we must be prepared to have a larger mailing. If we assume that 8 out of 10 full-time workers are expected to respond to such a survey (i.e., a rate of return of 80%), a total of 25 such employees must be contacted to obtain the desired 20 responses. Therefore, the questionnaire instrument was distributed in its final form to 25 full-time employees drawn from the personnel files of the company.

To draw the random sample, we use a table of random numbers. The population frame comprised a listing of the names and company mailbox numbers of all $N = 500$ full-time employees obtained from the company personnel files. Since the population size (500) is a three-digit number, each assigned code number must also be three digits so that every full-time worker has an equal chance for selection. Thus, a code of 001 is given to the first full-time employee in the population listing, a code of 002 is given to the second full-time employee in the population listing, …, a code of 352 to the three hundred fifty-second individual in the population listing, and so on until a code of 500 is given to the Nth full-time worker in the listing. Since $N = 500$ is the largest possible coded value, all three-digit code sequences greater than N (i.e., 501 through 999 and 000) are discarded.

To select the random sample, a random starting point for the table of random numbers must be established. One such method is to close one's eyes and strike the table of random numbers with a pencil. Suppose we use such a procedure and thereby select row 06, column 05, of Table 1.1 (a replica of Table E.1) as the starting point. Reading from left to right in Table 1.1 in sequences of three digits without skipping, the individual employees are selected for the survey.

The individual with code number 003 is the first full-time employee in the sample (row 06 and columns 05–07). The second individual selected has code number 364 (row 06 and columns 08–10). The third code number selected is 884. Since the highest code for any employee is 500, this number is discarded, as is the next value of 720. Individuals with code numbers 433, 463, 363, 109, 470, 432, 087, and 100 are selected third through tenth, respectively.

Table 1.1 Using a table of random numbers.

	Row	00000 12345	00001 67890	11111 12345	11112 67890	22222 12345	22223 67890	33333 12345	33334 67890
	01	49280	88924	35779	00283	81163	07275	89863	02348
	02	61870	41657	07468	08612	98083	97349	20775	45091
	03	43898	65923	25078	86129	78496	97653	91550	08078
	04	62993	93912	30454	84598	56095	20664	12872	64647
	05	33850	58555	51438	85507	71865	79488	76783	31708
Begin	06	97340	03364	~~88472~~	~~04334~~	~~63919~~	36394	~~11095~~	92470
Selection	07	~~70543~~	~~29776~~	~~10087~~	10072	55980	64688	68239	20461
(Row 06,	08	89382	93809	00796	95945	34101	81277	66090	88872
Column 05)	09	37818	72142	67140	50785	22380	16703	53362	44940
	10	60430	22834	14130	96593	23298	56203	92671	15925
	11	82975	66158	84731	19436	55790	69229	28661	13675
	12	39087	71938	40355	54324	08401	26299	49420	59208
	13	55700	24586	93247	32596	11865	63397	44251	43189
	14	14756	23997	78643	75912	83832	32768	18928	57070
	15	32166	53251	70654	92827	63491	04233	33825	69662
	16	23236	73751	31888	81718	06546	83246	47651	04877
	17	45794	26926	15130	82455	78305	55058	52551	47182
	18	09893	20505	14225	68514	46427	56788	96297	78822
	19	54382	74598	91499	14523	68479	27686	46162	83554
	20	94750	89923	37089	20048	80336	94598	26940	36858
	21	70297	34135	53140	33340	42050	82341	44104	82949
	22	85157	47954	32979	26575	57600	40881	12250	73742
	23	11100	02340	12860	74697	96644	89439	28707	25815
	24	36871	50775	30592	57143	17381	68856	25853	35041
	25	23913	48357	63308	16090	51690	54607	72407	55538
	⋮	⋮	⋮	⋮	⋮	⋮	⋮	⋮	⋮

Source: Partially extracted from The Rand Corporation, *A Million Random Digits with 100,000 Normal Deviates* (Glencoe, IL: The Free Press, 1955) and displayed in Table E.1 in Appendix E at the back of this text.

The selection process continues in a similar manner until the needed sample size of 25 full-time employees is obtained. During the selection process, if any three-digit coded sequence repeats, the employee corresponding to that coded sequence is included again as part of the sample if sampling with replacement; however, the repeating coded sequence is merely discarded if sampling without replacement.

Problems for Section 1.10

1.12 When is a sample random? What are some potential problems with using "fish bowl" methods to draw a simple random sample?

1.13 Suppose that I want to select a random sample of size 1 from a population of 3 items (which we can call A, B, and C). My rule for drawing the sample is: Flip a coin; if it is heads, pick item A; if it is tails, flip the coin again; this time, if it is heads, choose B; if tails, choose C. Explain why this is a random sample but not a simple random sample.

1.14 Suppose that a population has 4 members (call them A, B, C, and D). I would like to draw a random sample of size 2, which I decide to do in the following way: Flip a coin; if it is heads, my sample will be items A and B; if it is tails, the sample will be items C and D. Although this is a random sample, it is not a simple random sample. Explain why. (If you did Problem 1.13, compare the procedure described there with the procedure described in this problem.)

1.15 For a population list containing $N = 902$ individuals, what code number would you assign for
(a) The first person on the list?
(b) The fortieth person on the list?
(c) The last person on the list?

1.16 For a population of $N = 902$, verify that by starting in row 05 of the table of random numbers (Table E.1), only six rows are needed to draw a sample of size $n = 60$ *without replacement.*

1.11 ETHICAL AND OTHER ISSUES IN SURVEY RESEARCH

Data collection occurs at an early stage of a statistical analysis. After formulating a problem of interest, we would plan how to obtain the information that will be needed for decision making when attempting to solve the problem. Thus, the importance of obtaining good data can never be overstated. Always think of GIGO.

1.11.1 The Sample Survey

Every day, we read about the results of surveys or opinion polls in our newspaper, or we hear some interesting or titillating commentary on our radio and television. Clearly, advances in information technology have led to a proliferation of survey research. Not all this research is good, meaningful, or important (Reference 2). It becomes essential that we learn to evaluate critically what we read and hear and discount surveys that lack objectivity or credibility. In particular, we must examine the purpose of the survey, why it was conducted, and for whom. Remember that there are four main reasons for collecting data: (1) to provide input to a research study, (2) to measure performance, (3) to enhance decision making, or (4) to satisfy our curiosity. An opinion poll or survey conducted to satisfy curiosity is mainly for entertainment. Its result is an "end in itself" rather than a "means to an end." We should be more skeptical of such a survey because the result should not be put to further use.

The first step in evaluating the worthiness of a survey is to determine whether it was based on a probability or a nonprobability sample (as discussed in Section 1.9). You may recall that in an enumerative study, the only way for us to make correct statistical inferences from a sample to a population and interpret the results is through the use of a probability sample. Surveys employing nonprobability sampling methods are subject to serious, perhaps unintentional, interview biases that may render the results meaningless. In 1948, for example, each of the major pollsters employed quota sampling and incorrectly predicted the outcome of the presidential election (see Reference 17).

Even when surveys employ random probability sampling methods, they are subject to potential errors. There are four types of survey errors (Reference 9):

1. Coverage error or selection bias
2. Nonresponse error or nonresponse bias
3. Sampling error
4. Measurement error

A good survey research design attempts to reduce or minimize these various survey errors, often at considerable cost.

1.11.2 Coverage Error

The key to proper sample selection is an adequate population frame or up-to-date list of all the subjects from which the sample will be drawn. **Coverage error** results from the exclusion of certain groups of subjects from this population listing so that they have no chance of being selected in the sample. Coverage error results in a **selection bias.** If the listing is inadequate because certain groups of subjects in the population were not properly included, any random probability sample selected will provide an estimate of the characteristics of the *target* population, not the *actual* population.

Perhaps the most famous case of selection bias occurred in the 1936 *Literary Digest* poll. In that year, *Literary Digest,* a well-respected magazine, predicted that Governor Alf Landon of Kansas would receive 57% of the votes and overwhelmingly win the presidential election. When the actual votes were counted, Landon received only 38% and President Franklin Delano Roosevelt was easily reelected to a second term in office. The size of the error in the *Literary Digest* poll was considered enormous and unprecedented with respect to any major poll. Having lost its credibility, the magazine went bankrupt.

What went wrong? The prediction from the *Literary Digest* poll was based on the responses from 2.4 million individuals, a huge sample size. A major reason was selection bias. In 1936, the country was still suffering from the Great Depression. However, the *Literary Digest* compiled its population frame from such sources as telephone books, club membership lists, magazine subscriptions, and automobile registrations (Reference 7)—thereby catering to the rich and excluding from its listing the majority of the voting population who, during this period of economic hardship, could not afford such amenities as telephones, club memberships, magazine subscriptions, and automobiles. Thus, the 57% estimate for the Landon vote may have been very close to the *target* population, but certainly not the *actual* population.

1.11.3 Nonresponse Error

Not everyone will be willing to respond to a survey. In fact, research has indicated that individuals in the upper and lower economic classes tend to respond less frequently to surveys than do people in the middle class. **Nonresponse error** results from the failure to collect data on all subjects in the sample. And nonresponse error results in a nonresponse bias. Since it cannot be generally assumed that persons who do not respond to surveys are similar to those who do, it is extremely important to follow up on the nonresponses after a specified period of time. Several attempts should be made, either by mail or by telephone, to convince such individuals to change their minds. Based on these results, an effort is made to tie together the estimates obtained from the initial respondents with those obtained from the follow-ups so that we can be reasonably sure that the inferences made from the survey are valid (Reference 1).

1.11.4 Sampling Error

There are three main reasons for drawing a sample rather than taking a complete census—it is more expedient, less costly, and more efficient. However, when selecting the subjects using a random probability sample, depending on where one starts in the table of random numbers, chance dictates who in the population frame will or will not be included. Although only one sample is actually selected, if many different samples were to be drawn, hopefully each sample would be a miniature representation of the population and yield reasonable estimates of its characteristics. **Sampling error** reflects the heterogeneity or "chance differences" from sample to sample based on the probability of subjects being selected in the particular samples.

When we read about the results of surveys or polls in newspapers or magazines, there is often a statement regarding margin of error or precision—for example, "the results of this poll are expected to be within ±4 percentage points of the actual value." Sampling error can be reduced by taking larger sample sizes, although this will increase the cost of conducting the survey.

1.11.5 Measurement Error

In the practice of good survey research, a questionnaire is designed with the intent that it will enable meaningful information to be gathered. The obtained data must be *valid;* that is, the "right" responses must be assessed, and in a manner that will elicit meaningful measurements.

But there is a dilemma here—obtaining meaningful measurements is often easier said than done. Consider the following proverb:

> *A man with one watch always knows what time it is;*
> *a man with two watches always searches to identify the correct one;*
> *a man with ten watches is always reminded of the difficulty in measuring time.*

Unfortunately, the process of obtaining a measurement is often governed by what is convenient, not what is needed. And the measurements obtained are often only a proxy for the ones really desired.

Measurement error refers to inaccuracies in the recorded responses that occur because of a weakness in question wording, an interviewer's effect on the respondent, or the effort made by the respondent.

Much attention has been given to measurement error that occurs because of a weakness in question wording. A question should be clear, not ambiguous. Moreover, it should be objectively presented in a neutral manner; "leading questions" must be avoided.

As an example, in November 1993, the Labor Department reported that the unemployment rate in the United States has been underestimated for more than a decade because of poor questionnaire wording in the Current Population Survey. In particular, the wording led to a significant undercount of women in the labor force. Since unemployment rates are tied to benefit programs such as state unemployment compensation systems, it was imperative that government survey researchers rectify the situation by adjusting the questionnaire wording.

A second source of measurement error occurs in personal interviews and in telephone interviews—a "halo effect" in which the respondent feels obligated to please the interviewer. This type of error can be minimized by proper interviewer training.

A third source of measurement error occurs because of the effort (or lack thereof) on the part of the respondent. Sometimes the measurements obtained are gross exaggerations, either willful or owing to lack of recall on the part of the respondent. In either case, this hampers the utility of the survey—recall GIGO. This type of error can be minimized in two ways: (1) by carefully scrutinizing the data and calling back those individuals whose responses seem unusual and (2) by establishing a program of random callbacks in order to ascertain the reliability of the responses.

1.11.6 Ethical Issues

With respect to the proliferation of survey research (Reference 2), Eric Miller, editor of the newsletter *Research Alert,* stated that "there's been a slow sliding in ethics. The scary part is that people make decisions based on this stuff. It may be an invisible crime, but it's not a victimless one." Not all survey research is good, meaningful, or important, and not all survey research is ethical. We must become healthy skeptics and critically evaluate what we read and hear. We must examine the purpose of the survey, why it was conducted, and for whom, and then discard it if it is found to be lacking in objectivity or credibility.

Ethical considerations arise with respect to the four types of potential errors that may occur when designing surveys that use random probability samples. We must try to distinguish between poor survey design and unethical survey design. The key is *intent*. Coverage error or selection bias becomes an ethical issue only if particular groups or individuals are purposely excluded from the population frame so that the survey results would likely indicate a position more favorable to that of the survey's sponsor. In a similar vein, nonresponse error or nonresponse bias becomes an ethical issue only if particular groups or individuals are less likely to be available to respond to a particular survey format and the sponsor knowingly designs the survey in a manner aimed at excluding such groups or individuals. Sampling error becomes an ethical issue only if the findings are purposely presented without reference to sample size and margin of error so that the sponsor could promote a viewpoint that might otherwise be truly insignificant. Measurement error, however, becomes an ethical issue in any of three ways. A survey sponsor may purposely choose loaded, lead-in questions that would guide the responses in a particular direction. In addition, an interviewer, through mannerisms and tone, may purposely create a halo effect or otherwise guide the responses in a particular direction. Furthermore, a respondent having a disdain for the survey process may willfully provide false information.

Problems for Section 1.11

1.17 "A survey indicates that Americans overwhelmingly preferred a Chrysler to a Toyota after test-driving both." What information would you want to know before you accept the results of this survey?

1.18 "A survey indicates that the vast majority of college students picked Gap jeans as the most 'in' clothing." What information would you want to know before you accept the results of this survey?

1.12 INTRODUCTION AND DATA COLLECTION: A REVIEW AND A PREVIEW

As you can see in the following summary chart, this chapter provided an introduction to statistics and discussed data collection. You should be able to answer the following conceptual questions:

1. What is the difference between a sample and a population?
2. What is the difference between a statistic and a parameter?
3. What is the difference between an enumerative study and an analytical study?
4. What is the difference between a categorical and a numerical random variable?
5. What is the difference between discrete and continuous data?
6. What are the various levels of measurement?
7. What is an operational definition and why is it so important?
8. What are the main reasons for obtaining data and what methods can be used to accomplish this?

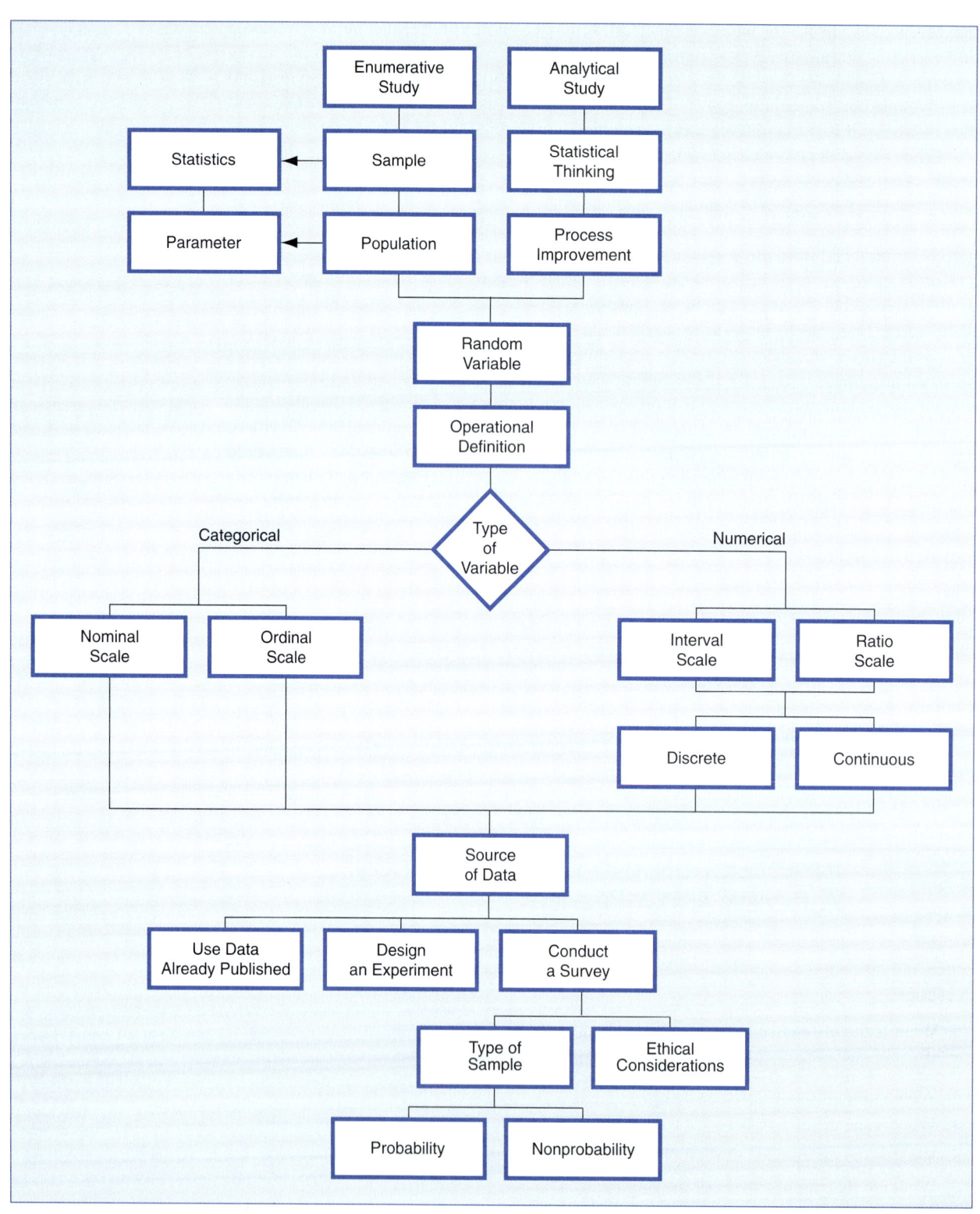

Chapter 1 summary chart.

22   **Chapter 1**   Introduction and Data Collection

9. What is the difference between probability and nonprobability sampling?
10. Why is the compilation of a complete population frame so important for survey research?
11. What is the difference between sampling with versus without replacement?
12. What distinguishes the four potential sources of error when dealing with surveys designed using probability sampling?

Check over the list of questions to see if you know the answers and could (1) explain your answers to someone who did not read this chapter and (2) reference specific readings or examples that support your answers. Also, reread any of the sections that may have seemed unclear to see whether they make more sense now.

Once data have been collected—be it from a published source, a designed experiment, or a survey—the data must be organized and prepared in order to assist us in making various analyses. In the next two chapters, methods of tabular and chart presentation will be demonstrated, various "exploratory data analysis" techniques will be described, and a variety of descriptive summary measures useful for data analysis and interpretation will be developed.

Getting It All Together

Key Terms

analytical studies 6
categorical random variable 9
continuous data 10
coverage error 19
data 8
descriptive statistics 5
discrete data 10
enumerative studies 6
GIGO 9
inferential statistics 5
interval scale 11
measurement error 20
nominal scale 10
nonprobability sample 14
nonresponse error 19
numerical random variable 9
operational definition 12
ordinal scale 10
parameter 5

population 5
population frame 15
primary and secondary sources 8
probability sample 14
random variable 9
ratio scale 11
sample 5
sample with replacement 15
sample without replacement 15
sampling error 20
selection bias 19
simple random sample 14
statistic 5
statistical thinking 4
table of random numbers 16
target population 15
Total Quality Management 4
universe 5

Chapter Review Problems

1.19 Determine whether each of the following random variables is categorical or numerical. If it is numerical, determine whether the phenomenon of interest is discrete or continuous. In addition, provide the level of measurement and an operational definition for each of the variables.
 (a) Brand of personal computer used
 (b) Cost of personal computer system

(c) Amount of time the personal computer is used per week
(d) Primary use for the personal computer
(e) Number of persons in the household who use the personal computer
(f) Number of computer magazine subscriptions
(g) Word processing package primarily used

1.20 Determine whether each of the following random variables is categorical or numerical. If it is numerical, determine whether the phenomenon of interest is discrete or continuous. In addition, provide the level of measurement and an operational definition for each of the variables.
(a) Amount of money spent on clothing in the last month
(b) Number of women's winter coats owned
(c) Favorite department store
(d) Amount of time spent shopping for clothing in the last month
(e) Most likely time period during which shopping for clothing takes place (weekday, weeknight, or weekend)
(f) Number of pairs of women's gloves owned
(g) Primary type of transportation used when shopping for clothing

1.21 Suppose the following information is obtained from Robert Keeler on his application for a home mortgage loan at the Metro County Savings and Loan Association:
(a) Place of Residence: Stony Brook, New York
(b) Type of Residence: Single-family home
(c) Date of Birth: April 9, 1962
(d) Monthly Payments: $1,427
(e) Occupation: Newspaper reporter/author
(f) Employer: Daily newspaper
(g) Number of Years at Job: 4
(h) Number of Jobs in Past Ten Years: 1
(i) Annual Family Salary Income: $66,000
(j) Other Income: $16,000
(k) Marital Status: Married
(l) Number of Children: 2
(m) Mortgage Requested: $120,000
(n) Term of Mortgage: 30 years
(o) Other Loans: Car
(p) Amount of Other Loans: $8,000

Classify each of the responses by type of data and level of measurement.

1.22 Suppose that the manager of the Customer Service Division of a consumer electronics company was interested in determining whether customers who had purchased a videocassette recorder over the past 12 months were satisfied with their products. Using the warranty cards submitted after the purchase, the manager was planning to survey 1,425 of these customers.
(a) Describe both the population and sample of interest to the manager.
(b) Describe the type of data that the manager primarily wishes to collect.
(c) Develop a first draft of the questionnaire by writing a series of three categorical questions and three numerical questions that you feel would be appropriate for this survey. Provide an operational definition for each variable.

1.23 In a political poll to try to predict the outcome of an election, what is the population to which we usually want to generalize? How might we get a random sample from that population? What might be some problems with the sampling in these polls?

1.24 Given a population of $n = 93$, draw a sample of size $n = 15$ *without replacement* by starting in row 29 of the table of random numbers (Table E.1). Reading across the row, list the 15 coded sequences obtained.

1.25 Do Problem 1.24 by sampling *with replacement*.

1.26 For a population of $N = 1,250$, a *two-stage* usage of the table of random numbers (Table E.1) may be recommended to avoid wasting time and effort. To obtain the sample by a two-stage approach, list the four-digit coded sequences after adjusting the first digit in each sequence as follows: If the first digit is a 0, 2, 4, 6, or 8, change the digit to 0. If the first digit is a 1, 3, 5, 7, or 9, change the digit to 1. Thus, starting in row 07 of the table of random numbers (Table E.1), the sequence 7054 becomes 1054, the sequence 3297 becomes 1297, and so on. Verify that only 10 rows are needed to draw a sample of size $n = 60$ *without replacement*.

1.27 For a population of $N = 2,202$, a *two-stage* usage of the table of random numbers (Table E.1) may be recommended to avoid wasting time and effort. To obtain the sample by a two-stage approach, list the four-digit coded sequences after adjusting the first digit in each sequence as follows: If the first digit is a 0, 3, or 6, change the digit to 0; if the first digit is a 1, 4, or 7, change the digit to 1; if the first digit is a 2, 5, or 8, change the digit to 2; if the first digit is a 9, discard the sequence. Thus, starting in row 07 of the table of random numbers (Table E.1), the sequence 7054 becomes 1054, the sequence 3297 becomes 0297, and so on. Verify that only 10 rows are needed to draw a sample of size $n = 60$ *without replacement*.

Team Projects

TP1.1 Suppose the following information is obtained for Frank Moore upon his admittance to the Brandwein College infirmary:
(a) Sex: Male
(b) Residence or Dorm: Mogelever Hall
(c) Class: Sophomore
(d) Temperature: 102.2°F (oral)
(e) Pulse: 70 beats per minute
(f) Blood Pressure: 130/80 mg/mm
(g) Blood Type: B Positive
(h) Known Allergies to Medicines: No
(i) Preliminary Diagnosis: Influenza
(j) Estimated Length of Stay: 3 days

Classify each of the 10 responses by type of data and level of measurement. Provide an operational definition for each variable. (*Hint:* Be careful with blood pressure; it's tricky.)

TP1.2 Provide an operational definition for each of the following:
(a) A good student
(b) The number of children per household
(c) An excellent movie
(d) A light-tasting beer
(e) A cute dress or outfit
(f) A quality product
(g) A smooth-riding automobile

TP1.3 Suppose that the Cat Fanciers Association (CFA) was planning to survey 1,500 of its members primarily to determine the percentage of its membership that currently own more than one cat.
(a) Describe both the population and sample of interest to the CFA.
(b) Describe the type of data that the CFA primarily wishes to collect.
(c) Develop a first draft of the questionnaire needed by writing a series of five categorical questions and five numerical questions that you feel would be appropriate for this survey. Provide an operational definition for each variable.
(d) **ACTION** Write a draft of the cover letter needed for this survey.

Endnotes

1. *Newsday,* April 25, 1988.
2. It is interesting to note that whether we are sampling with replacement from finite populations or sampling without replacement from infinite populations (such as some continuous, ongoing production process), the formulas used are the same.
3. Learning from the experiences of the 1969 draft lottery, the 1970 lottery attempted to correct for the possible mixing and selecting problems. Today, the mixing and selecting process used in televised state lotteries appears to be random—the only human involvement is the announcing of the chosen numbers.

References

1. Cochran, W. G., *Sampling Techniques,* 3d ed. (New York: Wiley, 1977).
2. Crossen, C., "Margin of Error: Studies and Surveys Proliferate, but Poor Methodology Makes Many Unreliable," *The Wall Street Journal,* November 14, 1991, pp. A1 and A9.
3. Deming, W. E., *Sample Design in Business Research* (New York: Wiley, 1960).
4. Deming, W. E.,"On Probability as a Basis for Action," *American Statistician,* Vol. 29, 1975, pp. 146–152.
5. Deming, W. E., *Out of the Crisis* (Cambridge, MA: Massachusetts Institute of Technology Center for Advanced Engineering Study, 1986).
6. Deming, W. E., *The New Economics for Industry, Government, Education* (Cambridge, MA: Massachusetts Institute of Technology Center for Advanced Engineering Study, 1993).
7. Gallup, G. H., *The Sophisticated Poll-Watcher's Guide* (Princeton, NJ: Princeton Opinion Press, 1972).
8. Goleman, D., "Pollsters Enlist Psychologists in Quest for Unbiased Results," *The New York Times,* September 7, 1993, pp. c1 and c11.
9. Groves, R. M., *Survey Errors and Survey Costs* (New York: Wiley, 1989).
10. Hansen, M. H., W. N. Hurwitz, and W. G. Madow, *Sample Survey Methods and Theory,* Vols. I and II (New York: Wiley, 1953).
11. Juran, J. M., *Juran on Leadership for Quality* (New York: The Free Press, 1989).
12. Juran, J. M., and F. M. Gryna, *Quality Planning and Analysis,* 2d ed. (New York: McGraw-Hill, 1980).
13. Kendall, M. G., and R. L. Plackett, eds., *Studies in the History of Statistics and Probability,* Vol. II (London: Charles W. Griffin, 1977).
14. Kirk, R. E., ed., *Statistical Issues: A Reader for the Behavioral Sciences* (Monterey, CA: Brooks/Cole, 1972).
15. Lotus 1-2-3 Release 5 (Cambridge, MA: Lotus Development Corporation, 1994).
16. *Microsoft Excel Version 7* (Redmond, WA: Microsoft Corporation, 1996).
17. Mosteller, F., and others, *The Pre-Election Polls of 1948* (New York: Social Science Research Council, 1949).
18. Norusis, M., *SPSS Guide to Data Analysis for SPSS-X: With Additional Instructions for SPSS/PC+* (Chicago, IL: SPSS Inc., 1986).
19. Pearson, E. S., ed., *The History of Statistics in the Seventeenth and Eighteenth Centuries* (New York: Macmillan, 1978).
20. Pearson, E. S., and M. G. Kendall, eds., *Studies in the History of Statistics and Probability* (Darien, CT: Hafner, 1970).
21. Rand Corporation, *A Million Random Digits with 100,000 Normal Deviates* (New York: Free Press, 1955).
22. Robbins, S. P., *Management,* 4th ed. (Englewood Cliffs, NJ: Prentice Hall, 1994).
23. Ryan, B. F., and B. L. Joiner, *Minitab Student Handbook,* 3d ed. (North Scituate, MA: Duxbury Press, 1994).
24. *SAS Language and Procedures Usage,* Version 6 (Raleigh, NC: SAS Institute, 1988).
25. Walker, H. M., *Studies in the History of the Statistical Method* (Baltimore, MD: Williams & Wilkins, 1929).
26. Wattenberg, B. E., ed., *Statistical History of the United States: From Colonial Times to the Present* (New York: Basic Books, 1976).

supplement
Introduction to Using MICROSOFT EXCEL

PART 1 ● Microsoft Excel Orientation

1S.1 WHAT ARE SPREADSHEET APPLICATION PROGRAMS?

In Section 1.2, when we discussed how software could help the manager use data to make decisions, we stated that a spreadsheet application program, Microsoft Excel, would be used as the software of choice in this text.

Spreadsheet application programs are the personal productivity programs best suited for the interactive manipulation of numerical data. These programs allow users to create electronic **spreadsheets,** or **worksheets,** which are rectangular arrays of (horizontal) rows and (vertical) columns into which entries are made.

Paper-based worksheets have long been used for the management and analysis of financial data by accountants. Not surprisingly, these professionals were among the first to understand the advantage of using *electronic* worksheets in which the effects of changing data could be immediately calculated and displayed.

Since their inception, spreadsheet programs have matured, gaining new functionality and growing more powerful over the years. In the case of Microsoft Excel, current versions of this program directly support most of the types of statistical analyses that are discussed in this text.

1S.1.1 Why Use Microsoft Excel?

Although other spreadsheet application programs can perform a variety of statistical analyses, Microsoft Excel's availability on a variety of personal computer operating systems and its dominant market share in business made it an easy choice for this text. Microsoft Excel also allows users to create electronic **workbooks,** collections of worksheets and other types of sheets. This feature can be used to develop solutions to statistical problems in a format that is the electronic equivalent of a typical (paper-based) management report. In this text, workbooks developed contain separate sheets that present the data and analysis for a problem as well as sheets that document and explain the spreadsheet methods used for the analysis. (The use of workbooks for this purpose is a pedagogical feature of this text.)

> ### *1S.1.2 Making the Most Effective Use of the Excel Sections*
>
> The Microsoft Excel sections in this text are designed to assist you in learning how to obtain and interpret pertinent statistics using features of Excel. Reviewing the worksheet designs and following the step-by-step instructions provided in the Excel sections while working with the Excel program is the most effective way of using this text. Through this active participation you will not only enhance your understanding of statistical concepts, but will also learn more about the specifics of both developing spreadsheet-based solutions and using the Microsoft Excel program.
>
> The workbooks stored on the diskette that accompanies this text include and explain all the Excel examples from this text (see Appendix F for details). To enhance your learning, we suggest that you construct and save your own versions of the workbooks developed in the Microsoft Excel sections, following the instructions provided in those sections, using the workbooks on the accompanying diskette to double-check your work.

1S.1.3 Organization of This Supplement

In this supplement we first provide an orientation to the basic concepts necessary to operate any program, such as Microsoft Excel, that runs in a windowing environment such as Windows and the Macintosh operating system. We then survey the basic features and components of Microsoft Excel before we describe the planning methodology used to create all workbooks discussed in this text. The supplement concludes with the walkthrough of a solution to a simple statistical problem. This problem introduces you to the basic features of Microsoft Excel that enable you to use Excel as a tool for statistical analysis.

Computer novices should read all sections of this supplement; more experienced users may wish to proceed directly to the discussion of the planning methodology (Section 1S.6). Novice or not, upon finishing this supplement, you should be ready to use Microsoft Excel in a productive manner.

1S.2 USING WINDOWING ENVIRONMENTS

All versions of Microsoft Excel (version 5 for Windows 3.1, version 5.0 for the Mac OS, and version 7 for Windows 95) load and run in a **windowing environment**, a user interface in which **windows**, or frames, are used as containers to subdivide the screen. In windowing environments you accomplish tasks by pointing to and choosing on-screen objects that represent components or functions of your computer system. Typically, this pointing and choosing is done with a **mouse pointer,** an on-screen pointer that moves when you move a **mouse** or similar input device. Three mouse operations are frequently used in windowing environments and throughout our discussion of Microsoft Excel in this text and are defined as follows:

1. To **select** an on-screen object, you move the mouse pointer (by moving the mouse) directly over an object and then **click,** or press, the left mouse button (or click the single button, if your mouse contains only one button).

2. To **drag** an object, you move the mouse pointer over an object and hold down the left (or single) mouse button while moving the mouse.

3. To **double-click** an object, you move the mouse pointer directly over an object and click the left (or single) mouse button twice in rapid succession.

4. To **right-click** an object, you move the mouse pointer directly over an object and press the right mouse button. (If your mouse contains only one mouse button, right-clicking is the same as simultaneously holding down the control key and pressing the mouse button.)

> Mouse operations can be practiced using the MOUSING.XLS workbook, stored on the diskette that accompanies this text. To use this file, you will need to learn how to run the Microsoft Excel application and also how to open Excel workbook files (see Section 1S.3).

MOUSING.XLS

In a windowing environment, mouse operations are applied to a variety of on-screen objects. Typically, users select **icons**, graphics that represent a specific program application or document, in order to begin their work. (Selecting the Microsoft Excel program icon or an icon representing an Excel workbook as displayed in Figure 1S.1 would be typical ways to load and run Excel.)

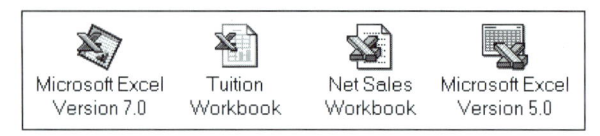

FIGURE 1S.1 Microsoft Excel program icon and Excel workbook icon from the Microsoft Windows environment.

1S.3 USING DIALOG BOXES IN MICROSOFT EXCEL

Selecting on-screen objects in a windowing environment frequently triggers the display of **dialog boxes**, special windows that accept user input or report status information. In Microsoft Excel, dialog boxes appear to guide the user through such common computing tasks such as opening, saving, and printing files and can appear to help the user complete various spreadsheet and data analysis tasks. This section uses common task dialog boxes for opening and printing files to introduce and define objects found in all types of dialog boxes. Because common task dialog boxes differ according to the windowing environment being used by Excel, this section uses dialog boxes from both Excel version 5 for Windows 3.1 and version 7 for Windows 95 to illustrate the type of variation one would encounter when executing the same common task in different versions of Excel.

> Dialog boxes for the spreadsheet and data analysis functions of Excel are standardized across the versions for the Mac OS, Windows 3.1, and Windows 95, differing only in cosmetic details such as typeface used or the styling of the window frame. When illustrating such dialog boxes, the authors have chosen to use one illustration that will be recognizable to users of either of the two version 5's or version 7.

Figure 1S.2 shows the File Open dialog box as it appears in the Microsoft Excel version 5 for Windows 3.1. On the left-hand side of the dialog box, under the heading File Name is an **edit box,** an area on the screen into which you can directly type or edit a value, in this case the name of a file. Below the edit box is a linked **scrollable list box** that allows you to select

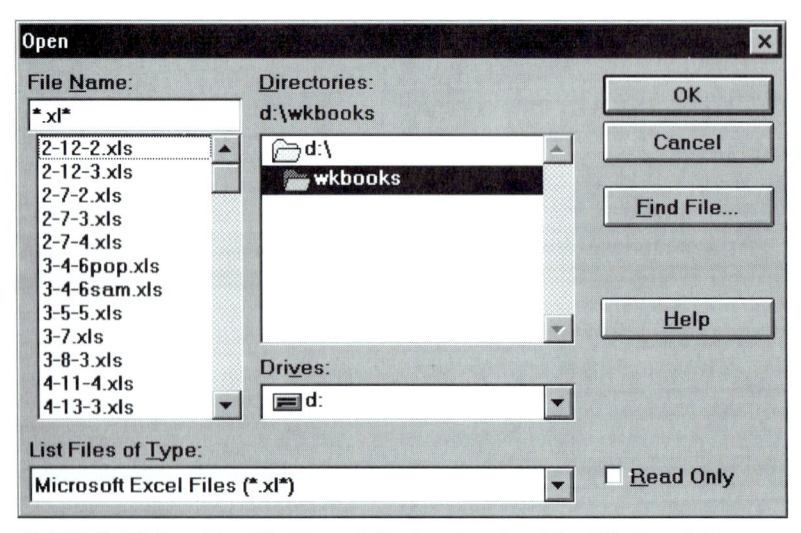

FIGURE 1S.2 The File Open dialog box (version 5 for Windows 3.1).

an item for the edit box by scrolling, or moving through, the list of choices. (Clicking the buttons on the **scroll bar,** immediately to the right of the list, is one way of scrolling through the list.)

In the center, under the label Directories, is another scrollable list box that appears without an accompanying edit box (a directory name must be picked from the list; it cannot be typed). Below the scrollable list boxes under the labels Drives and List Files of Type are two examples of a **drop-down list box.** These objects allow you to select an item from a non-scrollable list that appears when the drop-down button is clicked.

In the lower right is an example of a **check box,** a way of specifying an optional action. The optional action of this check box is to open a file as read only, meaning no changes can be saved to the file. (You may want to use this particular feature option when first opening and exploring a workbook file from the diskette that accompanies this text.)

Finally, on the right-hand side are four clickable buttons, two of which are common to many dialog boxes: the **OK button** (Open in version 7), that accepts all values, selections, and options as currently displayed in the dialog box and the **Cancel button,** that cancels the current operation (in this case, the opening of a file). Clicking these buttons will close the dialog box and cause it to disappear from the screen. (The version 7 File | Open dialog box, illustrated in Figure F.1 on page F–8, has a special properties button that can be used in conjunction with the Excel files on the diskette that accompanies this text. See Appendix F for details.)

Figure 1S.3 shows the File Print dialog box from version 7 of two additional objects commonly encountered, but which are not present in the File Open dialog boxes. In this File Print dialog box, notice that under the Print What and Page Range labels are two sets of **option buttons,** buttons that are used to present a set of mutually exclusive choices. Selecting an option button (also known as a radio button) always deselects, or clears, the other option buttons in the set, thereby preventing you from selecting two conflicting choices.

This dialog box also contains three sets of **spinner buttons** that appear to the right of the edit boxes associated with the copies:, from:, and to: labels. These pairs of buttons provide an alternate means of entering a value in the edit boxes by allowing you to increase or decrease the numeric value that appears in the associated edit box. (This version 7 dialog box, unlike the Print dialog boxes for the two version 5's, also allows the user to directly change the printer to be used.)

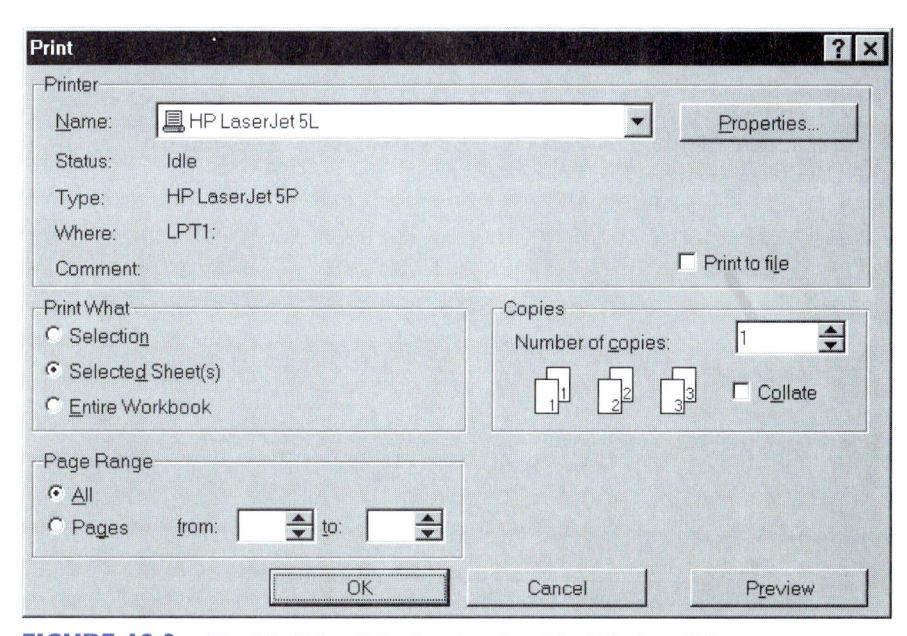

FIGURE 1S.3 The File Print dialog box (version 7 for Windows 95).

> A sharp-eyed reader will have noticed that many menu choices, dialog box labels, and button captions contain an underlined character and that many menus include combination keystrokes, such as Crtl+P, as part of their menu choices. These are all examples of keyboard shortcuts that can be useful to a skilled typist. As they can vary from one windowing environment to another, such shortcuts are neither described nor used in this text.

1S.4 USING THE MICROSOFT EXCEL APPLICATION WINDOW

Double-clicking a Microsoft Excel program or workbook icon causes a windowing environment to load Excel and display the initial Excel application window. Figure 1S.4 shows this initial window for version 7, using the settings described later in this section. (The application windows for the other versions are similar).

This application window is divided into two parts, the lower of which displays the first sheet from an open (and new) workbook. It is into this lower area that you type entries in order to implement solutions for the problems you seek to solve. For now, focus on the upper part of the window which contains several **bars**, or rows of objects, that display important information or that perform the various functions of the program.

The first bar, the **title bar**, displays the name of the program and the name of the currently opened workbook ("Book1" in Figure 1S.4). The title bar also contains **buttons**, clickable graphics that simulate the operation of push buttons, that perform system tasks to control the display of the Excel window. (As these buttons vary across the different versions of Excel they are not described in detail here.)

The second bar, the **menu bar**, contains words and symbols that when selected cause Excel to execute some task or cause Excel to display a **pull-down menu**, a vertical list of fur-

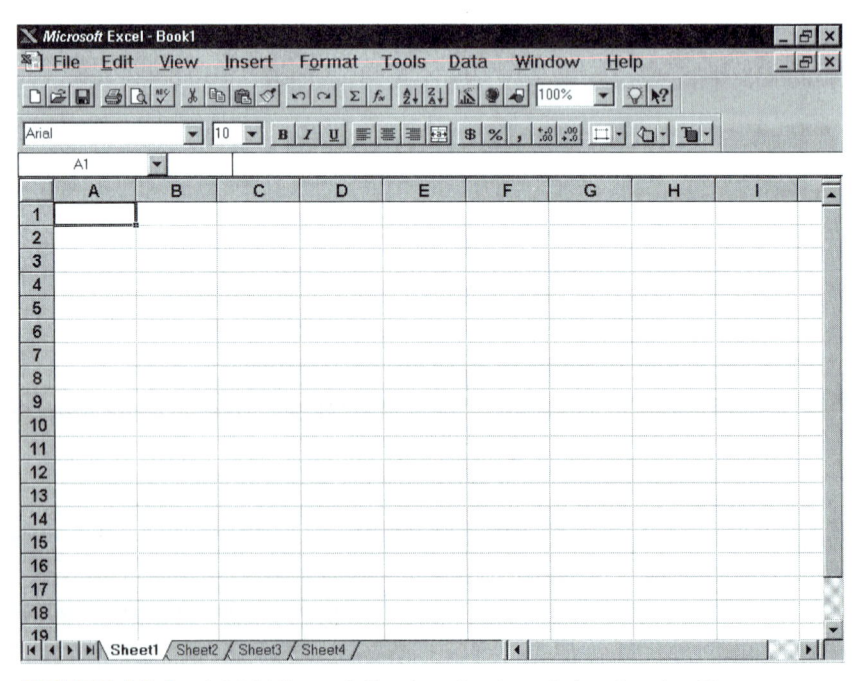

FIGURE 1S.4 Initial Microsoft Excel application window (version 7).

Supplement Introduction to Using Microsoft Excel

ther choices. For example, to display the File Open dialog box shown in Figure 1S.2, one would select the word File from the menu bar, and then from the subsequent pull-down menu select the word Open. Likewise, to display the File Print dialog box, one would again select File, but then select Print from the pull-down. (From this point on, we will abbreviate such selections by writing "File | Open" or "File | Print" using the vertical slash character | to separate menu choices.)

Immediately below the menu bar are two **toolbars**, containing buttons that perform common computing and worksheet formatting tasks. In Figure 1S.4, both the Standard and Formatting toolbars have been selected for viewing, and both have been snapped into position below the menu bar. (They can also appear as free-floating objects in their own mini-window.) Generally, the buttons on toolbars act as shortcuts, allowing the user to perform in a single mouse click what might otherwise require several menu selections. As you get proficient in using Excel, you may want to learn the shortcuts associated with these buttons.

Before beginning to use your copy of Microsoft Excel, you want to make your Excel windows appear as similar as is possible to Figure 1S.4 by following the boxed instructions that follow. (Again, as was noted in the discussion of dialog boxes in Section 1S.3, there may be some cosmetic differences to your display.) You may also want to verify that all of the spreadsheet features used in this text are properly installed and initialized in your copy of Excel.

To set up the Microsoft Excel application window as illustrated in Figure 1S.4, load the Excel program and issue the following commands:

1. **To display the two toolbars in the window.** Select View | Toolbars. The Toolbars dialog box appears. Select the Standard and Formatting check boxes and deselect all other toolbars on the list. Click the OK button.

2. **To display the formula and status bar and to standardize the display of the worksheet area.** Select Tools | Options. The Options dialog box appears containing two rows of labeled folder tabs. Select the View tab to make its view options visible if another tab's options are visible. (Normally, the view options will be visible when this dialog box is first displayed.) In the View Show groups, verify that the Formula bar and Status bar check boxes are selected. Under the Window Options group, make sure that Gridlines, Row & Column Headers, Horizontal scroll bar, Vertical scroll bar, and Sheet tabs check boxes are selected. Deselect the Formulas check box, if it has been selected. Click the OK button. (Other check boxes not mentioned can have the setting of your choice.)

To ensure that the Excel features used in this text are active and available in your copy of the program, load Excel and issue the following commands.

1. **To verify calculation and edit settings.** Select Tools | Options. In the Options dialog box that appears, select the Calculation tab. Verify that the Automatic calculation option button has been selected. Continue by selecting the Edit tab. Make sure that all check boxes *except* the Fixed Decimal check box have been selected. (Deselect the Fixed Decimal check box if necessary.) Click the OK box.

2. **To verify general Excel settings.** Select Tools | Options. In the Options dialog box that appears, select the General tab. Verify that the A1 option button has been selected. Change the value in the Sheets In New Workbook edit box to 4 if it is some other value. Verify that the Standard font is Arial and that the Size is 10. Use the drop-down list to select these values, if necessary. (Note: If Arial is not available on your system, use the font of your choice.) Click the OK button.

> 3. **To verify the installation of the Data Analysis Tools.** Select Tools. If Data Analysis appears as a menu choice, select it and verify that the dialog box that appears contains a scrollable list box in which the first item is "Anova: Single Factor" and the last choice is "z-Test: Two Samples for Means." (You will have to scroll the list to see the last choice.) Click the Cancel button. This verifies the installation of the tools.
>
> If Data Analysis does not appear as a Tools menu choice, select Add-Ins…from the Tools menu. The Add-Ins dialog box appears. Select the Analysis ToolPak check box—if there is such a check box—from the "Add-Ins Available:" list. Click the OK button and exit Excel (Select File | Exit.). Reload Excel and follow the instructions in the preceding paragraph to verify installation of the tools.
>
> If you cannot find an "Analysis ToolPak" check box in the "Add-Ins Available:" list, most likely the Data Analysis Tools component was not included when your copy of Excel was installed. (Consult with your technical support staff or review the chapter on installing Microsoft Excel in your Excel manual.)

IS.5   ORGANIZING EXCEL WORKBOOKS

Using spreadsheet application programs to develop accurate solutions to statistical problems requires careful planning and organization. Using these programs without first having a understanding of what is to be done and how it is to be done can cause us to produce worksheets—and in the case of Microsoft Excel, entire workbooks—that contain inaccurate or misleading results or that fail to clearly communicate the management information we seek.

To minimize such problems, we developed a standard format to help organize all the workbooks discussed in this text. This organization reflects the activities of analyzing the problem, designing a solution, implementing the solution, and verifying the solution that experienced users employ when developing any computer-based solution to a business problem. Understanding this format will help you make better use of the workbooks included on the diskette accompanying this text. (And because being able to properly organize solutions is such an important skill to master, a simple problem illustrating the use of our method is presented later in this supplement.)

Each of our workbooks contains an **Overview sheet** that serves as the table of contents for the workbook and that states the goals of the workbook, and the variables and the calculations needed to solve the problem. The overview sheet reflects our analysis of the problem and can be used by you as a guide to understanding the rest of the workbook.

The organization of each workbook also includes at least one **Data sheet** and one **Calculations sheet** that features the variables required by the problem and the calculations necessary to perform the statistical analyses required for the problem. The data and the calculations have been segregated in the workbooks to facilitate testing of the workbooks as well as to make it easier for you to modify the workbooks for use with other sets of data.

Where the results shown on a Calculations sheet are too cluttered for a clear and concise presentation, they have been reformatted onto an additional *Results sheet* that summarizes the solution to the problem. In a similar manner, when the programming techniques implemented to produce a solution are too complicated to easily summarize on the Overview sheet, they are explained separately in a *Behind the Scenes* sheet.

Many workbooks also contain additional sheets that have been created by Microsoft Excel as the result of using a feature of the program such as the Chart Wizard or the Data Analysis Tool. In addition, several workbooks contain "macro" procedures that have been placed in separate *module sheets* as required by the Excel program.

> ### Renaming Sheets
>
> All sheets in our workbooks have been given self-descriptive names, such as "Overview," "Sorted Data," and "Calculations," that are suggestive of their function or of the information they contain. To give the sheets you create similar names, you will need to **rename** your sheets, replacing the names in the form of Sheet1, Sheet2, Sheet3, and so on that Microsoft Excel automatically assigns to them.
>
> One way to rename a sheet is to double-click the **sheet tab** of the sheet you wish to rename. Sheet tabs appear near the bottom of the workbook window and display the current name of their sheets. Double-clicking a sheet tab causes a Rename Sheet dialog box containing the Name edit box to appear. By selecting the edit box and changing its contents and then clicking the OK button, you can rename the sheet. (If you do not see sheet tabs in your Excel application window, review the instructions given in the box at the end of Section 1S.4. If your workbook contains any sheets, you may need to use the horizontal scroll buttons to the left of the tabs to display the tab of the sheet you wish to rename.)

PART 2 • Getting Started with Microsoft Excel

The next 11 sections guide you through all the steps of analyzing, developing, implementing, and testing that we used to develop the workbooks of this text. As you read these sections, you will also be introduced to the basic concepts and operating procedures necessary to construct workbooks in Microsoft Excel. (This knowledge is assumed in Excel discussions that appear later in this text.)

1S.6 DESIGNING SHEETS FOR A WORKBOOK

To illustrate the development methodology described in the previous section, suppose that we have been asked to explore the variability among the costs that would be incurred attending any one of six colleges in the state of Pennsylvania. Assume that we have been provided data for the three variables (school, tuition, and room and board) that are shown in Table 1S.1 (These data, from a sample drawn from a population of 90 Pennsylvania colleges, will also be used when we discuss descriptive statistics in Chapter 3.)

As the first step in developing a Microsoft Excel workbook solution, we decide to express the problem as "What are the maximum and minimum values for the tuition and

Table 1S.1 Tuition and room and board expenses for a sample of six Colleges in Pennsylvania.

SCHOOL	TUITION (IN $000)	ROOM AND BOARD (IN $000)
University of Pittsburgh	10.3	4.1
East Stroudsburg University	4.9	2.9
Geneva College	8.9	4.1
Drexel University	11.7	3.8
California Univ. of Penn.	6.3	2.9
Slippery Rock University	7.7	3.4

Table 1S.2 Initial design of the Data sheet for the Pennsylvania college costs analysis data ($000).

SCHOOL	TUITION	ROOM AND BOARD	TOTAL COST
xxxxx	xx.x	x.x	xxx.x
xxxxx	xx.x	x.x	xxx.x
.	.	.	.
.	.	.	.
.	.	.	.
xxxxx	xx.x	x.x	xxx.x

room and board variables in this sample?" We also decide to ask the same question of a total cost variable, which we will define as the sum of the tuition and room and board expenses for each school.

Continuing on, we determine that a listing of the maximum and minimum values for these three variables, along with the range (the difference between the maximum and the minimum value) from each of the three will be the results. (See Section 3.5 for a complete discussion of the range statistic.) For our design for the Data Sheet, we develop Table 1S.2, using x's and ellipses as place holders for actual data that will be entered into the table. (We place a decimal point in the tuition, room and board, and total cost place holders to remind ourselves that the data were recorded to the nearest tenth in thousands of dollars.)

A Good Practice: Data Sheet Design

Note that the design of the data sheet reflects the arrangement of values as found in Table 1S.1 since this is the source of the data for our problem. Designing a data sheet in this manner is a good practice, one that will minimize your data entry errors later.

For this (relatively simple) problem, we decide that the Calculations sheet can serve as the result sheet as well. We produce Table 1S.3 as the initial design for this sheet; we will postpone the exact details of the sheet later in the development process.

Table 1S.3 Initial design of the Calculations sheet Pennsylvania college costs analysis.

	TUITION	ROOM AND BOARD	TOTAL COST
Maximum	xxxx	xxxx	xxxx
Minimum	xxxx	xxxx	xxxx
Range	xxxx	xxxx	xxxx

1S.7 DESIGNING THE DATA SHEET

Having completed our initial analysis and design of the sheets for this problem, we are now ready to implement them as worksheets in a Microsoft Excel workbook. Doing this will require mapping specific elements of the design such as the title and column headings to specific locations in a specific worksheet.

Recall that earlier in this chapter, worksheets were identified as rectangular arrays of (horizontal) rows and (vertical) columns into which entries are made. Mapping, therefore, requires associating the columns and lines (rows) of our initial designs with specific columns and rows in a worksheet. Using a standard notation for worksheets, we can label the columns of our initial design with letters and label our lines (rows) with numbers to identify into which **cells** (the intersections of the rows and columns) the entries should be made.

For our Data sheet, we decide that we should start with the column headings of Table 1S.1 and that the first column heading (School) should appear in cell A1; that is, it should be entered into the intersection of the first ("A") column and the first ("1") row. From this we quickly determine that the tuition and room and board headings should appear in cells B1 and C1, respectively, and that the values of each variable for the six colleges should appear in the next six rows (2–7). With these decisions made, we can revise Table 1S.2 to produce a design that can be directly implemented into a worksheet (see Table 1S.4). (Later in this supplement, the initial design for the Calculations sheet will be similarly revised.)

Table 1S.4 Revised design of the Data sheet for the Pennsylvania college costs analysis.

	A	B	C	D
1	School	Tuition	Room and Board	Total Cost
2	xxxxx	xx.x	x.x	xxx.x
3	xxxxx	xx.x	x.x	xxx.x
4	.	.	.	.
5	.	.	.	.
6	.	.	.	.
7	xxxxx	xx.x	x.x	xxx.x

Cells and Cell Addresses

In the preceding section, "A1" is an example of a **cell address.** Many times, a cell address in this column letter and row number format will be sufficient to specify the mapping of a cell entry. At other times—for example, when a design is implemented that references cells in other sheets or when certain features of Microsoft Excel are used—the address will need to include the name of the sheet. In such cases, the cell address is written in the form *sheetname!column-row* such as Data!A1 for the cell in the first column and row of the sheet named "Data." This form allows you to distinguish between two similarly located cells of two different sheets—for example, Data!A1 and Calculations!A1. (Should the name of the sheet include one or more spaces, then the sheet name needs to be enclosed in a pair of single quotation marks as in 'Sorted Data'!A1.)

Note that when a cell address does not contain the name of a sheet, the current sheet is implied. See discussion about formulas in Section 1S.9 to learn more about how this can minimize some typing when implementing a worksheet design.

1S.8 ENTERING VALUES INTO A DATA SHEET

Having specified cell addresses for the various parts of the Data worksheet in the previous section, we are now ready to enter values into the cells of the Data sheet. To do this, activate the Microsoft Excel application, and select File | New to create a new, blank worksheet window. Rename Sheet1 as Data, following the instructions given in Section 1S.5.

Select cell A1 by clicking its interior. A special border, the **cell highlight,** appears around the cell. This highlight indicates that cell A1 is now the **active cell**—that is, the cell into the next value to be typed will be entered. (Also note that A1, the address of the active cell, appears in the cell reference box.) Type the column heading School. As you type, notice that your keystrokes appear both in the edit box of the formula bar as well as in cell A1 itself. Press the Enter key (or click the check button to the left of the edit box) to complete the entry. (Users of keyboards that do not contain an Enter key, should press the Return key when instructions in this text call for pressing the Enter key.) If you made a mistake typing, you can use one of the methods described in the "Correcting Errors" box to correct your mistake.

Correcting Errors

At some point, as you enter text or numbers, you will probably make an entry that needs to be corrected. When you make a mistake while typing an entry, you can press the Escape key to cancel the entry, or you can use one of the following methods to edit your entry.

- Press the Backspace key to erase characters to the left of the cursor one character at a time.
- Press the Delete key to erase characters to the right of the edit cursor one character at a time.
- Replace the in-error text by clicking at the start of the error and dragging the mouse pointer over the rest of the error and then typing the replacement text.
- If you change your mind, you can undo your last edit by selecting the command: Edit | Undo. (There is also an Edit | Redo should you change your mind a second time and wish to keep the edit after all.)

As you complete this entry, notice that part of the heading extends into the next column. A way of overcoming this problem by having the columns automatically adjusted will be discussed in Section 1S.11 (Enhancing the Appearance of a Data Sheet).

Continue by selecting cell B1 and entering the column heading Tuition. Then select cell C1 to enter the Room and Board column heading. Finally, select cell D1 and enter the column heading Total Cost.

Now that all column headings have been entered, we can begin to enter the values that will appear underneath them. We will type in values by columns, using the feature of the Enter (Return) key to automatically advance the cell highlight down one row after each entry. (If we wished to enter values by rows, we could end each entry by pressing the Tab key, which would advance the cell highlight one column to the right.)

Select cell A2 and type the name University of Pittsburgh and press the Enter (or Return) key. Then type East Stroudsburg University in cell A3 and press the Enter key. Now type Geneva College in cell A4, press Enter; type Drexel University in cell A5, press Enter; type California Univ. of Penn. in cell A6, press Enter; and finally type Slippery Rock University in cell A7 and press Enter.

Select cell B2 and enter the tuition values from Table 1S.1 into cells B2:B7. Select cell C2 and enter the room and board values data from Table 1S.1 into cells C2:C7. Because the total cost variable was defined (unlike the other variables), and not taken from our source table, we momentarily defer until Section 1S.10 an explanation of how to enter the data for this column.

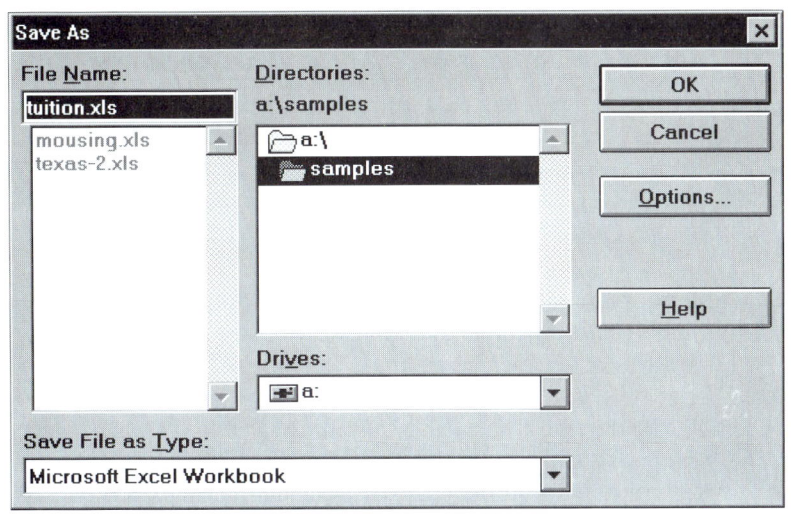

FIGURE 1S.5   The File Save As dialog box (version 5 for Windows 3.1).

Instead, having now entered all the data from our source table in the Data worksheet, we should **save,** or store a copy of our work on disk, before continuing. To save our work, select File | Save As. The File Save As dialog box which appears (see Figure 1S.5) will be similar to the File | Open dialog box for the version of Microsoft Excel being used. Specify the appropriate choices for drive, directory, and file name under which to store the workbook. (These choices vary based on the type of system used and whether you are saving your work to a diskette, local hard disk, or network disk. Ask your instructor if you are unsure as to the values you should choose.)

> *A Good Practice*
>
> After you have finished a single task, such as entering all the data from a source table, stop and save your work. Don't wait until you have finished all the tasks associated with developing a workbook. Unless otherwise directed by your instructor, you should always name your files using a name that *does not match* any of the files on the diskette that accompanies this text. This will prevent accidental alteration of the files on the diskette and avoid possible confusion.

1S.9   USING FORMULAS IN A DATA SHEET

Now that we have saved our work, we are ready to include the data for the total cost variable in cells D2:D7 of the Data worksheet. Recall that since this variable was not part of our original source table, we defined it in Section 1S.7 to be equal to the sum of the tuition and room and board expenses for each school. Therefore, one way to generate the total cost data would be to follow this definition and manually compute the values—for example, adding 10.3 and 4.1 to get 14.4 as the value for the total cost variable for the University of Pittsburgh.

Although it might be argued that for this very small and very simple problem, manual calculation would be the best method to generate the total cost data, it is generally wiser to have Microsoft Excel generate the values than to do it yourself. For Excel to generate these values, we will need to develop and enter **formulas,** or instructions to perform a calculation or some other task, in the appropriate cells of our Data worksheet (cells D2:D7 in this example).

Table 1S.5 Arithmetic Operators in Microsoft Excel

ARITHMETIC OPERATION	EXCEL OPERATOR
Addition	+
Subtraction	–
Multiplication	*
Division	/
Exponentiation (a number raised to a power)	^

To distinguish them from other types of cell entries, all formulas always begin with the = (equals sign) symbol. Creating formulas requires knowledge of the **operators,** or special symbols, used to express arithmetic operations. Operators used in formulas in this text are listed in Table 1S.5.

As our definition of the total cost variable calls for the addition of two quantities, we will use the + (plus sign) in our total cost formula, combining it with the cell addresses that contain the values we wish to add. In the case of calculating the total for the University of Pittsburgh, cells Data!B2 and Data!C2, which contain the tuition and room and board values for that school, are added together.

Assembling these pieces, we can form the formula =Data!B2+Data!C2 and enter it into cell D2. However, because we are entering a formula in the same sheet as the sheet to which it refers, we can write the formula using the shorthand notation =B2+C2, and Microsoft Excel will correctly interpret the addresses as referring to the current (Data) sheet.

Similar formulas can now be constructed and entered for the D column cells in rows 3–7. We adjust the original formula so as to refer to the tuition and room and board variables for the other five colleges. The formulas are as follows:

=B3+C3 for cell D3, the total cost variable for East Stroudsburg University

=B4+C4 for cell D4, the total cost variable for Geneva College

=B5+C5 for cell D5, the total cost variable for Drexel University

=B6+C6 for cell D6, the total cost variable for California Univ. of Penn.

=B7+C7 for cell D7, the total cost variable for Slippery Rock University

Again, note that the results of the formulas, and not the formulas themselves, are displayed in the worksheet.

1S.10 USING COPY TO FACILITATE THE ENTRY OF FORMULAS

As the construction and entry of similar formulas is a common task when developing workbooks, Microsoft Excel includes a copy shortcut feature that can facilitate this task.

To demonstrate this feature, let's pretend that we have not entered any of the formulas of the previous section. (If you have already entered those formulas, select cell D2 and press the Delete key to delete the formula in cell D2. Then do the same for cells D3–D7 to delete the formulas in those cells.) Begin by entering the first formula for the total cost variable, =B2+C2, into cell D2. Select cell D2 as shown in Figure 1S.6 Panel A. Then move the mouse pointer so that it is over the square handle in the lower right-hand corner of the cell highlight. The mouse pointer changes from an outlined plus sign to a plain plus sign (Panel B). Then drag the mouse pointer through the group of cells or **cell range** D3:D7 that will contain the similar formulas (see Panel C). Release the mouse button to complete the operation.

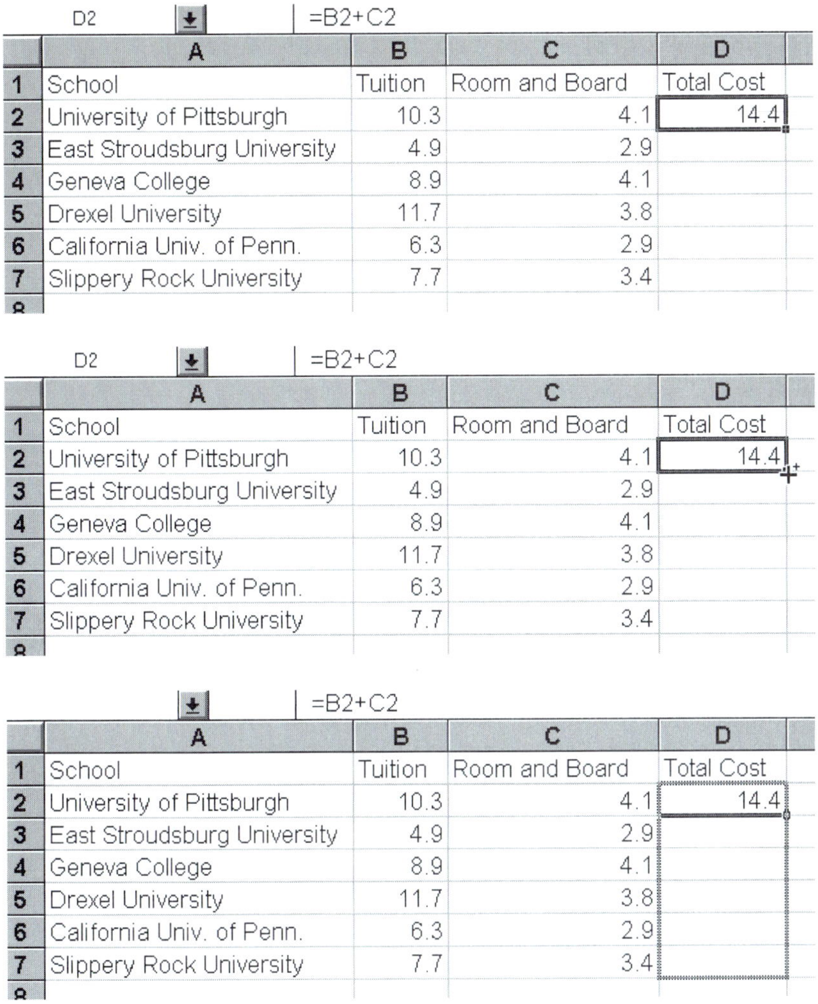

FIGURE 1S.6 The copying operation in Microsoft Excel. Panels A, B, and C.

Microsoft Excel pastes formulas adjusted to refer to the proper cells, and not exact copies of the formula. In this example, since we were copying down a column, Excel adjusted the row portion of all addresses in the formulas, just as we did earlier when we manually entered those formulas in the previous section. This can be verified if we select any of the cells in range D3:D7 and examine the cell's contents in the formula bar area. (We need to select a cell and examine the formula bar area because, as noted above, cells containing formulas display the result of the formula and not the formula itself.)

With the inclusion of these formulas, all the entries to the Data worksheet have now been made. (Before continuing, this would also be another good place to save our workbook using the procedure described in the Good Practices box at the end of the Section.1S.8.)

1S.11 ENHANCING THE APPEARANCE OF A DATA SHEET

Now that we have completed all the entries of the Data worksheet, we can make changes to enhance the appearance of the Data worksheet. The first improvement involves adjusting the width of the columns so that all column headings and column values can be easily read. This can be done by moving the mouse pointer to the column headings and clicking on the column heads for columns A–D and issuing the command Format | Column | Autofit. Notice that the column widths have automatically adjusted to the widest entry in each column.

Another improvement involves changing the column headings from a plain format to boldface. Perhaps the simplest way to do this is to first select the cell range to be reformatted (cells A3:D3) and then to click the B (boldface) button of the formatting toolbar. (To select the range A3:D3, first select cell A3 and then drag the mouse pointer across cells B3 through D3.) Then, with these three cells highlighted, click on the B (boldface) button on the formatting toolbar.

Another change we could consider is making the column headings appear centered over their columns. (Notice that they are currently left-justified in their columns.) If you wish to center the headings, select the range A3:D3 as before and click on the center button on the formatting toolbar. The column headings for columns A–D are now centered.

Having previously entered the numbers in columns B and C and the formulas in column D, you should observe that the numeric values in column B are not aligned on the decimal point. This problem may be fixed by first selecting the range B2:D7 and then clicking on the increase decimal point box on the formatting toolbar. You will now see that all values displayed have a single decimal point and are aligned on the decimal.

An additional improvement that is useful is centering a title over a range of columns. If we wanted to center a first-row worksheet title over a range of columns, we would do the following:

❶ Enter the title in the leftmost cell in the desired row.

❷ Select the range of cells in which the title is to be centered.

❸ Issue the command Format | Cells.

❹ Select the Alignment tab on the Cell Format dialog box.

❺ Select the center Across Selection button and click the OK button.

At this point, you should again save your work, using the procedure described in the Good Practices box at the end of the Section 1S.8.

1S.12 USING FUNCTIONS IN FORMULAS

In Section 1S.9, we used the plus sign arithmetic operator to construct our formula. We could have just as easily used the Sum **function,** one of many such pre-programmed instructions that can be used when solving a variety of common arithmetic, business, engineering, and statistical problems.

To use the Sum function, we would have typed the formula =SUM(B2:C2) into cell D2 instead of the formula =B2+C2. In the formula =SUM(B2:C2), the word SUM identifies the sum function, the pair of parentheses () bracket the cells of interest, and B2:C2 is the address of the cell range of interest, the cells whose values will be used by the function. (Since using the colon in this manner to express a range is so common in Excel, it will be the standard for indicating ranges in the rest of the text.)

Formulas that contain functions are otherwise identical to formulas that do not. This means that in this case we could individually type the formulas =SUM(B3:C3), =SUM(B4:C4), =SUM(B5:C5), =SUM(B6:C6), and =SUM(B7:C7), into cells D3:D7, respectively, or use the copy shortcut feature to place these formulas into the proper cells.

1S.13 IMPLEMENTING A RESULTS SHEET

Earlier in this supplement we decided that one worksheet would serve both for the calculations as well as for the results for the college costs analysis, and we developed Table 1S.2 as the design for this worksheet. In this section we implement that sheet, illustrating the use of additional functions as we go along.

Recall that for our results, we wished to present the maximum and minimum values, along with the range (defined as the difference between the maximum and the minimum value), for the three variables of interest. We can use the MAX and MIN functions in formulas to produce the maximum and minimum values for the variables and use the subtraction operator (the minus sign) in formulas to produce the range values.

We decide to map the title for the sheet to cell B1, the column headings Tuition, Room and Board, and Total Cost to cells B3, C3, and D3, and the labels Maximum, Minimum, and Range to cells A4, A5, and A6, respectively. We revise our initial design, replacing the place holders with the formulas to be entered into the worksheet. Table 1S.6 shows our revision.

Table 1S.6 Revised design of the Calculations sheet for the Pennsylvania college costs analysis.

	A	B	C	D
1		Pennsylvania College Costs Analysis		
2				
3		Tuition	Room and Board	Total Cost
4	Maximum	=MAX(DATA!B2:B7)	=MAX(DATA!C2:C7)	=MAX(DATA!D2:D7)
5	Minimum	=MIN(DATA!B2:B7)	=MIN(DATA!C2:C7)	=MIN(DATA!D2:D7)
6	Range	=B4 – B5	=C4 – C5	=D4 – D5

Note the inclusion of the worksheet name (Data) in the ranges used by the MAX and MIN formulas. This inclusion is necessary since we are referring to a range of cells on another sheet. Note that the range could also be entered as DATA!B2:DATA!B7, but we chose to use the same shorter way that the Excel program uses when it reports ranges in certain dialog boxes.

As for the range formulas, we could have constructed formulas in the form of =MAX(range) – MIN(range), for example, =MAX(DATA!B2:B7) – MIN(DATA!B2:B7) for the range of the tuition variable, but we chose the simpler alternative, which has the added advantage of more clearly communicating the relationships among the cells of this worksheet.

To implement this design, we would activate the Microsoft Excel application and open our Tuition workbook. We then would select the tab of a previously unused worksheet (e.g., Sheet2), rename the sheet "Calculations," and enter all the values and formulas in the proper cells.

A completed worksheet is shown in Figure 1S.7. After entering all values and formulas, you should review all entries for errors and test all calculations with simple numbers. Review

	A	B	C	D
1	School	Tuition	Room and Board	Total Cost
2	University of Pittsburgh	10.3	4.1	14.4
3	East Stroudsburg University	4.9	2.9	7.8
4	Geneva College	8.9	4.1	13
5	Drexel University	11.7	3.8	15.5
6	California Univ. of Penn.	6.3	2.9	9.2
7	Slippery Rock University	7.7	3.4	11.1

FIGURE 1S.7 Completed Calculations sheet for the Pennsylvania college costs analysis.

Customizing Your Printouts

The File | Page Setup command offers numerous options that control the way your printed pages look. Selecting this command produces the Page Setup dialog box illustrated in Figure 1S.8.

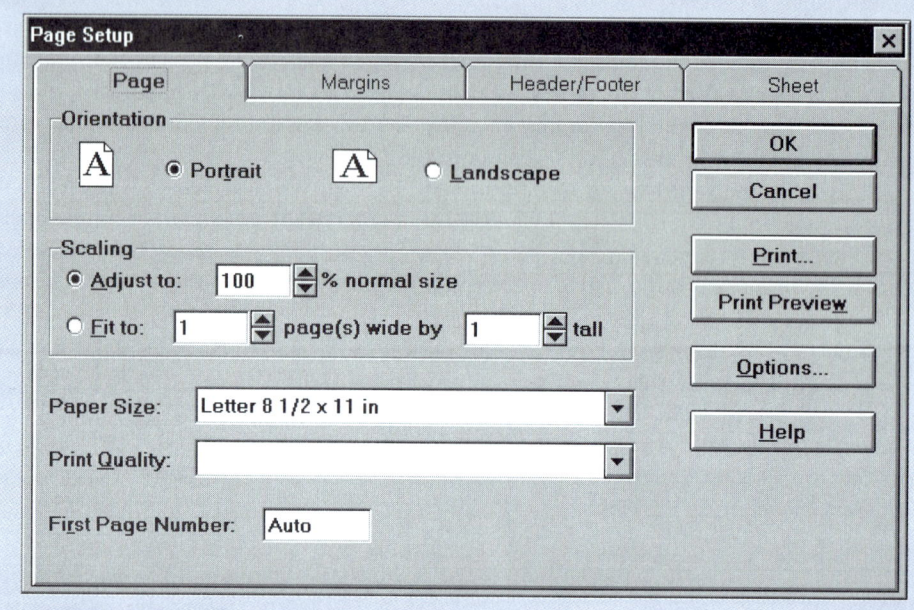

FIGURE 1S.8 The Page Setup dialog box.

The Page Setup dialog box contains tabs labeled Page, Margins, Header/Footer, and Sheet. Some of the choices available from these tabs are useful to know about.

- **Page tab choices.** The **Orientation option buttons** control whether sheets from your workbook are printed vertically (Portrait) or horizontally (Landscape). Landscape orientation is particularly useful for worksheets that have a greater width than length. The **Scaling option buttons** allow you either to reduce or enlarge the printed size of a worksheet (Adjust option) or to specify the number of pages to fit on one printed page (Fit option).

- **Margins tab choices.** Choices on this tab control the size of all margins on the printed page as well as whether the printed worksheets are centered on a page.

- **Header/Footer tab choices.** Choices here allow you to choose from many different styles of headers and footers that can be added to your printouts. You can also construct your own custom headers and footers.

- **Sheet Tab choices.** Most useful among the many Sheet choices are the Print check boxes that allow you to specify whether gridlines and row and column headings are included in the printouts of your worksheets. There is also a check box to specify draft quality printout; this is particularly useful when using older and slower types of printers.

of all formulas can be greatly aided by selecting the Tools | Options command and then, when the Options dialog box appears, selecting the Formulas checkbox of the View tab. This causes Excel to display the formulas—and not their results—on-screen. After this adjustment, you may need to issue the Format | Column | Autofit command discussed in Section 1S.11 to view all the formulas clearly. (You can then print the worksheet for later review by selecting the File | Print command and responding appropriately to the choices in the dialog box that appears. See the discussion of the File Print dialog box in Section 1S.3).

1S.14 INTRODUCTION TO WHAT IF ANALYSIS

As mentioned earlier in this supplement, one of the advantages of using spreadsheet applications is that the results of changing data can be immediately calculated and displayed. This dynamic nature of spreadsheet applications is especially valuable in statistical analysis, where it can be used to explore the sensitivity of the effect of changes in some portion of a data set on the results obtained. Frequently, we change the values of one variable and note the new results, a process that is commonly called "what if analysis."

With our completed and tested Tuition workbook, we could ask "What would happen to our results if the tuition of Geneva College was raised to 9.7 thousand dollars?" To see the effects of this change, we would open our Tuition workbook; select the Data worksheet tab; and select cell B4, the cell that contains the value for the tuition variable for Geneva; and enter the new value 9.7. Immediately, the value of the total cost variable changes from 13 to 13.8. Selecting the tab for the Results sheet, we notice no changes to the maximum, minimum, and range values. (You may wish to compare these results to the results produced by asking "What if the University of Pittsburgh's tuition, in cell B2, was changed from 10.3 to 13 thousand dollars?")

What if analyses are a valuable tool for understanding a variety of statistical methods. In Section 1S.15, we will learn about the scenario manager feature of Microsoft Excel, which facilitates the comparison of results among several, related what if analyses.

1S.15 USING THE SCENARIO MANAGER

When we wish to compare and save results produced by several related what ifs, or in cases in which we wish to preserve original values of our variables, we can use the **scenario manager** feature of Microsoft Excel. This feature allows us to associate a set of prestored values with a range of cells—called a scenario in Excel—and to save that set for later viewing or modification.

To illustrate the scenario manager, assume that we would like to determine and compare what happens to the value of the total cost range in each of the following four cases:

- Original data. Values as in Table 1S.1 on page 35.
- Pitt scholarship. We receive a scholarship that would pay for our tuition expenses for attending the University of Pittsburgh. (The tuition value for the University of Pittsburgh changes to zero.)
- ESU scholarship. We receive a scholarship that would pay for our tuition for attending East Stroudsburg University. (Tuition value for East Stroudsburg University changes to zero.)
- Drexel work/study. We receive a work/study grant that reduces the Drexel tuition to 3.0 thousand dollars. (The Drexel tuition value changes to 3.0.)

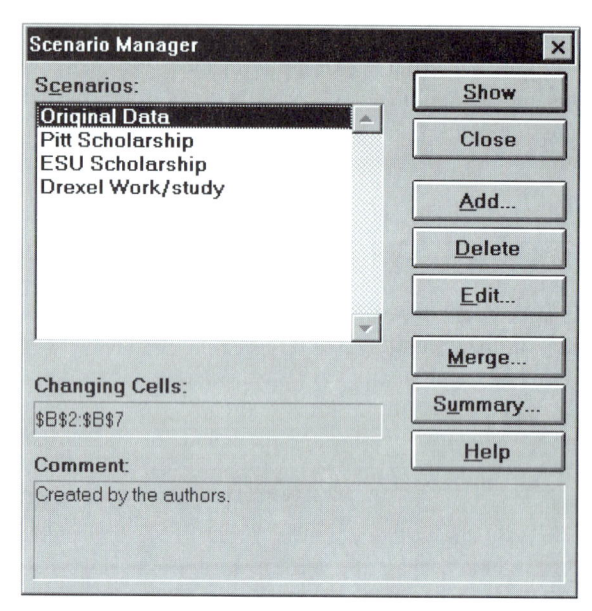

FIGURE IS.9 The Scenario Manager dialog box.

In this example we will need to define four scenarios, one for each of these cases, in order to compare results. To define the scenarios, we first select Tools | Scenarios, and the Scenario Manager dialog box appears (see Figure IS.9).

To define the four scenarios, we do the following:

❶ To define the original data scenario, click the Add button. The Add Scenario dialog box appears. Select the Scenario Name edit box and enter "Original Data" as the name for this scenario. Select the Changing Cells edit box and enter B2:B7. Click the OK button. The Scenario values dialog box appears, which contains edit boxes for each of the cells in the range. (Only the first five edit boxes are visible. Using the scroll bars would make the others visible.) As we want to store all the current values, immediately click the OK button. The scenario "original data" is now defined, and the Scenario Manager dialog box reappears.

❷ To define the Pitt scholarship scenario, click the Add button to get to the Add Scenario dialog box. Select the Scenario Name edit box and enter "Pitt Scholarship" as the name for this scenario. Verify that the Select the Changing Cells edit box contains B2:B7 (edit it, if necessary), and click the OK button. The Scenario Values dialog box appears. Locate the edit box for cell B2. Change the value in this box from 10.3 to 0 and then click the OK button. The scenario "Pitt scholarship" is now defined. Select Show the "Original Data" scenario to restore all the original data values.

❸ To define the ESU scholarship scenario, click the Add button to get to the Add Scenario dialog box. Select the Scenario Name edit box and enter "ESU Scholarship" as the name for this scenario. Verify that the Select the Changing Cells edit box contains B2:B7 (edit it, if necessary), and click the OK button. The Scenario Values dialog box appears. Locate the edit box for cell B3. Change the value in this box from 4.9 to 0 and then click the OK button. The scenario "ESU scholarship" is now defined. Select Show the "Original Data" scenario to restore all the original data values.

❹ To define the Drexel work/study scenario, again use a procedure similar to the previous two, naming the scenario "Drexel Work/study" and locating the edit box for cell

B5 in the Scenario Values dialog box. This time, as cell B7 is not one of the first five cells, we will have to scroll through the list of cells until it appears in the dialog box. Change the value in this box from 11.7 to 3 and then click the OK button. The scenario "Drexel work/study" is now defined. Select Show the "Original Data" scenario to restore all the original data values.

Having defined our scenarios, we can now use them in the following cycle:

① Select Tools | Scenarios if the Scenario Manager dialog box is not displayed.

② Select the scenario of interest and click the Show button, followed by the Close button.

③ Select the Results sheet tab to see the effects of the scenario.

④ Select Show the "Original Data" scenario to restore all the original data values.

⑤ Repeat steps 1–4 for each additional case.

All scenarios are saved when the workbook is saved, and they remain available until explicitly deleted (by pressing the Delete button in the Scenario Manager dialog box).

1S.16 IMPORTING DATA INTO A WORKSHEET

When performing statistical analyses throughout this text, we often encounter data sets with a large number of observations. When such sets have previously been entered and stored in a data file, it makes sense to try to import the contents of the file into our Data sheet to avoid having to reenter each observation one at a time as described in Section 1S.8.

Although Microsoft Excel can import data that have been stored in any one of various special file types used by other spreadsheet applications, in the field of statistics, data is more commonly found stored as **text files,** files that contain unlabeled and unformatted values that are separated by delimiters such as spaces, commas, or tab characters. When such a file is opened in Microsoft Excel, the program loads the **Text Import Wizard,** a series of linked dialog boxes that can be used to specify the organization of the data. As a user successfully steps through these dialog boxes, the Text Import Wizard will transfer the contents of the text file to a new worksheet in a new workbook. The new worksheet can then be transferred to another workbook using the File | Move or Copy Sheet command.

One text file found on the accompanying diskette is the TEXASC&U.TXT file, the out-of-state tuition data set of a population of 60 Texas colleges and universities that will be discussed extensively in Chapters 2 and 3. To import the data from this file, we would do the following:

TEXASC&U.TXT

① Select the command File | Open.

② Select the directory and drive that contains the text file TEXASC&U.TXT.

③ Enter the name TEXASC&U.TXT in the File Name edit box (or select it from the File Name list box) and click the OK button. The first dialog box of the Excel Text Wizard appears (see Figure 1S.10, Panel A).

④ Since the data values for the variables have been placed in aligned, fixed-width columns, select the Fixed Width option button. Press the Next button at the bottom of the dialog box to continue to the second dialog box of the wizard, illustrated in Figure 1S.10, Panel B. (Note that what the text calls variables are referred to as *fields* in the dialog box.)

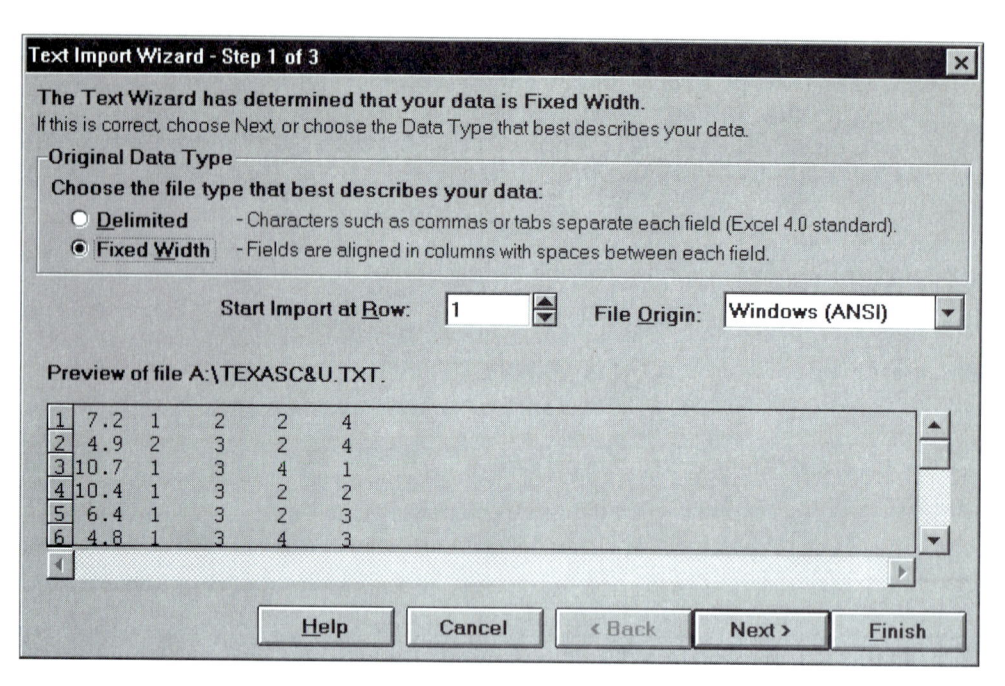

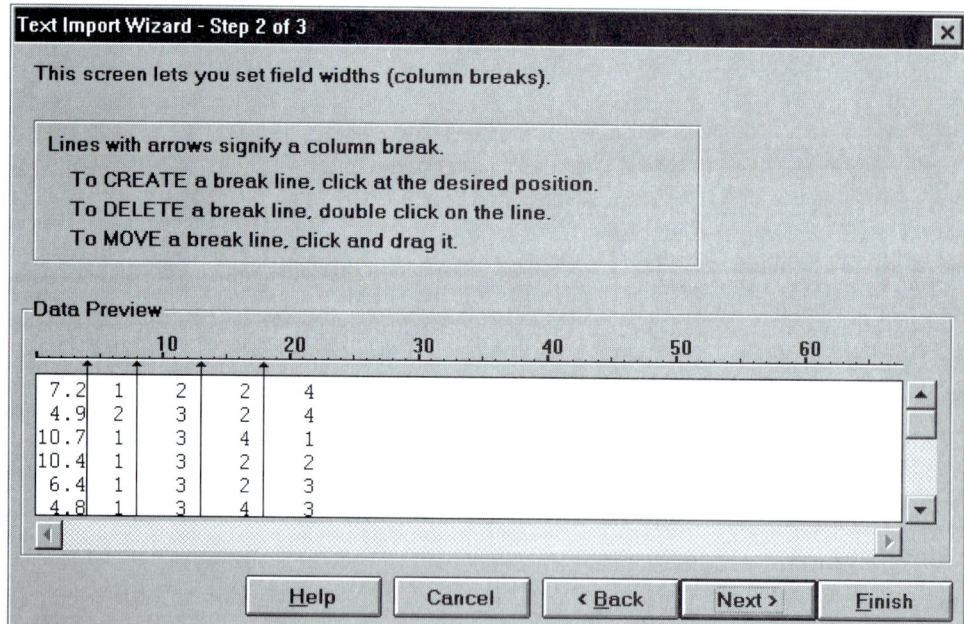

FIGURE 1S.10 Using the Text Import Wizard for the Texas out-of-state tuition data. Panel A and Panel B.

⑤ This second dialog box (Panel B) allows you to specify how to place the data from each line in the text file into columns. In this case, accept the Text Import Wizard's suggested placement by pressing the Next button to continue to the third and last dialog box. (To override suggestions, click on the vertical arrow break line and drag it.)

⑥ This third dialog box (Panel C) allows you to select the format of the columns for each variable in the new worksheet and to designate which columns of data to exclude from the new worksheet, if any. In this case—as in most cases— accepting the

Supplement Introduction to Using Microsoft Excel

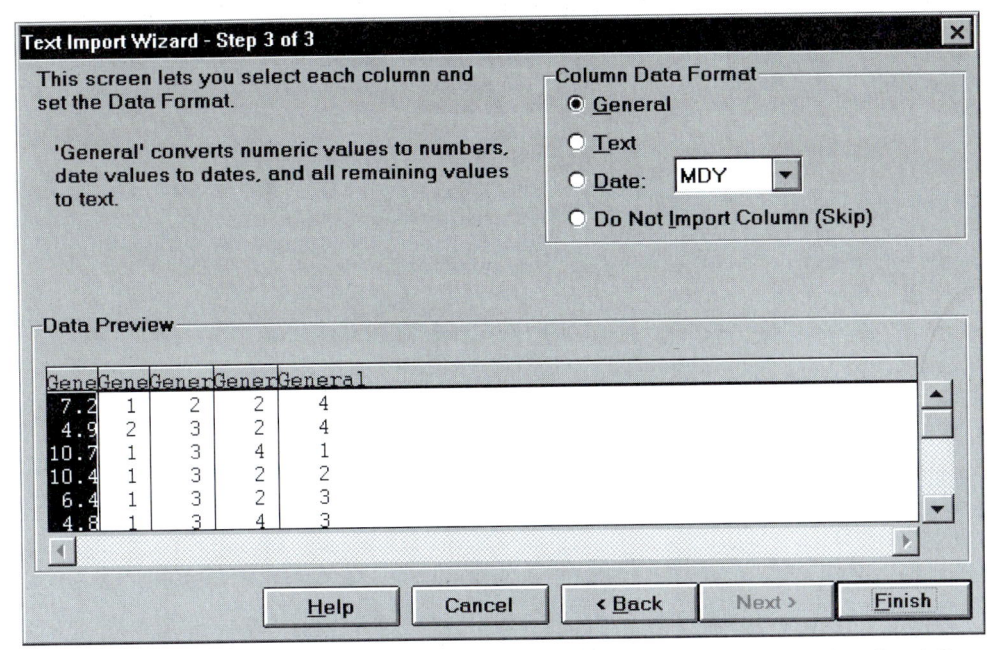

FIGURE 1S.10 Using the Text Import Wizard for the Texas out-of-state tuition data. Panel C.

Text Import Wizard's suggested "general" format for all columns, by clicking the Finish button, is all that needs to be done. (Otherwise, we would select a column and then select the option button corresponding to our choice for that column.)

Clicking the Finish button completes the text import process. At this point, you should review the new worksheet to check that the data have been correctly entered into all columns. After checking the data, select row 1 (by clicking on the row label 1 that is immediately to the left of cell A1) and select the command Insert | Rows. You will then have space to enter (and format) column headings as we did in Sections 1S.8 and 1S.11. Rename the worksheet "Data," save your workbook, and continue developing your workbook.

1S.17 SUMMARY

In this chapter we have provided an introduction to the basic concepts of spreadsheet applications in general and the features of the Microsoft Excel program in particular. We also used an example to illustrate the generation of a workbook-based solution to a simple statistical problem, by first analyzing the problem to be solved and then designing, implementing, and testing the solution.

Such a full discussion of all these steps for problems presented in later chapters is beyond the scope of this text. Instead, Microsoft Excel discussions in the remainder of this text will begin with detailed solution plans that need to be implemented. You should always remember, though, that thoughtful analysis preceded the development of those plans.

In the remainder of the text we will be learning about many additional aspects of Microsoft Excel in the context of specific statistical analyses. Our approach will be to integrate the discussion of Microsoft Excel with the statistical topics covered. In many instances, after covering a specific statistical procedure, we will demonstrate how to use Excel to analyze data that relate to the statistical methods just covered. In other situations, for pedagogical purposes, the discussion of Excel occurs after all the statistical topics in the chapter have been introduced.

Key Terms

Active cell 38
Bar 32
Buttons 32
Cancel button 31
Cell 37
Cell address 37
Cell range 40
Cell highlight 38
Check box 31
Clicking 29
Data sheet 34
Dialog box 30
Double-click 29
Dragging 29
Drop-down list box 31
Edit box 30
Formula 39
Function 42
Icons 29
Menu bar 32
Mouse 29
Mouse pointer 29
OK button 31

Operators 40
Option buttons 31
Orientation 44
Overview sheet 34
Pull-down menu 32
Renaming 35
Right-clicking 29
Scaling option buttons 44
Scenario manager 45
Scroll bar 31
Scrollable list box 31
Select 29
Sheet tab 35
Spinner buttons 31
Spreadsheets 28
Text files 47
Text Import Wizard 47
Title bar 32
Toolbars 33
Windows 29
Windowing environment 29
Workbook 28
Worksheet 28

References

1. Cobb Group, *Running Microsoft Excel 5* (Redmond, WA: Microsoft Press, 1994).
2. Grauer, R., and M. Barber, *Exploring Microsoft Excel 5.0* (Englewood Cliffs, NJ: Prentice-Hall, 1994).
3. *Microsoft Excel Version 7* (Redmond, WA: Microsoft Corp., 1996).
4. Parsons, J. J., Oja, D., and D. Auer, *Comprehensive Microsoft Excel 5.0 for Windows* (Cambridge, MA: Course Technology Inc., 1995).
5. Wells, E., *Developing Microsoft Excel 5 Solutions* (Redmond, WA: Microsoft Corp., 1995).

chapter 2

Presenting Data in Tables and Charts

CHAPTER OBJECTIVE To show how to organize and present collected data in tables and charts.

2.1 INTRODUCTION

In the preceding chapter we learned how to collect data through survey research. As pointed out in Section 1.9, since sampling saves time, money, and labor, we usually deal with sample information rather than data from an entire population. Nevertheless, regardless of whether we are dealing with a sample or a population, as a general rule, whenever a set of data that we have collected contains about 20 or more observations, the best way to examine such *mass data* is to present it in summary form by constructing appropriate tables and charts. We can then extract the important features of the data from these tables and charts.

Thus, this chapter is about data presentation. In particular, we will demonstrate how large sets of data can be organized and most effectively presented in the form of tables and charts in order to enhance data analysis and interpretation—two key aspects of the decision-making process.

2.2 ORGANIZING NUMERICAL DATA: THE ORDERED ARRAY AND STEM-AND-LEAF DISPLAY

In order to introduce the relevant ideas for Chapters 2 and 3, let us suppose that a company providing college advisory services to high school students throughout the United States has hired an analyst to compare the tuition rates charged to out-of-state residents by colleges and universities in different regions of the country. Table 2.1 displays the tuition rates charged to out-of-state residents by each of the 60 colleges and universities in the state of Texas (see Special Data Set 1 of Appendix D, pages D1–D2). When a set of data such as this one is collected, it is usually in **raw form**; that is, the numerical observations are not arranged in any particular order or sequence. As seen from Table 2.1, as the number of observations gets large, it becomes more and more difficult to focus on the major features in a set of data; thus, we need ways to organize the observations so that we can better understand what information the data are conveying. Two commonly used methods for accomplishing this are the *ordered array* and the *stem-and-leaf display*.

2.2.1 The Ordered Array

If we place the raw data in rank order, from the smallest to the largest observation, the ordered sequence obtained is called an **ordered array.** When the data are sorted into an ordered array,

Table 2.1 Raw data pertaining to tuition rates (in $000) for out-of-state residents at 60 colleges and universities in Texas.

TEXASC&U.TXT

7.2	4.9	10.7	10.4	6.4	4.8	4.7	4.6	6.0	5.4
4.8	4.7	8.3	3.8	4.8	8.3	6.4	6.6	4.5	8.0
3.6	2.4	8.5	8.8	7.7	4.9	8.6	12.0	4.9	7.0
11.0	4.9	3.9	4.9	4.4	4.9	4.9	8.0	3.6	7.4
7.9	4.9	5.8	3.9	11.6	10.3	3.4	3.9	5.0	3.9
8.0	3.5	4.9	5.8	4.1	3.9	3.5	4.8	5.9	3.6

Source: See Special Data Set 1, Appendix D, pages D1–D2, taken from "America's Best Colleges, 1994 College Guide," *U.S. News & World Report,* extracted from College Counsel 1993 of Natick, Mass. Reprinted by special permission, *U.S. News & World Report,* © 1993 by *U.S. News & World Report* and by College Counsel.

Table 2.2 Ordered array of out-of-state tuition rates (in $000) at 60 Texas colleges and universities.

2.4	3.4	3.5	3.5	3.6	3.6	3.6	3.8	3.9	3.9
3.9	3.9	3.9	4.1	4.4	4.5	4.6	4.7	4.7	4.8
4.8	4.8	4.8	4.9	4.9	4.9	4.9	4.9	4.9	4.9
4.9	4.9	5.0	5.4	5.8	5.8	5.9	6.0	6.4	6.4
6.6	7.0	7.2	7.4	7.7	7.9	8.0	8.0	8.0	8.3
8.3	8.5	8.6	8.8	10.3	10.4	10.7	11.0	11.6	12.0

Source: Table 2.1.

as in Table 2.2, our evaluation of their major features is facilitated. It becomes easier to pick out extremes, typical values, and concentrations of values.

Although it is useful to place the raw data into an ordered array prior to developing summary tables and charts or computing descriptive summary measures (see Chapter 3), the greater the number of observations present in a data set, the more useful it is to organize the data set into a stem-and-leaf display in order to study its characteristics (References 1, 13, and 14).

2.2.2 The Stem-and-Leaf Display

A **stem-and-leaf display** separates data entries into "leading digits" or "stems" and "trailing digits" or "leaves." For example, since the tuition rates (in $000) in the Texas data set all have one- or two-digit integer numbers, either the ones column or the tens column would be the leading digit, and the remaining column would be the trailing digit. Thus, an entry of 7.2 (corresponding to $7,200) has a leading digit of 7 and a trailing digit of 2.

Figure 2.1 depicts the stem-and-leaf display of the tuition rates for all 60 colleges and universities in Texas. The column of numbers to the left of the vertical line is called the "stem." These numbers correspond to the *leading digits* of the data. In each row the "leaves" branch out to the right of the vertical line, and these entries correspond to *trailing digits*.

- **Constructing the Stem-and-Leaf Display** The stem-and-leaf display can be constructed using the data from Table 2.1. Note that the first institution, Abilene Christian University, has a tuition rate of 7.2 thousand dollars. Therefore, the trailing digit of 2 is listed as the first leaf value next to the stem value of 7 (the leading digit). The second institution, Angelo State University, has a tuition rate of 4.9 thousand dollars. Here the trailing digit of 9

```
 2 | 4
 3 | 869694995956
 4 | 9876878599994999918
 5 | 48089
 6 | 4046
 7 | 27049
 8 | 33058600
 9 |
10 | 743
11 | 06
12 | 0
```
N = 60

FIGURE 2.1 Stem-and-leaf display of out-of-state tuition rates at the Texas schools.
Source: Table 2.1.

is listed as the first leaf value next to the stem value of 4. Continuing, the third institution, Austin College, has a tuition rate of 10.7 thousand dollars so that the trailing digit of 7 is listed as the first leaf value next to the stem value of 10. The fourth institution, Baylor University, has a tuition rate of 10.4 thousand dollars, so the trailing digit of 4 is listed as the second leaf value next to the stem value of 10.

At this point in its construction, our stem-and-leaf display appears as follows:

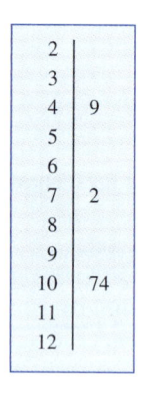

```
 2 |
 3 |
 4 | 9
 5 |
 6 |
 7 | 2
 8 |
 9 |
10 | 74
11 |
12 |
```

Note that two of the four schools have the same stem. As more and more schools are included, those possessing the same stems and, perhaps, even the same leaves within stems (i.e., the same tuition rates) will be observed. Such leaf values will be recorded adjacent to the previously recorded leaves, opposite the appropriate stem—resulting in Figure 2.1.

To assist us in further examining the data, we may wish to rearrange the leaves within each of the stems by placing the digits in ascending order, row by row. The **revised stem-and-leaf display** is presented in Figure 2.2.

```
 2 | 4
 3 | 455666899999
 4 | 1456778888999999999
 5 | 04889
 6 | 0446
 7 | 02479
 8 | 00033568
 9 |
10 | 347
11 | 06
12 | 0
```
N = 60

FIGURE 2.2 Revised stem-and-leaf display of out-of-state tuition rates at the Texas schools.

Another type of rearrangement is also useful. If we desire to alter the size of the stem-and-leaf display, it is flexible enough for such an adjustment. Suppose, for example, we want to increase the number of stems so that we can attain a lighter concentration of leaves on the remaining stems. This is accomplished in the stem-and-leaf display presented in Figure 2.3.

Note that each stem from Figure 2.2 has been split into two new stems—one for the *low*-unit digits 0, 1, 2, 3, or 4 and the other for the *high*-unit digits 5, 6, 7, 8, or 9. These are represented by L and H, respectively, as indicated in the stem listings of Figure 2.3.

However, some researchers would argue that the data displayed in Figure 2.3 are under-summarized since we are failing to capture how the data are truly clustering within various

2L	4
2H	
3L	4
3H	55666899999
4L	14
4H	56778888999999999
5L	04
5H	889
6L	044
6H	6
7L	024
7H	79
8L	00033
8H	568
9L	
9H	
10L	34
10H	7
11L	0
11H	6
12L	0

$N = 60$

FIGURE 2.3 Revised stem-and-leaf display of out-of-state tuition rates at the Texas schools.
Source: Figure 2.2.

2,3	4**455666899999**
4,5	14567788889999999999**04889**
6,7	044 6**02479**
8,9	00033 **568**
10,11	34 7**06**
12,13	0

$N = 60$

FIGURE 2.4 Revised stem-and-leaf display of out-of-state tuition rates at the Texas schools after condensing stems.
Source: Figure 2.2.

groupings. Hence, instead of expanding the display, as in Figure 2.3, we might wish to condense the data, as in Figure 2.4.

Note that consecutive pairs of stems from Figure 2.2 form the reduced set of stems in Figure 2.4 and the leaves corresponding to the *higher* member of each pair are **boldfaced**.

The (revised) stem-and-leaf display is, perhaps, the most versatile technique in descriptive statistics. It simultaneously organizes the data for further descriptive analyses (as we will see in Chapter 3), and it prepares the data for both tabular and chart form.

2.2.3 Using Microsoft Excel to Sort Data into an Ordered Array

In this section we began to study the out-of-state tuition rates of 60 Texas colleges by forming an ordered array and a stem-and-leaf display. Although the stem-and-leaf display is not available in Microsoft Excel,[1] we can use the Data | Sort command to obtain an ordered array. To obtain an ordered array similar to Table 2.2 on page 53, begin by opening the TEXAS-1.XLS workbook. Click on the Data worksheet tab. You will observe that the Texas data consists of six variables that have been labeled School, Tuition, Type, Setting, Calendar, and Focus. Since we want to sort by the tuition variable, click anywhere in Column B and select the command Data | Sort.

TEXAS-1.XLS

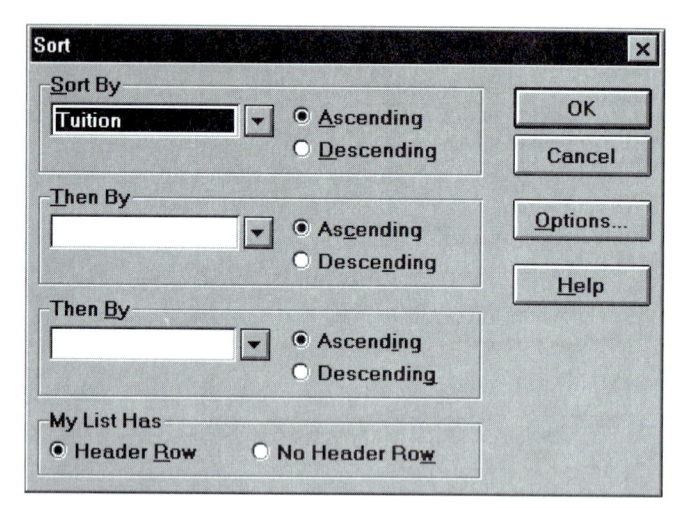

FIGURE 2.1.EXCEL Sort dialog box for the Texas out-of-state tuition data.

The Sort dialog box presented in Figure 2.1.Excel appears. In the upper left portion of this dialog box, Excel provides a list box that lists the names of all the variables in your data. These names were obtained from the column headings for each column. Excel allows you to sort the data according to one, two, or three variables. The first variable that the data are to be sorted by is entered at the top followed by (if desired) the second and third variables by which the data should be sorted. For each variable to be sorted, you have the choice of sorting in ascending order (lowest to highest) or descending order (highest to lowest). Since we want to obtain an ordered array for the tuition variable, verify that tuition appears in the Sort By box and that the first ascending option button has been selected. You also will notice that the Header Row button has been selected in this dialog box. This will allow the data to be sorted without moving the column headings. Click the OK button to sort the data. Figure 2.2.Excel represents the Texas data sorted in ascending order by the amount of tuition.

Problems for Section 2.2

- 2.1 Given the following stem-and-leaf display:

9	714
10	82230
11	561776735
12	394282
13	20

    (a) Rearrange the leaves and form the revised stem-and-leaf display.
    (b) Place the data into an ordered array.
    (c) Which of these two devices seems to give more information? Discuss.

- 2.2 Upon examining the monthly billing records of a mail-order book company, the auditor takes a sample of 20 of its unpaid accounts. The amounts owed the company are

    MAILORD.TXT

    $4, $18, $11, $7, $7, $10, $5, $33, $9, $12
    $3, $11, $10, $6, $26, $37, $15, $18, $10, $21

    (a) Develop the ordered array.
    (b) Form the stem-and-leaf display.

56 **Chapter 2** Presenting Data in Tables and Charts

	A	B	C	D	E	F
1	School	Tuition	Type	Setting	Calendar	Focus
2	Prairie View A&M U.	2.4	Public	Rural	Semester	RU
3	U. of Houston	3.4	Public	Urban	Semester	NU
4	U. of Texas at Arlington	3.5	Public	Suburban	Semester	NU
5	U. of Texas, San Antonio	3.5	Public	Urban	Semester	RU
6	Paul Quinn C.	3.6	Private	Urban	Semester	RLA
7	Texas C.	3.6	Private	Urban	Semester	RLA
8	Wiley C.	3.6	Private	Urban	Semester	RLA
9	Jarvis Christian C.	3.8	Private	Rural	Semester	RLA
10	Sul Ross State U.	3.9	Public	Rural	Semester	RU
11	Texas Women's U.	3.9	Public	Urban	Semester	RU
12	U. of Houston-Downtown	3.9	Public	Urban	Semester	RU
13	U. of North Texas	3.9	Public	Urban	Semester	RU
14	U. of Texas-Pan American	3.9	Public	Urban	Semester	RU
15	U. of Texas at El Paso	4.1	Public	Urban	Semester	RU
16	Texas A&I U.	4.4	Public	Rural	Semester	RU
17	Midwestern State U.	4.5	Public	Urban	Semester	RU
18	East Texas State U.	4.6	Private	Urban	Quarter	RU
19	East Texas Baptist U.	4.7	Private	Urban	4-1-4	RLA
20	Huston-Tillotson C.	4.7	Private	Urban	Semester	RLA
21	Dallas Baptist U.	4.8	Private	Urban	4-1-4	RLA
22	Howard Payne U.	4.8	Private	Rural	Semester	RLA
23	Lamar U.	4.8	Public	Urban	Semester	RU
24	Wayland Baptist U.	4.8	Private	Urban	4-1-4	RU
25	Angelo State U.	4.9	Public	Urban	Semester	RU
26	Sam Houston State U.	4.9	Public	Rural	Semester	RU
27	Southwest Texas State U.	4.9	Public	Urban	Semester	RU
28	Stephen F. Austin State U.	4.9	Public	Rural	Semester	RU
29	Tarleton State U.	4.9	Public	Rural	Semester	RU
30	Texas A&M U.	4.9	Public	Urban	Semester	NU
31	Texas A&M at Galveston	4.9	Public	Urban	Semester	RU
32	Texas Tech U.	4.9	Public	Urban	Semester	RU
33	U. of Texas at Austin	4.9	Public	Urban	Semester	NU
34	U. of Mary Hardin-Baylor	5	Private	Suburban	Semester	RLA
35	Houston Baptist U.	5.4	Private	Urban	Quarter	RU
36	Texas Wesleyan U.	5.8	Private	Urban	Semester	RU
37	U. of Texas at Dallas	5.8	Public	Suburban	Semester	RU
38	West Texas State U.	5.9	Public	Rural	Semester	RU
39	Hardin Simmons U.	6	Private	Urban	Semester	RU
40	Concordia Lutheran C.	6.4	Private	Urban	Semester	RLA
41	Lubbock Christian U.	6.4	Private	Urban	Semester	RLA
42	McMurry U.	6.6	Private	Urban	4-1-4	RLA
43	Southwestern Adventist	7	Private	Rural	Semester	RLA
44	Abilene Christian U.	7.2	Private	Suburban	Semester	RU
45	Texas Lutheran C.	7.4	Private	Urban	Semester	RLA
46	St. Mary's U.	7.7	Private	Urban	Semester	RU
47	Texas Southern U.	7.9	Public	Urban	Semester	RU
48	Our Lady of the Lake	8	Private	Urban	Semester	RU
49	Texas Christian U.	8	Private	Urban	Semester	NU
50	U. of St. Thomas	8	Private	Urban	Semester	RU
51	Incarnate Word C.	8.3	Private	Urban	Semester	RLA
52	LeTourneau U.	8.3	Private	Urban	Semester	RLA
53	Rice U.	8.5	Private	Urban	Semester	NU
54	Schreiner C.	8.6	Private	Rural	4-1-4	RLA
55	St. Edward's U.	8.8	Private	Urban	Semester	RU
56	U. of Dallas	10.3	Private	Suburban	Semester	NLA
57	Baylor U.	10.4	Private	Urban	Semester	NU
58	Austin C.	10.7	Private	Urban	4-1-4	NLA
59	Southwestern U.	11	Private	Suburban	Semester	RLA
60	Trinity U.	11.6	Private	Urban	Semester	RU
61	Southern Methodist U.	12	Private	Suburban	Semester	NU

FIGURE 2.2.EXCEL Sorted Texas out-of-state tuition data.

2.3 The following data represent the retail price (in dollars) of a sample of 39 different brands of bathroom scales:

SCALES.TXT

50	50	50	28	65	40	50	22	32	30
79	50	22	20	35	24	25	120	35	35
65	20	14	25	24	48	15	10	17	50
25	22	60	30	12	30	10	12	20	

Source: "Bathroom Scales," Copyright 1993 by Consumers Union of United States, Inc., Yonkers, N.Y. 10703. Adapted by permission from *Consumer Reports*, January 1993, pp. 34–35. Although these data sets originally appeared in *Consumer Reports*, the selective adaptation and resulting conclusions presented are those of the authors and are not sanctioned or endorsed in any way by Consumers Union, the publisher of *Consumer Reports*.

(a) Develop the ordered array.
(b) Form the stem-and-leaf display.

2.4 The following data are the book values (in dollars) (i.e., net worth divided by number of outstanding shares) for a random sample of 50 stocks from the New York Stock Exchange (NYSE):

STOCK1.TXT

7	9	8	6	12	6	9	15	9	16
8	5	14	8	7	6	10	8	11	4
10	6	16	5	10	12	7	10	15	7
10	8	8	10	18	8	10	11	7	10
7	8	15	23	13	9	8	9	9	13

(a) Develop the ordered array.
(b) Form the stem-and-leaf display.

2.5 A medical doctor speaking on a late-night television show conjectures that "cancer appears to be more prevalent in states with large urban populations and in states in the eastern part of the United States." The following data represent the incidence rate of cancer (i.e., reported incidence per 100,000 population) in all 50 states during a recent year:

CANCER.TXT

State	Incidence of Cancer per 100,000 Population	State	Incidence of Cancer per 100,000 Population
Alabama	433	Kentucky	414
Alaska	442	Louisiana	422
Arizona	360	Maine	391
Arkansas	383	Maryland	491
California	366	Massachusetts	443
Colorado	282	Michigan	454
Connecticut	434	Minnesota	366
Delaware	500	Mississippi	438
Florida	367	Missouri	390
Georgia	406	Montana	372
Hawaii	371	Nebraska	336
Idaho	307	Nevada	422
Illinois	402	New Hampshire	403
Indiana	438	New Jersey	464
Iowa	377	New Mexico	375
Kansas	345	New York	329

(continued)

(continued)

State	Incidence of Cancer per 100,000 Population	State	Incidence of Cancer per 100,000 Population
North Carolina	355	Tennessee	408
North Dakota	408	Texas	313
Ohio	463	Utah	229
Oklahoma	326	Vermont	376
Oregon	396	Virginia	440
Pennsylvania	442	Washington	364
Rhode Island	445	West Virginia	409
South Carolina	418	Wisconsin	398
South Dakota	348	Wyoming	238

Source: National Cancer Institute.

(a) Develop the ordered array.
(b) Form the stem-and-leaf display.

2.6 The following data represent the type (creamy versus chunky), score (0 = poor, 100 = excellent), cost (in cents), and amount of sodium (in mgs) of a sample of 37 brands of peanut butter:

PEANUT.XLS

Product	Type	Score	Cost (¢)	Sodium (mgs)
Jif	Creamy	68	22	220
Smucker's Natural	Creamy	65	27	15
Deaf Smith Arrowhead Mills	Creamy	62	32	0
Adams 100% Natural	Creamy	56	26	0
Adams	Creamy	56	26	168
Skippy	Creamy	56	19	225
Laura Scudder's All Natural	Creamy	53	26	165
Kroger	Creamy	50	14	240
Country Pure Brand (Safeway)	Creamy	50	21	225
NuMade (Safeway)	Creamy	45	20	187
Peter Pan	Creamy	44	21	225
Peter Pan (2)	Creamy	41	22	3
A&P	Creamy	40	12	225
Hollywood Natural	Creamy	40	32	15
Food Club	Creamy	39	17	225
Pathmark	Creamy	36	9	255
Lady Lee (Lucky Stores)	Creamy	30	16	225
Albertsons	Creamy	30	17	225
Shur Fine (Shurfine Central Corp.)	Creamy	22	16	225
Smucker's Natural	Chunky	80	27	15
Jif	Chunky	75	23	162
Skippy	Chunky	75	21	211
Adams 100% Natural	Chunky	62	26	0
Deaf Smith Arrowhead Mills	Chunky	62	32	0
Country Pure Brand (Safeway)	Chunky	62	21	195
Laura Scudder's All Natural	Chunky	56	24	165
Smucker's Natural	Chunky	53	26	188

(continued)

(continued)

Product	Type	Score	Cost (¢)	Sodium (mgs)
Food Club	Chunky	52	17	195
Kroger	Chunky	50	14	255
A&P	Chunky	47	11	225
Peter Pan	Chunky	47	22	180
NuMade (Safeway)	Chunky	42	21	208
Health Valley 100% Natural	Chunky	42	34	3
Lady Lee (Lucky Stores)	Chunky	40	16	225
Albertsons	Chunky	36	17	225
Pathmark	Chunky	34	9	210
Shur Fine (Shurfine Central Corp.)	Chunky	34	16	195

Source: "Peanut Butter," Copyright 1990 by Consumers Union of United States, Inc., Yonkers, N.Y. 10703. Adapted by permission of *Consumer Reports,* September 1990, p. 590. Although these data sets originally appeared in *Consumer Reports,* the selective adaptation and resulting conclusions presented are those of the authors and are not sanctioned or endorsed in any way by Consumers Union, the publisher of *Consumer Reports.*

For each of the three variables (score, cost, and sodium)
(a) Use the Sort command in Microsoft Excel to develop the ordered array.
(b) Form the stem-and-leaf display.

SHAMPOO.TXT

2.7 The following data are the cost per ounce (in cents) for random samples of 31 conventional shampoos labeled for "normal" hair and 29 shampoos labeled for "fine" hair:

Normal Hair					Fine Hair				
79	63	19	9	37	69	9	23	22	8
49	20	16	55	69	12	32	12	18	74
23	14	9	87	44	19	63	49	37	55
13	16	23	20	64	85	44	87	17	11
28	18	32	81	85	23	50	65	51	35
47	50	8	13	21	14	20	28	8	
9									

Source: "Shampoos," Copyright 1992 by Consumers Union of United States, Inc., Yonkers, N.Y. 10703. Adapted by permission from *Consumer Reports,* June 1992, pp. 400–401. Although these data sets originally appeared in *Consumer Reports,* the selective adaptation and resulting conclusions presented are those of the authors and are not sanctioned or endorsed in any way by Consumers Union, the publisher of *Consumer Reports.*

(a) Develop the ordered array for each data set.
(b) Form the stem-and-leaf display for each data set.

2.3 TABULATING NUMERICAL DATA: THE FREQUENCY DISTRIBUTION

Using either the raw data, an ordered array, or a revised stem-and-leaf display of out-of-state tuition rates for the 60 colleges and universities in Texas (see Tables 2.1 and 2.2 on pages 52 and 53 and Figure 2.1 on page 53), the analyst wishes to construct the appropriate tables and charts that will enhance the report she is preparing for the marketing manager of the college advisory service.

Regardless of whether an ordered array or a stem-and-leaf display is selected for *organizing* the data, as the number of observations gets large, it becomes necessary to further condense the data into appropriate summary tables. Thus, we may wish to arrange the data into **class groupings** (i.e., categories) according to conveniently established divisions of the range of the observations. Such an arrangement of data in tabular form is called a frequency distribution.

A **frequency distribution** is a summary table in which the data are arranged into conveniently established, numerically ordered class groupings or categories.

When the observations are *grouped* or condensed into frequency distribution tables, the process of data analysis and interpretation is made much more manageable and meaningful. In such summary form the major data characteristics can be approximated, thus compensating for the fact that when the data are so grouped, the initial information pertaining to individual observations that was previously available is lost through the grouping or condensing process.

In constructing the frequency distribution table, attention must be given to

1. Selecting the appropriate *number* of class groupings for the table
2. Obtaining a suitable *class interval* or *width* of each class grouping
3. Establishing the *boundaries* of each class grouping to avoid overlapping

2.3.1 Selecting the Number of Classes

The number of class groupings to be used is primarily dependent on the number of observations in the data. Larger numbers of observations require a larger number of class groups. In general, however, the frequency distribution should have at least five class groupings, but no more than 15. If there are not enough class groupings or if there are too many, little information would be obtained. As an example, a frequency distribution having but one class grouping that spans the entire range of tuition rates could be formed as follows:

Tuition Rates (in $000)	Number of Schools
2.0–13.0	60
Total	60

From such a summary table, however, no additional information is obtained that was not already known from scanning either the raw data or the ordered array. A table with too much data concentration is not meaningful. The same would be true at the other extreme—if a table had too many class groupings, there would be an underconcentration of data, and very little would be learned.

2.3.2 Obtaining the Class Intervals

When developing the frequency distribution table, it is desirable to have each class grouping of equal width. To determine the width of each class, the *range* of the data is divided by the number of class groupings desired:

$$\text{Width of interval} \cong \frac{\text{range}}{\text{number of desired class groupings}} \quad (2.1)$$

Since there are only 60 observations in our tuition rate data, we decided that six class groupings would be sufficient. From the ordered array in Table 2.2 (page 53), the range is computed as 12.0 − 2.4 = 9.6 thousand dollars, and, using Equation (2.1), the **width of the class interval** is approximated by

$$\text{Width of interval} \cong \frac{9.6}{6} = 1.6 \text{ thousand dollars}$$

For convenience and ease of reading, the selected interval or width of each class grouping is rounded to 2.0 thousand dollars.

2.3.3 Establishing the Boundaries of the Classes

To construct the frequency distribution table, it is necessary to establish clearly defined **class boundaries** for each class grouping so that the observations either in raw form or in an ordered array can be properly tallied. Overlapping of classes must be avoided.

Since the width of each class interval for the tuition rate data has been set at 2.0 thousand dollars, the boundaries of the various class groupings must be established so as to include the entire range of observations. Whenever possible, these boundaries should be chosen to facilitate the reading and interpreting of data. Thus, the first class interval ranges from 2.0 to under 4.0, the second from 4.0 to under 6.0, and so on. The data in their raw form (Table 2.1) or from the ordered array (Table 2.2) are then tallied into each class as shown:

Tuition Rates (in $000)	Tallies	Frequency
2.0 but less than 4.0	ℍℍ ℍℍ ///	13
4.0 but less than 6.0	ℍℍ ℍℍ ℍℍ ℍℍ ////	24
6.0 but less than 8.0	ℍℍ ////	9
8.0 but less than 10.0	ℍℍ ///	8
10.0 but less than 12.0	ℍℍ	5
12.0 but less than 14.0	/	1
Total		60

By establishing the boundaries of each class as above, all 60 observations have been tallied into six classes, each having an interval width of 2.0 thousand dollars, without overlapping. The frequency distribution is presented in Table 2.3.

Table 2.3 Frequency distribution of the Texas out-of-state tuition rates.

Tuition Rates (in $000)	Number of Schools
2.0 but less than 4.0	13
4.0 but less than 6.0	24
6.0 but less than 8.0	9
8.0 but less than 10.0	8
10.0 but less than 12.0	5
12.0 but less than 14.0	1
Total	60

Source: Table 2.1.

The main advantage of using this summary table is that the major data characteristics become immediately clear to the reader. For example, we see from Table 2.3 that the *approximate range* of the 60 tuition rates is from 2.0 to 14.0 thousand dollars, with out-of-state tuition at most Texas schools tending to cluster between 4.0 and 6.0 thousand dollars.

On the other hand, the major disadvantage of this summary table is that we cannot know how the individual values are distributed within a particular class interval without access to the original data. Thus, for the five schools with out-of-state tuition rates between 10.0 and 12.0 thousand dollars, it is not clear from Table 2.3 whether the values are distributed throughout the interval, cluster near 10.0 thousand dollars, or cluster near 12.0 thousand dollars. The class midpoint, however, is the value used to represent all the data summarized into a particular interval.

> The **class midpoint** is the point halfway between the boundaries of each class and is representative of the data within that class.

The class midpoint for the interval "2.0 but less than 4.0" is 3.0 thousand dollars. (The other class midpoints are, respectively, 5.0, 7.0, 9.0, 11.0, and 13.0 thousand dollars.)

2.3.4 Subjectivity in Selecting Class Boundaries

The selection of class boundaries for frequency distribution tables is highly subjective. Hence, for data sets that do not contain many observations, the choice of a particular set of class boundaries over another might yield a different picture to the reader. For example, for the tuition rate data, using a class-interval width of 2.5 thousand dollars instead of 2.0 (as was used in Table 2.3) may cause shifts in the way in which the observations distribute among the classes. This is particularly true if the number of observations in the data set is not very large.

Such shifts in data concentration do not occur only because the width of the class interval is altered. We may keep the interval width at 2.0 thousand dollars but choose different lower and upper class boundaries. Such manipulation may also cause shifts in the way in which the data distribute—especially if the size of the data set is not very large. Fortunately, as the number of observations in a data set increases, alterations in the selection of class boundaries affect the concentration of data less and less.

Problems for Section 2.3

Note: *To use Microsoft Excel to solve these problems, refer to Section 2.7.*

2.8 A random sample of 50 executive vice-presidents is selected from various public relations firms in the United States, and the annual salaries of these company officers are obtained. The salaries range from $52,000 to $137,000. Set up the class boundaries for a frequency distribution
(a) If 5 class intervals are desired
(b) If 6 class intervals are desired
(c) If 7 class intervals are desired
(d) If 8 class intervals are desired

2.9 If the asking price of one-bedroom cooperative and condominium apartments in Queens, a borough of New York City, varies from $103,000 to $295,000,
(a) Indicate the class boundaries of 10 classes into which these values can be grouped.
(b) What class-interval width did you choose?
(c) What are the 10 class midpoints?

• 2.10 The raw data displayed here are the electric and gas utility charges during the month of July 1995 for a random sample of 50 three-bedroom apartments in Manhattan:

UTILITY.TXT

Raw Data on Utility Charges ($)

96	171	202	178	147	102	153	197	127	82
157	185	90	116	172	111	148	213	130	165
141	149	206	175	123	128	144	168	109	167
95	163	150	154	130	143	187	166	139	149
108	119	183	151	114	135	191	137	129	158

(a) Form a frequency distribution
  (1) Having 5 class intervals
  (2) Having 6 class intervals
  (3) Having 7 class intervals
  [*Hint:* To help you decide how best to set up the class boundaries, you should first place the raw data either in a stem-and-leaf display (by letting the leaves be the trailing digits) or in an ordered array.]
(b) Form a frequency distribution having 7 class intervals with the following class boundaries: $80 but less than $100, $100 but less than $120, and so on.

2.11 Construct a frequency distribution from the book value data in Problem 2.4 on page 58.

2.12 Construct a frequency distribution from the cancer incidence data in Problem 2.5 on pages 58–59.

2.13 Construct separate frequency distributions for each of the three numerical variables (score, cost, and sodium) from the peanut butter data in Problem 2.6 on pages 59–60.

2.14 Given the ordered arrays in the accompanying table dealing with the lengths of life (in hours) of a sample of forty 100-watt light bulbs produced by Manufacturer A and a sample of forty 100-watt light bulbs produced by Manufacturer B:

Ordered arrays of length of life of two brands of 100-watt light bulbs (in hours).

Manufacturer A				
684	697	720	773	821
831	835	848	852	852
859	860	868	870	876
893	899	905	909	911
922	924	926	926	938
939	943	946	954	971
972	977	984	1005	1014
1016	1041	1052	1080	1093

Manufacturer B				
819	836	888	897	903
907	912	918	942	943
952	959	962	986	992
994	1004	1005	1007	1015
1016	1018	1020	1022	1034
1038	1072	1077	1077	1082
1096	1100	1113	1113	1116
1153	1154	1174	1188	1230

(a) Form the frequency distribution for each brand. (*Hint:* For purposes of comparison, choose class-interval widths of 100 hours for each distribution.)

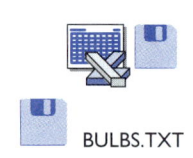

(b) For purposes of answering Problems 2.19, 2.24, and 2.30, form the frequency distribution for each brand according to the following schema [if you have not already done so in part (a) of this problem]:

Manufacturer A: 650 but less than 750, 750 but less than 850, and so on

Manufacturer B: 750 but less than 850, 850 but less than 950, and so on

(c) Change the class-interval width in (b) to 50 so that you have intervals from 650 to under 700, 700 to under 750, 750 to under 800, and so on.

2.4 TABULATING NUMERICAL DATA: THE RELATIVE FREQUENCY DISTRIBUTION AND PERCENTAGE DISTRIBUTION

The frequency distribution is a summary table into which the original data are grouped to facilitate data analysis. To enhance the analysis, however, it is almost always desirable to form either the relative frequency distribution or the percentage distribution, depending on whether we prefer proportions or percentages. These two equivalent distributions are shown as Tables 2.4 and 2.5, respectively.

The **relative frequency distribution** depicted in Table 2.4 is formed by dividing the frequencies in each class of the frequency distribution (Table 2.3 on page 62) by the total number of observations. A **percentage distribution** (Table 2.5) may then be formed by multiplying each relative frequency or proportion by 100.0. Thus, from Table 2.4 it is clear that the proportion of schools in Texas with out-of-state tuition rates from 12.0 to under 14.0 thousand dollars is .017, while from Table 2.5 it can be seen that 1.7% of the schools have such tuition rates.

Working with a base of 1 for proportions or 100.0 for percentages is usually more meaningful than using the frequencies themselves. Indeed, the use of the relative frequency distribution or percentage distribution becomes essential whenever one set of data is being compared with other sets of data, especially if the numbers of observations in each set differ.

As a case in point, let us suppose that a personnel manager wanted to compare daily absenteeism among the clerical workers in two department stores. If, on a given day, 6 clerical workers out of 50 in Store A were absent and 3 clerical workers out of 10 in Store B were absent, what conclusions can be drawn? It is inappropriate to say that *more* absenteeism occurred in Store A. Although there were twice as many absences in Store A as there were in

Table 2.4 Relative frequency distribution of the Texas out-of-state tuition rates.

Tuition Rates (in $000)	Proportion of Schools
2.0 but less than 4.0	.217
4.0 but less than 6.0	.400
6.0 but less than 8.0	.150
8.0 but less than 10.0	.133
10.0 but less than 12.0	.083
12.0 but less than 14.0	.017
Total	1.000

Source: Data are taken from Table 2.3 on page 62.

Table 2.5 Percentage distribution of the Texas out-of-state tuition rates.

Tuition Rates (in $000)	Percentage of Schools
2.0 but less than 4.0	21.7
4.0 but less than 6.0	40.0
6.0 but less than 8.0	15.0
8.0 but less than 10.0	13.3
10.0 but less than 12.0	8.3
12.0 but less than 14.0	1.7
Total	100.0

Source: Data are taken from Table 2.3 on page 62.

Store B, there were also five times as many clerical workers employed in Store A. Hence, in these types of comparisons, we must formulate our conclusions from the *relative rates* of absenteeism, not from the actual counts. Thus, we can state that the absenteeism rate is two and a half times higher in Store B (30.0%) than it is in Store A (12.0%).

Now suppose that, when developing her report for the marketing manager of the college advisory service, our analyst decides to compare the 60 out-of-state Texas tuition rates with those reported from 45 higher education institutions in the state of North Carolina. Table 2.6 displays information on the out-of-state resident tuition rate for *each* of the 45 North Carolina colleges and universities (see Special Data Set 1 of Appendix D on page D3).

To compare the out-of-state tuition rates from the 60 Texas schools with those from the 45 North Carolina schools, we develop a percentage distribution for the latter group. This new table will then be compared with Table 2.5.

Table 2.7 depicts both the frequency distribution and the percentage distribution of the tuition rates charged to out-of-state residents by the 45 North Carolina schools. This table has been constructed instead of two separate tables to save space. Note that the class groupings selected in Table 2.7 match, where possible, those selected in Table 2.3 for the Texas schools. The boundaries of the classes should match or be multiples of each other in order to facilitate comparisons.

Using the percentage distributions of Tables 2.5 and 2.7, it is now meaningful to compare the schools in the two states in terms of the tuition rates charged to out-of-state residents. From the

NCC&U-T.TXT

Table 2.6 Raw data pertaining to tuition rates (in $000) for out-of-state residents at 45 colleges and universities in North Carolina.

6.5	4.0	7.1	8.3	5.4	7.6	9.0	15.7	16.7
6.4	5.0	8.5	5.7	7.7	7.2	12.4	7.1	5.5
9.7	4.4	7.0	6.3	8.3	6.9	5.7	7.6	7.9
7.9	6.0	8.2	10.4	9.9	3.9	9.8	8.2	5.6
7.9	6.4	7.4	7.0	13.0	8.7	6.4	6.7	7.4

Source: See Special Data Set 1, Appendix D, page D3, taken from "America's Best Colleges, 1994 College Guide," *U.S. News & World Report*, extracted from College Counsel 1993 of Natick, Mass. Reprinted by special permission, *U.S. News & World Report*, © 1993 by *U.S. News & World Report* and by College Counsel.

Table 2.7 Frequency distribution and percentage distribution of North Carolina out-of-state tuition rates.

Tuition Rates (in $000)	Number of Schools	Percentage of Schools
2.0 but less than 4.0	1	2.2
4.0 but less than 6.0	8	17.8
6.0 but less than 8.0	21	46.7
8.0 but less than 10.0	10	22.2
10.0 but less than 12.0	1	2.2
12.0 but less than 14.0	2	4.4
14.0 but less than 16.0	1	2.2
16.0 but less than 18.0	1	2.2
Totals	45	99.9*

*Error due to rounding.
Source: Data are taken from Table 2.6.

two tables it is apparent that the tuition rates are generally lower in Texas than in North Carolina. For example, in Texas, tuition rates are typically clustering between 4.0 and 6.0 thousand dollars (i.e., 40.0% of the schools) while in North Carolina the tuition rates are typically clustering between 6.0 and 8.0 thousand dollars (or 46.7% of the schools). We can also approximate the *ranges* in tuition rates from the tables. In North Carolina, the range in tuition rates is approximated to be 16.0 thousand dollars (i.e., the difference between 18.0, the upper boundary of the last class, and 2.0, the lower boundary of the first class), while in Texas the range is approximated as 12.0 thousand dollars (or 14.0 – 2.0). Other descriptive summary measures that would enhance a comparative analysis of the tuition rates between the two states will be discussed in Chapter 3.

Problems for Section 2.4

Note: *To use Microsoft Excel to solve these problems, refer to Section 2.7.*

- 2.15 Form the percentage distribution from the frequency distribution developed in Problem 2.10(b) on page 64 regarding utility charges.
- 2.16 Form the percentage distribution from the frequency distribution developed in Problem 2.11 on page 64 regarding book values of companies listed on the NYSE.
  2.17 Form the percentage distribution from the frequency distribution developed in Problem 2.12 on page 64 regarding cancer incidence.
  2.18 Form the percentage distributions corresponding to the frequency distributions for each of the three numerical variables (score, cost, and sodium) developed in Problem 2.13 on page 64 regarding peanut butter characteristics.

S-2-18.XLS

  2.19 Form the percentage distributions from the frequency distributions developed in Problem 2.14(b) on page 65 concerning the life of light bulbs manufactured by two competing companies, A and B.

2.5 GRAPHING NUMERICAL DATA: THE HISTOGRAM AND POLYGON

It is often said that "one picture is worth a thousand words." Indeed, statisticians often employ graphic techniques to more vividly describe sets of data. In particular, histograms and polygons are used to describe numerical data that have been grouped into frequency, relative frequency, or percentage distributions.

2.5.1 Histograms

Histograms are vertical bar charts in which the rectangular bars are constructed at the boundaries of each class.

When plotting histograms, the random variable or phenomenon of interest is displayed along the horizontal axis; the vertical axis represents the number, proportion, or percentage of observations per class interval—depending on whether the particular histogram is, respectively, a frequency histogram, a relative frequency histogram, or a percentage histogram.

Vertical Axis Label	↔	Type of Chart
Number of observations	↔	Frequency histogram (or polygon)
Proportion of observations	↔	Relative frequency histogram (or polygon)
Percentage of observations	↔	Percentage histogram (or polygon)

A percentage histogram is depicted in Figure 2.5 for the out-of-state tuition rates at all 60 colleges and universities in Texas. It is interesting to note the close visual relationship portrayed by the stem-and-leaf display and the histogram. Look at Figure 2.4 on page 55 and our histogram in Figure 2.5. If we were to rotate the stem-and-leaf display 90° (i.e., hold our book sideways), a frequency histogram would be depicted so that its class groupings would be represented by the stems and its vertical bars would be represented by individual leaves on each stem.

When comparing two or more sets of data, neither stem-and-leaf displays nor histograms can be constructed on the same graph. Superimposing the vertical bars of one histogram on another would cause difficulty in interpretation. For such cases, it is necessary to construct relative frequency or percentage polygons.

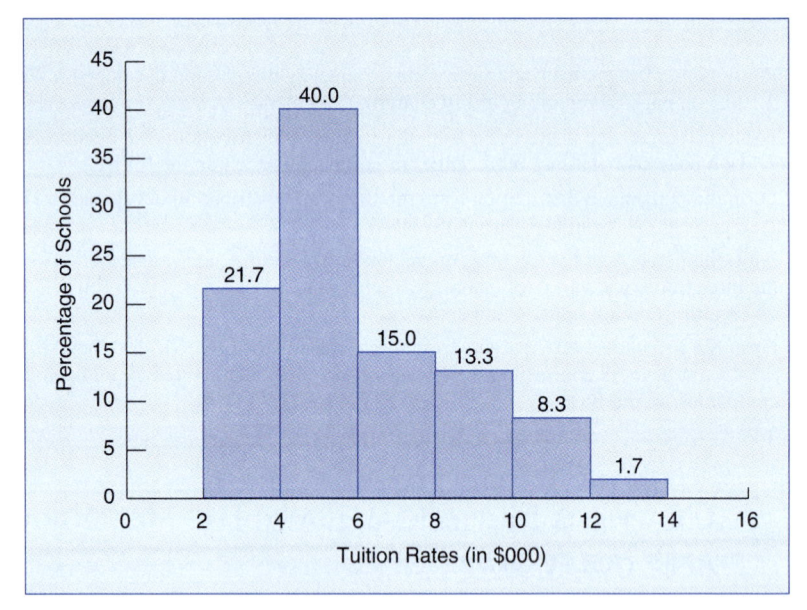

FIGURE 2.5 Percentage histogram of the Texas out-of-state tuition rates.
Source: Data are taken from Table 2.5 on page 65.

2.5.2 Polygons

As with histograms, when plotting polygons the phenomenon of interest is displayed along the horizontal axis, and the vertical axis represents the number, proportion, or percentage of observations per class interval.

The **percentage polygon** is formed by letting the midpoint of each class represent the data in that class and then connecting the sequence of midpoints at their respective class percentages.

Because consecutive midpoints are connected by a series of straight lines, the polygon is sometimes jagged in appearance. However, when dealing with a very large set of data, if we were to make the boundaries of the classes in its frequency distribution closer together (and thereby increase the number of classes in that distribution), the jagged lines of the polygon would "smooth out."

Figure 2.6 shows the percentage polygon for the out-of-state tuition rates at all 60 Texas schools, and Figure 2.7 compares the percentage polygons for the tuition rates at the 60 Texas schools versus the 45 North Carolina schools. The differences in the structure of the two distributions, previously discussed when comparing Tables 2.5 and 2.7, are clearly indicated here.

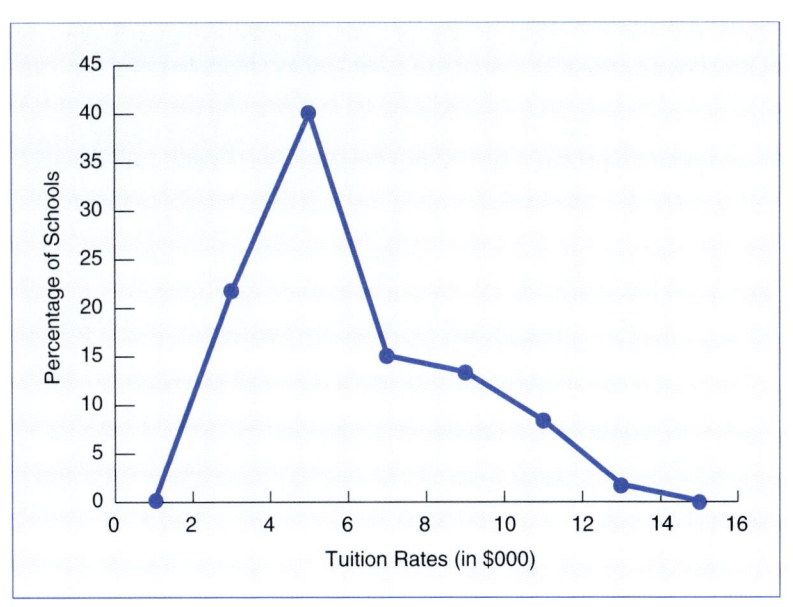

FIGURE 2.6 Percentage polygon of the Texas out-of-state tuition rates.
Source: Data are taken from Table 2.5.

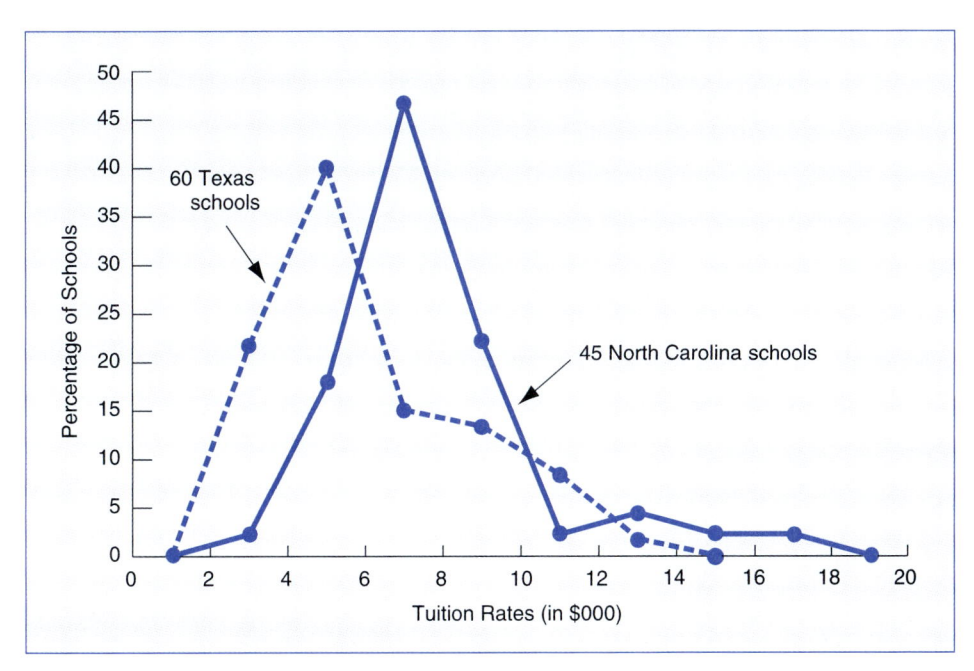

FIGURE 2.7 Percentage polygons of out-of-state tuition rates for Texas and North Carolina schools. *Source:* Data are taken from Tables 2.5 and 2.7.

- **Polygon Construction** Notice that the polygon is a representation of the shape of the particular distribution. Since the area under the percentage distribution (the entire curve) must be 100.0%, it is necessary to connect the first and last midpoints with the horizontal axis so as to enclose the area of the observed distribution. In Figure 2.6, this is accomplished by connecting the first observed midpoint with the midpoint of a "fictitious preceding" class (i.e., 1.0 thousand dollars) having 0.0% observations and by connecting the last observed midpoint with the midpoint of a "fictitious succeeding" class (i.e., 15.0 thousand dollars) having 0.0% observations.

2.5 Graphing Numerical Data: The Histogram and Polygon

Notice, too, that when polygons (Figure 2.6) or histograms (Figure 2.5) are constructed, the vertical axis must show the true zero or "origin" so as not to distort or otherwise misrepresent the character of the data. The horizontal axis, however, does not need to specify the zero point for the phenomenon of interest. For aesthetic reasons, the range of the random variable should constitute the major portion of the chart, and, when zero is not included, "breaks" (─⋀─) in the axis are appropriate.

Problems for Section 2.5

Note: To use Microsoft Excel to solve these problems, refer to Section 2.7.

UTILITY.TXT

2.20 From the percentage distribution developed in Problem 2.15 on page 67 regarding utility charges
(a) Plot the percentage histogram.
(b) Plot the percentage polygon.

STOCK1.TXT

2.21 From the percentage distribution developed in Problem 2.16 on page 67 regarding book values of companies listed on the NYSE
(a) Plot the percentage histogram.
(b) Plot the percentage polygon.

CANCER.TXT

2.22 From the percentage distribution developed in Problem 2.17 on page 67 regarding cancer incidence
(a) Plot the percentage histogram.
(b) Plot the percentage polygon.

S-2-23.XLS

2.23 From the percentage distributions developed in Problem 2.18 on page 67 for each of the three numerical variables (score, cost, and sodium) regarding peanut butter characteristics
(a) Plot the respective percentage histograms.
(b) Plot the respective percentage polygons.

BULBS.TXT

2.24 From the percentage distributions developed in Problem 2.19 on page 67 regarding life of light bulbs
(a) Plot the percentage histograms on separate graphs.
(b) Plot the percentage polygons on one graph.

2.6 CUMULATIVE DISTRIBUTIONS AND CUMULATIVE POLYGONS

Two other useful methods of data presentation that facilitate analysis and interpretation are the cumulative distribution tables and the cumulative polygon charts. Both of these may be developed from the frequency distribution table, the relative frequency distribution table, or the percentage distribution table.

2.6.1 The Cumulative Percentage Distribution

Depending on our individual preference for proportions or percentages, when comparing two or more sets of data of differing size, we select either the relative frequency distribution or the percentage distribution. Since we already have the percentage distributions of out-of-state tuition rates at 60 Texas schools and at 45 North Carolina schools in Tables 2.5 and 2.7 (pages 65 and 66), we can use these tables to construct the respective cumulative percentage distributions. See Tables 2.8 and 2.9.

A **cumulative percentage distribution table** is constructed by first recording the lower boundaries of each class from the percentage distribution and then inserting an extra boundary at the end. We compute the cumulative percentages in the *"less than"* column by determining the percentage of observations less than each of the stated boundary values. Thus from

Table 2.8 Cumulative percentage distribution of out-of-state tuition rates for the 60 Texas schools.

Tuition Rates (in $000)	Percentage of Schools "Less Than" Indicated Value
2.0	0.0
4.0	21.7
6.0	61.7
8.0	76.7
10.0	90.0
12.0	98.3
14.0	100.0

Source: Data are taken from Table 2.5.

Table 2.9 Cumulative percentage distribution of out-of-state tuition rates for the 45 North Carolina schools.

Tuition Rates (in $000)	Percentage of Schools "Less Than" Indicated Value
2.0	0.0
4.0	2.2
6.0	20.0
8.0	66.7
10.0	88.9
12.0	91.1
14.0	95.6
16.0	97.8
18.0	100.0

Source: Data are taken from Table 2.7.

Table 2.5, we see that 0.0% of the out-of-state tuition rates in Texas institutions are less than 2.0 thousand dollars; 21.7% of the tuition rates are less than 4.0 thousand dollars; 61.7% of the tuition rates are less than 6.0 thousand dollars; and so on until all (100.0%) of the tuition rates are less than 14.0 thousand dollars. This cumulating process is observed in Table 2.10.

Table 2.10 Forming the cumulative percentage distribution.

Out-of-State Tuition Rates (in $000)	Percentage of Schools in Class Interval	Percentage of Schools "Less Than" Lower Boundary of Class Interval
2.0 but less than 4.0	21.7	0.0
4.0 but less than 6.0	40.0	21.7
6.0 but less than 8.0	15.0	61.7 = 21.7 + 40.0
8.0 but less than 10.0	13.3	76.7 = 21.7 + 40.0 + 15.0
10.0 but less than 12.0	8.3	90.0 = 21.7 + 40.0 + 15.0 + 13.3
12.0 but less than 14.0	1.7	98.3 = 21.7 + 40.0 + 15.0 + 13.3 + 8.3
14.0 but less than 16.0	0.0	100.0 = 21.7 + 40.0 + 15.0 + 13.3 + 8.3 + 1.7

2.6.2 Cumulative Percentage Polygon

To construct a **cumulative percentage polygon** (also known as an **ogive**), we note that the phenomenon of interest—tuition rates—is again plotted on the horizontal axis, while the cumulative percentages (from the "*less than*" column) are plotted on the vertical axis. At each lower boundary, we plot the corresponding (cumulative) percentage value from the listing in the cumulative percentage distribution. We then connect these points with a series of straight-line segments.

Figure 2.8 on page 72 illustrates the cumulative percentage polygon of the Texas out-of-state tuition rates. The major advantage of the ogive over other charts is the ease with which we can interpolate between the plotted points.

● **Approximating the Percentages** As one example, the analyst at the college advisory service might wish to approximate the percentage of Texas colleges and universities

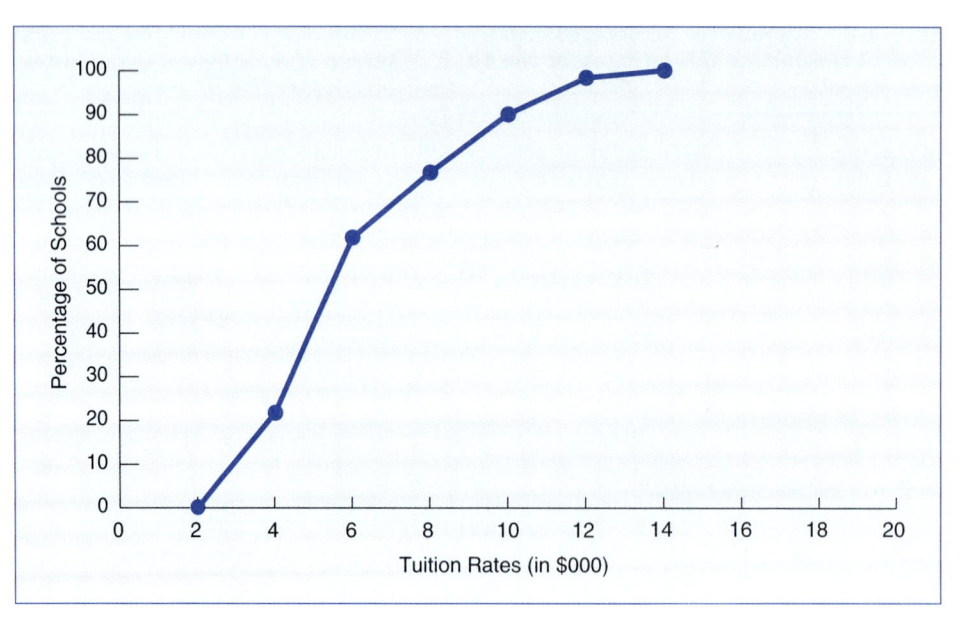

FIGURE 2.8 Cumulative percentage polygon of the Texas out-of-state tuition rates.
Source: Data are taken from Table 2.8.

that charge an out-of-state tuition rate below a specified amount, say 7.0 thousand dollars. To accomplish this, a vertical line is projected upward at 7.0 until it intersects the "less than" curve. The desired percentage is then approximated by reading horizontally from the point of intersection to the percentage indicated on the vertical axis. In this case, approximately 69.2% of the Texas schools have tuition rates under 7.0 thousand dollars. (This, of course, implies that about 30.8% of the schools have tuition rates of at least 7.0 thousand dollars.)

● **Approximating the Values** Even more important, the analyst, when preparing her report for the marketing manager of the college advisory service, may also wish to approximate various tuition rates that correspond to particular cumulative percentages. For example, 25.0% of all Texas schools have out-of-state tuition rates below what amount? To determine this, a horizontal line is drawn from the specified cumulative percentage point (25.0) until it intersects the "less than" curve. The desired tuition rate is then approximated by dropping a perpendicular (a vertical line) at the point of intersection to the horizontal axis. From Figure 2.8, we note that this rate is approximately 4.2 thousand dollars. Other percentage points commonly considered for such analysis (see Chapter 3) are the 50.0% value and the 75.0% value.

● **Comparing Two or More Cumulative Distributions** Such approximations as these are extremely helpful when comparing two or more sets of data. Figure 2.9 depicts the cumulative percentage polygons of out-of-state tuition rates for both the Texas schools and the North Carolina schools.

From Figure 2.9, we note that, in general, the Texas ogive is drawn to the left of the North Carolina ogive. For example, in Texas, 25% of all tuition rates are below 4.2 thousand dollars, while in North Carolina, we see that 25% of all tuition rates are below 6.1 thousand dollars. In Texas, 50% of all tuition rates are below 5.4 thousand dollars, while in North Carolina, 50% of all tuition rates are below 7.2 thousand dollars. Furthermore, in Texas, 75% of all tuition rates are below 7.7 thousand dollars, while in North Carolina, we see that 75% of all tuition rates are below 8.7 thousand dollars. These comparisons enable us to confirm our earlier impression that out-of-state tuition rates are lower in Texas than in North Carolina.

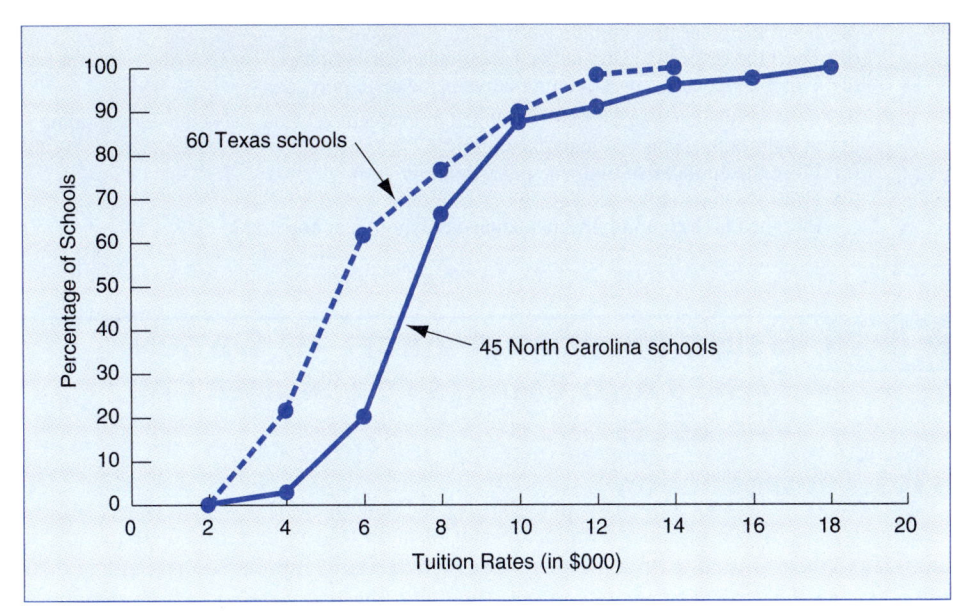

FIGURE 2.9 Cumulative percentage polygons of out-of-state tuition rates for the Texas and North Carolina schools. *Source:* Data are taken from Tables 2.5 and 2.7.

Problems for Section 2.6

Note: *To use Microsoft Excel to solve these problems, refer to Section 2.7.*

2.25 Examine Figure 2.9.
   (a) 10.0% of the out-of-state tuition rates in each state are below what amounts?
   (b) 40.0% of the out-of-state tuition rates in each state are below what amounts?
   (c) 60.0% of the out-of-state tuition rates in each state are below what amounts?
   (d) 90.0% of the out-of-state tuition rates in each state are below what amounts?
   (e) What percentage of the out-of-state tuition rates in each state are below 5.0 thousand dollars?
   (f) What percentage of the out-of-state tuition rates in each state are below 11.0 thousand dollars?
   (g) Discuss your findings.
   (h) How might your information be of assistance to the research analyst at the college advisory service? Discuss.

TEXAS-2.XLS
NCC&U.TXT

• 2.26 From the frequency distribution developed in Problem 2.10(b) on page 64 regarding utility charges
   (a) Form the cumulative frequency distribution.
   (b) Form the cumulative percentage distribution.
   (c) Plot the ogive (cumulative percentage polygon).

UTILITY.TXT

• 2.27 From the frequency distribution developed in Problem 2.11 on page 64 regarding book values of companies listed on the NYSE
   (a) Form the cumulative frequency distribution.
   (b) Form the cumulative percentage distribution.
   (c) Plot the ogive (cumulative percentage polygon).

STOCK1.TXT

2.28 From the frequency distribution developed in Problem 2.12 on page 64 regarding cancer incidence
   (a) Form the cumulative frequency distribution.
   (b) Form the cumulative percentage distribution.
   (c) Plot the ogive (cumulative percentage polygon).

CANCER.TXT

2.29 From the frequency distributions developed in Problem 2.13 on page 64 for each of the three numerical variables (score, cost, and sodium) regarding peanut butter characteristics

S-2-29.XLS

(a) Form the respective cumulative frequency distributions.
(b) Form the respective cumulative percentage distributions.
(c) Plot the respective ogives (cumulative percentage polygons).

2.30 From the frequency distributions developed in Problem 2.14 on pages 64–65 regarding life of light bulbs from two manufacturers

BULBS.TXT

(a) Form the cumulative frequency distributions.
(b) Form the cumulative percentage distributions.
(c) Plot the ogives (cumulative percentage polygons) on one graph.

2.7 USING MICROSOFT EXCEL TO OBTAIN TABLES AND CHARTS FOR NUMERICAL VARIABLES

2.7.1 Introduction

In Sections 2.3–2.6, we organized the Texas out-of-state tuition data into a frequency distribution and then formed relative frequency, percentage, and cumulative percentage distributions. Using these distributions, we then developed the histogram, percentage polygon, and cumulative percentage polygon. In this section we will discuss how Microsoft Excel can be used to obtain tables and charts for these data. To organize our workbook for this purpose, we need to design a Data sheet and a Calculations sheet. Additional sheets including those for the histogram and frequency polygon charts will be developed as a consequence of selecting the appropriate Excel command. Table 2.1.Excel represents the design of the Data sheet, while Table 2.2.Excel represents the design of the Calculations sheet.

TEXAS-1.XLS

2.7.2 Using the FREQUENCY Function to Obtain a Frequency Distribution

One way to obtain a frequency distribution in Excel is to use the special array formulas that contain the FREQUENCY function of Excel. We can develop formulas in the form:

=FREQUENCY (*cell range, upper class boundaries*)

Table 2.1.Excel Design of the Data sheet for the Texas out-of-state tuition workbook.

	A	B	C	D	E	F	G
1	School	Tuition	Type	Setting	Calendar	Focus	Upper Class
2	xx	xx	xx	xx	xx	xx	Boundaries
3	.	.	.	.	.	.	1.99
4	.	.	.	.	.	.	3.99
5	.	.	.	.	.	.	5.99
6	.	.	.	.	.	.	7.99
7	.	.	.	.	.	.	9.99
8	.	.	.	.	.	.	11.99
9	.	.	.	.	.	.	13.99
⋮	.	.	.	.	.	.	
61	xx	xx	xx	xx	xx	xx	

Chapter 2 Presenting Data in Tables and Charts

Table 2.2.Excel Design of the Calculations sheet for the Texas out-of-state tuition workbook.

	A	B	C	D
1		Texas Frequency Distribution		
2				
3	Class	Frequency	Relative Frequency	Percentage
4	=DATA!G3	{=FREQUENCY(DATA!B2:B61,DATA!G3:G9)}	=B4/$B$11	=C4
5	=DATA!G4	.	=B5/$B$11	=C5
6	=DATA!G5	.	=B6/$B$11	=C6
7	=DATA!G6	.	=B7/$B$11	=C7
8	=DATA!G7	.	=B8/$B$11	=C8
9	=DATA!G8	.	=B9/$B$11	=C9
10	=DATA!G9	.	=B10/$B$11	=C10
11	Total:	=SUM(B4:B10)		

where *cell range* = the cell range for the data to be analyzed

upper class boundaries = the cell range that contains the values that represent the upper class boundaries

To create a frequency distribution for the Texas tuition data similar to Table 2.3 on page 62, open your Texas tuition workbook or the TEXAS-1.XLS file and do the following:

1. Select the Data sheet and enter the upper class boundaries in the cell range G3:G9 that corresponds to the class intervals shown in Table 2.3. For this problem, they are the values 1.99, 3.99, 5.99, 7.99, 9.99, 11.99, and 13.99. We have written 1.99 as the approximation of less than 2, 3.99 as the approximation of less than 4, ..., and 13.99 as the approximation of less than 14. These limits need to be entered on the Data sheet when we use the Data Analysis tool in Section 2.7.3.

2. Issue the command Insert | Worksheet and name the worksheet as Calculations.

3. Enter title and column headings for this sheet.

4. In the cell range A4:A10, enter the formulas to copy the values of the upper class boundaries. As before, you can enter =Data!G3 into cell A4 and copy the formula to the remainder of the range (see Section 1S.10).

5. We are now ready to assemble the formula to calculate the frequencies. As the values for the Texas tuition data are located in cells B2:B61 on the Data sheet and the upper class boundaries are located in the range G3:G9 on the same sheet, we can use the formula =FREQUENCY(Data!B2:B61,Data!G3:G9). To enter this formula, we take advantage of a shortcut feature of Excel that will place this formula into a range of cells in one operation. To do this, first select the range for the formula (in this case, B4:B10 of the Calculations sheet). Then type the formula and, *while holding down the Control and Shift keys*, press the Enter key. The frequencies now appear on Column B on this sheet (see Figure 2.3.Excel). If you use the formula bar to review the formulas just entered in the range B4:B10, you will notice that the formulas are enclosed in curly braces { }. This indicates that these cells contain a special type of formula that cannot be individually edited.

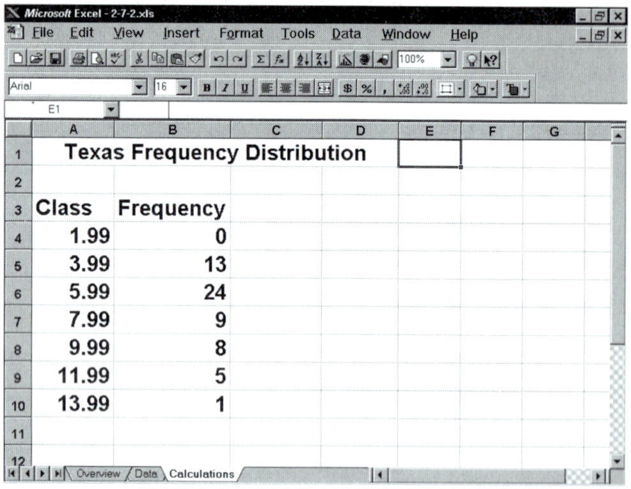

FIGURE 2.3.EXCEL
Frequency distribution for the Texas out-of-state tuition data obtained from the Excel FREQUENCY function.

Class	Frequency
1.99	0
3.99	13
5.99	24
7.99	9
9.99	8
11.99	5
13.99	1

Texas Frequency Distribution

▲ WHAT IF EXAMPLE

We can store alternative sets of upper class boundaries and compare the results using the Scenario Manager feature (see Section 1S.15). Define scenarios for the range Data!G3:G9 using the upper class boundaries of this chapter (1.99, 3.99, 5.99, 7.99, 9.99, 11.99, 13.99) as well as the set of values of .99, 2.99, 4.99, 6.99, 8.99, 10.99, and 12.99. Additional scenarios in which the class-interval width is some other value than two are also possible. However, these scenarios would require the editing of the Data and Calculations sheets. A frequency distribution with this different class-interval width would be obtained and could be compared to the one illustrated in Figure 2.3.Excel.

Having obtained the set of frequencies, we can now calculate the relative frequencies and percentages. As a first step, enter the formula =SUM(B4:B10) in cell B11 of the Calculations sheet to obtain the total frequency. (The formula is entered into cell B11 for presentation purposes.) This value can now be used as the denominator in a set of formulas in column C to calculate the relative frequencies. To enter the relative frequencies for the first class, enter the formula =B4/$B$11 in cell C4 and press Enter. In this formula, the address in the denominator has been entered as an **absolute address**. This is an address that will *not* be adjusted by Excel during the copying operation. (This makes sense here because we want to divide the cell frequencies by the same value, the total frequency, which is contained in cell B11.)

The relative frequencies are now displayed in cells C4:C10. They can be converted to percentages by doing the following:

❶ Copy cell C4 to cell D4 by entering the formula =C4 in cell D4.

❷ Copy this formula through the range D4:D10.

❸ Select the cell range D4:D10 and press the Percent format button.

❹ Adjust the decimal by pressing the Increase Decimal button once for each decimal place desired. In Figure 2.4.Excel, we pressed the Increase Decimal button twice to obtain two decimal places.

The frequencies, relative frequencies, and percentages are illustrated in Figure 2.4.Excel.

FIGURE 2.4.EXCEL
Frequency distribution, relative frequency distribution, and percentage distribution obtained from Excel for the Texas out-of-state tuition data.

2.7.3 Using the Data Analysis Tool to Obtain Frequency and Cumulative Frequency Distributions and Histograms

In addition to (or instead of) using the FREQUENCY function to obtain frequency distributions, we could use the Histogram option of the Data Analysis tool of Excel to obtain frequency and cumulative frequency distributions as well as a histogram and cumulative frequency polygon. The Data Analysis tool is a set of predefined routines that we can use to perform most of the statistical procedures we will discuss in this text. These routines allow us to obtain results that are often difficult or impossible to obtain using simple formulas. The Data Analysis tool also typically generates numerous statistics from a single-user operation. However, the statistics generated by this tool do not change when the data being analyzed changes. In such circumstances, it may be better to use formulas where possible. For this reason, we discuss both the use of the Data Analysis tool and the use of formulas.

The Histogram option of the Data Analysis tool can be used to obtain both a frequency distribution and charts such as a histogram and cumulative percentage polygon. To use the Histogram option of the Data Analysis tool, the upper class boundaries of the class intervals *must* be entered on the sheet that contains the data to be analyzed. Retrieve the TEXAS-2.XLS workbook file, which is the original Texas out-of-state tuition data as shown in Table 2.1.Excel, and select the Data Sheet by clicking on the Data Sheet tab. To create the histogram and cumulative percentage polygon, we need to do the following:

TEXAS-2.XLS

❶ Select the command Tools | Data Analysis and then select Histogram from the Analysis Tools list box that appears (see Figure 2.5.Excel). Click the OK button to display the Histogram dialog box. If Data Analysis is not a choice on your Tools menu, the Data Analysis component of Excel is probably not properly installed (review Section 1S.4 before continuing).

❷ In the Input area of the Histogram dialog box, enter the range Data!$B$2:$B$61 in the Input Range edit box. Enter the range Data!$G$3:$G$9 in the Bin Range edit box.

❸ In the Output area of the Histogram dialog box, click the New Worksheet Ply button and enter Histogram as the name of the new sheet in the edit box to the right of this button. Select the Cumulative Percentages and Chart Output check boxes and leave

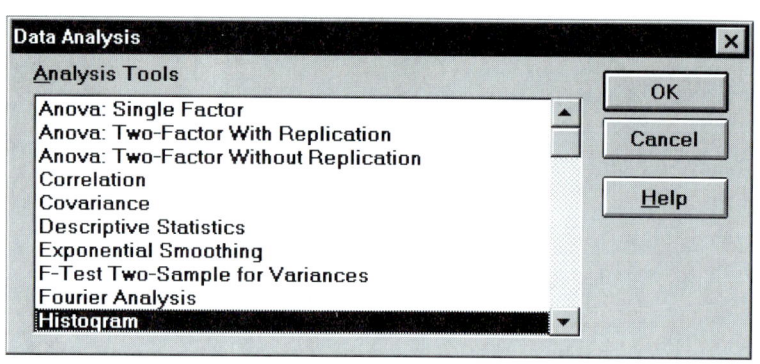

FIGURE 2.5.EXCEL Data Analysis dialog box.

FIGURE 2.6.EXCEL Histogram dialog box. Panel A.

the Pareto check box unselected. Click the OK button. The dialog box should look like the one shown in Figure 2.6.Excel, Panel A.

Excel will generate both a frequency distribution and cumulative percentage distribution and superimpose the cumulative percentage polygon onto the histogram (see Figure 2.6.Excel, Panel B). If only a histogram was desired, the Cumulative Percentages check box would not be selected in step 3.

Observe that as was the case with the FREQUENCY function, the frequencies and cumulative percentages provided refer to the upper boundaries of the class. This means that 21.67% of the schools have tuition less than 4 thousand dollars, 61.67% have tuition less than 6 thousand dollars, 76.67% have tuition less than 8 thousand dollars, 90% have tuition less than 10 thousand dollars, 98.33% have tuition less than 12 thousand dollars, and 100% have tuition less than 14 thousand dollars.

In Figure 2.6.Excel Panel B, a different vertical axis is included for each chart since the two graphs are superimposed. The vertical axis on the left side of the chart provides frequencies for the histogram, while the vertical axis on the right provides percentages for the cumulative percentage polygon.

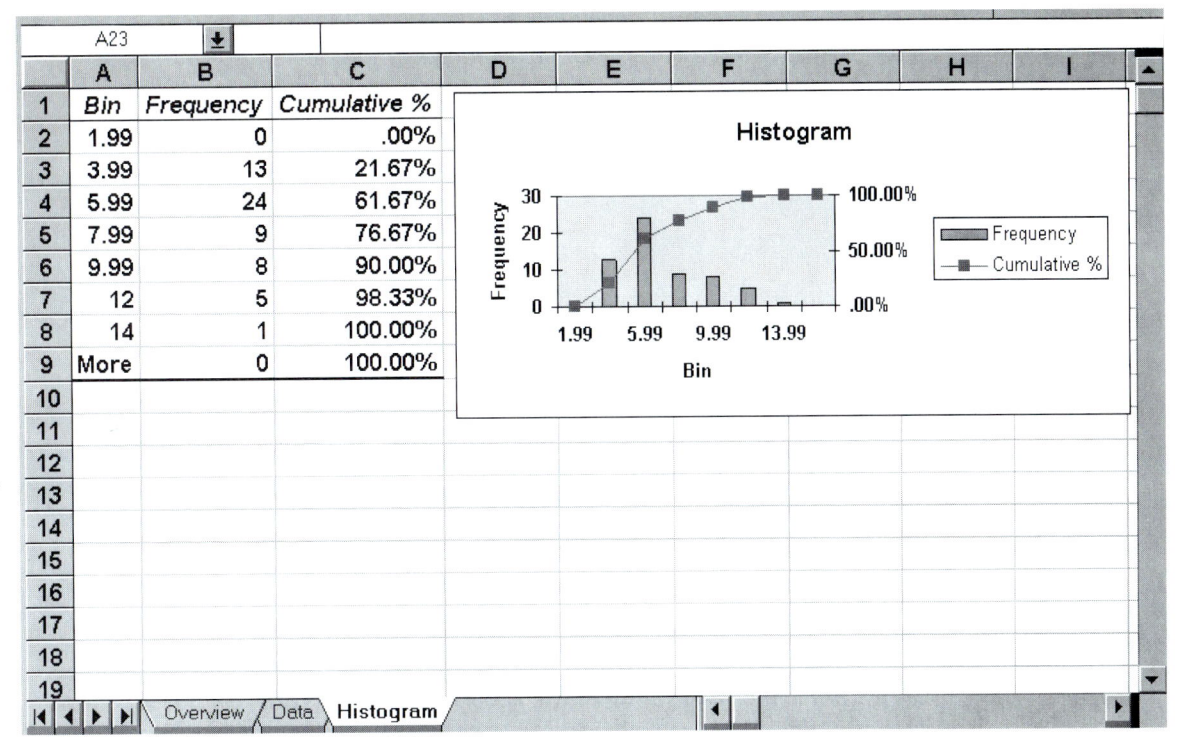

	A	B	C	D	E	F	G	H	I
1	Bin	Frequency	Cumulative %						
2	1.99	0	.00%						
3	3.99	13	21.67%						
4	5.99	24	61.67%						
5	7.99	9	76.67%						
6	9.99	8	90.00%						
7	12	5	98.33%						
8	14	1	100.00%						
9	More	0	100.00%						

FIGURE 2.6.EXCEL Histogram and cumulative percentage distribution obtained from Excel for the Texas out-of-state tuition data. Panel B.

Observe that this chart contains two errors. There are gaps between the bars that correspond to the class intervals, and there is an additional class, labeled More by Excel. To remove the gaps, we need to do the following:

❶ Double-click on the white area in the chart box.

❷ Select Format | I Column Group. This displays the Format Column Groups dialog box.

❸ In the Gap Width edit box of the Options tab, change the value to 0. Click the OK button. The Histogram now has continuous bars.

To remove the additional class, we need to do the following:

❶ First select a cell outside the chart. Then double-click on one of the bars displayed on the histogram. A halo appears around the chart box, and a second set of data points appears superimposed over the bars along with a formula that begins with the word Series in the edit box above the worksheet.

❷ Select the command Format | Selected Data Series to display the Format Data Series dialog box. Click the Name and Values tab. In the Y Values edit box, change the ending cell from $B$9 to $B$8. Click the OK button.

❸ Then click on one of the data points of the Cumulative Percentage chart. Select Format | Selected Data Series a second time, and the Name and Values tab appears; change the ending cell in the Y Values edit box, from $C$9 to $C$8. Click the OK button. Our resulting chart now has the proper number of classes (7).

In addition to changing these features, we will probably also want to enlarge the chart for greater clarity and display meaningful axis labels. To enlarge the chart:

2.7 Using Microsoft Excel to Obtain Tables and Charts for Numerical Variables

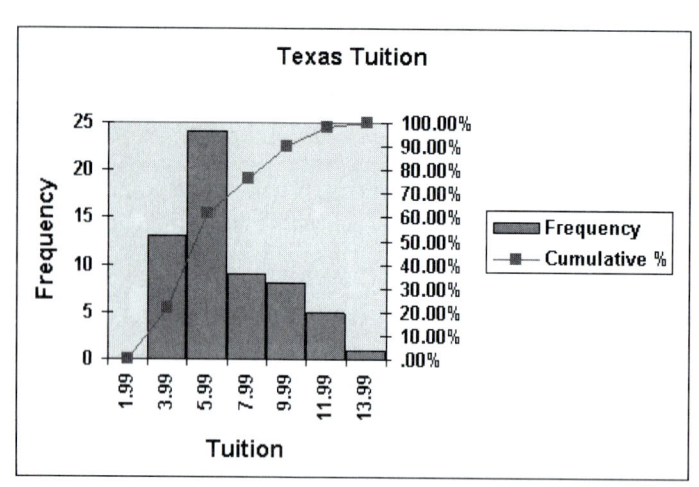

FIGURE 2.7.EXCEL Revised histogram obtained from Excel for the Texas out-of-state tuition data.

① First select a cell outside the chart and then single-click in the white area inside the chart box. A set of eight square handles appears on the border of the chart box.

② Move the mouse pointer directly over the lower left hand corner of the chart box. The mouse pointer changes to a small double-sided arrow. Drag the mouse pointer (recall that dragging requires holding down the mouse button while moving the mouse) toward cell D15. The mouse pointer changes to a simple plus sign, and as you drag it, the border of the chart box expands.

③ While the mouse pointer is over cell D15, release the mouse button. The chart is now enlarged. Note that additional tick marks are now displayed on the Y-axis for each chart.

To change the label on the X-axis, first select a cell outside the chart box and, with the mouse pointer directly over the current X-axis label (Bin), double-click the mouse. A haloed border with handles appears around the word Bin. Click inside the haloed border and replace the label bin with the label Tuition.

To change the title, first select a cell outside the chart box and with the mouse pointer directly over the current title Histogram, double-click the mouse. A haloed border with handles appears around the word Histogram. (If a haloed border appears around the X-axis label, click one more time on the current title.) Click inside the haloed border and replace the label Histogram with the label Texas Tuition. Click outside the chart box to complete the editing of these labels. Other parts of the chart could be reformatted by double-clicking them and then selecting the appropriate Format command.

The histogram that results from these changes will be similar to the one presented in Figure 2.7.Excel.

2.7.4 Using the Chart Wizard to Obtain Frequency Polygons and Histograms

Thus far, we have used the FREQUENCY function to obtain various tables and the Data Analysis tool to obtain a variety of tables and charts. We will now discuss how the Microsoft Excel Chart Wizard can be used to develop a frequency polygon and a histogram.

The Microsoft Excel Chart Wizard contains many different types of charts that can be developed for both numerical and categorical variables. The Chart Wizard can be used by clicking on the Chart Wizard icon on the top (Standard) toolbar (it is the third icon to the left of the zoom control on the right-hand side of the toolbar) or by issuing the command Insert | Chart.

The Chart Wizard presents five linked dialog boxes, one at a time, that allow you to create a variety of charts for both numerical and categorical variables. The first dialog box asks you to provide the range for your data. The second dialog box asks you to select the type of chart desired. The third dialog box gives you the choice of many options that relate to the chart that has been chosen in the second dialog box. The fourth dialog box allows you to specify the orientation of your data and identify names for categories and data sets. The fifth dialog box allows you to format the chart by entering such things as titles and a legend. Charts created by the Chart Wizard can be further edited and formatted as will be described later in this section.

To obtain a frequency polygon from the Chart Wizard, open the workbook you developed in Section 2.7.2 or open the workbook file TEXAS-3.XLS. Note that in the design of the Calculations sheet, the upper class boundaries appear to the left of the cells containing the frequencies. We must arrange these boundaries and frequencies in this manner in order to have the Chart Wizard prepare a frequency polygon or histogram with a properly labeled X-axis. To obtain the frequency polygon from the Chart Wizard, do the following:

TEXAS-3.XLS

❶ Select Insert | Chart | As New Sheet (since we will want to place the frequency polygon on a separate sheet). The first dialog box of the Chart Wizard, asking for the range of the data to be plotted, appears.

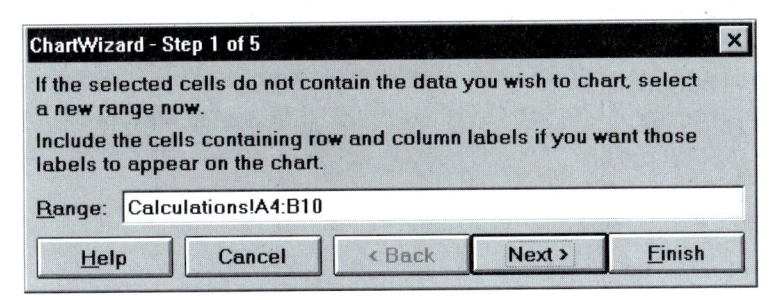

FIGURE 2.8.EXCEL Chart Wizard steps to obtain a line chart for the Texas out-of-state tuition data. Panel A.

❷ In the Range edit box, enter Calculations!A4:B10 (Figure 2.8.Excel, Panel A). If the upper class boundaries and the frequencies do not appear in adjacent columns, enter the range of the upper class boundaries followed by a comma followed by the range of the frequencies. (For example, if the frequencies appeared in column M, rows 4–10, we would write Calculations!A4:A10,Calculations!M4:M10. Click the Next button to continue to the second dialog box.

❸ In the second dialog box, select the Line Chart choice and click the Next button. (See Panel B.)

❹ In the third dialog box, select the first choice, a line chart that contains plotted points. (See Panel C.) Click the Next button.

❺ In the fourth dialog box, select the Columns option button under the Data Series in heading. Then enter 1 in the First Columns edit box and enter 0 in the First Rows edit box. (See Panel D.) Click the Next button to move to the next dialog box.

❻ In the fifth dialog box (Panel E), you can enter a chart legend for the chart as well as titles for the axes. To obtain the frequency polygon in Figure 2.9.Excel, (a) select the No option button for Add a legend, and (b) enter Texas Tuition as the title, enter

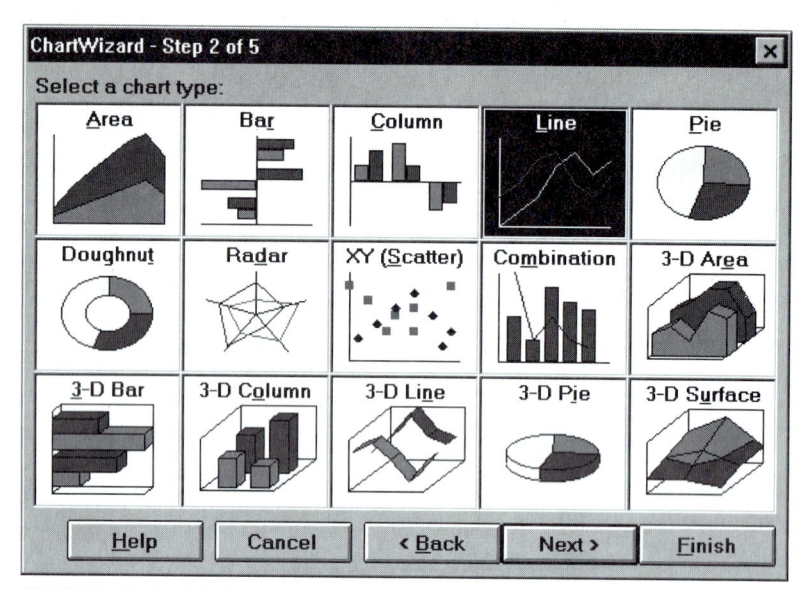

FIGURE 2.8.EXCEL Excel Chart Wizard Step 2. Panel B.

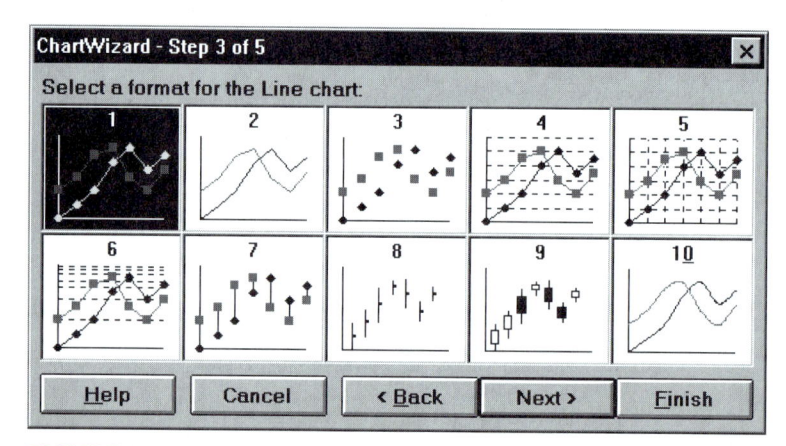

FIGURE 2.8.EXCEL Excel Chart Wizard Step 3. Panel C.

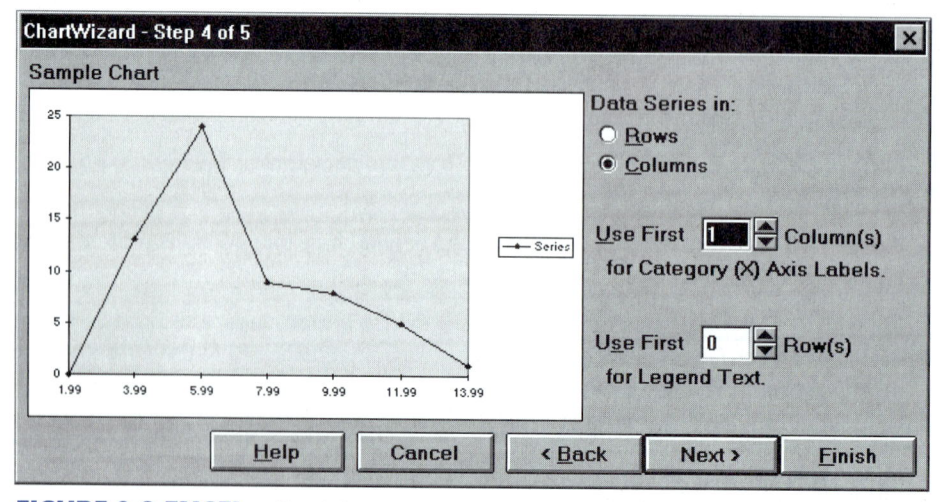

FIGURE 2.8.EXCEL Excel Chart Wizard Step 4. Panel D.

Chapter 2 Presenting Data in Tables and Charts

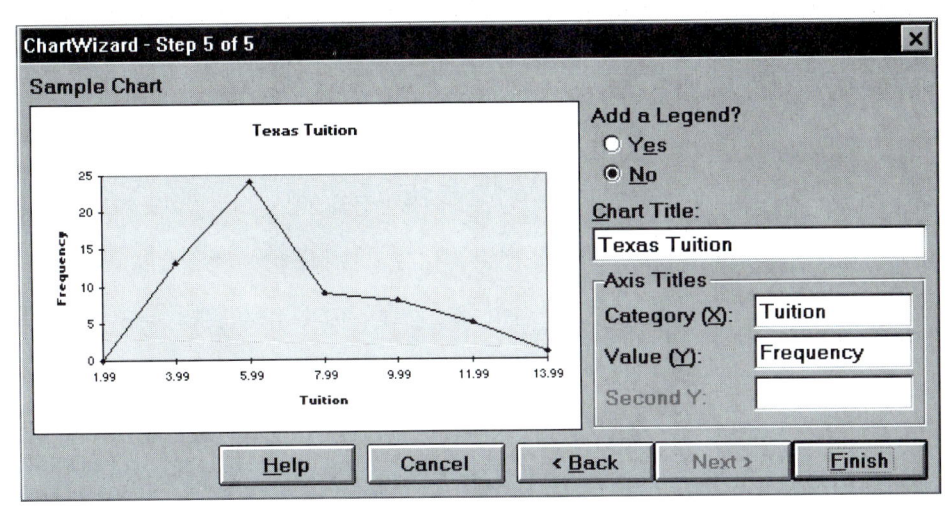

FIGURE 2.8.EXCEL   Excel Chart Wizard Step 5. Panel E.

Tuition for the category (X) edit box, and enter Frequency for the Value (Y) edit box. (See Panel E.) Click the Finish button.

Figure 2.9.Excel, Panel A, displays the frequency polygon we have obtained. Rename this sheet as Frequency Polygon.

If we examine this frequency polygon, we see that, as was the case in Figure 2.6.Excel on page 79, the category markings on the X-axis refer to the upper limits of the classes, not the class midpoints. To change these markings, double-click on the X-axis. The Format Axis dialog box appears. Select the Scale Tab and select the Value (Y) Axis crosses between categories check box. Click the OK button. Then change the class labels found in the range Calculations!A4:A10 to 1, 3, 5, 7, 9, 11, and 13. Figure 2.9.Excel, Panel B, displays the revised frequency polygon.

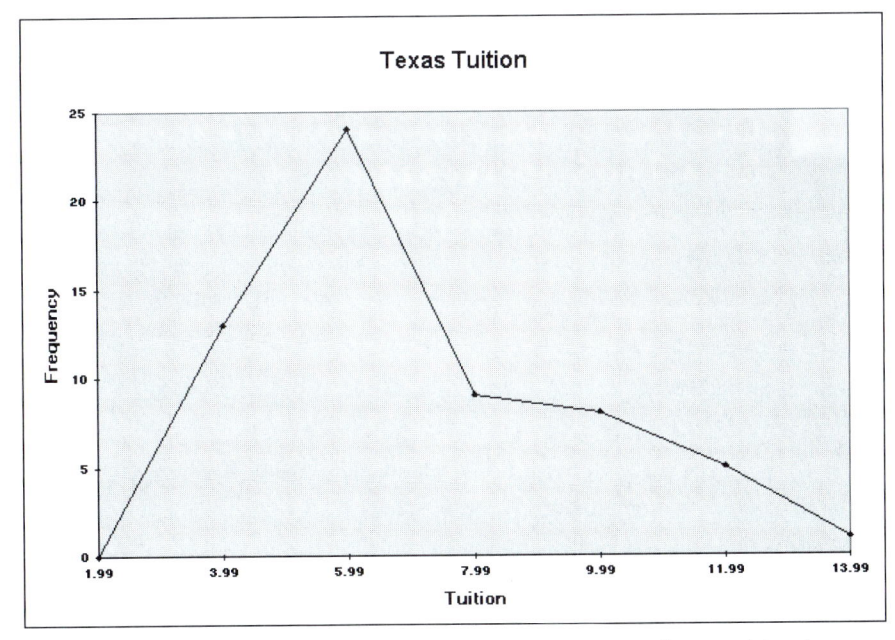

FIGURE 2.9.EXCEL   Frequency polygon for the Texas out-of-state tuition data. Panel A.

2.7   Using Microsoft Excel to Obtain Tables and Charts for Numerical Variables

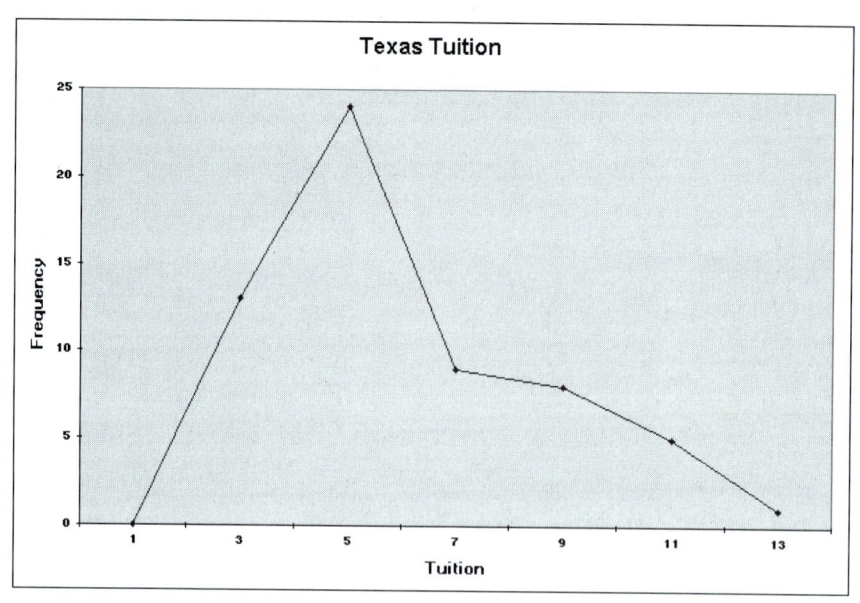

FIGURE 2.9.EXCEL Revised frequency polygon obtained from Excel for the Texas out-of-state tuition data. Panel B.

In addition, if we wanted to obtain a percentage polygon or a cumulative percentage polygon, we would use the column of percentages or cumulative percentages instead of the column of frequencies when we define the range for the data.

To obtain a histogram from the Chart Wizard, we would follow steps similar to those described for the polygon. Steps 1 and 2 would be the same, as would steps 5 and 6. In step 3, we would select the Column Chart, while in step 4, we would select choice 8, a series of bars attached to each other.

As was the case with the polygon, if we wanted a percentage histogram, we would use the column of percentages instead of the column of frequencies.

2.8 ORGANIZING AND TABULATING CATEGORICAL DATA: THE SUMMARY TABLE

Thus far in this chapter we have learned that when collecting a large set of numerical data, the best way to examine it is first to organize and present it in appropriate tabular and chart format. Often, however, the data we collect are categorical, not numerical. Thus, in the remainder of this chapter, we will demonstrate how categorical data can be organized and presented in the form of tables and charts.

In order to do this, let us suppose once again that our analyst at the college advisory service wants to evaluate various features pertaining to colleges and universities in the state of North Carolina. Special Data Set 1 in Appendix D on page D3 displays information on the out-of-state tuition rate, type of institution, setting of school, academic calendar, and institutional focus for each of the 45 North Carolina colleges and universities. We note that the tuition rate variable is *numerical* while the other variables are all *categorical*. Earlier in this chapter we were concerned only with the former; a detailed study of the responses to the categorical variables will be undertaken here.

When dealing with categorical phenomena, the observations may be tallied into *summary tables* and then graphically displayed as either *bar charts, pie charts,* or *Pareto diagrams*.

Table 2.11 Frequency and percentage summary table pertaining to institutional focus for 45 colleges and universities in North Carolina.

NCC&U.TXT

Institutional Focus	Number of Schools	Percentage of Schools
National Liberal Arts Schools (NLA)	2	4.4
National Universities (NU)	4	8.9
Regional Liberal Arts Schools (RLA)	16	35.6
Regional Universities (RU)	22	48.9
Specialty Schools (SS)	1	2.2
Totals	45	100.0

Source: Data are taken from Special Data Set 1, Appendix D, page D3.

To illustrate the development of a **summary table,** let us consider the data obtained by our analyst on institutional focus. From Special Data Set 1 in Appendix D, we see that, of the 45 colleges and universities in North Carolina, 2 are classified by the College Counsel as national liberal arts schools (NLA), 4 are classified as national universities (NU), 16 are regional liberal arts schools (RLA), 22 are regional universities (RU), and 1 is a specialty school (SS). This information is presented in Table 2.11.

2.9 GRAPHING CATEGORICAL DATA: BAR AND PIE CHARTS

To express the information provided in Table 2.11 graphically, the percentage bar chart (Figure 2.10) or percentage pie chart (Figure 2.11) can be displayed.

Figure 2.10 depicts a percentage bar chart for the North Carolina institutional focus data presented in Table 2.11. In **bar charts,** each category is depicted by a bar, the length of which represents the frequency or percentage of observations falling into a category.

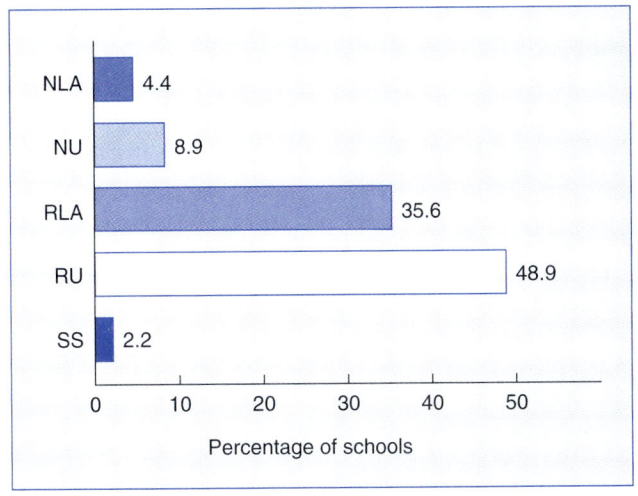

FIGURE 2.10 Percentage bar chart depicting the institutional focus of the North Carolina schools.
Source: Data are taken from Table 2.11.

To construct a bar chart, the following suggestions are made:

1. The bars should be constructed horizontally (as in Figure 2.10) when the categorized observations are the outcomes of a categorical variable. The bars should be constructed vertically when the categorized observations are the outcomes of a numerical variable.
2. All bars should have the same width (as in Figure 2.10) so as not to mislead the reader. Only the lengths should differ.
3. Spaces between bars should range from one-half the width of a bar to the width of a bar.
4. Scales and guidelines are useful aids in reading a chart and should be included. The zero point or origin should be indicated.
5. The axes of the chart should be clearly labeled.
6. Any "keys" to interpreting the chart may be included within the body of the chart or below the body of the chart.
7. Footnotes or source notes, when appropriate, are presented after the title of the chart or at the bottom edge of the chart's frame.

Figure 2.11 depicts a percentage pie chart for the North Carolina institutional focus data presented in Table 2.11.

If we had to construct a **pie chart** (when software such as Microsoft Excel was not available), we would use both the compass and the protractor—the former to draw the circle, the latter to measure off the appropriate pie sectors. Since the circle has 360°, the protractor would be used to divide up the pie based on the percentage "slices" desired. As an example, in Table 2.11, 8.9% of the North Carolina schools are classified by the College Counsel as national universities. Thus, we would multiply 360 by .089, mark off the resulting 32° with the protractor, and then connect the appropriate points to the center of the pie, forming a slice comprising 8.9% of the area of the pie.

Using this procedure with all the categories in Table 2.11 would enable us to construct the entire pie chart displayed in Figure 2.11. However, we certainly recommend that, when available, software be used for developing a pie chart.

The purpose of graphical presentation is to display data accurately and clearly. Figures 2.10 and 2.11 attempt to convey the same information with respect to institutional focus. Whether these charts succeed, however, has been a matter of much concern (see References 2–4, 11, 12). In particular, research in the human perception of graphs (Reference 4) concludes that the pie chart presents the weakest display. The bar chart is preferred to the pie chart, because it has been observed that the human eye can more accurately judge length comparisons against a fixed scale

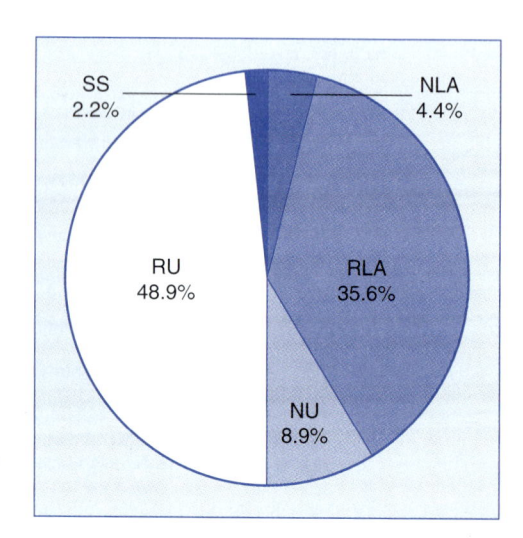

FIGURE 2.11 Percentage pie chart depicting the institutional focus of the North Carolina schools. *Source:* Data are taken from Table 2.11.

(as in a bar chart) than angular measures (as in a pie chart). Nevertheless, the pie chart has two distinct advantages: (1) it is aesthetically pleasing, and (2) it clearly shows that the total for all categories or slices of the pie adds to 100%. Thus, the selection of a particular chart is still highly subjective and often dependent on the aesthetic preferences of the user.

Problems for Section 2.9

Note: *To use Microsoft Excel to solve these problems, refer to Section 2.12.*

2.31 The board of directors of a large housing cooperative wishes to investigate the possibility of hiring a supervisor for an outdoor playground. All 616 households in the cooperative were polled, with each household having one vote, regardless of its size. The following data were collected:

Should the co-op hire a supervisor?	
Yes	146
No	91
Not sure	58
No response	321
Total	616

(a) Convert the data to percentages and construct
   (1) A bar chart
   (2) A pie chart
(b) Which of these charts do you prefer to use here? Why?
(c) Eliminating the "no response" group, convert the 295 responses to percentages and construct
   (1) A bar chart
   (2) A pie chart
(d) **ACTION** Based on your findings in (a) and (c), what would you recommend that the board of directors do? Write a letter to the president of the board.

2.32 The following data represent the market shares (in percent) held by manufacturers of portable, transportable, and mobile cellular phones sold during 1992:

Manufacturer	Market Share (in %)
Motorola	22
Nokia	14
Mitsubishi	10
NovAtel	9
Toshiba	8
All others	37
Total	110

Source: The New York Times, October 31, 1993, Sect. 3, p. 1.

(a) Construct a bar chart.
(b) Construct a pie chart.
(c) Which of these charts do you prefer to use here? Why?

2.33 The data at the top of page 88 represent the market share (in percent) held by manufacturers of Windows business applications software during 1992:

Manufacturer	Market Share (in %)
Aldus	4.0
Lotus	14.6
Microsoft	60.0
Software Publishing	2.9
Wordperfect	9.6
Others	8.8
Totals	99.9*

* Due to rounding.
Source: *The New York Times*, October 13, 1993, p. D1.

(a) Construct a bar chart.
(b) Construct a pie chart.
(c) Which of these charts do you prefer to use here? Why?

2.34 Imports to the United States from developing countries accounted for 41.4% of an estimated total of 575.9 billion dollars in the year 1993. On the other hand, exports from the United States to developing countries accounted for 40.7% of an estimated total of 459.6 billion dollars in that year. The following table presents a breakdown by country or region (in percent share) of U.S. imports and exports for the year 1993:

Country or Region	Imports into U.S. Percent Share	Exports from U.S. Percent Share
Africa	2.3	1.6
Asia (excluding Japan)	23.5	17.2
Canada	19.2	21.7
European Community	16.6	20.8
Japan	18.4	10.4
Latin America	12.9	16.8
Mideast	2.7	4.7
Other	4.4	6.8
Total	100.0	100.0

Source: *The New York Times*, December 19, 1993, p. F7.

(a) Construct separate bar charts for imports and exports.
(b) Construct separate pie charts for imports and exports.
(c) Which of these charts do you prefer to use here? Why?
(d) **ACTION** Analyze the data and write a memo to your economics professor based on your findings.

2.10 GRAPHING CATEGORICAL DATA: THE PARETO DIAGRAM

The **Pareto diagram** is a special type of vertical bar chart in which the categorized responses are plotted in the descending rank order of their frequencies and combined with a cumulative polygon on the same scale. The main principle behind this graphical device is its ability to sep-

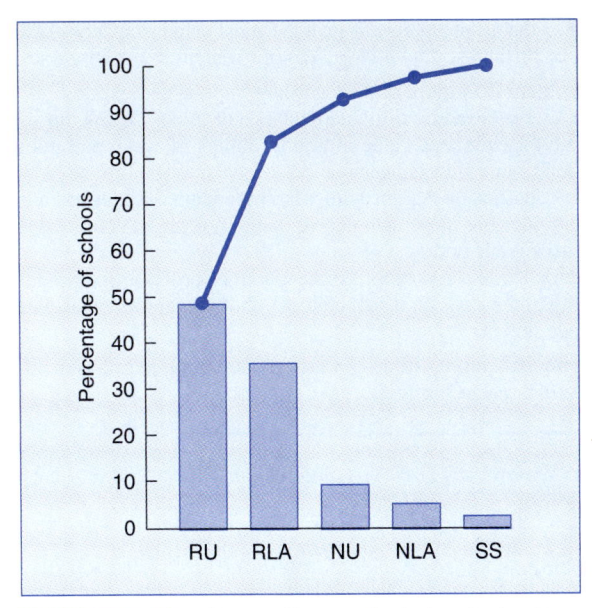

FIGURE 2.12 Pareto diagram depicting the institutional focus of the North Carolina schools.
Source: Data are taken from Table 2.11 on page 85.

arate the "vital few" from the "trivial many," enabling us to focus on the important responses. Hence the chart achieves its greatest utility when the categorical variable of interest contains many categories. The Pareto diagram is widely used in the statistical control of process and product quality (see Chapter 10).

To illustrate the Pareto diagram, we observe that in Figure 2.10, the bar chart pertaining to institutional focus presents the categories as national liberal arts schools, national universities, regional liberal arts schools, regional universities, and specialty schools. Since regional universities and regional liberal arts schools dominate the institutional focus in the state of North Carolina, a Pareto diagram may be formed by changing the ordering. Such a plot is depicted in Figure 2.12. From the lengths of the vertical bars, we observe that almost one out of every two of these schools is classified as a regional university. From the cumulative polygon, we note that 84.5% of these schools are classified as either a regional university or a regional liberal arts school.

In constructing the Pareto diagram, the vertical axis contains the percentages (from 100 on top to 0 on bottom) and the horizontal axis contains the categories of interest. The equally spaced bars must also be of equal width and, for visual impact (Reference 10), we suggest that the bars be of the same tint. The point on the cumulative percentage polygon for each category is centered at the midpoint of each respective bar. Hence, when studying a Pareto diagram, we should be focusing on two things—the magnitudes of the differences in bar lengths corresponding to adjacent descending categories and the cumulative percentages of these adjacent categories.

Problems for Section 2.10

Note: *To use Microsoft Excel to solve these problems, refer to Section 2.12.*

2.35 Refer to the data in Problem 2.33 on pages 87–88 regarding percent market share attained by manufacturers of Windows business applications software.
(a) Form a Pareto diagram.

(b) Which of the graphs seems to give the most visual impact—the Pareto diagram here or one of the charts drawn in (a)–(b) of Problem 2.33? Discuss.

2.36 Refer to Problem 2.34 on page 88 regarding imports and exports.
(a) Set up a table based on estimated balance of trade. That is, for each country or region, calculate the estimated value of import dollars minus export dollars—yielding either a trade deficit or a trade surplus.
(b) For the countries or regions with which the United States has a trade deficit (i.e., import dollars are more than export dollars), form a Pareto diagram.
(c) Summarize your findings.
(d) **ACTION** Write a report for your economics professor based on your findings in (c). List potential social, political, cultural, and/or economic reasons that may have led to this trade deficit.

2.37 The following data represent daily water consumption per household in a suburban water district during a recent summer:

Reason for Water Usage	Gallons per Day
Bathing and showering	99
Dish washing	13
Drinking and cooking	11
Laundering	33
Lawn watering	150
Toilet	88
Miscellaneous	20
Total	414

(a) Form a Pareto diagram.
(b) Summarize your findings.
(c) **ACTION** Since the town council is concerned about future water shortages, write a letter based on your findings in (b) pinpointing problem areas and proposing legislation that might conserve water through changes in personal habits.

2.38 The following data represent the oil production of OPEC members in December 1992, in millions of barrels per day:

Country	Daily Oil Production (in millions of barrels)
Algeria	0.77
Gabon	0.30
Indonesia	1.35
Iran	3.50
Iraq	0.55
Kuwait	1.30
Libya	1.45
Nigeria	1.90
Qatar	0.42
Saudi Arabia	8.20
United Arab Emirates	2.25
Venezuela	3.50
Total	25.49

Source: The New York Times, January 25, 1993, p. D2.

(a) Form a Pareto diagram.
(b) Summarize your findings.
(c) **ACTION** Write a letter to the business editor of your local newspaper on this matter.

2.39 A patient-satisfaction survey conducted for a sample of 210 individuals discharged from a large urban hospital during the month of June led to the following list of 384 complaints:

Reason for Complaint	Number
Anger with other patients/visitors	13
Failure to respond to buzzer	71
Inadequate answers to questions	38
Lateness for tests	34
Noise	28
Poor food service	117
Rudeness of staff	62
All others	21
Total	384

(a) Form a Pareto diagram.
(b) Summarize your findings.
(c) **ACTION** Write a memo to the chief executive officer of the hospital regarding your findings and offer suggestions for improvement.

2.40 The following table presents the number of stockholder meetings held outside the United States at which U.S. pension-fund clients of *Global Proxy Services Corporation* voted in the 1992–93 proxy season:

Country	Number of Meetings Held
Australia	49
Belgium	50
Canada	87
England	374
France	72
Germany	99
Holland	83
Hong Kong	116
Italy	115
Japan	1,249
Switzerland	61
Other	396
Total	2,751

Source: The New York Times, July 16, 1993, p. D1.

(a) Form a Pareto diagram.
(b) Summarize your findings.
(c) **ACTION** Write a letter to the business editor of your local newspaper on this matter.

2.11 TABULARIZING CATEGORICAL DATA USING CONTINGENCY TABLES

It is often desirable to examine the responses to two categorical variables simultaneously. For example, our analyst at the college advisory service might be interested in examining whether or not there is any pattern or relationship between type of institution (i.e., private or public) and the institutional focus. Using Special Data Set 1 in Appendix D on page D3, Table 2.12 depicts this information for all 45 colleges and universities in the state of North Carolina. Such two-way tables of cross-classification are known as **contingency tables.**

NCC&U.TXT

Table 2.12 Contingency table displaying type of institution and institutional focus of the North Carolina schools.

Type of Institution	NLA	NU	RLA	RU	SS	Totals
Private	2	1	16	11	0	30
Public	0	3	0	11	1	15
Totals	2	4	16	22	1	45

Source: Data are taken from Special Data Set 1, Appendix D, page D3.

To construct Table 2.12, for example, the joint responses for each of the 45 schools with respect to type of institution and institutional focus are tallied into one of the 10 possible "cells" of the table. Thus from Special Data Set 1 in Appendix D on page D3, the first school listed (Appalachian State University) is a public regional university. These joint responses were tallied into the cell composed of the second row and fourth column. The second institution (Barber Scotia College) is a private regional liberal arts school. These joint responses were tallied into the cell composed of the first row and third column. The remaining 43 joint responses were recorded in a similar manner.

In order to explore any possible pattern or relationship between type of institution and the institutional focus, it is useful to first convert these results into percentages based on

1. The overall total (i.e., the 45 colleges and universities in North Carolina)
2. The row totals (private or public)
3. The column totals [national liberal arts school (NLA), national university (NU), regional liberal arts school (RLA), regional university (RU), or specialty school (SS)]

This is accomplished in Tables 2.13, 2.14, and 2.15, respectively. Let us examine some of the many findings presented in these tables. From Table 2.13, we note that

1. 66.7% of the institutions in North Carolina are private.
2. 8.9% of the institutions in North Carolina are classified as national universities.
3. 2.2% of the institutions in North Carolina are private national universities.

From Table 2.14, we note that

1. 53.3% of the private institutions are classified as regional liberal arts schools.
2. 73.3% of the public institutions are classified as regional universities.

Table 2.13 Contingency table displaying type of institution and institutional focus of the North Carolina schools (percentages based on overall total).

Type of Institution	NLA	NU	RLA	RU	SS	Totals
Private	4.4	2.2	35.6	24.4	0.0	66.7
Public	0.0	6.7	0.0	24.4	2.2	33.3
Totals	4.4	8.9	35.6	48.9	2.2	100.0

Source: Data are taken from Table 2.12.

Table 2.14 Contingency table displaying type of institution and institutional focus of the North Carolina schools (percentages based on row totals).

Type of Institution	NLA	NU	RLA	RU	SS	Totals
Private	6.7	3.3	53.3	36.7	0.0	100.0
Public	0.0	20.0	0.0	73.3	6.7	100.0
Totals	4.4	8.9	35.6	48.9	2.2	100.0

Source: Data are taken from Table 2.12.

Table 2.15 Contingency table displaying type of institution and institutional focus of the North Carolina schools (percentages based on column totals).

Type of Institution	NLA	NU	RLA	RU	SS	Totals
Private	100.0	25.0	100.0	50.0	0.0	66.7
Public	0.0	75.0	0.0	50.0	100.0	33.3
Totals	100.0	100.0	100.0	100.0	100.0	100.0

Source: Data are taken from Table 2.12.

And, from Table 2.15, we note that

1. 25.0% of the institutions classified as national universities are private.
2. 50.0% of the institutions classified as regional universities are public.

The tables, therefore, indicate a pattern: North Carolina institutions of higher learning comprise mainly regional universities and regional liberal arts schools—the former are evenly split between public and private schools and the latter are all private schools.

Problems for Section 2.11

2.41 In a recent study, researchers were looking at the relationship between the type of college attended and the level of job that people who graduated in 1975 held at the time of the study. The researchers examined only graduates who went into industry. The cross-tabulation of the data is as follows:

	Type of College		
Management Level	**Ivy League**	**Other Private**	**Public**
High (Sr. V-P or above)	45	62	75
Middle	231	563	962
Low	254	341	732

(a) Construct a table with either row or column percentages, depending on which you think is more informative.
(b) Interpret the results of the study.
(c) What other variable or variables might you want to know before advising someone to attend an Ivy League or other private school if he or she wants to get to the top in business?

2.42 People returning from vacations in different countries were asked how they enjoyed their vacation. Their responses were as follows:

	Response to Country			
Country	**Yuck**	**So-So**	**Good**	**Great**
England	5	32	65	45
Italy	3	12	32	43
France	8	23	28	25
Guatemala	9	12	6	2

(a) Construct a table of row percentages.
(b) What would you conclude from this study?
(c) **ACTION** Write a letter to the travel editor of your local newspaper regarding your findings.

2.43 The defeat of the incumbent, George Bush, in the 1992 presidential election was attributed to poor economic conditions and high unemployment. Suppose that a survey of 800 adults taken soon after the election resulted in the following cross-classification of financial condition with education level:

	Education Level			
Financial Conditions	**H.S. Degree or Lower**	**Some College**	**College Degree or Higher**	**Totals**
Worse off now than before	261	48	38	347
No difference	104	73	41	218
Better off now than before	65	39	131	235
Totals	430	160	210	800

(a) Construct a table of column percentages.
(b) What would you conclude from this study?
(c) **ACTION** Write a letter to your political science professor regarding your findings.

2.12 USING MICROSOFT EXCEL TO OBTAIN TABLES AND CHARTS FOR CATEGORICAL VARIABLES

2.12.1 Introduction

In Sections 2.8–2.11, we studied the data for colleges and universities in North Carolina and formed a one-way summary table for the institutional focus of schools and a two-way contingency table for the type of school and its institutional focus. Using the summary table, we developed the bar chart, the pie chart, and the Pareto diagram. In this section we will discuss how we can use Excel to obtain these tables and charts for the North Carolina data. To organize our workbook for this purpose, we need to design and add a OneWayTable sheet in a workbook that contains a Data sheet similar to the one for the Texas tuition data. Table 2.3.Excel represents the design of the Data sheet.

Note that this design reserves space for a PivotTable, which is a special type of table that Microsoft Excel creates to summarize data. If summary data is already available, category labels could be entered in column A and their corresponding frequencies in column B.

Table 2.3.Excel Design of the One-Way Tables sheet for obtaining one-way summary tables and charts.

	A	B	C	D	E	F
1			One-Way Summary Table for N.C. schools			
2						
3		Reserved for				
4			Focus	Frequency	Percentage	Cumulative Percentage
5			=$A$5	=$B$5	=D5/$B$10	=E5
6			.	.	.	F5+E6
7			.	.	.	.
8			.	.	.	.
9			=$A$9	=$B$9	=D9/$B$10	F8+E9
10		Pivot Table				

2.12.2 Using the PivotTable Wizard to Obtain One-Way Summary Tables and Contingency Tables

The PivotTable is a type of table that summarizes information about different variables (called *fields* in Microsoft Excel). The results can be provided in various forms including counts, averages, or totals.

Selecting the command Data | PivotTable begins the PivotTable Wizard. Using this Wizard, we specify the variables of interest, the organization, and the type of data to be summarized. In this section, in developing one- and two-way tables for categorical variables, we will be interested in obtaining a count of the number of observations in a group or in a cell of a two-way contingency table.

To create a summary table for the institutional focus of North Carolina schools similar to Table 2.11 on page 85, open the NC-1.XLS file. Observe that the first row in the Data sheet contains the column headings Tuition, Type, Setting, Calendar, and Focus. Providing column headings allows the PivotTable Wizard to form an on-screen button for each variable.

NC-1.XLS

❶ Since we want the output of the PivotTable to appear on a separate sheet, click on Insert | Worksheet, rename the new sheet as OneWayTable, and click on the Data Sheet tab to return to the Data sheet.

❷ Select the command Data | PivotTable. The first of four PivotTable Wizard dialog boxes appears. (See Figure 2.10.Excel, Panel A.) Click on the Microsoft Excel List or Database button and click the Next button to move to the next dialog box.

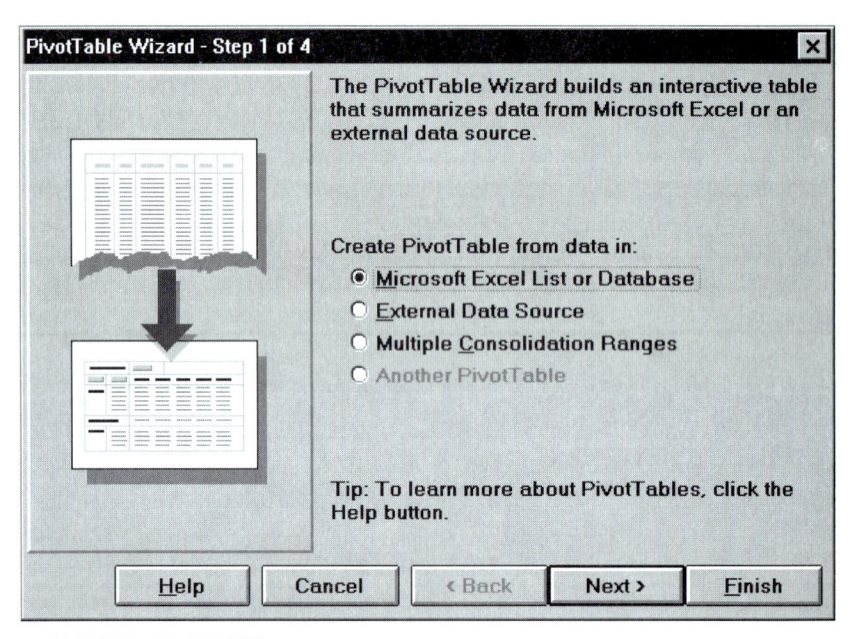

FIGURE 2.10.EXCEL PivotTable Wizard steps for the North Carolina institutional focus data. Step 1. Panel A.

❸ This second dialog box (Panel B) asks you to enter the range of your data. Enter Data!B1:F46 and click the Next button.

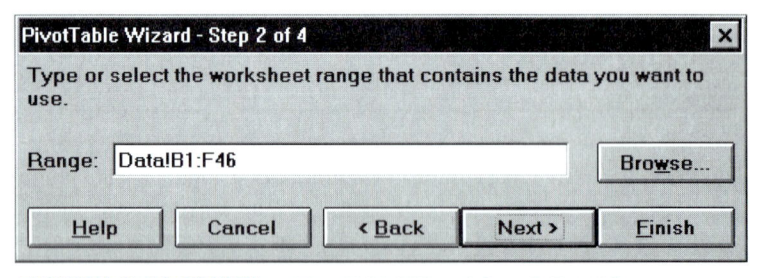

FIGURE 2.10.EXCEL PivotTable Wizard Step 2. Panel B.

❹ The third of four PivotTable Wizard dialog boxes, illustrated in Panel C, allows you to indicate the variables to be included in the PivotTable and the type of data to be analyzed. On the right side of the dialog box appear draggable labels containing the column headings for the five variables of the data sheet, tuition, type, setting, calendar, and focus. Since we wish to obtain a summary table for the variable Focus, we need to drag the Focus label and drop it (by releasing the mouse button) in the rectangular box named Row. Once you have dragged and dropped, you will observe that Focus appears at the top of

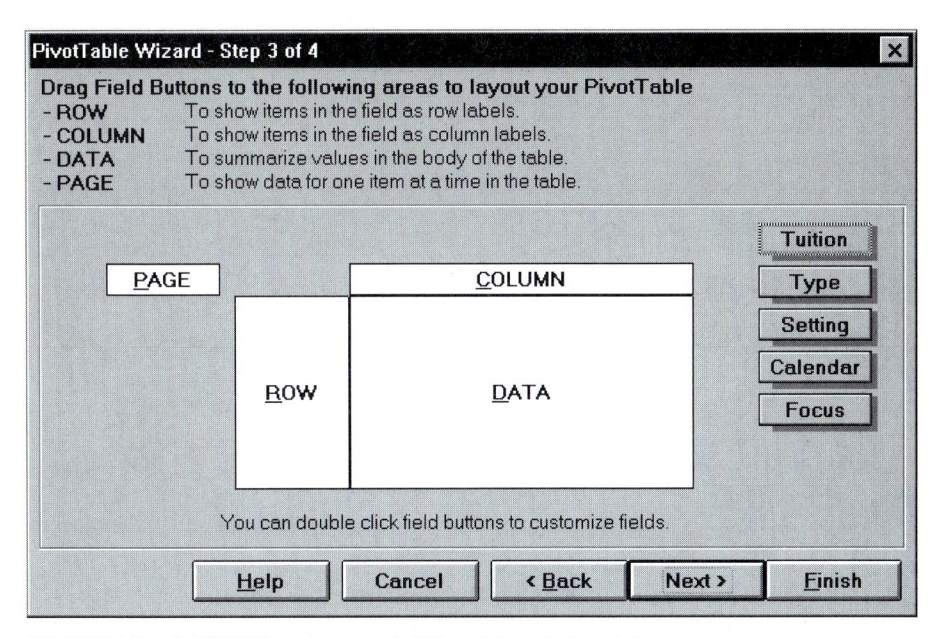

FIGURE 2.10.EXCEL   PivotTable Wizard Step 3. Panel C.

the Row box. Now, to obtain a count of the number of schools in each class, we select Focus again and drag and drop it to the Data area. The name in the box titled Data should change to Count of Focus. If it does not, double-click on the name indicated. This will present you with a PivotTable Field dialog box. In the Summarize by list box, select Count and then click the Options button to display the extended PivotTable Field dialog box. In the Show Data as drop down list box shown in Figure 2.10.Excel, Panel D, select

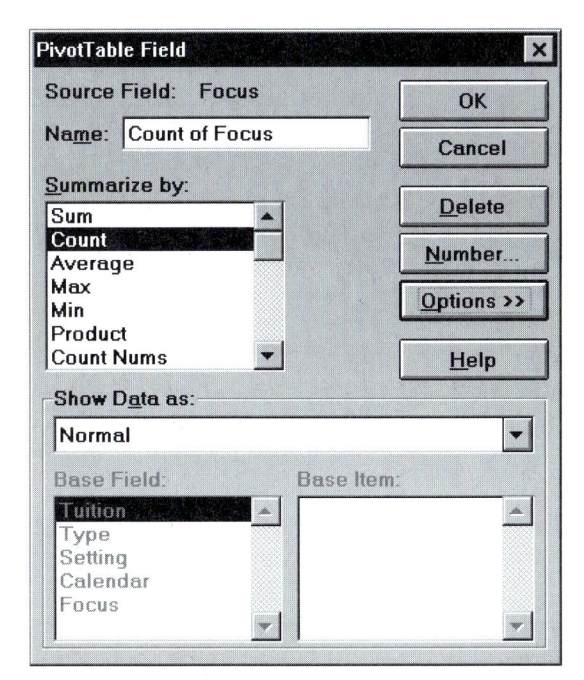

FIGURE 2.10.EXCEL   Extended PivotTable Field dialog box. Panel D.

2.12 Using Microsoft Excel to Obtain Tables and Charts for Categorical Variables

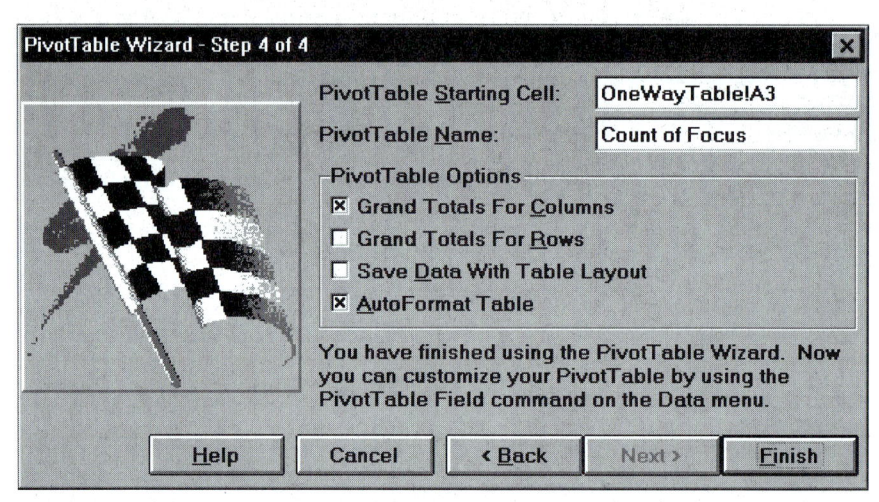

FIGURE 2.10.EXCEL PivotTable Wizard Step 4. Panel E.

Normal to obtain frequencies. Click the OK button to return to dialog box 3 of the PivotTable Wizard and then click the Next button. The fourth dialog box appears.

❺ The fourth dialog box in this Wizard allows you to place and format your table. You enter OneWayTable!A3 in the PivotTable Starting Cell edit box and then enter Count of Focus in the PivotTable Name edit box. Continue by selecting the Grand Total for Columns check box to verify the count of the sample size and then the AutoFormat Table check box. (See Panel E.) Click the Finish button to have Microsoft Excel produce a PivotTable similar to the one illustrated in Figure 2.11.Excel.

FIGURE 2.11.EXCEL
Summary table of North Carolina institutional focus data.

Having used the PivotTable Wizard for the one-way summary table, we can use the Wizard a second time to obtain a two-way contingency table such as Table 2.11, on page 85, which cross-classifies the type of institution with the focus of the institution. Repeat steps 1–3 as before, this time renaming the new sheet TwoWayTable. Continue with the following steps 4 and 5:

❹ In the third PivotTable Wizard dialog box, since we wish to obtain a two-way contingency table for the variables Type (in the rows) and Focus (in the columns), we need to select and drag and drop the Type label in the row box on the left side of the dialog box and select and drag and drop the Focus label in the Column area. Select the Focus

98   **Chapter 2**   Presenting Data in Tables and Charts

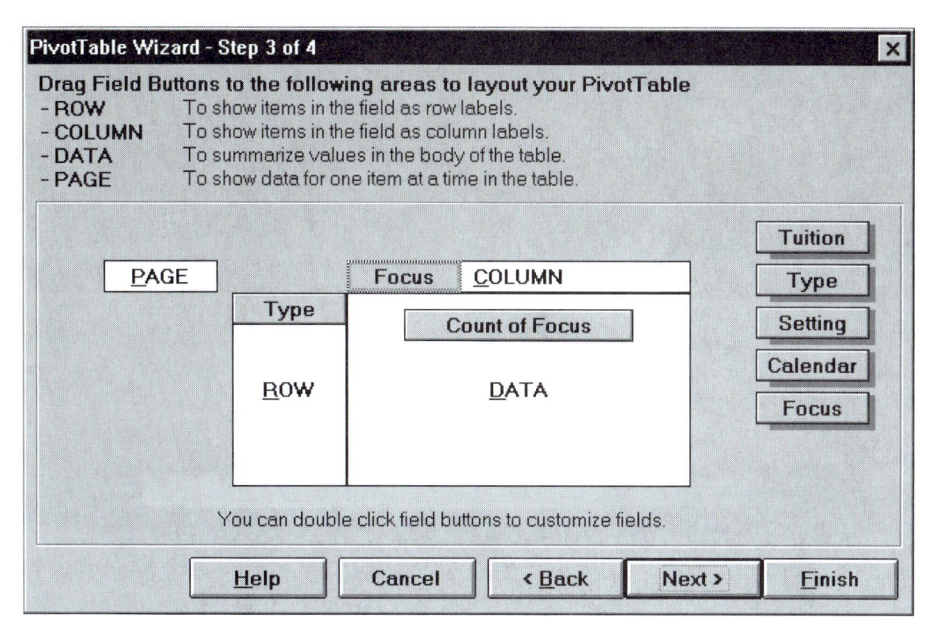

FIGURE 2.12.EXCEL Complete PivotTable Wizard step 3 dialog box for the North Carolina institutional focus data.

label a second time and drag and drop it to the Data area. (See Figure 2.12.Excel.) Click the Next button. The fourth dialog box appears.

⑤ In the fourth and final dialog box (Panel E), enter TwoWayTable!A3 in the PivotTable Starting Cell edit box and then enter Type X Focus in the PivotTable Name edit box. Continue by selecting the Grand Total for Columns check box and the Grand Total for Rows check box to verify the count of the sample size and then the AutoFormat Table check box. Click the Finish button to have Microsoft Excel produce a PivotTable similar to the one illustrated in Figure 2.13.Excel.

Now suppose we wish to obtain contingency tables similar to Tables 2.13, 2.14, and 2.15 on page 93 that express the results in terms of total percentages, row percentages, and column percentages. Then in step 3 of the PivotTable Wizard from the Show Data drop-down list box in the extended PivotTable Field dialog box, we need to select % of Total, % of Row Total, or % of Column Total, respectively.

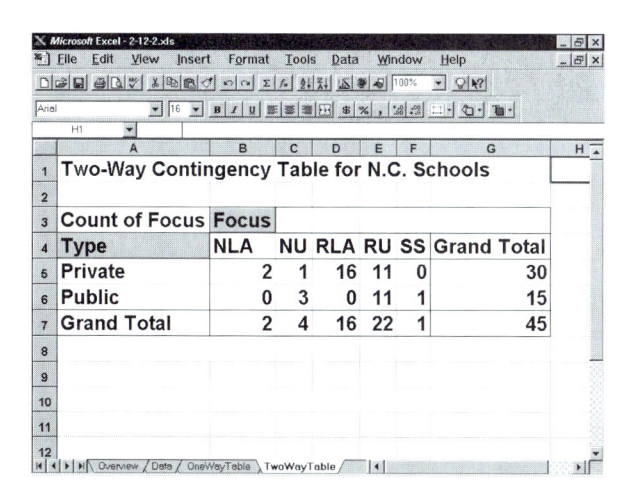

FIGURE 2.13.EXCEL
Contingency table obtained from Excel for the North Carolina institutional focus data.

2.12 Using Microsoft Excel to Obtain Tables and Charts for Categorical Variables

2.12.3 Using the Chart Wizard to Obtain Bar Charts, Pie Charts, and Pareto Diagrams

In Section 2.7.4, we used the line chart option of the Chart Wizard to obtain a frequency polygon for the Texas out-of-state tuition rates. Now we will use the Chart Wizard in conjunction with the output of the one-way PivotTable to obtain a bar chart, pie chart, and Pareto diagram for the North Carolina institutional focus data summarized in Table 2.11 on page 85 and in Figure 2.11.Excel. To obtain these charts from the Chart Wizard, open the workbook file you developed in Section 2.12.2 or open the file NC-2.XLS and do the following:

NC-2.XLS

❶ Select Insert | Chart | As New Sheet (since we will want to place the bar chart on a separate sheet). The first dialog box of the Chart Wizard, asking for the range of the data to be plotted, appears.

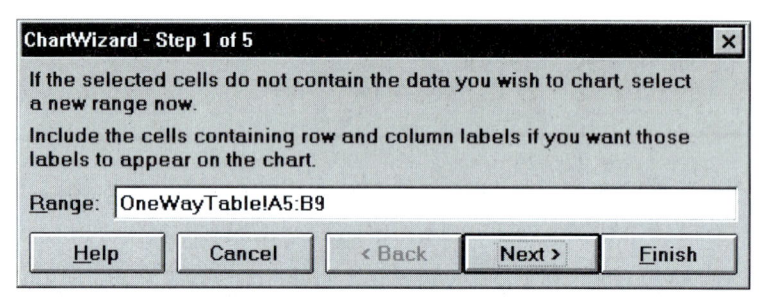

FIGURE 2.14.EXCEL Chart Wizard steps for obtaining a bar chart for the North Carolina institutional focus data. Step 1. Panel A.

❷ In the Range edit box, enter OneWay Table!A5:B9. (See Figure 2.14.Excel, Panel A.) Note that to obtain a chart with a properly labeled *X*-axis, the cells containing the class categories must be located in a column to the left of the cells containing the frequencies (as is the case with this PivotTable). Click the Next button to continue to the second dialog box.

❸ In the second dialog box select the Bar Chart choice. (See Panel B.) Click the Next button.

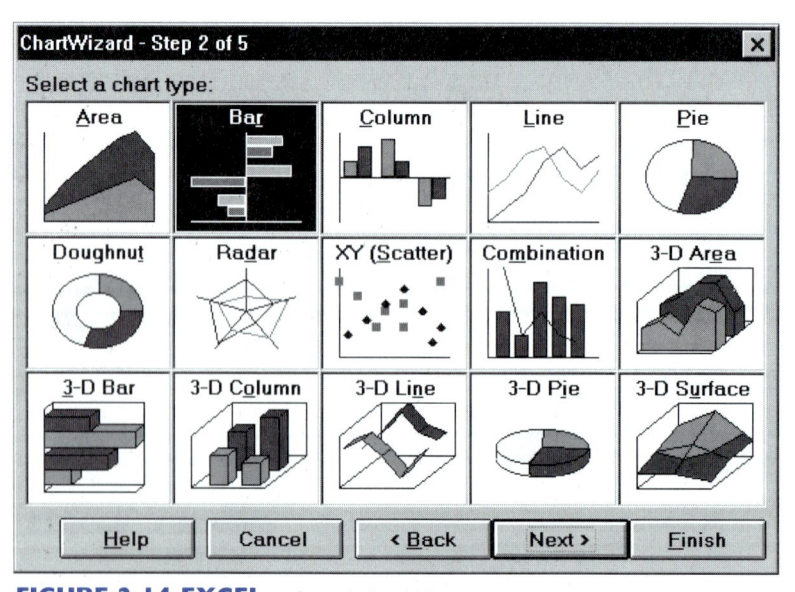

FIGURE 2.14.EXCEL Step 2. Panel B.

100 Chapter 2 Presenting Data in Tables and Charts

④ In the third dialog box select the third choice, a bar chart that has bars going out to the right from the vertical axis. (See Panel C.) Click the Next button.

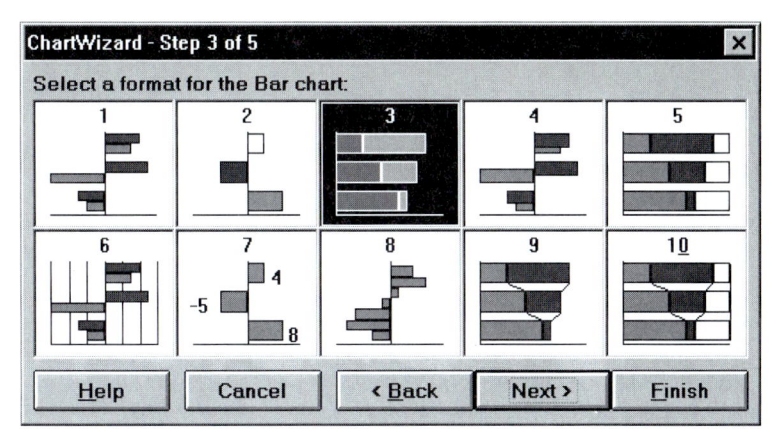

FIGURE 2.14.EXCEL   Step 3. Panel C.

⑤ In the fourth dialog box select the Columns option button under the Data Series in heading. Then enter 1 in the First Columns edit box and enter 0 in the First Rows edit box. (See Figure 2.14.Excel, Panel D.) Click the Next button to move to the next dialog box.

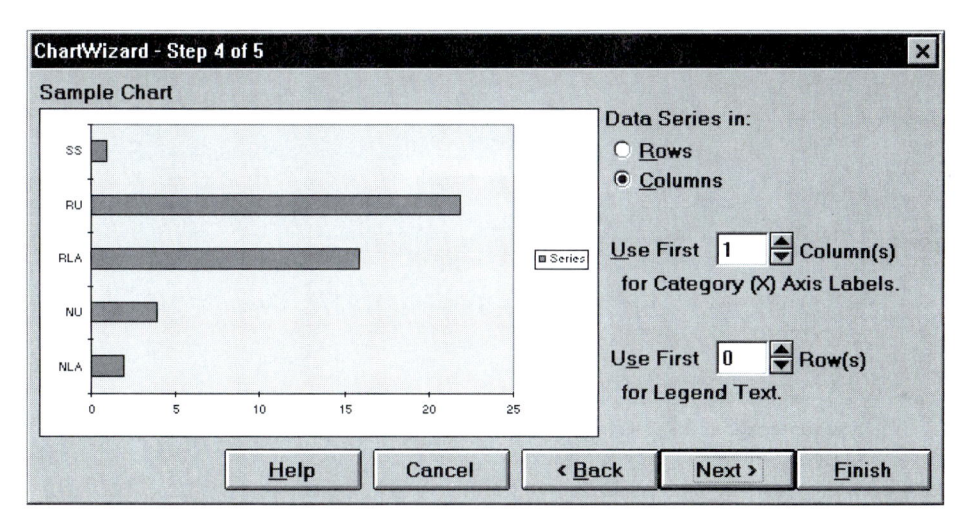

FIGURE 2.14.EXCEL   Step 4. Panel D.

⑥ In the fifth dialog box you can enter a chart legend for the chart as well as titles for the axes. To obtain the bar chart in Figure 2.15.Excel, select the No option button for Add a Legend and enter the following: N.C. Schools Focus as the title, Focus for the Category (X) edit box, and Frequency for the Value (Y) edit box. (See Panel E.) Click the Finish button.

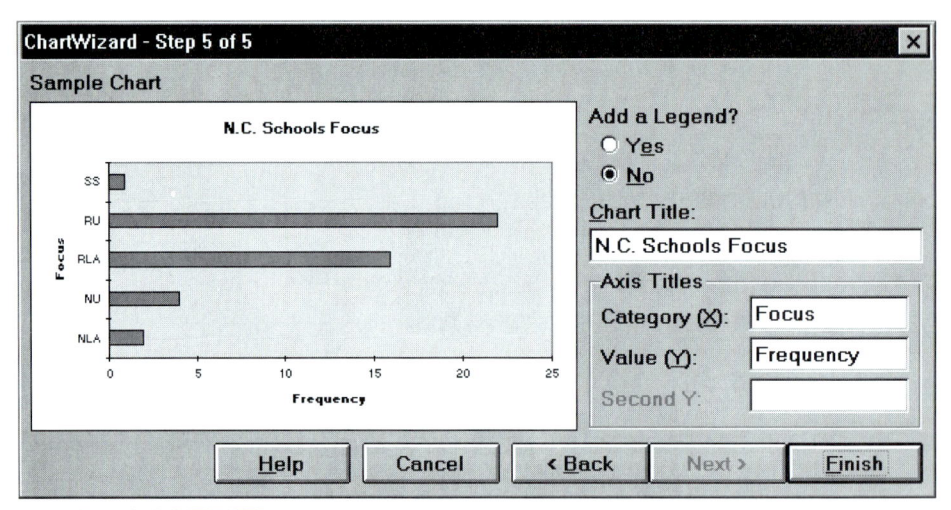

FIGURE 2.14.EXCEL   Step 5. Panel E.

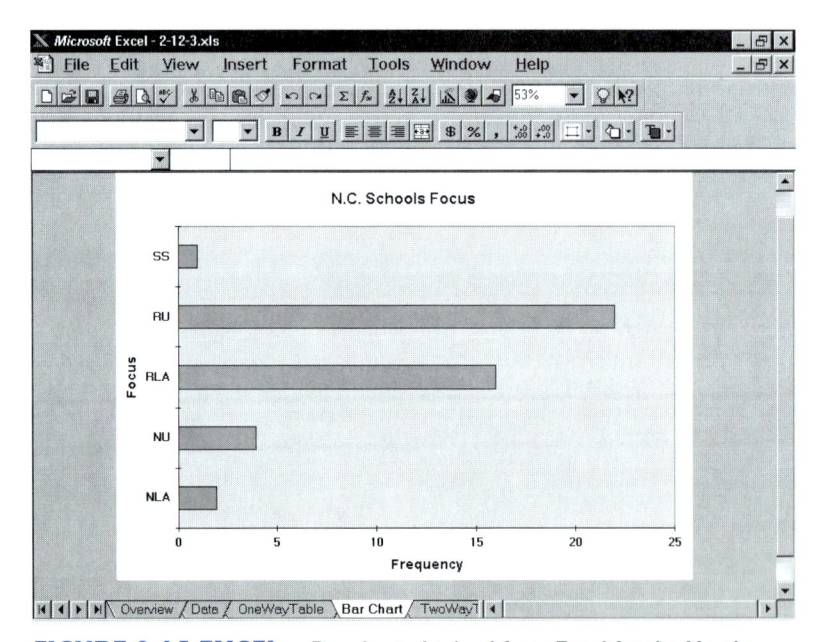

FIGURE 2.15.EXCEL   Bar chart obtained from Excel for the North Carolina institutional focus data.

Figure 2.15.Excel displays the bar chart we have obtained. Rename this sheet as BarChart. Now that we have obtained the bar chart for these data, we can return to the Chart Wizard to obtain a pie chart. Repeat steps 1 and 2 of the previous description for the bar chart and continue with steps 3–6 as follows:

❸ Click the chart titled Pie, which is the last chart on the right in the first row (Figure 2.14.Excel, Panel B). Then click the Next button.

❹ Click on Option 7, which shows the percentages corresponding to each section of the pie. Then click the Next button.

❺ Click the Columns button for Data Series. Since the categories are in the first column, enter 1 for First Columns and enter 0 for First Rows and then click the Next button.

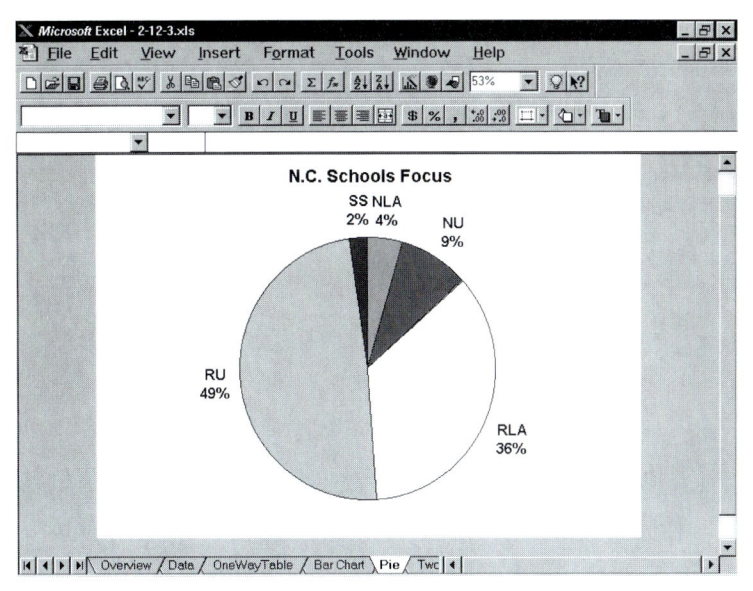

FIGURE 2.16.EXCEL Pie chart obtained from Excel for the North Carolina institutional focus data.

❻ Click the No button for Add a Legend. Select the Chart Title edit box and enter N. C. Schools Focus. Click on the Finish button.

Figure 2.16.Excel displays the pie chart we have obtained. Change the name of this sheet to Pie Chart.

Now that we have obtained the bar chart and the pie chart for these data, we would like to also develop a Pareto diagram. The Pareto diagram is an option that is not available in current versions of the Chart Wizard, so we need to use Excel formulas to develop the cumulative percentages and then use the combination chart selection on the Chart Wizard.

Assuming that we have the summary PivotTable shown in Figure 2.11.Excel, we need to copy the results of the PivotTable in columns A and B to columns C and D to be able to properly sort the categories in order using the Sort command. Then we sort the categories and frequencies in columns C and D and obtain the percentages for each category in descending order in column E and the cumulative percentages in column F. This may be accomplished by implementing the design of the OneWay Tables sheet shown in Table 2.3.Excel on page 95 and doing the following:

❶ First copy the contents of the PivotTable to columns C and D, making sure to use absolute addresses.

❷ Now we need to sort the categories in descending order according to the frequency in each category. Select the range C4:D9, and select the command Data | Sort. The Sort dialog box appears. In the Sort By drop-down list box, select Frequency and then select the Descending option button. Then click the OK button.

❸ Enter the formula =D5/$B$10 in cell E5 and copy this formula to cells E6 through E9. To change the results to percentage format, select the range E5:E9 and click the % button and then the Increase Decimal button twice (to get two decimal places).

❹ To obtain the cumulative percentage distribution, enter the formula =E5 in cell F5 and then enter =F5+E6 into cell F6. Copy the contents of cell F6 to cells F7:F9.

2.12 Using Microsoft Excel to Obtain Tables and Charts for Categorical Variables

Now that we have obtained the percentage in each class and the cumulative percentages, we can use the Chart Wizard to obtain the Pareto diagram as we did with the bar chart and the pie chart. With the OneWayTable sheet active, select Insert | Chart | As New Sheet, and enter the following at each step of the Chart Wizard:

Wizard Step 1. Enter the range of the data as C5:C9,E5:F9. (The comma is necessary here since the data are located in two separate *nonadjacent* areas.)

Wizard Step 2. Select the Combination chart. This chart will provide a bar chart and a line chart.

Wizard Step 3. Select the second choice that will plot a separate bar and line chart with two *Y*-axes.

Wizard Step 4. Select the Columns option button for Data Series, enter 1 in the First Columns edit box, and enter 0 in the First Rows edit box.

Wizard Step 5. Select the No option button for Add a Legend and enter Pareto Diagram as the Chart Title and Focus in the Category X edit box.

Figure 2.17.Excel illustrates the frequency and cumulative distributions along with the Pareto diagram. Rename this sheet as Pareto.

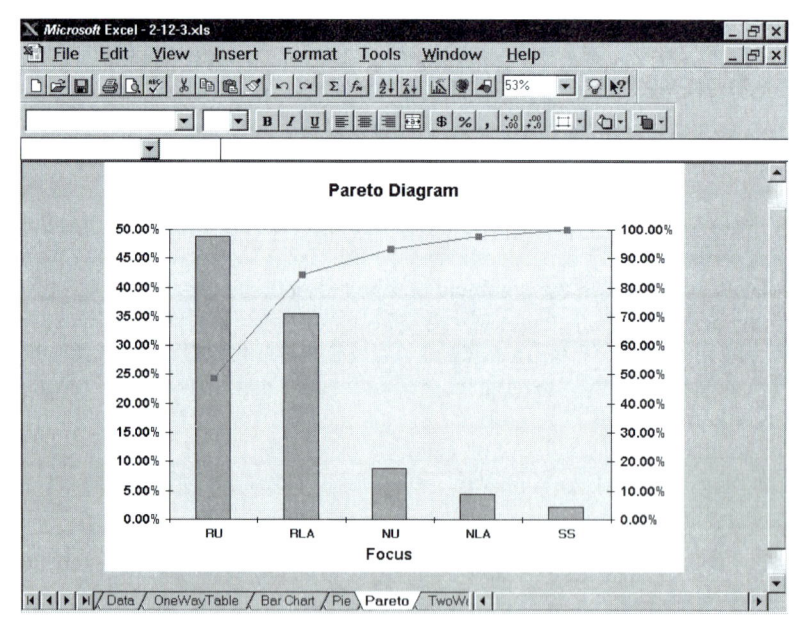

FIGURE 2.17.EXCEL Pareto diagram obtained from Excel for the North Carolina institutional focus data.

2.13 PROPER TABULAR AND CHART PRESENTATION AND ETHICAL ISSUES

To this point we have studied how a collected set of data is presented in tabular and chart form. If our analysis is to be enhanced by a visual display of data, it is essential that the tables and charts be presented clearly and carefully. Tabular frills and other **"chart junk"** must be eliminated so as not to cloud the message given by the data with unnecessary adornments (References 7, 12, 13, and 15). In addition to eliminating chart junk when we display charts, we must also avoid common errors that distort our visual impression (References 6, 9, and 10). Two such errors are

1. Failing to compare two or more sets of data on a relative basis
2. Failing to indicate the zero point at the bottom of the vertical axis

2.13.1 Failing to Compare Data Sets on a Relative Basis

In Section 2.4, we demonstrated why it is necessary to compare two or more sets of data on a relative basis, and Figures 2.7 (page 69) and 2.9 (page 73) respectively displayed the proper percentage polygons and percentage ogives comparing the out-of-state tuition rates for 60 Texas schools and 45 North Carolina schools. Using frequency counts rather than percentages or proportions would be misleading. To see this, look at Figures 2.13 and 2.14 where frequency polygons and frequency ogives improperly "compare" the Texas and North Carolina out-of-state tuition rates. In addition, to dramatize the visual distortion, we include the out-of-state tuition rates charged by the 90 Pennsylvania schools (see Special Data Set 1 of Appendix D on pages D4–D5). As can be seen from Figures 2.13 and 2.14, the frequency polygons and frequency ogives for the Texas and North Carolina schools are overwhelmed by those for the Pennsylvania schools, and no meaningful comparisons can be made from such distorted charts.

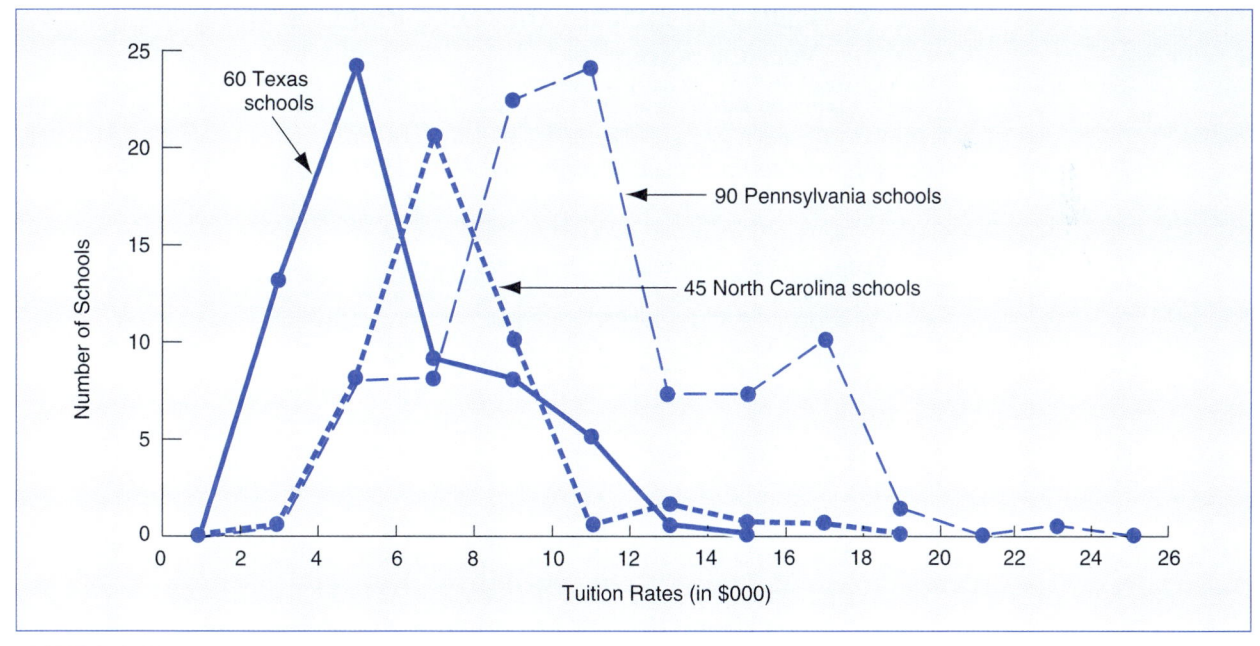

FIGURE 2.13 "Improper" frequency polygons for the out-of-state tuition rates for schools in Texas, North Carolina, and Pennsylvania. *Source:* Data are taken from Tables 2.3, 2.7, and "America's Best Colleges, 1994 College Guide," *U.S. News & World Report*, extracted from College Counsel 1993 of Natick, Mass. Reprinted by special permission, *U.S. News & World Report*, © 1993 by *U.S. News & World Report* and by College Counsel.

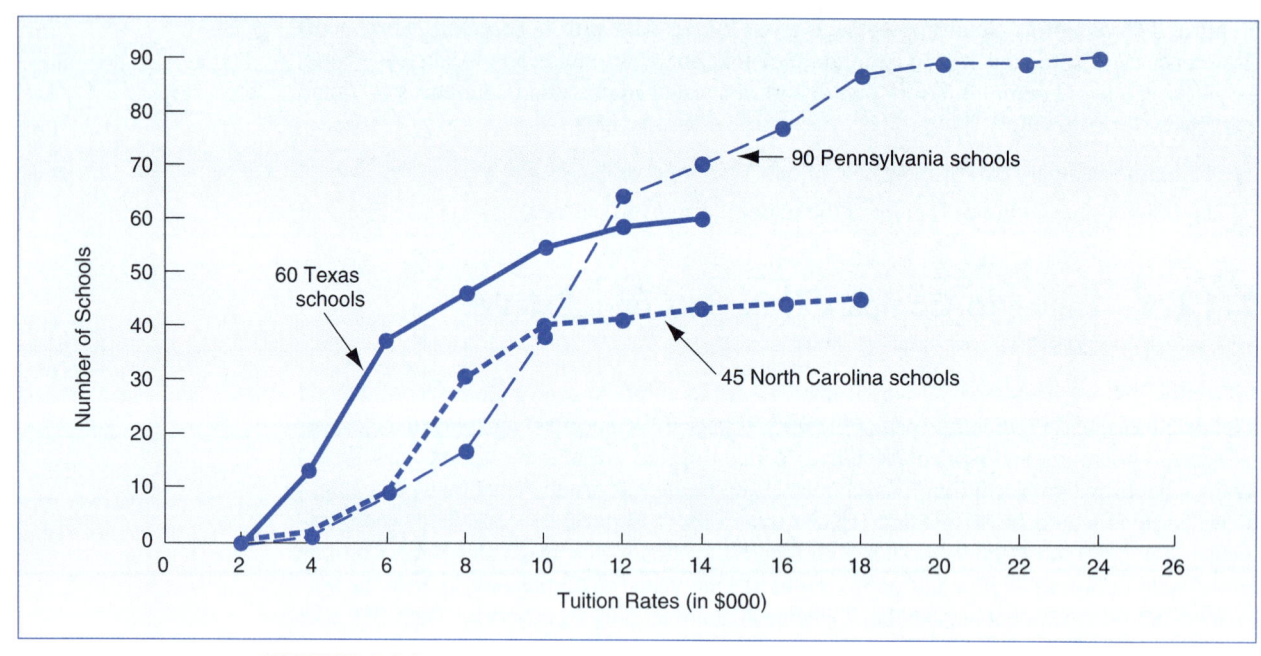

FIGURE 2.14 "Improper" cumulative frequency polygons for the out-of-state tuition rates for schools in Texas, North Carolina, and Pennsylvania. *Source:* Data are taken from Tables 2.3, 2.7, and "America's Best Colleges, 1994 College Guide," *U.S. News & World Report,* extracted from College Counsel 1993 of Natick, Mass. Reprinted by special permission, *U.S. News & World Report,* © 1993 by *U.S. News & World Report* and by College Counsel.

2.13.2 Failing to Indicate the Zero Point on the Vertical Axis

The starting point on the vertical axis must be indicated with a zero so as not to distort the visual impression regarding the magnitude of changes occurring in the chart. By taking only a slice of the vertical axis, such changes can be exaggerated. Figure 2.15 displays such a visual distortion.

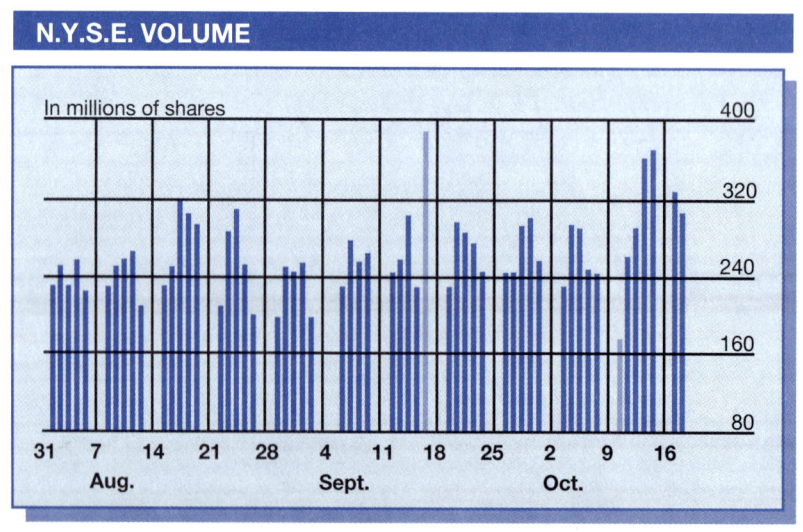

FIGURE 2.15 "Improper" display of New York Stock Exchange sales volume (in millions of shares traded) over time. *Source: The New York Times,* October 20, 1993, p. D7.

106 Chapter 2 Presenting Data in Tables and Charts

Note that in this chart, zero was omitted from the vertical axis. Because of this the reader gets a distorted view of the magnitude of the differences in the daily transactions. For example, over the portrayed time period, the most active trading session occurred on Friday, September 17, while the least active trading session occurred on Monday, October 11 (Columbus Day). However, from the incorrectly drawn graph, the vertical bar representing the most active trading session is three times longer than the vertical bar representing the least active session—leaving the impression that three times as many shares were traded on September 17 than on October 11. Had the zero point been properly displayed on the vertical axis, the graph would have accurately portrayed the fact that only twice as many shares were traded on September 17 than on October 11.

2.13.3 Ethical Issues

Ethical considerations arise when we are deciding what data to present in tabular and chart format and what not to present. It is vitally important when presenting data to document both good and bad results. When making oral presentations and presenting written reports, it is essential that the results be given in a fair, objective, and neutral manner. Thus, we must try to distinguish between poor data presentation and unethical presentation. Again, as in our discussion of ethical considerations in data collection (Section 1.11), the key is *intent*. Often, when fancy tables and chart junk are presented or pertinent information is omitted, it is simply done out of ignorance. However, unethical behavior occurs when an individual willfully hides the facts by distorting a table or chart or by failing to report pertinent findings.

Problem for Section 2.13

2.44 **(Student Project)** Bring to class a chart from a newspaper or magazine that you believe to be a poorly drawn representation of some numerical variable. Be prepared to submit the chart to the instructor with comments as to why you feel it is inappropriate. Also, be prepared to present and comment on this in class.

2.14 DATA PRESENTATION: A REVIEW AND A PREVIEW

As you can see in the following summary chart, this chapter was about data presentation. To be sure you understand the material covered in this chapter, you should be able to answer the following conceptual questions:

1. Why is it necessary to organize a set of numerical data that we collect?
2. What are the main differences between an ordered array and a stem-and-leaf display?
3. Under what conditions is it most appropriate to construct and use frequency distributions and percentage distributions?
4. How do histograms and polygons differ with respect to their construction and use?
5. When should a percentage ogive (i.e., cumulative percentage polygon) be constructed and how should it be used?
6. Why is the percentage ogive such a useful tool?
7. Why would you construct a frequency and percentage summary table?
8. How do you construct a frequency and percentage summary table?
9. How do you construct a bar chart, pie chart, and Pareto diagram?
10. What are the advantages and/or disadvantages for using a bar chart, a pie chart, or a Pareto diagram?

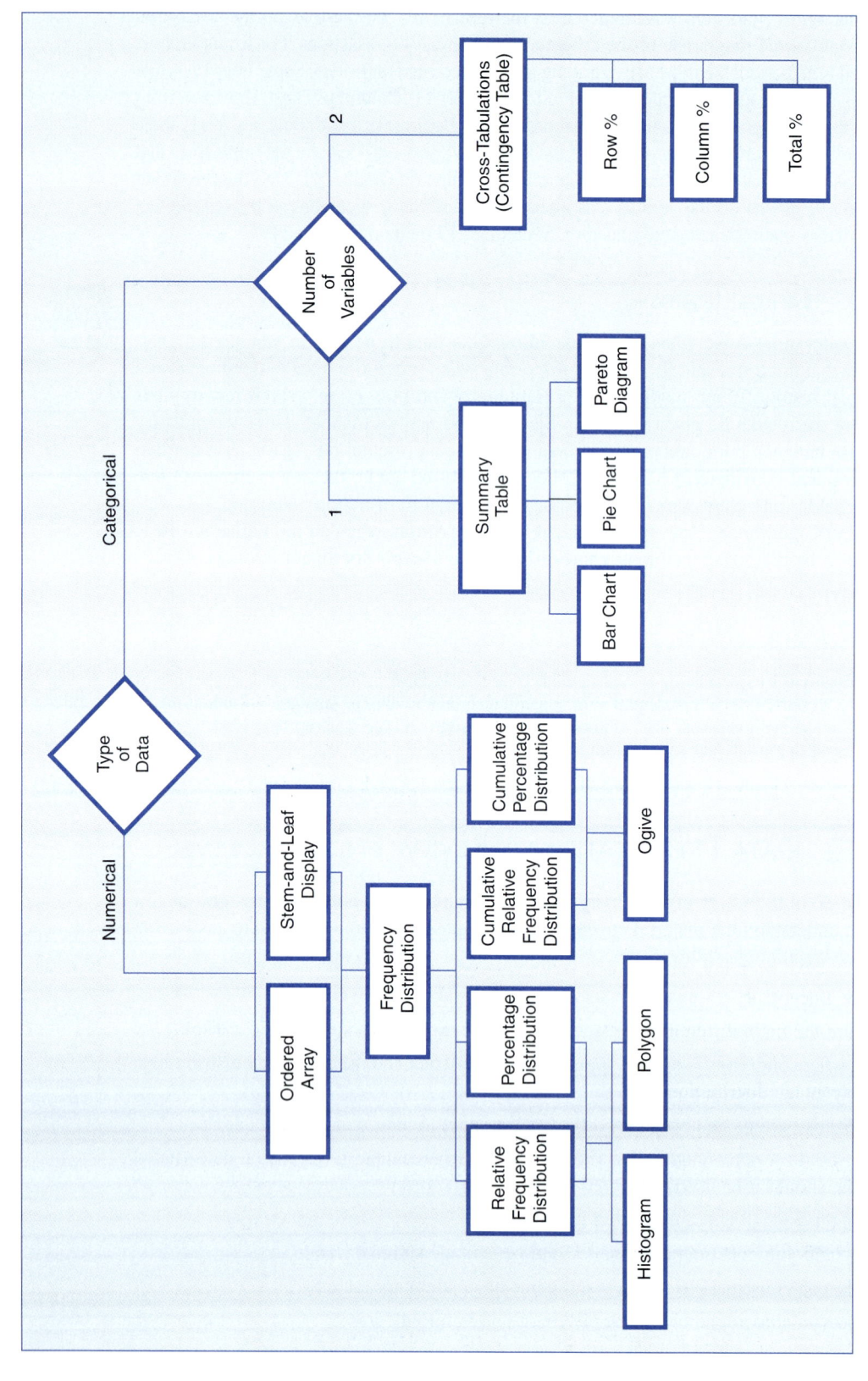

Chapter 2 summary chart.

11. What kinds of percentage breakdowns can assist you in interpreting the results found through the cross-classification of data based on two categorical variables?

12. What are some of the ethical issues to be concerned with when presenting data in tabular or chart format?

Check over the list of questions to see if you know the answers and could (1) explain your answers to someone who did not read this chapter and (2) reference specific readings or examples that support your answers. Also, reread any of the sections that may have seemed unclear to see whether they make sense now.

Once the collected data have been presented in tabular and chart format, we are ready to make various analyses. In the following chapter, a variety of descriptive summary measures useful for data analysis and interpretation will be developed.

Getting It All Together

Key Terms

absolute address   76
bar chart   85
"chart junk"   104
class boundaries   62
class groupings   61
class midpoint   63
contingency table   92
cumulative percentage distribution table   70
frequency distribution   61
histogram   67
ogive (cumulative percentage polygon)   71

ordered array   52
Pareto diagram   88
percentage distribution   65
pie chart   86
percentage polygon   68
raw form   52
relative frequency distribution   65
revised stem-and-leaf display   54
stem-and-leaf display   53
summary table   85
width of the class interval   62

Chapter Review Problems

Note: *You may use Microsoft Excel to solve these problems.*

2.45 The following data are the retail prices for a random sample of 30 pencil tire pressure gauge models:

GAUGE.TXT

4.50	6.50	2.00	2.50	4.00	3.50	5.00	3.00	5.00	5.50
1.00	7.50	3.00	2.00	3.00	3.50	3.50	5.00	6.00	4.50
2.00	3.00	3.50	3.50	3.00	3.00	4.00	1.50	1.50	2.50

Source: "Tire-pressure Gauges," Copyright 1993 by Consumers Union of United States, Inc., Yonkers, N.Y. 10703. Adapted by permission from *Consumer Reports,* February 1993, pp. 98–99. Although these data sets originally appeared in *Consumer Reports,* the selective adaptation and resulting conclusions presented are those of the authors and are not sanctioned or endorsed in any way by Consumers Union, the publisher of *Consumer Reports.*

(a) Place the raw data in a stem-and-leaf display. (*Hint:* Let the leaves be the tenths digits.)
(b) Place the raw data in an ordered array.
(c) Form the frequency distribution and percentage distribution.
(d) Plot the percentage histogram.

(e) Plot the percentage polygon.
(f) Form the cumulative percentage distribution.
(g) Plot the ogive (cumulative percentage polygon).
(h) **ACTION** If you were considering purchasing a pencil tire pressure gauge, what else would you want to know? Write a list of questions you would ask in an automobile supply store.

2.46 TOOTHPST.TXT The following data represent the cost per month of use (in dollars) and the cleaning test score (0–100) for a random sample of 39 brands of tubed toothpaste:

Toothpaste	Cost per Month	Score
Ultra Brite Original	.58	86
Gleem	.66	79
Caffree Regular	1.02	77
Crest Tartar Control Fresh Mint Gel	.53	75
Colgate Tartar Control Gel	.57	74
Crest Tartar Control Original	.53	72
Ultra Brite Gel Cool Mint	.52	72
Colgate Clear Blue Gel	.71	71
Crest Cool Mint Gel	.55	70
Crest Regular	.59	69
Crest Sparkle	.51	64
Close-Up Tartar Control Gel	.67	63
Close-Up Anti-Plaque	.62	62
Colgate Tartar Control Paste	.66	62
Tom's of Maine Cinnamint	1.07	62
Aquafresh Tartar Control	.80	60
Aim Anti-Tartar Gel	.79	58
Aim Extra-Strength Gel	.44	57
Slimer Gel	1.04	57
Arm & Hammer Baking Soda Fresh Mint Gel	1.12	55
Aquafresh	.79	56
Aquafresh Extra Fresh	.81	53
Close-Up Paste	.64	85
Topol Spearmint Gel	1.77	82
Topol Spearmint	1.32	76
Close-Up Mint Gel	.64	72
Aim Regular-Strength Gel	.55	70
Pepsodent	.39	58
Colgate Baking Soda	1.22	51
Colgate Regular	.74	50
Colgate Junior Gel	.44	39
Colgate Peak	.97	29
Arm & Hammer Baking Soda Fresh Mint	1.26	28
Rembrandt	4.73	53
Sensodyne Original	1.29	80
Sensodyne Gel	1.34	48
Viadent Original Anti-Plaque	1.40	53
Denquel	1.77	37
Butler Protect Gel	1.11	20

Source: "Toothpastes," Copyright 1992 by Consumers Union of United States, Inc., Yonkers, N.Y. 10703. Adapted by permission from *Consumer Reports,* September 1992, pp. 604–605. Although these data sets originally appeared in *Consumer Reports,* the selective adaptation and resulting conclusions presented are those of the authors and are not sanctioned or endorsed in any way by Consumers Union, the publisher of *Consumer Reports.*

For each of the two numerical variables:
(a) Form the stem-and-leaf display.
(b) Form a combined frequency and percentage distribution table.
(c) Plot the percentage polygon.
(d) Form the cumulative percentage distribution.
(e) Plot the percentage ogive.
(f) **ACTION** Write a report for your marketing professor summarizing your findings and characterizing this product.

2.47 The following sets of data are based on closing stock price for random samples of 25 issues traded on the American Exchange and 50 issues traded on the New York Exchange:

American Exchange (25 issues)	New York Exchange (50 issues)	
$ 6.88	$36.50	$26.00
.75	23.50	19.00
3.88	8.25	46.00
4.12	57.50	23.50
11.88	27.12	22.62
15.88	3.75	12.88
16.50	25.00	5.50
8.75	15.50	37.50
9.25	36.12	9.88
7.50	6.00	59.12
5.38	9.12	35.25
14.38	33.38	20.62
2.50	22.50	24.00
4.88	8.75	80.50
6.38	8.62	29.38
33.62	5.75	3.75
4.88	21.88	64.75
9.00	6.12	14.25
2.00	25.00	46.38
20.00	15.88	4.75
14.25	24.00	25.00
4.00	10.88	35.00
15.25	18.75	9.00
2.38	53.88	12.38
49.50	20.38	31.00

(a) Using interval widths of $10, form the frequency distribution and percentage distribution for each exchange.
(b) Plot the frequency histogram for each exchange.
(c) On one graph, plot the percentage polygon for each exchange.
(d) Form the cumulative percentage distribution for each exchange.
(e) On one graph, plot the ogive (cumulative percentage polygon) for each exchange.
(f) **ACTION** Write a brief report to your finance professor comparing and contrasting the two exchanges.

2.48 A wholesale appliance distributing firm wished to study its accounts receivable for two successive months. Two independent samples of 50 accounts were selected for each of the two months. The results are summarized in the table at the top of page 112.

Frequency distributions for accounts receivable.

Amount	March Frequency	April Frequency
$0 to under $2,000	6	10
$2,000 to under $4,000	13	14
$4,000 to under $6,000	17	13
$6,000 to under $8,000	10	10
$8,000 to under $10,000	4	0
$10,000 to under $12,000	0	3
Totals	50	50

(a) Plot the frequency histogram for each month.
(b) On one graph, plot the percentage polygon for each month.
(c) Form the cumulative percentage distribution for each month.
(d) On one graph, plot the ogive (cumulative percentage polygon) for each month.
(e) **ACTION** Write a brief report to your accounting professor comparing and contrasting the accounts receivable of the two months.

2.49 The following table contains the cumulative distributions and cumulative percentage distributions of braking distance (in feet) at 80 mph for a sample of 25 U.S.-manufactured automobile models and for a sample of 72 foreign-made automobile models obtained in a recent year:

Cumulative frequency and percentage distributions for the braking distance (in feet) at 80 mph for U.S.-manufactured and foreign-made automobile models.

Braking Distance (in ft)	U.S.-Made Automobile Models "Less Than" Indicated Values Number	Percentage	Foreign-Made Automobile Models "Less Than" Indicated Values Number	Percentage
210	0	0.0	0	0.0
220	1	4.0	1	1.4
230	2	8.0	4	5.6
240	3	12.0	19	26.4
250	4	16.0	32	44.4
260	8	32.0	54	75.0
270	11	44.0	61	84.7
280	17	68.0	68	94.4
290	21	84.0	68	94.4
300	23	92.0	70	97.2
310	25	100.0	71	98.6
320	25	100.0	72	100.0

Based on these data, answer the following questions:
(a) How many models of U.S.-made automobiles have braking distances of 240 feet or more?
(b) What is the percentage of U.S.-made automobiles with braking distances less than 260 feet?

(c) Which group of car models—U.S. made or foreign made—have the wider *range* in braking distance?
(d) How many foreign-made automobile models have braking distances between 260 feet and 269.9 feet (inclusive)?
(e) Use the cumulative distributions to construct the frequency distributions and percentage distributions for each group of car models.
(f) On one graph, plot the two percentage ogives.
(g) **ACTION** Write a brief report comparing and contrasting the braking distance information for the two groups of car models.

2.50 Environmental conservation is a national issue of major importance. It has been said that Americans threw out 227.1 million tons of garbage in a recent year, enough to fill one of the World Trade Center's Twin Towers from bottom to top every day. Typically, disposal of garbage is achieved through landfills (87%), incineration (7%), and recycling (5%).

Suppose that the consulting firm at which you are employed provides the following table showing the percentage breakdown of the sources of waste:

Source	Percentage
Paper and paperboard	37.1
Yard waste	17.9
Glass	9.7
Metals	9.6
Food waste	8.1
Plastics	7.2
Wood	3.8
Rubber and leather	2.5
Textiles	2.1
All other	2.0
Total	100.0

(a) Form the appropriate graph to pinpoint the "vital few" from the "trivial many."
(b) Analyze the data and summarize your findings.
(c) **ACTION** Write a letter to the Environmental Protection Agency based on your findings and request government information on the potential for recycling for each of the items.

2.51 Serious cervical injuries to professional football players in the last several years have heightened interest in reducing spinal injuries. The following summary tables respectively display the percentage of spinal injuries categorized by cause and the percentage of sports injuries categorized by particular sport:

Spinal Injury Causes	Percentage
Falls	20.8
Motor vehicles	47.7
Sports	14.2
Violence	14.6
Other	2.7
Total	100.0

Sports Spinal Injury Causes	Percentage
Diving	66.0
Football	6.1
Gymnastics	2.2
Horseback riding	2.0
Non-skiing winter sports	2.3
Snow skiing	3.8
Surfing	3.1
Trampoline	2.6
Wrestling	2.3
Other	9.6
Total	100.0

Source: The New York Times, November 20, 1991, p. B11.

(a) For the data on overall spinal injury causes, construct
   (1) A bar chart.
   (2) A pie chart.
(b) Which chart do you prefer for purposes of presentation? Why?
(c) For the data on sports spinal injuries, develop the appropriate graph to pinpoint the "vital few" from the "trivial many."
(d) Analyze the data and summarize your findings.
(e) **ACTION** Write a letter to the sports editor of your local newspaper explaining your findings.

2.52 The following data represent the market share of all drinks and a stratification of the carbonated soft drink market share based on supermarket sales:

Type of Carbonated Soft Drink	Market Share (in %)
Caffeinated cola	48.0
Caffeine-free cola	10.4
Citrus	3.4
Club soda	0.4
Cream	1.4
Dr. Pepper	3.9
Ginger ale	3.5
Grape	1.2
Grapefruit	1.0
Lemon-lime	9.8
Mineral water	1.0
Orange	3.7
Root beer	3.7
Sweetened seltzer	0.4
Tonic water	0.7
Unsweetened seltzer	2.2
Other	5.3
Total	100.0

Source: The New York Times, May 2, 1992, p. 19.

Type of Drink	Market Share (in %)
Beer	12
Carbonated soft drinks	25
Coffee	11
Juice	6
Milk	15
Tap water	19
Other	12
Total	100

(a) For the data on market share of all types of drinks, construct
   (1) A bar chart
   (2) A pie chart
(b) Which chart do you prefer for purposes of presentation? Why?
(c) For the data on carbonated soft drink market share, develop the appropriate graph to pinpoint the "vital few" from the "trivial many."
(d) Analyze the data and summarize your findings.
(e) **ACTION** Write a letter to the food editor of your local newspaper explaining your findings.
(f) **(Class Project)** Let each student in the class respond to the question: "Which type of carbonated soft drink do you most prefer?" so that the teacher may tally the results into a summary table on the blackboard.
   (1) Convert the data to percentages and construct a Pareto diagram.
   (2) Compare and contrast the findings from the class with those obtained nationally based on market shares. What do you conclude? Discuss.

2.53 The following data represent the global market sales of all products manufactured by Motorola, Inc., in 1992 and a stratification of its net sales by business segment:

Region	Market Sales (in %)
Asia-Pacific	15
Europe	21
Japan	7
United States	48
Other	9
Total	100

Business Segment	Net Sales (in %)
Communications	29
General systems	26
Government electronics	5
Information systems	4
Semiconductor	32
Other	4
Total	100

Source: The New York Times, October 31, 1993, Sect. 3, p. 6.

S-2-53.XLS

(a) For the data on global market sales of all products, construct
   (1) A bar chart
   (2) A pie chart
(b) Which chart do you prefer for purposes of presentation? Why?
(c) For the data on net sales by business segment, develop the appropriate graph to pinpoint the "vital few" from the "trivial many."
(d) Analyze the data and summarize your findings.
(e) **ACTION** Write a letter to your marketing professor explaining your findings.

2.54 The following table provides a percentage breakdown of where personal computers were sold in 1987 and in 1993:

	Percentage of Sales	
Type	1987	1993
Direct response	0	14
Direct sellers	17	4
Mail order	4	3
Mass merchants	3	8
Superstores	0	6
Trade dealers	60	44
Value-added resellers	11	13
Other	5	8
Totals	100	100

Source: The New York Times, May 30, 1993, p. F5.

(a) For each year construct an appropriate graph and analyze the data.
(b) **ACTION** Write a letter to your marketing professor discussing the implications of your analysis of these shifting trends.

2.55 **(Class Project)** Let each student in the class be cross-classified based on gender (male, female) and current employment status (yes, no) with the results tallied on the blackboard.
(a) Construct a table with either row or column percentages, depending on which you think is more informative.
(b) What would you conclude from this study?
(c) What other variables would you want to know regarding employment in order to enhance your findings?

2.56 Go to the World Wide Web page for this text (http://www.prenhall.com/phbusiness) for additional exercises.

Team Projects

TP2.1 The analyst at the college advisory service requires help to finish her report regarding tuition rates charged to out-of-state residents by colleges and universities in different regions of the country. In order to meet the deadline for a presentation to the board of directors, the marketing manager decides to hire your group, the _____ Corporation, to assist the analyst in her endeavors. Given Special Data Set 1 of Appendix D on pages D4–D5 regarding the out-of-state tuition rates in the Pennsylvania schools, the _____ Corporation is ready to

116 Chapter 2 Presenting Data in Tables and Charts

(a) Outline how the group members will proceed with their tasks.
(b) Form the frequency and percentage distribution in the same table.
(c) Plot the percentage polygon.
(d) Form the cumulative percentage distribution.
(e) Plot the percentage ogive.
(f) Perform a descriptive analysis comparing the tuition rates in Pennsylvania to those in Texas and North Carolina.
(g) In addition, the analyst would like to form a separate contingency table (based on row percentages) for type of institution (private or public) and setting (rural, suburban, or urban) for each of the three states.
(h) Write and submit an executive summary, attaching all tables and charts.
(i) Prepare and deliver a 10-minute oral presentation to the marketing manager.

PENNC&U.TXT

C&U.TXT

TP2.2 A popular family magazine interested in publishing an article on the dietary virtues (or lack thereof) of ready-to-eat cereals hires your group, the _____ Corporation, to study their cost and nutritional characteristics. The theme the article is intending to present is that "ready-to-eat cereals are a quick and efficient way to get the family started each weekday." Armed with Special Data Set 2 of Appendix D on pages D6–D7 displaying useful information on 84 such cereals, the _____ Corporation is ready to
(a) Outline how the group members will proceed with their tasks.
(b) Form the frequency and percentage distribution in the same table.
(c) Plot the percentage polygon.
(d) Form the cumulative percentage distribution.
(e) Plot the percentage ogive.
(f) Perform a descriptive analysis.
(g) Additionally, form a contingency table cross-classifying type of ready-to-eat cereal (high fiber, moderate fiber, low fiber) with level of calories per serving (below 155, at or above 155).
(h) Write and submit an executive summary, attaching all tables and charts.
(i) Prepare and deliver a 10-minute oral presentation to the food editor of the magazine.

CEREAL.TXT

TP2.3 The manufacturer of well-known men's and women's fragrances is planning to develop a new product line to be marketed for the upcoming holiday season. The marketing director hires your group, the _____ Corporation, to study the characteristics of currently available fragrances so the manufacturer will be in a better position to price its newly developed product line. Armed with Special Data Set 3 of Appendix D on pages D8–D9 displaying useful information on 83 such fragrances, the _____ Corporation is ready to
(a) Outline how the group members will proceed with their tasks.
(b) Form the frequency and percentage distribution in the same table.
(c) Plot the percentage polygon.
(d) Form the cumulative percentage distribution.
(e) Plot the percentage ogive.
(f) Perform a descriptive analysis.
(g) Additionally form a contingency table cross-classifying type of fragrance (perfume, cologne, or "other") with intensity of fragrance (very strong, strong, medium, or mild).
(h) Construct a table based on total percentages.
(i) Construct a table based on row percentages.
(j) Construct a table based on column percentages.
(k) Repeat (g)–(j) for women's fragrances only.
(l) Repeat (g)–(j) for men's fragrances only.
(m) Compare and contrast the results in (k) and (l).
(n) Write and submit an executive summary, attaching all tables and charts.
(o) Prepare and deliver a 10-minute oral presentation to the marketing director.

FRAGRANC.TXT

Endnote

1. The diskette that accompanies this text includes a Visual Basic for Applications module (STEMLEAF.XLS) that aids in the creation of a stem-and-leaf display.

References

1. Chambers, J. M., W. S. Cleveland, B. Kleiner, and P. A. Tukey, *Graphical Methods for Data Analysis* (Boston, MA: Duxbury Press, 1983).
2. Cleveland, W. S., "Graphs in Scientific Publications," *The American Statistician*, Vol. 38 (November 1984), pp. 261–269.
3. Cleveland, W. S., "Graphical Methods for Data Presentation: Full Scale Breaks, Dot Charts, and Multibased Logging," *The American Statistician*, Vol. 38 (November 1984), pp. 270–280.
4. Cleveland, W. S., and R. McGill, "Graphical Perception: Theory, Experimentation, and Application to the Development of Graphical Methods," *Journal of the American Statistical Association*, Vol. 79 (September 1984), pp. 531–554.
5. Cobb Group, *Running Microsoft Excel 5* (Redmond, WA: Microsoft Press, 1994).
6. Croxton, F., D. Cowden, and S. Klein, *Applied General Statistics,* 3d ed. (Englewood Cliffs, NJ: Prentice-Hall, 1967).
7. Ehrenberg, A. S. C., "Rudiments of Numeracy," *Journal of the Royal Statistical Society*, Series A, Vol. 140 (1977), pp. 277–297.
8. *Microsoft Excel Version 7* (Redmond, WA: Microsoft Press, 1996).
9. Huff, D., *How to Lie with Statistics* (New York: W. W. Norton, 1954).
10. Kimble, G. A., *How to Use (and Misuse) Statistics* (Englewood Cliffs, NJ: Prentice-Hall, 1978).
11. Tufte, E. R., *The Visual Display of Quantitative Information* (Cheshire, CT: Graphics Press, 1983).
12. Tufte, E. R., *Envisioning Information* (Cheshire, CT: Graphics Press, 1990).
13. Tukey, J., *Exploratory Data Analysis* (Reading, MA: Addison-Wesley, 1977).
14. Velleman, P. F., and D. C. Hoaglin, *Applications, Basics, and Computing of Exploratory Data Analysis* (Boston, MA: Duxbury Press, 1981).
15. Wainer, H., "How to Display Data Badly," *The American Statistician*, Vol. 38 (May 1984), pp. 137–147.

chapter 3

Summarizing and Describing Numerical Data

CHAPTER OBJECTIVE To provide an understanding of the characteristics of numerical data (central tendency, variation, shape) and their corresponding descriptive summary measures as an aid to data analysis.

3.1 INTRODUCTION: WHAT'S AHEAD

In the preceding chapters we learned how to present numerical data in both tabular and chart format. Now, how do we make sense out of such information? Although presenting data is an essential component of descriptive statistics, it doesn't tell the whole story. *Good data analysis* not only involves *presenting* the numerical data and *observing* what the data are trying to convey, it also involves *computing* and *summarizing* the key features and *analyzing* the findings. In this chapter we will examine these aspects: the summarization, description, and, ultimately, interpretation of the data.

3.2 EXPLORING THE DATA

To introduce the relevant ideas of this chapter, let us return to the analyst at the college advisory service, who (in Chapter 2) wanted to study the out-of-state tuition rates of colleges and universities in different regions of the country. In particular, three states have been selected for immediate evaluation—Texas, North Carolina, and Pennsylvania. Suppose, for exploratory purposes, our analyst was to start by selecting a random sample of six schools from the population listing of 90 colleges and universities in the state of Pennsylvania (see Special Data Set 1 of Appendix D on pages D4–D5).

Pennsylvania School	Tuition (in $000)
University of Pittsburgh	10.3
East Stroudsburg University	4.9
Geneva College	8.9
Drexel University	11.7
California Univ. of Pa.	6.3
Slippery Rock University	7.7

We note that the six schools (recorded in the order in which they were selected) are presented along with their out-of-state tuition rates (in thousands of dollars). What can be learned from such data that will assist our analyst with her evaluation? Based on this sample, we observe the following:

1. The data are in **raw form.** That is, the collected data seem to be in a random sequence with no apparent pattern to the manner in which the individual observations are listed.

2. Each of the tuition rates occurs only once. That is, no one of them is observed more frequently than any other.

3. The spread in the tuition rates ranges from 4.9 to 11.7 thousand dollars.

4. There do not appear to be any unusual or extraordinary tuition rates in this sample. Arranged in numerical order (i.e., the *ordered array*), the tuition rates here (in thousands of dollars) are 4.9, 6.3, 7.7, 8.9, 10.3, 11.7. (If the tuition rates, in thousands of dollars, had been 4.9, 6.3, 7.7, 8.9, 10.3, and 28.0, then 28.0 thousand dollars would be considered an **extreme value** or an **outlier.**)

If our analyst were to ask us to examine the data and present a short summary of our findings, then comments similar to the four above are basically all that we could be expected to make without more formal statistical training. However, when making such comments, we have both analyzed and interpreted what the data are trying to convey. An analysis is *objective;* we should all agree with these findings. On the other hand, an interpretation is *subjective;* we may form different conclusions when interpreting our analytical findings. From the above, points 2–4 are based on analysis, whereas point 1 is an interpretation. With respect to the latter, no formal analytical test was made—it is simply our conjecture that there is no pattern to the sequence of collected data. Our conjecture would seem to be appropriate if the sample of six schools was randomly and independently drawn from the population listing using sampling methods described in Chapter 1.

Let us now see how we can add to our understanding of what the data are telling us by more formally examining three properties of numerical data.

3.3 PROPERTIES OF NUMERICAL DATA

The three major properties that describe a set of numerical data are

1. Central tendency
2. Variation
3. Shape

In any analysis and/or interpretation, a variety of descriptive measures representing the properties of central tendency, variation, and shape may be used to extract and summarize the major features of the data set. If these descriptive summary measures are computed from a sample of data, they are called *statistics;* if they are computed from an entire population of data, they are called *parameters.* Since statisticians usually take samples rather than use entire populations, our primary emphasis in this text is on statistics rather than parameters.

3.4 MEASURES OF CENTRAL TENDENCY

Most sets of data show a distinct tendency to group or cluster about a certain central point. Thus, for any particular set of data, it usually becomes possible to select some typical value or **average** to describe the entire set. Such a descriptive typical value is a measure of **central tendency** or **location.**

Five types of averages often used as measures of central tendency are the arithmetic mean, the median, the mode, the midrange, and the midhinge.

3.4.1 The Arithmetic Mean

The **arithmetic mean** (also called the **mean**) is the most commonly used *average*[1] or measure of central tendency. It is calculated by summing all the observations in a set of data and then dividing the total by the number of items involved.

• **Introducing Algebraic Notation** Thus, for a sample containing a set of n observations $X_1, X_2, \ldots, X_n$, the arithmetic mean (given by the symbol $\bar{X}$—called "X bar") can be written as

$$\bar{X} = \frac{X_1 + X_2 + \cdots + X_n}{n}$$

To simplify the notation, for convenience the term

$$\sum_{i=1}^{n} X_i$$

(meaning the *summation of all the X_i values*) is conventionally used whenever we wish to add together a series of observations. That is,

$$\sum_{i=1}^{n} X_i = X_1 + X_2 + \cdots + X_n$$

Rules pertaining to summation notation are presented in Appendix B on pages B1–B5. Using this summation notation, the arithmetic mean of the sample can be expressed as the following.

> The arithmetic mean is the sum of the values divided by the number of values:
>
> $$\bar{X} = \frac{\sum_{i=1}^{n} X_i}{n} \qquad (3.1)$$

where

$\bar{X}$ = sample arithmetic mean

n = sample size

X_i = ith observation of the random variable X

$\sum_{i=1}^{n} X_i$ = summation of all X_i values in the sample (see Appendix B)

For our analyst's sample, the out-of-state tuition rates (in thousands of dollars) are

X_1 = 10.3 at University of Pittsburgh

X_2 = 4.9 at East Stroudsburg University

X_3 = 8.9 at Geneva College

X_4 = 11.7 at Drexel University

X_5 = 6.3 at California Univ. of Pa.

X_6 = 7.7 at Slippery Rock University

The arithmetic mean for this sample is calculated as

$$\bar{X} = \frac{\sum_{i=1}^{n} X_i}{n} = \frac{10.3 + 4.9 + 8.9 + 11.7 + 6.3 + 7.7}{6} = 8.30 \text{ thousand dollars}$$

We observe here that the mean is computed as 8.3 thousand dollars even though not one particular school in the sample actually had that tuition rate. In addition, we see from the **dot scale** of Figure 3.1 that for this set of data, three observations are smaller than the mean and three are larger. The mean acts as a *balancing point* so that observations that are smaller balance out those that are larger.

Note that the calculation of the mean is based on all the observations ($X_1, X_2, \ldots, X_n$) in the set of data. No other commonly used measure of central tendency possesses this charac-

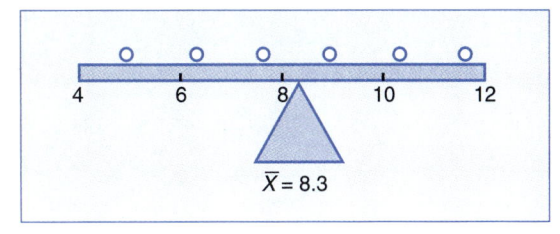

FIGURE 3.1 Dot scale representing out-of-state tuition rates (in $000) at six Pennsylvania schools.

teristic. Since its computation is based on every observation, the arithmetic mean is greatly affected by any extreme value or values. In such instances, the arithmetic mean presents a distorted representation of what the data are conveying; hence the mean would not be the best average to use for describing or summarizing such a set of data.

To further demonstrate the characteristics of the mean, suppose that our analyst takes a random sample of $n = 6$ schools from the population frame of 60 in the state of Texas and another random sample of $n = 6$ schools from the listing of 45 in the state of North Carolina. The out-of-state tuition rates (in thousands of dollars) are reported as follows:

Texas School	Tuition (in $000)
Concordia Lutheran College	6.4
Southern Methodist University	12.0
Texas A & M University	4.9
Lubbock Christian University	6.4
Rice University	8.5
Trinity University	11.6

North Carolina School	Tuition (in $000)
Methodist College	8.3
Warren Wilson College	8.7
Campbell University	7.6
Belmont Abbey College	8.3
Catawba College	9.0
North Carolina State University	7.9

The respective dot scales are displayed in Figures 3.2 and 3.3.

$$\sum_{i=1}^{n} X_i$$

FIGURE 3.2 Dot scale representing out-of-state tuition rates (in $000) at six Texas schools.

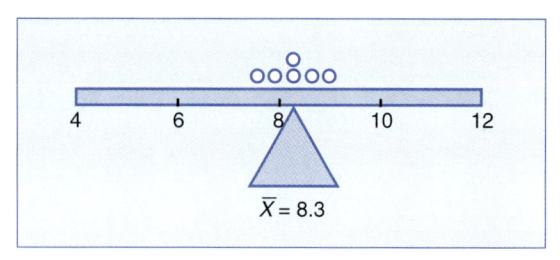

FIGURE 3.3 Dot scale representing out-of-state tuition rates (in $000) at six North Carolina schools.

Note that the mean tuition rate for each of these samples is also 8.3 thousand dollars. Nevertheless, as can be observed from Figures 3.2 and 3.3, the two samples drawn here have distinctly different features—with respect to each other and with respect to the sample of six Pennsylvania schools depicted in Figure 3.1. For example, three of the six Texas schools have tuition rates quite different from those of Trinity University and Southern Methodist University. For this sample, the arithmetic mean is presenting a somewhat distorted representation of what the data are conveying, and it may not be the best average to use. On the other hand, for the North Carolina and Pennsylvania samples, the mean is the appropriate descriptive measure for characterizing and summarizing the respective data sets because outliers are not present. In fact, the North Carolina data are quite *homogeneous*. Two of the six schools in this sample have tuition rates equivalent to the mean. From Figures 3.1–3.3, it is clear that the rates charged by these six North Carolina schools contain the least amount of scatter or variability among the three samples. In addition, it can also be observed that the out-of-state tuition data in each of the Pennsylvania and North Carolina samples possess the property of symmetry, while the data for the Texas sample do not. (The properties of variation and shape will be addressed further in Sections 3.5 and 3.6.)

3.4.2 The Median

The **median** is the middle value in an ordered sequence of data. If there are no ties, half of the observations will be smaller and half will be larger. The median is unaffected by any extreme observations in a set of data. Thus, whenever an extreme observation is present, it is appropriate to use the median rather than the mean to describe a set of data.

To calculate the median from a set of data collected in its raw form, we must first place the data into an ordered array. We then use the *positioning point formula*

$$\frac{n+1}{2}$$

to find the place in the ordered array that corresponds to the median value. One of two rules is followed:

Rule 1: If the size of the sample is an *odd* number, the median is represented by the numerical value corresponding to the positioning point—the $(n + 1)/2$ ordered observation.

Rule 2: If the size of the sample is an *even* number, then the positioning point lies between the two middle observations in the ordered array. The median is the *average* of the numerical values corresponding to these two middle observations.

● **Even-Sized Sample** For our analyst's six-school sample from Pennsylvania, the raw data (in thousands of dollars) are

            10.3   4.9   8.9   11.7   6.3   7.7

The ordered array becomes

                  4.9   6.3   7.7   8.9   10.3   11.7
Ordered
Observation   1   2   3   4   5   6
                              Median = 8.30 thousand dollars

For these data, the positioning point is $(n + 1)/2 = (6 + 1)/2 = 3.5$. Therefore, the median is obtained by averaging the third and fourth ordered observations:

$$\frac{7.7 + 8.9}{2} = 8.30 \text{ thousand dollars}$$

As can be seen from the ordered array, the median is unaffected by extreme observations. Regardless of whether the largest tuition rate is 11.7, 21.7, or 31.7 thousand dollars, the median is still 8.3 thousand dollars.

● **Odd-Sized Sample** Had the sample size been an odd number, the median would be represented by the numerical value given to the $(n + 1)/2$ observation in the ordered array. Thus, in the following ordered array for $n = 5$ students' GMAT scores, the median is the value of the third [that is, $(5 + 1)/2$] ordered observation—590:

                  500   570   590   600   690
                                Median
Ordered
Observation   1   2   3   4   5

● **Ties in the Data** When computing the median, we ignore the fact that tied values may be present in the data. Suppose, for example, that the following set of data represents the starting salaries (in thousands of dollars) for a sample of $n = 7$ marketing majors who have recently graduated from your college:

            24.1   22.6   27.0   19.8   21.5   23.7   22.6

The ordered array becomes

              19.8   21.5   22.6   22.6   23.7   24.1   27.0
                                     Median
Ordered
Observation   1   2   3   4   5   6   7

For this odd-sized sample the median positioning point is the $(n + 1)/2 =$ fourth ordered observation. Thus, the median is 22.6 thousand dollars, the middle value in the ordered sequence, even though the third ordered observation is also 22.6 thousand dollars.

● **Characteristics of the Median** To summarize, the calculation of the median value is affected by the number of observations, not by the magnitude of any extreme(s). Any observation selected at random is just as likely to exceed the median as it is to be exceeded by it.

3.4.3 The Mode

Sometimes, when summarizing or describing a set of data, the mode is used as a measure of central tendency. The **mode** is the value in a set of data that appears most frequently. It may be obtained from an ordered array. Unlike the arithmetic mean, the mode is not affected by the occurrence of any extreme values. However, the mode is used only for descriptive purposes because it is more variable from sample to sample than other measures of central tendency.

Using the ordered array for out-of-state tuition rates for the six-school sample from Pennsylvania,

$$4.9 \quad 6.3 \quad 7.7 \quad 8.9 \quad 10.3 \quad 11.7$$

we see that there is no mode. None of the tuition rates is "most typical."

Note that there is a difference between *no mode* and a mode of 0, as illustrated in the following ordered array of noontime temperatures (°F) in Duluth during the first week of January:

Ordered Array
(Duluth, Minnesota) $-4°$ $-2°$ $-1°$ $-1°$ $0°$ $0°$ $0°$

Mode = 0°

In addition, a data set can have more than one mode, as illustrated in the following ordered array of noontime temperatures (°F) in Richmond during the first week of January:

Ordered Array
(Richmond, Virginia)   21°   28°   28°   35°   41°   43°   43°

In Richmond, we see there are two modes—28° and 43°. Such data are described as *bimodal*.

3.4.4 The Midrange

The **midrange** is the average of the *smallest* and *largest* observations in a set of data. This can be written as follows:

> The midrange is obtained by adding the smallest and largest values and dividing by 2.
>
> $$\text{Midrange} = \frac{X_{smallest} + X_{largest}}{2} \tag{3.2}$$

Using the ordered array of out-of-state tuition rates from our analyst's six-school sample from Pennsylvania,

$$4.9 \quad 6.3 \quad 7.7 \quad 8.9 \quad 10.3 \quad 11.7$$

the midrange is computed from Equation (3.2) as

$$\text{Midrange} = \frac{X_{smallest} + X_{largest}}{2}$$

$$= \frac{4.9 + 11.7}{2} = 8.30 \text{ thousand dollars}$$

The midrange is often used as a summary measure both by financial analysts and by weather reporters, since it can provide an adequate, quick, and simple measure to characterize the *entire* data set—be it a series of daily closing stock prices over a whole year or a series of recorded hourly temperature readings over a whole day.

In dealing with data such as daily closing stock prices or hourly temperature readings, an extreme value is not likely to occur. Nevertheless, in most applications, despite its simplicity, the midrange must be used cautiously. Since it involves only the smallest and largest observations in a data set, the midrange becomes distorted as a summary measure of central tendency if an outlier is present. In such situations, the midrange is inappropriate; a summary measure somewhat similar in format to the midrange that is always appropriate because it is totally unaffected by outliers is the *midhinge*.

3.4.5 The Midhinge

The **midhinge** is the average of the *first* and *third quartiles* in a set of data. This means that

> The midhinge is obtained by adding the first and third quartiles and dividing by 2.
>
> $$\text{Midhinge} = \frac{Q_1 + Q_3}{2} \quad (3.3)$$

where
Q_1 = first quartile
Q_3 = third quartile

It is a summary measure used to overcome potential problems introduced by extreme values in the data.

- **Quartiles: Measures of "Noncentral" Location** Aside from the measures of central tendency, there also exist some useful measures of "noncentral" location that are employed particularly when summarizing or describing the properties of large sets of numerical data.[2] The most widely used of these measures are the **quartiles.**

Whereas the median is a value that splits the ordered array in half (50.0% of the observations are smaller and 50.0% of the observations are larger), the quartiles are descriptive measures that split the ordered data into four quarters.

The **first quartile, Q_1,** is the value such that 25.0% of the observations are smaller and 75.0% of the observations are larger.

The **second quartile, Q_2,** is the median—50.0% of the observations are smaller and 50.0% are larger.

The **third quartile, Q_3,** is the value such that 75.0% of the observations are smaller and 25.0% of the observations are larger.

To approximate the quartiles, the following *positioning point formulas* are used:

Q_1 = value corresponding to the $\frac{n+1}{4}$ ordered observation

Q_2 = median, the value corresponding to the $\frac{2(n+1)}{4} = \frac{n+1}{2}$ ordered observation

Q_3 = value corresponding to the $\frac{3(n+1)}{4}$ ordered observation

The following rules are used for obtaining the quartile values:

1. If the resulting positioning point is an integer, the particular numerical observation corresponding to that positioning point is chosen for the quartile.
2. If the resulting positioning point is halfway between two integers, the average of their corresponding values is selected.

3. If the resulting positioning point is neither an integer nor a value halfway between two integers, a simple rule used to approximate the particular quartile is to *round off* to the nearest integer positioning point and select the numerical value of the corresponding observation.

To compute our midhinge [see Equation (3.3)], we first need to compute Q_1 and Q_3. For example, from the ordered array pertaining to the out-of-state tuition rates (in thousands of dollars) at the six Pennsylvania schools, we have

$$Q_1 = \frac{n+1}{4} \text{ ordered observation}$$

$$= \frac{6+1}{4} = 1.75\text{th} \cong 2\text{nd ordered observation}$$

Therefore, Q_1 can be approximated as 6.30 thousand dollars.

$$Q_3 = \frac{3(n+1)}{4} \text{ ordered observation}$$

$$= \frac{3(6+1)}{4} = 5.25\text{th} \cong 5\text{th ordered observation}$$

Therefore, Q_3 can be approximated as 10.30 thousand dollars.

Returning to Equation (3.3), we may now compute the midhinge as

$$\text{Midhinge} = \frac{Q_1 + Q_3}{2}$$

$$= \frac{6.3 + 10.3}{2} = 8.30 \text{ thousand dollars}$$

This result, the average of Q_1 and Q_3 (two measures of noncentral location), cannot be affected by potential outliers, since no observation smaller than Q_1 or larger than Q_3 is considered. Summary measures such as the midhinge and the median, which cannot be affected by outliers, are called **resistant measures.**

3.4.6 Using Microsoft Excel Functions to Obtain Measures of Location

● **Introduction** In this section we have discussed a variety of measures of central tendency (arithmetic mean, median, mode, midrange, and midhinge) and two measures of noncentral location, the first quartile and the third quartile. We will now illustrate how Excel functions can be used to compute each of these measures of central tendency. In addition, we will also discuss how we can subdivide a set of data into subgroups so that various descriptive statistics can be computed for each subgroup.

● **Excel Functions for Descriptive Statistics** You may recall that in Section 1.S.12, we used the SUM function to add a set of values, while in Section 2.7, we used the FREQUENCY function to develop a frequency distribution. Among the functions that are available to obtain measures of central tendency are

AVERAGE (*range*)	Computes the arithmetic mean of the values in the cell *range*.
MEDIAN (*range*)	Computes the median of the values in the cell *range*.
MODE (*range*)	Computes the mode of the values in the cell *range*.

MIN (*range*)		Computes the smallest value of the values in the cell *range*.
MAX (*range*)		Computes the largest value of the values in the cell *range*.
SMALL (*range, k*)		Computes the *k*th smallest value of the values in the cell *range*.

We note that although Microsoft Excel contains a quartile function, it is not considered to be a reliable measure, so we need to calculate our own quartiles.

To illustrate the use of these functions for obtaining measures of central tendency, let us return to the six-school sample randomly selected from the population of 90 schools in the state of Pennsylvania. The data pertaining to this sample are presented in Table 1S.2 on page 36 and also on page 35. To obtain the measures of central tendency from Microsoft Excel that we have studied in this section, we need to design a Calculations sheet as shown in Table 3.1.Excel.

To implement this Calculations sheet, open the workbook file PA-SAM-1.XLS and enter the labels and formulas in columns A and B of a new worksheet named Calculations. In doing this, there are several things that we need to do and to be aware of:

3-4-6 SAM.XLS

❶ Be aware that when you enter the formula for the mode in cell B6, instead of a numerical value, cell B6 displays the error message #N/A or #NUM! The reason for this is that since each of the values occurs once, there is no mode. Whenever this occurs, Excel simply displays a message in the cell instead of the modal value.

❷ To obtain the Midrange, we average the minimum and maximum values we obtained in cells B7 and B8 since Excel does not have a Midrange function. Thus we enter =(B7+B8)/2 in cell B9.

❸ We can compute the quartiles by using the COUNT and SMALL functions of Excel. We first need to determine the ranked values that correspond to the first and third quartiles. Once we have determined these values, we can use the SMALL function to determine the quartiles. To do this, we enter =(COUNT(Data!B2:B7)+1)/4 in cell B10. This will compute the rank for the first quartile according to the rule discussed on page 127. In this example, the value computed is 1.75, indicating that the first quartile

Table 3.1.Excel Design of the Calculations sheet for descriptive statistics of the six-school sample from Pennsylvania for out-of-state tuition rates.

	A	B
1	Pennsylvania Schools Data Analysis	
2		
3	Sample Statistics	Tuition
4	Mean	=AVERAGE(Data!B2:B7)
5	Median	=MEDIAN(Data!B2:B7)
6	Mode	=MODE(Data!B2:B7)
7	Minimum	=MIN(Data!B2:B7)
8	Maximum	=MAX(Data!B2:B7)
9	Midrange	=(B7+B8)/2
10	First Quartile rank	=(COUNT(Data!B2:B7) + 1)/4
11	First Quartile	=SMALL(Data!B2:B7,*first quartile rank*)
12	Third Quartile rank	=3*(COUNT(Data!B2:B7)+1)/4
13	Third Quartile	=SMALL(Data!B2:B7,*third quartile rank*)
14	Midhinge	=(B11+B13)/2

Note: The first and third quartile ranks will be determined after the formulas for cells B10 and B12 are entered.

corresponds to the 1.75 ranked value. According to our rule for finding quartiles, we may round this to the nearest integer value, which is 2 for this data set.

❹ To obtain the first quartile, we use the SMALL function to find the second smallest value that corresponds to the value for the first quartile for these data. Thus, in cell B11, we enter =SMALL(Data!B2:B7,2) and press the Return key. Note that we have entered the numerical value of 2 as the first quartile rank.

❺ To obtain the third quartile, we first have to determine the rank for the third quartile according to the rule discussed on page 127. To do this, we move to cell B12 and enter =3*(COUNT(Data!B2:B7)+1)/4. We can now observe that the value displayed in cell B12 is 5.25, indicating that the third quartile corresponds to the 5.25 ranked value. According to our rule for finding quartiles, we may round this to the nearest integer value, which is 5.

❻ To obtain the third quartile, we need to use the SMALL function to find the fifth smallest value that corresponds to the value for the third quartile for these data. We thus click on cell B13 and enter =SMALL(Data!B2:B7,5). Note that we have entered the numerical value of 5 as the third quartile rank.

> The rules for obtaining the quartiles used in steps 4 and 6 can be implemented as a set of Microsoft Excel formulas that use the IF function. A complete discussion of this approach is beyond the scope of this text. However, this approach is illustrated and explained in the 3-4-6-SAM.XLS workbook included on the diskette that accompanies this text. The approach used in this workbook eliminates the manual calculation of rank values.

❼ To obtain the midhinge, we can average the quartiles in cells B11 and B13. You should click on cell B14 and then enter =(B11+B13)/2.

Figure 3.1.Excel represents the descriptive statistics obtained by using these Excel functions.

	A	B
1	Sample Statistics for Pennsylv:	
2		
3	Tuition Variable	
4	Mean	8.3
5	Median	8.3
6	Mode	#N/A
7	Minimum	4.9
8	Maximum	11.7
9	Midrange	8.3
10	First Quartile rank	1.75
11	First Quartile	6.3
12	Third Quartile rank	5.25
13	Third Quartile	10.3
14	Midhinge	8.3

FIGURE 3.1.EXCEL Measures of central tendency obtained from Excel for the out-of-state tuition rates for the six-school sample from Pennsylvania.

▲ WHAT IF EXAMPLE

Although implementing formulas using the functions described in this section takes longer than using the Data Analysis tool, we gain the ability to change our data and view the results without having to reissue any new commands. We can thus study the effect of extreme values on the various measures of central tendency by pretending that the tuition at Slippery Rock University is 19.7 instead of 7.7. Change the value on the Data sheet in cell B9 to 19.7, and then go to the Calculations sheet to see the updated set of descriptive statistics shown in Figure 3.2.Excel.

	A	B
1	Sample Statistics for Pennsylv:	
2		
3	Tuition Variable	
4	Mean	10.3
5	Median	9.6
6	Mode	#N/A
7	Minimum	4.9
8	Maximum	19.7
9	Midrange	12.3
10	First Quartile rank	1.75
11	First Quartile	6.3
12	Third Quartile rank	5.25
13	Third Quartile	11.7
14	Midhinge	9

FIGURE 3.2.EXCEL Revised measures of central tendency obtained from Excel for the out-of-state tuition rates for the six-school sample from Pennsylvania.

By comparing these results with those of Figure 3.1.Excel, we see that the arithmetic mean and the midrange are greatly affected by this extreme value of 19.7, while the median and midhinge are much less affected. The arithmetic mean changed to 10.3 and the midrange to 12.3, while the median became 9.6 and the midhinge became 9.

As before, to see the effects of many different changes, use the Scenario Manager (see Section 1S.15) to store and use sets of alternative data values.

● **Using Microsoft Excel to Obtain Statistics Classified by Group** We are often interested in obtaining statistics for a numerical variable broken down by a categorical variable of interest. For example, we might want to determine the average tuition for private colleges and also for public colleges for the population of 90 Pennsylvania schools. This can be accomplished by using the PivotTable feature selection on the Data menu that was discussed in Section 2.12. To obtain the average tuition for public colleges and also for private colleges, we open the workbook for the tuition data for the population of 90 Pennsylvania schools PA-POP-1.XLS and select Insert | Worksheet and rename the worksheet as Pivot. Then we click on Data | PivotTable.

You now see the first of four PivotTable Wizard dialog boxes as illustrated in Figure 2.10.Excel on page 96. Click on the Microsoft Excel List or Database Option button; then click the Next button to move to step 2 of the PivotTable Wizard.

3-4-6 POP.XLS

This second of four PivotTable Wizard dialog boxes asks you to enter the range of your data. You should enter Data!B1:F91 and click the Next button to move to step 3.

The third of four PivotTable Wizard dialog boxes asks you to indicate the variables to be included in the PivotTable and the type of data to be analyzed. On the right side of the dialog box, you will see boxes that correspond to the names of the variables tuition, type, setting, calendar, and focus. Since we wish to obtain a summary table for the variable Type, drag and drop the Type label to the row Area. Next drag and drop the Tuition label into the Data area. The label just dropped into the Data area needs to read Average of Tuition. If the name does not change to Average of Tuition, double-click on the dropped label. This will present you with a PivotTable Field dialog box. In the Summarize By list box, select Average. (Notice that other statistics such as minimum, maximum, and standard deviation are also available.) Click the OK button to return to dialog box 3 of the PivotTable Wizard and then click the Next button to move to step 4 of the PivotTable Wizard.

At this last of the four PivotTable Wizard dialog boxes you need to first enter the starting cell for the PivotTable (Pivot!A3) and then provide a name for the table such as Average of Tuition. Select the Grand Total for Columns check box to verify the count of the sample size and then the AutoFormat Table box so Excel can format the table provided. Click the Finish button.

Figure 3.3.Excel illustrates the PivotTable we have obtained.

FIGURE 3.3.EXCEL Average tuition by type of school for 90 colleges and universities in Pennsylvania.

	A	B
3	Average of Tuition	
4	Type	Total
5	Private	12.13676471
6	Public	7.022727273
7	Grand Total	10.88666667

Problems for Section 3.4

Note: These problems may be solved using the Microsoft Excel functions discussed in Section 3.4.6 or the Data Analysis tool discussed in Section 3.7.

- 3.1 Given the following two sets of data—each with samples of size $n = 7$:

    Set 1:   10   2   3   2   4   2   5
    Set 2:   20   12   13   12   14   12   15

    (a) For each set, compute the mean, median, mode, midrange, and midhinge.
    (b) Compare your results and summarize your findings.
    (c) Compare the first sampled item in each set, compare the second sampled item in each set, and so on. Briefly describe your findings here in light of your summary in part (b).

- 3.2 A track coach must decide on which one of two sprinters to select for the 100-meter dash at an upcoming meet. The coach will base the decision on the results of five races between the two athletes run with 15-minute rest intervals between. The times (in seconds) for the five races are shown at the top of page 133.
    (a) Based on these data, which of the two sprinters should the coach select? Why?
    (b) Should the selection be different if the coach knew that Sharyn had fallen at the start of the fourth race? Why?
    (c) Discuss the differences in the concepts of the mean and the median as measures of central tendency and how this relates to (a) and (b).

	Race				
Athlete	1	2	3	4	5
Sharyn	12.1	12.0	12.0	16.8	12.1
Tamara	12.3	12.4	12.4	12.5	12.4

3.3 Suppose that, owing to an error, a data set containing the price-to-earnings (PE) ratios from nine companies traded on the American Stock Exchange was recorded as 13, 15, 14, 17, 13, 16, 15, 16, and 61, where the last value should have been 16 instead of 61. Using Excel, show how much the mean, median, and midrange are affected by the error (i.e., compute these statistics for the "bad" and "good" data sets, and compare the results of using different estimators of central tendency).

S-3-3.XLS

3.4 A manufacturer of flashlight batteries took a sample of 13 from a day's production and burned them continuously until they failed. The numbers of hours they burned were

342, 426, 317, 545, 264, 451, 1049,

631, 512, 266, 492, 562, 298

FLASHBAT.TXT

(a) Compute the mean, median, mode, midrange, and midhinge. Looking at the distribution of times, which descriptive measures seem best and which worst? (And why?)
(b) In what ways would this information be useful to the manufacturer? Discuss.
(c) Suppose that the first value was 1,342 instead of 342. Repeat (a) using this value. Comment on the difference in the results.

3.5 It is stated in Section 3.5.3 that one important property of the arithmetic mean is

$$\sum_{i=1}^{n}\left(X_i - \overline{X}\right) = 0$$

(a) Using the out-of-state tuition rates from the six-school sample for Texas (see page 123), verify that this property holds.
(b) Using the out-of-state tuition rates from the six-school sample for North Carolina (see page 123), verify that this property holds.

3.6 The following data are the amount of calories in a 30-gram serving for a random sample of 10 types of fresh-baked chocolate chip cookies:

COOKIES.TXT

Product	Calories	Product	Calories
Hillary Rodham Clinton's	153	David's	146
Original Nestle Toll House	152	David's Chocolate Chunk	149
Mrs. Fields	146	Great American Cookie Company	138
Stop & Shop	138	Pillsbury Oven Lovin'	168
Duncan Hines	130	Pillsbury	147

Source: "Chocolate-chip Cookies" Copyright 1993 by Consumers Union of United States, Inc., Yonkers, N.Y. 10703. Adapted by permission from *Consumer Reports,* October 1993, pp. 646–647. Although this data set originally appeared in *Consumer Reports,* the selective adaptation and resulting conclusions presented are those of the authors and are not sanctioned or endorsed in any way by Consumers Union, the publisher of *Consumer Reports.*

(a) What is the mean amount of calories? The median amount of calories?
(b) How would this information be of use to a company about to market a new fresh-baked chocolate chip cookie? Discuss.

TAPES.TXT

3.7 The following data are the list prices (in dollars) for a random sample of 17 Type IV (metal) audiotapes:

4.95	18.99	5.29
5.29	3.49	3.99
3.99	5.50	9.95
11.00	5.99	14.99
5.99	3.49	3.50
4.59	2.99	

Source: "AudioCassette Tapes" Copyright 1993 by Consumers Union of United States, Inc., Yonkers, N.Y. 10703. Adapted by permission from *Consumer Reports*, January 1993, pp. 38–39. Although this data set originally appeared in *Consumer Reports*, the selective adaptation and resulting conclusions presented are those of the authors and are not sanctioned or endorsed in any way by Consumers Union, the publisher of *Consumer Reports*.

(a) What is the mean price? The median price?
(b) **ACTION** What other information would you want to know before purchasing such a tape? Prepare a list of questions that you would ask the salesperson.

ADAMHA.TXT

3.8 The following data represent the amount of funding (in millions of dollars) provided by the Alcohol, Drug Abuse, and Mental Health Administration through grants to a sample of 21 institutions during a recent year:

Name of Institution	Amount Funded ($000,000)
Johns Hopkins University	14.9
University of California at San Francisco	14.1
University of Washington	6.8
Yale University	13.1
Stanford University	7.6
University of California at Los Angeles	13.2
Harvard University	5.1
University of Michigan	11.9
University of Pennsylvania	8.5
Columbia University	7.1
Washington University at St. Louis	5.1
Duke University	5.0
University of Minnesota	6.2
University of California at San Diego	5.5
University of North Carolina	3.8
University of Wisconsin	3.4
University of Rochester	3.5
Yeshiva University	2.8
University of Chicago	4.1
University of Pittsburgh	15.9
Cornell University	5.7

Source: U.S. Department of Health and Human Services, Public Health Services.

(a) Organize the data into an ordered array or a stem-and-leaf display.
(b) Compute the mean, median, mode, midrange, and midhinge.
(c) Describe the property of central tendency for these data.

3.9 For the last 10 days in June, the "Shore Special" train was late arriving at its destination by the following times (in minutes; a negative number means that the train was early by that number of minutes):

TRAIN1.TXT

    −3, 6, 4, 10, −4, 124, 2, −1, 4, 1

(a) If you were hired by the railroad to show that the railroad is providing good service, what are some of the summary measures you would use to accomplish this?
(b) If you were hired by a TV station that was producing a documentary to show that the railroad is providing bad service, what summary measures would you use?
(c) If you were trying to be objective and unbiased in assessing the railroad's performance, which summary measures would you use? (This is the hardest part, because you cannot answer without making additional assumptions about the relative costs of being late by various amounts of time.)
(d) What would be the effect on your conclusions if the value of 124 had been incorrectly recorded and should have been 12?

3.10 In order to estimate how much water will be needed to supply the community of Falling Rock in the next decade, the town council asked the city manager to find out how much water a sample of families currently uses. The sample of 15 families used the following number of gallons (in thousands) in the past year:

WATER.TXT

    11.2, 21.5, 16.4, 19.7, 14.6, 16.9, 32.2, 18.2,

    13.1, 23.8, 18.3, 15.5, 18.8, 22.7, 14.0

(a) What is the mean amount of water used per family? The median? The midrange? The midhinge?
(b) Suppose that 10 years from now, the town council expects that there will be 45,000 families living in Falling Rock. How many gallons of water will be needed annually, if the rate of consumption per family stays the same?
(c) In what ways would the information provided in (a) and (b) be useful to the town council? Discuss.
(d) Why might the town council have used the data from a survey rather than just measuring the total consumption in the town? (Think about what types of users are not yet included in the estimation process.)

3.5 MEASURES OF VARIATION

A second important property that describes a set of numerical data is variation. **Variation** is the amount of **dispersion** or "spread" in the data. Two sets of data may differ in both central tendency and variation; or, as shown in Figure 3.1 and 3.3, two sets of data may have the same measures of central tendency but differ greatly in terms of variation. The data set depicted in Figure 3.3 is much less variable than that depicted in Figure 3.1 (see pages 124 and 123).

Five measures of variation are the range, the interquartile range, the variance, the standard deviation, and the coefficient of variation.

3.5.1 The Range

The **range** is the difference between the largest and smallest observations in a set of data:

$$\text{Range} = X_{largest} - X_{smallest} \quad (3.4)$$

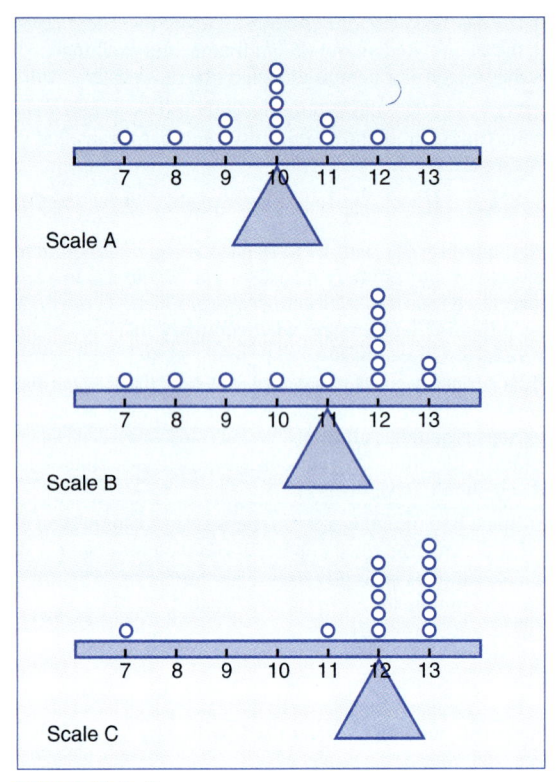

FIGURE 3.4 Comparing three data sets with the same range.

Using the ordered array of out-of-state tuition rates (in thousands of dollars) from our sample of six Pennsylvania schools,

4.9   6.3   7.7   8.9   10.3   11.7

the range is 11.7 – 4.9 = 6.80 thousand dollars.

The range measures the *total spread* in the set of data. Although the range is a simple measure of total variation in the data, its distinct weakness is that it does not take into account *how* the data are actually distributed between the smallest and largest values. This can be observed from Figure 3.4. As evidenced in scale C, it would be improper to use the range as a measure of variation when at least one of its components are extreme observations.

3.5.2 The Interquartile Range

The **interquartile range** (also called **midspread**) is the difference between the *third* and *first quartiles* in a set of data:

$$\text{Interquartile range} = Q_3 - Q_1 \qquad (3.5)$$

This measure considers the spread in the middle 50% of the data and thus is in no way influenced by extreme values.

For the Pennsylvania out-of-state tuition data, we have

$$\text{Interquartile range} = Q_3 - Q_1 = 10.3 - 6.3 = 4.0 \text{ thousand dollars}$$

This is the range in tuition rates for the *middle group* of Pennsylvania schools.

3.5.3 The Variance and the Standard Deviation

Although the range is a measure of the total spread and the interquartile range is a measure of the middle spread, neither of these measures of variation takes into consideration *how* the observations distribute or cluster. Two commonly used measures of variation that do take into account *how* all the values in the data are distributed are the **variance** and its square root, the **standard deviation.** These measures evaluate how the values fluctuate about the mean.

- **Defining the Sample Variance** The sample variance is *roughly* (or *almost*) the average of the squared differences between each of the observations in a set of data and the mean. Thus, for a sample containing n observations, $X_1, X_2, \ldots, X_n$, the sample variance (given by the symbol S^2) can be written as

$$S^2 = \frac{(X_1 - \bar{X})^2 + (X_2 - \bar{X})^2 + \cdots + (X_n - \bar{X})^2}{n - 1}$$

Using our summation notation, the above formulation can be expressed as the following:

> The variance is the sum of the squared differences around the arithmetic mean divided by the sample size minus 1.
>
> $$S^2 = \frac{\sum_{i=1}^{n}(X_i - \bar{X})^2}{n - 1} \qquad (3.6)$$

where
$\bar{X}$ = sample arithmetic mean

n = sample size

X_i = ith value of the random variable X

$\sum_{i=1}^{n}(X_i - \bar{X})^2$ = summation of all the squared differences between the X_i values and $\bar{X}$

Had the denominator been n instead of $n - 1$, the average of the squared differences around the mean would have been obtained. However, $n - 1$ is used here because of certain desirable mathematical properties possessed by the statistic S^2 that make it appropriate for statistical inference (see Chapter 5). As the sample size increases, the difference in division by n or $n - 1$ becomes smaller and smaller.

- **Defining the Sample Standard Deviation** The **sample standard deviation** (given by the symbol S) is the square root of the sample variance, expressed as follows:

> The standard deviation is the square root of the sum of the squared differences around the arithmetic mean divided by the sample size minus 1.
>
> $$S = \sqrt{\frac{\sum_{i=1}^{n}(X_i - \overline{X})^2}{n-1}} \tag{3.7}$$

● **Computing S^2 and S** To compute the variance, we

1. Obtain the difference between each observation and the mean
2. Square each difference
3. Add the squared results together
4. Divide the summation by $n - 1$

To compute the standard deviation, we simply take the square root of the variance.

For our six-school sample from Pennsylvania, the raw (out-of-state tuition rate) data (in thousands of dollars) are

$$10.3 \quad 4.9 \quad 8.9 \quad 11.7 \quad 6.3 \quad 7.7$$

and $\overline{X} = 8.30$ thousand dollars.

The sample variance is computed as

$$S^2 = \frac{\sum_{i=1}^{n}(X_i - \overline{X})^2}{n-1}$$

$$= \frac{(10.3 - 8.3)^2 + (4.9 - 8.3)^2 + \cdots + (7.7 - 8.3)^2}{6-1}$$

$$= \frac{31.84}{5}$$

$$= 6.368 \text{ (in squared thousands of dollars)}$$

and the sample standard deviation is computed as

$$S = \sqrt{S^2} = \sqrt{\frac{\sum_{i=1}^{n}(X_i - \overline{X})^2}{n-1}} = \sqrt{6.368} = 2.52 \text{ thousands of dollars}$$

● **Obtaining S^2 and S** Since in the preceding computations we are squaring the differences, *neither the variance nor the standard deviation can ever be negative.* The only time S^2 and S can be zero is when there is no variation at all in the data—when each observation in the sample is exactly the same. In such an unusual case the range would also be zero.

But numerical data are inherently variable—not constant. Any random phenomenon of interest that we can think of usually takes on a variety of values. For example, colleges and universities charge different rates of tuition for out-of-state residents just as people have different IQs, incomes, weights, heights, ages, pulse rates, and so on. It is because numerical data inherently vary that it becomes so important to study not only measures (of central tendency) that summarize the data, but also measures (of variation) that reflect how the numerical data are dispersed.

- **What the Variance and the Standard Deviation Indicate** The variance and the standard deviation measure the "average" scatter around the mean—that is, how larger observations fluctuate above it and how smaller observations distribute below it.

  The variance possesses certain useful mathematical properties. However, its computation results in squared units—squared thousands of dollars, squared dollars, squared inches, and so on. Thus, our primary measure of variation will be the standard deviation, whose value is in the original units of the data—thousands of dollars, dollars, or inches.

  In the six-school sample, the standard deviation is 2.52 thousand dollars. This tells us that the *majority* of the tuition rates in this sample are clustering within 2.52 thousand dollars around the mean of 8.30 thousand dollars (i.e., between 5.78 and 10.82 thousand dollars).

- **Why We Square the Deviations** The formulas for variance and standard deviation could not merely use

$$\sum_{i=1}^{n}\left(X_i - \overline{X}\right)$$

as a numerator, because, as you may recall, the mean acts as a *balancing point* for observations larger and smaller than it. Therefore, the sum of the deviations about the mean is always zero[3]; that is,

$$\sum_{i=1}^{n}\left(X_i - \overline{X}\right) = 0$$

To demonstrate this, let us again refer to the out-of-state tuition data for the six Pennsylvania schools:

$$10.3 \quad 4.9 \quad 8.9 \quad 11.7 \quad 6.3 \quad 7.7$$

Therefore,

$$\sum_{i=1}^{n}\left(X_i - \overline{X}\right) = (10.3 - 8.3) + (4.9 - 8.3) + (8.9 - 8.3)$$
$$+ (11.7 - 8.3) + (6.3 - 8.3) + (7.7 - 8.3)$$
$$= 0$$

This is depicted in the accompanying dot scale diagram displayed in Figure 3.5.

As already noted, three of the observations are smaller than the mean and three are larger. Although the sum of the six deviations (2.0, −3.4, 0.6, 3.4, −2.0, and −0.6) is zero, the sum of the squared deviations allows us to study the variation in the data. Hence we use

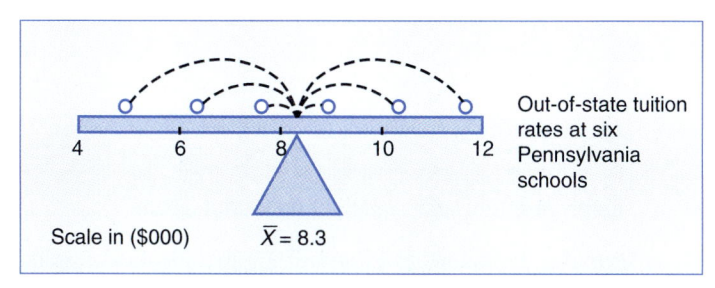

FIGURE 3.5 The mean as a *balancing point*.

$$\sum_{i=1}^{n}(X_i - \overline{X})^2$$

when computing the variance and standard deviation. In the squaring process, observations that are farther from the mean get more weight than observations closer to the mean.

The respective squared deviations for the Pennsylvania out-of-state tuition data are

$$4.00 \quad 11.56 \quad 0.36 \quad 11.56 \quad 4.00 \quad 0.36$$

We note that the fourth observation (X_4 = 11.7 thousand dollars) is 3.4 thousand dollars higher than the mean and the second observation (X_2 = 4.9 thousand dollars) is 3.4 thousand dollars lower. In the squaring process both these values contribute substantially more to the calculation of S^2 and S than do the other observations in the sample, which are closer to the mean.

Therefore, we may generalize as follows:

1. The more spread out or dispersed the data are, the larger will be the range, the interquartile range, the variance, and the standard deviation.

2. The more concentrated or homogeneous the data are, the smaller will be the range, the interquartile range, the variance, and the standard deviation.

3. If the observations are all the same (so that there is no variation in the data), the range, interquartile range, variance, and standard deviation will all be zero.

● **Computing S^2 and S: Calculator Formulas** The formulas for variance and standard deviation, Equations (3.6) and (3.7), are *definitional formulas*, but they may not be practical to use when we use a calculator. For our Pennsylvania out-of-state tuition data, the mean, 8.30 thousand dollars, is not an integer. For these more typical situations, where the observations and the mean are unlikely to be integers, the following *calculator formulas* for the variance and the standard deviation are given:

The variance is equal to the sum of the squared X's minus the sample size times the square of the arithmetic mean, divided by the sample size minus 1.

$$S^2 = \frac{\sum_{i=1}^{n} X_i^2 - n\overline{X}^2}{n - 1} \tag{3.8}$$

$$S = \sqrt{\frac{\sum_{i=1}^{n} X_i^2 - n\overline{X}^2}{n - 1}} \tag{3.9}$$

where $\sum_{i=1}^{n} X_i^2$ = summation of the squares of the individual observations

$n\overline{X}^2$ = sample size times the square of the sample mean

The calculator formulas, Equations (3.8) and (3.9), give identical results to the definitional formulas, Equations (3.6) and (3.7). Since the denominators are the same, it can be shown through expansion and the use of summation rules (see Appendix B) that

$$\sum_{i=1}^{n}(X_i - \bar{X})^2 = \sum_{i=1}^{n} X_i^2 - n\bar{X}^2$$

Returning to the Pennsylvania out-of-state tuition data, the variance and standard deviation are recomputed using Equations (3.8) and (3.9) as follows:

$$S^2 = \frac{\sum_{i=1}^{n} X_i^2 - n\bar{X}^2}{n-1}$$

$$= \frac{(10.3^2 + 4.9^2 + \cdots + 7.7^2) - 6(8.3^2)}{6-1}$$

$$= \frac{(106.09 + 24.01 + \cdots + 59.29) - 6(68.89)}{5}$$

$$= \frac{445.18 - 413.34}{5}$$

$$= \frac{31.84}{5} = 6.368 \text{ (in squared thousands of dollars)}$$

and

$$S = \sqrt{6.368} = 2.52 \text{ thousand dollars}$$

3.5.4 The Coefficient of Variation

Unlike the previous measures we have studied, the **coefficient of variation** is a *relative measure* of variation. It is expressed as a percentage rather than in terms of the units of the particular data.

The coefficient of variation, denoted by the symbol CV, measures the scatter in the data relative to the mean. It may be computed as follows:

> The coefficient of variation is equal to the standard deviation divided by the arithmetic mean, multiplied by 100%.
>
> $$\text{CV} = \left(\frac{S}{\bar{X}}\right)100\% \qquad (3.10)$$

where
S = standard deviation in a set of numerical data
$\bar{X}$ = arithmetic mean in a set of numerical data

Returning to the out-of-state tuition rates for the six-school sample from Pennsylvania, the coefficient of variation is

$$\text{CV} = \left(\frac{S}{\bar{X}}\right)100\% = \left(\frac{2.52}{8.30}\right)100\% = 30.4\%$$

For this sample the relative size of the "average spread around the mean" to the mean is 30.4%.

As a relative measure, the coefficient of variation is particularly useful when comparing the variability of two or more sets of data that are expressed in different units of measurement.

The coefficient of variation is also very useful when comparing two or more sets of data that are measured in the same units but differ to such an extent that a direct comparison of the respective standard deviations is not very helpful. As an example, suppose that a potential investor is considering purchasing shares of stock in one of two companies, A or B, which are listed on the American Stock Exchange. If neither company offers dividends to its stockholders and if both companies are rated equally high (by various investment services) in terms of potential growth, the potential investor might want to consider the volatility (variability) of the two stocks to aid in the investment decision. Now suppose that each share of stock in company A has averaged $50 over the past months with a standard deviation of $10. In addition, suppose that in this same time period, the price per share for company B stock averaged $12 with a standard deviation of $4. In terms of the actual standard deviations, the price of company A shares seems to be more volatile than that of company B shares. However, since the average prices per share for the two stocks are so different, it would be more appropriate for the potential investor to consider the variability in price relative to the average price in order to examine the volatility/stability of the two stocks. For company A, the coefficient of variation is $CV_A = (\$10/\$50)100\% = 20.0\%$; for company B, the coefficient of variation is $CV_B = (\$4/\$12)100\% = 33.3\%$. Thus, relative to the mean, the price of stock B is much more variable than the price of stock A.

3.5.5 Using Microsoft Excel Functions to Obtain Measures of Variation

In this section we have discussed a variety of measures of variation (range, interquartile range, standard deviation, variance, and coefficient of variation). We will now illustrate how Excel formulas (including functions) can be used to compute each of these measures of variation. To obtain these measures for the sample of six Pennsylvania schools, we need to allocate rows on the Calculations sheet as illustrated in Table 3.1.Excel on page 129. Additional rows are indicated in Table 3.2.Excel.

PA-SAM-2.XLS

Open the workbook developed in Section 3.4.6 or PA-SAM-2.XLS. Insert a new worksheet named Calculations, and enter the names of the statistics in cells A15:A19 and the formulas for the statistics in cells B15:B19. In doing this, we need to remember that the data is on the Data sheet and the calculations are on the Calculations sheet.

First, to compute the range, we subtract the minimum value in cell B7 from the maximum value in cell B8. Then, to compute the interquartile range, we subtract the first quartile in cell B11 from the third quartile in cell B13.

Once these are computed, we use the **STDEV function** to compute the standard deviation of the data in cells B2:B7 of the Data sheet and the **VAR function** to compute the variance of the same data. Finally, to compute the coefficient of variation, we divide the standard deviation in cell B17 by the arithmetic mean in cell B4. Since the coefficient of variation is usually expressed

Table 3.2.Excel Additional rows for measures of variation on the Calculations sheet for the out-of-state tuition data for the six-school sample from Pennsylvania.

	A	B
15	Range	=B8–B7
16	Interquartile Range	=B13–B11
17	Standard Deviation	=STDEV(Data!B2:B7)
18	Variance	=VAR(Data!B2:B7)
19	Coefficient of Variation	=B17/B4

as a percentage, we move the mouse pointer to cell B19, click the % button on the formatting toolbar, and then click the Increase Decimal button twice to provide a two-decimal-place result.

Figure 3.4.Excel represents the descriptive statistics obtained by using these Excel functions.

	A	B
1	Sample Statistics for Pennsylvania	
2		
3	Tuition Variable	
4	Mean	8.3
5	Median	8.3
6	Mode	#N/A
7	Minimum	4.9
8	Maximum	11.7
9	Midrange	8.3
10	First Quartile rank	1.75
11	First Quartile	6.3
12	Third Quartile rank	5.25
13	Third Quartile	10.3
14	Midhinge	8.3
15	Range	6.8
16	Interquartile Range	4
17	Standard Deviation	2.52349
18	Variance	6.368
19	Coefficient of Variation	30.40%

FIGURE 3.4.EXCEL Descriptive statistics obtained from Excel for the out-of-state tuition data for the six-school sample from Pennsylvania.

▲ WHAT IF EXAMPLE

As we did in Section 3.4.6 with the measures of central tendency, we can observe the effect of extreme values on the various measures of variation by pretending that the tuition at Slippery Rock University was incorrectly entered as 19.7 instead of 7.7. Change the value on the Data sheet in cell B7 to 19.7, and then go to the Calculations sheet to see the updated set of descriptive statistics shown in Figure 3.5.Excel.

By comparing these two sets of results, we can see that the range and the standard deviation were greatly affected by this extreme value of 19.7, while the interquartile range was much less affected. The range moved up to 14.8 from 6.8 and the standard deviation from 2.52 to 5.24, while the interquartile range became 5.4 instead of 4.0.

As before, if we were interested in seeing the effects of many different changes, we could use the Scenario Manager (see Section 1S.15) to store and use sets of alternative data values.

	A	B
1	Sample Statistics for Pennsylvania	
2		
3	Tuition Variable	
4	Mean	10.3
5	Median	9.6
6	Mode	#N/A
7	Minimum	4.9
8	Maximum	19.7
9	Midrange	12.3
10	First Quartile rank	1.75
11	First Quartile	6.3
12	Third Quartile rank	5.25
13	Third Quartile	11.7
14	Midhinge	9
15	Range	14.8
16	Interquartile Range	5.4
17	Standard Deviation	5.2429
18	Variance	27.488
19	Coefficient of Variation	50.90%

FIGURE 3.5.EXCEL Revised descriptive statistics obtained from Excel for the out-of-state tuition data for the six-school sample from Pennsylvania.

Problems for Section 3.5

Note: These problems may be solved using the Microsoft Excel functions discussed in Section 3.5.5 or the Data Analysis tool discussed in Section 3.7.

3.11 Verify that the computation of the standard deviation is identical, regardless of whether the definitional formula (3.7) or the calculator formula (3.9) is used, for the following:
   (a) The out-of-state tuition data from the six-school sample from Texas (see page 123).
   (b) The out-of-state tuition data from the six-school sample from North Carolina (see page 123).

3.12 For each data set in Problem 3.1 on page 132:
   (a) Compute the range, interquartile range, variance, standard deviation, and coefficient of variation.
   (b) Compare your results and discuss your findings.
   (c) Based on your answers to Problem 3.1 and (a) and (b) above, what can you generalize about the properties of central tendency and variation?

3.13 Using the data from Problem 3.2 on page 132, compute the range, interquartile range, variance, standard deviation, and coefficient of variation for each of the two athletes' race-time trials.

3.14 Using the PE ratio data from Problem 3.3 on page 133, compute the range, interquartile range, variance, standard deviation, and coefficient of variation for the data set with the error (61) and then recompute these statistics after the PE ratio is corrected to 16.
   (a) Discuss the differences in your findings for each measure of dispersion.
   (b) Which measure seems to be affected most by the error?

3.15 For the following, refer to the battery life data in Problem 3.4 on page 133:
   (a) Calculate the range, variance, and standard deviation.
   (b) For many sets of data, the range is about six times the standard deviation. Is this true here? (If not, why do you think it is not?)
   (c) Using the information above, what would you advise the manufacturer to do if he wanted to be able to say in advertisements that these batteries "should last 400 hours"? (*Note*: There is no right answer to this question; the point is to consider how to make such a statement precise.)
   (d) Do (a)–(c) with the first value equal to 1342 instead of 342.

3.16 Using the grants data from Problem 3.8 on page 134:
   (a) Compute the range, interquartile range, variance, standard deviation, and coefficient of variation for amount funded (in millions of dollars).
   (b) Discuss the property of variation for these data.

3.17 Using the train time data from Problem 3.9 on page 135:
   (a) Compute the range, interquartile range, variance, standard deviation, and coefficient of variation for "lateness" (in minutes).
   (b) Discuss the property of variation for these data.
   (c) Repeat (a) using the value of 12 instead of 124 and compare the results with those obtained in (a).

3.18 Using the water usage data from Problem 3.10 on page 135:
   (a) Compute the range, interquartile range, variance, standard deviation, and coefficient of variation in water consumption.
   (b) Discuss the property of variation for these data.

3.6 SHAPE

A third important property of a set of data is its **shape**—the manner in which the data are distributed. Either the distribution of the data is **symmetrical** or it is not. If the distribution of data is not symmetrical, it is called asymmetrical or **skewed.**

To describe the shape, we need only compare the mean and the median. If these two measures are equal, we may generally consider the data to be symmetrical (or *zero-skewed*). On the other hand, if the mean exceeds the median, the data may generally be described as *positive* or **right-skewed.** If the mean is exceeded by the median, those data can generally be called *negative* or **left-skewed.** That is,

Mean > median: positive or right-skewness

Mean = median: symmetry or zero-skewness

Mean < median: negative or left-skewness

Positive skewness arises when the mean is increased by some unusually high values; negative skewness occurs when the mean is reduced by some extremely low values. Data are symmetrical when there are no really extreme values in a particular direction so that low and high values balance each other out.

Figure 3.6 depicts the shapes of three data sets: The data on scale L are negative or left-skewed (since the distortion to the left is caused by extremely small values); the data on scale R are positive or right-skewed (since the distortion to the right is caused by extremely large values); and the data on scale S are symmetrical (the low and high values on the scale balance, and the mean equals the median).

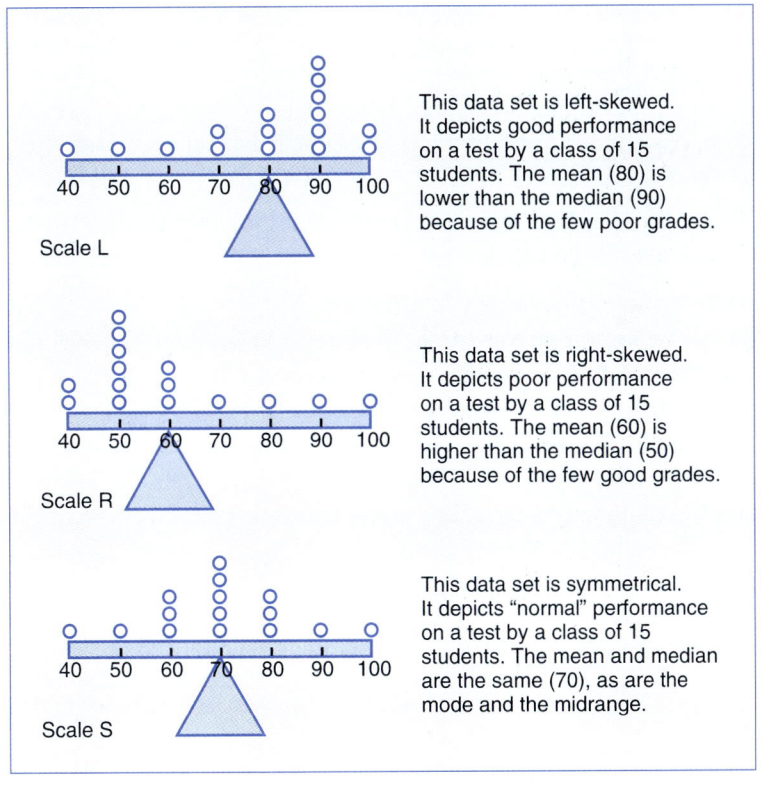

FIGURE 3.6 A comparison of three data sets differing in shape.

For our sample of six Pennsylvania schools, the out-of-state tuition data are displayed along the dot scale in Figure 3.1 (see page 123). The mean and the median are equal to 8.3 thousand dollars, and the data appear to be symmetrically distributed around these measures of central tendency.

Problems for Section 3.6

FLASHBAT.TXT

- 3.19 For each data set in Problem 3.1 on page 132:
  (a) Describe the shape.
  (b) Compare your results and discuss your findings.
- 3.20 Using the battery life data from Problem 3.4 on page 133, describe the shape.
- 3.21 Using the grants data from Problem 3.8 on page 134, describe the shape.
- 3.22 Using the train "lateness" data from Problem 3.9 on page 135, describe the shape.
- 3.23 Using the water consumption data from Problem 3.10 on page 135, describe the shape.

3.7 USING THE DATA ANALYSIS TOOL TO OBTAIN DESCRIPTIVE STATISTICS

In addition to (or instead of) using various functions to compute descriptive statistics, we could use the **Data Analysis tool** of Excel to simultaneously obtain a set of descriptive statistics. The advantage of the Data Analysis tool is that it is easier to use because numerous statistics can be obtained from a single-user operation. A disadvantage is that if data is changed after the Data Analysis tool has been used, the entire sequence of steps would have to be repeated with the new data.

PA-SAM-1.XLS

With the workbook for the out-of-state tuition data for the sample of six Pennsylvania schools opened, click on the Data sheet tab. The Descriptive Statistics Data Analysis tool is then used as follows:

❶ Select Tools | Data Analysis. Select Descriptive Statistics from the Analysis Tools list box and click the OK button.

❷ The Descriptive Statistics dialog box appears.
  (a) The input range represents the location of the data. Enter the range of the cells pertaining to the data (Data!B1:B7).
  (b) Since the data are located in a column, click the Grouped By Columns option button.
  (c) Since the label for the tuition variable is in cell B1, the first cell of the input range, select the Labels in First Row check box.
  (d) Leave the Confidence Interval box unchecked since it will be discussed in Chapter 6.
  (e) To obtain the minimum and maximum values, select the Kth Largest and the Kth Smallest check boxes and enter 1 in their edit boxes. If we want to obtain other measures of position, by entering the value for k (such as 3 for the third smallest or largest), we can indicate which ranked value we desire to obtain.

❸ Select the New Worksheet Ply option button and enter the name Descriptive in its edit box.

❹ Select the Summary Statistics check box. The Descriptive Statistics dialog box should look like the one depicted in Figure 3.6.Excel. Click the OK button.

Figure 3.7.Excel represents the output obtained from the Descriptive Statistics option of the Data Analysis tool for the out-of-state tuition data for the six-school sample from Pennsylvania.

Although this output is similar to the one illustrated in Figure 3.4.Excel on page 143, there are some minor differences. There are three statistics included here that are not part of Figure 3.4.Excel. These are standard error, skewness, and kurtosis. The standard error is the standard deviation divided by the square root of the sample size and will be discussed in

146 **Chapter 3** Summarizing and Describing Numerical Data

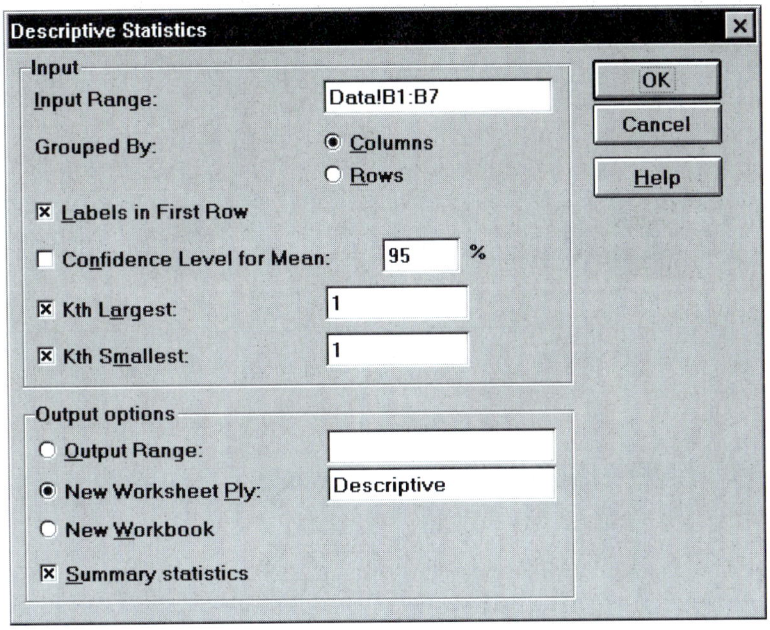

FIGURE 3.6.EXCEL Descriptive Statistics dialog box for the out-of-state tuition worksheet for the six-school sample from Pennsylvania.

	A	B
1	Tuition	
2		
3	Mean	8.3
4	Standard Error	1.030210335
5	Median	8.3
6	Mode	#N/A
7	Standard Deviation	2.523489647
8	Sample Variance	6.368
9	Kurtosis	-1.079565541
10	Skewness	2.1743E-15
11	Range	6.8
12	Minimum	4.9
13	Maximum	11.7
14	Sum	49.8
15	Count	6
16	Largest(1)	11.7
17	Smallest(1)	4.9
18	Confidence Level(95.000%)	2.019172162

FIGURE 3.7.EXCEL Descriptive statistics obtained from the Excel Data Analysis tool for the out-of-state tuition rates for the six-school sample from Pennsylvania.

Chapter 5. Skewness is a measure of the lack of symmetry in the data set and is based on a statistic that is a function of the cubed differences around the arithmetic mean. Kurtosis is a measure of the relative concentration of values in the center of the distribution as compared to the tails and is based on the differences around the arithmetic mean raised to the fourth power and is not discussed in this text (see Reference 4).

> Note that the value for skewness is formatted as 2.1743E–15 using a numeric format known as scientific notation. Excel uses this type of format to display very small or very large values. The number after the letter E represents the number of digits that the decimal point needs to be moved to the left (for a negative number) or to the right (for a positive number). For example, the number 3.7431E02 would mean that the decimal point should be moved two places to the right, producing the number 374.31. The number 3.7431E–02 would mean that the decimal point should be moved two places to the left, producing the number .0037431. Thus, the result of the skewness statistic, equal to 2.1743E–15 would mean that the decimal point should be moved 15 places to the left, producing the number 0.000000000000021743.

3.8 FIVE-NUMBER SUMMARY AND BOX-AND-WHISKER PLOT

Now that we have studied the three major properties of numerical data (central tendency, variation, and shape), it is important that we identify and describe the major features of the data in a summarized format. One approach to this "exploratory data analysis" is to develop a *five-number summary* and to construct a *box-and-whisker plot* (References 5 and 6).

3.8.1 Five-Number Summary

A **five-number summary** consists of

$$X_{smallest} \quad Q_1 \quad median \quad Q_3 \quad X_{largest}$$

It combines three measures of central tendency (the median, midhinge, and midrange) and two measures of variation (the interquartile range and range) to provide us with a better idea as to the shape of the distribution.

If the data were perfectly symmetrical, the following would be true:

1. The distance from Q_1 to the median would equal the distance from the median to Q_3.
2. The distance from $X_{smallest}$ to Q_1 would equal the distance from Q_3 to $X_{largest}$.
3. The median, the midhinge, and the midrange would all be equal. (These measures would also equal the mean in the data.)

On the other hand, for nonsymmetrical distributions the following would be true:

1. In right-skewed distributions the distance from Q_3 to $X_{largest}$ greatly exceeds the distance from $X_{smallest}$ to Q_1.
2. In right-skewed distributions, median < midhinge < midrange.
3. In left-skewed distributions, the distance from $X_{smallest}$ to Q_1 greatly exceeds the distance from Q_3 to $X_{largest}$.
4. In left-skewed distributions, midrange < midhinge < median.

For our Pennsylvania out-of-state tuition data, the five-number summary is

$$4.9 \quad 6.3 \quad 8.3 \quad 10.3 \quad 11.7$$

We may now use the five-number summary to study the shape of this distribution. From the preceding rules, it is clear that the out-of-state tuition data for our sample of six Pennsylvania schools are perfectly symmetrical.

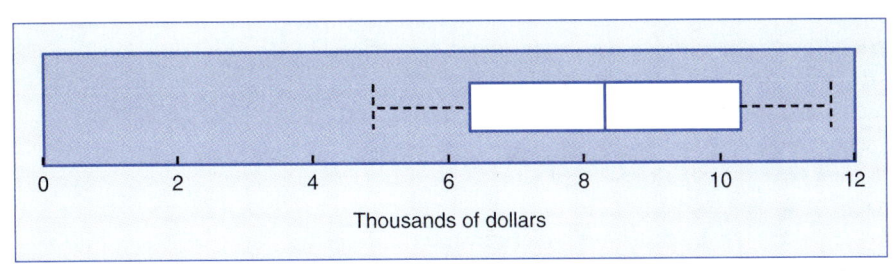

FIGURE 3.7 Box-and-whisker plot of out-of-state tuition rates at six Pennsylvania schools.

3.8.2 Box-and-Whisker Plot

In its simplest form, a **box-and-whisker plot** provides a graphical representation of the data through its five-number summary. Such a plot is depicted in Figure 3.7 for the out-of-state tuition rates at the six Pennsylvania schools.

The vertical line drawn within the box represents the location of the median value in the data. Note further that the vertical line at the left side of the box represents the location of Q_1 and the vertical line at the right side of the box represents the location of Q_3. Therefore, we see that the box contains the middle 50% of the observations in the distribution. The lower 25% of the data are represented by a dashed line (i.e., a *whisker*) connecting the left side of the box to the location of the smallest value, $X_{smallest}$. Similarly, the upper 25% of the data are represented by a dashed line connecting the right side of the box to $X_{largest}$.

This visual representation of the tuition rates depicted in Figure 3.7 indicates the symmetrical shape of the data. Not only do we observe that the vertical median line is centered in the box, we also see that the whisker lengths are clearly the same.

To summarize what we have learned about graphical representation of our data, Figure 3.8 on page 150 demonstrates the differences between "exploratory data analysis" and polygons by depicting four different types of distributions through their box-and-whisker plots and corresponding polygons.

When a data set is perfectly symmetrical, as is the case in Figure 3.8(a) and (d), the mean, median, midrange, and midhinge will be the same. In addition, the length of the left whisker will equal the length of the right whisker, and the median line will divide the box in half. In practice, it is unlikely that we will observe a data set that is perfectly symmetrical. However, we should be able to state that our data set is approximately symmetrical if the lengths of the two whiskers are almost equal and the median line almost divides the box in half.

On the other hand, when our data set is left-skewed as in Figure 3.8(b), the few small observations distort the midrange and mean toward the left tail. In this case,

$$\text{midrange} < \text{mean} < \text{midhinge} < \text{median} < \text{mode}$$

For this left-skewed distribution, we may observe from Figure 3.8(b) that the skewed (i.e., distorted) nature of the data set indicates that there is a heavy clustering of observations at the high end of the scale (i.e., the right side); 75% of all data values are found between the left edge of the box (Q_1) and the end of the right whisker ($X_{largest}$). Therefore, the long left whisker contains the distribution of only the smallest 25% of the observations, demonstrating the distortion from symmetry in this data set.

If the data set is right-skewed as in Figure 3.8(c), the few large observations distort the midrange and mean toward the right tail. In this case,

$$\text{mode} < \text{median} < \text{midhinge} < \text{mean} < \text{midrange}$$

For the right-skewed data set in Figure 3.8(c), the concentration of data points will be on the low end of the scale (i.e., the left side of the box-and-whisker plot). Here, 75% of all data val-

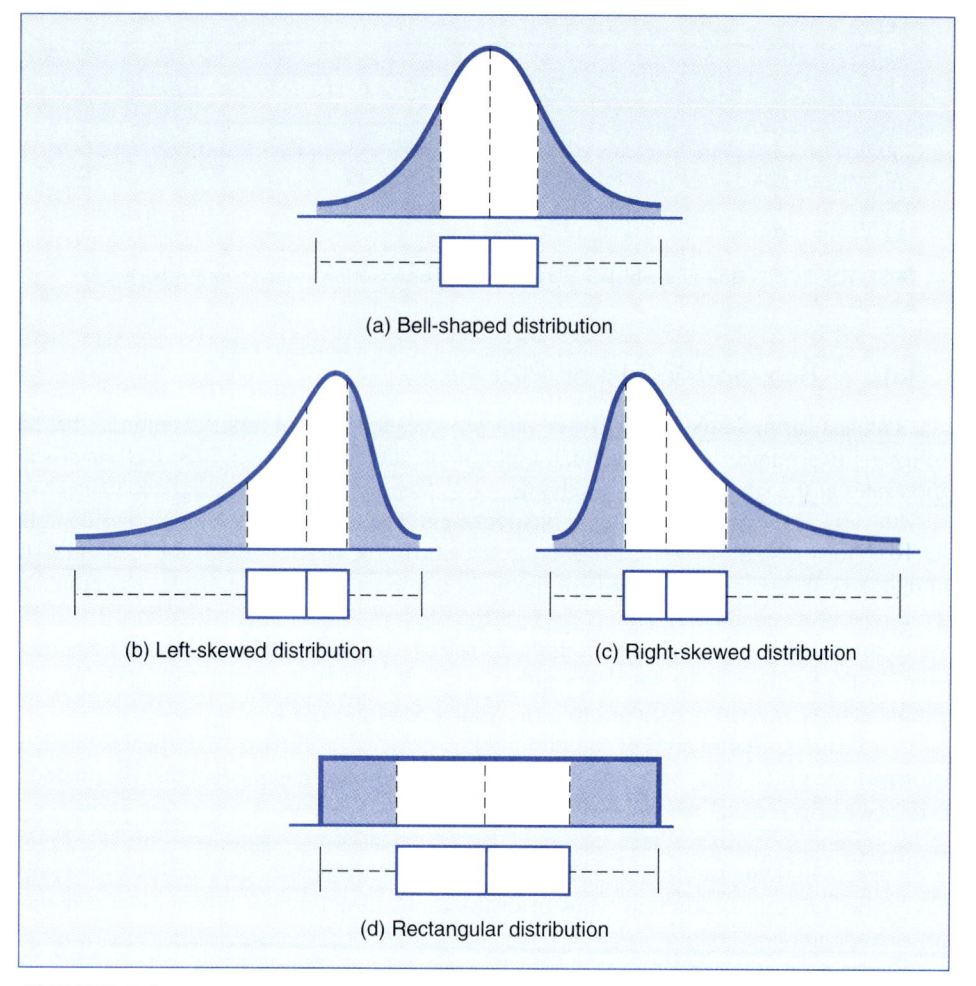

FIGURE 3.8 Four hypothetical distributions examined through their box-and-whisker plots and their corresponding polygons. *Note:* Areas under polygon are split into quartiles corresponding to the five-number summary for the box-and-whisker plots.

(a) Bell-shaped distribution
(b) Left-skewed distribution
(c) Right-skewed distribution
(d) Rectangular distribution

ues are found between the beginning of the left whisker ($X_{smallest}$) and the right edge of the box (Q_3), and the remaining 25% of the observations are dispersed along the long right whisker at the upper end of the scale.

3.8.3 Simulating a Box-and-Whisker Plot Using Microsoft Excel

Although the Chart Wizard does not currently contain an option for a Box-and-Whisker plot, it is possible to format a Line chart choice to obtain an approximation of such a plot. To illustrate this, open the workbook file PA-SAM-2.XLS containing the tuition data and statistics for the sample of six Pennsylvania schools.

3-8-3.XLS

To simplify the task, first consolidate the display of the five statistics needed for a box-and-whisker plot by implementing the following formulas on the Calculations sheet:

Cell E4: enter=$B$7 to display the minimum value.
Cell E5: enter=$B$11 to display the first quartile.
Cell E6: enter=$B$5 to display the median.

Cell E7:   enter=$B$13 to display the third quartile.

Cell E8:   enter=$B$8 to display the maximum value.

(Note: these formulas must be entered using absolute addresses.)

Next copy the formulas in the range E4:E8 to columns G, H, and I in the same rows. One way this can be accomplished is through these four steps:

(a) Select the range E4:E8.

(b) Select the command Edit | Copy.

(c) Select the range G4:I8.

(d) Select the command Edit | Paste.

After this copy operation, *delete* the formulas in cells G4, G8, I4, and I8 by moving the cell highlight over each of these cells and pressing the Delete key. Add a heading as is shown in Figure 3.8.Excel Panel A. At this point, we are ready to insert a chart that will simulate a box-and-whisker plot.

To create a simulated box-and-whisker plot, select Insert | Chart | As New Sheet and step through the Chart Wizard following these directions:

Step 1: Enter Calculations!F4:J8 as the range.

Step 2: Select the Line chart choice.

Step 3: Select the first format choice. If a dialog box with the message "Number must be between 0 and 1" appears, click the OK button to proceed to step 4.

Step 4: Select the Rows option button for Data Series in and enter 0 for First Columns and 0 for First Rows.

Step 5: Select the No option button for Add a Legend and enter Tuition Box-and-Whisker Plot as the Chart Title. Enter Thousands of Dollars in the Value (Y) edit box.

The Chart Wizard produces a new chart on a new sheet which now needs to be modified to look more like a box-and-whisker plot. To do this, select any point on the chart and then select the command Format | 1 Line Group. Select the Options tab from the dialog box that appears and then select the High-Low Lines check box. Click the OK button. The chart begins to look like a box-and-whisker plot.

Continue by selecting the X axis. Select Format | Selected Axis. In the Format Axis dialog box, select the Patterns tab and then select the None option button for both Tick-Mark Labels and Tick mark Type. Click the OK button. This produces an approximation of a box-and-whisker plot. Note: Further refinements are possible by changing the settings for Line and Marker in the Patterns tabs of the Format dialog box that appears when selecting points

F	G	H	I	J
		Values for Box-and-Whisker Plot		
			4.9	
	6.3	6.3	6.3	
	8.3	8.3	8.3	
	10.3	10.3	10.3	
		11.7		

FIGURE 3.8.EXCEL
Completed additions to the Calculations Sheet. Panel A.

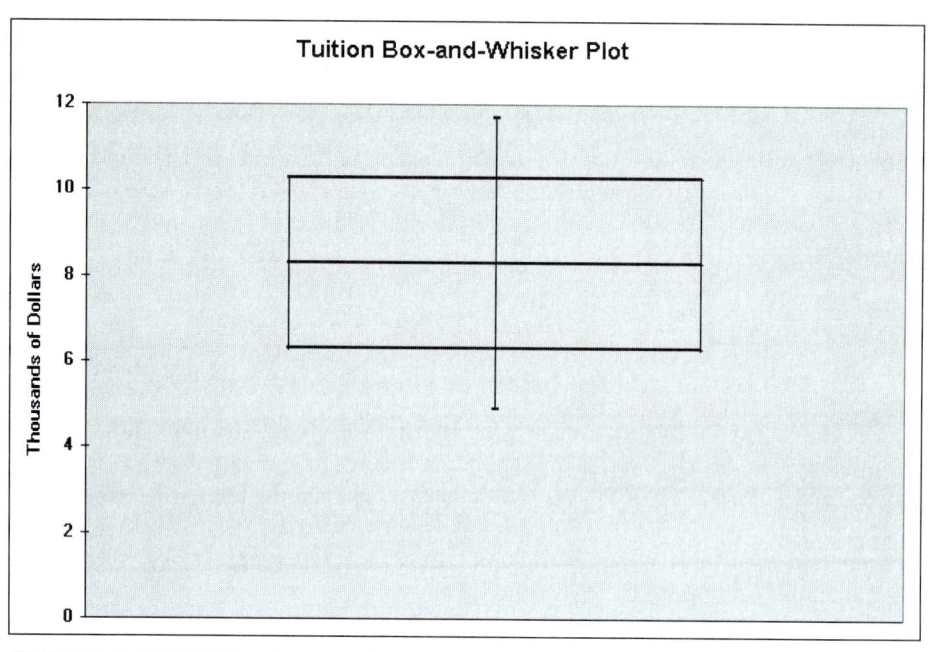

FIGURE 3.8.EXCEL Simulated Box-and-Whisker plot. Panel B.

or lines in the chart and then choosing the command Format | Selected Data Series (or Data Point). Figure 3.8.Excel Panel B shows a simulated box-and-whisker plot after line weight and color, and marker style and color has been changed.

Problems for Section 3.8

FLASHBAT.TXT

3.24 Using the battery life data from Problem 3.4 on page 133:
 (a) List the five-number summary.
 (b) Form the box-and-whisker plot and describe the shape.
 (c) Compare your answer in (b) with that from Problem 3.20 on page 146. Discuss.

ADAMHA.TXT

• 3.25 Using the grants data from Problem 3.8 on page 134:
 (a) List the five-number summary.
 (b) Form the box-and-whisker plot and describe the shape.
 (c) Compare your answer in (b) with that from Problem 3.21 on page 146. Discuss.

TRAIN1.TXT

3.26 Using the train "lateness" data from Problem 3.9 on page 135:
 (a) List the five-number summary.
 (b) Form the box-and-whisker plot and describe the shape.
 (c) Compare your answer in (b) with that from Problem 3.22 on page 146. Discuss.

WATER.TXT

3.27 Using the water consumption data from Problem 3.10 on page 135:
 (a) List the five-number summary.
 (b) Form the box-and-whisker plot and describe the shape.
 (c) Compare your answer in (b) with that from Problem 3.23 on page 146. Discuss.

3.9 CALCULATING DESCRIPTIVE SUMMARY MEASURES FROM A POPULATION

In Sections 3.4–3.6, we examined various *statistics* that summarize or describe numerical information from a *sample*. In particular, we used these statistics to describe the properties of central tendency, variation, and shape for the out-of-state tuition data obtained from the sample of $n = 6$ Pennsylvania schools. Suppose, however, that our analyst at the college advisory service now wishes to conduct a more thorough investigation of the tuition rates charged (in thousands of dollars) to out-of-state residents in each of the 90 colleges and universities in the state of Pennsylvania (i.e., the *population*). The resulting measures (the *parameters*) computed from the population of $N = 90$ Pennsylvania schools to summarize and describe the properties of central tendency, variation, and shape could then be used by our analyst in her report comparing and contrasting regional differences in out-of-state tuition rates.

3.9.1 Population Measures of Central Tendency

• **The Population Mean** The **population mean** is given by the symbol μ, the Greek lowercase letter mu. It is obtained from the following:

The population mean is equal to the sum of all the values in the population divided by the population size.

$$\mu = \frac{\sum_{i=1}^{N} X_i}{N} \tag{3.11}$$

where
N = population size

X_i = ith value of the random variable X

$\sum_{i=1}^{N} X_i$ = summation of all X_i values in the population

• **The Population Median, Mode, Midrange, and Midhinge** The median, mode, midrange, and midhinge for a population of size N are obtained as previously described in Sections 3.4.2–3.4.5 by replacing n with N.

3.9.2 Population Measures of Variation

• **The Population Range and Interquartile Range** The range and interquartile range for a population of size N are obtained as previously described in Sections 3.5.1 and 3.5.2 by replacing n with N.

• **The Population Variance and Standard Deviation** The **population variance** is given by the symbol σ^2, the Greek lowercase letter sigma squared, and the **population standard deviation** is given by the symbol σ. It is obtained from the following:

The population variance is equal to the sum of the squared differences around the population mean, divided by the population size.

$$\sigma^2 = \frac{\sum_{i=1}^{N}(X_i - \mu)^2}{N} \tag{3.12}$$

where
N = population size
X_i = ith value of the random variable X

$\sum_{i=1}^{N} X_i$ = summation of all X_i values in the population

$\sum_{i=1}^{N} (X_i - \mu)^2$ = summation of all the squared differences between the X_i values and μ

and the population standard deviation is the square root of the population variance:

$$\sigma = \sqrt{\frac{\sum_{i=1}^{N}(X_i - \mu)^2}{N}} \qquad (3.13)$$

We note that the formulas for the population variance and standard deviation differ from those for the sample in that $(n - 1)$ in the denominator of S^2 and S [see Equations (3.6) and (3.7)] is replaced by N in the denominator of σ^2 and σ.

- **The Population Coefficient of Variation** The **population coefficient of variation**, given by the symbol CV_{pop}, measures the scatter in the data relative to the mean. It may be computed by

$$CV_{pop} = \left(\frac{\sigma}{\mu}\right) 100\% \qquad (3.14)$$

where
σ = standard deviation in the population
μ = arithmetic mean in the population

3.9.3 Results

The raw data of out-of-state tuition rates charged (in thousands of dollars) at all $N = 90$ colleges and universities in the state of Pennsylvania are presented in Special Data Set 1 of Appendix D on pages D4–D5. From these data, the revised stem-and-leaf display shown in Figure 3.9 is obtained.

Using the raw data or the data arranged in a stem-and-leaf display, the following summary measures are obtained:

- **Mean**

$$\mu = \frac{\sum_{i=1}^{N} X_i}{N} = \frac{979.8}{90} = 10.89 \text{ thousand dollars}$$

```
 2 | 7
 3 |
 4 | 048999
 5 | 05
 6 | 011113
 7 | 77
 8 | 3334449
 9 | 113334455667777
10 | 001222233346677
11 | 244567779
12 | 36
13 | 02357
14 | 1239
15 | 246
16 | 144
17 | 0177899
18 | 39
19 |
20 |
21 |
22 | 3
```

FIGURE 3.9 Revised stem-and-leaf display of out-of-state tuition rates at 90 Pennsylvania colleges and universities. *Source:* Special Data Set 1 of Appendix D, pages D4–D5.

PENNC&U.TXT

- **Median**

$$\text{Positioning point} = \frac{N+1}{2} \text{ ordered observation}$$

$$= \frac{90+1}{2} = 45.5\text{th ordered observation}$$

To obtain the median, we simply count (left to right, row by row) to the 45th and 46th ordered observations and take the average. In our data, these observations are found in the row with a "stem" of 10. The respective "leaves" are 2 and 2—corresponding to tuition rates of 10.2 and 10.2 thousand dollars. Thus, the median is (10.2 + 10.2)/2 = 10.20 thousand dollars.

- **Mode** The most frequently observed out-of-state tuition rates in Pennsylvania are 6.1, 9.7, and 10.2 thousand dollars. The data are multimodal.

- **Midrange**

$$\frac{X_{smallest} + X_{largest}}{2} = \frac{2.7 + 22.3}{2} = 12.50 \text{ thousand dollars}$$

- Q_1

$$\text{Positioning point} = \frac{N+1}{4} \text{ ordered observation}$$

$$= \frac{90+1}{2} = 22.75\text{th ordered observation}$$

$$\cong 23\text{rd ordered observation}$$

To obtain Q_1, we simply count (left to right, row by row) to the 23rd ordered observation. In our data, the "leaf" is 4, branching off the "stem" of 8. Therefore, $Q_1 = 8.40$ thousand dollars.

- Q_3

$$\text{Positioning point} = \frac{3(N+1)}{4} \text{ ordered observation}$$

$$= \frac{273}{4} = 68.25\text{th ordered observation}$$

$$\cong 68\text{th ordered observation}$$

To obtain Q_3, we simply count (left to right, row by row) to the 68th ordered observation. In our data, the "leaf" is 3, branching off the "stem" of 13. Therefore, $Q_3 = 13.30$ thousand dollars.

- **Midhinge**

$$\frac{Q_1 + Q_3}{2} = \frac{8.4 + 13.3}{2} = 10.85 \text{ thousand dollars}$$

- **Range**

$$X_{largest} - X_{smallest} = 22.3 - 2.7 = 19.60 \text{ thousand dollars}$$

- **Interquartile Range**

$$Q_3 - Q_1 = 13.3 - 8.4 = 4.90 \text{ thousand dollars}$$

- **Variance**

$$\sigma^2 = \frac{\sum_{i=1}^{N}(X_i - \mu)^2}{N} = \frac{(14.9 - 10.89)^2 + (16.4 - 10.89)^2 + \cdots + (4.8 - 10.89)^2}{90}$$

$$= 15.594 \text{ (in squared thousands of dollars)}$$

- **Standard Deviation**

$$\sigma = \sqrt{\sigma^2} = \sqrt{15.594} = 3.95 \text{ thousand dollars}$$

- **Coefficient of Variation**

$$CV_{pop} = \left(\frac{\sigma}{\mu}\right)100\% = \left(\frac{3.95}{10.89}\right)100\% = 36.3\%$$

3.9.4 Shape

The shape of the population is obtained through a *relative* comparison of the mean and the median, supported by an evaluation of the five-number summary and box-and-whisker plot.

The five-number summary is

$X_{smallest}$	Q_1	median	Q_3	$X_{largest}$
2.70	8.40	10.20	13.30	22.30

and the corresponding box-and-whisker plot is displayed in Figure 3.10.

Among the 90 colleges and universities in the state of Pennsylvania, the population of out-of-state tuition rates can be considered as right-skewed in shape because the mean

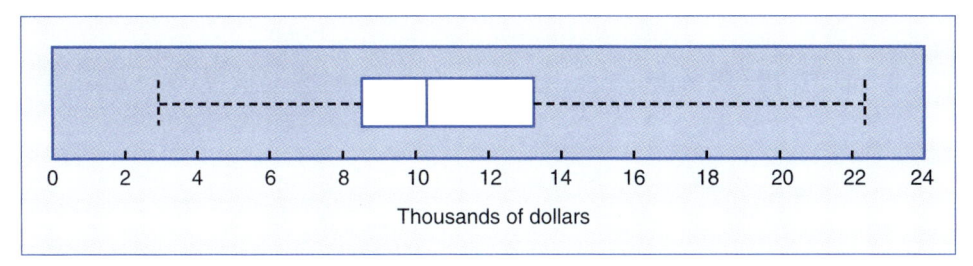

FIGURE 3.10 Box-and-whisker plot of out-of-state tuition rates at 90 Pennsylvania schools.

(10.89 thousand dollars) exceeds the median (10.20 thousand dollars). Similar conclusions can be drawn from the analysis of the box-and-whisker plot depicted in Figure 3.10.

3.9.5 Summarizing the Findings from the Sample and Population

Table 3.1 summarizes the results of utilizing the various descriptive measures we have investigated in this chapter. We observe that the various statistics computed from the sample of 6 schools appear to differ from the corresponding characteristics obtained from the population of 90 schools. Why this has happened, however, is simply a function of chance. When drawing the random sample, our analyst at the college advisory service properly used a table of random numbers (Table E.1), as discussed in Section 1.10. Unfortunately, due to the small size of the sample and purely by chance, the out-of-state tuition rates charged by the selected colleges and universities are fairly homogeneous and fail to account for the range in rates that exists in the entire population of 90 schools. This is clearly depicted on the dot scale shown in Figure 3.11 on page 158. The sample data are not right-skewed because not one of the six

Table 3.1 Using descriptive measures on the two data sets from Pennsylvania.

Descriptive Measure	Out-of-State Tuition Rates Sample ($n = 6$)	Population ($N = 90$)
Mean	8.30	10.89
Median	8.30	10.20
Mode	No mode	Multimodal
$X_{smallest}$	4.90	2.70
$X_{largest}$	11.70	22.30
Midrange	8.30	12.50
Q_1	6.30	8.40
Q_3	10.30	13.30
Midhinge	8.30	10.85
Range	6.80	19.60
Interquartile range	4.00	4.90
Variance	6.368	15.594
Standard deviation	2.52	3.95
Coefficient of variation	30.4%	36.3%
Shape	Symmetrical	Right-skewed

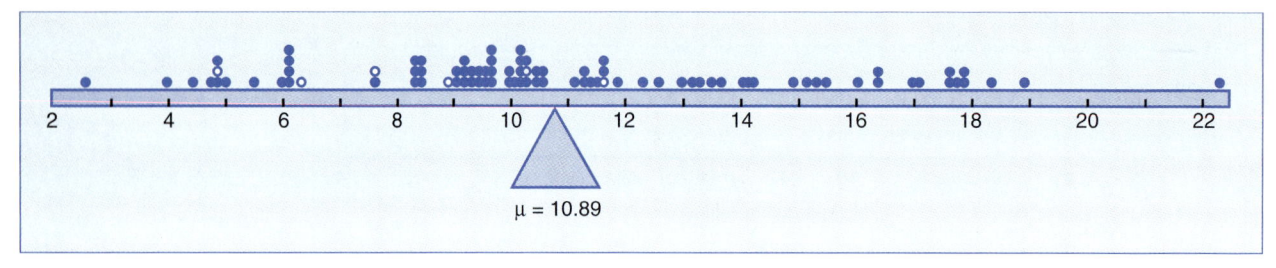

FIGURE 3.11 Dot scale showing the out-of-state tuition rates (in $000) at 90 Pennsylvania schools. Source: Figure 3.9.

schools has a tuition rate for out-of-state residents (light dots) that was among the highest 30% in the population of schools.

3.9.6 Using the Standard Deviation: The Empirical Rule

In most data sets, a large portion of the observations tend to cluster somewhat near the median. In right-skewed data sets, this clustering occurs to the left of (i.e., below) the median, and in left-skewed data sets, the observations tend to cluster to the right of (i.e., above) the median. In symmetrical data sets, where the median and mean are the same, the observations tend to distribute equally around these measures of central tendency. When extreme skewness is not present and such clustering is observed in a data set, we can use the so-called *empirical rule* to examine the property of data variability and get a better sense of what the standard deviation is measuring.

> The **empirical rule** states that for most data sets, we will find that roughly two out of every three observations (i.e., 67%) are contained within a distance of 1 standard deviation around the mean and roughly 90–95% of the observations are contained within a distance of 2 standard deviations around the mean.

Hence, the standard deviation, as a measure of average variation around the mean, helps us to understand how the observations distribute above and below the mean and helps us to focus on and flag unusual observations (i.e., outliers) when analyzing a set of numerical data.

From Table 3.1, we recall that for the population of 90 Pennsylvania schools, the mean tuition rate charged to out-of-state residents, μ, is 10.89 thousand dollars and the standard deviation, σ, is 3.95 thousand dollars. From the revised stem-and-leaf display (Figure 3.9), we note that 58 out of 90 schools (64.4%) had a tuition rate between $\mu - 1\sigma$ and $\mu + 1\sigma$ (i.e., between 6.94 and 14.84 thousand dollars). Moreover, we see that 87 out of 90 schools (96.7%) had a tuition rate between $\mu - 2\sigma$ and $\mu + 2\sigma$ (i.e., between 2.99 and 18.79 thousand dollars). Finally, we note that all 90 schools (100%) had a tuition rate within $\mu - 3\sigma$ and $\mu + 3\sigma$ (i.e., between 0 and 22.74 thousand dollars).[4] It is interesting to note that even though the out-of-state tuition data are right-skewed in shape, the percentages of colleges and universities with tuition rates falling within 1 or more standard deviations about the mean are not very different from what would be expected had the data been distributed as a symmetrical, bell-shaped distribution.

3.9.7 Obtaining the Population Standard Deviation and Variance from Microsoft Excel

The only measures covered in this section that differ from those discussed in previous sections of this chapter are the population standard deviation and variance. To compute these population parameters, we would simply use the STDEVP function instead of the STDEV function, and the VARP function instead of the VAR function.

Problems for Section 3.9

3.28 Given the following set of data for a population of size $N = 10$:

        7   5   11   8   3   6   2   1   9   8

(a) Compute the mean, median, mode, midrange, and midhinge.
(b) Compute the range, interquartile range, variance, standard deviation, and coefficient of variation.
(c) Are these data skewed? If so, how?

3.29 Given the following set of data for a population of size $N = 10$:

        7   5   6   6   6   4   8   6   9   3

(a) Compute the mean, median, mode, midrange, and midhinge.
(b) Compute the range, interquartile range, variance, standard deviation, and coefficient of variation.
(c) Are these data skewed? If so, how?
(d) Compare the measures of central tendency to those of Problem 3.28(a). Discuss.
(e) Compare the measures of variation to those of Problem 3.28(b). Discuss.

3.30 The following data represent the quarterly sales tax receipts (in $000) submitted to the comptroller of Gmoserville Township for the period ending March 1995 by all 50 business establishments in that locale:

10.3	11.1	9.6	9.0	14.5
13.0	6.7	11.0	8.4	10.3
13.0	11.2	7.3	5.3	12.5
8.0	11.8	8.7	10.6	9.5
11.1	10.2	11.1	9.9	9.8
11.6	15.1	12.5	6.5	7.5
10.0	12.9	9.2	10.0	12.8
12.5	9.3	10.4	12.7	10.5
9.3	11.5	10.7	11.6	7.8
10.5	7.6	10.1	8.9	8.6

(a) Organize the data into an ordered array or stem-and-leaf display.
(b) Compute the mean, median, mode, midrange, and midhinge for this population.
(c) Compute the range, interquartile range, variance, standard deviation, and coefficient of variation for this population.
(d) Form the box-and-whisker plot and describe the shape of these quarterly sales tax receipts data.
(e) What proportion of these businesses have quarterly sales tax receipts
   (1) Within ±1 standard deviation of the mean?
   (2) Within ±2 standard deviations of the mean?
   (3) Within ±3 standard deviations of the mean?
(f) Are you surprised at the results in (e)? (*Hint*: Compare and contrast your findings versus what would be expected based on the empirical rule.)
(g) **ACTION** Assist the comptroller of this township by writing a draft of the memo that will be sent to the governor regarding the collected receipts.
(h) How would this information be of use to the governor? Discuss.

3.10 Recognizing and Practicing Proper Descriptive Summarization and Exploring Ethical Issues

In this chapter we have studied how a set of numerical data is characterized through the computation of various descriptive summary measures regarding the properties of central tendency, variation, and shape. The next step is data analysis and interpretation; the former is *objective*, the latter is *subjective*. How *are* we going to make use of our results and how *should* we make use of our results? Do we use our results primarily to subjectively support a prior position or claim that we have made, or do we use our findings to objectively illuminate what the data are trying to convey?

Since a major role of the statistician is to analyze and interpret results, the computed summary measures should be used primarily to enhance data analysis and interpretation. We must avoid errors that may arise either in the objectivity of what is being analyzed or in the subjectivity of what is being interpreted (References 2 and 3).

3.10.1 Avoiding Errors in Analysis and Interpretation

You may recall that, at the beginning of this chapter (see Section 3.2), prior to learning the descriptive summary measures that characterize the three properties of numerical data (central tendency, variation, and shape), we were asked to examine and describe a set of numerical data pertaining to tuition rates charged to out-of-state residents from a sample of six Pennsylvania colleges and universities. Thus, without a knowledge of the contents of this chapter, we attempted to analyze and interpret what the data were trying to convey.

Our analysis was *objective;* we should all have agreed with our limited visual findings: There was no typical tuition rate value; the spread in the tuition rates ranged from 4.9 to 11.7 thousand dollars; and there were no outliers present in the data. On the other hand, having now read the chapter and thus gained a knowledge about various descriptive summary measures and their strengths and weaknesses, how could we improve on our previously objective analysis? Since the data are not skewed, shouldn't we report the mean or median? Doesn't the standard deviation provide more information about the property of variation than the range? Shouldn't we describe the data set as symmetrical in shape?

Objectivity in data analysis means reporting the most appropriate summary measures for a given data set—those that best meet the assumptions about the given data set. In our example we properly assumed that the data were in raw form; that is, there was no pattern to the sequence of collected data. Had this assumption been violated, we would still have been able to make objective descriptive comments as indicated here, but we would not have been able to draw inferences on the population of colleges and universities in the state of Pennsylvania; such inferences depend on the assumption that the sampled schools were randomly and independently selected. Thus, only through knowledge and awareness can good, objective data analysis take place.

On the other hand, our data interpretation was *subjective;* we could have formed different conclusions when interpreting our analytical findings. We all see the world from different perspectives. Some of us will look at the ordered array of tuition rates in thousands of dollars (4.9, 6.3, 7.7, 8.9, 10.3, 11.7) and conclude that out-of-state residents attending schools in Pennsylvania pay too much; others, attending more expensive private institutions, will look at the same data set and conclude that out-of-state residents pay too little. Thus, since data interpretation is subjective, it must be done in a fair, neutral, and clear manner.

3.10.2 Ethical Issues

Ethical issues are vitally important to all data analysis. As daily consumers of information, we owe it to ourselves to question what we read in newspapers and magazines and what we hear on the radio or television. Over time, much skepticism has been expressed about the purpose, the focus, and the objectivity of published studies. Perhaps no comment was ever more telling than a quip often attributed[5] to the famous nineteenth-century British statesman Benjamin Disraeli—*"There are three kinds of lies: lies, damned lies, and statistics."*

Again, as was mentioned in Section 2.13.3, ethical considerations arise when we are deciding what results to present in a report and what not to present. It is vitally important to document both good and bad results. In addition, when making oral presentations and presenting written reports, it is essential that the results be given in a fair, objective, and neutral manner. Thus, we must try to distinguish between poor presentation of results and unethical presentation. Once more, as in our prior discussions on ethical considerations, the key is *intent*. When pertinent information is omitted, it is often simply done out of ignorance. However, unethical behavior occurs when one willfully chooses an inappropriate summary measure (e.g., the mean or midrange for a very skewed set of data) to distort the facts in order to support a particular position. In addition, unethical behavior occurs when one selectively fails to report pertinent findings because it would be detrimental to the support of a particular position.

Problems for Section 3.10

3.31 You receive a telephone call from a friend who is also studying statistics this semester. Your friend has just used Excel to obtain descriptive summary measures for several numerical variables pertaining to a survey concerning student life on campus. He says "I've been asked to write a report and prepare a 5-minute classroom presentation on student life on campus. I'm looking at my computer printout—I've got all these descriptive summary measures for each of my seven numerical variables. There's so much information here, I just can't get started. Do you have any suggestions?" You think for a moment, and then reply………

3.32 An arbitrator is asked to examine a dispute over salaries paid to professional baseball players. The owner of a particular team claims that the average salary per annum is too high. The agent for the players argues that the average salary for the players on this team is too low. How should the arbitrator evaluate these two conflicting statements?

3.11 SUMMARIZING AND DESCRIBING NUMERICAL DATA: A REVIEW

As you can see on the following summary chart, this chapter was about data summarization and description. You should now be able to answer the following conceptual questions:

1. What should we be looking for when we attempt to characterize and describe the properties of a set of numerical data?
2. What do we mean by the property of location or central tendency?
3. What are the differences among the various measures of central tendency such as the mean, median, mode, midrange, and midhinge, and what are the advantages and disadvantages to each?
4. What is the difference between measures of central tendency and noncentral tendency?

5. What do we mean by the property of variation?
6. What are the differences among the various measures of variation such as the range, interquartile range, variance, standard deviation, and coefficient of variation, and what are the advantages and disadvantages to each?
7. How does the empirical rule help explain the ways in which the observations in a set of numerical data cluster, congregate, and distribute?
8. What do we mean by the property of shape?
9. Why are such exploratory data analysis techniques as the five-number summary and box-and-whisker plot so useful?
10. What are some of the ethical issues to be concerned with when distinguishing between the use of appropriate and inappropriate descriptive summary measures reported in newspapers and magazines?

Check over the list of questions to see if you know the answers and could (1) explain your answers to someone who did not read this chapter and (2) reference specific readings or examples that support your answers. Also, reread any of the sections that may have seemed unclear to see whether they make sense now.

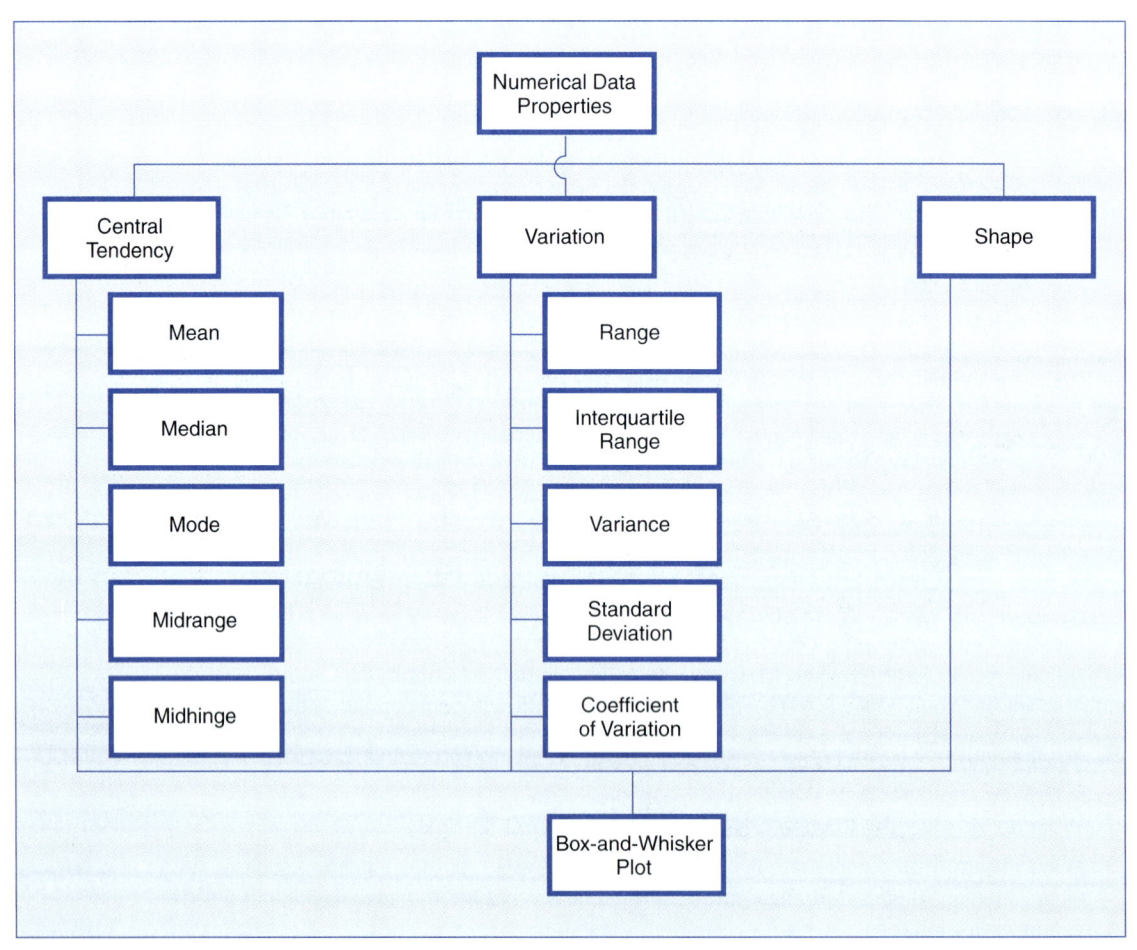

Chapter 3 summary chart.

Getting It All Together

Key Terms

arithmetic mean 121
average 121
AVERAGE function 128
box-and-whisker plot 149
central tendency or location 121
coefficient of variation 141
Data Analysis tool 146
dot scale 122
empirical rule 158
five-number summary 148
interquartile range 136
left-skew 145
MAX function 129
mean 121
median 124
MEDIAN function 128
midhinge 127
midrange 126
midspread 136
MIN function 129
mode 125
MODE function 128
outlier or extreme value 120

population coefficient of variation 154
population mean 153
population standard deviation 153
population variance 153
properties of numerical data 121
Q_1: first quartile 127
Q_2: second quartile (or median) 127
Q_3: third quartile 127
quartiles 127
range 135
raw form 120
resistant measures 128
right-skew 145
shape 144
skewness 144
SMALL function 129
standard deviation 137
STDEV function 142
symmetry 144
VAR function 142
variance 137
variation or dispersion 135

Chapter Review Problems

Note: *The Chapter Review Problems can be solved using Microsoft Excel.*

- 3.33 The following data are the retail prices (in dollars) for a random sample of 32 corded telephone models:

  CORDPHON.TXT

44	35	55	54	78	107	45	63
45	22	36	44	50	50	60	30
39	60	25	25	25	24	46	71
60	40	22	10	20	30	12	10

  Source: "Corded Phones" Copyright 1992 by Consumers Union of United States, Inc., Yonkers, N.Y. 10703. Adapted by permission from *Consumer Reports*, December 1992, pp. 780–781. Although this data set originally appeared in *Consumer Reports,* the selective adaptation and resulting conclusions presented are those of the authors and are not sanctioned or endorsed in any way by Consumers Union, the publisher of *Consumer Reports.*

  (a) Completely analyze the data.
  (b) **ACTION** Write an article for a local newspaper dealing with consumer affairs in order to enlighten its readership on this matter.

STUDIO.TXT

3.34 The following data are the monthly rental prices for a sample of 10 unfurnished studio apartments in Manhattan and a sample of 10 unfurnished studio apartments in Brooklyn Heights:

Manhattan									
$955	$1000	$985	$980	$940	$975	$965	$999	$1247	$1119
Brooklyn Heights									
$750	$775	$725	$705	$694	$725	$690	$745	$575	$800

(a) For each set of data compute the mean, median, midhinge, range, interquartile range, standard deviation, and coefficient of variation.
(b) What can be said about unfurnished studio apartments renting in Manhattan versus those renting in Brooklyn Heights?
(c) **ACTION** How might this information be of use to an individual desiring to move into the New York City area? Write an article about this for the real estate column of your local newspaper.

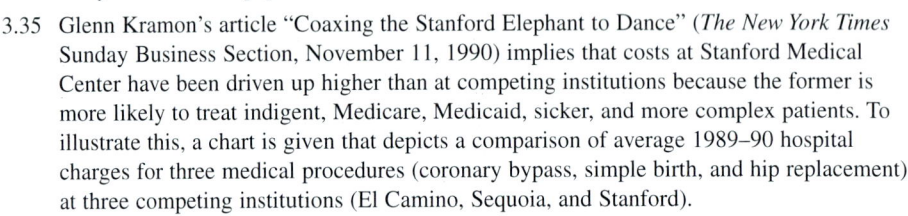

3.35 Glenn Kramon's article "Coaxing the Stanford Elephant to Dance" (*The New York Times* Sunday Business Section, November 11, 1990) implies that costs at Stanford Medical Center have been driven up higher than at competing institutions because the former is more likely to treat indigent, Medicare, Medicaid, sicker, and more complex patients. To illustrate this, a chart is given that depicts a comparison of average 1989–90 hospital charges for three medical procedures (coronary bypass, simple birth, and hip replacement) at three competing institutions (El Camino, Sequoia, and Stanford).

Your CEO knows you are currently taking a course in statistics and calls you in to discuss the article. She tells you that as she was leaving a meeting of hospital CEOs last night, one of them mentioned that this chart is totally meaningless and asked her opinion. She now requests that you prepare her response. You smile, take a deep breath, and reply………

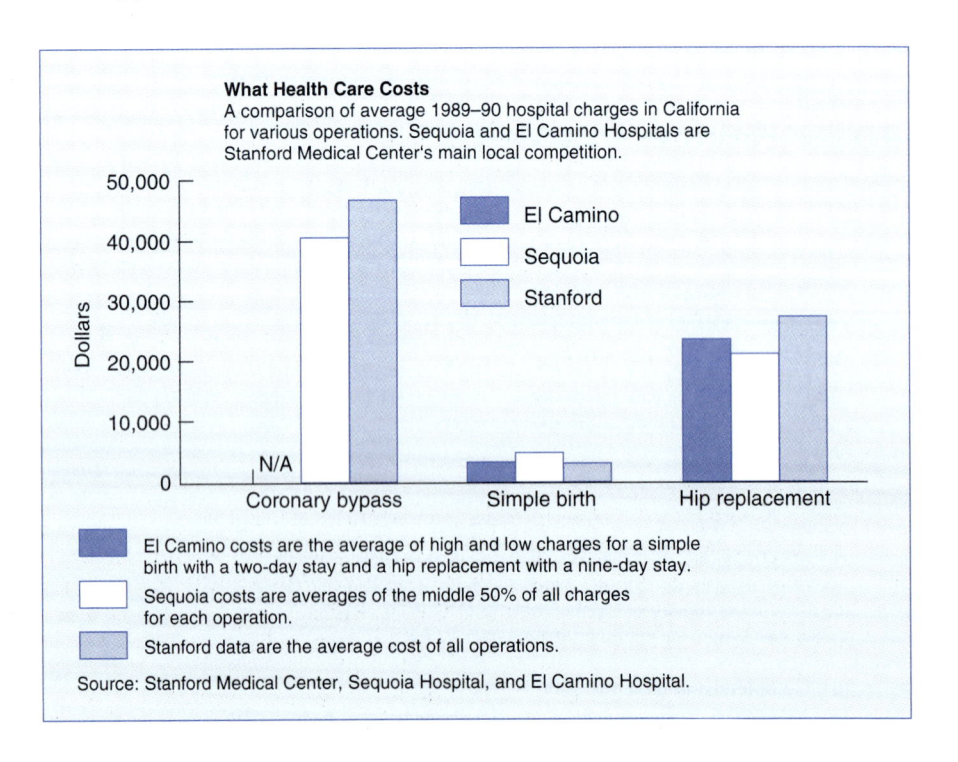

What Health Care Costs
A comparison of average 1989–90 hospital charges in California for various operations. Sequoia and El Camino Hospitals are Stanford Medical Center's main local competition.

El Camino costs are the average of high and low charges for a simple birth with a two-day stay and a hip replacement with a nine-day stay.
Sequoia costs are averages of the middle 50% of all charges for each operation.
Stanford data are the average cost of all operations.
Source: Stanford Medical Center, Sequoia Hospital, and El Camino Hospital.

3.36 A college was conducting a Phonothon to raise money for the building of an Arts Center. The provost hoped to obtain one-half million dollars for this purpose. The following data represent the amounts pledged (in thousands of dollars) by all alumni who were called during the first nine nights of the campaign.

FUNRAISE.TXT

$$16, 18, 11, 17, 13, 10, 22, 15, 16$$

(a) Compute the mean, median, and standard deviation.
(b) Describe the shape of this set of data.
(c) Estimate the total amount that will be pledged (in thousands of dollars) by all alumni if the campaign is to last 30 nights. (*Hint*: Total = $N\bar{X}$.)
(d) **ACTION** Write a memo to the provost summarizing your findings to date and, if necessary, offering any needed recommendations.
(e) How might this information assist the provost? Discuss.

3.37 The following data represent the tuition charged (in thousands of dollars) at a sample of 15 preparatory schools in the northeast and at a sample of 15 preparatory schools in the midwest during a recent academic year.

PREP.TXT

Northeast Prep Schools			Midwest Prep Schools		
10.5	8.9	9.6	7.9	10.6	8.4
10.1	9.3	9.1	8.2	10.1	9.2
10.0	9.7	11.2	9.1	8.5	10.7
11.0	10.4	10.5	9.3	7.5	9.5
9.8	10.0	9.9	8.8	9.3	9.8

(a) For each set of data compute the mean, median, midhinge, range, interquartile range, standard deviation, and coefficient of variation.
(b) For each set of data form the stem-and-leaf display and box-and-whisker plot.
(c) List the five-number summary and interpret the shape of each data set.
(d) Summarize your findings.
(e) **ACTION** Suppose you have a cousin who seeks your advice regarding the cost of attending a preparatory school in the northeast versus those in the midwest. Write him a letter based on your summary in (d).

3.38 Using Special Data Set 1 of Appendix D on pages D1–D5 regarding the out-of-state tuition rates at the schools in Texas, North Carolina, and Pennsylvania:
(a) Obtain various descriptive summary measures for each of these populations.
(b) **ACTION** Write and submit an executive summary comparing and contrasting the results across the three states.

3.39 Go to the World Wide Web page for this text (http://www.prenhall.com/phbusiness) for additional exercises.

Team Projects

TP3.1 Refer to TP2.2 on page 117. Your group, the _____ Corporation, has been hired by the food editor of a popular family magazine to study the cost and nutritional characteristics of ready-to-eat cereals. Having prepared the appropriate tables and charts (see TP2.2), the _____ Corporation is ready to enhance its preliminary analysis. Armed with Special Data Set 2 of Appendix D on pages D6–D7 displaying useful information on 84 such cereals:

CEREAL.TXT

(a) Outline how the group members will proceed with their tasks.
(b) Obtain various descriptive summary measures on cost, weight, calories, and sugar (in grams per serving)—broken down by type of cereal.
(c) Write and submit an executive summary describing the results.
(d) Prepare and deliver a 10-minute oral presentation to the food editor of the magazine.

FRAGRAN.TXT

TP3.2 Refer to TP2.3 on page 117. Your group, the _____ Corporation, has been hired by the marketing director of a manufacturer of well-known men's and women's fragrances to study the characteristics of currently available fragrances. The results of your group's endeavors are to enable the manufacturer to make pricing decisions regarding a new product line that is being planned for distribution in the upcoming holiday season. Having prepared the appropriate tables and charts (see TP2.3), the _____ Corporation is ready to enhance its preliminary analysis. Armed with Special Data Set 3 of Appendix D on pages D8–D9 displaying useful information on the cost per ounce of 83 such fragrances:
(a) Outline how the group members will proceed with their tasks.
(b) Obtain various descriptive summary measures on the cost per ounce based on:
   (1) Product gender (women's versus men's)
   (2) Type of fragrance (perfume, cologne, or "other")
   (3) Intensity (very strong, strong, medium, or mild)
(c) Write and submit an executive summary describing the results.
(d) Prepare and deliver a 10-minute oral presentation to the marketing director.

Case A—Kalosha Industries Employee Satisfaction Survey

The vice-president for human resources of Kalosha Industries, a diversified manufacturer of consumer products, wants to develop a profile of its managerial, technical, and sales staff that will measure job satisfaction, longevity, and attitudes. The following questionnaire was developed with the assistance of a survey research firm, and a random sample of 122 members of the managerial, technical, and sales staff were randomly selected and surveyed.

EMPSTAT.TXT

Employee Satisfaction Survey

1. What is your age (as of last birthday)? _____
2. What is your gender? [1] Male [2] Female
3. On the whole, how satisfied are you with your job?
   [1] Very Satisfied [2] Moderately Satisfied
   [3] A Little Dissatisfied [4] Very Dissatisfied
4. Which *one* of the following job characteristics is *most important* to you?
   [1] High Income [2] No Danger of Being Fired
   [3] Flexible Hours [4] Opportunities for Advancement
   [5] Enjoying the Work
5. How many years altogether have you worked for your present employer? _____
6. In the next five years, how likely are you to be promoted?
   [1] Very Likely [2] Likely [3] Not Sure
   [4] Unlikely [5] Very Unlikely
7. Does your job allow you to take part in making decisions that affect your work?
   [1] Always [2] Much of the Time
   [3] Sometimes [4] Never
8. As part of your job, do you participate in budgetary decisions?
   [1] Yes [2] No

9. How proud are you to be working for this organization?
   - [1] Very Proud
   - [2] Somewhat Proud
   - [3] Indifferent
   - [4] Not At All Proud
10. Would you turn down another job for more pay in order to stay with this organization?
    - [1] Very Likely
    - [2] Likely
    - [3] Not Sure
    - [4] Unlikely
    - [5] Very Unlikely
11. In general, how would you describe relations in your workplace between coworkers and colleagues?
    - [1] Very Good
    - [2] Good
    - [3] So So
    - [4] Bad
    - [5] Very Bad

The data file EMPSAT.TXT contains the results of the questionnaire for the sample of 122 employees. Your task is to write a report that summarizes the characteristics of these employees. You want to make sure to study the longevity of the employees in the organization and their attitudes and experiences on the job. Appended to your report should be all appropriate tables, charts, and descriptive statistical information obtained from the survey results.

Case B—Campus Cafeteria Nutrition Study

Ann Foster, Vice-President for Student Services at a rural liberal arts college, held a meeting with Camille Neller, the newly appointed Director of Food Services, and Dr. Diane Barry, Professor of Nutrition, regarding a series of student and parent complaints over the menu offered by the college's cafeteria. Since freshmen are obligated to purchase a meal plan requiring a minimum of two meals per day at the college's cafeteria, there has been some concern that the menu does not always offer an inexpensive, quick, and wholesome meal. When asked by Vice-President Foster to respond to these comments, Mrs. Neller stated that she has been on campus only 3 weeks and has been primarily following the menu provided by her predecessor while experimenting with a gourmet meal selection each day. "Now that these concerns have been called to my attention, I intend to pursue another course," she said. "Considering the fact that the college is situated in a rural area and that, in particular, canned food products need to be stored for the winter months when deliveries from the nearest town might be delayed, I wish to study the nutritional content of canned soup because this item could easily be made available for all lunch and dinner meals and may even provide the nutrients for the wholesome, inexpensive, quick meal that was requested." Dr. Barry agreed that such a study would be useful and should provide the necessary information to make a decision regarding implementation. Vice-President Foster then asked Dr. Barry to direct the study and to report her findings to Mrs. Neller in 2 weeks. The Vice-President offered to support the effort: "Do what it takes," she said, "we must show the students and their parents that we are responsive to their needs." Dr. Barry requested that a student assistant be provided. "Whomever you want to hire," responded the Vice-President.

Dr. Barry has hired you to assist her in this study and has provided you with the following data on 47 different canned soup products to investigate their nutritional value. These data represent the cost, calories, fat, calories from fat, and sodium of 47 different canned soups.

Nutritional characteristics of 47 different canned soups.

Brand	Product	Type	Cost	Calories	Fat	Calories From Fat	Sodium
Campbell's Homestyle	CN	CC	.35	60	2	30	880
Progresso	CN	CR	.66	75	2	24	730
Campbell's	CN	CC	.18	60	2	30	870
Nissin Cup O'Noodles	CN	DI	.33	170	8	42	970
Progresso Healthy Class	CN	CR	.77	80	2	23	460
Lipton Soup Mix	CN	DC	.21	80	2	23	700
Campbell's Ramen Noodle	CN	DC	.09	190	8	38	970
Nissin Tip Ramen	CN	DC	.11	200	9	41	960
Campbell's Soup Mix	CN	DC	.26	100	2	18	700
Pathmark	CN	CC	.17	60	2	30	840
ShopRite	CN	CC	.19	60	2	30	840
Maruchan Ramen	CN	DC	.09	190	9	43	780
Lady Lee	CN	CC	.19	60	2	30	840
Weight Watchers	CN	CR	.76	60	1	15	790
Knorr Chicken Flavor	CN	DC	.54	110	2	16	800
Campbell's Home Cookin'	CN	CR	.74	105	3	26	860
Hain	CN	CR	.96	110	4	33	800
Mrs. Grass Noodle Soup	CN	DC	.12	70	2	26	900
Campbell's Cup Instant	CN	DI	.48	105	3	26	1190
Lipton Cup-A-Soup Instant	CN	DI	.36	65	1	14	890
Campbell's Chunky Classic	CN	CR	.74	120	4	30	810
Campbell's Healthy Request	CN	CR	.70	80	2	23	470
Pritikin Chicken Soup	CN	CR	.97	80	1	11	180
Campbell's Low Sodium	CN	CR	.80	125	4	29	65
Healthy Choice	CN	CR	.78	95	2	19	580
Hain Vegetarian	V	CR	.83	125	3	22	670
Campbell's Home Cookin'	V	CR	.53	110	2	16	680
Campbell's Chunky	V	CR	.53	120	3	23	800
Healthy Choice Garden	V	CR	.71	105	1	9	600
Progresso Tomato	V	CR	.46	75	2	24	940
Progresso Vegetable	V	CR	.44	75	1	12	680
Healthy Choice Tomato	V	CR	.73	140	3	19	540
Campbell's Homestyle	V	CC	.34	60	2	30	880
Campbell's Home Cookin'	V	CR	.53	110	1	8	640
Campbell's Made With Beef	V	CC	.23	90	2	20	830
Health Valley Fat-Free	V	CR	.92	55	1	6	280
Campbell's Healthy Request	V	CR	.55	90	1	10	480
Pritikin	V	CR	.94	90	1	10	160
Campbell's	T	CC	.15	90	2	20	670
Campbell's Healthy Request	T	CC	.20	90	2	20	410
ShopRite	T	CC	.13	100	1	9	710
Kroger	T	CC	.14	100	1	9	630
Lady Lee	T	CC	.16	80	0	0	700
Pathmark	T	CC	.15	100	1	9	630
Vons	T	CC	.18	100	1	9	710
Health Valley Org.	T	CR	.87	75	1	12	300
Campbell's Italian	T	CC	.28	90	0	0	740

Notes: For product: CN = chicken noodle, V = vegetable, T = tomato
For type: CC = canned/condensed, CR = canned, ready to serve,
DC = dry/cook-up, DI = dry/instant
Cost in cents per 8-oz serving

Calories per 8-oz serving
Fat in grams per 8-oz serving
Calories as a percentage of fat per 8-oz serving
Sodium level in milligrams per 8-oz serving

Source: "Soup's On" Copyright 1993 by Consumers Union of United States, Inc., Yonkers, N.Y. 10703. Adapted by permission from *Consumer Reports*, November 1993, pp. 698–699. Although this data set originally appeared in *Consumer Reports,* the selective adaptation and resulting conclusions presented are those of the authors and are not sanctioned or endorsed in any way by Consumers Union, the publisher of *Consumer Reports*.

SOUP.DAT

(a) Undertake a complete descriptive evaluation of all numerical variables given in the table.

(b) Perform a similar evaluation comparing and contrasting each of these numerical variables based on whether the product is a chicken noodle soup or either a vegetable or tomato soup.

(c) Perform a similar evaluation comparing and contrasting each of these numerical variables based on type of soup—canned/condensed, canned and ready to serve, dry/cookup, or dry/instant.

(d) Make recommendations regarding your findings.

Endnotes

1. Although the word *average* refers to any summary measure of central tendency, it is most often used as a synonym for the mean.

2. These measures are called **quantiles.** Some of the more widely used quantiles are the **deciles** (which split the ordered data into *tenths*) and the **percentiles** (which split the ordered data into *hundredths*). For further information on these measures, see Reference 5.

3. Using the summation rules from Appendix B, we show the following proof:

$$\sum_{i=1}^{n}(X_i - \overline{X}) = 0$$

$$\sum_{i=1}^{n} X_i - \sum_{i=1}^{n} \overline{X} = 0$$

$$\sum_{i=1}^{n} X_i - n\overline{X} = 0$$

$$\sum_{i=1}^{n} X_i - \sum_{i=1}^{n} X_i = 0$$

4. Here $\mu \pm 3\sigma$ yields the interval –0.96 to 22.74 thousand dollars; however, a negative tuition rate is not meaningful, and we record the interval as 0 to 22.74 thousand dollars.

5. The quip is most often attributed to Benjamin Disraeli (1804–81), twice Prime Minister of England. However, a recent report [Woerner, D., "Who Really Said It?" *Chance*, Vol. 6 (Fall 1993), p. 37] indicates that it might have been first stated by someone else.

References

1. Cobb Group, *Running Microsoft Excel Version 5* (Redmond, WA: Microsoft Press, 1994).
2. Huff, D., *How to Lie with Statistics* (New York: W.W. Norton, 1954).
3. Kimble, G. A., *How to Use (and Misuse) Statistics* (Englewood Cliffs, NJ: Prentice-Hall, 1978).
4. *Microsoft Excel Version 7* (Redmond, WA: Microsoft Corporation, 1996).
5. Tukey, J., *Exploratory Data Analysis* (Reading, MA: Addison-Wesley, 1977).
6. Velleman, P. F., and D. C. Hoaglin, *Applications, Basics, and Computing of Exploratory Data Analysis* (Boston, MA: Duxbury Press, 1981).

chapter 4

Basic Probability and Discrete Probability Distributions

CHAPTER OBJECTIVE To develop an understanding of the basic concepts of probability, characteristics of probability distributions, and the binomial and other discrete distributions.

4.1 INTRODUCTION

In this chapter we begin by discussing three different approaches to determine the probability of occurrence of different phenomena. We will then learn how to compute a variety of different types of probabilities and use these concepts to develop the probability distribution and two of its characteristics, the expected value and the variance. We will then discuss the binomial distribution and other discrete probability distributions.

4.2 OBJECTIVE AND SUBJECTIVE PROBABILITY

What do we mean by the word probability? **Probability** is the likelihood or chance that a particular event will occur. It could refer to

1. The chance of picking a black card from a deck of cards
2. The chance that an individual prefers one product over another
3. The chance that a new consumer product on the market will be successful

In each of these examples, the probability involved is a proportion or fraction whose values range between 0 and 1 inclusively. We note that an event that has no chance of occurring (i.e., the **null event**) has a probability of 0, while an event that is sure to occur (i.e., the **certain event**) has a probability of 1.

Each of the above examples refers to one of three approaches to the subject of probability. The first is often called the **a priori classical probability** approach. Here the probability of success is based on prior knowledge of the process involved. In the simplest case, where each outcome is equally likely, this chance of occurrence of the event may be defined as follows:

(4.1)

where X = number of outcomes in which the event we are looking for occurs

   T = total number of possible outcomes

A standard deck of cards is presented in Figure 4.1. If we want to find the probability of picking a black card (where we are defining black as a "success") the correct answer would be 26/52 or 1/2, since there are 26 black cards in a standard deck of cards.

What does this probability tell us? If we replace each card after it is drawn, does it mean that one out of the next two cards selected will be black? No, because we cannot say for sure what will happen on the next several selections. However, we can say that in the long run, if this selection process is continually repeated, the proportion of black cards selected will approach .50.

In this first example, the number of successes and the number of outcomes are known from the composition of the deck of cards. However, in the second approach to probability, called the **empirical classical probability** approach, although the probability is still defined as the ratio of the number of favorable outcomes to the total number of outcomes, these outcomes are based on observed data, not upon prior knowledge of a process. This type of probability could refer to the proportion of individuals in a survey who prefer a particular brand of soft drink, a certain political candidate, or have a part-time job while attending school.

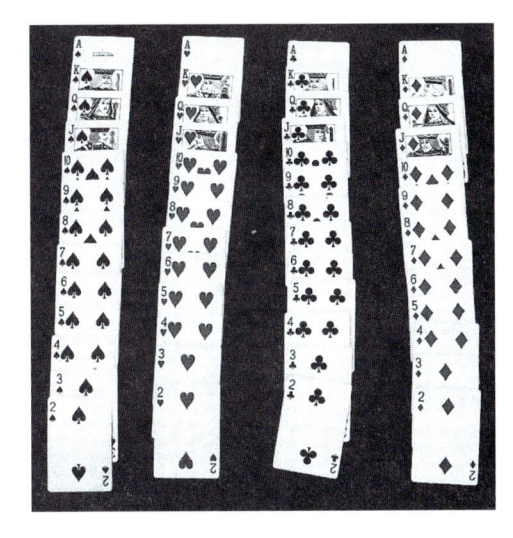

FIGURE 4.1 Standard deck of 52 playing cards.

The third approach to probability is called the **subjective probability** approach. Whereas in the previous two approaches the probability of a favorable event was computed *objectively*—either from prior knowledge or from actual data—*subjective* probability refers to the chance of occurrence assigned to an event by a particular individual. This chance may be quite different from the subjective probability assigned by another individual. For example, the inventor of a new toy may assign quite a different probability to the chance of success for the toy than the president of the company that is considering marketing the toy. The assignment of subjective probabilities to various events is usually based on a combination of an individual's past experience, personal opinion, and analysis of a particular situation. Subjective probability is especially useful in making decisions in situations in which the probability of various events cannot be determined empirically.

Problems for Section 4.2

4.1 For each of the following, indicate whether the type of probability involved is an example of a priori classical probability, empirical classical probability, or subjective probability.
(a) That the next toss of a fair coin will land on heads
(b) That the New York Mets will win next year's baseball World Series
(c) That the sum of the faces of two dice will be 7
(d) That the train taking a commuter to work will be more than 10 minutes late
(e) That a Republican will win the next presidential election in the United States

4.3 BASIC PROBABILITY CONCEPTS

4.3.1 Sample Spaces and Events

The basic elements of probability theory are the outcomes of the process or phenomenon under study. Each possible type of occurrence is referred to as an event.

A **simple event** can be described by a single characteristic. The collection of all the possible events is called the **sample space**.

We can achieve a better understanding of these terms by referring to the following example. A standard deck of 52 playing cards (see Figure 4.1) contains four suits (spades, hearts, clubs, and diamonds), each of which has 13 different cards (ace, king, queen, jack, 10, 9, 8, 7, 6, 5, 4, 3, 2).

If we randomly select a card from the deck,

1. What is the probability the card is black?
2. What is the probability the card is an ace?
3. What is the probability the card is a black ace?
4. What is the probability the card is black *or* an ace?
5. If we knew that the card selected was black, what is the probability that it is also an ace?

In the case of the deck of cards, the sample space consists of the entire deck of 52 cards, made up of various events, depending on how they are classified. For example, if the events are classified by suit, there are four events: spade, heart, club, and diamond. If the events are classified by card value, there are 13 events: ace, king, ..., and 2. The manner in which the sample space is subdivided depends on the types of probabilities that are to be determined. With this in mind, it is of interest to define both the complement of an event and a joint event as follows:

The **complement** of event A includes all events that are not part of event A. It is given by the symbol A'.

The complement of the event black would consist of all those cards that were not black (i.e., all the red cards). The complement of spade would contain all cards that were not spades (i.e., diamonds, hearts, and clubs).

A **joint event** is an event that has two or more characteristics.

The event black ace is a joint event, since the card must be both black *and* ace to qualify as a black ace.

4.3.2 Contingency Tables

There are several ways in which a particular sample space can be viewed. In this text, the method we will use involves assigning the appropriate events to a **table of cross-classifications**. Such a table is also called a **contingency table** (see Section 2.11).

If the two variables of interest for the card example were "presence of ace" and "color of card," the contingency table would look as shown in Table 4.1. The values in each cell of the table were obtained by subdividing the sample space of 52 cards according to the number of aces and the color of the card. It can be noted that if the row and column (margin) totals are known, only one cell entry in this 2×2 table is needed to obtain the entries in the remaining three cells.

Table 4.1 Contingency table for face-color variables.

	Red	Black	Totals
Ace	2	2	4
Non-ace	24	24	48
Totals	26	26	52

Problems for Section 4.3

4.2 In the past several years, credit card companies have made an aggressive effort to solicit new accounts from college students. Suppose that a sample of 200 students at your college indicated the following information in terms of whether the student possessed a bank credit card and/or a travel and entertainment credit card:

	Travel and Entertainment Credit Card	
Bank Credit Card	Yes	No
Yes	60	60
No	15	65

(a) Give an example of a simple event.
(b) Give an example of a joint event.
(c) What is the complement of having a bank credit card?
(d) Why is "having a bank credit card *and* having a travel and entertainment credit card" a joint event?

4.3 Numerous intensive studies have been conducted of consumer planning for the purchase of durable goods such as television sets, refrigerators, washing machines, stoves, and automobiles. In one such study, 1,000 individuals in a randomly selected sample were asked whether they were planning to buy a new television in the next 12 months. A year later the same persons were interviewed again to find out whether they actually bought a new television. The response to both interviews is cross-tabulated as follows:

	Buyers	Nonbuyers	Totals
Planned to buy	200	50	250
Did not plan to buy	100	650	750
Totals	300	700	1,000

(a) Give an example of a simple event.
(b) Give an example of a joint event.
(c) What is the complement of "planned to buy"?

4.4 A sample of 500 respondents was selected in a large metropolitan area to determine various information concerning consumer behavior. Among the questions asked was "Do you enjoy shopping for clothing?" Of 240 males, 136 answered yes. Of 260 females, 224 answered yes.
(a) Set up a 2 × 2 table to evaluate the probabilities.
(b) Give an example of a simple event.
(c) Give an example of a joint event.
(d) What is the complement of "enjoy shopping for clothing"?

4.5 A company has made available to its employees (without charge) extensive health club facilities that may be used before work, during the lunch hour, after work, and on weekends. Records for the last year indicate that of 250 employees, 110 used the facilities at some time. Of 170 males employed by the company, 65 used the facilities.
(a) Set up a 2 × 2 table to evaluate the probabilities of using the facilities.
(b) Give an example of a simple event.
(c) Give an example of a joint event.
(d) What is the complement of "used the health club facilities"?

4.6 Each year, ratings are compiled concerning the performance of new cars during the first 90 days of use. Suppose that the cars have been categorized according to two attributes, whether or not the car needs warranty-related repair (Yes or No) and the country in which the company manufacturing the car is based (United States, not United States). Based on

the data collected, the probability that the new car needs a warranty repair is .04, the probability that the car is manufactured by an American-based company is .60, and the probability that the new car needs a warranty repair *and* was manufactured by an American-based company is .025.
(a) Set up a 2 × 2 table to evaluate the probabilities of a warranty-related repair.
(b) Give an example of a simple event.
(c) Give an example of a joint event.
(d) What is the complement of "manufactured by an American-based company"?

4.4 SIMPLE (MARGINAL) PROBABILITY

Thus far, we have focused on the meaning of probability and on defining and illustrating sample spaces. We will now begin to answer some of the questions posed in the previous sections by developing rules for obtaining different types of probability.

The most obvious rule for probabilities is that they range in value from 0 to 1. An impossible event has probability 0 of occurring, and a certain event has probability 1 of occurring. **Simple probability** refers to the probability of occurrence of a simple event, $P(A)$, such as

- The probability of selecting a black card
- The probability of selecting an ace

We have already noted that the probability of selecting a black card is 26/52 or 1/2, since there are 26 black cards in the 52-card deck.

How would we find the probability of picking an ace from the deck? We would find the number of aces in the deck by totaling the black aces and the red aces in the deck:

$$P(\text{ace}) = \frac{\text{number of aces in deck}}{\text{number of cards in deck}}$$

$$= \frac{\text{number of red aces} + \text{number of black aces}}{\text{total number of cards}}$$

$$= \frac{2+2}{52} = \frac{4}{52}$$

Simple probability is also called **marginal probability**, since the total number of successes (aces in this case) can be obtained from the appropriate margin of the contingency table (see Table 4.1 on page 174).

Problems for Section 4.4

- **4.7** Referring to Problem 4.2 on page 175, if a student is selected at random, what is the probability that
  (a) The student has a bank credit card?
  (b) The student does not have a bank credit card?
  (c) The student has a travel and entertainment credit card?
  (d) The student does not have a travel and entertainment credit card?

- **4.8** Referring to Problem 4.3 on page 175, if an individual is selected at random, what is the probability that in the last year he or she
  (a) Has bought a new television?
  (b) Planned to buy a new television?
  (c) Did not plan to buy a new television?
  (d) Has not bought a new television?

- **4.9** Referring to Problem 4.4 on page 175, what is the probability that a respondent chosen at random
  - (a) Is a male?
  - (b) Enjoys shopping for clothing?
  - (c) Is a female?
  - (d) Does not enjoy shopping for clothing?
- **4.10** Referring to Problem 4.5 on page 175, what is the probability that an employee chosen at random
  - (a) Is a male?
  - (b) Has used the health club facilities?
  - (c) Is a female?
  - (d) Has not used the health club facilities?
- **4.11** Referring to Problem 4.6 on page 175, what is the probability that a new car selected at random
  - (a) Needs a warranty-related repair?
  - (b) Is not manufactured by an American-based company?
  - (c) Does not need a warranty-related repair?
  - (d) Is manufactured by an American-based company?

4.5 JOINT PROBABILITY

Whereas marginal probability refers to the occurrence of simple events, **joint probability** refers to phenomena containing two or more events, such as the probability of a black ace or a red queen.

Recall that a joint event A and B means that both event A and event B must occur simultaneously. Referring to Table 4.1 on page 174, those cards that are black *and* aces consist only of the outcomes in the single cell "black ace." Since there are two black aces, the probability of picking a card that is a black ace is

$$P(\text{black and ace}) = \frac{\text{number of black aces}}{\text{number of cards in deck}}$$

$$= \frac{2}{52}$$

Now that we have discussed the concept of joint probability, the marginal probability of a particular event can be viewed in an alternative manner. We have already shown that the marginal probability of an event consists of a set of joint probabilities. For example, if B consists of two events, B_1 and B_2, then we can observe that $P(A)$, the probability of event A, consists of the joint probability of event A occurring with event B_1 and the joint probability of event A occurring with event B_2. Thus, in general,

$$P(A) = P(A \text{ and } B_1) + P(A \text{ and } B_2) + \cdots + P(A \text{ and } B_k) \qquad (4.2)$$

where $B_1, B_2, \ldots, B_k$ = *mutually exclusive* and *collectively exhaustive* events

Two events are **mutually exclusive** if both events cannot occur at the same time.

Two events are **collectively exhaustive** if one of the events must occur.

For example, being male *and* being female are mutually exclusive and collectively exhaustive events. No one is both (they are mutually exclusive), and everyone is one or the other (they are collectively exhaustive).

Therefore, returning to our card example, the probability of an ace can be expressed as follows:

$$P(\text{ace}) = P(\text{ace and red}) + P(\text{ace and black})$$
$$= \frac{2}{52} + \frac{2}{52}$$
$$= \frac{4}{52}$$

This is, of course, the same result that we would obtain if we added up the number of outcomes that made up the simple event "ace."

Problems for Section 4.5

- 4.12 Referring to Problem 4.2 on page 175, what is the probability that if a student is selected at random
  - (a) The student has a bank credit card *and* a travel and entertainment card?
  - (b) The student does not have a bank credit card *and* has a travel and entertainment card?
  - (c) The student has neither a bank credit card *nor* a travel and entertainment card?

- 4.13 Referring to Problem 4.3 on page 175, what is the probability that in the last year he or she
  - (a) Planned to buy *and* actually bought a new television?
  - (b) Planned to buy *and* actually did not buy a new television?
  - (c) Did not plan to buy *and* actually did not buy a new television?

- 4.14 Referring to Problem 4.4 on page 175, what is the probability that a respondent chosen at random
  - (a) Is a female *and* enjoys shopping for clothing?
  - (b) Is a male *and* does not enjoy shopping for clothing?
  - (c) Is a male *and* enjoys shopping for clothing?

- 4.15 Referring to Problem 4.5 on page 175, what is the probability that an employee chosen at random
  - (a) Is a female *and* has used the health club facilities?
  - (b) Is a male *and* has not used the health club facilities?
  - (c) Is a female *and* has not used the health club facilities?

- 4.16 Referring to Problem 4.6 on page 175, what is the probability that a new car chosen at random
  - (a) Needs a warranty repair *and* is manufactured by a company based in the United States?
  - (b) Needs a warranty repair *and* is not manufactured by a company based in the United States?
  - (c) Does not need a warranty repair *and* is not manufactured by a company based in the United States?

4.6 ADDITION RULE

Having developed a means of finding the probability of event *A* and the probability of event "*A* and *B*," we should like to examine a rule that is used for finding the probability of event "*A* or *B*." This rule, the **additon rule**, considers the occurrence of either event *A* or event *B* or both *A* and *B*.

Suppose we refer to the event "black *or* ace" which includes all cards that are black, are aces, or are black aces. Each cell of the contingency table (Table 4.1 on page 174) can be examined to determine whether it is part of the event in question. If we want to study the event "ace *or* black" from Table 4.1, the cell "red ace" is part of the event, since it includes all cards that are aces. The cell "black non-ace" is included because it contains black cards. Finally, the cell "black ace" has both the characteristics of interest. Therefore, the probability can be obtained as follows:

$$P(\text{ace or black}) = P(\text{ace and red}) + P(\text{ace and black}) + P(\text{non-ace and black})$$

$$= \frac{2}{52} + \frac{2}{52} + \frac{24}{52}$$

$$= \frac{28}{52}$$

The computation of $P(A \text{ or } B)$, the probability of the event A or B, can be expressed in the following **general addition rule:**

> The probability of A or B is equal to the probability of A plus the probability of B minus the probability of A and B.
>
> $$P(A \text{ or } B) = P(A) + P(B) - P(A \text{ and } B) \qquad (4.3)$$

Applying this addition rule to the preceding example, we obtain the following result:

$$P(\text{ace or black}) = P(\text{ace}) + P(\text{black}) - P(\text{ace and black})$$

$$= \frac{4}{52} + \frac{26}{52} - \frac{2}{52}$$

$$= \frac{28}{52}$$

The addition rule consists of taking the probability of A and adding it to the probability of B; the probability of A and B must then be subtracted from this total because it has already been included twice in computing the probability of A and the probability of B. This can be demonstrated by referring to the contingency table. If the outcomes of the event "ace" are added to those of the event "black," then the joint event "ace and black" (the intersection) has been included in each of these simple events. Therefore, since this has been "double counted," it must be subtracted to provide the correct result.

4.6.1 Mutually Exclusive Events

In certain circumstances, however, the joint probability need not be subtracted because it is equal to zero. Such situations occur when no outcomes exist for a particular event. For example, suppose we want to know the probability of picking either a heart *or* a spade if we select only one card from a standard deck of 52 playing cards. Using the addition rule, we have the following:

$$P(\text{heart or spade}) = P(\text{heart}) + P(\text{spade}) - P(\text{heart and spade})$$

$$= \frac{13}{52} + \frac{13}{52} - \frac{0}{52} = \frac{26}{52}$$

We realize that the probability that a card will be both a heart *and* a spade simultaneously is zero, since in a standard deck each card may take on only one particular suit. The joint occurrence in this case is nonexistent (called the **null set**) because it contains no outcomes, since a card cannot be a heart and spade simultaneously.

As mentioned previously, whenever the joint probability does not contain any outcomes, the events involved are considered to be *mutually exclusive*. This refers to the fact that the occurrence of one event (a heart) means that the other event (a spade) cannot occur. Thus, the addition rule for mutually exclusive events reduces to

$$P(A \text{ or } B) = P(A) + P(B) \tag{4.4}$$

4.6.2 Collectively Exhaustive Events

Now consider what the probability would be of selecting a card that was red *or* black. Since red and black are mutually exclusive events, using Equation (4.4) we would have

$$P(\text{red } or \text{ black}) = P(\text{red}) + P(\text{black})$$
$$= \frac{26}{52} + \frac{26}{52} = \frac{52}{52} = 1.0$$

The probability of red *or* black adds up to 1.0. This means that the card selected must be red or black, since they are the only colors in a standard deck. Since one of these events must occur, they are considered to be *collectively exhaustive* events.

Problems for Section 4.6

Note: *Microsoft Excel may be used to solve the problems in this section. See Section 4.9.*

4.17 For each of the following, state whether the events that are created are mutually exclusive and/or collectively exhaustive. If they are not, either reword the categories to make them mutually exclusive and collectively exhaustive or explain why this would not be useful.
 (a) Registered voters were asked whether they registered as Republican or Democrat.
 (b) Respondents were classified on car ownership into the categories American, European, Japanese, none.
 (c) People were asked, "Do you currently live in (i) an apartment, (ii) a house?"
 (d) A product was classified as defective or not defective.
 (e) People were asked, "Do you intend to purchase a color television in the next 6 months?" (i) Yes, (ii) No

4.18 The probability of each of the following events is zero. For each, state why. Tell what common characteristic of these events makes their probability zero.
 (a) A person who is registered as a Republican and a Democrat
 (b) A product that is defective and not defective
 (c) A house that is a ranch style and split-level style.

4.19 Referring to Problem 4.2 on page 175, if a student is selected at random, what is the probability that
 (a) The student has a bank credit card *or* has a travel and entertainment card?
 (b) The student does not have a bank credit card *or* has a travel and entertainment card?
 (c) The student has a bank credit card *or* does not have a bank credit card?

4.20 Referring to Problem 4.3 on page 175, if an individual is selected at random, what is the probability that in the last year he or she
 (a) Planned to buy a new television *or* actually bought a new television?
 (b) Did not plan to buy a new television *or* did not actually buy a new television?
 (c) Planned to buy a new television *or* did not plan to buy a new television?

4.21 Referring to Problem 4.4 on page 175, what is the probability that a respondent chosen at random

(a) Is a female *or* enjoys shopping for clothing?
(b) Is a male *or* does not enjoy shopping for clothing?
(c) Is a male *or* a female?

4.22 Referring to Problem 4.5 on page 175, what is the probability that an employee chosen at random
(a) Is a female *or* has used the health club facilities?
(b) Is a male *or* has not used the health club facilities?
(c) Has used the health club facilities *or* has not used the health club facilities?

4.23 Referring to Problem 4.6 on page 175, what is the probability that a new car chosen at random
(a) Needs a warranty repair *or* was manufactured by an American-based company?
(b) Needs a warranty repair *or* was not manufactured by an American-based company?
(c) Needs a warranty repair *or* does not need a warranty repair?

4.7 CONDITIONAL PROBABILITY

Each example we have studied thus far in this chapter has involved the probability of a particular event when sampling from the entire sample space. However, how would we find various probabilities if certain information about the events involved is already known? For example, if we were told that a card is black, what would be the probability that the card is an ace?

When we are computing the probability of a particular event A, given information about the occurrence of another event B, this probability is referred to as **conditional probability**, $P(A|B)$. The conditional probability $P(A|B)$ can be defined as follows:

> The probability of A given B is equal to the probability of A and B divided by the probability of B.
>
> $$P(A|B) = \frac{P(A \text{ and } B)}{P(B)} \quad (4.5)$$

where

$P(A \text{ and } B)$ = joint probability of A and B

$P(B)$ = marginal probability of B

Rather than using Equation (4.5) for finding conditional probability, we can use the contingency table. In our example, we wish to find $P(\text{ace}|\text{black})$. Here the information is given that the card is black. Therefore, the sample space does not consist of all 52 cards in the deck; it consists only of the black cards. Of the 26 black cards, two are aces. Therefore, the probability of an ace, given that we know the card is black, is

$$P(\text{ace} | \text{black}) = \frac{\text{number of black aces}}{\text{number of black cards}}$$

$$= \frac{2}{26}$$

This result (2/26) can also be obtained by using Equation (4.5) as follows:

If

$$P(A|B) = \frac{P(A \text{ and } B)}{P(B)}$$

where

$$\text{event } A = \text{ace}$$
$$\text{event } B = \text{black}$$

then

$$P(\text{ace} \mid \text{black}) = \frac{2/52}{26/52}$$

$$= \frac{2}{26}$$

- **Statistical Independence** In the first example, we observed that the probability that the card picked is an ace, given that we know it is black, is 2/26. We may remember that the probability of picking an ace out of the deck, $P(\text{ace})$, was 4/52, which reduces to 2/26. This result reveals some important information. The prior knowledge that the card was black did not affect the probability that the card was an ace. This characteristic is called **statistical independence** and can be defined as follows:

$$P(A \mid B) = P(A) \tag{4.6}$$

where

$$P(A \mid B) = \text{conditional probability of } A \text{ given } B$$
$$P(A) = \text{marginal probability of } A$$

Thus, we may note that two events A and B are statistically independent if and only if $P(A \mid B) = P(A)$. In a 2×2 contingency table, once this holds for one combination of A and B, it will be true for all others.[1] Here, "color of the card" and "being an ace" are statistically independent events. Knowledge of one event in no way affects the probability of the second event.

Problems for Section 4.7

Note: *Microsoft Excel may be used to solve the problems in this section. See Section 4.9.*

- **4.24** Referring to Problem 4.2 on page 175
  - (a) Assume that we know that the student has a bank credit card. What is the probability then that he or she has a travel and entertainment card?
  - (b) Assume we know that the student does not have a travel and entertainment card. What then is the probability that he or she has a bank credit card?
  - (c) Are the two events, having a bank credit card and having a travel and entertainment card, statistically independent? Explain.

- **4.25** Referring to Problem 4.3 on page 175
  - (a) If the respondent planned to buy a new television, what is the probability that he or she actually bought one?
  - (b) If the respondent did not plan to buy a new television, what is the probability that he or she did not buy a new television?
  - (c) Are planning to buy a new television and actually buying one statistically independent? Explain.

- **4.26** Referring to Problem 4.4 on page 175
  - (a) Suppose the respondent chosen is a female. What then is the probability that she does not enjoy shopping for clothing?
  - (b) Suppose the respondent chosen enjoys shopping for clothing. What then is the probability that the individual is a male?
  - (c) Are enjoying shopping for clothing and the gender of the individual statistically independent? Explain.

- **4.27** Referring to Problem 4.5 on page 175

(a) Suppose we select a female employee of the company. What then is the probability that she has used the health club facilities?
(b) Suppose we select a male employee of the company. What then is the probability that he has not used the health club facilities?
(c) Are gender of the individual and use of the health club facilities statistically independent? Explain.

4.28 Referring to Problem 4.6 on page 175
(a) Suppose we know that the car was manufactured by a company based in the United States. What then is the probability that the car needs a warranty repair?
(b) Suppose we know that the car was not manufactured by a company based in the United States. What then is the probability that the car needs a warranty repair?
(c) Are need for a warranty repair and location of the car manufacturer statistically independent?

4.8 MULTIPLICATION RULE

The formula for conditional probability can be manipulated algebraically so that the joint probability $P(A \text{ and } B)$ can be determined from the conditional probability of an event. Using Equation (4.5)

$$P(A|B) = \frac{P(A \text{ and } B)}{P(B)}$$

and solving for the joint probability $P(A \text{ and } B)$, we have the **general multiplication rule:**

> The probability of A and B is equal to the probability of A given B times the probability of B.
>
> $$P(A \text{ and } B) = P(A|B)P(B) \qquad (4.7)$$

To demonstrate the use of this multiplication rule, we turn to an example. Suppose that 20 marking pens are displayed in a stationery store. Six are red and 14 are blue. We are to select 2 markers randomly from the set of 20. What is the probability that both markers selected are red? Here the multiplication rule can be used in the following way:

$$P(A \text{ and } B) = P(A|B)P(B)$$

Therefore if

$$A_R = \text{second marker selected is red}$$
$$B_R = \text{first marker selected is red}$$

we have

$$P(A_R \text{ and } B_R) = P(A_R|B_R)P(B_R)$$

The probability that the first marker is red is 6/20, since 6 of the 20 markers are red. However, the probability that the second marker is also red depends on the result of the first selection. If the first marker is not returned to the display after its color is determined (sampling *without* replacement), then the number of markers remaining will be 19. If the first marker is red, the probability that the second is also red is 5/19, since 5 red markers remain in the display. Therefore, using Equation (4.7), we have the following:

$$P(A_R \text{ and } B_R) = \left(\frac{5}{19}\right)\left(\frac{6}{20}\right)$$

$$= \frac{30}{380} = .079$$

However, what if the first marker selected is returned to the display after its color is determined? Then the probability of picking a red marker on the second selection is the same as on the first selection (sampling *with* replacement), since there are 6 red markers out of 20 in the display. Therefore, we have the following:

$$P(A_R \text{ and } B_R) = P(A_R|B_R)P(B_R)$$

$$= \left(\frac{6}{20}\right)\left(\frac{6}{20}\right)$$

$$= \frac{36}{400} = .09$$

This example of sampling *with* replacement illustrates that the second selection is independent of the first, since the second probability is not influenced by the first selection. Therefore, the **multiplication rule for independent events** can be expressed as follows [by substituting $P(A)$ for $P(A|B)$]:

> If A and B are independent, the probability of A and B is equal to the probability of A times the probability of B.
>
> $$P(A \text{ and } B) = P(A)P(B) \qquad (4.8)$$

If this rule holds for two events, A and B, then A and B are statistically independent. Therefore, there are two ways to determine statistical independence.

1. Events A and B are statistically independent if and only if $P(A|B) = P(A)$.
2. Events A and B are statistically independent if and only if $P(A \text{ and } B) = P(A)P(B)$.

It should be noted that for a 2 × 2 contingency table, if this is true for one joint event, it will be true for all joint events.[1] For example, if the probability of a card being an ace is independent of it being black, then the probability of it being an ace is independent of it being red, the probability of not being an ace is independent of it being black, and the probability of not being an ace is independent of it being red.

Now that we have discussed the multiplication rule, we can write the formula for marginal probability [Equation (4.2)] as follows. If

$$P(A) = P(A \text{ and } B_1) + P(A \text{ and } B_2) + \ldots + P(A \text{ and } B_k)$$

then, using the multiplication rule, we have

> $$P(A) = P(A|B_1)P(B_1) + P(A|B_2)P(B_2) + \cdots + P(A|B_k)P(B_k) \qquad (4.9)$$

where $B_1, B_2, \ldots, B_k = k$ mutually exclusive and collectively exhaustive events

We illustrate this formula by referring to Table 4.1 on page 174. Using Equation (4.9), we may compute the probability of an ace as follows:

$$P(A) = P(A|B_1)P(B_1) + P(A|B_2)P(B_2)$$

Since

$P(A)$ = probability of an ace

$P(B_1)$ = probability of a red card

$P(B_2)$ = probability of a black card

$$P(A) = \left(\frac{2}{26}\right)\left(\frac{26}{52}\right) + \left(\frac{2}{26}\right)\left(\frac{26}{52}\right)$$

$$= \frac{2}{52} + \frac{2}{52}$$

$$= \frac{4}{52}$$

Problems for Section 4.8

- 4.29 Referring to Problem 4.2 on page 175, use Equation (4.8) to determine whether having a bank credit card is statistically independent of having a travel and entertainment credit card.

4.30 Referring to Problem 4.3 on page 175, use Equation (4.8) to determine whether planning to buy a new television and actually buying one are statistically independent.

- 4.31 Referring to Problem 4.4 on page 175, use Equation (4.8) to determine whether enjoying shopping for clothing is statistically independent of the gender of the individual.

4.32 Referring to Problem 4.5 on page 175, use Equation (4.8) to determine whether using the health club facilities is statistically independent of the gender of the individual.

4.33 Referring to Problem 4.6 on page 175, use Equation (4.8) to determine whether need for warranty repair is statistically independent of the location of the company manufacturing the car.

4.34 Suppose you believe that the probability that you will get an A in Statistics is .6, and the probability that you will get an A in Organizational Behavior is .8. If these events are independent, what is the probability that you will get an A in both Statistics and Organizational Behavior? Give some plausible reasons why these events may not be independent, even though the teachers of these two subjects may not communicate about your work.

4.35 A standard deck of cards is being used to play a game. There are four suits (hearts, diamonds, clubs, and spades), each having 13 cards (ace, 2, 3, 4, 5, 6, 7, 8, 9, 10, jack, queen, and king), making a total of 52 cards. This complete deck is thoroughly mixed, and you will receive the first two cards from the deck *without* replacement.
   (a) What is the probability that both cards are queens?
   (b) What is the probability that the first card is a 10 *and* the second card is a 5 or 6?
   (c) If we were sampling *with* replacement, what would be the answer in (a)?
   (d) In the game of blackjack, the picture cards (jack, queen, king) count as 10 points, and the ace counts as either 1 or 11 points. All other cards are counted at their face value. Blackjack is achieved if your two cards total 21 points. What is the probability of getting blackjack in this problem?

4.36 A box of nine golf gloves contains two left-handed gloves and seven right-handed gloves.
   (a) If two gloves are randomly selected from the box *without* replacement, what is the probability that
      (1) both gloves selected will be right-handed?
      (2) there will be one right-handed glove *and* one left-handed glove selected?
   (b) If three gloves are selected, what is the probability that all three will be left-handed?
   (c) If we were sampling *with* replacement, what would be the answers to (a)(1) and (b)?

4.9 USING MICROSOFT EXCEL TO OBTAIN PROBABILITIES

In Sections 4.4–4.8, we studied a variety of different types of probability, including marginal probability, joint probability, the addition rule, conditional probability, and the multiplication rule. Microsoft Excel can be used to obtain these probabilities from data that have been cross-classified into a contingency table. You may recall that in Section 2.12 we used the PivotTable Wizard to develop a contingency table from raw data. By using various options of the PivotTable Wizard, the contingency table obtained could provide cell frequencies, row percentages, column percentages, or total percentages.

However, as in the case of the playing card example of Table 4.1 on page 174, the values in each cell of the table may already have been summarized in a contingency table. For this situation, Microsoft Excel formulas can be used to compute the desired probabilities. To illustrate these calculations, we can design a workbook that contains a Calculations worksheet as illustrated in Table 4.1.Excel.

4–9.XLS

If we wish to obtain the marginal probability of an ace, we need to divide the number of aces by the number of cards in the deck. From Table 4.1.Excel, this involves dividing cell D4 by cell D6. To obtain the joint probability of black and ace, we need to divide the number of black aces by the number of cards in the deck. Using Table 4.1.Excel, we observe that this involves dividing cell C4 by cell D6.

To use the addition rule, we need to add the marginal probabilities and subtract the corresponding joint probability. Thus, to obtain the probability of ace or black, we need to add the probability of ace (cell B8) to the probability of black (cell B9) and subtract the joint probability of a black ace that is contained in cell B10. In a similar manner, we can obtain the probability of red or black by adding the probability of red (computed in cell B12) to the probability of black (cell B9). The joint probability of these two events is zero, so the probability of red or black is the sum of cells B12 and B9.

Finally, to obtain a conditional probability, we divide the joint probability of A and B by the probability of B. To find the probability of ace given black (ace|black), we divide the number of black aces contained in cell C4 by the number of black cards computed in cell C6. Figure 4.1.Excel represents these calculations for the data of Table 4.1.Excel.

Table 4.1.Excel Design for Calculations sheet for the playing card example.

	A	B	C	D
1		Calculating Probabilities		
2				
3		Red	Black	Total
4	Ace	2	2	= B4+C4
5	Non-Ace	24	24	= B5+C5
6	Total	=B4+B5	=C4+C5	= B6+C6
7				
8	P(ace)	=D4/D6		
9	P(black)	=C6/D6		
10	P(black and ace)	=C4/D6		
11	P(ace or black)	= B8+B9–B10		
12	P(red)	=B6/D6		
13	P(red or black)	= B12+B9		
14	P(ace | black)	=C4/C6		

	A	B	C	D
1	Calculating Probabilities			
2				
3		Red	Black	Total
4	Ace	2	2	4
5	Non-Ace	24	24	48
6	Total	26	26	52
7				
8	P(ace)	0.076923		
9	P(black)	0.5		
10	P(black and ace)	0.038462		
11	P(ace or black)	0.538462		
12	P(red)	0.5		
13	P(red or black)	1		
14	P(ace | black)	0.076923		

FIGURE 4.1.EXCEL Calculating probabilities for the playing card example using Microsoft Excel.

4.10 THE PROBABILITY DISTRIBUTION FOR A DISCRETE RANDOM VARIABLE

As discussed in Section 1.8, a numerical random variable is some phenomenon of interest whose responses or outcomes may be expressed numerically. Such a random variable may also be classified as discrete or continuous—the former arising from a counting process and the latter from a measuring process. This chapter deals with some probability distributions that represent discrete random variables.

We may define the probability distribution for a discrete random variable as follows:

A probability distribution for a discrete random variable is a mutually exclusive listing of all possible numerical outcomes for that random variable such that a particular probability of occurrence is associated with each outcome.

Assuming that a fair six-sided die will not stand on edge or roll out of sight (*null events*), Table 4.2 represents the probability distribution for the outcomes of a single roll of the fair die. Since all possible outcomes are included, this listing is complete (or *collectively exhaustive*), and thus the probabilities must sum to 1. We can then use this table to obtain various probabilities for the rolling of a fair die.

The probability of a face ⚂ is

$$P(\boxdot) = 1/6$$

Using the addition rule for mutually exclusive events, the probability of an odd face is

$$P(\text{odd}) = P(\boxdot) + P(\boxdot) + P(\boxdot)$$
$$= 1/6 + 1/6 + 1/6 = 3/6$$

Moreover, the probability of a face of ⚁ or less is

Table 4.2 Theoretical probability distribution of the results of rolling one fair die.

Face of Outcome	Probability
1 ⚀	1/6
2 ⚁	1/6
3 ⚂	1/6
4 ⚃	1/6
5 ⚄	1/6
6 ⚅	1/6
Total	1

$$P(⚁ \text{ or less}) = P(⚀) + P(⚁)$$
$$= 1/6 + 1/6 = 2/6$$

And the probability of a face larger than ⚅ is

$$P(> ⚅) = 0$$

4.11 MATHEMATICAL EXPECTATION AND EXPECTED MONETARY VALUE

To summarize a discrete probability distribution, we shall compute its major characteristics—the mean and the standard deviation.

4.11.1 Expected Value of a Discrete Random Variable

The mean (μ) of a probability distribution is the expected value of its random variable.

> The **expected value of a discrete random variable** is a weighted average over all possible outcomes—the weights being the probability associated with each of the outcomes.

This summary measure can be obtained by multiplying each possible outcome X_i by its corresponding probability $P(X_i)$ and then summing the resulting products. Thus, the expected value of the discrete random variable X, symbolized as $E(X)$, may be expressed as follows:

$$\mu = E(X) = \sum_{i=1}^{N} X_i\, P(X_i) \qquad (4.10)$$

where X = discrete random variable of interest

X_i = ith outcome of X

$P(X_i)$ = probability of occurrence of the ith outcome of X

$i = 1, 2, \ldots, N$

For the theoretical probability distribution of the results of rolling one fair die (Table 4.2), the expected value of the roll may be computed as

$$\mu = E(X) = \sum_{i=1}^{N} X_i P(X_i)$$

$$= (1)(1/6) + (2)(1/6) + (3)(1/6) + (4)(1/6) + (5)(1/6) + (6)(1/6)$$

$$= 1/6 + 2/6 + 3/6 + 4/6 + 5/6 + 6/6$$

$$= 21/6 = 3.5$$

Notice that the expected value of the results of rolling a fair die is not "literally meaningful," since we can never obtain a face of 3.5. However, we can expect to observe the six different faces with equal likelihood, so we should have about the same number of ones, twos, ..., and sixes. In the long run, over many rolls, the average value would be 3.5.

To make this particular situation meaningful, however, we introduce the following carnival game: How much money should we be willing to put up to have the opportunity of rolling a fair die if we were to be paid, in dollars, the amount on the face of the die? Since the expected value of a roll of a fair die is 3.5, the expected long-run payoff is $3.50 per roll. This means that, on any particular roll, our payoff will be $1.00, $2.00, ..., or $6.00, but over many, many rolls the payoff can be expected to average out to $3.50 per roll. Now if we want the game to be fair, neither we nor our opponent (the "house") should have an advantage. Thus, we should be willing to pay $3.50 per roll to play. If the house wants to charge us $4.00 per roll, we can expect to lose from such gambling, on the average, $0.50 per roll over time, and unless we derive some intrinsic satisfaction (costing on the average $0.50 per roll), we should refrain from participating in such a game.

Usually, though, in any casino or carnival-type game the expected long-run payoff to the participant is negative; otherwise the house would not be in business (References 6 and 7). Since games such as Craps, Under-or-over-seven, Chuck-a-luck, and Roulette (see Reference 6), attract large numbers of participants, something other than expected monetary value undoubtedly is the ultimate criterion used by participants. (The concept of the *expected utility of money* is discussed in Reference 2.) It is the criterion that rational participants are considering, implicitly or explicitly, when they partake in such games. On the other hand, however, the house uses the expected-monetary-value criterion when it participates in such games.

4.11.2 Variance and Standard Deviation of a Discrete Random Variable

The **variance (σ^2) of a discrete random variable** may be defined as the weighted average of the squared differences between each possible outcome and its mean—the weights being the probabilities of each of the respective outcomes.

This summary measure can be obtained by multiplying each possible squared difference $(X_i - \mu)^2$ by its corresponding probability $P(X_i)$ and then summing up the resulting products. Hence, the variance of the discrete random variable X may be expressed as follows:

$$\sigma^2 = \sum_{i=1}^{N}(X_i - \mu)^2 P(X_i) \qquad (4.11)$$

where
X = discrete random variable of interest
X_i = ith outcome of X
$P(X_i)$ = probability of occurrence of the ith outcome of X
$i = 1, 2, \ldots, N$

Moreover, the **standard deviation (σ) of a discrete random variable** is given by

$$\sigma = \sqrt{\sum_{i=1}^{N}(X_i - \mu)^2 P(X_i)} \qquad (4.12)$$

For the theoretical probability distribution of the results of rolling one fair die (Table 4.2), the variance and the standard deviation may be computed by

$$\begin{aligned}\sigma^2 &= \sum_{i=1}^{N}(X_i - \mu)^2 P(X_i) \\ &= (1 - 3.5)^2(1/6) + (2 - 3.5)^2(1/6) + (3 - 3.5)^2(1/6) + (4 - 3.5)^2(1/6) \\ &\quad + (5 - 3.5)^2(1/6) + (6 - 3.5)^2(1/6) \\ &= 2.9166\end{aligned}$$

and

$$\sigma = 1.71$$

For our carnival game, the mean payoff per roll is $3.50 with a standard deviation of $1.71.

4.11.3 Expected Monetary Value

As prospective participants in the carnival game, the most important question that we had to address was whether or not it was *profitable* for us to play the game. To answer this question, we had to realize that the random variable of interest from the "gambling viewpoint" was not really X, the *outcome on the face* of the die (as in Table 4.2 on page 188), but rather V, the *dollar value* associated with the resulting outcome from rolling the die. Thus, for participating in the game, the values for V ranged from –$3.00 to +$2.00 since it cost $4.00 for every roll of the die (see Table 4.3).

For purposes of decision making, the objective is to compare the **expected monetary values** (denoted by EMV) of alternative strategies (e.g., "play the carnival game" versus "don't play"). The expected monetary value indicates the average profit that would be gained if a particular strategy were selected in many decision-making situations (e.g., "play the game many times"). Hence, to play the game

$$\text{EMV}(\text{play}) = E(V) = \sum_{i=1}^{N} V_i P(V_i)$$

Table 4.3 Theoretical probability distribution representing the dollar value for participating in carnival game.

Result (X)	Dollar Value (V)	Probability
1 ⚀	−3	1/6
2 ⚁	−2	1/6
3 ⚂	−1	1/6
4 ⚃	0	1/6
5 ⚄	1	1/6
6 ⚅	2	1/6
		1

$$= (-3)(1/6) + (-2)(1/6) + (-1)(1/6) + (0)(1/6) + (1)(1/6) + 2(1/6)$$
$$= -.50$$

while not playing the game

$$\text{EMV(not play)} = E(V) = 0$$

Thus, our long-run expected payoff for participating is negative. On the average, we'd be losing 50 cents each time we decide to roll the die. This means that if we played for an evening in which the die is rolled 100 times, we would expect to collect $350 but pay out $400. Thus, at the end of play we would expect to lose $50 on the 100 rolls—an average of 50 cents per roll (or 12½ cents per dollar wagered).

- **Assigning Probabilities**   To compute the expected monetary value for the various strategies, a set of probabilities must be assigned to the mutually exclusive and collectively exhaustive listing of outcomes and events. In many cases, no information is available about the probability of occurrence of the various events and, thus, equal probabilities are assigned. In other instances, the probabilities of the events can be estimated in several ways. First, information may be available from past experience that can be used for estimating the probabilities. Second, a subjective assessment of the likelihood of the various events may be given by managers or other supervisory personnel. Third, the probabilities of the events could follow a particular discrete probability distribution. These probability distributions will be the subject of the remainder of this chapter.

4.11.4 Using Microsoft Excel to Obtain Expected Values and Variances

In this section we studied the expected value and variance of a discrete probability distribution. Although Microsoft Excel does not have any functions that can compute these characteristics, we can use formulas to calculate both the expected value and variance (and standard deviation of the discrete probability distribution). Table 4.2.Excel shows the design of a

4-11-4.XLS

Table 4.2.Excel Design for Calculations sheet for the fair die example.

	A	B	C
1		Calculating Expected Values and Variances	
2			
3	X	P(X)	[(X − E(X))]^2
4	1	= 1/6	=(A4 − $B$11)^2
5	2	= 1/6	=(A5 − $B$11)^2
6	3	= 1/6	=(A6 − $B$11)^2
7	4	= 1/6	=(A7 − $B$11)^2
8	5	= 1/6	=(A8 − $B$11)^2
9	6	= 1/6	=(A9 − $B$11)^2
10	Total	= SUM (B4:B9)	
11	Expected Value	=SUMPRODUCT(A4:A9,B4:B9)	
12	Variance	=SUMPRODUCT(C4:C9,B4:B9)	
13	Standard Deviation	=SQRT(B12)	

Calculations sheet that computes the expected value, variance, and standard deviation of the probability distribution of rolling one fair die.

To compute the expected value, we need to multiply each X value by its probability of occurrence $P(X)$ and sum these products. This may be accomplished in Microsoft Excel by using the SUMPRODUCT function. The SUMPRODUCT function computes the sum of the products between a set of two variables, X and Y, $\sum_{i=1}^{N} X_i Y_i$. The format of the function is

=SUMPRODUCT (*range of X variable, range of Y variable*)

Referring to Table 4.2.Excel, since the range of X is from A4 through A9, and the range of $P(X)$ is from B4 through B9, we may enter the formula =SUMPRODUCT(A4:A9,B4:B9) in cell B11 to obtain the expected value.

Once the expected value has been computed, we are ready to obtain the variance and the standard deviation. Referring to Equation (4.11) on page 190, for each X, we need to obtain the squared differences between X and the expected value, and then multiply this by the probability of occurrence $P(X)$. This can be done by entering the formula =(A4−$B$11)^2 in Cell C4 and copying the formula into cells C5 through cell C9. Since we always wish to subtract the expected value, notice that in the formula we use the absolute reference $B$11 for the expected value. Once the values in cells C4:C9 have been obtained, we can use the SUMPRODUCT function to multiply these squared differences by the probability of occurrence using the formula =SUMPRODUCT(C4:C9,B4:B9) in cell B12. The standard deviation can be computed in cell B13 by simply taking the square root of the variance in cell B12. Figure 4.2.Excel represents these calculations for the data of Table 4.2.Excel.

> ▲ **WHAT IF EXAMPLE**
>
> Now that we have set up this Calculations sheet for the expected value and variance, we can evaluate the effect of changes in the probability of occurrence of the different events on the expected value and variance. Although this would not be appropriate for the fair die example, it would be of importance in situations where we were determining the expected monetary value based on subjective probabilities (see Problems 4.37–4.39,

	A	B	C
1	Calculating Expected Values and Variances		
2			
3	X	P(X)	[(X-E(X)]^2
4	1	0.166667	6.25
5	2	0.166667	2.25
6	3	0.166667	0.25
7	4	0.166667	0.25
8	5	0.166667	2.25
9	6	0.166667	6.25
10	Total	1	
11	Expected Value	3.5	
12	Variance	2.916667	
13	Standard Deviation	1.707825	

FIGURE 4.2.EXCEL Calculating expected values and variances for the fair die example using Microsoft Excel.

4.43 and 4.44). We can illustrate this by changing the probabilities in cells B4:B9 to .1, .1, .1, .1, .3, and .3, respectively. Notice the effect on the expected value in cell B11, the variance in cell B12, and the standard deviation in cell B13 of these changes in the probabilities. As before, if we are interested in seeing the effects of many different changes, we could use the Scenario Manager (see Section 1S.15) to store and use sets of alternative probability values. (Because the Scenario Manager cannot store formulas, the original values of one-sixth will be converted to a decimal approximation if they are stored in a scenario.)

Problems for Section 4.11

Note: *Microsoft Excel may be used to solve the problems in this section.*

- 4.37 Given the following probability distributions:

Distribution A		Distribution B	
X	P(X)	X	P(X)
0	.50	0	.05
1	.20	1	.10
2	.15	2	.15
3	.10	3	.20
4	.05	4	.50

  (a) Compute the mean for each distribution.
  (b) Compute the standard deviation for each distribution.
  (c) **ACTION** Compare and contrast the results in (a) and (b). Discuss what you have learned.

4.38 Given the following probability distributions:

Distribution C		Distribution D	
X	P(X)	X	P(X)
0	.20	0	.10
1	.20	1	.20
2	.20	2	.40
3	.20	3	.20
4	.20	4	.10

(a) Compute the mean for each distribution.
(b) Compute the standard deviation for each distribution.
(c) **ACTION** Compare and contrast the results in (a) and (b). Discuss what you have learned.

4.39 An employee for a vending concession at a baseball stadium must choose between working behind the hot dog counter and receiving a fixed sum of $50 for the evening and walking around the stands selling beer on a commission basis. If the latter is chosen, the employee can make $90 on a warm night, $70 on a moderate night, $45 on a cool night, and $15 on a cold night. At this time of year the probabilities of a warm, moderate, cool, or cold night are .1, .3, .4, and .2, respectively.
(a) Determine the mean or expected value to be earned by selling beer that evening.
(b) Compute the standard deviation.
(c) Which product should the employee sell? Why?
(d) What are the results in (a)–(c) if the probabilities are .3, .2, .3, and .2, respectively?

4.40 A state lottery is conducted in which 10,000 tickets are sold for $1 each. Six winning tickets are randomly selected: one grand-prize winner of $5,000, one second-prize winner of $2,000, one third-prize winner of $1,000, and three other winners of $500 each.
(a) Compute the expected value of playing this game.
(b) **ACTION** Would you play this game? Why?

4.41 Let us consider rolling a pair of six-sided dice. The random variable of interest represents the total of the two numbers (i.e., faces) that occurs when the pair of fair dice are rolled. The probability distribution is as follows:

X	P(X)
2	1/36
3	2/36
4	3/36
5	4/36
6	5/36
7	6/36
8	5/36
9	4/36
10	3/36
11	2/36
12	1/36
	1

(a) Determine the mean or expected sum from rolling a pair of the fair dice.
(b) Compute the variance and the standard deviation.

The game of Craps involves the rolling of a pair of fair dice. A *field bet* in the game of Craps is a one-roll bet and is based on the outcome of the pair of dice. For every $1.00 bet you make, you lose the $1.00 if the sum is 5, 6, 7, or 8; you win $1.00 if the sum is 3, 4, 9, 10, or 11; or you win $2.00 if the sum is either 2 or 12.
(c) Form the probability distribution function representing the different outcomes that are possible in a field bet.
(d) Determine the mean of this probability distribution.
(e) What is the player's expected long-run profit (or loss) from a $1.00 field bet? Interpret.
(f) What is the expected long-run profit (or loss) to the house from a $1.00 field bet? Interpret.
(g) **ACTION** Would you play this game and make a field bet?

4.42 In the carnival game Under-or-over-seven, a pair of fair dice are rolled once, and the resulting sum determines whether or not the player wins or loses his or her bet. For example, the player can bet $1.00 that the sum is under 7—that is, 2, 3, 4, 5, or 6. For such a bet the player will lose $1.00 if the outcome equals or exceeds 7 or will win $1.00 if the result is under 7. Similarly, the player can bet $1.00 that the sum is over 7—that is, 8, 9, 10, 11, or 12. Here the player wins $1.00 if the result is over 7 but loses $1.00 if the result is 7 or under. A third method of play is to bet $1.00 on the outcome 7. For this bet the player will win $4.00 if the result of the roll is 7 and lose $1.00 otherwise.
(a) Form the probability distribution function representing the different outcomes that are possible for a $1.00 bet on being under 7.
(b) Form the probability distribution function representing the different outcomes that are possible for a $1.00 bet on being over 7.
(c) Form the probability distribution function representing the different outcomes that are possible for a $1.00 bet on 7.
(d) Prove that the expected long-run profit (or loss) to the player is the same—no matter which method of play is used.
(e) **ACTION** Would you prefer to play Under-or-over-seven or make a field bet in Craps (Problem 4.41)? Why?

- 4.43 The Islander Fishing Company purchases clams for $1.50 per pound from Peconic Bay fishermen for sale to various New York restaurants for $2.50 per pound. Any clams not sold to the restaurants by the end of the week can be sold to a local soup company for $0.50 per pound. The probabilities of various levels of demand are as follows:

Demand (pounds)	Probability
500	.2
1,000	.4
2,000	.4

(*Hint*: The company can purchase 500, 1,000, or 2,000 pounds.)

(a) For each possible purchase level (500, 1000, or 2000 pounds), compute the profit (or loss) for each level of demand.
(b) **ACTION** Using the expected-monetary-value criterion, determine the optimal number of pounds of clams that the company should purchase from the fishermen. Discuss.
(c) What would be the effect on the results in (a) and (b) if the clams can be sold for $3 per pound?
(d) What would be the effect on the results if the probabilities of demand were .4, .4, and .2, respectively?

4.44 An investor has a certain amount of money available to invest now. Three alternative portfolio selections are available. The estimated profits of each portfolio under each economic condition are indicated in the following payoff table at the top of page 196.

	Portfolio Selection		
Event	A	B	C
Economy declines	$500	–$2,000	–$7,000
No change	$1,000	$2,000	–$1,000
Economy expands	$2,000	$5,000	$20,000

Based on his own past experience, the investor assigns the following probabilities to each economic condition:

$$P(\text{economy declines}) = .30$$

$$P(\text{no change}) = .50$$

$$P(\text{economy expands}) = .20$$

S-4-44.XLS

(a) **ACTION** Determine the best portfolio selection for the investor according to the expected-monetary-value criterion. Discuss.
(b) What would be the effect on the results if the probabilities of the economic conditions were:
  (1) .1, .6, and .3?
  (2) .1, .3, and .6?
  (3) .4, .4, and .2?
  (4) .6, .3, and .1?

4.12 DISCRETE PROBABILITY DISTRIBUTION FUNCTIONS

The probability distribution for a discrete random variable may be

1. A *theoretical* listing of outcomes and probabilities (as in Table 4.2), which can be obtained from a mathematical model representing some phenomenon of interest.
2. An *empirical* listing of outcomes and their observed relative frequencies.
3. A *subjective* listing of outcomes associated with their subjective probabilities representing the degree of conviction of the decision maker as to the likelihood of the possible outcomes (as discussed in Section 4.2).

In the remainder of this chapter we will be concerned mainly with the first kind of probability distribution—the listing obtained from a mathematical model representing some phenomenon of interest.

A **model** is considered to be a miniature representation of some underlying phenomenon. In particular, a mathematical model is a mathematical expression representing some underlying phenomenon. For discrete random variables, this mathematical expression is known as a **probability distribution function**.

When such mathematical expressions are available, the exact probability of occurrence of any particular outcome of the random variable can be computed. In such cases, then, the entire probability distribution can be obtained and listed. For example, the probability distribution function represented in Table 4.2 is one in which the discrete random variable of interest is said to follow the **uniform probability distribution**. The essential characteristic of the uniform distribution is that all outcomes of the random variable are equally likely to occur.

Thus, the probability that the face ⚄ of the fair die turns up is the same as that for any other result—1/6—since there are six possible outcomes.

In addition, other types of mathematical models have been developed to represent various discrete phenomena that occur in the social and natural sciences, in medical research, and in business. One of the most useful of these represents data characterized by the binomial probability distribution. This distribution will now be developed.

4.13 BINOMIAL DISTRIBUTION

The **binomial distribution** is a discrete probability distribution function that is extremely useful for describing many phenomena.

The binomial distribution possesses four essential properties:

1. The possible observations may be obtained by two different sampling methods. Either each observation may be considered as having been selected from an *infinite population without replacement* or from a *finite population with replacement.*
2. Each observation may be classified into one of two mutually exclusive and collectively exhaustive categories, usually called *success* and *failure.*
3. The probability of an observation's being classified as success, p, is constant from observation to observation. Thus, the probability of an observation's being classified as failure, $1 - p$, is constant over all observations.
4. The outcome (i.e., success or failure) of any observation is independent of the outcome of any other observation.

The discrete random variable or phenomenon of interest that follows the binomial distribution is the number of successes obtained in a sample of n observations. Thus, the binomial distribution has enjoyed numerous applications:

- In games of chance:
  What is the probability that red will come up 15 or more times in 19 spins of the roulette wheel?
- In product quality control:
  What is the probability that in a sample of 20 tires of the same type, none will be defective if 8% of all such tires produced at a particular plant are defective?
- In education:
  What is the probability that a student can pass a 10-question multiple-choice exam (each question containing 4 choices) if the student guesses on each question? (Passing is defined as getting 60% of the items correct—that is, getting at least 6 out of 10 items correct.)
- In finance:
  What is the probability that a particular stock will show an increase in its closing price on a daily basis over the next 10 (consecutive) trading sessions if stock market price changes really are random?

In each of these examples the four properties of the binomial distribution are clearly satisfied. For the roulette example, a particular set of spins may be construed as the sample taken from an infinite population of spins without replacement. When spinning the roulette wheel, each observation is categorized as red (success) or not red (failure). The probability of spinning red on an American roulette wheel, p, is 18/38 and is assumed to remain stable over all

FIGURE 4.2 American roulette wheel.

observations. Thus, the probability of failure (spinning black or green), $1 - p$, is 20/38 each and every time the roulette wheel spins. Moreover, the roulette wheel has no memory—the outcome of any one spin is independent of preceding or following spins so that, for example, the probability of obtaining red on the 32nd spin, given that the previous 31 spins were all red, remains equal to p, 18/38, if the roulette wheel is a fair one (see Figure 4.2).

In the product quality-control example, the sample of tires is also selected without replacement from an ongoing production process, an infinite population of manufactured tires.[2] As each tire in the sample is inspected, it is categorized as defective or nondefective according to the operational definition of conformance to specifications that has been previously developed. Over the entire sample of tires the probability of any particular tire being classified as defective, p, is .08 so that the probability of any tire being categorized as nondefective, $1 - p$, is .92. (Note that when we are looking for defective tires, the discovery of such an event is deemed a success. This is one of the instances referred to previously where, for statistical purposes, the term *success* may refer to business failures, deaths due to a particular illness, and other phenomena that, in nonstatistical terminology, would be deemed unsuccessful.) The production process is assumed to be stable. In addition, for such a production process, the probability of one tire being classified as defective or nondefective is independent of the classification for any other tire.

Similar statements pertaining to the binomial distribution's characteristics in the education example and in the finance example can be made as well. This is left to the reader. (See Problems 4.45 and 4.46 on page 206.)

The four examples of binomial probability models previously described are distinguished by the parameters n and p. Each time a set of parameters—the number of observations in the sample, n, and the probability of success, p—is specified, a particular binomial probability distribution can be generated.

4.13.1 Development of the Model

As another example of a phenomenon that satisfies the conditions of the binomial distribution, and one that is convenient for intuitively deriving an expression for the probabilities that arise in binomial problems, we shall return to the fair die example discussed in Section 4.10. Here, however, we consider success to be the outcome face ⚃ and failure to be any other outcome. Suppose we are now interested in three rolls of this same die in order to determine how frequently the face ⚃ is obtained.[3] What might occur? None of the rolls might land on ⚃; one of the rolls may be a ⚃; two of the rolls may land on ⚃; or all three rolls may land on ⚃. Can the binomial random variable, the number of ⚃ faces occurring on three rolls of a fair die, take on any other value? That would be impossible since, if we roll the same die three times and are interested in how often a particular value (face ⚃) occurs, that value cannot exceed the number of rolls, n, nor can it be lower than zero. Hence, the range of a binomial random variable is from 0 to n.

Suppose then, for example, that we roll a fair die three times and observe the following result:

First Roll	Second Roll	Third Roll
⚃	Not 5	⚃

We now wish to determine the probability of this occurrence; that is, what is the probability of obtaining two successes (face ⚃) in three rolls in this *particular sequence*? Since it may be assumed that rolling dice is a stable process, the probability that each roll occurs as noted is

First Roll	Second Roll	Third Roll
$p = 1/6$	$1 - p = 5/6$	$p = 1/6$

Since each outcome is independent of the others, the probability of obtaining the given sequence is

$$p(1-p)p = p^2(1-p)^1 = p^2(1-p) = (1/6)^2(5/6) = 5/216$$

Thus, out of 216 possible and equally likely outcomes from rolling a fair die three times, five will have the face ⚃ as the first and last roll, with a face other than ⚃ (i.e., ⚀, ⚁, ⚂, ⚃, or ⚅) as the middle roll, and the particular sequence noted above will be obtained.

Now, however, we may ask how many different sequences are there for obtaining two faces of ⚃ out of $n = 3$ rolls of the die? This involves using the **rule of combinations:**

> The number of ways of selecting X objects out of n objects, irrespective of order, is equal to
>
> $$\frac{n!}{X!(n-X)!} \tag{4.13}$$

4.13 Binomial Distribution

where $n! = n(n-1)\ldots(1)$ is called *n factorial* and $0! = 1$.

This expression may be denoted by the symbol $\binom{n}{X}$. Therefore, with $n = 3$ and $X = 2$, we have

$$\binom{n}{X} = \frac{n!}{X!(n-X)!} = \frac{3!}{2!(3-2)!} = \frac{3 \times 2 \times 1}{(2 \times 1)(1)} = 3$$

such sequences. These three possible sequences are

Sequence 1 = [⚄] [Not 5] [⚄] with probability $p(1-p)p = p^2(1-p)^1 = 5/216$

Sequence 2 = [⚄] [⚄] [Not 5] with probability $pp(1-p) = p^2(1-p)^1 = 5/216$

Sequence 3 = [Not 5] [⚄] [⚄] with probability $(1-p)pp = p^2(1-p)^1 = 5/216$

Therefore, the probability of obtaining exactly two faces of [⚄] from three rolls of a die is equal to

(number of possible sequences) × (probability of a particular sequence)

$(3) \times (5/216) = 15/216 = .0694$

A similar, intuitive derivation can be obtained for the other three possible outcomes of the random variable—no face [⚄], one face [⚄], or all three faces [⚄]. However, as n, the number of obser-vations, gets large, this type of intuitive approach becomes quite laborious, and a mathematical model is more appropriate. In general, the following mathematical model represents the binomial probability distribution for obtaining the number of successes (X), given a knowledge of the parameters n and p:

$$P(X) = \frac{n!}{X!(n-X)!} p^X (1-p)^{n-X} \qquad (4.14)$$

where $P(X)$ = the probability of X successes given a knowledge of n and p

n = sample size

p = probability of success

$1 - p$ = probability of failure

X = number of successes in the sample ($X = 0, 1, 2, \ldots, n$)

We note, however, that the generalized form shown in Equation (4.14) is merely a restatement of what we had intuitively derived.

The binomial random variable X can have any integer value X from 0 through n. In Equation (4.14) the product

$$p^X(1-p)^{n-X}$$

tells us the probability of obtaining exactly X successes out of n observations *in a particular sequence*, while the term

$$\frac{n!}{X!(n-X)!}$$

tells us *how many combinations* of the X successes out of n observations are possible. Hence, given the number of observations n and the probability of success p, we may determine the probability of X successes:

$$P(X) = \text{(number of possible sequences)}$$
$$\times \text{(probability of a particular sequence)}$$
$$= \frac{n!}{X!(n-X)!} p^X (1-p)^{n-X}$$

by substituting the desired values for n, p, and X and computing the result.

Thus, as previously shown, the probability of obtaining exactly two faces of ⚅ from three rolls of a die is

$$P(X-2) = \frac{3!}{2!(3-2)!}\left(\frac{1}{6}\right)^2 \left(1-\frac{1}{6}\right)^{3-2}$$

$$= \frac{3!}{2!\,1!}\left(\frac{1}{6}\right)^2 \left(\frac{5}{6}\right)^1$$

$$= 3\left(\frac{1}{6}\right)\left(\frac{1}{6}\right)\left(\frac{5}{6}\right) = \frac{15}{216} = .0694$$

Such computations may become quite tedious, especially as n gets large. However, we may obtain the probabilities by using Excel (see Section 4.13.3) and thereby avoid any computational drudgery.

4.13.2 Characteristics of the Binomial Distribution

Each time a set of parameters—n and p—is specified, a particular binomial probability distribution can be generated.

- **Shape** We note that a binomial distribution may be symmetric or skewed. Whenever p = .5, the binomial distribution will be symmetric regardless of how large or small the value of n. However, when p ≠ .5, the distribution will be skewed. The closer p is to .5 and the larger the number of observations, n, the less skewed the distribution will be. Thus, the distribution of the number of occurrences of ⚅ in three rolls of a fair die is skewed to the right, since p = 1/6.

We leave it to the reader to verify the effect of n and p on the shape of the distribution by plotting the histogram in Problem 4.51(c) on page 207. However, to summarize the preceding characteristics, three binomial distributions are depicted in Figure 4.3. Panel A represents the probability of obtaining the face ⚅ on three rolls of a fair die; Panel B represents the probability of obtaining "heads" on three tosses of a fair coin; and Panel C represents the probability of obtaining "heads" on four tosses of a fair coin. Thus, a comparison of Panels A and B demonstrates the effect on shape when the sample sizes are the same but the probabilities for success differ. A comparison of Panels B and C shows the effect on shape when the probabilities for successes are the same but the sample sizes differ.

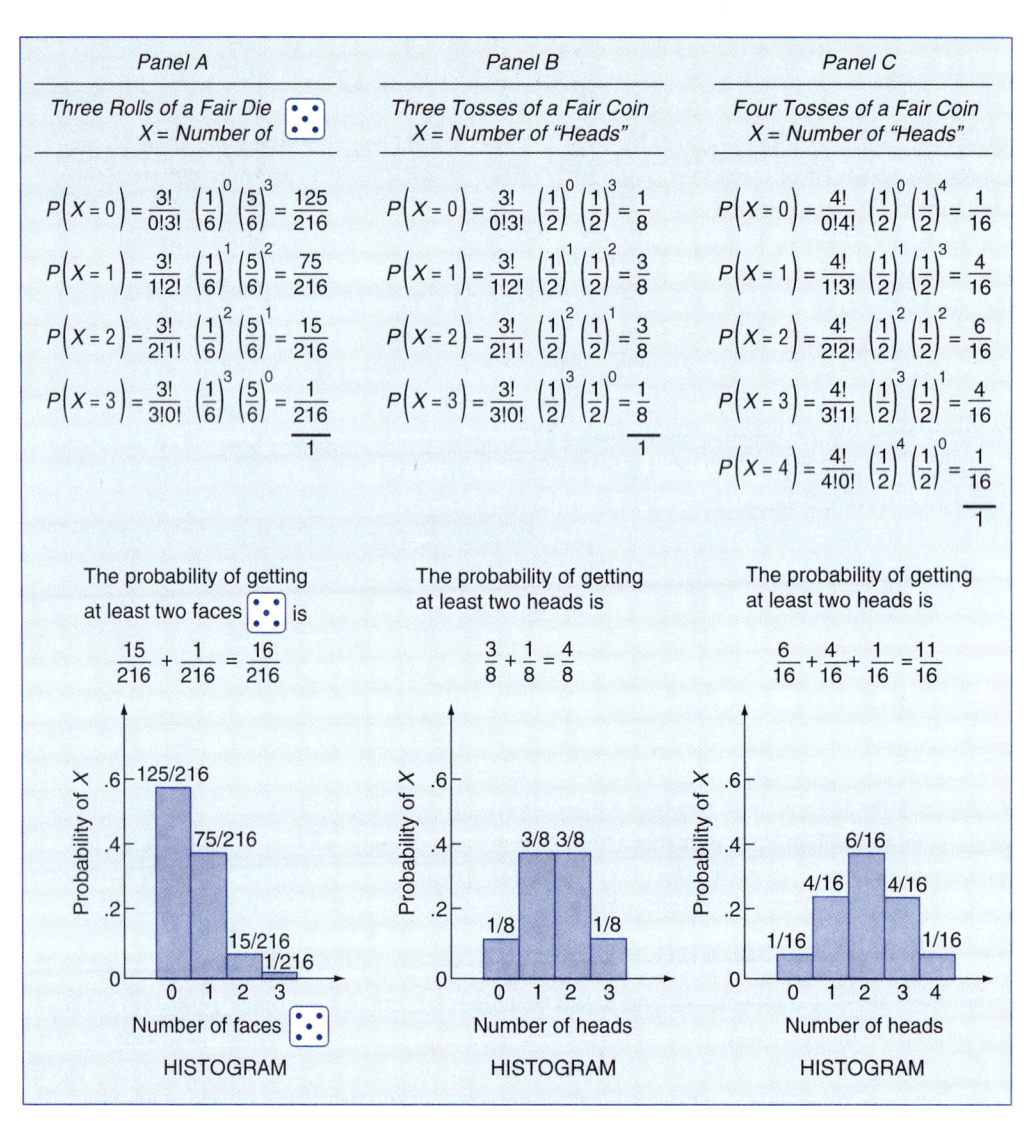

Panel A	Panel B	Panel C
Three Rolls of a Fair Die ⚂ X = Number of ⚂	Three Tosses of a Fair Coin X = Number of "Heads"	Four Tosses of a Fair Coin X = Number of "Heads"
$P(X=0) = \frac{3!}{0!3!}\left(\frac{1}{6}\right)^0\left(\frac{5}{6}\right)^3 = \frac{125}{216}$	$P(X=0) = \frac{3!}{0!3!}\left(\frac{1}{2}\right)^0\left(\frac{1}{2}\right)^3 = \frac{1}{8}$	$P(X=0) = \frac{4!}{0!4!}\left(\frac{1}{2}\right)^0\left(\frac{1}{2}\right)^4 = \frac{1}{16}$
$P(X=1) = \frac{3!}{1!2!}\left(\frac{1}{6}\right)^1\left(\frac{5}{6}\right)^2 = \frac{75}{216}$	$P(X=1) = \frac{3!}{1!2!}\left(\frac{1}{2}\right)^1\left(\frac{1}{2}\right)^2 = \frac{3}{8}$	$P(X=1) = \frac{4!}{1!3!}\left(\frac{1}{2}\right)^1\left(\frac{1}{2}\right)^3 = \frac{4}{16}$
$P(X=2) = \frac{3!}{2!1!}\left(\frac{1}{6}\right)^2\left(\frac{5}{6}\right)^1 = \frac{15}{216}$	$P(X=2) = \frac{3!}{2!1!}\left(\frac{1}{2}\right)^2\left(\frac{1}{2}\right)^1 = \frac{3}{8}$	$P(X=2) = \frac{4!}{2!2!}\left(\frac{1}{2}\right)^2\left(\frac{1}{2}\right)^2 = \frac{6}{16}$
$P(X=3) = \frac{3!}{3!0!}\left(\frac{1}{6}\right)^3\left(\frac{5}{6}\right)^0 = \frac{1}{216}$	$P(X=3) = \frac{3!}{3!0!}\left(\frac{1}{2}\right)^3\left(\frac{1}{2}\right)^0 = \frac{1}{8}$	$P(X=3) = \frac{4!}{3!1!}\left(\frac{1}{2}\right)^3\left(\frac{1}{2}\right)^1 = \frac{4}{16}$
		$P(X=4) = \frac{4!}{4!0!}\left(\frac{1}{2}\right)^4\left(\frac{1}{2}\right)^0 = \frac{1}{16}$
1	1	1
The probability of getting at least two faces ⚂ is $\frac{15}{216} + \frac{1}{216} = \frac{16}{216}$	The probability of getting at least two heads is $\frac{3}{8} + \frac{1}{8} = \frac{4}{8}$	The probability of getting at least two heads is $\frac{6}{16} + \frac{4}{16} + \frac{1}{16} = \frac{11}{16}$

FIGURE 4.3 Comparison of three binomial distributions.

● **The Mean** The mean of the binomial distribution can be obtained as the product of its two parameters, *n* and *p*. Instead of using Equation (4.10), which holds for all discrete probability distributions, for data that are binomially distributed, we use the following to compute the mean:

> The mean of the binomial distribution (μ) is equal to the sample size *n* multiplied by the probability of success *p*.
>
> $$\mu = E(X) = np \tag{4.15}$$

Intuitively, this makes sense. For example, if we toss a fair die three times, how frequently should we "expect" the face ⚂ to come up? On the average, over the long run, we would theoretically expect

202 Chapter 4 Basic Probability and Discrete Probability Distributions

$$\mu = E(X) = np = 3\left(\frac{1}{6}\right) = \frac{3}{6} = 0.50$$

occurrences of face ⚃ in three tosses—the same result that would be obtained from the more general expression shown in Equation (4.10).

- **The Standard Deviation** The standard deviation of the binomial distribution is calculated using the formula

$$\sigma = \sqrt{np(1-p)} \tag{4.16}$$

Referring to our die example, we compute

$$\sigma = \sqrt{3\left(\frac{1}{6}\right)\left(\frac{5}{6}\right)} = \sqrt{\frac{15}{36}} = 0.646$$

This is the same result that would be obtained from the more general expression shown in Equation (4.12).

- **Summary** In this section we have developed the binomial model as a useful discrete probability distribution in its own right. However, the binomial distribution plays an even more important role when it is used in statistical inference problems regarding estimating or testing proportions (as will be discussed in Chapters 6, 7, and 9).

4.13.3 Using Microsoft Excel to Obtain Binomial Probabilities

In this section we studied the binomial distribution and used Equation (4.14) on page 200 to compute binomial probabilities. Rather than using this equation, which involves tedious calculations, particularly for large sample sizes, we can use the BINOMDIST function of Microsoft Excel. The format of this function is as follows:

BINOMDIST (*X, n, p, cumulative*)

where
X = the number of successes

n = the sample size

p = the probability of success

cumulative = False if you wish to compute the probability of *exactly* X successes or True if you wish to compute the probability of *X or fewer successes*

To illustrate the use of this formula, suppose we return to the fair die example in Section 4.13. On page 199, we computed the probability of obtaining two occurrences of the face ⚃ in three rolls as .0694. To obtain this result using Excel, we would enter the formula =BINOMDIST(2,3,1/6,False) in a cell. If we wanted the probability of two or fewer occurrences of the face ⚃, the cumulative selection would have been set equal to true =BINOMDIST(2,3,1/6,True).

Table 4.3.Excel Design for Calculations sheet for computing binomial probabilities for the fair die example (where illustrated for n = 3 and p = .5).

	A	B	C	D	E	F
1			Calculating Binomial Probabilities			
2						
3	n	p	Mean	Variance	Std. Deviation	
4	3	.5	=A4*B4	=A4*B4*(1 – B4)	=SQRT(D4)	
5						
6	X	P(X)	P(<=X)	P(<X)	P(>X)	P(X>=X)
7	0	=BINOMDIST (A7,$A$4,$B$4,False)	=BINOMDIST (A7,$A$4,$B$4,True)	=C7 – B7	=1 – C7	=1 – D7
8	1	=BINOMDIST (A8,$A$4,$B$4,False)	=BINOMDIST (A8,$A$4,$B$4,True)	=C8 – B8	=1 – C8	=1 – D8
9	2	=BINOMDIST (A9,$A$4,$B$4,False)	=BINOMDIST (A9,$A$4,$B$4,True)	=C9 – B9	=1 – C9	=1 – D9
10	3	=BINOMDIST (A10,$A$4,$B$4,False)	=BINOMDIST (A10,$A$4,$B$4,True)	=C10 – B10	=1 – C10	=1 – D10

4-13-3.XLS

Now that we have introduced the BINOMDIST function and illustrated its use in computing probabilities, we are ready to develop a design for generating an entire set of probabilities similar to those displayed in Figure 4.3 on page 202. Table 4.3Excel contains the design for a Calculations sheet that provides for the computation of the probabilities for all values of X for a given value of n and p.

We may begin the plan for obtaining the binomial probabilities by opening a new workbook and renaming the active sheet Calculations. After the appropriate labels have been typed in cells A3:E3, we need to enter the value for n in cell A4 and the value for p in cell B4. Once these values have been entered, the formula for the expected value (np) can be entered in cell C4 as =A4*B4, the formula for the variance [$np(1 - p)$] can then be entered in cell D4 as =A4*B4*(1 – B4), and the formula for the standard deviation $\left[\sqrt{np(1-p)}\right]$ can be entered in cell E4 as =SQRT(D4).

The possible outcomes for the number of successes (X) are then entered in column A beginning in row 7 and continuing through the value of n. If $n = 3$, values of 0, 1, 2, and 3 need to be entered in cells A7:A10. Once this has been completed, we can create formulas using the BINOMDIST funtion to compute the binomial probabilities for each value of X in cells B7:B10. Enter the formula =BINOMDIST(A7,$A$4,$B$4,False) in cell B7 and then copy the formula down the rest of the range (B8:B10). Since we want to obtain the probability of exactly zero successes ($X = 0$), we enter False as the fourth argument in the function.

To obtain the cumulative probability of less than or equal to X successes, substitute True as the fourth argument in the BINOMDIST function, and enter =BINOMDIST (A7,$A$4,$B$4,True) in cell C7. Then copy this formula to the subsequent rows containing the remainder of the X values (cells C8:C10). To obtain the probability of less than X successes [$P(<X)$] and to obtain the probability of more than X successes [$P(>X)$], enter the formulas as shown for columns D and E. Finally, to include the probability of at least X successes ($P \geq X$), enter the formulas in cells F7:F10. Figure 4.3.Excel represents these calculations for the data of Table 4.3.Excel.

Now that the binomial probabilities have been calculated for each X value, we can use Microsoft Excel's Chart Wizard (see Section 2.12) to obtain a bar chart similar to the one displayed in Panel B of Figure 4.3. The selections in the five dialog boxes for the Chart Wizard are as follows:

	A	B	C	D	E	F
1	Calculating Binomial Probabilities					
2						
3	n	p	Mean	Variance	Std. Deviation	
4	3	0.5	1.5	0.75	0.866025404	
5						
6	X	P(x)	P(<=X)	P(<X)	P(>X)	P(>=X)
7	0	0.125	0.125	0	0.875	1
8	1	0.375	0.5	0.125	0.5	0.875
9	2	0.375	0.875	0.5	0.125	0.5
10	3	0.125	1	0.875	0	0.125

FIGURE 4.3.EXCEL Calculating binomial probabilities for the fair die example using Microsoft Excel.

First dialog box: Enter Calculations!A7:B10 as the range.
Second dialog box: Select Column as the chart type.
Third dialog box: Select the first format choice that contains bars above and below the X axis.
Fourth dialog box: Select the Columns option button for Data Series, enter 1 in the First Columns edit box and 0 in the First Rows edit box.
Fifth dialog box: Select the No Option button for Add a Legend, enter the chart title Binomial Probabilities, the Category (X) title Number of Successes (X), and the value (Y) title P(X).

With the gaps between X values removed (see Section 2.7.3), rename the sheet containing this chart as Bar Chart. Figure 4.4.Excel displays the chart of the binomial distribution for the data of Figure 4.3.Excel.

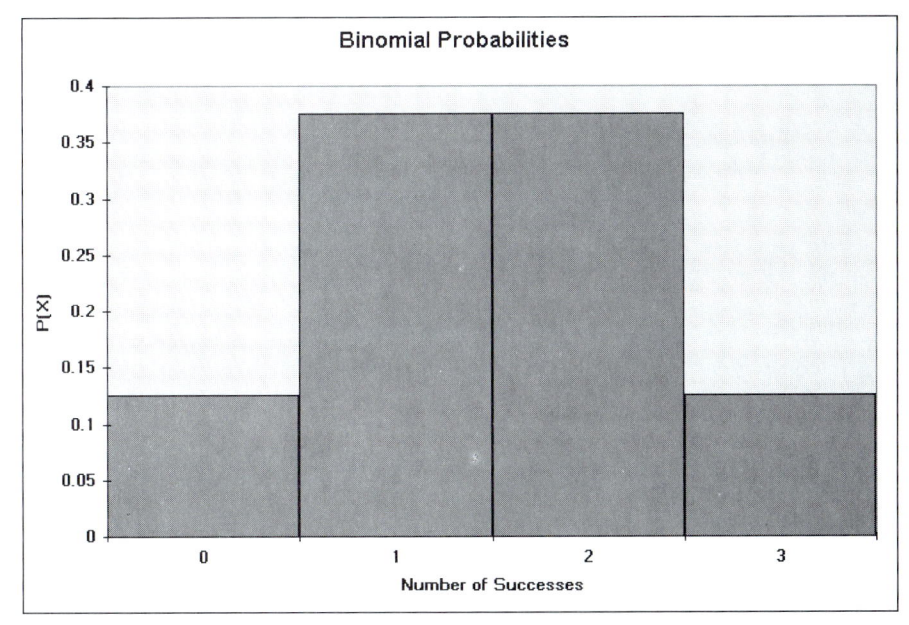

FIGURE 4.4.EXCEL Histogram of binomial probabilities obtained from Microsoft Excel for the fair die example.

> ▲ WHAT IF EXAMPLE
>
> The design of the Calculations worksheet presented in Table 4.3.Excel allows us to study the effect of changes in n and p on the probability of X successes and the shape of the bar chart of the binomial probabilities obtained from the Chart Wizard. For example, referring to Figure 4.3, if we wanted to obtain the results displayed in Panel A for $p = 1/6$, we would substitute = 1/6 for 0.5 in cell B4. Our results would be the same as those in Panel A of Figure 4.3. If we wanted to see the effect of increasing the sample size to 4 while keeping $p = .5$ as in Panel B, we would type 4 into cell A11, and then copy cells B10:F10 to cells B11:F11. The results obtained would be the same as those presented in Panel C of Figure 4.3.

Problems for Section 4.13

Note: *Microsoft Excel may be used to solve the problems in this section.*

4.45 Describe how the four properties of the binomial distribution could be satisfied in the education example on page 197.

4.46 Describe how the four properties of the binomial distribution could be satisfied in the finance example on page 197.

4.47 Determine the following:
(a) If $n = 4$ and $p = .12$, then what is $P(X = 0)$?
(b) If $n = 10$ and $p = .40$, then what is $P(X = 9)$?
(c) If $n = 10$ and $p = .50$, then what is $P(X = 8)$?
(d) If $n = 6$ and $p = .83$, then what is $P(X = 5)$?
(e) If $n = 10$ and $p = .90$, then what is $P(X = 9)$?

4.48 In the carnival game of Chuck-a-luck, three fair dice are rolled after the player has placed a bet on the occurrence of a particular face of the dice, say ⚄. For every $1.00 bet that you place, you lose the $1.00 if none of the three dice shows the face ⚄; you win $1.00 if one die shows the face ⚄; you win $2.00 if two of the dice show the face ⚄; or you win $3.00 if all three dice show the face ⚄.
(a) Form the probability distribution function representing the different monetary values (winnings or losses) that are possible (from one roll of the three dice). (*Hint:* Review Section 4.13.1 and see Figure 4.3, Panel A).
(b) Determine the mean of this probability distribution.
(c) What is the player's expected long-run profit (or loss) from a $1.00 bet? Interpret.
(d) What is the expected long-run profit (or loss) to the house? Interpret.
(e) **ACTION** Would you play Chuck-a-luck and make a bet? Discuss.

4.49 Suppose that warranty records show that the probability that a new car needs a warranty repair in the first 90 days is .05. If a sample of three new cars is selected,
(a) What is the probability that
    (1) none needs a warranty repair?
    (2) at least one needs a warranty repair?
    (3) more than one needs a warranty repair?
(b) What assumptions are necessary in (a)?
(c) What are the mean and standard deviation of the probability distribution in (a)?
(d) What would be your answers to (a)–(c) if the probability of needing a warranty repair was .10?

4.50 The probability that a salesperson will sell a magazine subscription to someone who has been randomly selected from the telephone directory is .20.
   (a) If the salesperson calls 10 individuals this evening, what is the probability that
      (1) No subscriptions will be sold?
      (2) Exactly two subscriptions will be sold?
      (3) At least two subscriptions will be sold?
      (4) At most two subscriptions will be sold?
   (b) If the salesperson made calls to 20 individuals, what would be your answers to (a)?
   (c) If the probability of selling a magazine subscription was only .10, what would be your answers to (a)?

• 4.51 An important part of the customer-service responsibilities of a natural gas utility company concerns the speed with which calls relating to no heat in a house can be serviced. Suppose that one service variable of importance refers to whether or not the repair person reaches the home within a 2-hour period. Past data indicate that the likelihood is .60 that the repair person reaches the home within a 2-hour period.
   (a) If a sample of five service calls for "no heat" is selected, what is the probability that the repair person will arrive at
      (1) All five houses within a 2-hour period?
      (2) At least three houses within the 2-hour period?
   (b) Find the probability that the repair person will arrive at zero, one, and two houses and plot the histogram for the probability distribution.
   (c) What is the shape of the distribution plotted in (c)? Explain.
   (d) If the company was able to increase the probability to .80, what would be your answers to (a)–(c)?
   (e) If the company was able to increase the probability to .90, what would be your answers to (a)–(c)?

S-4-51.XLS

4.14 OTHER DISCRETE PROBABILITY DISTRIBUTIONS

Now that we have discussed the binomial distribution, in this section we will introduce several other discrete probability distributions that are useful in a variety of circumstances in business.

4.14.1 Hypergeometric Distribution

Both the binomial distribution and the **hypergeometric distribution** are concerned with the same thing—the number of successes in a sample containing n observations. What differentiates these two discrete probability distributions is the manner in which the data are obtained. For the binomial model, the sample data are drawn with replacement from a finite population or *without* replacement from an infinite population (as discussed in Section 1.10). On the other hand, for the hypergeometric model, the sample data are drawn without replacement from a finite population. Hence, while the probability of success, p, is constant over all observations of a binomial experiment, and the outcome of any particular observation is independent of any other, the same cannot be said for a hypergeometric experiment; here the outcome of one observation is affected by the outcomes of the previous observations.

In general, a mathematical expression of the hypergeometric distribution for obtaining X successes, given a knowledge of the parameters n, N, and A, is

$$P(X) = \frac{\binom{A}{X}\binom{N-A}{n-X}}{\binom{N}{n}}$$

$$= \frac{\dfrac{A!}{X!(A-X)!} \times \dfrac{(N-A)!}{(n-X)![(N-A)-(n-X)]!}}{\dfrac{N!}{n!(N-n)!}}$$

(4.17)

where $P(X)$ = the probability of X successes given a knowledge of n, N, and A

n = sample size

N = population size

A = number of successes in the population

$N - A$ = number of failures in the population

X = number of successes in the sample

The number of successes in the sample, X, cannot exceed the number of successes in the population, A, or the sample size, n. Thus, the range of the hypergeometric random variable is limited to the sample size (as was the range for the binomial random variable) or to the number of successes in the population—whichever is smaller.

To illustrate the hypergeometric distribution, let us consider the following problem. Suppose we wish to determine the probability of obtaining two clubs in five draws made without replacement from a thoroughly shuffled standard deck of 52 cards. Here the population, $N = 52$, is finite.

Using the Equation (4.17), we would have

$$P(X = 2) = \frac{\binom{13}{2}\binom{39}{3}}{\binom{52}{5}}$$

$$= \frac{\dfrac{13!}{2!\,11!} \times \dfrac{39!}{3!\,36!}}{\dfrac{52!}{5!\,47!}}$$

$$= .2743$$

4.14.2 Negative Binomial Distribution

The **negative binomial distribution** is used to determine the probability that a certain number of failures occurs before there are a certain number of successes. The mathematical expression for the negative binomial distribution for obtaining the probability that the Xth success occurs on the nth trial is

$$P(n) = \binom{n-1}{X-1} p^X (1-p)^{n-X} \qquad (4.18)$$

where p = probability of success

n = number of trials until the Xth success is observed

We may illustrate the negative binomial distribution with the following example. Suppose an organization is trying to create a team of eight people with backgrounds in the delivery of a particular service process. Past experience has shown that 40% of the candidates for the team have such a background. What is the probability that the selection is completed when the tenth candidate is considered (i.e., only two candidates are rejected before the team of eight people is selected)?

For this example, $n = 10$, $X = 8$, and $p = .4$, so that we have

$$P(n = 10) = \binom{10-1}{8-1}(.4)^8 (1-.4)^{10-8}$$

$$P(n = 10) = \left(\frac{9!}{7!(2!)}\right)(.4)^8 (.6)^{10-8}$$

$$P(n = 10) = (36)(.4)^8 (.6)^2$$

$$P(n = 10) = .00849347$$

Thus, the probability that the selection of the team would be completed with the tenth candidate considered (with only two candidates rejected) is .00849347.

4.14.3 Poisson Distribution

The Poisson distribution is another discrete probability distribution that has many important practical applications. Not only are numerous discrete phenomena represented by a Poisson process, but the Poisson model is also used for providing approximations to the binomial distribution.

A **Poisson process** is said to exist if we can observe discrete events in an *area of opportunity*—a continuous interval (of time, length, surface area, etc.)—in such a manner that if we shorten the area of opportunity or interval sufficiently,

1. The probability of observing exactly one success in the interval is stable.
2. The probability of observing more than one success in the interval is zero.
3. The occurrence of a success in any one interval is statistically independent of that in any other interval.

To better understand the Poisson process, suppose we examine the number of customers arriving during the 12 noon to 1 P.M. lunch hour at a bank located in the central business district in a large city. Any arrival of a customer is a discrete event at a particular point over the continuous 1-hour interval. Over such an interval of time, there might be an average of 180 arrivals. Now if we were to break up the 1-hour interval into 3,600 consecutive 1-second intervals,

1. The expected (or average) number of customers arriving in any 1-second interval would be .05.
2. The probability of having more than one customer arriving in any 1-second interval approaches 0.
3. The arrival of one customer in any 1-second interval has no effect on (i.e., is statistically independent of) the arrival of any other customer in any other 1-second interval.

The **Poisson distribution** has one parameter, called λ (the Greek lowercase letter lambda), which is the average or expected number of successes per unit. The number of successes X of the Poisson random variable ranges from 0 to ∞.

The mathematical expression for the Poisson distribution for obtaining X successes, given that λ successes are expected, is

$$P(X) = \frac{e^{-\lambda} \lambda^X}{X!} \qquad (4.19)$$

where

$P(X)$ = the probability of X successes given a knowledge of λ

λ = expected number of successes

e = mathematical constant approximated by 2.71828

X = number of successes per unit

To demonstrate Poisson model applications, let us return to the example of the bank lunch-hour customer arrivals: If, on average, three customers arrive per minute, what is the probability that in a given minute exactly two customers will arrive? What is the chance that more than two customers will arrive in a given minute?

Using Equation (4.19), we have, for the first question

$$P(X) = \frac{e^{-3.0}(3.0)^2}{2!} = \frac{9}{(2.71828)^3(2)} = .2240$$

To answer the second question—the probability that in any given minute more than two customers will arrive—we have

$$P(X > 2) = P(X = 3) + P(X = 4) + \cdots + P(X = \infty)$$

Since all the probabilities in a probability distribution must sum to 1, the terms on the right side of the equation, $P(X > 2)$, also represent the complement of the probability that X is less than or equal to 2 $[1 - P(X \le 2)]$.
Thus

$$P(X > 2) = 1 - P(X \le 2) = 1 - [P(X = 0) + P(X = 1) + P(X = 2)]$$

Now, using Equation (4.19), we have

$$P(X > 2) = 1 - \left[\frac{e^{-3.0}(3.0)^0}{0!} + \frac{e^{-3.0}(3.0)^1}{1!} + \frac{e^{-3.0}(3.0)^2}{2!} \right]$$

$$P(X > 2) = 1 - [.0498 + .1494 + .2240]$$

$$= 1 - .4232 = .5768$$

Thus, we see that there is roughly a 42.3% chance that two or fewer customers will arrive at the bank per minute. Therefore, a 57.7% chance exists that three or more customers will arrive.

4.14.4 Using Microsoft Excel to Obtain Hypergeometric, Negative Binomial, and Poisson Probabilities

In this section we studied the hypergeometric, negative binomial, and Poisson distributions, and used Equations (4.17), (4.18), and (4.19) on pages 208, 209, and 210, respectively, to compute their respective probabilities. Rather than using these equations, which involve tedious calculations, we can use the appropriate Microsoft Excel functions.

- **The Hypergeometric Distribution** The HYPGEOMDIST function can be used to obtain the probability of X successes in the hypergeometric distribution. The format of this function is

$$\text{HYPGEOMDIST}(X, n, A, N)$$

where
X = the number of successes in the sample

n = the sample size

A = the number of successes in the population

N = the population size

As an example of computing a hypergeometric probability using this function, suppose we return to the two-club/five-draw example in Section 4.14.1. On page 208, we computed this probability to be .2743. To obtain this result using Excel, we would enter the formula =HYPGEOMDIST(2,5,13,52) in a cell, and we would obtain the value .27427971.

Now that we have introduced the HYPGEOMDIST function and illustrated its use in computing probabilities, we are ready to develop a plan for generating an entire set of probabilities for $X = 0, 1, \ldots, n$. Table 4.4 Excel contains the design for an Excel worksheet named

4-14-4.XLS

Table 4.4.Excel Design for computing hypergeometric probabilities for the two-club/five-draw example (where illustrated for $n = 5, A = 13, N = 52$).

	A	B	C
1		Calculating Hypergeometric Probabilities	
2			
3	n	A	N
4	5	13	52
5			
6	X	P(X)	
7	0	=HYPGEOMDIST(A7,$A$4,$B$4,$C$4)	
8	1	=HYPGEOMDIST(A8,$A$4,$B$4,$C$4)	
9	2	=HYPGEOMDIST(A9,$A$4,$B$4,$C$4)	
10	3	=HYPGEOMDIST(A10,$A$4,$B$4,$C$4)	
11	4	=HYPGEOMDIST(A11,$A$4,$B$4,$C$4)	
12	5	=HYPGEOMDIST(A12,$A$4,$B$4,$C$4)	

Hypergeometric, which provides for the computation of the probabilities for all values of X for a given value of X, n, A, and N.

Open a new workbook and rename the active sheet Hypergeometric. Type the appropriate labels in cells A3:C3. Enter the value for n in cell A4, A in cell B4, and the value for N in cell C4.

The possible outcomes for the number of successes (X) are then entered in column A beginning in row 7 and continuing through the value of n. If $n = 5$, values of 0, 1, 2, 3, 4, and 5 need to be entered in cells A7:A12. Once this has been completed, we are able to compute the hypergeometric probabilities for each value of X. Starting in cell B7, we enter the formula =HYPGEOMDIST(A7,$A$4,$B$4,$C$4). Notice that each of the arguments in the function refer to other cells of the worksheet so that if these values change, they will change accordingly. We then use the copy command to copy this formula to the subsequent rows containing the remainder of the X values (in this case, cells B8:B12). Figure 4.5.Excel represents these calculations for the data of Table 4.4.Excel.

	A	B	C	D
1	Calculating Hypergeometric Probabilities			
2				
3	n	A	N	
4	5	13	52	
5				
6	X	P(x)		
7	0	0.221533613		
8	1	0.411419568		
9	2	0.274279712		
10	3	0.081542617		
11	4	0.010729292		
12	5	0.000495198		

FIGURE 4.5.EXCEL Calculating hypergeometric probabilities for the two-club/five-draw example using Microsoft Excel.

Now that the hypergeometric probabilities have been calculated for each X value, we can use Microsoft Excel's Chart Wizard (see Section 2.12) to obtain a bar chart. The selections in the five dialog boxes for the Chart Wizard would be as follows:

First dialog box: Enter Hypergeometric! A7:B12 as the range.
Second dialog box: Select Column as the chart type.
Third dialog box: Select the first format choice that contains bars above and below the X axis.
Fourth dialog box: Select the Columns options button for Data Series, enter 1 in the First Columns edit box, and 0 in the First Rows edit box.
Fifth dialog box: Select the No option button for Add a Legend. Enter the chart title Hypergeometric Probabilities, the Category (X) title Number of Successes (X), and the value Y title P(X).

With the gaps between the X values removed (see Section 2.7.3), rename the chart sheet Hypergeometric chart. Figure 4.6.Excel displays the chart of the hypergeometric distribution for the data of Figure 4.5.Excel.

Hypergeometric Probabilities

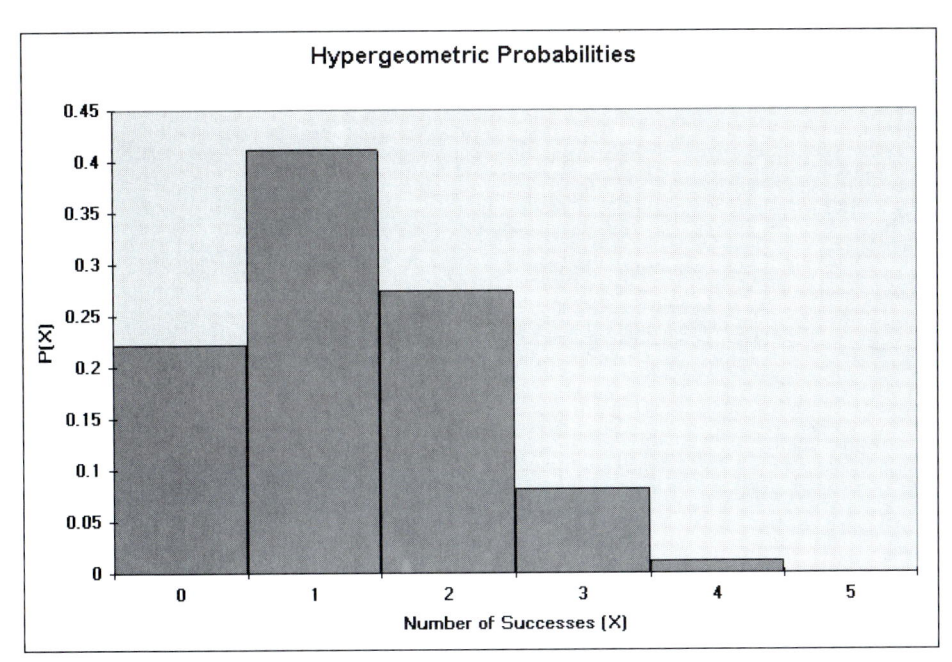

FIGURE 4.6.EXCEL Histogram of hypergeometric probabilities obtained from Microsoft Excel for the two-club/five-draw example.

> ▲ **WHAT IF EXAMPLE**
>
> The layout of the Hypergeometric worksheet presented in Table 4.4.Excel allows us to study the effect of changes in n, A, and N on the probability of X successes and the shape of the bar chart of the hypergeometric probabilities obtained from the Chart Wizard. For example, if we wanted to examine the effect of changing the sample size on the probability of obtaining a certain number of clubs, we could change the value in cell A4 of Table 4.4.Excel, and either remove or add rows in column B as necessary to have the correct set of X values.

● **The Negative Binomial Distribution** The negative binomial distribution is used to determine the probability that a certain number of failures occurs before there are a certain number of successes. Using Microsoft Excel, the NEGBINOMDIST function can be used to obtain negative binomial probabilities. The format of this function is

$$\text{NEGBINOMDIST}(n - X, X, p)$$

where
  $n - X$ = the number of failures
  X = the number of successes
  p = the probability of success

To illustrate the NEGBINOMDIST function, suppose we return to the team-selection example in Section 4.14.2 concerning the probability of obtaining the eighth success on the tenth selection. On page 209, we computed this probability to be .00849347. To obtain this result using Excel, we would enter the formula =NEGBINOMDIST(2,8,.4) in a cell. In general, if

4.14 Other Discrete Probability Distributions   **213**

Table 4.5.Excel Design for computing negative binomial probabilities for the team-selection example (illustrated for $n - X = 2, X = 8, p = .4$).

	A	B
1	Negative Binomial Calculations	
2		
3	Number of Successes	8
4	Number of Trials	10
5	p	0.4
6	Number of Failures	=B4 – B3
7	Probability	=NEGBINOMDIST(B6,B3,B5)

FIGURE 4.7.EXCEL Obtaining negative binomial probabilities for the team-selection example using Microsoft Excel.

	A	B
1	Negative Binomial Calculations	
2		
3	Number of Successes	8
4	Number of Trials	10
5	p	0.4
6	Number of Failures	2
7	Probability	0.008493

4-14-4.XLS

we want to compute any negative binomial probability using Microsoft Excel, we can use the design illustrated in Table 4.5.Excel to develop a worksheet named Negative Binomial.

The results are displayed in Figure 4.7.Excel.

● **The Poisson Distribution** The Poisson distribution is used to determine the probability that a certain number of successes occurs in an area of opportunity, given the expected or average number of successes λ. Using Microsoft Excel, the POISSON function can be used to obtain Poisson probabilities. The format of this function is

$$\text{POISSON}(X, \lambda, cumulative)$$

where $\quad X$ = the number of successes

λ = the expected or average number of successes in an area of opportunity

cumulative = False if you wish to compute the probability of *exactly X* successes or True if you wish to compute the probability of *X or fewer* successes

To illustrate the POISSON function, suppose we return to the example concerning the bank lunch-hour customer arrivals in Section 4.14.3. On page 210, we computed the probability that two customers would arrive at the bank in the next minute if the average arrivals were three per minute as .2240. To obtain this result using Excel, we would enter the formula =POISSON(2,3,False) in a cell. If we wanted the probability of two or fewer arrivals, we would enter =POISSON(2,3,True).

Now that we have introduced the POISSON function and illustrated its use in computing probabilities, we are ready to develop a plan for generating an entire set of probabilities. Table 4.6.Excel contains the design for an Excel worksheet titled Poisson that provides for the computation of the probabilities for all values of X for a given value of λ.

4-14-4.XLS

Begin by inserting a new worksheet and renaming it Poisson. After the appropriate labels have been typed, enter the value for the mean in cell D3.

Table 4.6.Excel Layout for computing Poisson probabilities for the bank lunch-hour customer arrivals data (illustrated for $\lambda = 3$).

	A	B	C	D	E	F
1			Calculating Poisson Probabilities			
2						
3			Mean:	3		
4						
5						
6	X	P(X)	P(<=X)	P(<X)	P(>X)	P(>=X)
7	0	=POISSON(A7,$D$3,False)	=POISSON(A7,$D$3,True)	=C7 – B7	=1 – C7	=1 – D7
8	1	=POISSON(A8,$D$3,False)	=POISSON(A8,$D$3,True)	=C8 – B8	=1 – C8	=1 – D8
9	2	=POISSON(A9,$D$3,False)	=POISSON(A9,$D$3,True)	=C9 – B9	=1 – C9	=1 – D9
10	3	=POISSON(A10,$D$3,False)	=POISSON(A10,$D$3,True)	=C10 – B10	=1 – C10	=1 – D10
⋮	⋮	⋮	⋮	⋮	⋮	⋮
22	15	=POISSON(A22,$D$3,False)	=POISSON(A22,$D$3,True)	=C22 – B22	=1 – C22	=1 – D22

The possible outcomes for the number of successes (X) are then entered in column A beginning in row 7. The maximum value for X is theoretically ∞, but as a general rule, the probability of an X value of more than 15–20 units above its average is less than .0001. Thus, with $\lambda = 3$, we enter X values of 0–15 in cells A7:A22. Once this has been completed, we are able to compute the Poisson probabilities for each value of X. Starting in cell B7, we enter the formula =POISSON(A7,$D$3,False). Since we want to obtain the probability of exactly 0 successes (X = 0), we enter False for the third argument in the formula. We then use the copy command to copy this formula to the subsequent rows containing the remaining X values (cells B8:B22).

If we wanted to obtain the cumulative probability of less than or equal to X successes, as displayed in cell C7, we would substitute True as the third argument in the POISSON formula, enter =POISSON(A7,$D$3,True) in cell C7, and then copy this formula to the subsequent rows containing the remaining X values (in this case, cells C8:C22). In a similar manner, we could also obtain the probability of less than X successes [P(<X)] by entering the formula =C7–B7 in cell D7 and then copying this formula to the subsequent rows of X values (in this case, cells D8:D22). We could obtain the probability of more than X successes [P(>X)] by entering the formula =1–C7 in cell E7 and then copying this formula to the pertinent rows (in this case, cells E8:E22). Finally, if we wanted the probability of at least X successes, (P≥X), we would enter the formula = 1–D7 in cell F7 and then copy this formula to the pertinent rows (in this case, cells F8:F22). Figure 4.8.Excel represents these calculations for the data of Table 4.6.Excel.

Now that the Poisson probabilities have been calculated for each X value, we can use Microsoft Excel's Chart Wizard (see Section 2.12) to obtain a bar chart similar to the one displayed in Panel B of Figure 4.3. The selections in the five dialog boxes for the Chart Wizard would be as follows:

First dialog box: Enter Poisson!A7:B22 as the range.
Second dialog box: Select Column as the chart type.

	A	B	C	D	E	F
1		Calculating Poisson Probabilities				
2						
3			Mean:	3		
4						
5						
6	X	P(X)	P(<=X)	P(<X)	P(>X)	P(>=X)
7	0	0.049787068	0.049787068	0	0.950212932	1
8	1	0.149361205	0.199148273	0.049787068	0.800851727	0.950212932
9	2	0.224041808	0.423190081	0.199148273	0.576809919	0.800851727
10	3	0.224041808	0.647231889	0.423190081	0.352768111	0.576809919
11	4	0.168031356	0.815263245	0.647231889	0.184736755	0.352768111
12	5	0.100818813	0.916082058	0.815263245	0.083917942	0.184736755
13	6	0.050409407	0.966491465	0.916082058	0.033508535	0.083917942
14	7	0.021604031	0.988095496	0.966491465	0.011904504	0.033508535
15	8	0.008101512	0.996197008	0.988095496	0.003802992	0.011904504
16	9	0.002700504	0.998897512	0.996197008	0.001102488	0.003802992
17	10	0.000810151	0.999707663	0.998897512	0.000292337	0.001102488
18	11	0.00022095	0.999928613	0.999707663	7.13866E-05	0.000292337
19	12	5.52376E-05	0.999983851	0.999928613	1.6149E-05	7.13866E-05
20	13	1.27471E-05	0.999996598	0.999983851	3.40191E-06	1.6149E-05
21	14	2.73153E-06	0.99999933	0.999996598	6.70386E-07	3.40191E-06
22	15	5.46306E-07	0.999999876	0.99999933	1.2408E-07	6.70386E-07

FIGURE 4.8.EXCEL Obtaining Poisson probabilities for the bank lunch-hour customer arrivals using Microsoft Excel.

Third dialog box: Select the first format choice that contains bars above and below the X axis.

Fourth dialog box: Select the Columns option button for Data Series, enter 1 in the First Columns edit box and 0 in the First Rows edit box.

Fifth dialog box: Select the No options button for Add a Legend, enter the chart title Poisson Chart, the Category (X) title Number of Successes (X), and the value Y title P(X).

With the gaps between the X values removed (see Section 2.7.3), rename the chart sheet Poisson chart. Figure 4.9.Excel displays the chart of the Poisson distribution for the data of Figure 4.8.Excel.

> ▲ **WHAT IF EXAMPLE**
>
> The layout of the Poisson worksheet presented in Table 4.6.Excel allows us to study the effect of changes in λ on the probability of X successes and the shape of the bar chart of the Poisson probabilities obtained from the Chart Wizard. For example, we could change the value of λ to 5, and determine the probabilities of the various number of successes followed by a bar chart from the Chart Wizard. This could then be compared to the results obtained from $\lambda = 3$.

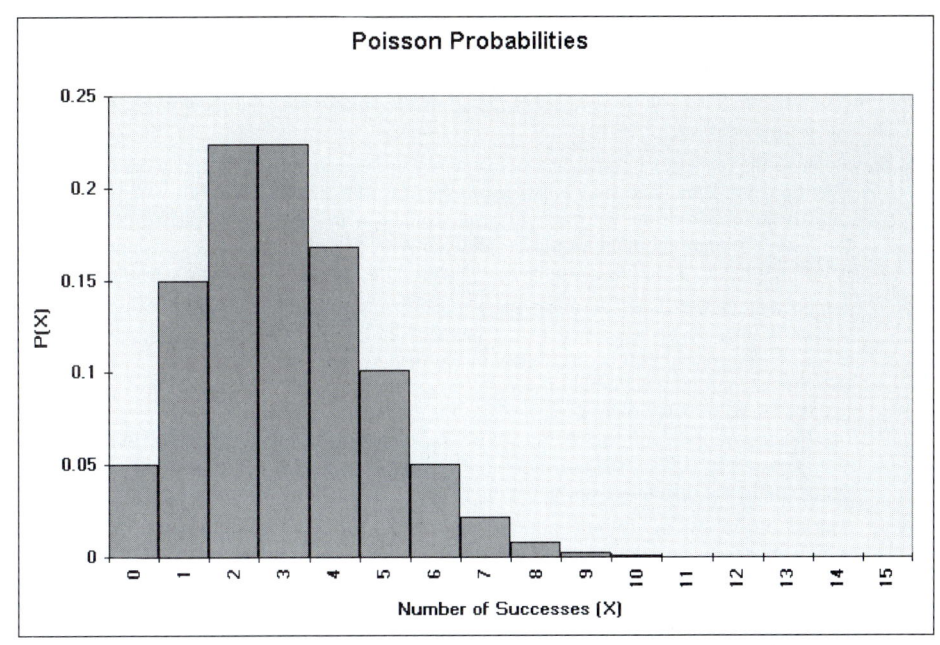

FIGURE 4.9.EXCEL Histogram of Poisson probabilities obtained from Microsoft Excel for the bank lunch-hour customer arrival data.

Problems for Section 4.14

Note: *Microsoft Excel may be used to solve the problems in this section.*

4.52 An auditor for the Internal Revenue Service is selecting a sample of six tax returns from persons in a particular profession for possible audit. If two or more of these indicate "improper" deductions, the entire group (population) of 100 tax returns will be audited.
 (a) What is the probability that the entire group will be audited if the true number of improper returns in the population is
   (1) 25? (3) 5?
   (2) 30? (4) 10?
 (b) Discuss the differences in your results depending on the true number of improper returns in the population.

4.53 The dean of a business school wishes to form an executive committee of 5 from among the 40 tenured faculty members at the school. The selection is to be random, and at the school there are 8 tenured faculty members in accounting.
 (a) What is the probability that the committee will contain
   (1) none of them?
   (2) at least one of them?
   (3) not more than one of them?
 (b) What would be your answers to (a) if the committee consisted of seven members?

4.54 From an inventory of 48 cars being shipped to local automobile dealers, suppose 12 have had defective radios installed.
 (a) What is the probability that one particular dealership receiving eight cars
   (1) obtains all with defective radios?
   (2) obtains none with defective radios?
   (3) obtains at least one with a defective radio?
 (b) What would be your answer to (a) if six cars have had defective radios installed?

4.55 A state lottery is conducted in which six winning numbers are selected from a total of 54 numbers. What is the probability that if six numbers are randomly selected,
 (a) All six numbers will be winning numbers?
 (b) Five numbers will be winning numbers?

(c) Four numbers will be winning numbers?
(d) Three numbers will be winning numbers?
(e) None of the numbers will be winning numbers?
(f) What would be your answers to (a)–(e) if the six winning numbers were selected from a total of 40 numbers?

4.56 Refer to the team-selection example discussed in Section 4.14.2. Suppose the probability that the candidates have such a background is .60. What is the probability that
(a) The selection of the team would be completed with the tenth candidate considered?
(b) The first team member would be selected with the third candidate considered?
(c) The third team member would be selected with the sixth candidate considered?

4.57 Suppose an auditor is evaluating a set of vouchers to investigate a particular type of transaction that has occurred in 10% of the vouchers in the past. What is the probability that
(a) The first transaction is discovered in the fifth voucher selected?
(b) The second transaction is discovered in the tenth voucher selected?
(c) The third transaction is discovered in the twentieth voucher selected?
(d) What would be your answers to (a)–(c) if the particular type of transaction has occurred in 20% of the vouchers in the past?
(e) Explain the difference in the results in (a)–(c) and (d) of this problem.

4.58 Marketing surveys are often conducted with the use of quota sampling procedures in which an interviewer is instructed to conduct surveys with respondents who have certain characteristics. Suppose an interviewer is conducting a survey at a location in a shopping mall and has been instructed to obtain five interviews with females who have brown hair. From past experience, it is estimated that 30% of all female shoppers who pass this location will be females who have brown hair. What is the probability that
(a) The first interview is conducted with the fourth female to pass this location?
(b) The second interview is conducted with the seventh female to pass this location?
(c) The set of five interviews is completed with the fifteenth female to pass this location?
(d) What would be your answers to (a)–(c) if the percentage of females with brown hair was 20%?
(e) Explain the difference in the results in (a)–(c) and (d) of this problem.

4.59 Determine the following:
(a) If $\lambda = 2.5$, then what is $P(X = 2)$?
(b) If $\lambda = 8.0$, then what is $P(X = 8)$?
(c) If $\lambda = 0.5$, then what is $P(X = 1)$?
(d) If $\lambda = 3.7$, then what is $P(X = 0)$?
(e) If $\lambda = 4.4$, then what is $P(X = 7)$?

4.60 The average number of claims per hour made to the Gnecco & Trust Insurance Company for damages or losses incurred in moving is 3.1. What is the probability that in any given hour
(a) Fewer than three claims will be made?
(b) Exactly three claims will be made?
(c) Three or more claims will be made?
(d) More than three claims will be made?

4.61 Based on past records, the average number of two-car accidents in a New York City police precinct is 3.4 per day. What is the probability that there will be
(a) At least six such accidents in this precinct on any given day?
(b) Not more than two such accidents in this precinct on any given day?
(c) Fewer than two such accidents in this precinct on any given day?
(d) At least two but no more than six such accidents in this precinct on any given day?
(e) What would be your answers to (a)–(d) if the average was five such accidents per day?

• 4.62 The quality-control manager of Marilyn's Cookies is inspecting a batch of chocolate-chip cookies that has just been baked. If the production process is in control, the average number of chip parts per cookie is 6.0. What is the probability that in any particular cookie being inspected
(a) Fewer than five chip parts will be found?
(b) Exactly five chip parts will be found?
(c) Five or more chip parts will be found?
(d) Four or five chip parts will be found?
(e) What would be your answer to (a)–(d) if the average number of chip parts per cookie is 5.0?

4.63 Refer to Problem 4.62. How many cookies in a batch of 100 being sampled should the manager expect to discard if company policy requires that all chocolate-chip cookies sold must have at least four chocolate-chip parts?

4.15 ETHICAL ISSUES AND PROBABILITY

Ethical issues may arise when any statements relating to probability are being presented for public consumption, particularly when these statements are part of an advertising campaign for a product or service. Unfortunately, a substantial portion of the population is not very comfortable with any type of numerical concept (see Reference 5) and misinterprets the meaning of the probability. In some instances, the misinterpretation is not intended, but in other cases advertisements may unethically try to mislead potential customers.

One example of a potentially unethical application of probability relates to the sales of tickets to a state lottery in which the customer typically selects a set of numbers (such as six) from a larger list of numbers (such as 54). Although virtually all participants know that they are unlikely to win the lottery, they also have very little idea of how unlikely it is for them to select, for example, all six winning numbers out of the list of 54 numbers [see Problem 4.55(a) on page 217]. In addition, they have little idea how likely it is that they can win a consolation prize by selecting either four winning numbers or five winning numbers [see Problem 4.55(b) and (c)]. Given this background, it seems to us that a recent advertising campaign in which a commercial for a state lottery said "We won't stop until we have made everyone a millionaire" is at the very least deceptive and at the worst unethical. Actually, given the fact that the lottery brings millions of dollars in revenue into the state treasury, the state is never going to stop running it, although in anyone's lifetime no one can be sure of becoming a millionaire by winning the lottery.

A second example of a potentially unethical application of probability relates to an investment newsletter that promises a 20% annual return on investment with a 90% probability. In such a situation, it seems imperative that the investment service needs to (1) explain the basis on which this probability estimate rests, (2) provide the probability statement in another format such as 9 chances in 10, and (3) explain what happens to the investment in the 10% of the cases in which a 20% return is not achieved.

Problems for Section 4.15

4.64 **ACTION** Write an advertisement for the state lottery that describes the probability for winning in an ethical fashion.

4.65 **ACTION** Write an advertisement for the investment newsletter that states the probability for a 20% annual return in an ethical fashion.

4.16 BASIC PROBABILITY: A REVIEW AND A PREVIEW

As shown in the chapter summary chart on page 221, this chapter was about basic probability, probability distributions, and the binomial, hypergeometric, negative binomial, and Poisson distributions. To be sure you understand the material covered in this chapter, you should be able to answer the following conceptual questions:

1. What are the differences among a priori classical probability, empirical classical probability, and subjective probability?

2. What is the difference between a simple event and a joint event?
3. How can the addition rule be used to find the probability of occurrence of event *A or B*?
4. What is the difference between mutually exclusive events and collectively exhaustive events?
5. How does conditional probability relate to the concept of statistical independence?
6. How does the multiplication rule differ for events that are and are not independent?
7. What is the meaning of the expected value of a probability distribution?
8. What are the assumptions of the binomial distribution?
9. Under what circumstances should the hypergeometric distribution be used instead of the binomial distribution?
10. What is the difference between the binomial distribution and the negative binomial distribution?
11. What are the assumptions of the Poisson distribution, and how do they differ from those of binomial distribution?

Getting It All Together

Key Terms

addition rule   178
a priori classical probability   172
binomial distribution   197
certain event   172
collectively exhaustive   177
complement   174
conditional probability   181
contingency table   174
empirical classical probability   172
expected monetary value (EMV)   190
expected value of a discrete random variable   188
general addition rule   179
general multiplication rule   183
hypergeometric distribution   207
joint event   174
joint probability   177
marginal probability   176
model   196
multiplication rule for independent events   184

mutually exclusive   177
negative binomial distribution   208
null event   172
Poisson distribution   210
Poisson process   209
probability   172
probability distribution for a discrete random variable   187
probability distribution function   196
rule of combinations   199
sample space   173
simple event   173
simple probability   176
standard deviation σ of a discrete random variable   190
statistical independence   182
subjective probability   173
table of cross-classifications   174
uniform probability distribution   196
variance of a discrete random variable   189

Chapter Review Problems

Note: *The Chapter Review Problems may be solved using Microsoft Excel.*

4.66 When rolling a die once, what is the probability that
(a) The face of the die is odd?

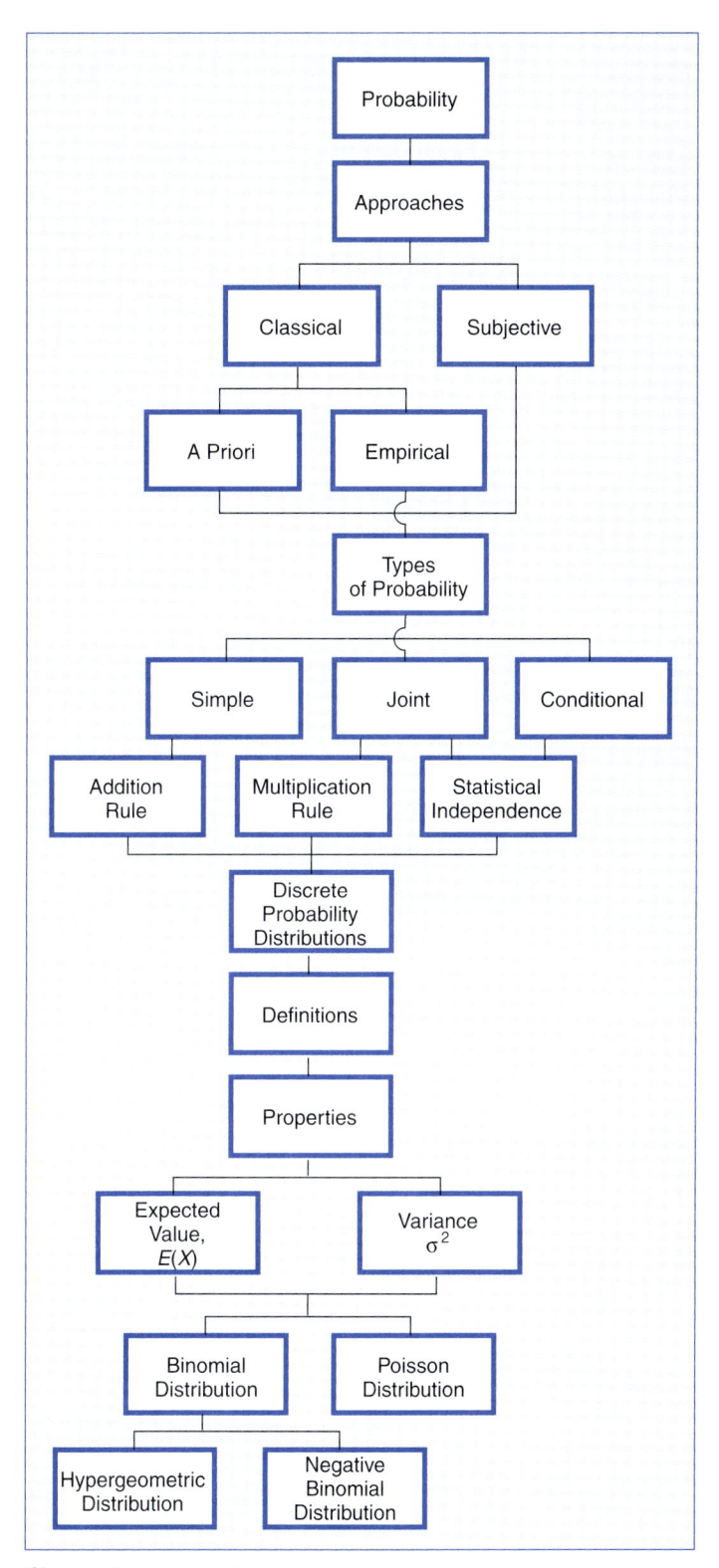

Chapter 4 summary chart.

(b) The face is even *or* odd?
(c) The face is even *or* a 1?
(d) The face is odd *or* a 1?
(e) The face is both even and a 1?
(f) Given the face is odd, it is a 1?

- 4.67 The director of a large employment agency wishes to study various characteristics of its job applicants. A sample of 200 applicants has been selected for analysis. Seventy applicants have had their current jobs for at least 5 years; 80 of the applicants are college graduates; 25 of the college graduates have had their current jobs at least 5 years.
    (a) What is the probability that an applicant chosen at random
        (1) is a college graduate?
        (2) is a college graduate *and* has held the current job less than 5 years?
        (3) is a college graduate *or* has held the current job at least 5 years?
    (b) Given that a particular employee is a college graduate, what is the probability that he or she has held the current job less than 5 years?
    (c) Determine whether being a college graduate *and* holding the current job for at least 5 years are statistically independent.
    (*Hint:* Set up a 2 × 2 table to evaluate the probabilities.)

4.68 A soft drink bottling company maintains records concerning the number of unacceptable bottles of soft drink obtained from the filling and capping machines. Based on past data, the probability that a bottle came from machine I *and* was nonconforming is .01 and the probability that a bottle came from machine II *and* was nonconforming is .025. Half the bottles are filled on machine I and the other half are filled on machine II.
    (a) Give an example of a simple event.
    (b) Give an example of a joint event.
    (c) If a filled bottle of soft drink is selected at random, what is the probability that
        (1) it is a nonconforming bottle?
        (2) it was filled on machine II?
        (3) it is filled on machine I *and* is a conforming bottle?
        (4) it is filled on machine II *and* is a conforming bottle?
        (5) it is filled on machine I *or* is a conforming bottle?
    (d) Suppose we know that the bottle was produced on machine I. What is the probability that it is nonconforming?
    (e) Suppose we know that the bottle is nonconforming. What is the probability that it was produced on machine I?
    (f) Explain the difference in the answers to (d) and (e).
    (*Hint*: Set up a 2 × 2 table to evaluate the probabilities.)

4.69 The producer of a nationally distributed brand of potato chips would like to determine the feasibility of changing the product package from a cellophane bag to an unbreakable container. The product manager believes that there would be three possible national market responses to a change in product package: weak, moderate, and strong. The projected payoffs, in increased or decreased profit as compared to the current product package, are the following:

	Strategy	
Event	Use New Package	Keep Old Package
Weak national response	−$4,000,000	0
Moderate national response	+$1,000,000	0
Strong national response	+$5,000,000	0

Based on past experience, the product manager assigns the following probabilities to the different levels of national response:

$$P(\text{weak national response}) = .30$$

$$P(\text{moderate national response}) = .60$$

$$P(\text{strong national response}) = .10$$

  (a) Using the expected-monetary-value criterion, determine whether the new product package should be adopted.
  (b) What would be your answer in (a) if the probabilities were .6, .3, and .1, respectively?
  (c) What would be your answer in (a) if the probabilities were .1, .3, and .6, respectively?

4.70 Based on past experience, 15% of the bills of a large mail-order book company are incorrect. A random sample of three current bills is selected.
  (a) What is the probability that
    (1) exactly two bills are incorrect?
    (2) no more than two bills are incorrect?
    (3) at least two bills are incorrect?
  (b) What assumptions about the probability distribution are necessary to solve this problem?
  (c) What would be your answers to (a) if the percentage of incorrect bills was 10%?

4.71 The quality-control manager of Ruby's Gambling Equipment Company, which manufactures dice for sale to gambling casinos, must ensure the "fairness" of the dice prior to shipment. Suppose that a particular die is rolled 20 times.
  (a) What is the probability that the die lands on an odd face (i.e., ⚀, ⚂, or ⚄)
    (1) exactly 17 times?
    (2) at least 17 times?
    (3) at most 17 times?
    (4) more than 17 times?
    (5) fewer than 17 times?
  (b) How many times can the die be expected to land on an odd face if the die is truly a fair one?

4.72 Based on past experience, printers in a university computer lab are available 90% of the time. If a random sample of 10 time periods are selected
  (a) What is the probability that printers are available
    (1) exactly nine times?
    (2) at least nine times?
    (3) at most nine times?
    (4) more than nine times?
    (5) fewer than nine times?
  (b) How many times can the printers be expected to be available?
  (c) What would be your answers to (a) and (b) if the printers were available 95% of the time?

• 4.73 Suppose that on a lengthy mathematics test, Lauren gets 70% of the items right.
  (a) For a 10-item quiz, calculate the probability that Lauren will get
    (1) at least seven items right.
    (2) less than six items right (and therefore fail the quiz).
    (3) nine or 10 items right (and get an A on the quiz).
  (b) What is the expected number of items that Lauren will get right? What proportion of the time will she get that number right?
  (c) What is the standard deviation of the number of items that Lauren will get right?
  (d) What would be your answers to (a)–(c) if she got 80% correct?

• 4.74 The manufacturer of the disk drives used in one of the well-known brands of microcomputers expects 2% of the disk drives to malfunction during the microcomputer's warranty period. In a sample of 10 disk drives, what is the probability that

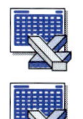

(a) None will malfunction during the warranty period?
(b) Exactly one will malfunction during the warranty period?
(c) At least two will malfunction during the warranty period?
(d) What would be your answers to (a)–(c) if 1% of the disk drives were expected to malfunction?

Endnotes

1. In a contingency table with R rows and C columns, the rule would have to be examined for $(R-1)(C-1)$ separate combinations of A and B.
2. As an example of a binomial random variable arising from sampling with replacement from a finite population, consider the probability of obtaining two clubs in five draws from a randomly shuffled deck of cards, where the selected card is replaced and the deck well shuffled after each draw.
3. Three rolls of the same die is equivalent to one roll of each of three dice. See Problem 4.48 on the game of Chuck-a-luck (page 206).

References

1. Berenson, M. L., and D. M. Levine, *Basic Business Statistics: Concepts and Applications,* 6th. ed. (Englewood Cliffs, NJ: Prentice-Hall, 1996).
2. Mathur, K., and D. Solow, *Management Science: The Art of Decision Making* (Englewood Cliffs, NJ: Prentice-Hall, 1994).
3. Mendenhall, W., and T. Sincich, *Statistics for Engineering and the Sciences*, 4th ed. (Englewood Cliffs, NJ: Prentice-Hall, 1995).
4. *Microsoft Excel Version 7* (Redmond, WA: Microsoft Corp., 1996).
5. Paulos, J. A. *Innumeracy* (New York: Hill and Wang, 1988).
6. Scarne, J., *Scarne's New Complete Guide to Gambling* (New York: Simon and Schuster, 1974).
7. Thorp, E. O., *Beat the Dealer* (New York: Random House, 1962).

chapter 5

The Normal Distribution and Sampling Distributions

CHAPTER OBJECTIVE To show how the normal and other probability distributions can be used to represent certain types of continuous phenomena and to develop the concept of a sampling distribution along with the central limit theorem.

5.1 INTRODUCTION

In Chapter 4, we developed the concept of a probability distribution for a discrete random variable, and we studied the binomial and other discrete distributions. In this chapter we will turn our discussion to the most important probability distribution in statistics, the normal distribution. We will begin by discussing the properties of the normal distribution and then develop various applications. We will then study a simple graphical tool, the normal probability plot, which can be used to evaluate whether a set of data appears to be normally distributed. After a brief discussion of other continuous distributions, we will then describe the properties of sample estimators, develop the concept of a sampling distribution, and discuss the central limit theorem.

5.2 MATHEMATICAL MODELS OF CONTINUOUS RANDOM VARIABLES

Now that we have studied some discrete probability distributions, we turn our attention to **continuous probability density functions**—those that arise due to some measuring process on various phenomena of interest. Continuous models have important applications in engineering and the physical sciences as well as in business and the social sciences. Some examples of continuous random phenomena are height, weight, time between arrivals (of customers at a bank), and customer servicing times.

When a mathematical expression is available to represent some underlying continuous phenomenon, the probability that various values of the random variable occur within certain ranges or intervals may be calculated. However, the *exact* probability of a *particular value* from a continuous distribution is zero.

As an example, the probability distribution represented in Table 5.1 is obtained by categorizing a distribution in which the continuous random phenomenon of interest is said to follow

Table 5.1 Thickness of 10,000 brass washers manufactured by a large company.

Thickness (inches)	Relative Frequency or Probability
Under .0180	48/10,000 = .0048
.0180 < .0182	122/10,000 = .0122
.0182 < .0184	325/10,000 = .0325
.0184 < .0186	695/10,000 = .0695
.0186 < .0188	1198/10,000 = .1198
.0188 < .0190	1664/10,000 = .1664
.0190 < .0192	1896/10,000 = .1896
.0192 < .0194	1664/10,000 = .1664
.0194 < .0196	1198/10,000 = .1198
.0196 < .0198	695/10,000 = .0695
.0198 < .0200	325/10,000 = .0325
.0200 < .0202	122/10,000 = .0122
.0202 or above	48/10,0000 = 0.0048
Total	1.0000

the Gaussian or "bell-shaped" **normal probability density function.** If the nonoverlapping (mutually exclusive) listing contains all possible class intervals (is collectively exhaustive), the probabilities will again sum to 1. This is demonstrated in Table 5.1. Such a probability distribution may be considered as a relative frequency distribution (see Section 2.4), where, except for the two open-ended classes, the midpoint of every other class interval represents the data in that interval.

Unfortunately, obtaining probabilities or computing expected values and standard deviations for continuous phenomena involves mathematical expressions that require a knowledge of integral calculus and are beyond the scope of this book. Nevertheless, one continuous probability density function that we shall focus on has been deemed so important for applications that special probability tables (such as Table E.2 of Appendix E) have been devised to eliminate the need for what otherwise would require laborious mathematical computations. This particular continuous probability density function is known as the Gaussian or **normal distribution.**

5.3 THE NORMAL DISTRIBUTION

5.3.1 Importance of the Normal Distribution

The normal distribution is vitally important in statistics for three main reasons:

1. Numerous continuous phenomena seem to follow it or can be approximated by it.
2. We can use it to approximate various discrete probability distributions.
3. It provides the basis for *classical statistical inference* because of its relationship to the *central limit theorem.*

5.3.2 Properties of the Normal Distribution

The normal distribution has several important theoretical properties. Among these are

1. It is bell-shaped and symmetrical in its appearance.
2. Its measures of central tendency (mean, median, mode, midrange, and midhinge) are all identical.
3. Its "middle spread" is equal to 1.33 standard deviations. This means that the interquartile range is contained within an interval of two-thirds of a standard deviation below the mean to two-thirds of a standard deviation above the mean.
4. Its associated random variable has an infinite range ($-\infty < X < +\infty$).

In practice, some of the variables we observe may only approximate these theoretical properties. This occurs for two reasons: (1) The underlying population distribution may be only approximately normal, and (2) any actual sample may deviate from the theoretically expected characteristics. For some phenomenon that may be approximated by the normal distribution model,

1. Its polygon may be only approximately bell-shaped and symmetrical in appearance.
2. Its measures of central tendency may differ slightly from each other.
3. The value of its interquartile range may differ slightly from 1.33 standard deviations.
4. Its *practical range* will not be infinite but will generally lie within 3 standard deviations above and below the mean (i.e., range ≈ 6 standard deviations).

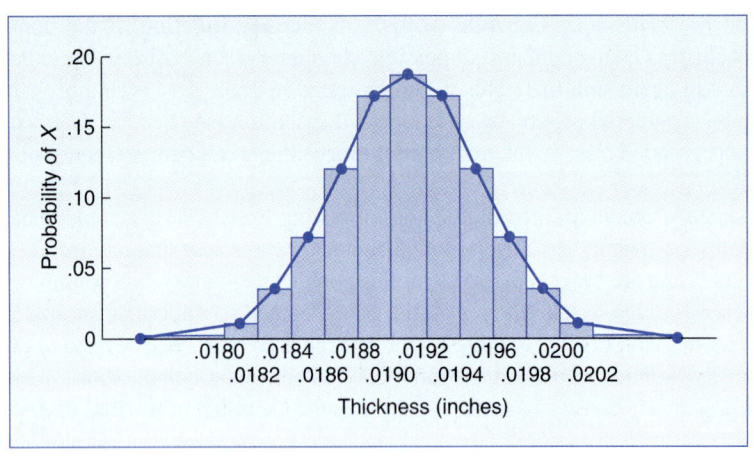

FIGURE 5.1 Relative frequency histogram and polygon of the thickness of 10,000 brass washers. *Source:* Data are taken from Table 5.1.

As a case in point, let us refer to Figure 5.1, which depicts the relative frequency histogram and polygon for the distribution of the thickness of 10,000 brass washers presented in Table 5.1 on page 226. For these data, the first three theoretical properties of the normal distribution seem to have been satisfied; however, the fourth does not hold. The random variable of interest, thickness, cannot possibly take on values of zero or below, nor can a washer be so thick that it becomes unusable. From Table 5.1, we note that only 48 out of every 10,000 brass washers manufactured can be expected to have a thickness of 0.0202 inch or more, while an equal number can be expected to have a thickness under 0.0180 inch. Thus, the chance of randomly obtaining a washer so thin or so thick is .0048 + .0048 = .0096—or almost 1 in 100.

5.3.3 The Mathematical Model

The mathematical model or expression representing a probability density function is denoted by the symbol $f(X)$. For the normal distribution, the model used to obtain the desired probabilities is

$$f(X) = \frac{1}{\sqrt{2\pi}\,\sigma} e^{-(1/2)[(X-\mu)/\sigma]^2} \tag{5.1}$$

where
- e = the mathematical constant approximated by 2.71828
- π = the mathematical constant approximated by 3.14159
- μ = the population mean
- σ = the population standard deviation
- X = any value of the continuous random variable, where $-\infty < X < +\infty$

Let us examine the components of the normal probability density function in Equation (5.1). Since e and π are mathematical constants, the probabilities of the random variable X are dependent only upon the two parameters of the normal distribution—the population mean μ and the population standard deviation σ. Every time we specify a *particular combination* of μ and σ, a *different* normal probability distribution will be generated. We illustrate this in Figure 5.2, where three different normal distributions are depicted. Distributions A and B

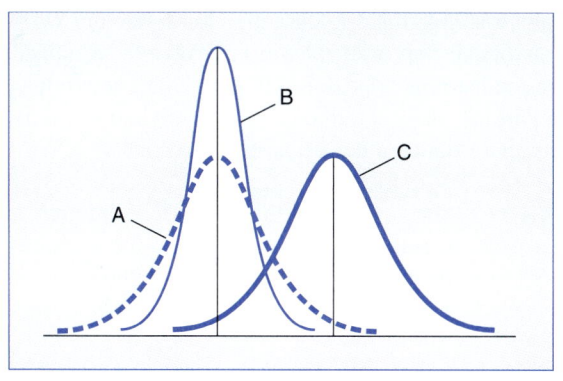

FIGURE 5.2 Three normal distributions having differing parameters μ and σ.

have the same mean (μ) but have different standard deviations. On the other hand, distributions A and C have the same standard deviation (σ) but have different means. Furthermore, distributions B and C depict two normal probability density functions that differ with respect to both μ and σ.

Unfortunately, the mathematical expression in Equation (5.1) is computationally tedious. To avoid having to make such computations, it would be useful to have a set of tables that would provide the desired probabilities. However, since an infinite number of combinations of the parameters μ and σ exists, an infinite number of such tables would be required.

5.3.4 Standardizing the Normal Distribution

Fortunately, by *standardizing* the data, we will only need one table. (See Table E.2.) By using the **transformation formula,** any normal random variable X is converted to a standardized normal random variable Z.

The Z value is equal to the difference between X and the population mean μ, divided by the standard deviation σ.

$$Z = \frac{X - \mu}{\sigma} \tag{5.2}$$

While the original data for the random variable X had mean μ and standard deviation σ, the standardized random variable Z will always have mean $\mu = 0$ and standard deviation $\sigma = 1$.

A **standardized normal distribution** is one whose random variable Z always has a mean $\mu = 0$ and a standard deviation $\sigma = 1$.

Substituting in Equation (5.1), we see that the probability density function of a standard normal variable Z is

$$f(Z) = \frac{1}{\sqrt{2\pi}} e^{-(1/2)Z^2} \tag{5.1a}$$

Thus, we can always convert any set of normally distributed data to its standardized form and then determine any desired probabilities from a table of the standardized normal distribution.

Suppose a consultant is investigating the time it takes factory workers in an automobile plant to assemble a particular part after the workers have been trained to perform the task using an individual-based learning approach. The consultant determines that the time in seconds to assemble the part for workers trained with this method is normally distributed with a mean μ of 75 seconds and a standard deviation σ of 6 seconds.

● **Transforming the Data** We see from Figure 5.3 that every measurement X has a corresponding standardized measurement Z obtained from the transformation formula (5.2). Hence, from Figure 5.3, it is clear that the 81 seconds required for a factory worker to complete the task is equivalent to 1 standardized unit (i.e., 1 *standard deviation*) above the mean, since

$$Z = \frac{81 - 75}{6} = +1$$

and that the 57 seconds required for a worker to assemble the part is equivalent to 3 standardized units (i.e., 3 *standard deviations*) below the mean because

$$Z = \frac{57 - 75}{6} = -3$$

Thus, the standard deviation has become the unit of measurement. In other words, a time of 81 seconds is 6 seconds (i.e., 1 standard deviation) higher, or *slower* than the average time of 75 seconds, and a time of 57 seconds is 18 seconds (i.e., 3 standard deviations) lower, or *faster* than the average time.

Suppose now that the consultant conducts the same study at another automobile plant, where the workers are trained to assemble the part by using a team-based learning method. Suppose that at this plant the consultant determines that the time to perform the task is normally distributed with mean μ of 60 seconds and a standard deviation σ of 3 seconds. The data are depicted in Figure 5.4. In comparison with the results for workers who have individual-based training, we note, for example, that for the workers whose training is team-based, an assembly time of 57 seconds is only 1 standard deviation below the mean for the group, since

$$Z = \frac{57 - 60}{3} = -1$$

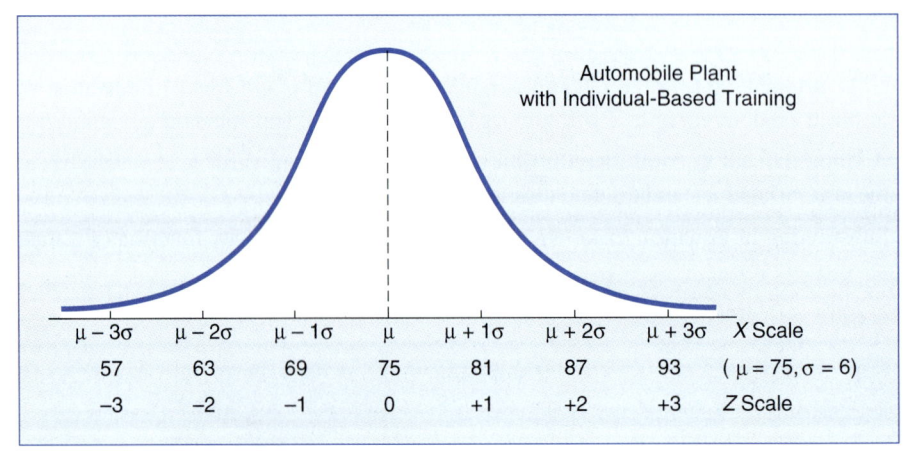

FIGURE 5.3 Transformation of scales.

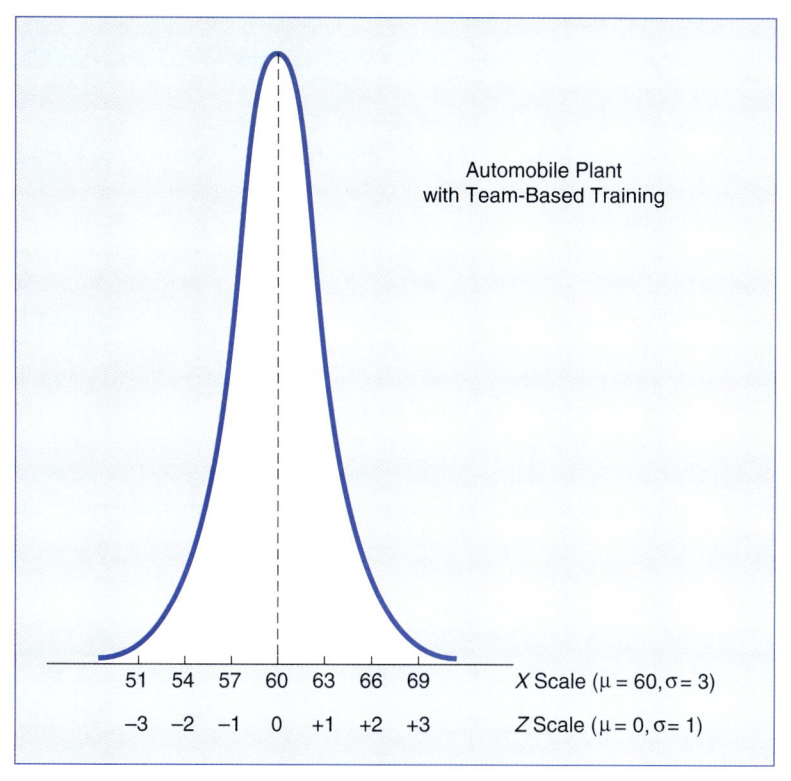

![Figure 5.4: Normal curve for Automobile Plant with Team-Based Training, showing X Scale (μ = 60, σ = 3) with values 51, 54, 57, 60, 63, 66, 69 and Z Scale (μ = 0, σ = 1) with values -3, -2, -1, 0, +1, +2, +3.]

FIGURE 5.4 A different transformation of scales.

We may also note that a time of 63 seconds is 1 standard deviation above the mean time for assemblage, since

$$Z = \frac{63 - 60}{3} = +1$$

and a time of 51 seconds is 3 standard deviations below the group mean because

$$Z = \frac{51 - 60}{3} = -3$$

5.3.5 Using the Normal Probability Tables

The two bell-shaped curves in Figures 5.3 and 5.4 depict the relative frequency polygons for the normal distributions representing the assembly time (in seconds) for all factory workers at the two automobile plants, one for the individual-based training and one for the team-based training. Since at each plant the times to assemble the part are known for every factory worker, the data represent the entire population at a particular plant, and therefore, the *probabilities* or proportion of area under the entire curve must add up to 1. Thus, the area under the curve between any two reported time values represents only a portion of the total area possible.

Now suppose the consultant wishes to determine the probability that a factory worker selected at random from those who underwent individual-based training should require between 75 and 81 seconds to complete the task. That is, what is the likelihood that the worker's time is between the plant mean and 1 standard deviation above this mean? This answer is found by using Table E.2.

Table E.2 represents the probabilities or areas under the normal curve calculated from the mean μ to the particular values of interest X. Using Equation (5.2), this corresponds to the probabilities or areas under the standardized normal curve from the mean ($\mu = 0$) to the transformed values of interest Z. Only positive entries for Z are listed in the table, since for a symmetrical distribution having a mean of 0, the area from the mean to +Z (i.e., Z standard deviations above the mean) must be identical to the area from the mean to –Z (i.e., Z standard deviations below the mean).

To use Table E.2, we note that all Z values must first be recorded to two decimal places. Thus, our particular Z value of interest is recorded as +1.00. To read the probability or area under the curve from the mean to Z = +1.00, we scan down the Z column from Table E.2 until we locate the Z value of interest (in tenths). Hence, we stop in the row Z = 1.0. Next we read across this row until we intersect the column that contains the hundredths place of the Z value. Therefore, in the body of the table the tabulated probability for Z = 1.00 corresponds to the intersection of the row Z = 1.0 with the column Z = .00 as shown in Table 5.2 (which is a replica of Table E.2). This probability is .3413. As depicted in Figure 5.5, there is a 34.13% chance that a factory worker selected at random who has had individual-based training will require between 75 and 81 seconds to assemble the part.

Table 5.2 Obtaining an area under the normal curve.

Z	.00	.01	.02	.03	.04	.05	.06	.07	.08	.09
0.0	.0000	.0040	.0080	.0120	.0160	.0199	.0239	.0279	.0319	.0359
0.1	.0398	.0438	.0478	.0517	.0557	.0596	.0636	.0675	.0714	.0753
0.2	.0793	.0832	.0871	.0910	.0948	.0987	.1026	.1064	.1103	.1141
0.3	.1179	.1217	.1255	.1293	.1331	.1368	.1406	.1443	.1480	.1517
0.4	.1554	.1591	.1628	.1664	.1700	.1736	.1772	.1808	.1844	.1879
0.5	.1915	.1950	.1985	.2019	.2054	.2088	.2123	.2157	.2190	.2224
0.6	.2257	.2291	.2324	.2357	.2389	.2422	.2454	.2486	.2518	.2549
0.7	.2580	.2612	.2642	.2673	.2704	.2734	.2764	.2794	.2823	.2852
0.8	.2881	.2910	.2939	.2967	.2995	.3023	.3051	.3078	.3106	.3133
0.9	.3159	.3186	.3212	.3238	.3264	.3289	.3315	.3340	.3365	.3389
1.0	.3413	.3438	.3461	.3485	.3508	.3531	.3554	.3577	.3599	.3621
1.1	.3643	.3665	.3686	.3708	.3729	.3749	.3770	.3790	.3810	.3830

Source: Extracted from Table E.2.

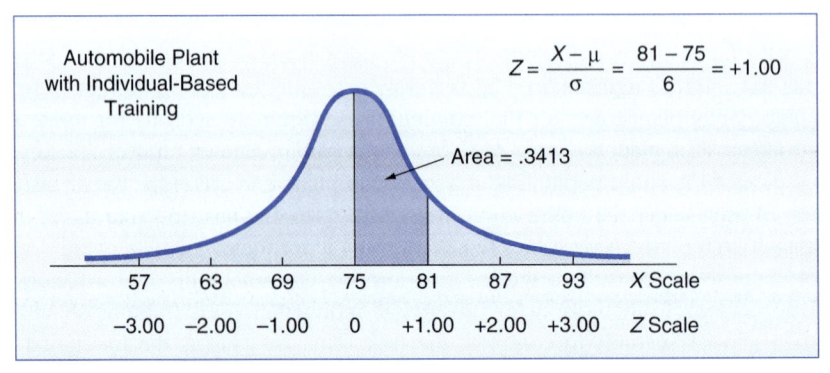

FIGURE 5.5 Determining the area between the mean and Z from a standardized normal distribution.

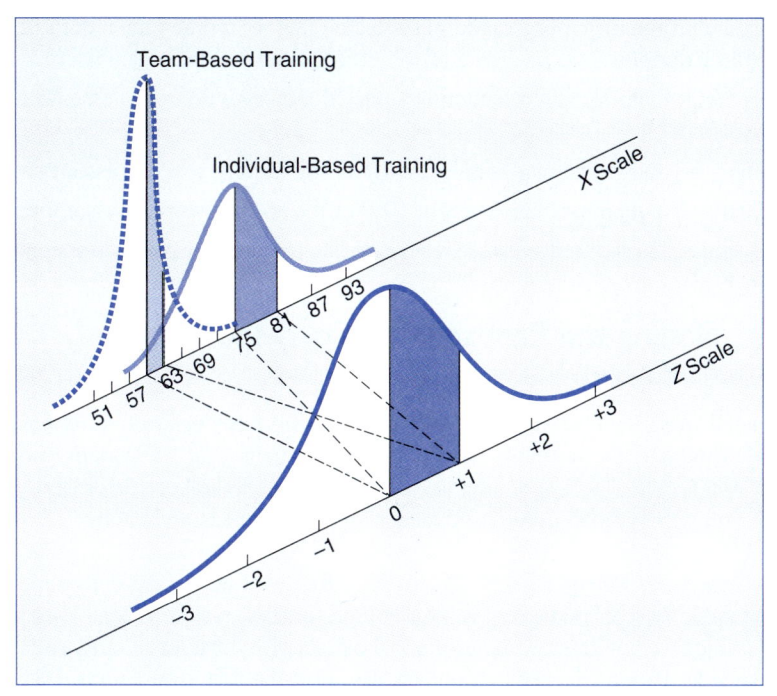

FIGURE 5.6 Demonstrating a transformation of scales for corresponding portions under two normal curves.

On the other hand, we know from Figure 5.4 that at the automobile plant where the workers received team-based training, a time of 63 seconds is 1 standardized unit above the mean time of 60 seconds. Thus, the likelihood that a randomly selected factory worker who receives team-based training will complete the assemblage in between 60 and 63 seconds is also .3413.[1] These results are clearly illustrated in Figure 5.6, which demonstrates that regardless of the value of the mean μ and standard deviation σ of a particular set of normally distributed data, a transformation to a standardized scale can always be made from Equation (5.2), and, by using Table E.2, any probability or portion of area under the curve can be obtained. From Figure 5.6, we see that the probability or area under the curve from 60 to 63 seconds for the workers who had team-based training is identical to the probability or area under the curve from 75 to 81 seconds for the workers who had individual-based training.

5.4 APPLICATIONS

Now that we have learned to use Table E.2 in conjunction with Equation (5.2), many different types of probability questions pertaining to the normal distribution can be resolved. To illustrate this, let us suppose that the consultant raises the following questions with regard to assembling a particular part by workers who have had individual-based training:

1. What is the probability that a randomly selected factory worker can assemble the part in under 75 seconds or in over 81 seconds?
2. What is the probability that a randomly selected factory worker can assemble the part in 69 – 81 seconds?

3. What is the probability that a randomly selected factory worker can assemble the part in under 62 seconds?
4. What is the probability that a randomly selected factory worker can assemble the part in 62–69 seconds?
5. How many seconds must elapse before 50% of the factory workers assemble the part?
6. How many seconds must elapse before 10% of the factory workers assemble the part?
7. What is the interquartile range (in seconds) expected for factory workers to assemble the part?

5.4.1 Finding the Probabilities Corresponding to Known Values

You may recall from Section 5.3.4 that for workers who have the individual-based training the assembly time data are normally distributed with a mean μ of 75 seconds and a standard deviation σ of 6 seconds. In responding to questions 1–4, we shall use this information as we seek to determine the probabilities associated with various measured values.

● **Question 1: Finding $P(X < 75 \text{ or } X > 81)$** How can we determine the probability that a randomly selected factory worker will perform the task in under 75 seconds or over 81 seconds? Since we have already determined the probability that a randomly selected factory worker will need between 75 and 81 seconds to assemble the part, from Figure 5.5 on page 232 we observe that our desired probability must be its *complement,* that is, 1 − .3413 = .6587.

Another way to view this problem, however, is to separately obtain both the probability of assembling the part in under 75 seconds and the probability of assembling the part in over 81 seconds and then use the *addition rule for mutually exclusive events* [Equation (4.4)] to obtain the desired result. This is depicted in Figure 5.7.

Since the mean and median are theoretically the same for normally distributed data, it follows that 50% of the workers can assemble the part in under 75 seconds.[2] To show this, from Equation (5.2) we have

$$Z = \frac{X - \mu}{\sigma} = \frac{75 - 75}{6} = 0.00$$

Using Table E.2, we see that the area under the normal curve from the mean to $Z = 0.00$ is .0000. Hence the area under the curve less than $Z = 0.00$ must be .5000 − .0000 = .5000 (which happens to be the area for the entire left side of the distribution from the mean to $Z = -\infty$, as shown in Figure 5.7).

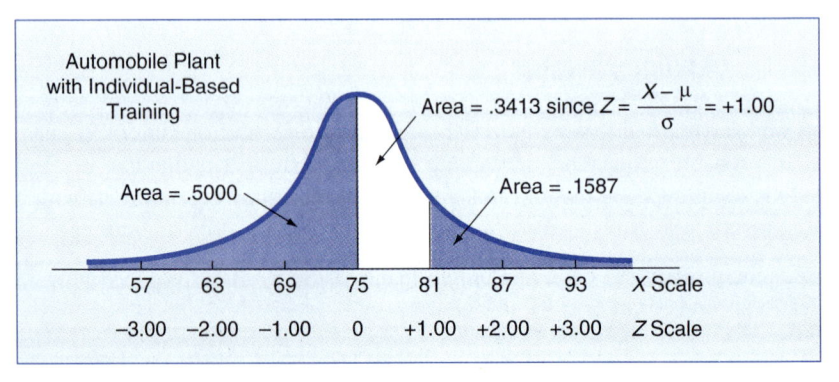

FIGURE 5.7 Finding $P(X < 75 \text{ or } X > 81)$.

Now we wish to obtain the probability of assembling the part in over 81 seconds. But Equation (5.2) only gives the areas under the curve from the mean to Z, not from Z to +∞. Thus, we find the probability from the mean to Z and subtract this result from .5000 to obtain the desired answer. Since we know that the area or portion of the curve from the mean to Z = +1.00 is .3413, the area from Z = +1.00 to Z = +∞ must be .5000 − .3413 = .1587. Hence, the probability that a randomly selected factory worker will perform the task is under 75 or over 81 seconds, $P(X < 75 \text{ or } X > 81)$, is .5000 + .1587 = .6587.

- **Question 2: Finding $P(69 \leq X \leq 81)$**   Suppose we are now interested in determining the probability that a randomly selected factory worker can complete the part in 69–81 seconds, that is, $P(69 \leq X \leq 81)$. We note from Figure 5.8 that one of the values of interest is above the mean assembly time of 75 seconds and the other value is below it. Since our transformation formula (5.2) permits us only to find probabilities from a particular value of interest to the mean, we can obtain our desired probability in three steps:

  1. Determine the probability from the mean to 81 seconds.
  2. Determine the probability from the mean to 69 seconds.
  3. Sum up the two mutually exclusive results.

For this example, we already completed step 1; the area under the normal curve from the mean to 81 seconds is .3413. To find the area from the mean to 69 seconds (step 2), we have

$$Z = \frac{X - \mu}{\sigma} = \frac{69 - 75}{6} = -1.00$$

Table E.2 shows only positive entries for Z. Because of symmetry, it is clear that the area from the mean to Z = −1.00 must be identical to the area from the mean to Z = +1.00. Discarding the negative sign then, we look up (in Table E.2) the value Z = 1.00 and find the probability to be .3413. Hence, from step 3, the probability that the part can be assembled in between 69 and 81 seconds is .3413 + .3413 = .6826. This is displayed in Figure 5.8.

- **Generalizing from the Standard Normal Distribution**   The preceding result is rather important. If we may generalize for a moment, we can see that for any normal distribution there is a .6826 chance that a randomly selected item will fall within ±1 standard deviation above or below the mean. We leave it to the reader to verify from Table E.2 (see Problem 5.4 on page 242) that there is a .9544 chance that any randomly selected normally distributed observation will fall within ±2 standard deviations above or below the mean and a .9973 chance that the observation will fall between ±3 standard deviations above or below the mean.

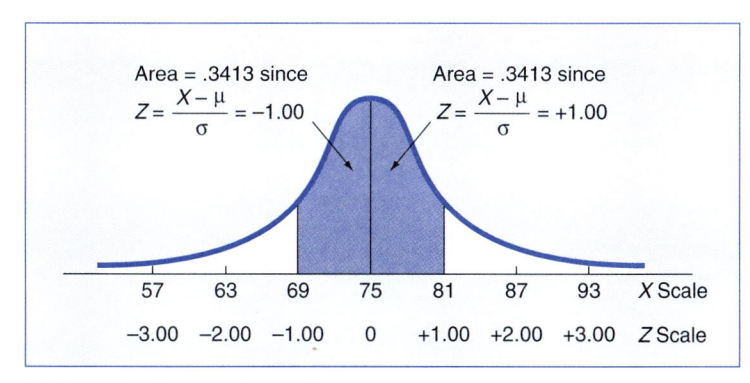

FIGURE 5.8   Finding $P(69 \leq X \leq 81)$.

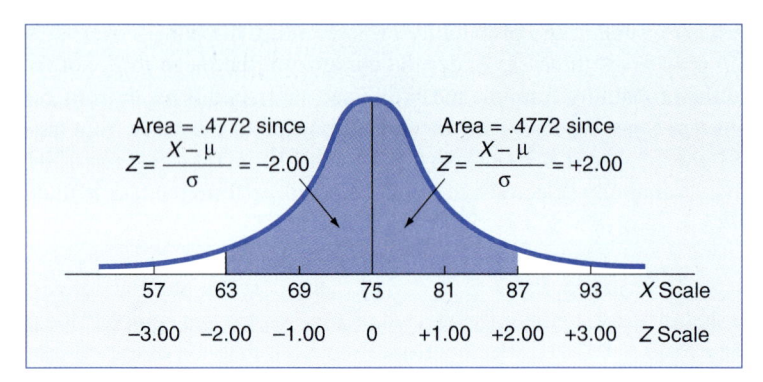

FIGURE 5.9 Finding $P(63 \leq X \leq 87)$.

For the workers who receive individual-based training, this tells us that slightly more than two out of every three of these factory workers (68.26%) can be expected to complete the task within ±1 standard deviation from the mean. Moreover, from Figure 5.9, slightly more than 19 out of every 20 factory workers (95.44%) can be expected to complete the assembly within ±2 standard deviations from the mean (i.e., between 63 and 87 seconds), and, from Figure 5.10, practically all factory workers (99.73%) can be expected to assemble the part within ±3 standard deviations from the mean (i.e., between 57 and 93 seconds).

From Figure 5.10, it is indeed quite unlikely (.0027 or only 27 factory workers in 10,000) that a randomly selected factory worker will be so fast or so slow that he or she could be expected to complete the assembly of the part in under 57 seconds or over 93 seconds. Thus, it is clear why 6 σ (i.e., 3 standard deviations above the mean to 3 standard deviations below the mean) is often used as a *practical approximation of the range* for normally distributed data.

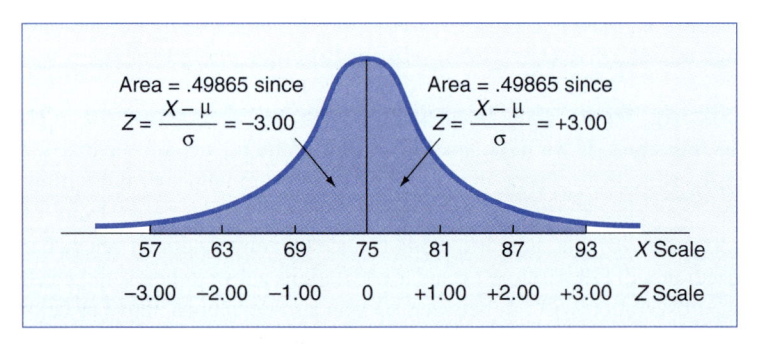

FIGURE 5.10 Finding $P(57 \leq X \leq 93)$.

● **Question 3: Finding $P(X < 62)$** To obtain the probability that a randomly selected factory worker can assemble the part in under 62 seconds, we should examine the shaded lower left-tailed region of Figure 5.11. The transformation formula (5.2) only permits us to find areas under the standardized normal distribution from the mean to Z, not from Z to $-\infty$. Thus, we must find the probability from the mean to Z and subtract this result from .5000 to obtain the desired answer.

To determine the area under the curve from the mean to 62 seconds, we have

$$Z = \frac{X - \mu}{\sigma} = \frac{62 - 75}{6} = \frac{-13}{6} = -2.17$$

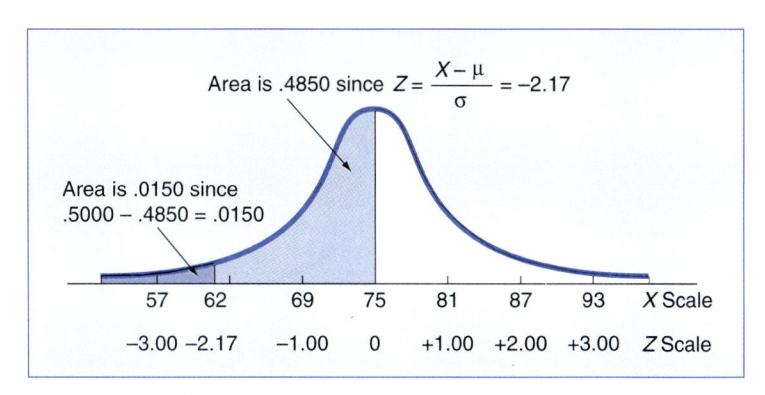

FIGURE 5.11 Finding $P(X < 62)$.

Neglecting the negative sign, we look up the Z value of 2.17 in Table E.2 by matching the appropriate Z row (2.1) with the appropriate Z column (.07) as shown in Table 5.3 (a replica of Table E.2). Therefore, the resulting probability or area under the curve from the mean to 2.17 standard deviations below it is .4850. Hence, the area from $Z = -2.17$ to $Z = -\infty$ must be $.5000 - .4850 = .0150$. This is indicated in Figure 5.11.

Table 5.3 Obtaining an area under the normal curve.

Z	.00	.01	.02	.03	.04	.05	.06	.07	.08	.09
0.0	.0000	.0040	.0080	.0120	.0160	.0199	.0239	.0279	.0319	.0359
0.1	.0398	.0438	.0478	.0517	.0557	.0596	.0636	.0675	.0714	.0753
0.2	.0793	.0832	.0871	.0910	.0948	.0987	.1026	.1064	.1103	.1141
0.3	.1179	.1217	.1255	.1293	.1331	.1368	.1406	.1443	.1480	.1517
0.4	.1554	.1591	.1628	.1664	.1700	.1736	.1772	.1808	.1844	.1879
⋮	⋮	⋮	⋮	⋮	⋮	⋮	⋮	⋮	⋮	⋮
2.0	.4772	.4778	.4783	.4788	.4793	.4798	.4803	.4808	.4812	.4817
2.1	.4821	.4826	.4830	.4834	.4838	.4842	.4846	.4850	.4854	.4857
2.2	.4861	.4864	.4868	.4871	.4875	.4878	.4881	.4884	.4887	.4890
2.3	.4893	.4896	.4898	.4901	.4904	.4906	.4909	.4911	.4913	.4916
2.4	.4918	.4920	.4922	.4925	.4927	.4929	.4931	.4932	.4934	.4936

Source: Extracted from Table E.2.

● **Question 4: Finding $P(62 \leq X \leq 69)$** As a final illustration of determining probabilities from the standardized normal distribution, suppose we wish to find how likely it is that a randomly selected factory worker can complete the task in 62–69 seconds. Since both values of interest are below the mean, we see from Figure 5.12 that the desired probability (or area under the curve between the two values) is less than .5000. Since our transformation formula (5.2) only permits us to find probabilities from a particular value of interest to the mean, we can obtain our desired probability in three steps:

1. Determine the probability or area under the curve from the mean to 62 seconds.
2. Determine the probability or area under the curve from the mean to 69 seconds.
3. Subtract the smaller area from the larger (to avoid double counting).

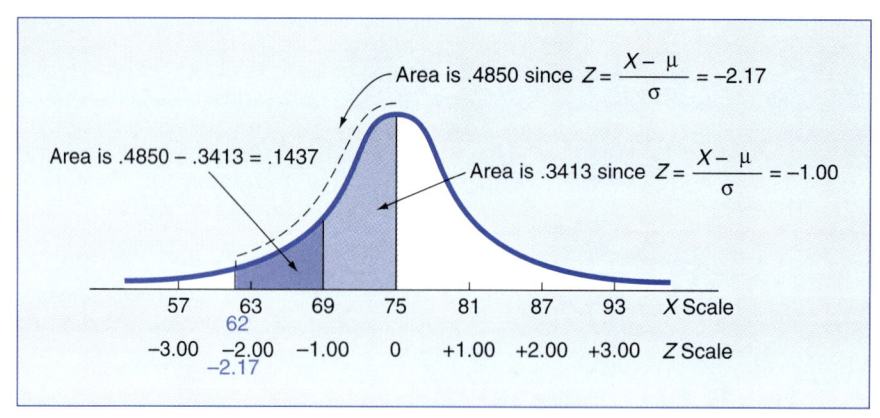

FIGURE 5.12 Finding $P(62 \leq X \leq 69)$.

For this example, we have already completed steps 1 and 2 in answering questions 3 and 2, respectively. The area from the mean to 62 seconds is .4850, and the area from the mean to 69 seconds is .3413. Hence, from step 3, by subtracting the smaller area from the larger one, we determine that there is only a .1437 probability of randomly selecting a factory worker who could be expected to complete the task in between 62 and 69 seconds. That is,

$$P(62 \leq X \leq 69) = P(62 \leq X \leq 75) - P(69 \leq X \leq 75)$$
$$= .4850 - .3413 = .1437$$

5.4.2 Finding the Values Corresponding to Known Probabilities

In our previous applications regarding normally distributed data, we have sought to determine the probabilities associated with various measured values. Now, however, suppose we wish to determine particular numerical values of the variables of interest that correspond to known probabilities. To illustrate, we will respond to questions 5–7.

● **Question 5** To determine how many seconds elapse before 50% of the factory workers assemble the part, we should examine Figure 5.13. Since this time value corresponds to the median, and the mean and median are equal in all symmetrical distributions, the median must be 75 seconds.

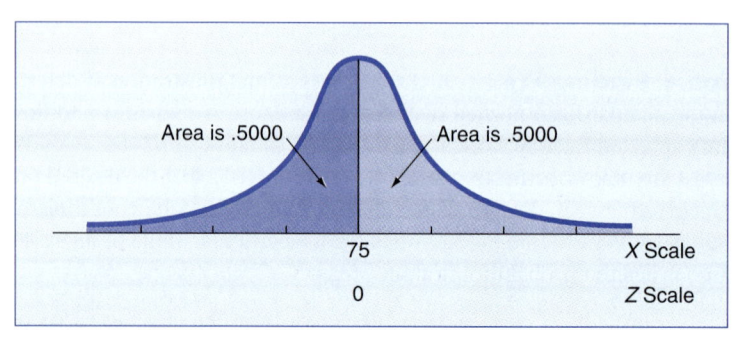

FIGURE 5.13 Finding X.

238 **Chapter 5** The Normal Distribution and Sampling Distributions

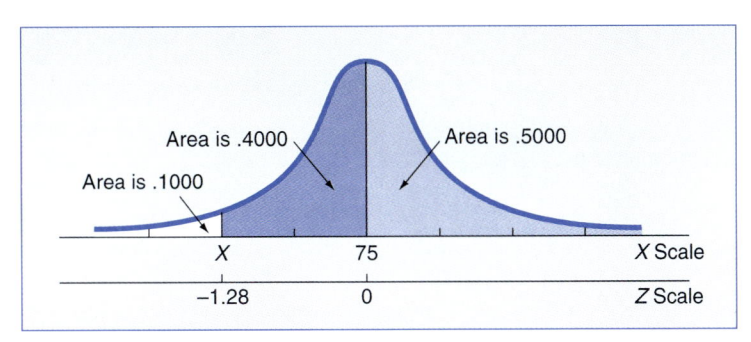

FIGURE 5.14 Finding Z to determine X.

- **Question 6** To determine how many seconds elapse before 10% of the factory workers assemble the part, we should focus on Figure 5.14. Since 10% of the factory workers are expected to complete the task in under X seconds, then 90% of the workers would be expected to require X seconds or more to do the job. From Figure 5.14, we may observe that this 90% can be broken down into two parts—times (in seconds) above the mean (i.e., 50% of the workers) and times between the mean and the desired value X (i.e., 40% of the workers). While we do not know X, we can determine the corresponding standardized value Z, since the area under the normal curve from the standardized mean 0 to this Z must be .4000. Using the body of Table E.2, we search for the area or probability .4000. The closest result is .3997, as shown in Table 5.4 (a replica of Table E.2).

Working from this area to the margins of the table, we see that the Z value corresponding to the particular Z row (1.2) and Z column (.08) is 1.28. However, from Figure 5.14, the Z value must be recorded as a negative (i.e., $Z = -1.28$), since it is below the standardized mean of 0.

Once Z is obtained, we can now use the transformation formula (5.2) to determine the value of interest, X. Since

$$Z = \frac{X - \mu}{\sigma}$$

Table 5.4 Obtaining a Z value corresponding to a particular area under the normal curve.

Z	.00	.01	.02	.03	.04	.05	.06	.07	.08	.09
0.0	.0000	.0040	.0080	.0120	.0160	.0199	.0239	.0279	.0319	.0359
0.1	.0398	.0438	.0478	.0517	.0557	.0596	.0636	.0675	.0714	.0753
0.2	.0793	.0832	.0871	.0910	.0948	.0987	.1026	.1064	.1103	.1141
0.3	.1179	.1217	.1255	.1293	.1331	.1368	.1406	.1443	.1480	.1517
0.4	.1554	.1591	.1628	.1664	.1700	.1736	.1772	.1808	.1844	.1879
:	:	:	:	:	:	:	:	:	:	:
1.0	.3413	.3438	.3461	.3485	.3508	.3531	.3554	.3577	.3599	.3621
1.1	.3643	.3665	.3686	.3708	.3729	.3749	.3770	.3790	.3810	.3830
1.2	.3849	.3869	.3888	.3907	.3925	.3944	.3962	.3980	.3997	.4015
1.3	.4032	.4049	.4066	.4082	.4099	.4115	.4131	.4147	.4162	.4177
1.4	.4192	.4207	.4222	.4236	.4251	.4265	.4279	.4292	.4306	.4319

Source: Extracted from Table E.2.

then the following is true.

> The X value is equal to the population mean μ plus the product of the Z value and the standard deviation σ.
>
> $$X = \mu + Z\sigma \qquad (5.3)$$

Substituting, we compute

$$X = 75 + (-1.28)(6) = 67.32 \text{ seconds}$$

Thus, we could expect that 10% of the workers will be able to complete the task in less than 67.32 seconds.

As a review, to find a *particular* value associated with a known probability, we must take the following steps:

1. Sketch the normal curve and then place the values for the means on the respective X and Z scales.
2. Split the appropriate half of the normal curve into two parts—the portion from the desired X to the mean and the portion from the desired X to the tail.
3. Shade the area of interest.
4. Using Table E.2, determine the appropriate Z value corresponding to the area under the normal curve from the desired X to the mean μ.
5. Using Equation (5.3), solve for X; that is,

$$X = \mu + Z\sigma$$

● **Question 7** To obtain the interquartile range, we must first find the value for Q_1 and the value for Q_3; then we must subtract the former from the latter.

To find the first quartile value, we must determine the time (in seconds) for which only 25% of the factory workers can be expected to assemble the part in less time. This is depicted in Figure 5.15.

Although we do not know Q_1, we can obtain the corresponding standardized value Z, since the area under the normal curve from the standardized mean 0 to this Z must be .2500. Using the body of Table E.2, we search for the area or probability .2500. The closest result is .2486, as shown in Table 5.5 (which is a replica of Table E.2).

Working from this area to the margins of the table, we see that the Z value corresponding to the particular Z row (0.6) and Z column (.07) is 0.67. However, from Figure 5.15, the

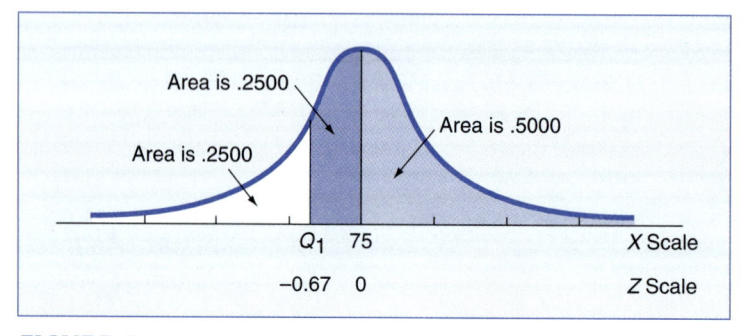

FIGURE 5.15 Finding Q_1.

Table 5.5 Obtaining a Z value corresponding to a particular area under the normal curve.

Z	.00	.01	.02	.03	.04	.05	.06	.07	.08	.09
0.0	.0000	.0040	.0080	.0120	.0160	.0199	.0239	.0279	.0319	.0359
0.1	.0398	.0438	.0478	.0517	.0557	.0596	.0636	.0675	.0714	.0753
0.2	.0793	.0832	.0871	.0910	.0948	.0987	.1026	.1064	.1103	.1141
0.3	.1179	.1217	.1255	.1293	.1331	.1368	.1406	.1443	.1480	.1517
0.4	.1554	.1591	.1628	.1664	.1700	.1736	.1772	.1808	.1844	.1879
0.5	.1915	.1950	.1985	.2019	.2054	.2088	.2123	.2157	.2190	.2224
0.6	.2257	.2291	.2324	.2357	.2389	.2422	.2454	.2486	.2518	.2549
0.7	.2580	.2612	.2642	.2673	.2704	.2734	.2764	.2794	.2823	.2852
0.8	.2881	.2910	.2939	.2967	.2995	.3023	.3051	.3078	.3106	.3133
0.9	.3159	.3186	.3212	.3238	.3264	.3289	.3315	.3340	.3365	.3389
1.0	.3413	.3438	.3461	.3485	.3508	.3531	.3554	.3577	.3599	.3621
1.1	.3643	.3665	.3686	.3708	.3729	.3749	.3770	.3790	.3810	.3830

Source: Extracted from Table E.2.

Z value must be recorded as a negative (i.e., Z = –0.67), since it lies to the left of the standardized mean of 0.

Once Z is obtained, the final step is to use Equation (5.3). Hence,

$$Q_1 = X = \mu + Z\sigma$$
$$= 75 + (-0.67)(6)$$
$$= 75 - 4$$
$$= 71 \text{ seconds}$$

To find the third quartile, we must determine the time (in seconds) for which 75% of the factory workers can be expected to assemble the part in less time (and 25% could complete the task in more time). This is displayed in Figure 5.16.

From the symmetry of the normal distribution, our desired Z value must be +0.67 (since Z lies to the right of the standardized mean of 0). Therefore, using Equation (5.3), we compute

$$Q_3 = X = \mu + Z\sigma$$
$$= 75 + (+0.67)(6)$$
$$= 75 + 4$$
$$= 79 \text{ seconds}$$

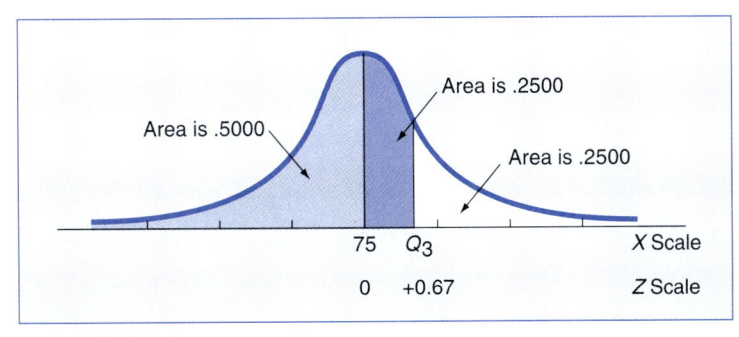

FIGURE 5.16 Finding Q_3.

The interquartile range or middle spread of the distribution is

$$\text{Interquartile range} = Q_3 - Q_1$$
$$= 79 - 71$$
$$= 8 \text{ seconds}$$

Problems for Section 5.4

Note: *Microsoft Excel can be used to solve the problems in this section (see Section 5.5).*

5.1 Given a standardized normal distribution with a mean of 0 and a standard deviation of 1 (Table E.2)
  (a) What is the probability that
    (1) Z is less than 1.57?
    (2) Z exceeds 1.84?
    (3) Z is between 1.57 and 1.84?
    (4) Z is less than 1.57 or greater than 1.84?
    (5) Z is between -1.57 and 1.84?
    (6) Z is less than -1.57 or greater than 1.84?
  (b) What is the value of Z if 50.0% of all possible Z values are larger?
  (c) What is the value of Z if only 2.5% of all possible Z values are larger?
  (d) Between what two values of Z (symmetrically distributed around the mean) will 68.26% of all possible Z values be contained?

5.2 Given a standardized normal distribution (with a mean of 0 and a standard deviation of 1), determine the following probabilities:
  (a) $P(Z > +1.34)$
  (b) $P(Z \leq +1.17)$
  (c) $P(0 \leq Z + 1.17)$
  (d) $P(Z < -1.17)$
  (e) $P(-1.17 \leq Z \leq +1.34)$
  (f) $P(-1.17 \leq Z \leq -0.50)$

5.3 Given a standardized normal distribution with a mean of 0 and a standard deviation of 1
  (a) What is the probability that
    (1) Z is between the mean and $+1.08$?
    (2) Z is less than the mean or greater than $+1.08$?
    (3) Z is between -0.21 and the mean?
    (4) Z is less than -0.21 or greater than the mean?
    (5) Z is at most $+1.08$?
    (6) Z is at least -0.21?
    (7) Z is between -0.21 and $+1.08$?
    (8) Z is less than -0.21 or greater than $+1.08$?
  (b) Determine the following probabilities:
    (1) $P(Z > +1.08)$
    (2) $P(Z < -0.21)$
    (3) $P(-1.96 \leq Z \leq -0.21)$
    (4) $P(-1.96 \leq Z \leq +1.08)$
    (5) $P(+1.08 \leq Z \leq +1.96)$
  (c) What is the value of Z if 50.0% of all possible Z values are smaller?
  (d) What is the value of Z if only 15.87% of all possible Z values are smaller?
  (e) What is the value of Z if only 15.87% of all possible Z values are larger?

5.4 Verify the following:
  (a) The area under the normal curve between the mean and 2 standard deviations above and below it is .9544.
  (b) The area under the normal curve between the mean and 3 standard deviations above and below it is .9973.

5.5 Monthly food expenditures for families of four in a large city average $420 with a standard deviation of $80. Assuming that the monthly food expenditures are normally distributed:

(a) What percentage of these expenditures is less than $350?
(b) What percentage of these expenditures is between $250 and $350?
(c) What percentage of these expenditures is between $250 and $450?
(d) What percentage of these expenditures is less than $250 or greater than $450?
(e) Determine Q_1 and Q_3 from the normal curve.
(f) What will your answers be to (a)–(e) if the standard deviation is $100?

5.6 Toby's Trucking Company determined that on an annual basis, the distance traveled per truck is normally distributed with a mean of 50.0 thousand miles and a standard deviation of 12.0 thousand miles.
(a) What proportion of trucks can be expected to travel between 34.0 and 50.0 thousand miles in the year?
(b) What is the probability that a randomly selected truck travels between 34.0 and 38.0 thousand miles in the year?
(c) What percentage of trucks can be expected to travel either below 30.0 or above 60.0 thousand miles in the year?
(d) How many of the 1,000 trucks in the fleet are expected to travel between 30.0 and 60.0 thousand miles in the year?
(e) How many miles will be traveled by at least 80% of the trucks?
(f) What will your answers be to (a)–(e) if the standard deviation is 10.0 thousand miles?

5.7 Plastic bags used for packaging produce are manufactured so that the breaking strength of the bag is normally distributed with a mean of 5 pounds per square inch and a standard deviation of 1.5 pounds per square inch.
(a) What proportion of the bags produced have a breaking strength
  (1) between 5 and 5.5 pounds per square inch?
  (2) between 3.2 and 4.2 pounds per square inch?
  (3) at least 3.6 pounds per square inch?
  (4) less than 3.17 pounds per square inch?
(b) Between what two values symmetrically distributed around the mean will 95% of the breaking strengths fall?
(c) What will your answers be to (a) and (b) if the standard deviation is 1.0 pound per square inch?

5.8 A set of final examination grades in an introductory statistics course was found to be normally distributed with a mean of 73 and a standard deviation of 8.
(a) What is the probability of getting at most a grade of 91 on this exam?
(b) What percentage of students scored between 65 and 89?
(c) What percentage of students scored between 81 and 89?
(d) What is the final exam grade if only 5% of the students taking the test scored higher?
(e) If the professor "curves" (gives A's to the top 10% of the class regardless of the score), are you better off with a grade of 81 on this exam or a grade of 68 on a different exam where the mean is 62 and the standard deviation is 3? Show statistically and explain.

5.9 A statistical analysis of 1,000 long-distance telephone calls made from the headquarters of Johnson & Shurgot Corporation indicates that the length of these calls is normally distributed with $\mu = 240$ seconds and $\sigma = 40$ seconds.
(a) What percentage of these calls lasted less than 180 seconds?
(b) What is the probability that a particular call lasted between 180 and 300 seconds?
(c) How many calls lasted less than 180 seconds or more than 300 seconds?
(d) What percentage of the calls lasted between 110 and 180 seconds?
(e) What is the length of a particular call if only 1% of all calls are shorter?

5.10 A building contractor claims he can renovate a 200-square-foot kitchen and dining room in 40 work hours, plus or minus 5 hours (i.e., the mean and standard deviation, respectively). The work includes plumbing, electrical installation, cabinets, flooring, painting, and the installation of new appliances. Assuming, from past experience, that times to complete similar projects are normally distributed with mean and standard deviation as estimated above
(a) What is the likelihood the project will be completed in less than 35 hours?
(b) What is the likelihood the project will be completed between 28 hours and 32 hours?
(c) What is the likelihood the project will be completed between 35 hours and 48 hours?
(d) 10% of such projects require more than how many hours?
(e) Determine the midhinge for completion time.

(f) Determine the interquartile range for completion time.
(g) What will your answers be to (a)–(f) if the standard deviation is 10 hours?

5.11 Wages for workers in a particular industry average $11.90 per hour and the standard deviation is $0.40. If the wages are assumed to be normally distributed:
(a) What percentage of workers receive wages between $10.90 and $11.90?
(b) What percentage of workers receive wages between $10.80 and $12.40?
(c) What percentage of workers receive wages between $12.20 and $13.10?
(d) What percentage of workers receive wages less than $11.00?
(e) What percentage of workers receive wages more than $12.95?
(f) What percentage of workers receive wages less than $11.00 or more than $12.95?
(g) What must the wage be if only 10% of all workers in this industry earn more?
(h) What must the wage be if 25% of all workers in this industry earn less?
(i) Determine the midhinge and interquartile range of the wages in this industry.

5.5 USING MICROSOFT EXCEL TO OBTAIN NORMAL PROBABILITIES

In Sections 5.1–5.4, we studied the bell-shaped normal distribution and examined numerous applications in which we computed the probability or area under the normal curve. Rather than using Equations (5.2) and (5.3) and Table E.2 to compute these probabilities, we can use several Microsoft Excel functions. There are five functions that are useful for obtaining probabilities under the normal curve.

The first function related to the normal distribution is the STANDARDIZE function, which computes the Z value for given values of X, μ, and σ. The format of this function is

$$=\text{STANDARDIZE}(X, \mu, \sigma)$$

Referring to the example in Section 5.4 in which $\mu = 75$ and $\sigma = 6$, if we wanted to find the Z value corresponding to $X = 81$, we would enter =STANDARDIZE(81,75,6) in a cell, and the value computed would be 1.0.

The second function related to the normal distribution is the NORMSDIST function, which computes the area or probability less than a given Z value. The format of this function is

$$=\text{NORMSDIST}(Z)$$

Referring to the example in Section 5.4 in which $\mu = 75$ and $\sigma = 6$, if we wanted to find the area below a Z value of 1.0, we would enter =NORMSDIST(1.0) in a cell, and the value computed would be .8413.

The third function related to the normal distribution is the NORMSINV function, which computes the Z value corresponding to a given cumulative area under the normal curve. The format of this function is

$$=\text{NORMSINV}(Probability < X)$$

where   Probability < X = the area under the curve less than X

If we wanted to find the Z value corresponding to a cumulative area of .025, we would enter =NORMSINV(.025), in a cell and the value computed would be –1.96.

The fourth function related to the normal distribution is the NORMDIST function, which computes the area or probability less than a given X value. The format of this function is

$$=\text{NORMDIST}(X, \mu, \sigma, \text{True})$$

Refer to the example in Section 5.4 in which $\mu = 75$ and $\sigma = 6$. If we wanted to find the area below an X value of 81, we would enter =NORMDIST(81,75,6,True) in a cell, and the value computed would be .8413.

The fifth function is the NORMINV function, which computes the X value corresponding to a given cumulative area under the normal curve. The format of this function is

$$=\text{NORMINV}(Probability < X, \mu, \sigma)$$

where $\quad Probability < X =$ the area under the curve less than X

Referring to the example in Section 5.4 in which $\mu = 75$ and $\sigma = 6$, if we wanted to find the X value corresponding to a cumulative area of .025, we would enter =NORMINV(.025,75,6) in a cell, and the value computed would be 63.24.

Now that we have introduced these functions and illustrated their use in computing probabilities under the normal curve and finding unknown Z and X values, we are ready to develop a plan for generating the answers to the questions discussed in Section 5.4. Table 5.1.Excel contains the design for a Calculations sheet that computes the probabilities under the normal curve and finds unknown Z and X values. Cells A8, A10, A13, and A14 contain formulas that use the concatenation operator (&) to form labels that dynamically change as the X values in cells B6 and B12 change.

5-5.XLS

To implement this design, open a new workbook, rename the active sheet as Calculations, and enter the labels and formulas found in column A. Using the example discussed in Section 5.4, since $\mu = 75$ and $\sigma = 6$, enter 75 in cell B3 and 6 in cell B4. In contrast to Table E.2, which provides the area between X and μ, the NORMDIST and NORMSDIST functions provide the cumulative area under the normal curve that is *less than* some value of X or Z. Working from this approach, we can obtain either the cumulative values in the left tail or the complement, the cumulative values in the right tail. To obtain an area between two values, we would compute the area less than the larger value and then subtract the area less than the smaller value.

Suppose we first want to find the probability that a worker assembles the part in less than 69 seconds. We enter 69 in cell B6, and then in cell B7, we use the STANDARDIZE function

Table 5.1.Excel Design for computing normal probabilities illustrated for the individually trained factory worker (with $\mu = 75$ and $\sigma = 6$).

	A	B
1	Calculating Normal Probabilities	
2		
3	Arithmetic Mean	75
4	Standard Deviation	6
5	Left Tail Probability	
6	First X Value	69
7	Z value	=STANDARDIZE(B6,B3,B4)
8	="P(X <= "&B6&")"	=NORMDIST(B6,B3,B4,True)
9	Right Tail Probability	
10	="P(X >= "&B6&")"	=1–B8
11	Interval Probability	
12	Second X Value	81
13	="P(X <= "&B12&")"	=NORMDIST(B12,B3,B4,True)
14	="P("&B6&" < X < "&B12&")"	=ABS(B13–B8)
15	Finding an X Value	
16	Cumulative Percent	.10
17	Z Value	=NORMSINV(B16)
18	X Value	=NORMINV(B16,B3,B4)

Note: The formula in cell B14 uses the ABS function to ensure that the difference is expressed as a nonnegative value.

with $X = 69$, $\mu = 75$, $\sigma = 6$, and enter =STANDARDIZE(B6,B3,B4). The resulting value is −1.0, the Z value representing the difference between 69 seconds and the arithmetic mean of 75 seconds, in standard deviation units. Then, to find the probability of a worker assembling a part in less than 69 seconds, we use the NORMDIST function with $X = 69$, $\mu = 75$, $\sigma = 6$, and enter =NORMDIST(B6,B3,B4,True) in cell B8. The resulting value is .1587, the area below 69 seconds. If we wanted to obtain the probability of the worker assembling the part in 69 or more seconds (the complement of less than 69 seconds), we would enter =1−B8 in cell B10. This result would be .8413.

Now that we have found the area below 69 seconds, we are ready to find the area between 69 and 81 seconds. We need to find the area below 81 seconds and then subtract the area below 69 seconds to obtain the desired result. We first enter 81 in cell B12. Then, using the NORMDIST function with $X = 81$, $\mu = 75$, $\sigma = 6$, we enter =NORMDIST(B12,B3,B4,True) in cell B13. The result obtained in cell B13 is .8413. If we subtract the value of .1587 in cell B8 from the value of .8413 in cell B13 by entering =ABS(B13−B8) in cell B14, the resulting value of .6826 will represent the area between 69 and 81 seconds (see page 235).

Having used Microsoft Excel to compute several different types of probabilities, we can now use the NORMSINV and NORMINV functions to obtain the Z and X values corresponding to a given lower tail area or quantile value. Suppose (as in Question 6 on page 239) we wanted to determine the amount of time it would take the fastest 10% of the workers to assemble a part. To find this, we need to enter this lower tail area of .10 in cell B16. We can find the corresponding Z value (see page 239) by entering =NORMSINV(B16) in cell B17. The resulting value will be −1.282. Then, in cell B18, we can determine the number of seconds (the X value) that corresponds to this lower tail area of .10 by entering =NORMINV(B16,B3,B4). The resulting value will be 67.31 seconds. (Note these values differ slightly from the ones calculated in Section 5.4.2 due to the rounding of the Z value to two decimal places.) Figure 5.1.Excel displays these calculations for the data of Table 5.1.Excel.

	A	B
1	Calculating Normal Probabilities	
2		
3	Arithmetic Mean	75
4	Standard Deviation	6
5	Left Tail Probability	
6	First X Value	69
7	Z Value	-1
8	P(X<=69)	0.15865526
9	Right Tail Probability	
10	P(X>=69)	0.84134474
11	Interval Probability	
12	Second X value	81
13	P(X<=81)	0.84134474
14	P(69<X<81)	0.68268948
15	Finding a X Value	
16	Cumulative Percent	0.1
17	Z Value	-1.281550794
18	X Value	67.31069523

FIGURE 5.1.EXCEL Obtaining normal probabilities using Microsoft Excel.

> ▲ **WHAT IF EXAMPLE**
>
> The design of the Calculations sheet presented in Table 5.1.Excel allows us to study the effect of changes in μ and σ on the probability of obtaining values below or above a particular X value or between two X values. For example, we could change the mean μ and/or the standard deviation σ and observe the effect on all the probabilities and X values computed on the Calculations sheet. As one example of this, change the standard deviation in cell B4 of the Calculations sheet from 6 to 8, and observe the effect on the Z value in cell B7, the $P(X \leq 69)$ in cell B8, the $P(X \geq 69)$ in cell B10, the $P(69 < X < 81)$ in cell B14, and the X value in cell B18. As before, if we are interested in seeing the effects of many different changes, we could use the Scenario Manager (see Section 1S.15) to store and use sets of alternative probability values.

5.6 ASSESSING THE NORMALITY ASSUMPTION: EVALUATING PROPERTIES AND CONSTRUCTING PROBABILITY PLOTS

Now that we have discussed the importance of the normal distribution and described its properties (Section 5.3) as well as demonstrated how it may be applied (Section 5.4), a very practical question must be considered. That is, we must be able to assess the likelihood that a particular data set can be assumed as coming from an underlying normal distribution or can be adequately approximated by it.

5.6.1 Exploring the Data—The Art of Data Analysis

The reader must be cautioned—*not all continuous random variables are normally distributed!* Often the continuous random phenomenon that we may be interested in studying will neither follow the normal distribution nor be adequately approximated by it (see Section 5.7). While some methods for studying such continuous phenomena are outside the scope of this text (see Reference 4), *nonparametric* techniques (see Reference 3) that do not depend on the particular form of the underlying random variable will be discussed in Chapter 8.

Hence, for a descriptive analysis of any particular set of data, the practical question remains: How can we decide whether our data set seems to follow or at least approximate the normal distribution sufficiently to permit it to be examined using the methodology of this chapter? Two descriptive *exploratory* approaches will be taken here to evaluate the *goodness-of-fit*.

1. A comparison of the data set's characteristics with the properties of an underlying normal distribution.
2. The construction of a normal probability plot.

More formal *confirmatory* approaches to the *goodness-of-fit* of a normal distribution can be found in References 3 and 6.

5.6.2 Evaluating the Properties

In Section 5.3.2, we noted that the normal distribution has several theoretical properties. We recall that it is bell-shaped and symmetrical in appearance; its measures of central tendency are all identical; its interquartile range is equal to 1.33 standard deviations; and its random variable is continuous in form and has an infinite range.

We also noted that, in actual practice, some of the continuous random phenomena we observe may only approximate these theoretical properties, either because the underlying population distribution may be only approximately normal or because any obtained sample data set may deviate from the theoretically expected characteristics. In such circumstances, the data may not be perfectly bell-shaped and symmetrical in appearance. The measures of central tendency will differ slightly, and the interquartile range will not be exactly equal to 1.33 standard deviations. In addition, in practice, the range of the data will not be infinite—it will be approximately equal to 6 standard deviations.

However, many continuous phenomena are neither normally distributed nor approximately normally distributed. For such phenomena, the descriptive characteristics of the respective data sets will not match well with these four properties of a normal distribution.

What then should we do to investigate the assumption of normality in our data? One approach is to compare and contrast the actual data characteristics against the corresponding properties from an underlying normal distribution. To accomplish this, the following three steps are suggested:

1. Make some tallies and plots and observe their appearance.
   - For small- or moderate-sized data sets, construct a stem-and-leaf display and box-and-whisker plot.
   - In addition, for large data sets, construct the frequency distribution and plot the histogram and polygon.
2. Compute descriptive summary measures and compare the characteristics of the data with the theoretical and practical properties of the normal distribution.
   - Obtain the mean, median, mode, midrange, and midhinge and note the similarities or differences in these five measures of central tendency.
   - Obtain the interquartile range and standard deviation. Note how well the interquartile range can be approximated by 1.33 times the standard deviation.
   - Obtain the range and note how well it can be approximated by 6 times the standard deviation.
3. Make some tallies to evaluate how the observations in the data set distribute themselves.
   - Determine whether approximately two-thirds of the observations lie between the mean ±1 standard deviation.
   - Determine whether approximately four-fifths of the observations lie between the mean ±1.28 standard deviations.
   - Determine if approximately 19 out of every 20 observations lie between the mean ±2 standard deviations.

As good data analysts, these are the kinds of things we should always be thinking of doing: *plotting, observing, computing,* and *describing.* Many of the descriptive statistical techniques we have studied so far come into play here. Nothing is new. Compared with other distributional forms, we know what a normal distribution is supposed to look like [see Figure 3.8(a) on page 150 comparing the polygon and box-and-whisker plot].

A second approach to evaluating the assumption of normality in our data is through the construction of a normal probability plot.

5.6.3 Constructing the Normal Probability Plot

You may recall that **quantiles** are defined as measures of "noncentral" location that are usually computed for summarizing large sets of numerical data [see Endnote 2 (page 169) pertaining to Section 3.4]. In Section 3.4 we stressed the median (which splits the ordered observations in half) and the quartiles (which split the ordered observations in fourths) and in that Endnote we mentioned other measures of position or quantiles such as the deciles (which split the ordered observations in tenths) and the percentiles (which split the ordered observations in hundredths). With this in mind, we may define a normal probability plot as follows:

> A **normal probability plot** is a two-dimensional plot of the observed data values on the *vertical* axis with their corresponding quantile values from a standardized normal distribution on the *horizontal* axis (see References 2 and 7).

If the plotted points seem to lie either on or close to an imaginary straight line rising from the lower left corner of the graph to the upper right corner, we would have evidence to believe that the obtained data set is (at least approximately) normally distributed. On the other hand, if the plotted points appear to deviate from this imaginary straight line in some patterned fashion, then we would have reason to believe that the obtained data set is not normally distributed and that the methodology presented in this chapter may not be appropriate.

To construct and use a normal probability plot, the following steps must be taken:

1. Place the values in the data set into an ordered array.
2. Find the corresponding standard normal quantile values.
3. Plot the corresponding pairs of points using the observed data values on the vertical axis and the associated standard normal quantile values on the horizontal axis.
4. Assess the likelihood that the random variable of interest is (at least approximately) normally distributed by inspecting the plot for evidence of linearity (i.e., a straight line).

These steps will be described in detail.

● **Obtaining the Ordered Array**   Since the original data set is likely to be obtained in raw form, the observations must be rearranged from smallest to largest to facilitate a matching with the corresponding standard normal quantile values. Thus, the original data are placed into an ordered array.

● **Finding the Standard Normal Quantile Values**   We know that a standard normal distribution is characterized by a mean of 0 and standard deviation of 1. Owing to its symmetry, the median or middle quantile value from a standard normal distribution must also be 0. Therefore, when dealing with a standard normal distribution the quantile values below the median will be negative and the quantile values above the median will be positive. However, the question we must still answer is: How can we obtain the quantile values from this distribution? The process by which we can accomplish this is known as an **inverse normal scores transformation** (see Reference 3).

The following is noted: Given a data set containing n observations from a standardized normal distribution, let the symbol O_1 represent the first (and smallest) ordered or quantile value, let the symbol O_2 represent its second smallest ordered value, let the symbol O_i represent the ith smallest ordered value, and let the symbol O_n represent the largest ordered value. Because of symmetry, the standard normal quantiles O_1 and O_n will have the same numerical value—except for sign. O_1 will be negative and O_n will be positive.

> The **first standard normal quantile**, O_1, is the Z value on a standard normal distribution below which the proportion $1/(n + 1)$ of the area under the curve is contained.
>
> The **second standard normal quantile**, O_2, is the Z value on a standard normal distribution below which the proportion $2/(n + 1)$ of the area under the curve is contained.
>
> The **ith standard normal quantile**, O_i, is the Z value on a standard normal distribution below which the proportion $i/(n + 1)$ of the area under the curve is contained.

The *n*th (and largest) standard normal quantile, O_n, is the Z value on a standard normal distribution below which the proportion $n/(n + 1)$ of the area under the curve is contained.

● **Making the Inverse Normal Scores Transformation** As in Section 5.4.2, once we know the probability or area under the curve, we may use the body of Table E.2 to locate the appropriate area and then its corresponding standard normal ordered value in the margins of this table. Thus, in general, to find the *i*th standard normal ordered value from a data set containing *n* observations, we sketch the standard normal distribution and locate the value O_i such that the proportion $i/(n + 1)$ of the area under the curve is contained below that value. By subtraction, we then compute the area under the curve from O_i to the mean μ of 0. We then find this area in the body of Table E.2 and, working to the margins of that table, we locate the corresponding standard normal ordered value.

To demonstrate this, let us suppose we wish to obtain the set of standard normal ordered values corresponding to a sample of 19 observations. The first standard normal ordered value, O_1, is that value below which the proportion $\dfrac{1}{n+1} = \dfrac{1}{19+1} = \dfrac{1}{20} = .05$ of the area under the normal curve is contained. From Figure 5.17, we see that the area from O_1 to the mean is .45 so that, from the body of Table 5.6, O_1 would fall halfway between -1.65 and -1.64. Since the standard normal ordered values are usually reported with two decimal places, the value -1.65 is chosen here.

The second standard normal ordered value, O_2, is that value below which the proportion $\dfrac{2}{n+1} = \dfrac{2}{19+1} = \dfrac{2}{20} = .10$ of the area under the normal curve is obtained. From Figure 5.18 and Table 5.7, O_2 would fall between -1.29 and -1.28 but closer to the latter. Hence, -1.28 is selected here.

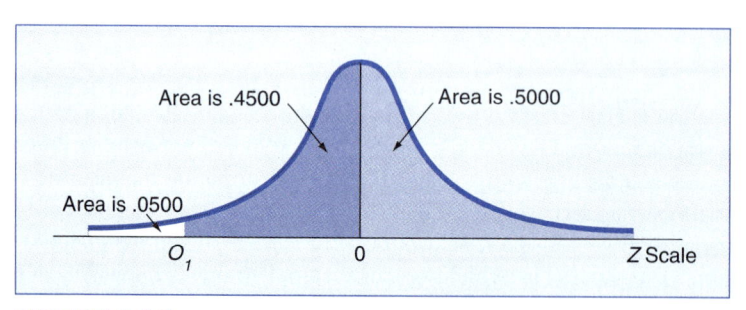

FIGURE 5.17 Finding the first standard normal ordered value from a data set with 19 observations.

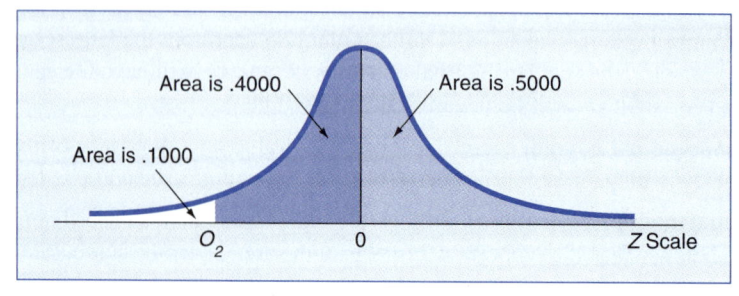

FIGURE 5.18 Finding the second standard normal ordered value from a data set with 19 observations.

Table 5.6 Obtaining a standard normal ordered value corresponding to a particular area under the normal curve.

Z	.00	.01	.02	.03	.04	.05	.06	.07	.08	.09
0.0	.0000	.0040	.0080	.0120	.0160	.0199	.0239	.0279	.0319	.0359
0.1	.0398	.0438	.0478	.0517	.0557	.0596	.0636	.0675	.0714	.0753
0.2	.0793	.0832	.0871	.0910	.0948	.0987	.1026	.1064	.1103	.1141
0.3	.1179	.1217	.1255	.1293	.1331	.1368	.1406	.1443	.1480	.1517
0.4	.1554	.1591	.1628	.1664	.1700	.1736	.1772	.1808	.1844	.1879
⋮	⋮	⋮	⋮	⋮	⋮	⋮	⋮	⋮	⋮	⋮
1.0	.3413	.3438	.3461	.3485	.3508	.3531	.3554	.3577	.3599	.3621
1.1	.3643	.3665	.3686	.3708	.3729	.3749	.3770	.3790	.3810	.3830
1.2	.3849	.3869	.3888	.3907	.3925	.3944	.3962	.3980	.3997	.4015
1.3	.4032	.4049	.4066	.4082	.4099	.4115	.4131	.4147	.4162	.4177
1.4	.4192	.4207	.4222	.4236	.4251	.4265	.4279	.4292	.4306	.4319
1.5	.4332	.4345	.4357	.4370	.4382	.4394	.4406	.4418	.4429	.4441
1.6	.4452	.4463	.4474	.4484	.4495	.4505	.4515	.4525	.4535	.4545
1.7	.4554	.4564	.4573	.4582	.4591	.4599	.4608	.4616	.4625	.4633

Source: Extracted from Table E.2.

Table 5.7 Obtaining a standard normal ordered value corresponding to a particular area under the normal curve.

Z	.00	.01	.02	.03	.04	.05	.06	.07	.08	.09
0.0	.0000	.0040	.0080	.0120	.0160	.0199	.0239	.0279	.0319	.0359
0.1	.0398	.0438	.0478	.0517	.0557	.0596	.0636	.0675	.0714	.0753
0.2	.0793	.0832	.0871	.0910	.0948	.0987	.1026	.1064	.1103	.1141
0.3	.1179	.1217	.1255	.1293	.1331	.1368	.1406	.1443	.1480	.1517
0.4	.1554	.1591	.1628	.1664	.1700	.1736	.1772	.1808	.1844	.1879
⋮	⋮	⋮	⋮	⋮	⋮	⋮	⋮	⋮	⋮	⋮
1.0	.3413	.3438	.3461	.3485	.3508	.3531	.3554	.3577	.3599	.3621
1.1	.3643	.3665	.3686	.3708	.3729	.3749	.3770	.3790	.3810	.3830
1.2	.3849	.3869	.3888	.3907	.3925	.3944	.3962	.3980	.3997	.4015
1.3	.4032	.4049	.4066	.4082	.4099	.4115	.4131	.4147	.4162	.4177
1.4	.4192	.4207	.4222	.4236	.4251	.4265	.4279	.4292	.4306	.4319

Source: Extracted from Table E.2.

Continuing in a similar manner, for example, the tenth standard normal ordered value, O_{10}, is that value below which the proportion $\frac{10}{n+1} = \frac{10}{19+1} = \frac{10}{20} = .50$ of the area under the normal curve is contained. Since we have located the median, this standard normal ordered value must be 0.00. We leave it as an exercise for the reader to show that the second largest standard normal ordered value, O_{18}, is +1.28 and the largest standard normal ordered value, O_{19}, is +1.65 (see Problem 5.12 on page 257).

5.6.4 Interpreting the Results

● **Hypothetical Applications** Table 5.8 presents ordered arrays of hypothetical midterm test scores from 19 students in each of four sections (sections I–IV) of a course in Introductory Finance. Also shown in Table 5.8 are the corresponding standard normal ordered values obtained from the previously described inverse normal scores transformation. If we were to construct normal probability plots for these four distinct data sets, what would they show us and how would we interpret the plots?

The normal probability plots for the four class sections are depicted in parts (a)–(d) of Figure 5.19. From (a), we observe that the points appear to deviate from a straight line in a random manner, so we may conclude that the data set from class section I is approximately normally distributed. [Note the corresponding polygon and box-and-whisker plot from Figure 3.8(a) on page 150.]

On the other hand, in Figure 5.19(b), we observe a nonlinear pattern to the plot. The points seem to rise somewhat more steeply at first and then seem to increase at a decreasing rate. This pattern is an example of a left-skewed data set. The steepness of the left side of the plot is indicative of the elongated left tail of the distribution of test scores from class section II. [Note the corresponding polygon and box-and-whisker plot from Figure 3.8(b).]

In Figure 5.19(c), we observe the opposite nonlinear pattern. The points here seem to rise more slowly at first and then seem to increase at an increasing rate. This pattern is an example of a right-skewed data set. The steepness of the right side of the plot is indicative

Table 5.8 Ordered arrays of hypothetical midterm test scores obtained from 19 students in each of four sections (I–IV) of a course in Introductory Finance and the corresponding standard normal ordered values.

(I) Bell-Shaped Normal Distribution	(II) Left-Skewed Distribution	(III) Right-Skewed Distribution	(IV) Rectangular-Shaped Distribution	O_i
48	47	47	38	−1.65
52	54	48	41	−1.28
55	58	50	44	−1.04
57	61	51	47	−0.84
58	64	52	50	−0.67
60	66	53	53	−0.52
61	68	53	56	−0.39
62	71	54	59	−0.25
64	73	55	62	−0.13
65	74	56	65	0.00
66	75	57	68	0.13
68	76	59	71	0.25
69	77	62	74	0.39
70	77	64	77	0.52
72	78	66	80	0.67
73	79	69	83	0.84
75	80	72	86	1.04
78	82	76	89	1.28
82	83	83	92	1.65

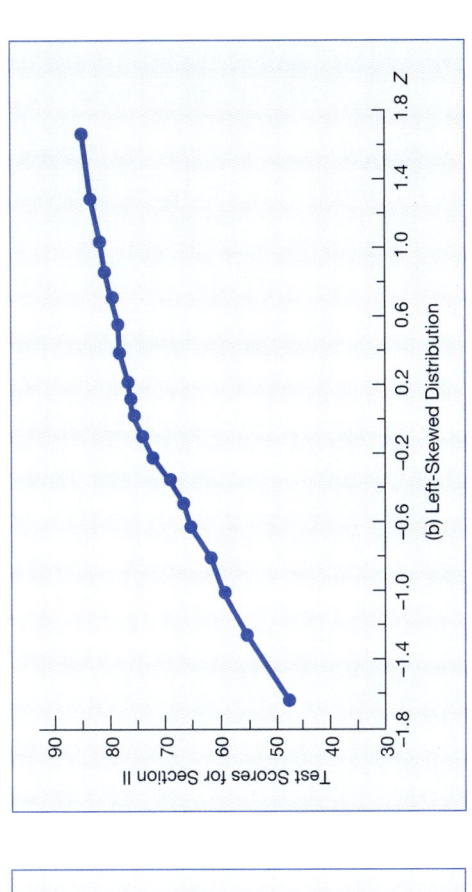

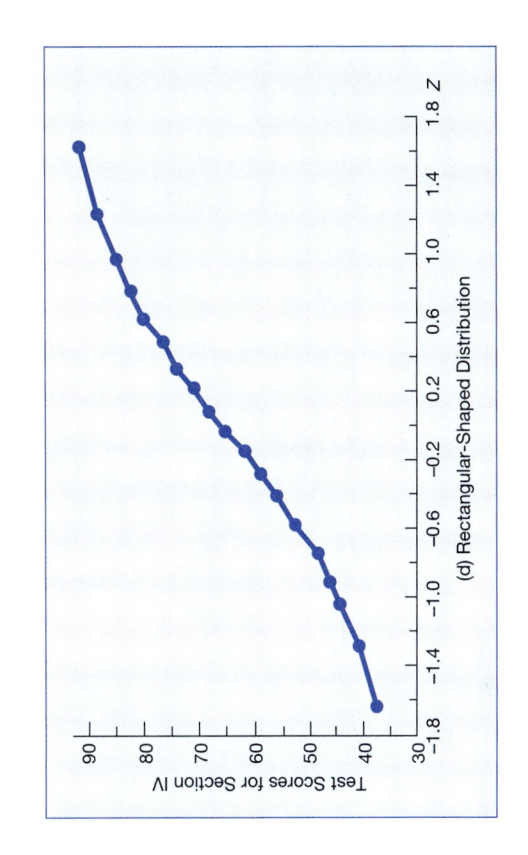

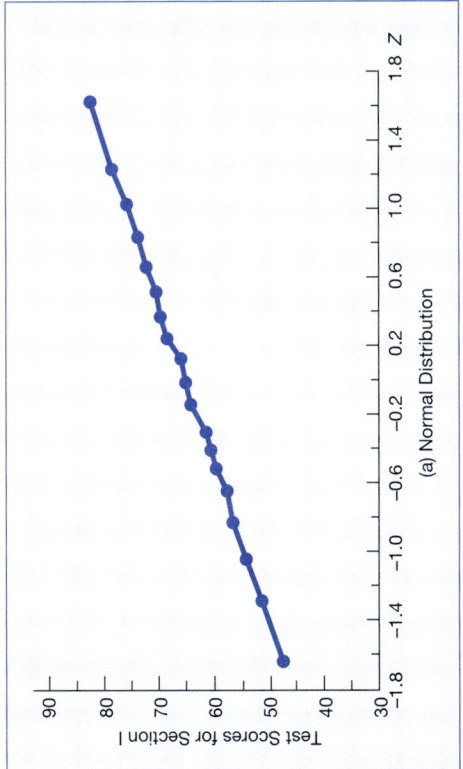

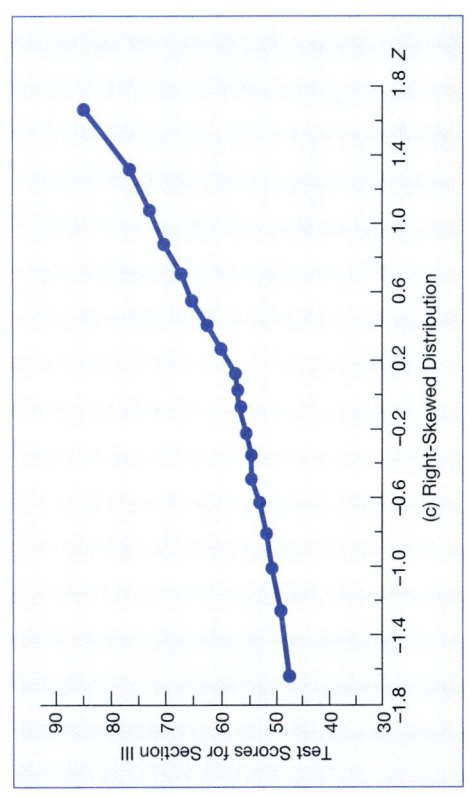

FIGURE 5.19 Normal probability plots for four hypothetical data sets.

of the elongated right tail of the distribution of test scores from class section III. [Note the corresponding polygon and box-and-whisker plot from Figure 3.8(c).]

From Figure 5.19(d), we observe a symmetrical plot with a pattern—that is, the pattern is linear over a large middle portion of the plot. However, on each side of the plot the curve seems to flatten out. This flattening out shows the opposite effect to what was observed in parts (b) and (c) as a result of skewness. Here there are no elongated tails. In fact, there are really no tails—the test scores in class section IV are rectangularly distributed. [Note the respective corresponding polygon and box-and-whisker plot from Figure 3.8(d).]

- **Real Application: Examining Texas Out-of-State Tuition Rates** Now that we have seen how to interpret normality or lack thereof from a set of hypothetical applications, we can demonstrate the normal probability plot using real applications. Using Special Data Set 1 from Appendix D, Figure 5.20 depicts the normal probability plot obtained from Microsoft Excel of the Texas out-of-state tuition rates (see Table 2.1 on page 52). (The characteristics of this data set has been described in Chapters 2 and 3.) From Figure 5.20, we observe that the Texas out-of-state tuition rates seem to be rising slowly at first and then seem to be increasing at an increasing rate, confirming our belief that the data set is right-skewed.

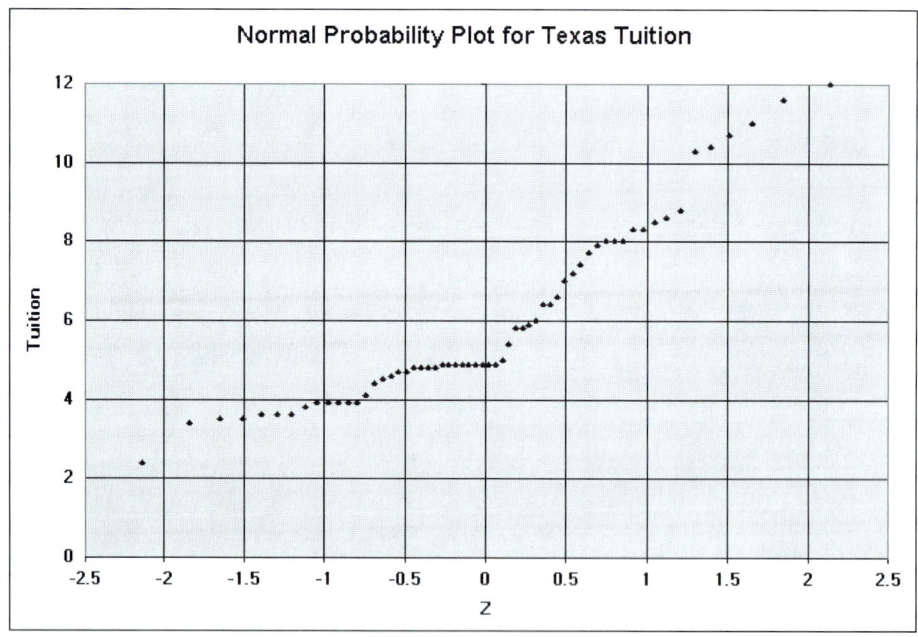

FIGURE 5.20 Normal probability plot obtained from Microsoft Excel of the Texas out-of-state tuition rates data. *Source:* Data are taken from Table 2.1.

5.6.5 Using Microsoft Excel to Obtain a Normal Probability Plot

In Sections 5.6.1–5.6.4, we developed the normal probability plot to evaluate whether a given set of data was normally distributed. Although current versions of Microsoft Excel do not have the normal probability plot included as part of the Data Analysis tool, we can use Excel functions and the Chart Wizard to develop a normal probability plot.

Table 5.2.Excel Design for a Calculations worksheet for Obtaining a Normal Probability Plot for the Texas out-of-state tuition data.

	A	B	C	D
1		Normal Probability Plot for Texas		
2				
3	Rank	Ordered Value	Z Value	Tuition
4	1	=A4/61	=NORMSINV(B4)	=Data!B2
5	2	=A5/61	=NORMSINV(B5)	=Data!B3
⋮	⋮	⋮	⋮	⋮
63	60	=A63/61	=NORMSINV(B63)	=Data!B61

Figure 5.20 on page 254 represents a normal probability plot for out-of-state tuition rates at 60 colleges and universities in Texas. To obtain this normal probability plot, open the TEXAS-1.XLS workbook to the Data sheet. A design for a Calculations sheet that can be used to obtain a normal probability plot is presented in Table 5.2.Excel. Note that the tuition data have been previously sorted in ascending order and therefore is ready to be used to obtain a normal probability plot.

TEXAS-1.XLS

To implement this design insert a new worksheet and rename it Calculations. Enter all column headings and the ranks in column A.

5-6-5.XLS

Although we could type the numbers 1 through 60 in the cell range A4:A63, a shortcut for preparing this column is to enter 1 in cell A4 and with A4 as the active cell, select Edit | Fill | Series. In the Series dialog box that appears, select the Columns and Linear option buttons, and enter 60 for the Stop Value as shown in Figure 5.2.Excel. Click the OK button. The ranks from 1 to 60 are now displayed in cells A4:A63 respectively.

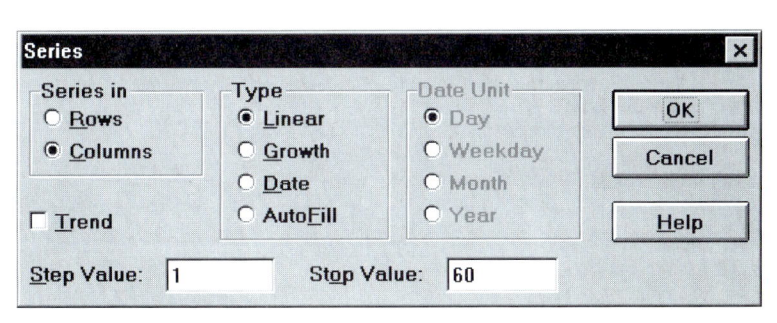

FIGURE 5.2.EXCEL Series dialog box.

The proportions $1/(n+1)$, $2/(n+1)$, ..., $n/(n+1)$ corresponding to each ordered value are entered in column B so that the inverse normal values can be obtained in column C. These are obtained by entering the rank divided by the sample size plus 1. Since the sample size plus 1 equals 61 for these data, enter =A4/61 in cell B4 and copy the formula through cell B63. Obtain the inverse normal values by using the NORMSINV function. Enter =NORMSINV(B4) in cell C4 and copy this formula through cell C63. Finally, duplicate the sorted tuition data from the Data sheet by entering =DATA!B2 in cell D4 and copying this formula through cell D63. Figure 5.3.Excel displays the completed Calculations sheet.

Now that the inverse normal values have been calculated for each tuition value, we can use Microsoft Excel's Chart Wizard (see Section 2.12) to obtain the normal probability plot depicted in Figure 5.20. After clicking on Insert | Chart | As New Sheet, the selections in the five dialog boxes for the Chart Wizard are as follows:

FIGURE 5.3.EXCEL Ranks, ordered values, and Z values obtained from Microsoft Excel for the Texas out-of-state tuition data.

	A	B	C	D
1		Normal Probability Plot for Texas		
2				
3	Rank	Ordered Value	Z Value	Tuition
4	1	0.016393443	-2.134675015	2.4
5	2	0.032786885	-1.841326593	3.4
6	3	0.049180328	-1.652852006	3.5
7	4	0.06557377	-1.509592948	3.5
8	5	0.081967213	-1.391958904	3.6
9	6	0.098360656	-1.290950422	3.6
10	7	0.114754098	-1.201626674	3.6
11	8	0.131147541	-1.120984052	3.8
12	9	0.147540984	-1.047037586	3.9
13	10	0.163934426	-0.97841621	3.9
14	11	0.180327869	-0.914117209	3.9
15	12	0.196721311	-0.853390247	3.9
16	13	0.213114754	-0.795660071	3.9
17	14	0.229508197	-0.740467385	4.1
18	15	0.245901639	-0.687443844	4.4
19	16	0.262295082	-0.636284767	4.5
20	17	0.278688525	-0.586742317	4.6
21	18	0.295081967	-0.538598215	4.7
22	19	0.31147541	-0.491672836	4.7
23	20	0.327868852	-0.445805881	4.8
24	21	0.344262295	-0.400857516	4.8
25	22	0.360655738	-0.356706096	4.8
26	23	0.37704918	-0.313240207	4.8
27	24	0.393442623	-0.270357532	4.9
28	25	0.409836066	-0.227967121	4.9
29	26	0.426229508	-0.185982572	4.9
30	27	0.442622951	-0.14432203	4.9
31	28	0.459016393	-0.102911599	4.9
32	29	0.475409836	-0.061677383	4.9
33	30	0.491803279	-0.020547759	4.9
34	31	0.508196721	0.020547759	4.9
35	32	0.524590164	0.061677383	4.9
36	33	0.540983607	0.102911599	5
37	34	0.557377049	0.14432203	5.4
38	35	0.573770492	0.185982572	5.8
39	36	0.590163934	0.227967121	5.8
40	37	0.606557377	0.270357532	5.9
41	38	0.62295082	0.313240207	6
42	39	0.639344262	0.356706096	6.4
43	40	0.655737705	0.400857516	6.4
44	41	0.672131148	0.445805881	6.6
45	42	0.68852459	0.491672836	7
46	43	0.704918033	0.538598215	7.2
47	44	0.721311475	0.586742317	7.4
48	45	0.737704918	0.636284767	7.7
49	46	0.754098361	0.687443844	7.9
50	47	0.770491803	0.740467385	8
51	48	0.786885246	0.795660071	8
52	49	0.803278689	0.853390247	8
53	50	0.819672131	0.914117209	8.3
54	51	0.836065574	0.97841621	8.3
55	52	0.852459016	1.047037586	8.5
56	53	0.868852459	1.120984052	8.6
57	54	0.885245902	1.201626674	8.8
58	55	0.901639344	1.290950422	10.3
59	56	0.918032787	1.391958904	10.4
60	57	0.93442623	1.509592948	10.7
61	58	0.950819672	1.652852006	11
62	59	0.967213115	1.841326593	11.6
63	60	0.983606557	2.134675015	12

Dialog box 1: Enter the range Calculations!C4:D63 and click the Next button.

Dialog box 2: Select the XY Scatter chart and click the Next button.

Dialog box 3: Choose format 3, which shows unconnected points plotted on a square grid, and click the Next button.

Dialog box 4: Select the Columns Option button for Data Series, enter 1 for First Columns, and enter 0 for First Rows. Click the Next button.

Dialog box 5: Select No for Add a Legend, enter Normal Probability Plot for Texas Tuition as the chart title, enter Z in the Category (X) edit box, and Tuition in the Value (Y) edit box. Click the Finish button.

You will notice that the tick marks for the Y-axis (the Tuition variable) are placed midway across the chart along the vertical line that corresponds to the X-axis value 0. To move the Y-axis tick marks to the left edge of the chart, click on the current tick marks and select Format | Selected Axis. In the Format Axis dialog box that appears, select the Patterns tab and then select the Low option button for the Tick Mark Labels. Click the OK button. The chart will now appear as in Figure 5.20 on page 254.

Problems for Section 5.6

Note: *Microsoft Excel can be used to solve the problems in this section.*

5.12 Show that for a sample of 19 observations, the 18th smallest (i.e., second largest) standard normal ordered value obtained from the inverse normal scores transformation is +1.28 and the 19th (i.e., largest) standard normal ordered value is +1.65.

5.13 Show that for a sample of 39 observations, the smallest and largest standard normal ordered values obtained from the inverse normal scores transformation are, respectively, −1.96 and +1.96, and the middle (i.e., 20th) standard normal ordered value is 0.00.

• 5.14 Using the inverse normal scores transformation on a sample of 6 observations, list the 6 expected proportions or areas under the standardized normal curve along with their corresponding standard normal ordered values.

• 5.15 Given the following ordered array (from left to right) of the amount of money (in dollars) withdrawn from a cash machine by 25 customers at a local bank:

ATM1.TXT

40	50	50	70	70	80	80	90	100	100
100	100	100	100	110	110	120	120	130	140
140	150	160	160	200					

Decide whether or not the data appear to be approximately normally distributed by
(a) Evaluating the actual versus theoretical properties.
(b) Constructing a normal probability plot.
(c) Discussing the results obtained in (a) and (b).

5.16 Given the following data on grocery bills (in dollars) paid by a random sample of 28 customers in a local supermarket:

GROCERY.TXT

44.24	35.56	45.93	49.92	38.94	41.16	44.84
27.28	50.66	50.97	45.93	46.58	28.73	25.93
24.21	23.84	54.58	52.62	47.36	30.84	48.62
31.15	38.58	34.96	45.32	53.81	40.22	37.19

Decide whether or not the data appear to be approximately normally distributed by
(a) Evaluating the actual versus theoretical properties.
(b) Constructing a normal probability plot.
(c) Discussing the results obtained in (a) and (b).

5.17 Given the following ordered array (from left to right) of final examination results obtained from 19 students in an Introductory Marketing class:

64	66	66	69	70	71	71	73	75	77
78	79	79	81	83	83	88	89	92	

Decide whether or not the data appear to be approximately normally distributed by
(a) Evaluating the actual versus theoretical properties.
(b) Constructing a normal probability plot.
(c) Discussing the results of (a) and (b).

5.18 Given the following data on amount of gasoline (in gallons) purchased at a highway gasoline station for a random sample of 24 automobiles:

12.78	8.89	10.09	10.64	15.98	13.95	9.48	10.84
10.88	9.93	7.74	5.80	11.84	10.29	10.89	6.68
12.09	8.28	8.83	7.95	7.33	12.56	8.86	9.15

Decide whether or not the data appear to be approximately normally distributed by
(a) Evaluating the actual versus theoretical properties.
(b) Constructing a normal probability plot.
(c) Discussing the results of (a) and (b).

5.7 OTHER CONTINUOUS PROBABILITY DISTRIBUTIONS

Now that we have discussed the normal distribution, in this section we will introduce other continuous probability distributions that are useful in a variety of circumstances in business.

5.7.1 The Exponential Distribution

The **exponential distribution** is widely used in waiting line or queuing theory to model the length of time between arrivals at a service facility such as a toll bridge crossing, a banking automatic teller machine (ATM), or a hospital emergency room. This distribution is defined by a single parameter, its mean, λ, the average number of arrivals per unit of time. The probability that the length of time before the next arrival is less than or equal to X is given by the following:

> The probability of an arrival in less than or equal to X amount of time is equal to one minus the mathematical constant e raised to a power equal to minus one times the product of the average number of arrivals λ and the value of X.
>
> $$P(\text{arrival time} \leq X) = 1 - e^{-\lambda X} \quad (5.4)$$

We may illustrate the exponential distribution by referring to the following example. Suppose customers arrive at a bank's ATM at the rate of 20 per hour. If a customer has just arrived, what is the probability that the next customer arrives within 6 minutes (0.1 hour)?

For this example we have $\lambda = 20$ and $X = 0.1$ so that using Equation (5.4), we have

$$P(\text{arrival time} \leq 0.1) = 1 - e^{-20(0.1)}$$
$$= 1 - e^{-2}$$

$$P(\text{arrival time} \leq 0.1) = 1 - .1353 = .8647$$

Thus, the probability that a customer will arrive within 6 minutes is .8647 or 86.47%.

A second continuous distribution that is often used in manufacturing is the Weibull distribution. This distribution is used to model the time until failure of industrial products. It is defined by two parameters, often denoted by α and β. (For further discussion of the Weibull and other continuous distributions, see Reference 4.)

5.7.2 Using Microsoft Excel to Obtain Exponential Probabilities

In Section 5.7.1, we discussed the exponential distribution, which is widely used in waiting line or queuing theory to model the length of time between arrivals. Rather than using Equation (5.4) on page 258 to compute probabilities from the exponential distribution, we can use the appropriate Microsoft Excel function.

The EXPONDIST function can be used to compute the probability of obtaining a value that is less than or equal to X in the exponential distribution. The format of this function is

$$\text{EXPONDIST}(X, \lambda, \text{True})$$

where λ = the mean of the exponential distribution

To illustrate this function, suppose we return to the ATM customer-arrival example in Section 5.7.1. In that example, with an arrival rate of 20 per hour, given that a customer has just arrived, we wanted to determine the probability that the next customer arrives within 6 minutes (0.1 hour).

For this example, we have $\lambda = 20$ and $X = 0.1$, and we computed this probability to be .8647. To obtain this result using Excel, we would enter the formula =EXPONDIST(.1,20,True) in a cell. In general, if we wanted to compute any exponential probability using Microsoft Excel, we could use the plan illustrated in Table 5.3.Excel to develop a Calculations worksheet. The results are displayed in Figure 5.4.Excel.

5-7-2.XLS

Table 5.3.Excel Design for computing exponential probabilities for the ATM customer-arrival example (where $X = .1$ and $\lambda = 20$).

	A	B
1		Exponential Calculations
2		
3	Value	.1
4	Mean	20
5	Probability	=EXPONDIST(B3,B4,True)

	A	B
1	Exponential Calculations	
2		
3	X Value	0.1
4	Mean	20
5	Probability	0.864664717

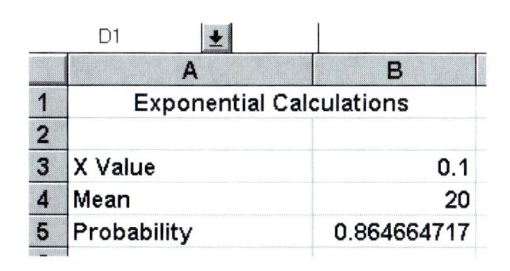

FIGURE 5.4.EXCEL Obtaining exponential probabilities using Microsoft Excel for the ATM customer-arrival example.

Problems for Section 5.7

Note: *Microsoft Excel can be used to solve the problems in this section.*

5.19 Suppose autos arrive at a toll booth located at the entrance to a bridge at the rate of 50 per minute during the 5–6 P.M. hour. If an auto has just arrived
 (a) What is the probability that the next auto arrives within 3 seconds (0.05 minute)?
 (b) What is the probability that the next auto arrives within 1 second (0.0167 minute)?
 (c) What will your answers be to (a) and (b) if the rate of arrival of autos is 60 per minute?
 (d) What will your answers be to (a) and (b) if the rate of arrival of autos is 30 per minute?

5.20 Customers arrive at the drive-up window of a fast-food restaurant at the rate of 2 per minute during the lunch hour.
 (a) What is the probability that the next customer arrives within 1 minute?
 (b) What is the probability that the next customer arrives within 5 minutes?
 (c) During the dinner time period, the arrival rate is 1 per minute. What would be your answer to (a) and (b) for the dinner time period?

5.21 Telephone calls arrive at the information desk of a large computer software company at the rate of 15 per hour.
 (a) What is the probability that the next call arrives within 3 minutes (0.05 hour)?
 (b) What is the probability that the next call arrives within 15 minutes (0.25 hour)?
 (c) Suppose the company has just introduced an updated version of one of its software programs and telephone calls are now arriving at the rate of 25 per hour. What would be your answers to (a) and (b)?

5.22 Suppose golfers arrive at the starter's booth of a public golf course at the rate of 8 per hour during the Monday-to-Friday midweek period. If a golfer has just arrived
 (a) What is the probability that the next golfer arrives within 15 minutes (0.25 hour)?
 (b) What is the probability that the next golfer arrives within 3 minutes (0.05 hour)?
 (c) Suppose the actual arrival rate on Fridays is 15 per hour. Then what would be your answers for (a) and (b) on Fridays?

5.8 INTRODUCTION TO SAMPLING DISTRIBUTIONS

A major goal of data analysis is to use statistics such as the sample mean and the sample proportion to estimate the corresponding parameters in the respective populations. We should realize that in *enumerative studies,* one is concerned with drawing conclusions about a population, not about a sample. As examples, a political pollster would be interested in the sample results only as a way of estimating the actual proportion of the votes that each candidate will receive from the population of voters. Likewise, an auditor, in selecting a sample of vouchers, is interested only in using the sample mean for estimating the population average amount.

In practice, a single sample of a predetermined size is selected at random from the population. The items that are to be included in the sample are determined through the use of a random number generator, such as a table of random numbers (see Section 1.10) or by using random number functions in Microsoft Excel (see Section 5.11). Hypothetically, to use the sample statistic to estimate the population parameter, we should examine every possible sample that could occur. If this selection of all possible samples were actually to be done, the distribution of the results would be referred to as a **sampling distribution**. The process of generalizing these sample results to the population is referred to as **statistical inference**.

5.9 SAMPLING DISTRIBUTION OF THE MEAN

5.9.1 Properties of the Arithmetic Mean

In Chapter 3, we discussed several measures of central tendency. Undoubtedly, the most widely used measure of central tendency is the arithmetic mean. It is also the best measure if the population can be assumed to be normally distributed.

Among several important mathematical properties of the arithmetic mean for a normal distribution are

1. Unbiasedness
2. Efficiency
3. Consistency

The first property, **unbiasedness**, involves the fact that the average of all the possible sample means (of a given sample size n) will be equal to the population mean μ.

This property can be demonstrated empirically by looking at the following example: Suppose that each of the four typists comprising a population of secretarial support service in a company are asked to type the same page of a manuscript. The number of errors made by each typist is as follows:

Typist	Number of Errors
A	3
B	2
C	1
D	4

This population distribution is shown in Figure 5.21.

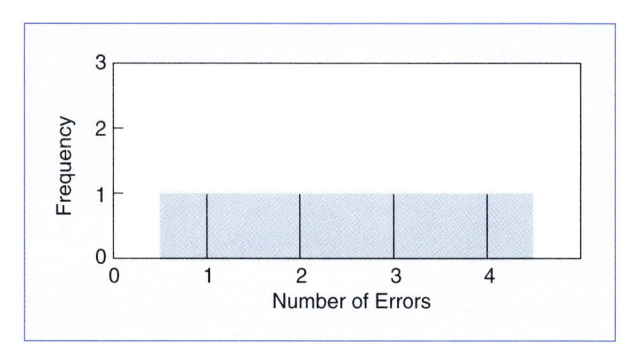

FIGURE 5.21 Number of errors made by a population of four typists.

You may recall from Section 3.9 that when the data from a population are available, the mean can be computed as follows:

The population mean μ is equal to the sum of the X values in the population divided by the population size N.

$$\mu = \frac{\sum_{i=1}^{N} X_i}{N} \qquad (5.5)$$

And the standard deviation can be computed as follows:

$$\sigma = \sqrt{\frac{\sum_{i=1}^{N}(X_i - \mu)^2}{N}} \qquad (5.6)$$

Thus,

$$\mu = \frac{3+2+1+4}{4} = 2.5 \text{ errors}$$

and

$$\sigma = \sqrt{\frac{(3-2.5)^2 + \cdots + (4-2.5)^2}{4}} = 1.12 \text{ errors}$$

If samples of two typists are selected *with* replacement from this population, there are 16 possible samples that could be selected ($N^n = 4^2 = 16$). These possible sample outcomes are shown in Table 5.9.

If all these 16 sample means are averaged, the mean of these values ($\mu_{\bar{X}}$) is equal to 2.5, which is the mean of the population μ.

On the other hand, if sampling was being done *without* replacement, there would be six possible samples of two typists:

$$\frac{N!}{n!(N-n)!} = \frac{4!}{2!\,2!} = 6$$

These six possible samples are listed in Table 5.10.

In this case, also, the average of all sample means ($\mu_{\bar{X}}$) is equal to the population mean, 2.5. Therefore, we have shown that the sample arithmetic mean is an unbiased estimator of the population mean. This tells us that although we don't know how close the average of any particular sample selected comes to the population mean, we are at least assured that the average of all the possible sample means that could have been selected will be equal to the population mean.

The second property possessed by the mean, **efficiency**, refers to the precision of the sample statistic as an estimator of the population parameter. For distributions such as the normal, the arithmetic mean is considered to be more stable from sample to sample than are other measures of central tendency. For a sample of size n, the sample mean will, on the average, come closer to the population mean than any other unbiased estimator so that the sample mean is a better estimator of the population mean.

The third property, **consistency**, refers to the effect of the sample size on the usefulness of an estimator. As the sample size increases, the variation of the sample mean from the pop-

Table 5.9 All 16 samples of n = 2 typists from a population of N = 4 typists when sampling *with* replacement.

Sample	Typists	Sample Outcomes	Sample Mean $\bar{X}_i$
1	A, A	3, 3	$\bar{X}_1 = 3$
2	A, B	3, 2	$\bar{X}_2 = 2.5$
3	A, C	3, 1	$\bar{X}_3 = 2$
4	A, D	3, 4	$\bar{X}_4 = 3.5$
5	B, A	2, 3	$\bar{X}_5 = 2.5$
6	B, B	2, 2	$\bar{X}_6 = 2$
7	B, C	2, 1	$\bar{X}_7 = 1.5$
8	B, D	2, 4	$\bar{X}_8 = 3$
9	C, A	1, 3	$\bar{X}_9 = 2$
10	C, B	1, 2	$\bar{X}_{10} = 1.5$
11	C, C	1, 1	$\bar{X}_{11} = 1$
12	C, D	1, 4	$\bar{X}_{12} = 2.5$
13	D, A	4, 3	$\bar{X}_{13} = 3.5$
14	D, B	4, 2	$\bar{X}_{14} = 3$
15	D, C	4, 1	$\bar{X}_{15} = 2.5$
16	D, D	4, 4	$\bar{X}_{16} = 4$
			$\mu_{\bar{X}} = 2.5$

Table 5.10 All six possible samples of n = 2 typists from a population of N = 4 typists when sampling *without* replacement.

Sample	Typists	Sample Outcomes	Sample Mean $\bar{X}_i$
1	A, B	3, 2	$\bar{X}_1 = 2.5$
2	A, C	3, 1	$\bar{X}_2 = 2$
3	A, D	3, 4	$\bar{X}_3 = 3.5$
4	B, C	2, 1	$\bar{X}_4 = 1.5$
5	B, D	2, 4	$\bar{X}_5 = 3$
6	C, D	1, 4	$\bar{X}_6 = 2.5$
			$\mu_{\bar{X}} = 2.5$

ulation mean becomes smaller so that the sample arithmetic mean becomes a better estimator of the population mean.

5.9.2 Standard Error of the Mean

The fluctuation in the average number of typing errors that was obtained from all 16 possible samples when sampling *with* replacement is illustrated in Figure 5.22 on page 264.

In this small example, although we can observe a good deal of fluctuation in the sample mean—depending on which typists were selected—there is not nearly as much fluctuation as in the actual population itself. The fact that the sample means are less variable than the population data follows directly from the **law of large numbers**. A particular sample mean averages together all the values in the sample. A population may consist of individual outcomes that can take on a wide range of values from extremely small to extremely large. However, if

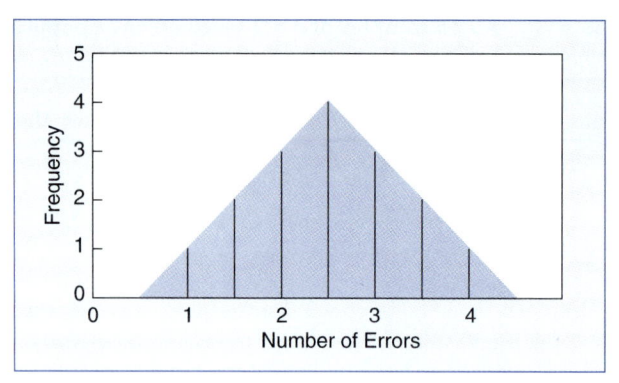

FIGURE 5.22 Sampling distribution of the average number of errors for samples of two typists.

an extreme value falls into the sample, although it will have an effect on the mean, the effect will be reduced since it is being averaged in with all the other values in the sample. As the sample size increases, the effect of a single extreme value gets even smaller, since it is being averaged with more observations.

This phenomenon is expressed statistically in the value of the standard deviation of the sample mean. This is the measure of variability of the mean from sample to sample and is referred to as the **standard error of the mean**, $\sigma_{\bar{X}}$. When sampling *with* replacement, the standard error of the mean is defined as follows:

> The standard error of the mean $\sigma_{\bar{X}}$ is equal to the standard deviation in the population σ divided by the square root of the sample size n.
>
> $$\sigma_{\bar{X}} = \frac{\sigma}{\sqrt{n}} \quad (5.7)$$

Therefore, as the sample size increases, the standard error of the mean will decrease by a factor equal to the square root of the sample size. This relationship between the standard error of the mean and the sample size will be further examined in Chapter 6 when we address the issue of sample size determination.

5.9.3 Sampling from Normal Populations

Now that we have introduced the idea of a sampling distribution and mentioned the standard error of the mean, we need to explore the question of what distribution the sample mean $\bar{X}$ will follow. It can be shown that if we sample *with* replacement from a population that is normally distributed with mean μ and standard deviation σ, the **sampling distribution of the mean** will also be normally distributed for *any size n* with mean $\mu_{\bar{X}} = \mu$ and have a standard error of the mean $\sigma_{\bar{X}}$.

In the most elementary case, if we draw samples of size $n = 1$, each possible sample mean is a single observation from the population, since

$$\bar{X} = \frac{\sum_{i=1}^{n} X_i}{n} = \frac{X_i}{1} = X_i$$

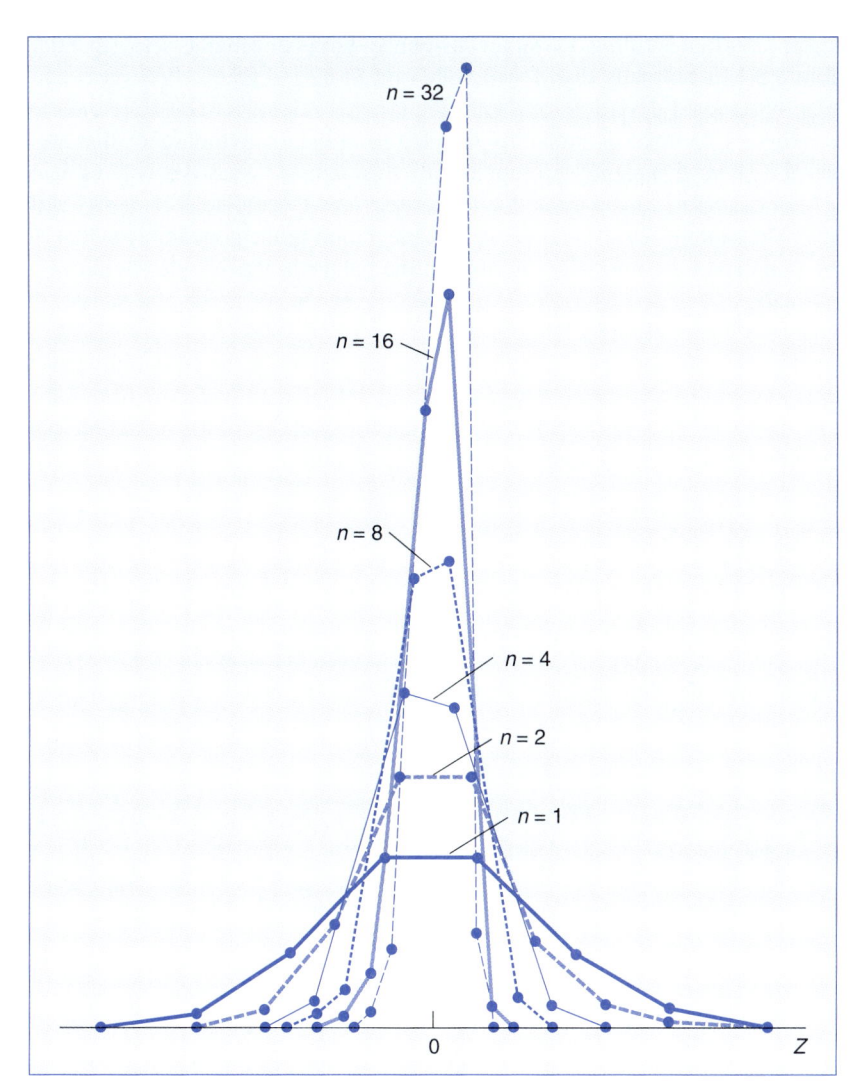

FIGURE 5.23 Sampling distributions of the mean from 500 samples of size n = 1, 2, 4, 8, 16, and 32 selected from a normal population.

If we know that the population is normally distributed with mean μ and standard deviation σ, then the sampling distribution of $\bar{X}$ for sample $n = 1$ must also follow the normal distribution with mean $\mu_{\bar{X}} = \mu$, and standard error of the mean $\sigma_{\bar{X}} = \sigma/\sqrt{1} = \sigma$. In addition, we note that as the sample size increases, the sampling distribution of the mean still follows a normal distribution with mean $\mu_{\bar{X}} = \mu$. However, as the sample size increases, the standard error of the mean decreases, so that a larger proportion of sample means are closer to the population mean. This can be observed by referring to Figure 5.23 where 500 samples of size 1, 2, 4, 8, 16, and 32 were randomly selected from a normally distributed population. We can see clearly from the polygons in Figure 5.23 that while the sampling distribution of the mean is approximately[3] normal for each sample size, the sample means are distributed more tightly around the population mean as the sample size is increased.

● **Application** We may obtain deeper insight into the concept of the sampling distribution of the mean if we examine the following: Suppose the packaging equipment in a manufacturing process that is filling 368-gram (13-ounce) boxes of cereal is set so that the amount of cereal in a box is normally distributed with a mean of 368 grams. From past experience, the population standard deviation for this filling process is known to be 15 grams.

If a sample of 25 boxes is randomly selected from the many thousands that are filled in a day and the average weight is computed for this sample, what type of result could be expected? For example, do you think that the sample mean would be 368 grams? 200 grams? 365 grams? The sample acts as a miniature representation of the population so that if the values in the population are normally distributed, the values in the sample should be approximately normally distributed. Thus, if the population mean is 368 grams, the sample mean has a good chance of being close to 368 grams.

To explore this problem even further, how can we determine the probability that the sample of 25 boxes will have a mean between 365 and 368 grams? We know from our study of the normal distribution (Section 5.3) that the area between any value X and the population mean μ can be found by converting to standardized Z units

$$Z = \frac{X - \mu}{\sigma}$$

and finding the appropriate value in the table of the normal distribution (Table E.2). In the examples in Section 5.4, we were studying how any single value X deviates from the mean. Now, in the cereal-fill example, the value involved is a sample mean, μ, and we wish to determine the likelihood of obtaining a sample mean between 365 and the population mean of 368. Thus, by substituting $\bar{X}$ for X, $\mu_{\bar{X}}$ for μ, and $\sigma_{\bar{X}}$ for σ, we have the following:

> The Z value is equal to the difference between the sample mean $\bar{X}$ and the population mean μ, divided by the standard error of the mean $\sigma_{\bar{X}}$.
>
> $$Z = \frac{\bar{X} - \mu_{\bar{X}}}{\sigma_{\bar{X}}} = \frac{\bar{X} - \mu}{\frac{\sigma}{\sqrt{n}}} \quad (5.8)$$

Note that, based on the property of unbiasedness, it is always true that $\mu_{\bar{X}} = \mu$. To find the area between 365 and 368 grams (Figure 5.24), we have

$$Z = \frac{\bar{X} - \mu}{\frac{\sigma}{\sqrt{n}}} = \frac{365 - 368}{\frac{15}{\sqrt{25}}} = \frac{-3}{3} = -1.00$$

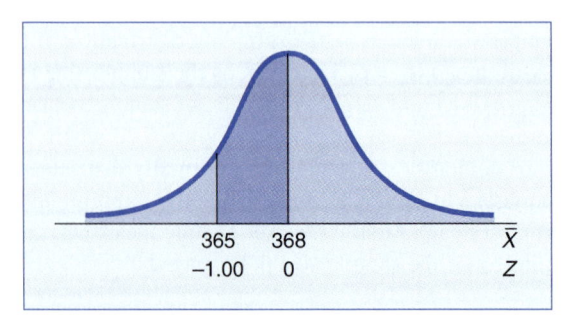

FIGURE 5.24 Diagram of normal curve needed to find area between 365 and 368 grams.

Looking up 1.00 in Table E.2, we find an area of .3413. Therefore, 34.13% of all the possible samples of size 25 would have a sample mean between 365 and 368 grams.

We must realize that this is not the same as saying that a certain percentage of *individual* boxes will have between 365 and 368 grams. In fact, that percentage can be computed from Equation (5.2) as follows:

$$Z = \frac{X - \mu}{\sigma} = \frac{365 - 368}{15} = \frac{-3}{15} = -0.20$$

The area corresponding to $Z = -0.20$ in Table E.2 is .0793. Therefore, 7.93% of the *individual* boxes are expected to contain between 365 and 368 grams. Comparing these results, we may observe that many more *sample means* than *individual boxes* will be between 365 and 368 grams. This result can be explained by the fact that each sample consists of 25 different values, some small and some large. The averaging process dilutes the importance of any individual value, particularly when the sample size is large. Thus, the chance that the mean of a sample of 25 will be close to the population mean is greater than the chance that a *single individual* value will be.

How would our results be affected by using a different sample size, such as 100 boxes instead of 25? Here we would have the following:

$$Z = \frac{\overline{X} - \mu}{\frac{\sigma}{\sqrt{n}}} = \frac{365 - 368}{\frac{15}{\sqrt{100}}} = \frac{-3}{1.5} = -2.00$$

From Table E.2, the area under the normal curve from the mean to $Z = -2.00$ is .4772. Therefore, 47.72% of the samples of size 100 would be expected to have means between 365 and 368 grams, as compared with only 34.13% for samples of size 25.

Instead of determining the proportion of sample means that are expected to fall within a certain interval, we might be more interested in finding out the interval within which a fixed proportion of the samples (means) would fall. For example, suppose we want to find an interval around the population mean that will include 95% of the sample means based on samples of 25 boxes. The 95% could be divided into two equal parts, half below the mean and half above the mean (see Figure 5.25). Analogous to Section 5.4, we are determining a distance below and above the population mean containing a specific area of the normal curve. From Equation (5.8), we have

$$Z_L = \frac{\overline{X}_L - \mu}{\frac{\sigma}{\sqrt{n}}}$$

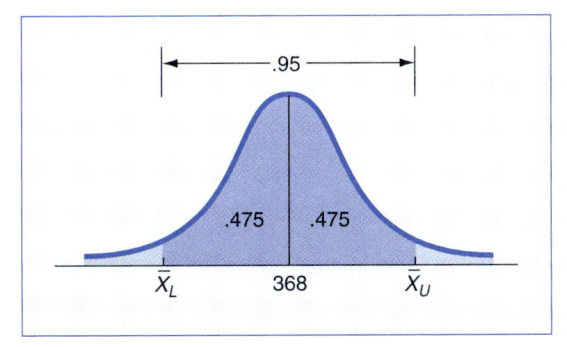

FIGURE 5.25 Diagram of normal curve needed to find upper and lower limits to include 95% of sample means.

where

$$Z_L = -Z$$

and

$$Z_U = \frac{\overline{X}_U - \mu}{\frac{\sigma}{\sqrt{n}}}$$

where

$$Z_U = +Z$$

Therefore, the lower value of $\overline{X}$ is

$$\overline{X}_L = \mu - Z\frac{\sigma}{\sqrt{n}} \qquad (5.9a)$$

and the upper value of $\overline{X}$ is

$$\overline{X}_U = \mu + Z\frac{\sigma}{\sqrt{n}} \qquad (5.9b)$$

Since $\sigma = 15$ and $n = 25$ and the value of Z corresponding to an area of .475 from the center of the normal curve is 1.96, the lower and upper values of $\overline{X}$ can be found as follows:

$$\overline{X}_L = 368 - (1.96)\frac{15}{\sqrt{25}} = 368 - 5.88 = 362.12$$

$$\overline{X}_U = 368 + (1.96)\frac{15}{\sqrt{25}} = 368 + 5.88 = 373.88$$

Our conclusion would be that 95% of all sample means based on samples of 25 boxes should fall between 362.12 and 373.88 grams.

5.9.4 Sampling from Nonnormal Populations

In the preceding section we explored the sampling distribution of the mean for the case in which the population itself was normally distributed. However, we should realize that in many instances, either we will know that the population is not normally distributed, or we may believe that it is unrealistic to assume a normal distribution. Thus, we need to examine the sampling distribution of the mean for populations that are not normally distributed. This issue brings us to an important theorem in statistics, the *central limit theorem*.

Central Limit Theorem: As the sample size (number of observations in each sample) gets *large enough,* the sampling distribution of the mean can be approximated by the normal distribution. This is true regardless of the shape of the distribution of the individual values in the population.

What sample size is large enough? A great deal of statistical research has gone into this issue. As a general rule, statisticians have found that for many population distributions, once the sample size is at least 30, the sampling distribution of the mean will be approximately normal. However, we may be able to apply the central limit theorem for even smaller sample sizes if some knowledge of the population is available (e.g., if the distribution is symmetrical).[4]

The application of the central limit theorem to different populations can be illustrated by referring to Figures 5.26 to 5.28 on pages 269–271. Each of the depicted sampling distributions has been obtained by using the computer to select 500 different samples from their respective population distributions. These samples were selected for varying sizes ($n = 2, 4, 8, 16, 32$) from three different continuous distributions (normal, uniform, and exponential).

Figure 5.26 illustrates the sampling distribution of the mean selected from a normal population. In the preceding section we stated that if the population is normally distributed, the sampling distribution of the mean will be normally distributed regardless of the sample size. An examination of the sampling distributions shown in Figure 5.26 gives empirical evidence

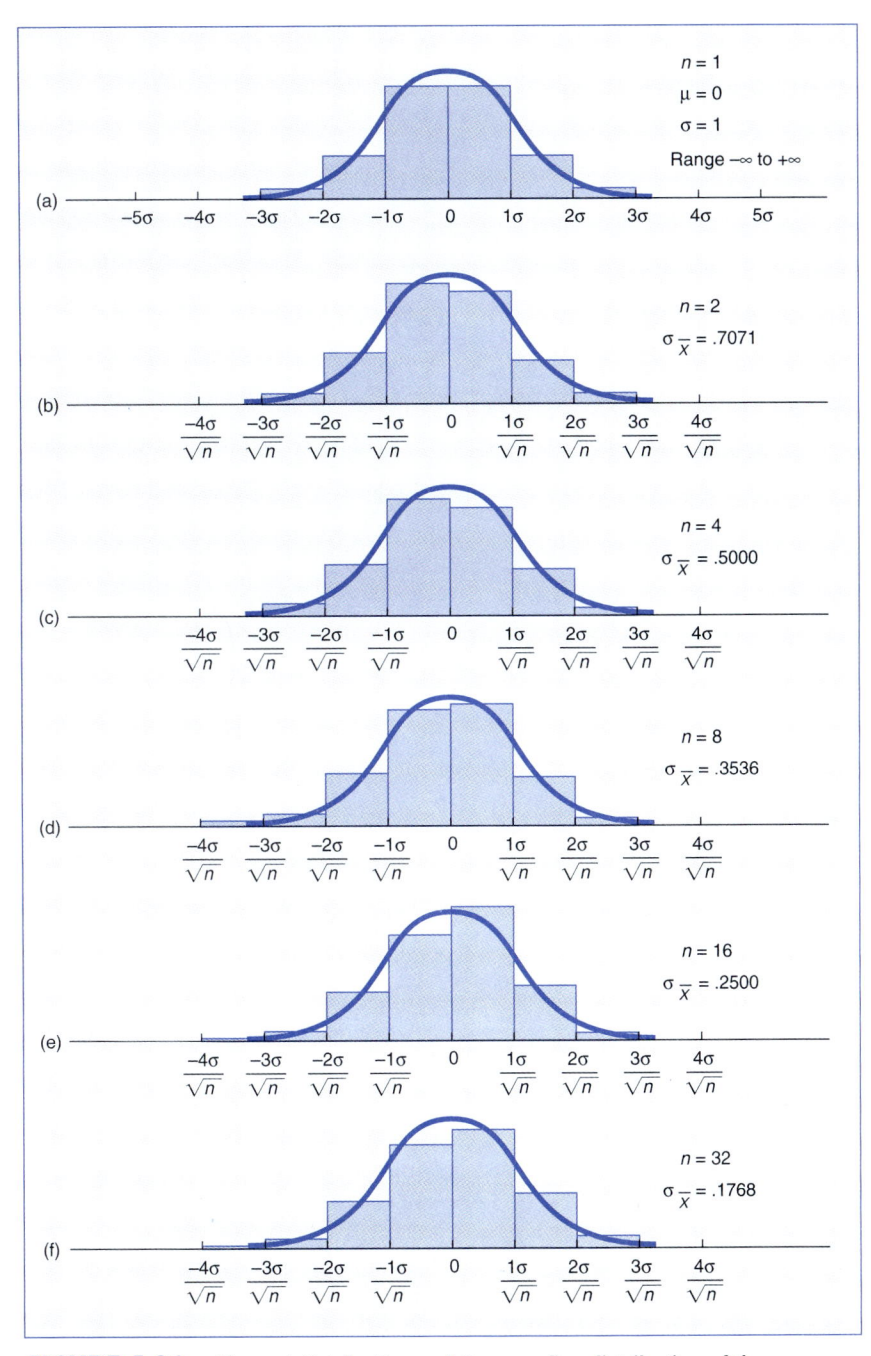

FIGURE 5.26 Normal distribution and the sampling distribution of the mean from 500 samples of size $n = 2, 4, 8, 16, 32$.

5.9 Sampling Distribution of the Mean

for this statement. For each sample size studied, the sampling distribution of the mean is *close* to the normal distribution that has been superimposed.

The second figure, Figure 5.27, presents the sampling distribution of the mean based on a population that follows a continuous uniform (rectangular) distribution. As depicted in part (a), for samples of size $n = 1$, each value in the population is equally likely. However, when samples of only two are selected, there is a peaking or *central limiting* effect already work-

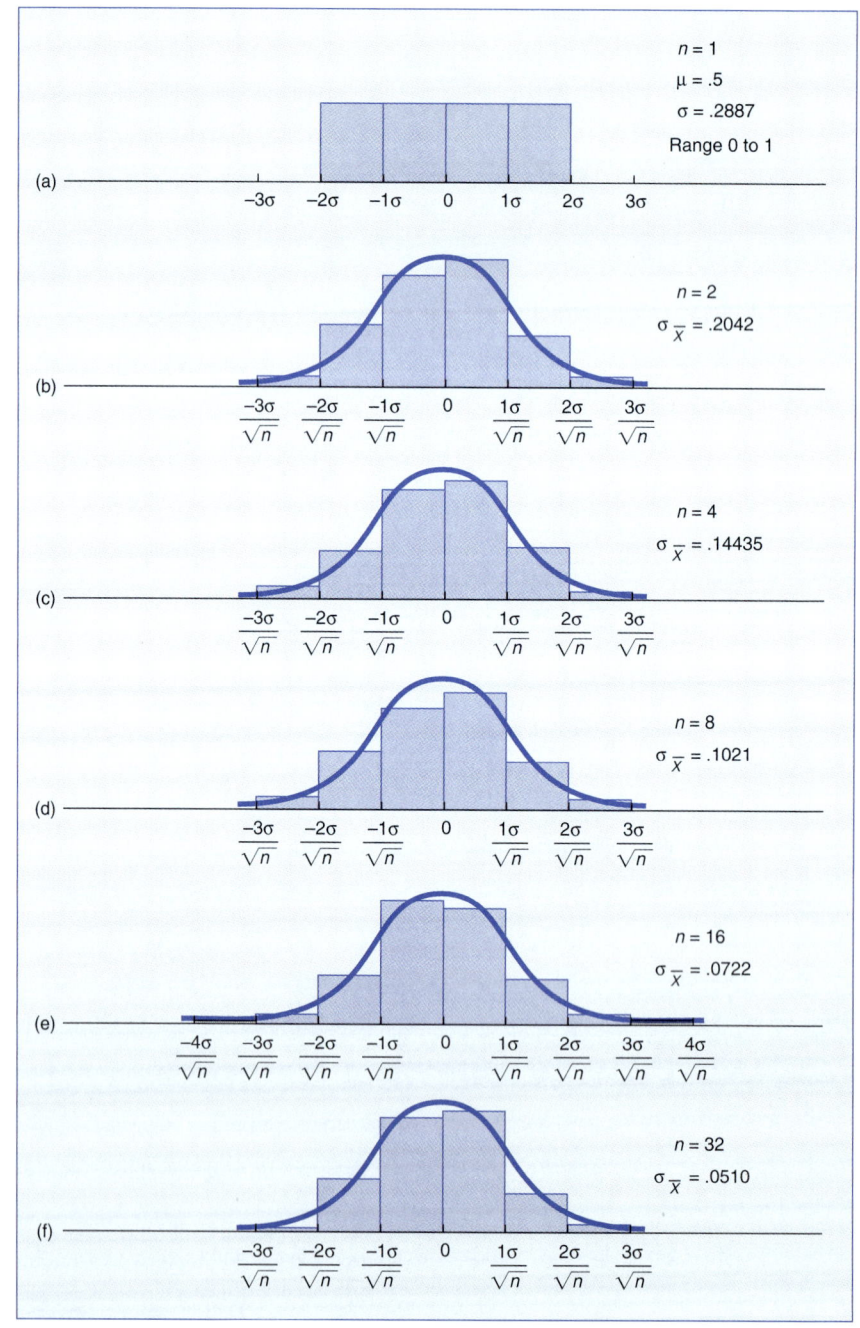

FIGURE 5.27 Continuous uniform (rectangular) distribution and the sampling distribution of the mean from 500 samples of size $n = 2, 4, 8, 16, 32$.

270 Chapter 5 The Normal Distribution and Sampling Distributions

ing. In this case, we can observe more values close to the mean of the population than far out at the extremes. As the sample size increases, the sampling distribution of the mean rapidly approaches a normal distribution. Once there are samples of at least eight observations, the sample mean approximately follows a normal distribution.

Finally, the third figure, Figure 5.28, depicts the sampling distribution of the mean obtained from an exponential distribution (see Section 5.7). From Figure 5.28, we note that

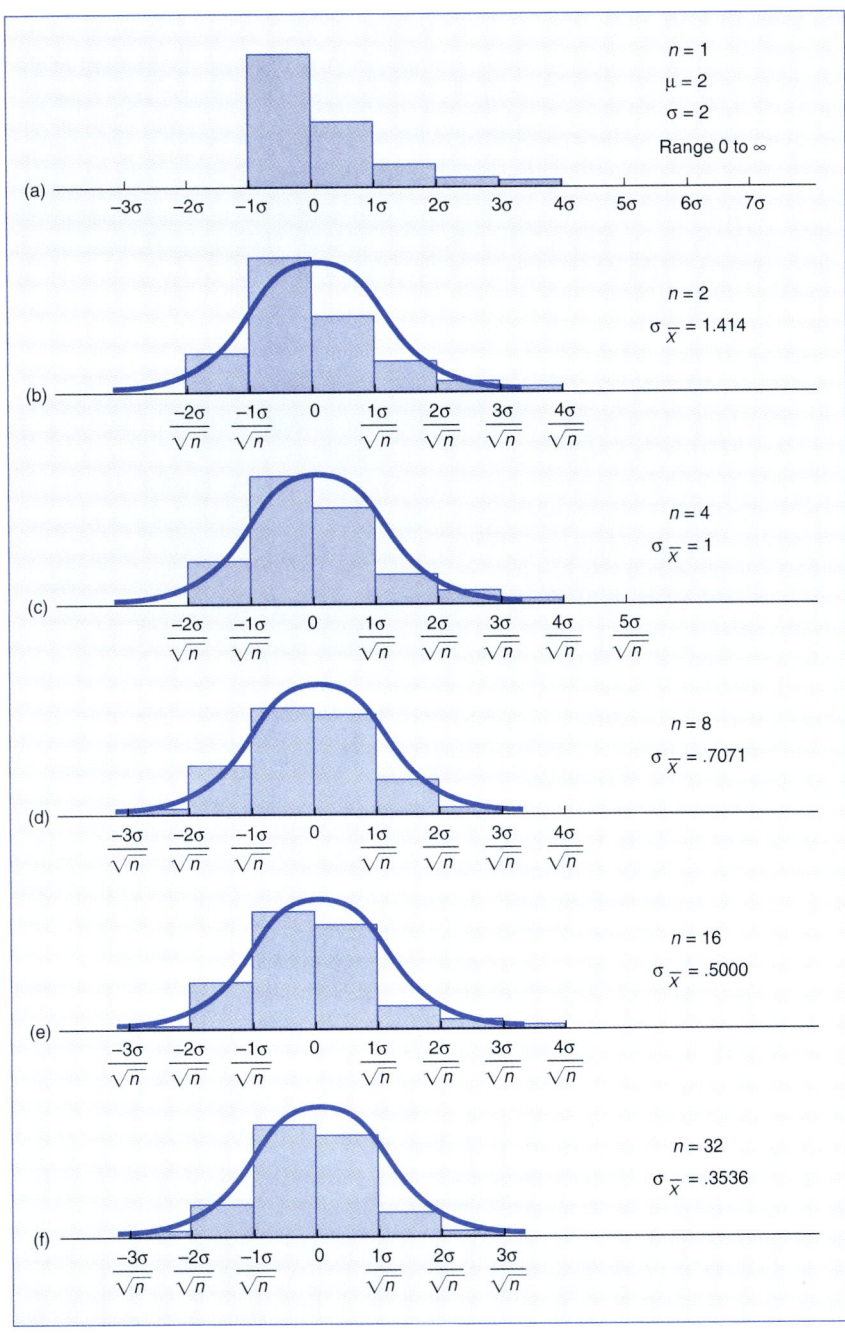

FIGURE 5.28 Exponential distribution and the sampling distribution of the mean from 500 samples of size $n = 2, 4, 8, 16, 32$.

5.9 Sampling Distribution of the Mean

as the sample size increases, the sampling distribution becomes less skewed. When samples of size 16 are taken, the distribution of the mean is slightly skewed, while for samples of size 32 the sampling distribution of the mean appears to be normally distributed.

We may now use the results obtained from our well-known statistical distributions (normal, uniform, exponential) to conclude as follows:

1. For most population distributions, regardless of shape, the sampling distribution of the mean will be approximately normally distributed if samples of at least 30 observations are selected.

2. If the population distribution is fairly symmetric, the sampling distribution of the mean will be approximately normal if samples of at least 15 observations are selected.

3. If the population is normally distributed, the sampling distribution of the mean will be normally distributed regardless of the sample size.

The central limit theorem, then, is of crucial importance in using statistical inference to draw conclusions about a population. It allows us to make inferences about the population mean without having to know the specific shape of the population distribution.

Problems for Section 5.9

5.23 Explain why a statistician would be interested in drawing conclusions about a population rather than merely describing the results of a sample.

5.24 Distinguish between a probability distribution and a sampling distribution.

5.25 For each of the following three populations, indicate what the sampling distribution for samples of 25 would consist of.
(a) Travel expense vouchers for a university in an academic year
(b) Absentee records (days absent/year) in 1995 for employees of a large manufacturing company
(c) Yearly sales (in gallons) of unleaded gasoline at service stations located in a particular county

5.26 The following data represent the number of days absent per year in a population of six employees of a small company:

1, 3, 6, 7, 7, 12

(a) Assuming that you sample *without* replacement
(1) Select all possible samples of size 2 and set up the sampling distribution of the mean.
(2) Compute the mean of all the sample means and also compute the population mean. Are they equal? What is this property called?
(3) Do parts (1) and (2) for all possible samples of size 3.
(4) Compare the shape of the sampling distribution of the mean obtained in parts (1) and (3). Which sampling distribution seems to have the least variability? Why?
(b) Assuming that you sample with replacement, do parts (1)–(4) of (a) and compare the results. Which sampling distributions seem to have the least variability, those in (a) or (b)? Why?

5.27 Referring to Table 2.6 on page 66 (North Carolina out-of-state tuition rates) and assuming that you sample *without* replacement

NCC&U.TXT

(a) Select all possible samples of size 2 and set up the sampling distribution of the mean.
(b) Compute the mean of all the sample means and also compute the population mean. Are they equal? What is this property called?

5.28 The diameter of Ping-Pong balls manufactured at a large factory is expected to be approximately normally distributed with a mean of 1.30 inches and a standard deviation of 0.04 inch. What is the probability that a randomly selected Ping-Pong ball will have a diameter
(a) Between 1.28 and 1.30 inches?
(b) Between 1.31 and 1.33 inches?

(c) Between what two values (symmetrically distributed around the mean) will 60% of the Ping-Pong balls fall (in terms of the diameter)?
(d) If many random samples of 16 Ping-Pong balls are selected
 (1) What would the mean and standard error of the mean be expected to be?
 (2) What distribution would the sample means follow?
 (3) What proportion of the sample means would be between 1.28 and 1.30 inches?
 (4) What proportion of the sample means would be between 1.31 and 1.33 inches?
 (5) 60% of the sample means will be between what two values?
(e) Compare the answers of (a) with (d)(3) and (b) with (d)(4). Discuss.
(f) Explain the difference in the results of (c) and (d)(5).
(g) Which is more likely to occur—an individual ball above 1.34 inches, a sample mean above 1.32 inches in a sample of size 4, or a sample mean above 1.31 inches in a sample of size 16? Explain.

5.29 Long-distance telephone calls are normally distributed with $\mu = 8$ minutes and $\sigma = 2$ minutes. If random samples of 25 calls were selected
(a) Compute $\sigma_{\bar{X}}$.
(b) What proportion of the sample means would be between 7.8 and 8.2 minutes?
(c) What proportion of the sample means would be between 7.5 and 8 minutes?
(d) If random samples of 100 calls were selected, what proportion of the sample means would be between 7.8 and 8.2 minutes?
(e) Explain the difference in the results of (b) and (d).
(f) Which is more likely to occur—an individual value above 11 minutes, a sample mean above 9 minutes in a sample of 25 calls, or a sample mean above 8.6 minutes in a sample of 100 calls? Explain.

5.30 The amount of time a bank teller spends with each customer has a population mean $\mu = 3.10$ minutes and standard deviation $\sigma = 0.40$ minute. If a random sample of 16 customers is selected
(a) What is the probability that the average time spent per customer will be at least 3 minutes?
(b) There is an 85% chance that the sample mean will be below how many minutes?
(c) What assumption must be made in order to solve (a) and (b)?
(d) If a random sample of 64 customers is selected, there is an 85% chance that the sample mean will be below how many minutes?
(e) What assumption must be made in order to solve (d)?
(f) Which is more likely to occur—an individual time below 2 minutes, a sample mean above 3.4 minutes in a sample of 16 customers, or a sample mean below 2.9 minutes in a sample of 64 customers? Explain.

5.10 SAMPLING DISTRIBUTION OF THE PROPORTION

When dealing with a categorical variable where each individual or item in the population can be classified as either possessing or not possessing a particular characteristic such as male or female, or prefer Brand A or do not prefer Brand A, the two possible outcomes could be assigned scores of 1 or 0 to represent the presence or absence of the characteristic. If only a random sample of n individuals were available, the sample mean for such a categorical variable would be found by summing all the 1 and 0 scores and then dividing by n. For example, if in a sample of five individuals, three preferred Brand A and two did not, there would be three 1's and two 0's. Summing the three 1's and two 0's and dividing by the sample size of 5 would give us a mean of 0.60, which is also the proportion of individuals in the sample who prefer Brand A. Therefore, when dealing with categorical data, the sample mean $\bar{X}$ (of the 1 and 0 scores) is the sample proportion p_s having the characteristic of interest. Thus, the sample proportion p_s can be defined as

$$p_s = \frac{X}{n} = \frac{\text{number of successes}}{\text{sample size}} \qquad (5.10)$$

The sample proportion p_s has the special property that it must be between 0 and 1. If all individuals possessed the characteristic, they would each be assigned a score of 1, and p_s would be equal to 1. If half the individuals possessed the characteristic, half would be assigned a score of 1, the other half would be assigned a score of 0, and p_s would be equal to .5. If none of the individuals possessed the characteristic, they would each be assigned a score of 0, and p_s would be equal to 0.

While the sample mean $\bar{X}$ is an estimator of the population mean μ, the statistic p_s is an estimator of the population proportion p. By analogy to the sampling distribution of the mean, the standard error of the proportion σ_{p_s} would be

$$\sigma_{p_s} = \sqrt{\frac{p(1-p)}{n}} \tag{5.11}$$

The **sampling distribution of the proportion** would actually follow the binomial distribution discussed in Section 4.13. However, the normal distribution can be used to approximate the binomial distribution when np and $n(1-p)$ are each at least 5. In most cases in which inferences are being made about the proportion, the sample size is substantial enough to meet the conditions for using the normal approximation (see Reference 1). Thus, in many instances, we may use the normal distribution to evaluate the sampling distribution of the proportion.

To illustrate the sampling distribution of the proportion, we will refer to the following example.

● **Application** The manager of the local branch of a savings bank has determined that 40% of all depositors have multiple accounts at the bank. If a random sample of 200 depositors is selected, what is the probability that the sample proportion of depositors with multiple accounts will be between .40 and .43?

Since $np = 200(.40) = 80$ and $n(1-p) = 200(.60) = 120$, the sampling distribution of the proportion can be assumed to be normally distributed. Thus, we have

$$Z = \frac{\bar{X} - \mu_{\bar{X}}}{\sigma_{\bar{X}}} = \frac{\bar{X} - \mu}{\frac{\sigma}{\sqrt{n}}}$$

and because we are dealing with sample proportions (not sample means), we have

$$p_s = \text{sample proportion}$$

$$p = \text{population proportion}$$

$$\sigma_{p_s} = \sqrt{\frac{p(1-p)}{n}}$$

and, substituting p_s for $\bar{X}$, p for $\mu_{\bar{X}}$, and $\sigma_{p_s} = \sqrt{p(1-p)/n}$ for $\sigma_{\bar{X}}$, we have

$$Z \cong \frac{p_s - p}{\sqrt{\frac{p(1-p)}{n}}} \tag{5.12}$$

Substituting,

$$Z \cong \frac{p_s - p}{\sqrt{\dfrac{p(1-p)}{n}}}$$

$$Z \cong \frac{.43 - .40}{\sqrt{\dfrac{(.40)(.60)}{200}}} = \frac{.03}{\sqrt{\dfrac{.24}{200}}}$$

$$= \frac{.03}{.0346}$$

$$= 0.87$$

Using Table E.2, the area under the normal curve from $Z = 0$ to $Z = 0.87$ is .3078. Therefore, the probability of obtaining a sample proportion between .40 and .43 is .3078. This means that if the true proportion of successes in the population were .40, then 30.78% of the samples of size 200 would be expected to have sample proportions between .40 and .43. (See Figure 5.29.)

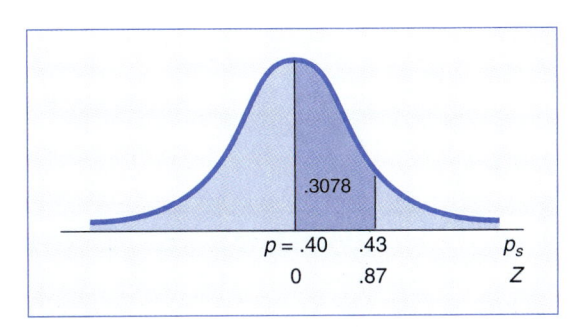

FIGURE 5.29 Diagram of normal curve needed to find the area between the proportions .40 and .43.

Problems for Section 5.10

5.31 Historically, 10% of a large shipment of machine parts are defective. If random samples of 400 parts are selected, what proportion of the samples will have
  (a) Between 9% and 10% defective parts?
  (b) Less than 8% defective parts?
  (c) If a sample size of only 100 was selected, what would your answers have been in (a) and (b)?
  (d) Which is more likely to occur—a percent defective above 13% in a sample of 100 or a percent defective above 10.5% in a sample of 400? Explain.

5.32 A political pollster is conducting an analysis of sample results in order to make predictions on election night. Assuming a two-candidate election, if a specific candidate receives at least 55% of the vote in the sample, then that candidate will be forecast as the winner of the election. If a random sample of 100 voters is selected, what is the probability that a candidate will be forecast as the winner when
  (a) The true percentage of his vote is 50.1%?
  (b) The true percentage of his vote is 60%?
  (c) The true percentage of his vote is 49% (and he will actually lose the election)?
  (d) If the sample size is increased to 400, what will your answer be to (a), (b), and (c)? Discuss.

5.33 Based on past data, 30% of the credit card purchases at a large department store are for amounts above $100. If random samples of 100 credit card purchases are selected
  (a) What proportion of samples are likely to have between 20% and 30% of the purchases over $100?

(b) Within what symmetrical limits of the population percentage will 95% of the sample percentages fall?

5.34 Suppose a marketing experiment is to be conducted in which students are to taste two different brands of soft drink. Their task is to correctly identify the brand tasted. If random samples of 200 students are selected and it is assumed that the students have no ability to distinguish between the two brands
(a) What proportion of the samples will have between 50% and 60% of the identifications correct?
(b) Within what symmetrical limits of the population percentage will 90% of the sample percentages fall?
(c) What is the probability of obtaining a sample percentage of correct identifications in excess of 65%?
(d) Which is more likely to occur—more than 60% correct identifications in a sample of 200 or more than 55% correct identifications in a sample of 1,000? Explain.

(*Hint*: If an individual has no ability to distinguish between the two soft drinks, then each one is equally likely to be selected.)

5.35 Historically, 93% of the deliveries of an overnight mail service arrive before 10:30 on the following morning. If random samples of 500 deliveries are selected, what proportion of the samples will have
(a) Between 93% and 95% of the deliveries arriving before 10:30 on the following morning?
(b) More than 95% of the deliveries arriving before 10:30 on the following morning?
(c) If samples of size 1,000 are selected, what will your answers be in (a) and (b)?
(d) Which is more likely to occur—more than 95% of the deliveries in a sample of 500 or less than 90% in a sample of 1,000 arriving before 10:30 on the following morning? Explain.

5.11 USING MICROSOFT EXCEL TO SELECT RANDOM SAMPLES AND SIMULATE SAMPLING DISTRIBUTIONS

In Sections 5.9 and 5.10, we developed the concept of the sampling distribution, first for the mean and then for the proportion. We observed how the central limit theorem can be applied to several different populations and also that under certain circumstances, the normal distribution can be used to approximate the binomial distribution. In this section we will use Microsoft Excel to select a random sample of values and to generate distributions of random numbers from a variety of populations.

5.11.1 Using Microsoft Excel Functions to Select a Simple Random Sample

In Section 1.10, we learned how to use a table of random numbers to select a simple random sample. Now we will use Microsoft Excel to obtain a sample of random numbers of any desired size.

There are two Excel functions, RANDBETWEEN and RAND, that provide random numbers based on simple random sampling. The RANDBETWEEN function produces *integer* values, while the RAND function provides values that are between 0 and 1.

The RANDBETWEEN Function takes the form

$$\text{RANDBETWEEN(lower limit, upper limit)}$$

The RAND Function takes the form

$$\text{RAND()}$$

and provides a random number between 0 and 1. If we wish to change the range of the numbers, we may do so by multiplying the RAND function by the value of the desired upper limit. For example, to generate a random number between 0 and 1000, we would enter -RAND()*1000 into the appropriate cell.

To use the RANDBETWEEN function to select a set of 25 employees from a population of 500 employees as we did in Section 1.10, open a new workbook and rename the active sheet Sample. After first entering the heading SAMPLE in cell A1, we need to enter the formula =RANDBETWEEN(1,500) in cell A2 and copy this formula down an additional 24 cells, from cells A3 through A26. You will observe that the cell range A2:A26 now contains random numbers between 001 and 500. However, depending on your results, you may also observe that some random numbers have been repeated. This is due to the fact that the RANDBETWEEN function generates random numbers by sampling *with* replacement. If you wish to sample *without* replacement, you need to remove any random numbers that have been repeated and, for those cells, perform the random selection again until none of the random numbers obtained represent repeating numbers.

5.11.2 Using the Data Analysis Tool to Select a Random Sample from a Population

If a population of values is already stored on a worksheet, the Sampling option of the Data Analysis tool can be used to obtain a sample of size n from the population of N items *with replacement*. This may be illustrated by referring to the population of 90 colleges and universities in Pennsylvania. Suppose we wanted to select a random sample of 10 schools from this population. We would open the PA-POP-1.XLS workbook and click on Tools | Data Analysis | Sampling. We would see the Sampling dialog box illustrated in Figure 5.5.Excel.

PA-POP-1.XLS

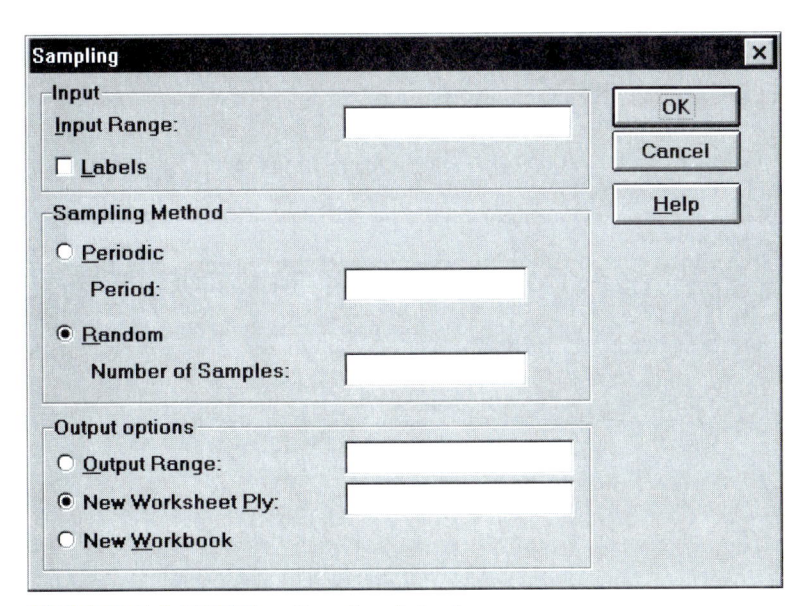

FIGURE 5.5.EXCEL Sampling dialog box.

We would then proceed as follows:

① Enter Data!B2:B91 in the Input Range edit box.

② Select the Random option button and enter 10 in the Number of Samples edit box.

③ Select the New Worksheet Ply option button and enter the name Sample in the edit box next to this choice, and then click the OK button.

FIGURE 5.6.EXCEL Random sample of the out-of-state tuition rates for 10 Pennsylvania colleges using Microsoft Excel.

	A
1	8.4
2	6.1
3	9.1
4	9.3
5	18.3
6	4.4
7	10
8	17.9
9	6
10	9.3

Figure 5.6.Excel represents the *particular* random sample that was obtained.

5.11.3 Using the Data Analysis Tool to Generate Distributions of Random Numbers

5-11-3.XLS

The Random Number Generation option of the Data Analysis tool generates sets of random numbers selected from a variety of distributions including the uniform and normal. In addition, discrete distributions stored on a worksheet can be used as the basis of generating random numbers.

To generate random numbers, open a new workbook, rename the sheet Uniform, select Tools | Data Analysis, and then select Random Number Generation from the Analysis Tools list box. Click the OK button. You will now see the Random Number Generation dialog box as displayed in Figure 5.7.Excel.

At the top of this dialog box, the Number of Variables edit box is used to indicate the number of different samples of size *n* that are to be selected. The Number of Random

FIGURE 5.7.EXCEL Random Number dialog box.

278 Chapter 5 The Normal Distribution and Sampling Distributions

Numbers edit box allows us to input the sample size n for each sample. The Distribution dropdown list box contains a set of choices. Of particular interest are the choices Uniform, Normal, and Discrete. If the Uniform probability distribution check box is selected, a Parameters check box appears that asks for lower and upper limits for the random numbers to be generated. If the Normal check box is selected, a Parameters check box appears that asks for the value of the mean and standard deviation. If the Discrete check box is selected, a Value and Probability Input Range appears. The cell location of all the X values and the probability of X values for the discrete probability distribution of interest would now be entered.

To illustrate the use of the Random Number Generator tool for simulating sampling distributions, we will develop sampling distributions of the mean based on a uniform population, a normal population, and a skewed population.

To develop a simulation of the sampling distribution of the mean with a uniformly distributed population of 100 samples of $n = 30$, select Tools | Data Analysis, select Random Number Generation and click OK as previously described. Then proceed as follows:

1. In the Number of Variables edit box, enter 100.
2. In the Number of Random Numbers edit box, enter 30.
3. Choose Uniform from the Distribution drop-down list.
4. Under Parameters, enter Between 0 and 1.
5. Select the New Worksheet Ply option button, and in the edit box to its right, enter Uniform. Click the OK button. (The resulting sheet will take a while for Microsoft Excel to produce.)

The data you obtain on the Uniform worksheet will display the 30 values selected from a uniform distribution for each of 100 samples. To obtain the sampling distribution of the mean, you need to do the following:

1. Enter the label Sample Means: in cell A31.
2. Compute the sample means by entering =AVERAGE(A1:A30) in cell A32 and copy this across all 100 columns (for the 100 samples) through cell CV32.
3. Enter the label Overall Average in cell A33 and compute the average of all 100 sample averages by entering =AVERAGE(A32:CV32) in cell A34. Note the closeness of the overall average obtained to the population average of 0.5 (the average of 0 and 1).
4. With the Uniform worksheet active, select Tools | Data Analysis, select Histogram from the Analysis Tools list box and click OK.
5. For the Input Range, enter A32:CV32.
6. Select the New Worksheet Ply option button and, in the edit box to its right, enter Uniform Histogram. Select the Chart Output check box and click the OK button. (By not providing an entry for Bin Range, Excel will create class intervals of equal width.)

The resulting bar chart can be changed into a histogram using the procedure first discussed in Section 2.7.3. A histogram similar to the one displayed in Figure 5.8.Excel will be obtained.

Now that we have obtained the sampling distribution of the mean for a uniformly distributed population, we can turn to a normally distributed population. To develop a simulation of the sampling distribution of the mean with a normally distributed population, insert a new worksheet into your workbook, rename it Normal and select Tools | Data Analysis. Then select Random Number Generation and click OK. If we wanted to select 100 samples of $n = 30$, we would do the following:

1. In the Number of Variables edit box, enter 100.
2. In the Number of Random Numbers edit box, enter 30.

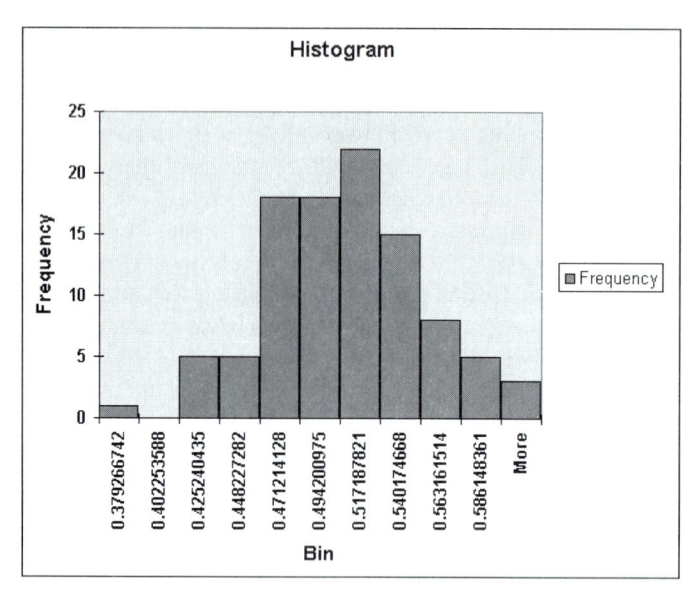

FIGURE 5.8.EXCEL Sampling distribution of the mean for samples of n = 30 from a uniform population obtained from Microsoft Excel.

③ Choose Normal from the Distribution drop-down list box.

④ Under Parameters, enter 0 in the Mean= edit box and 1 in the Standard Deviation= edit box to obtain the standardized normal distribution.

⑤ Select the New Worksheet Ply option button, and enter Normal as the new worksheet name. Click the OK button.

The data you obtain on the Normal worksheet will display the 30 values selected from a normal distribution for each of 100 samples. (The resulting sheet will take a while for Microsoft Excel to produce.) To obtain the sampling distribution of the mean, you need to do the following:

① Enter the label Sample Means: in cell A31.

② Compute the sample means by entering =AVERAGE(A1:A30) in cell A32 and copy this across all 100 columns (for the 100 samples) through cell CV32.

③ Enter the label Overall Average in cell A33 and compute the average of all 100 sample averages by entering =AVERAGE(A32:CV32) in cell A34. Note the closeness of the overall average obtained to the population average of 0.

④ Enter the label Standard Error of the Mean: in cell A35 and compute the standard deviation of all 100 sample means by entering =STDEV(A32:CV32) in cell A36. Note the closeness of this standard error of the mean to the population standard error of the mean equal to $1/\sqrt{30} = .18257$.

⑤ With the same sheet still active, select Tools | Data Analysis, then select Histogram and click OK.

⑥ For the Input Range, enter A32:CV32.

⑦ Select the New Worksheet Ply option button, enter Normal Histogram as the new name, select the Chart Output check box, and click the OK button. (By not providing an entry for Bin Range, Excel will create class intervals of equal width.)

Histogram

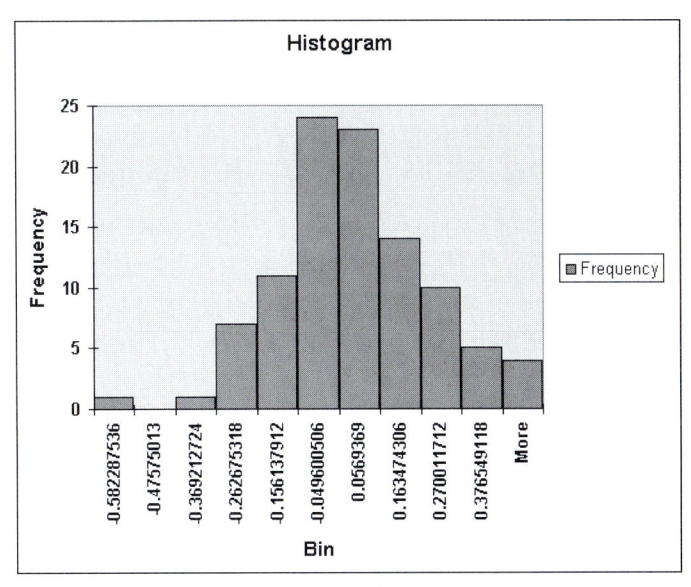

FIGURE 5.9.EXCEL Sampling distribution of the mean for Samples of $n = 30$ from a normal population obtained from Microsoft Excel.

The bar chart can be changed into a histogram using the procedure first discussed in Section 2.7.3. A histogram similar to the one displayed in Figure 5.9.Excel will be obtained.

The Random Number Generation option of the Data Analysis Histogram tool can also be used to develop sampling distributions based on any discrete distribution. To illustrate the sampling distribution of the mean from a right-skewed distribution, suppose the X values and probability of X values are entered and the expected value, variance, and standard deviation are calculated. Using the approach discussed in Section 4.11.4, the expected value or average of this distribution is .575 and the standard deviation is .8969.

To develop a simulation of the sampling distribution of the mean with the right-skewed population displayed in Table 5.4.Excel, continue with the same workbook (or open a new one), insert a worksheet and rename the sheet RightSkewedData. Select Tools | Data Analysis, then select Random Number Generation and click OK. If we wanted to select 100 samples of $n = 30$, we would do the following:

Table 5.4.Excel Design for the RightSkewed Data worksheet.

	A	B	C
1	Calculating Expected Values and Variances		
2			
3	X	P(X)	[(X–E(X)]2
4	0	0.6	=(A4–$B$11)^2
5	1	0.3	=(A5–$B$11)^2
6	2	.05	=(A6–$B$11)^2
7	3	.03	=(A7–$B$11)^2
8	4	.015	=(A8–$B$11)^2
9	5	.005	=(A9–$B$11)^2
10		SUM(B4:B9)	
11	Expected Value	=SUMPRODUCT(A4:A9,B4:B9)	
12	Variance	=SUMPRODUCT(C4:C9,B4:B9)	
13	Standard Deviation	=SQRT(B12)	

1. In the Number of variables edit box, enter 100.

2. In the Number of Random Numbers edit box, enter 30.

3. Choose Discrete from the Distribution drop-down list.

4. In the Value and Probability Input Range edit box, enter RightSkewedData!A4:B9.

5. Select the New Worksheet Ply option button. Enter RightSkewed as the new worksheet name, and click the OK button.

The output you will obtain on the RightSkewed worksheet will display the 30 values selected from the right-skewed distribution for each of 100 samples. To obtain the sampling distribution of the mean, you need to do the following:

1. Enter the label Sample Means: in cell A31.

2. Compute the arithmetic mean for each sample by entering =AVERAGE(A1:A30) in cell A32 and copy this across all 100 columns (for the 100 samples) through cell CV32.

3. Enter the label Overall Average: in cell A33 and compute the average of all 100 sample averages by entering =AVERAGE(A32:CV32) in cell A34. Note the closeness of the overall average obtained to the population average of .575.

4. Enter the label Standard Error of the Mean: in cell A35 and compute the standard deviation of all 100 sample means by entering =STDEV(A32:CV32) in cell A36. Note the closeness of this standard error of the mean to the population standard error of the mean equal to $.8969/\sqrt{30} = .1638$.

5. Select Tools | Data Analysis, then select Histogram and click OK.

6. For the Input Range, enter RightSkewed!A32:CV32.

7. Select the New Worksheet Ply option button, enter RightSkewed Histogram as the new worksheet name, select the Chart Output check box, and click the OK button. (By not providing an entry for Bin Range, Excel will create class intervals of equal width.)

The resulting bar chart can be changed into a histogram using the procedure first discussed in Section 2.7.3. A histogram similar to the one displayed in Figure 5.10.Excel will be obtained.

The use of a discrete population to simulate the sampling distribution of the mean can also be applied to illustrate the use of the normal distribution to approximate the binomial distribution. In Section 5.10, we noted that the normal distribution can be used to approximate the binomial distribution when np and $n(1 - p)$ are each at least 5. Thus, to study the accuracy of the approximation, we could study the binomial distribution for various values of n and p. This can be done by first developing the probabilities for each number of successes by following the approach used in Section 4.13.3 and then using the Random Number Generation option of the Data Analysis tool (with Discrete selected) to develop the sampling distribution.

▲ WHAT IF EXAMPLE

The Random Number Generation option of the Data Analysis tool offers a multitude of possibilities for exploring the effects of changes in the sampling distribution of the mean. Certainly, the most frequently investigated situation represents the effect of sample size on the shape of the sampling distribution of the mean. This can be studied for uniformly and normally distributed populations, and for skewed populations. For a given sample size, we can also study the effect of the shape of the population on the sampling distribution of the mean. In addition, the accuracy of the normal approximation to the binomial distribution can be studied for varying values of n and p.

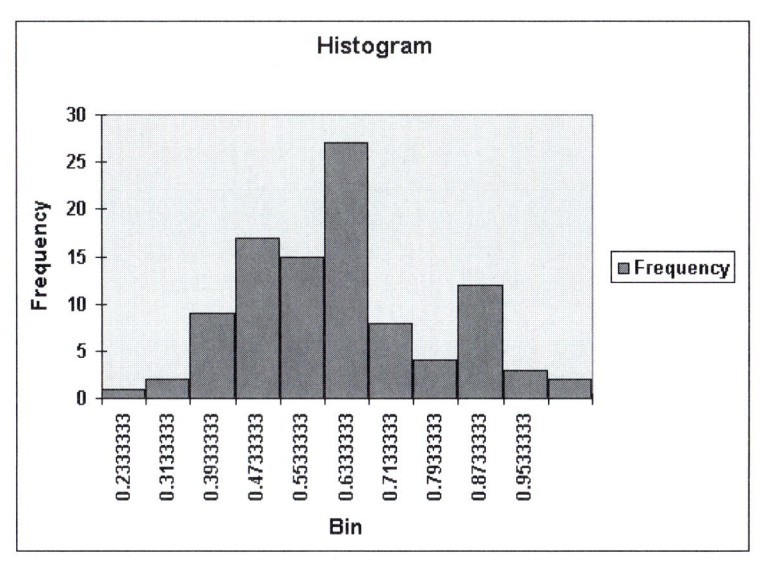

FIGURE 5.10.EXCEL Sampling distribution of the mean for $n = 30$ for a right-skewed population obtained from Microsoft Excel.

5.12 SAMPLING FROM FINITE POPULATIONS

The central limit theorem and the standard errors of the mean and the proportion were based on the premise that the samples selected were chosen *with* replacement. However, in virtually all survey research, sampling is conducted *without* replacement from populations that are of a finite size N. In these cases, particularly when the sample size n is not small as compared with the population size N (i.e., more than 5% of the population is sampled) so that $n/N > .05$, a **finite population correction factor (fpc)** should be used in defining both the standard error of the mean and the standard error of the proportion. The finite population correction factor may be expressed as

$$\text{fpc} = \sqrt{\frac{N-n}{N-1}} \qquad (5.13)$$

where
n = sample size

N = population size

Therefore, when dealing with means, we have

$$\sigma_{\bar{X}} = \frac{\sigma}{\sqrt{n}} \sqrt{\frac{N-n}{N-1}} \tag{5.14}$$

and when we are referring to proportions, we have

$$\sigma_{p_s} = \sqrt{\frac{p(1-p)}{n}} \sqrt{\frac{N-n}{N-1}} \tag{5.15}$$

Examining the formula for the finite population correction factor [Equation (5.13)], we see that the numerator will always be smaller than the denominator, so the correction factor will be less than 1. Since this finite population correction factor is multiplied by the standard error, the standard error becomes smaller when corrected. This means that we get more accurate estimates because we are sampling a large segment of the population.

● **Application**   We may illustrate the application of the finite population correction factor using two examples previously discussed in this chapter. In Section 5.9.3 on page 266, a sample of 25 cereal boxes was selected from a filling process. Suppose that a population of 2,000 boxes were filled on this particular day. Using the finite population correction factor, we would have

$$\sigma = 15, \quad n = 25, \quad N = 2{,}000$$

$$\sigma_{\bar{X}} = \frac{\sigma}{\sqrt{n}} \sqrt{\frac{N-n}{N-1}}$$

$$= \frac{15}{\sqrt{25}} \sqrt{\frac{2{,}000 - 25}{2{,}000 - 1}}$$

$$= 3\sqrt{.988} = 2.982$$

The probability of obtaining a sample whose mean is between 365 and 368 grams is computed as follows:

$$Z = \frac{\bar{X} - \mu_{\bar{X}}}{\sigma_{\bar{X}}} = \frac{-3}{2.982} = -1.01$$

From Table E.2, the approximate area under the normal curve is .3438.

It is evident in this example that the use of the finite population correction factor had a very small effect on the standard error of the mean and the subsequent area under the normal curve, since the sample was only 1.25% of the population size.

In the multiple banking accounts example on page 274, suppose there were a total of 1,000 different depositors at the bank. The previous sample of size 200 out of this finite population results in the following:

$$\sigma_{p_s} = \sqrt{\frac{p(1-p)}{n}} \sqrt{\frac{N-n}{N-1}}$$

$$= \sqrt{\frac{(.4)(.6)}{200}} \sqrt{\frac{1{,}000 - 200}{1{,}000 - 1}}$$

$$= \sqrt{\frac{24}{200}} \sqrt{\frac{800}{999}} = \sqrt{\frac{24}{200}} \sqrt{.801}$$

$$= (.0346)(.895) = .031$$

Using $\sigma_p = .031$ as the standard error of the sample proportion in Equation (5.12), then $Z = .03/.031 = 0.97$, and, from Table E.2, the appropriate area under the normal curve is .3340. In this example, the use of the finite population correction factor had a moderate effect on the standard error of the proportion and on the area under the normal curve, since the sample size is 20% (i.e., $n/N = .20$) of the population.

Problems for Section 5.12

- **5.36** Referring to Problem 5.28 on page 273, if the population consisted of a box of 200 Ping-Pong balls, what would be your answer to part (d)(4) of that problem?

- **5.37** Referring to Problem 5.30 on page 273, if there were a population of 500 customers, what would be your answers to (a) and (b) of that problem?

- **5.38** Referring to Problem 5.31 on page 275, if the shipment included 5,000 machine parts, what would be your answers to (a) and (b) of that problem?

- **5.39** Referring to Problem 5.35 on page 276, if the population consisted of 10,000 deliveries, what would be your answers to (a) and (b) of that problem?

5.13 CONTINUOUS DISTRIBUTIONS AND SAMPLING DISTRIBUTIONS: A REVIEW

As seen in the summary chart in this chapter we have studied the normal distribution, the exponential distribution, the sampling distribution of the sample mean, and the sampling distribution of the sample proportion. The importance of the normal distribution in statistics has been further emphasized by examining the central limit theorem. We have seen that knowledge of a population distribution is not always necessary in drawing conclusions from a sampling distribution of the mean or proportion.

To be sure you understand this material, you should be able to answer the following conceptual questions:

1. Why is it that only one table of the normal distribution is needed to find any probability under the normal curve?

2. How would you find the area between two values under the normal curve when both values are on the same side of the mean?

3. How would you find the X value that corresponds to a given percentile of the normal distribution?

4. Why do the individual observations have to be converted to standard normal ordered values in order to develop a normal probability plot?

5. Why is the sample arithmetic mean an unbiased estimator of the population arithmetic mean?

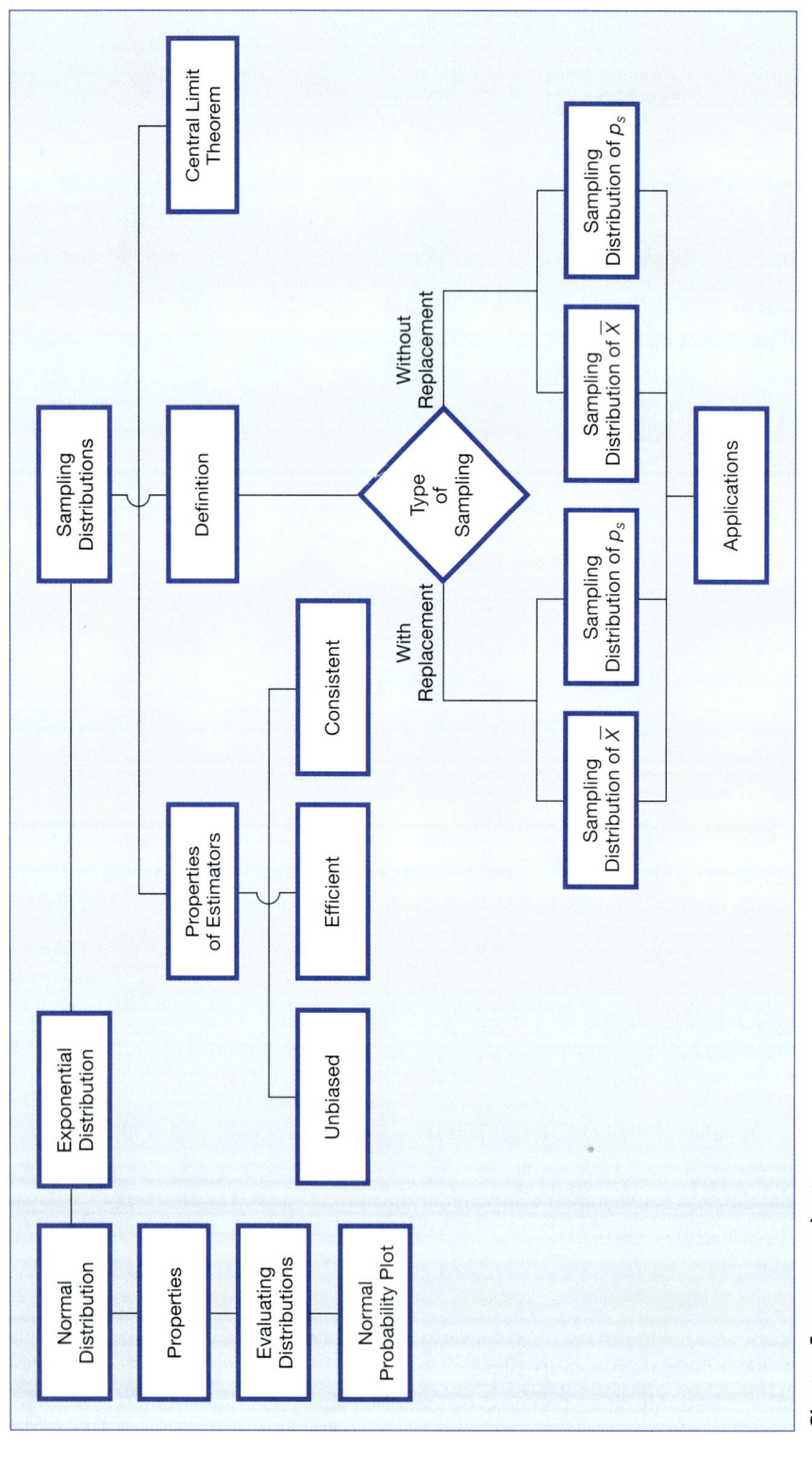

Chapter 5 summary chart.

6. Why does the standard error of the mean decrease as the sample size *n* increases?
7. Why does the sampling distribution of the mean follow a normal distribution for a *large enough* sample size even though the population may not be normally distributed?
8. Under what circumstances does the sampling distribution of the proportion approximately follow the normal distribution?
9. What is the effect on the standard error of using the finite population correction factor?

The concepts concerning sampling distributions are central to the development of statistical inference. The main objective of statistical inference is to take information based only on a sample and use this information to draw conclusions and make decisions about various population values. The statistical techniques developed to achieve these objectives are discussed fully in the next four chapters (confidence intervals and tests of hypotheses).

Getting It All Together

Key Terms

central limit theorem 268
consistency 262
continuous probability density function 226
efficiency 262
exponential distribution 258
finite population correction factor (fpc) 283
inverse normal scores transformation 249
law of large numbers 263
normal distribution 227
normal probability density function 227
normal probability plot 249

quantiles 248
sampling distribution 260
sampling distribution of the mean 264
sampling distribution of the proportion 274
standard error of the mean 264
standard normal quantile 249
standardized normal distribution 229
statistical inference 260
transformation formula 229
unbiasedness 261

Chapter Review Problems

Note: *Microsoft Excel may be used to solve the Chapter Review Problems.*

5.40 An industrial sewing machine uses ball bearings that are targeted to have a diameter of 0.75 inch. The specification limits under which the ball bearing can operate are 0.74 inch (lower) and 0.76 inch (upper). Past experience has indicated that the actual diameter of the ball bearings is approximately normally distributed with a mean of 0.753 inch and a standard deviation of 0.004 inch. What is the probability that a ball bearing will be
 (a) Between the target and the actual mean?
 (b) Between the lower specification limit and the target?
 (c) Above the upper specification limit?
 (d) Below the lower specification limit?
 (e) Above which value in diameter will 93% of the ball bearings be?

5.41 Suppose that the content of bottles of soft drink has been found to be normally distributed with a mean of 2.0 liters and a standard deviation of 0.05 liter. Bottles that contain less than 95% of the listed net content (1.90 liters in this case) can make the manufacturer subject to penalty by the state office of consumer affairs, while bottles that have a net content above 2.10 liters may cause excess spillage upon opening.
 (a) What proportion of the bottles will contain:

S-5-41.XLS

(1) Between 1.90 and 2.0 liters?
(2) Between 1.90 and 2.10 liters?
(3) Below 1.90 liters?
(4) Below 1.90 liters or above 2.10 liters?
(5) Above 2.10 liters?
(6) Between 2.05 and 2.10 liters?

(b) 99% of the bottles would be expected to contain at least how much soft drink?

(c) 99% of the bottles would be expected to contain an amount that is between which two values (symmetrically distributed)?

(d) Explain the difference in the results in (b) and (c).

(e) Suppose that in an effort to reduce the number of bottles that contain less than 1.90 liters, the bottler sets the filling machine so that the mean is 2.02 liters. Under these circumstances, what would be your answers in (a), (b), and (c)?

5.42 A city agency that processes building renovation permits has a policy that states that the permit is free if it is not ready at the end of 5 business days from when the application is made. Processing time is measured from when the permit is received (the time is stamped) to when the application has been fully processed.

(a) If the process has a mean of 3 days and a standard deviation of 1 day, what proportion of the permits will be free?

(b) If the process has a mean of 2 days and a standard deviation of 1.5 days, what proportion of the permits will be free?

(c) Which process [(a) or (b)] will result in more free permits? Explain.

(d) For the process described in (a), would it be better to focus on reducing the average to 2 days, or the standard deviation to 0.75 day? Explain.

5.43 Sally D. is 67 inches tall and weighs 135 pounds. If the heights of women are normally distributed with $\mu = 65$ inches and $\sigma = 2.5$ inches and if the weights of women are normally distributed with $\mu = 125$ pounds and $\sigma = 10$ pounds, determine whether Sally's more unusual characteristic is her height or her weight. Discuss.

5.44 A soft-drink machine is regulated so that the amount dispensed is normally distributed with $\mu = 7$ ounces and $\sigma = 0.5$ ounce. If samples of nine cups are taken, what value will be exceeded by 95% of the sample means?

5.45 The life of a type of transistor battery is normally distributed with $\mu = 100$ hours and $\sigma = 20$ hours.

(a) What proportion of the batteries will last between 100 and 115 hours?

(b) If random samples of 16 batteries are selected
(1) What proportion of the sample means will be between 100 and 115 hours?
(2) What proportion of the sample means will be more than 90 hours?
(3) Within what limits around the population mean will 90% of sample means fall?

(c) Is the central limit theorem necessary to answer (b)(1), (2), and (3)? Explain.

5.46 An orange juice producer buys all his oranges from a large orange orchard. The amount of juice squeezed from each of these oranges is approximately normally distributed with a mean of 4.70 ounces and a standard deviation of 0.40 ounce.

(a) What is the probability that a randomly selected orange will contain
(1) Between 4.70 and 5.00 ounces?
(2) Between 5.00 and 5.50 ounces?

(b) 77% of the oranges will contain at least how many ounces of juice? Suppose a sample of 25 oranges is selected:

(c) What is the probability that the sample mean will be at least 4.60 ounces?

(d) Between what two values symmetrically distributed around the population mean will 70% of the sample means fall?

(e) 77% of the sample means will be above what value?

(f) Are the results of (b) and (e) different? Explain why.

5.47 **(Class Project)**

(a) The table of random numbers is an example of a uniform distribution since each digit is equally likely to occur. Starting in the row corresponding to the day of the month in which you were born, use the table of random numbers (Table E.1) to select *one digit* at a time.

Select samples of size $n = 2$, $n = 5$, $n = 10$. Compute the sample mean $\bar{X}$ of each sample. For each sample size, each student should select five different samples so that a frequency distribution of the sample means can be developed for the results of the entire class. What can be said about the shape of the sampling distribution for each of these sample sizes?

(b) Use Microsoft Excel to develop the sampling distribution of the mean.

5.48 **(Class Project)** A coin having one side heads and the other side tails is to be tossed 10 times and the number of heads obtained is to be recorded. If each student performs this experiment five times, a frequency distribution of the number of heads can be developed from the results of the entire class. Does this distribution seem to approximate the normal distribution?

5.49 **(Class Project)** The number of cars waiting in line at a car wash is distributed as follows:

Length of Waiting Line (number of cars)	Probability
0	.25
1	.40
2	.20
3	.10
4	.04
5	.01

(a) The table of random numbers can be used to select samples from this distribution by assigning numbers as follows:
  (1) Start in the row corresponding to the day of the month in which you were born.
  (2) *Two-digit* random numbers are to be selected.
  (3) If a random number between 00 and 24 is selected, record a length of 0; if between 25 and 64, record a length of 1; if between 65 and 84, record a length of 2; if between 85 and 94, record a length of 3; if between 95 and 98, record a length of 4; if it is 99, record a length of 5.

Select samples of size $n = 2$, $n = 10$, $n = 25$. Compute the sample mean for each sample. For example, if a sample of size 2 results in random numbers 18 and 46, these would correspond to lengths of 0 and 1, respectively, producing a sample mean of 0.5. If each student selects five different samples for each sample size, a frequency distribution of the sample means (for each sample size) can be developed from the results of the entire class. What conclusions can you draw about the sampling distribution of the mean as the sample size is increased?

(b) Use Microsoft Excel to develop the sampling distribution of the mean.

5.50 **(Class Project)** The table of random numbers can be used to simulate the selection of different colored balls from a bowl as follows:
  (1) Start in the row corresponding to the day of the month in which you were born.
  (2) *One-digit* numbers are to be selected.
  (3) If a random digit between 0 and 6 is selected, consider the ball to be white; if a random digit is a 7, 8, or 9, consider the ball to be red.

Select samples of 10, 25, and 50 digits. In each sample, count the number of white balls and compute the proportion of white balls in the sample. If each student in the class selects five different samples for each sample size, a frequency distribution of the proportion of white balls (for each sample size) can be developed from the results of the entire class. What conclusions can be drawn about the sampling distribution of the proportion as the sample size is increased?

5.51 **(Class Project)** Suppose that step 3 of Problem 5.50 uses the following rule: If a random digit between 0 and 8 is selected, consider the ball to be white; if a random digit of 9 is selected, consider the ball to be red. Compare and contrast the results obtained in this problem and in Problem 5.50.

Team Projects

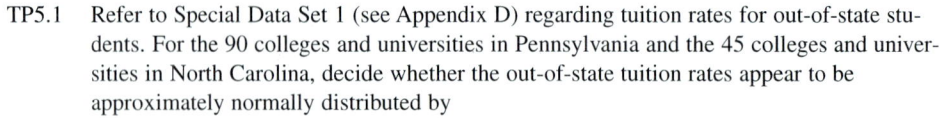

PENNC&U.TXT
NCC&U-T.TXT

PA-POP-I.XLS

TP5.1 Refer to Special Data Set 1 (see Appendix D) regarding tuition rates for out-of-state students. For the 90 colleges and universities in Pennsylvania and the 45 colleges and universities in North Carolina, decide whether the out-of-state tuition rates appear to be approximately normally distributed by
(a) Evaluating the actual versus theoretical properties.
(b) Constructing a normal probability plot.
(c) Stating your conclusions based on (a) and (b).
(d) Compare the conclusions reached in (c) with those for Texas reached in Section 5.6 and in Chapters 2 and 3.

CEREAL.TXT

TP5.2 Refer to Special Data Set 2 (see Appendix D) regarding ready-to-eat cereals and the following variables: cost, weight, and sugar. For each of these numerical variables, decide whether or not the data appear to be approximately normally distributed by
(a) Evaluating the actual versus theoretical properties.
(b) Constructing a normal probability plot.
(c) Stating your conclusions based on (a) and (b).

FRAGRANC.TXT

TP5.3 Refer to Special Data Set 3 (see Appendix D) regarding men's and women's fragrances. Decide whether or not the cost per ounce for men's fragrances and women's fragrances each appear to be approximately normally distributed by
(a) Evaluating the actual versus theoretical properties.
(b) Constructing a normal probability plot.
(c) Stating your conclusions based on (a) and (b).
(d) Compare the results for men's and women's fragrances. What conclusions can you reach?

Endnotes

1. Mathematically, this may be expressed as
$$P(60 \leq X \leq 63) = P(0 \leq Z \leq 1)$$
$$= .3413$$

2. Unlike the case of discrete random variables where the wording of the problem is so essential, we note that for continuous random variables there is much more flexibility in the wording. Hence, there are two ways to state our result: We can say that 50% of the workers can assemble the part in *under 75 seconds,* or we can say that 50% of the workers can assemble the part in 75 *seconds or less.* Semantics are unimportant, because with continuous random variables the probability of assembling the part in exactly 75 seconds (or any other exactly specified time) is 0.

3. We must remember that "only" 500 samples out of an infinite number of samples have been selected, so the sampling distributions shown are only approximations of the true distributions.

4. In some cases, sample sizes of 300 may not be enough to ensure normality.

References

1. Cochran, W. G., *Sampling Techniques,* 3d ed. (New York: Wiley, 1977).
2. Gunter, B., "Q-Q Plots," *Quality Progress* (February 1994), pp. 81–86.
3. Marascuilo, L. A., and M. McSweeney, *Nonparametric and Distribution-Free Methods for the Social Sciences* (Monterey, CA: Brooks/Cole, 1977).
4. Mendenhall, W., and T. Sincich, *Statistics for Engineering and the Sciences*, 4th ed. (Englewood Cliffs, NJ: Prentice-Hall, 1995).
5. *Microsoft Excel Version 7* (Redmond, WA: Microsoft Corp., 1996).
6. Ramsey, P. P., and P. H. Ramsey, "Simple Tests of Normality in Small Samples," *Journal of Quality Technology,* Vol. 22, 1990, pp. 299–309.
7. Sievers, G. L., "Probability Plotting," in Kotz, S., and N. L. Johnson, Eds., *Encyclopedia of Statistical Sciences,* Vol. 7 (New York: Wiley, 1986), pp. 232–237.

chapter 6

Estimation

CHAPTER OBJECTIVE To develop confidence interval estimates for means and proportions and to determine the sample size necessary to obtain a desired confidence interval.

6.1 INTRODUCTION

Statistical inference is the process of using sample results to draw conclusions about the characteristics of a population. In this chapter we shall examine statistical procedures that will enable us to *estimate* either a population mean, a population proportion, or a population total.

There are two major types of estimates: point estimates and interval estimates. A **point estimate** consists of a single sample statistic that is used to estimate the true value of a population parameter. For example, the sample mean $\bar{X}$ is a point estimate of the population mean μ and the sample variance S^2 is a point estimate of the population variance σ^2. Recall from Section 5.9.1 that the sample mean $\bar{X}$ possessed the highly desirable properties of unbiasedness and efficiency. Although in practice only one sample is selected, we know that the average value of all possible sample means is μ, the true population parameter.[1] Since the sample statistic ($\bar{X}$) varies from sample to sample (i.e., it depends on the elements selected in the sample), we need to take this into consideration to provide for a more informative estimate of the population characteristic. To accomplish this, we shall develop an **interval estimate** of the true population mean by taking into account the sampling distribution of the mean. The interval that we construct will have a specified confidence or probability of correctly estimating the true value of the population parameter μ. Similar interval estimates will also be developed for the population proportion, p, and the population total. We will then discuss how we can determine the size of the sample to be selected and demonstrate how a finite population can affect the width of the confidence interval developed and the sample size selected.

6.2 CONFIDENCE INTERVAL ESTIMATION OF THE MEAN (σ KNOWN)

In Section 5.9, we observed that from either the central limit theorem or knowledge of the population distribution, we can determine the percentage of sample means that fall within certain distances of the population mean. For instance, in Section 5.9.3, in the cereal filling-process example (in which $\mu = 368$, $\sigma = 15$, and $n = 25$) we observed that 95% of all sample means fall between 362.12 and 373.88 grams. The type of reasoning in this statement *(deductive reasoning)* is exactly opposite to the type of reasoning that is needed here *(inductive reasoning)*.

In statistical inference, we must take the results of a single sample and draw conclusions about the population, not vice versa. In practice, the population mean is the unknown quantity that is to be estimated. Suppose in the cereal filling-process example that the true population mean μ is unknown but the true population standard deviation σ is known to be 15 grams. Thus, rather than taking $\mu \pm (1.96)(\sigma/\sqrt{n})$ to find the upper and lower limits around μ as in Section 5.9.3, we will determine the consequences of substituting the sample mean $\bar{X}$ for the unknown μ and using $\bar{X} \pm (1.96)(\sigma/\sqrt{n})$ as an interval within which we estimate the unknown μ. Although in practice a single sample of size n is selected and the mean $\bar{X}$ is computed, we need to obtain a hypothetical set of all possible samples, each of size n, in order to understand the full meaning of the interval estimate that will be obtained.

Suppose, for example, that our sample of size $n = 25$ has a mean of 362.3 grams. The interval developed to estimate μ would be $362.3 \pm (1.96)(15)/(\sqrt{25})$ or 362.3 ± 5.88. That is, the estimate of μ would be

$$356.42 \le \mu \le 368.18$$

Since the population mean μ (equal to 368) is included *within* the interval, we observe that this sample has led to a correct statement about μ (see Figure 6.1).

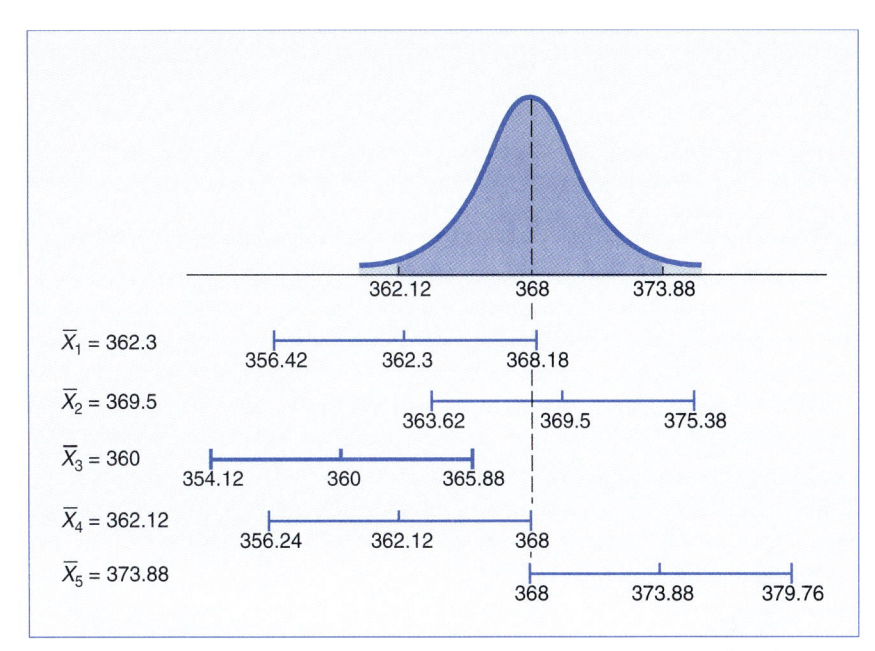

FIGURE 6.1 Confidence interval estimates from five different samples of size n = 25 taken from a population where μ = 368 and σ = 15.

To continue our hypothetical example, suppose that for a different sample of $n = 25$, the mean is 369.5. The interval developed from this sample would be $369.5 \pm (1.96)(15)/(\sqrt{25})$ or 369.5 ± 5.88. That is, the estimate of μ would be

$$363.62 \leq \mu \leq 375.38$$

Since the true population mean μ (equal to 368) is also included within this interval, we conclude that this statement about μ is correct.

Now, before we begin to think that we will *always* make correct statements about μ from the sample $\bar{X}$, suppose we draw a third hypothetical sample of size $n = 25$ in which the sample mean is equal to 360 grams. The interval developed here would be $360 \pm (1.96)(15)/(\sqrt{25})$ or 360 ± 5.88. In this case, the estimate of μ is

$$354.12 \leq \mu \leq 365.88$$

Observe that this estimate is *not* a correct statement, since the population mean μ is not included in the interval developed from this sample (see Figure 6.1). Thus, we are faced with a dilemma. For some samples the interval estimate of μ will be correct, and for others it will be incorrect. In addition, we must realize that in practice we select only *one* sample, and since we do not know the true population mean, we cannot determine whether our particular statement is correct.

What we can do to resolve this dilemma is to determine the proportion of samples producing intervals that result in correct statements about the population mean μ. To do this, we need to examine two other hypothetical samples: the case in which $\bar{X} = 362.12$ grams and the case in which $\bar{X} = 373.88$ grams. If $\bar{X} = 362.12$, the interval will be $362.12 \pm (1.96)(15)/(\sqrt{25})$ or 362.12 ± 5.88. That is,

$$356.24 \leq \mu \leq 368.00$$

Since the population mean of 368 is at the upper limit of the interval, the statement is a correct one (see Figure 6.1).

Finally, if $\bar{X} = 373.88$, the interval will be $373.88 \pm (1.96)(15)/(\sqrt{25})$ or 373.88 ± 5.88. That is,

$$368.00 \leq \mu \leq 379.76$$

In this case, since the population mean of 368 is included at the lower limit of the interval, the statement is a correct one.

Thus, from these examples (see Figure 6.1), we can determine that if the sample mean based on a sample of $n = 25$ falls anywhere between 362.12 and 373.88 grams, the population mean will be included *somewhere* within the interval. However, we know from our discussion of the sampling distribution in Section 5.9.3 that 95% of the sample means fall between 362.12 and 373.88 grams. Therefore, 95% of all sample means will include the population mean within the interval developed. The interval from 362.12 to 373.88 is referred to as a 95% confidence interval.

> In general, a 95% **confidence interval estimate** can be interpreted to mean that if all possible samples of the same size n were taken, 95% of them would include the true population mean somewhere within the interval around their sample means, and only 5% of them would not.

Since only one sample is selected in practice and μ is unknown, we never know for sure whether the specific interval obtained includes the population mean. However, we can state that we have 95% confidence that we have selected a sample whose interval does include the population mean.

In our examples we had 95% confidence of including the population mean within the interval. In some situations, we might desire a higher degree of assurance (such as 99%) of including the population mean within the interval. In other cases, we might be willing to accept less assurance (such as 90%) of correctly estimating the population mean.

In general, the level of confidence is symbolized by $(1 - \alpha) \times 100\%$, where α is the proportion in the tails of the distribution that is outside the confidence interval. Therefore, to obtain the $(1 - \alpha) \times 100\%$ confidence interval estimate of the mean with σ known, we have

$$\bar{X} \pm Z \frac{\sigma}{\sqrt{n}}$$

or   (6.1)

$$\bar{X} - Z \frac{\sigma}{\sqrt{n}} \leq \mu \leq \bar{X} + Z \frac{\sigma}{\sqrt{n}}$$

where Z = the value corresponding to an area of $(1 - \alpha)/2$ from the center of a standardized normal distribution

To construct a 95% confidence interval estimate of the mean, the Z value corresponding to an area of $.95/2 = .4750$ from the center of the standard normal distribution is 1.96. The value of Z selected for constructing such a confidence interval is called the **critical value** for the distribution.

There is a different critical value for each level of confidence, $1 - \alpha$. A **level of confidence** of 95% leads to a Z value of ± 1.96 (see Figure 6.2). If a level of confidence of 99% is desired, the area of .99 would be divided in half, leaving .495 between each limit and μ (see Figure 6.3). The Z value corresponding to an area of .495 from the center of the normal curve is approximately 2.58.

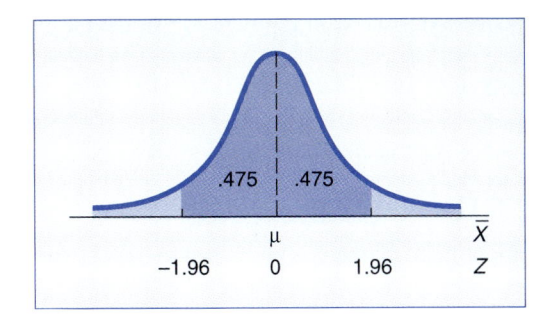

FIGURE 6.2 Normal curve for determining the Z value needed for 95% confidence.

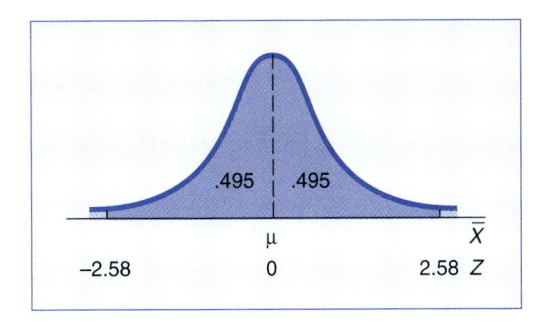

FIGURE 6.3 Normal curve for determining the Z value needed for 99% confidence.

Now that we have considered various levels of confidence, one might wonder why we wouldn't want to make the confidence level as close to 100% as possible. But any increase in the level of confidence is achieved only by simultaneously widening (and making less precise and less useful) the confidence interval obtained. Thus, we would have more confidence that the population mean is within a broader range of values. This trade-off between the width of the confidence interval and the level of confidence will be discussed in greater depth when we investigate how the sample size n is determined (see Section 6.6).

● **Application** To illustrate the application of the confidence interval estimate, we will use the following example. A manufacturer of computer paper has a production process that operates continuously throughout an entire production shift. The paper is expected to have an average length of 11 inches, and the standard deviation is known to be 0.02 inch. At periodic intervals, samples are selected to determine whether the average paper length is still equal to 11 inches or whether something has gone wrong in the production process to change the length of the paper produced. If such a situation has occurred, corrective action must be contemplated. A random sample of 100 sheets has been selected, and the average paper length is found to be 10.998 inches. If a 95% confidence interval estimate of the population average paper length is desired, using Equation (6.1), with $Z = 1.96$ for 95% confidence, we would have

$$\bar{X} \pm Z \frac{\sigma}{\sqrt{n}} = 10.998 \pm (1.96) \frac{.02}{\sqrt{100}}$$

$$= 10.998 \pm .00392$$

$$10.99408 \leq \mu \leq 11.00192$$

Thus, we would estimate, with 95% confidence, that the population mean is between 10.99408 and 11.00192 inches. Since 11, the value that indicates the production process is working properly, is included within the interval, there is no reason to believe that anything is wrong with the production process. There is 95% confidence that the sample selected is one where the true population mean is included somewhere within the interval developed.

If 99% confidence is desired, then using Equation (6.1), with $Z = 2.58$, we would have

$$\bar{X} \pm Z \frac{\sigma}{\sqrt{n}} = 10.998 \pm (2.58) \frac{.02}{\sqrt{100}}$$

$$= 10.998 \pm .00516$$

$$10.99284 \leq \mu \leq 11.00316$$

Once again, since 11 is included within this wider interval, there is no reason to believe that anything is wrong with the production process.

Problems for Section 6.2

Note: The problems in this section can be solved using Microsoft Excel (see Section 6.10).

6.1 A market researcher states that she has 95% confidence that the true average monthly sales of a product will be between $170,000 and $200,000. Explain the meaning of this statement.

6.2 Why can't the production manager in the example on page 295 have 100% confidence? Explain.

6.3 Is it true in the computer paper production example that 95% of the sample means will fall between 10.99408 and 11.00192 inches? Explain.

6.4 Is it true in the computer paper production example that we do not know for sure whether the true population mean is between 10.99408 and 11.00192 inches? Explain.

6.5 Suppose that the manager of a paint supply store wants to estimate the actual amount of paint contained in 1-gallon cans purchased from a nationally known manufacturer. It is known from the manufacturer's specifications that the standard deviation of the amount of paint is equal to 0.02 gallon. A random sample of 50 cans is selected, and the average amount of paint per 1-gallon can is 0.995 gallon.
(a) Set up a 99% confidence interval estimate of the true population average amount of paint included in a 1-gallon can.
(b) Based on your results, do you think that the store owner has a right to complain to the manufacturer? Why?
(c) Does the population amount of paint per can have to be normally distributed here? Explain.
(d) Explain why an observed value of 0.98 gallon for an individual can would not be unusual, even though it is outside the confidence interval you calculated.
(e) Suppose that you used a 95% confidence interval estimate. What would be your answers to (a) and (b)?

6.6 The quality control manager at a light bulb factory needs to estimate the average life of a large shipment of light bulbs. The process standard deviation is known to be 100 hours. A random sample of 50 light bulbs indicated a sample average life of 350 hours.
(a) Set up a 95% confidence interval estimate of the true average life of light bulbs in this shipment.
(b) Does the population of light bulb life have to be normally distributed here? Explain.
(c) Explain why an observed value of 320 hours would not be unusual, even though it is outside the confidence interval you calculated.
(d) Suppose that the process standard deviation changed to 80 hours. What would be your answer in (a)?

6.7 The inspection division of the Lee County Weights and Measures Department is interested in estimating the actual amount of soft drink that is placed in 2-liter bottles at the local bottling plant of a large nationally known soft-drink company. The bottling plant has informed the inspection division that the standard deviation for 2-liter bottles is 0.05 liter. A random sample of 100 2-liter bottles obtained from this bottling plant indicates a sample average of 1.99 liters.
(a) Set up a 95% confidence interval estimate of the true average amount of soft drink in each bottle.

(b) Does the population of soft-drink fill have to be normally distributed here? Explain.
(c) Explain why an observed value of 2.02 liters would not be unusual, even though it is outside the confidence interval you calculated.
(d) Suppose that the sample average changed to 1.97 liters. What would be your answer to (a)?

6.3 CONFIDENCE INTERVAL ESTIMATION OF THE MEAN (σ UNKNOWN)

Just as the mean of the population μ is usually not known, the actual standard deviation of the population σ is also not likely to be known. Therefore, we need to obtain a confidence interval estimate of μ using only the sample statistics of $\bar{X}$ and S. To achieve this, we turn to the work of William S. Gosset.

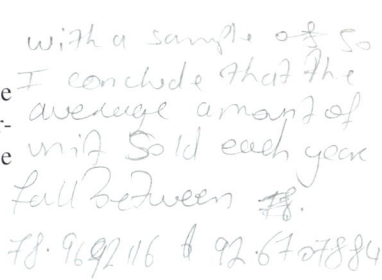

6.3.1 Student's t Distribution

At the turn of this century, a statistician named William S. Gosset, an employee of Guinness Breweries in Ireland (see Reference 6), was interested in making inferences about the mean when σ was unknown. Since Guinness employees were not permitted to publish research work under their own names, Gosset adopted the pseudonym "Student." The distribution that he developed has come to be known as **Student's t distribution.**

If the random variable X is normally distributed, then the statistic

$$\frac{\bar{X} - \mu}{\frac{S}{\sqrt{n}}}$$

has a *t* distribution with *n* − 1 *degrees of freedom.* Notice that this expression has the same form as Equation (5.8) on page 266, except that S is used to estimate σ, which is presumed unknown in this case.

6.3.2 Properties of the t Distribution

In appearance, the *t* distribution is very similar to the normal distribution. Both distributions are bell-shaped and symmetrical. However, the *t* distribution has more area in the tails and less in the center than does the normal distribution (see Figure 6.4). This is because σ is unknown and we are using S to estimate it. Since we are uncertain of the value σ, the values of *t* that we observe will be more variable than for Z.

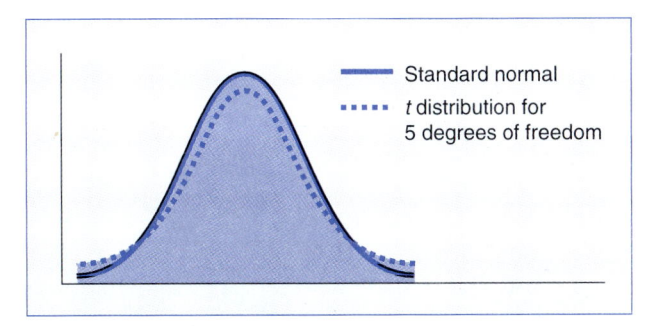

FIGURE 6.4 Standard normal distribution and t distribution for 5 degrees of freedom.

However, as the number of degrees of freedom increases, the t distribution gradually approaches the normal distribution until the two are virtually identical. This happens because, as the sample size gets larger, S becomes a better estimate of σ. With a sample size of about 120 or more, S estimates σ precisely enough that there is little difference between the t and Z distributions. For this reason, most statisticians will use Z instead of t when the sample size is over 120.

In practice, as long as the sample size is large enough and the population is not very skewed, the t distribution can be used to estimate the population mean when σ is unknown. The critical values of t for the appropriate degrees of freedom can be obtained from the table of the t distribution (see Table E.3). The top of each column of the t table indicates the area in the right tail of the t distribution (since positive entries for t are supplied, the values are for the upper tail); each row represents the particular t value for each specific degree of freedom. For example, with 34 degrees of freedom, if 95% confidence is desired, the appropriate value of t would be found in the manner shown in Table 6.1. The 95% confidence level indicates that there would be an area of .025 in each tail of the distribution. Looking in the column for an upper-tail area of .025 and in the row corresponding to 34 degrees of freedom results in a value for t of 2.0332. Since t is a symmetrical distribution, with a mean of 0, if the upper-tail value is +2.0332, the value for the lower-tail area (lower .025) will be −2.0332. A t value of 2.0332 means that the probability that t would exceed +2.0332 is .025 or 2.5% (see Figure 6.5).

Table 6.1 Determining the critical value from the t table for an area of .025 in each tail with 34 degrees of freedom.

	Upper-Tail Areas					
Degrees of Freedom	.25	.10	.05	.025	.01	.005
1	1.0000	3.0777	6.3138	12.7062	31.8207	63.6574
2	0.8165	1.8856	2.9200	4.3027	6.9646	9.9248
3	0.7649	1.6377	2.3534	3.1824	4.5407	5.8409
4	0.7407	1.5332	2.1318	2.7764	3.7469	4.6041
5	0.7267	1.4759	2.0150	2.5706	3.3649	4.0322
⋮	⋮	⋮	⋮	⋮	⋮	⋮
31	0.6825	1.3095	1.6955	2.0395	2.4528	2.7440
32	0.6822	1.3086	1.6939	2.0369	2.4487	2.7385
33	0.6820	1.3077	1.6924	2.0345	2.4448	2.7333
34	0.6818	1.3070	1.6909	2.0322	2.4411	2.7284
35	0.6816	1.3062	1.6896	2.0301	2.4377	2.7238

Source: Extracted from Table E.3.

FIGURE 6.5 t distribution with 34 degrees of freedom.

298 Chapter 6 Estimation

6.3.3 The Concept of Degrees of Freedom

You may recall from Chapter 3 that the sample variance S^2 requires the computation of

$$\sum_{i=1}^{n}(X_i - \bar{X})^2$$

Thus, in order to compute S^2, we first need to know $\bar{X}$. Therefore, we can say that only $n - 1$ of the sample values are free to vary. That is, there are $n - 1$ **degrees of freedom.**

We may illustrate this concept as follows: Suppose we have a sample of five values that has a mean of 20. How many distinct values do we need to know before we can obtain the remainder? The fact that $n = 5$ and $\bar{X} = 20$ also tells us that $\sum_{i=1}^{n} X_i = 100$, since $\sum_{i=1}^{n} X_i/n = \bar{X}$.

Thus, once we know four of the values, the fifth one will not be *free* to vary, since the sum must add to 100. For example, if four of the values are 18, 24, 19, and 16, the fifth value can only be 23 so that the sum equals 100.

6.3.4 The Confidence Interval Statement

The $(1 - \alpha) \times 100\%$ confidence interval estimate for the mean with σ unknown is expressed as follows:

$$\bar{X} \pm t_{n-1} \frac{S}{\sqrt{n}}$$

or (6.2)

$$\bar{X} - t_{n-1} \frac{S}{\sqrt{n}} \leq \mu \leq \bar{X} + t_{n-1} \frac{S}{\sqrt{n}}$$

where t_{n-1} is the critical value of the t distribution with $n - 1$ degrees of freedom for an area of $\alpha/2$ in the upper tail.

To see how confidence intervals for a mean can be constructed when the population standard deviation is unknown, we will use the following application.

● **Application** Suppose the marketing manager for a company that supplies home heating oil wants to estimate the average annual usage (in gallons) of single-family homes in a particular geographical area. A random sample of 35 single-family homes is selected, and the annual usage for these homes is summarized in Table 6.2. For these data we may use Excel to obtain the sample average $\bar{X} = 1{,}122.7$ gallons and the sample standard deviation $S = 295.72$ gallons.

Table 6.2 Annual amount of heating oil consumed (in gallons) in a sample of 35 single-family houses.

OILUSE.TXT

1150.25	1352.67	983.45	1365.11	942.71	1577.77	330.00
872.37	1126.57	1184.17	1046.35	1110.50	1050.86	851.60
1459.56	1252.01	373.91	1047.40	1064.46	1018.23	996.92
941.96	767.37	1598.57	1598.66	1343.29	1617.73	1300.76
1013.27	1402.59	1069.32	1108.94	1326.19	1074.86	975.86

If the marketing manager would like to have 95% confidence that the interval obtained includes the population average amount of heating oil consumed per year, using $\bar{X} = 1{,}122.7$, $S = 295.72$, $n = 35$, and $t_{34} = 2.0322$, we have

$$\bar{X} \pm t_{n-1} \frac{S}{\sqrt{n}} = 1{,}122.7 \pm (2.0322)\frac{295.72}{\sqrt{35}}$$

$$= 1{,}122.7 \pm 101.58$$

$$1{,}021.12 \leq \mu \leq 1{,}224.28$$

We would thus conclude with 95% confidence that the average amount of heating oil consumed per year is between 1,021.12 and 1,224.28 gallons. The 95% confidence interval states that we are 95% sure that the sample we have selected is one in which the population mean μ is located within the interval. This 95% confidence means that if all possible samples of size 35 were selected (something that would never be done in practice), 95% of the intervals developed would include the true population mean *somewhere* within the interval.

Problems for Section 6.3

Note: *The problems in this section can be solved using Microsoft Excel (see Section 6.10).*

6.8 Determine the critical value of t in each of the following circumstances:
(a) $1 - \alpha = .95$, $n = 10$.
(b) $1 - \alpha = .99$, $n = 10$.
(c) $1 - \alpha = .95$, $n = 32$.
(d) $1 - \alpha = .95$, $n = 65$.
(e) $1 - \alpha = .90$, $n = 16$.

6.9 A new breakfast cereal is test-marketed for 1 month at stores of a large supermarket chain. The results for a sample of 16 stores indicate average sales of $1,200 with a sample standard deviation of $180. Set up a 99% confidence interval estimate of the true average sales of this new breakfast cereal.

6.10 The manager of a branch of a local savings bank wants to estimate the average amount held in passbook savings accounts by depositors at the bank. A random sample of 30 depositors is selected, and the results indicate a sample average of $4,750 and a sample standard deviation of $1,200.
(a) Set up a 95% confidence interval estimate of the average amount held in all passbook savings accounts.
(b) If an individual had $4,000 in a passbook savings account, would this be considered unusual? Explain your answer.

6.11 A stationery store would like to estimate the average retail value of greeting cards that it has in its inventory. A random sample of 20 greeting cards indicates an average value of $1.67 and a standard deviation of $0.32. Set up a 95% confidence interval estimate of the average value of all greeting cards in the store's inventory.

6.12 The personnel department of a large corporation would like to estimate the family dental expenses of its employees to determine the feasibility of providing a dental insurance plan. A random sample of 10 employees reveals the following family dental expenses (in dollars) for the preceding year:

DENTAL.TXT

110, 362, 246, 85, 510, 208, 173, 425, 316, 179

(a) Set up a 90% confidence interval estimate of the average family dental expenses for all employees of this corporation.
(b) What assumption about the population distribution must be made in (a)?
(c) Give an example of a family dental expense that would be outside the confidence interval but would not be unusual for an individual family, and explain why this is not a contradiction.
(d) Suppose you used a 95% confidence interval in (a). What would be your answer to (a)?

300 **Chapter 6** Estimation

(e) Suppose the fourth value was $585 instead of $85. What would be your answer to (a)? What effect does this change have on the confidence interval?

6.13 The customer service department of a local gas utility would like to estimate the average length of time between the entry of the service request and the connection of service. A random sample of 15 houses is selected from the records available during the past year. The results recorded in number of days are as follows:

| 114 | 78 | 96 | 137 | 78 | 103 | 117 |
| 126 | 86 | 99 | 114 | 72 | 104 | 73 | 86 |

GASERVE.TXT

(a) Set up a 95% confidence interval estimate of the population average waiting time in the past year.
(b) What assumption about the population distribution must be made in (a)?
(c) Suppose the last value was 286 days instead of 86 days. What would be your answer to (a)? What effect does this change have on the confidence interval?

6.14 The director of quality of a large health maintenance organization wants to evaluate patient waiting time at a local facility. A random sample of 25 patients is selected from the appointment book. The waiting time is defined as the time from when the patient signs in to when he or she is seen by the doctor. The following data represent the waiting times (in minutes):

19.5	30.5	45.6	39.8	29.6
25.4	21.8	28.6	52.0	25.4
26.1	31.1	43.1	4.9	12.7
10.7	12.1	1.9	45.9	42.5
41.3	13.8	17.4	39.0	36.6

HMO.TXT

(a) Set up a 95% confidence interval estimate of the population average waiting time.
(b) What assumption about the population distribution must be made in (a)?
(c) Suppose that the value of 1.9 minutes was actually 101.9 minutes. What would be your answer to (a)? What effect does this change have on the confidence interval?

6.15 Set up a 95% confidence interval estimate for each of the following sets of data:

Set 1: 1, 1, 1, 1, 8, 8, 8, 8
Set 2: 1, 2, 3, 4, 5, 6, 7, 8

Explain why they have different confidence intervals even though they have the same mean and range.

6.16 Compute a 95% confidence interval for the numbers 1, 2, 3, 4, 5, 6, 20. Change the number 20 to 7, and recalculate the confidence interval. Using these results, describe the effect of an outlier (or extreme value) on the confidence interval.

6.4 ESTIMATION THROUGH BOOTSTRAPPING

Although confidence interval estimates have been widely applied in making inferences about population parameters, the estimating procedure used is based on assumptions that do not always hold. For example, the confidence interval estimate of μ discussed in Section 6.3 assumes that the underlying population is normally distributed. Although the confidence interval procedure is fairly insensitive to moderate departures from this assumption, if there is substantial nonnormality in the population, particularly if a small sample size is used, the confidence interval estimate for the mean may not be precise. In this section we will consider an alternative estimation approach called bootstrapping.

Bootstrapping estimation procedures involve selecting an initial sample and then repeated resampling from the initial sample. These procedures, developed by Efron (see

References 2, 3, and 5), are heavily computer intensive. What makes bootstrapping estimation useful is that the procedures are based on the initial sample and make no assumptions regarding the shape of the underlying population. In addition, the procedures do not require knowledge of any population parameters.

The steps in bootstrapping estimation of the mean are as follows:

1. Draw a random sample of size *n without replacement* from a population frame of size N.
2. Resample the initial sample by selecting *n* observations *with replacement* from the *n* observations in the initial sample.
3. Compute $\bar{X}$, the statistic of interest, from this *resample*.
4. Repeat steps 2 and 3 *m* different times (where *m* is typically selected as a value between 100 and 1,000, depending on the speed of the computer being used).
5. Form the **resampling distribution** of the statistic of interest (i.e., the distribution of the sample mean obtained from the *m* resamples) using a stem-and-leaf display or an ordered array.
6. To form a $(1 - \alpha) \times 100\%$ bootstrap confidence interval of the population mean μ, use the stem-and-leaf display or ordered array for the resampling distribution and find the value that cuts off the smallest $\alpha/2 \times 100\%$ and the value that cuts off the largest $\alpha/2 \times 100\%$ of the statistic. These values provide the lower and upper limits for the bootstrap confidence interval estimate of the unknown parameter.

To demonstrate the bootstrap confidence interval estimation procedure, we return to the heating oil usage example discussed on page 299. You may recall that the marketing manager wants to estimate the average annual consumption of heating oil of customers residing in single-family homes. A random sample of 35 customers is selected without replacement, and the *t* distribution is used to develop a 95% confidence interval estimate of μ. The estimate obtained is based on the assumption that the underlying population of oil usage is approximately normally distributed, an assumption that is not necessary for the bootstrap procedure.

Following the six-step procedure just described and using the sample data of Table 6.2 (page 299), a random resample of 35 observations is selected with replacement using Excel. The first resample is listed in Table 6.3. The mean of this sample is 1,003.26 gallons.

Notice that in this first resample, some values from the original sample such as 330.00 are repeated, and others such as 1,352.67 do not occur. If this resampling process is performed *m* = 200 times, the resampling distribution containing 200 resampling means can be developed. Table 6.4 presents the ordered array of the resampling distribution. To form a 95% bootstrap confidence interval estimate of μ, the smallest 2.5% and the largest 2.5% of the resample means need to be identified (step 6).

When 200 resample means are obtained, the fifth smallest value (i.e., $200 \times .025$) will cut off the lowest 2.5%, while the fifth largest value will cut off the highest 2.5%. From Table 6.4, we obtain the values of 1,006.53 and 1,227.02 gallons as the fifth smallest and fifth largest,

Table 6.3 First resample of size 35 from the sample of 35 single-family homes (annual amount of heating oil consumed).

1326.19	1150.25	1184.17	1013.27	330.00	1126.57	1343.29
330.00	1013.27	996.92	1047.40	1018.23	1064.46	1598.66
941.96	767.37	1050.86	1459.56	975.86	373.91	1018.23
373.91	1050.86	1074.86	1326.19	1110.50	1402.59	1343.29
1110.50	373.91	1069.32	330.00	1365.11	1110.50	941.96

Table 6.4 Ordered array of 200 resampled means obtained from Excel used to form 95% bootstrap confidence interval estimate of μ for the heating oil usage example.

988.50	994.49	1003.26	1004.25	1006.53	1014.38	1018.62
1030.21	1030.73	1032.04	1032.77	1034.17	1035.21	1041.45
1042.23	1046.16	1050.01	1050.97	1052.73	1053.00	1054.58
1055.08	1055.39	1055.76	1059.57	1059.88	1060.51	1061.70
1061.79	1062.46	1062.86	1063.31	1066.91	1068.84	1070.53
1071.00	1071.38	1074.99	1077.18	1077.63	1078.62	1079.41
1079.46	1081.45	1083.40	1084.25	1085.21	1085.98	1086.22
1086.83	1089.45	1089.95	1092.37	1092.53	1094.48	1095.25
1095.35	1095.40	1095.88	1096.26	1097.59	1100.06	1100.19
1100.20	1101.59	1103.41	1104.07	1104.92	1105.19	1106.67
1106.84	1111.14	1112.46	1112.47	1112.53	1113.58	1115.09
1115.71	1116.46	1116.51	1117.94	1118.67	1119.17	1119.71
1119.98	1120.31	1121.01	1121.46	1121.68	1121.92	1122.40
1123.34	1123.73	1124.38	1124.98	1125.63	1126.01	1127.00
1127.15	1127.30	1127.55	1127.86	1128.06	1129.03	1129.66
1129.84	1132.30	1132.78	1133.39	1134.03	1134.43	1135.07
1136.67	1136.90	1138.24	1138.52	1139.53	1139.81	1140.60
1142.72	1142.73	1143.03	1143.32	1143.35	1143.70	1144.47
1145.72	1146.21	1149.77	1150.87	1152.88	1153.13	1153.25
1153.35	1156.35	1156.43	1158.49	1158.62	1159.14	1159.45
1159.55	1161.01	1161.16	1161.50	1162.12	1163.15	1163.24
1164.06	1164.39	1165.45	1167.87	1168.50	1170.06	1171.11
1171.21	1172.03	1172.19	1172.90	1173.00	1173.37	1174.25
1176.32	1176.73	1178.39	1178.63	1182.16	1182.37	1183.12
1183.99	1184.27	1184.68	1185.80	1185.97	1186.86	1187.01
1187.27	1188.03	1188.25	1188.53	1189.39	1190.20	1191.10
1191.28	1191.86	1192.90	1193.21	1196.66	1199.30	1206.23
1206.28	1209.61	1216.85	1225.20	1226.46	1226.67	1227.02
1230.11	1233.43	1233.97	1251.88			

respectively. Therefore, the 95% bootstrap confidence interval for the population average amount of heating oil consumed is 1,006.53 to 1,227.02 gallons. This estimate is fairly close to the traditional confidence interval estimate of 1,021.12 to 1,224.28 obtained in Section 6.3. However, the bootstrap estimate requires less stringent assumptions than does the traditional confidence interval estimate.

Problems for Section 6.4

Note: *The problems in this section can be solved using the Microsoft Excel (see Section 6.10).*

6.17 Referring to Problem 6.12 on page 300
    (a) Generate 200 resamples each of size 10 and establish a 95% bootstrap confidence interval estimate of the average yearly family dental expenses.
    (b) Compare the results of (a) to those of Problem 6.12(a).

DENTAL.TXT

6.18 Referring to Problem 6.13 on page 301
    (a) Generate 200 resamples each of size 15 and establish a 95% bootstrap confidence interval estimate of the average waiting time for the connection of gas service.
    (b) Compare the results of (a) to those of Problem 6.13(a).

GASERVE.TXT

6.19 Referring to Problem 6.14 on page 301

HMO.TXT

(a) Generate 200 resamples each of size 25 and establish a 95% bootstrap confidence interval estimate of average waiting time for patients at the HMO.
(b) Compare the results of (a) to those of Problem 6.14(a).

6.5 CONFIDENCE INTERVAL ESTIMATION FOR THE PROPORTION

In this section we extend the concept of the confidence interval to categorical data to estimate the population proportion p from the sample proportion $p_s = X/n$. Recall from Chapter 5 that when both np and $n(1-p)$ are at least 5, the binomial distribution can be approximated by the normal distribution. Hence, we can set up the following $(1 - \alpha) \times 100\%$ confidence interval estimate for the population proportion p:

$$p_s \pm Z\sqrt{\frac{p_s(1-p_s)}{n}}$$

or (6.3)

$$p_s - Z\sqrt{\frac{p_s(1-p_s)}{n}} \leq p \leq p_s + Z\sqrt{\frac{p_s(1-p_s)}{n}}$$

where
p_s = sample proportion
p = population proportion
Z = critical value from the normal distribution
n = sample size

To see how this confidence interval estimate of the proportion can be utilized, we will examine the following application.

● **Application** The production manager for a large city newspaper wants to determine the proportion of newspapers printed that have a nonconforming attribute, such as excessive ruboff, improper page set-up, missing pages, duplicate pages, and so on. Past practice has involved the detailed examination of the first newspaper that comes off the press, but no further evaluation of the thousands of papers printed. The production manager has determined that a random sample of 200 newspapers should be selected for analysis. Suppose that, of this sample of 200, 35 contain some type of nonconformance. If the production manager wants to have 90% confidence in estimating the true population proportion, the confidence interval would be computed as follows:

$p_s = 35/200 = .175$, with a 90% level of confidence $Z = 1.645$

Using Equation (6.3), we have

$$p_s \pm Z\sqrt{\frac{p_s(1-p_s)}{n}} = .175 \pm (1.645)\sqrt{\frac{(.175)(.825)}{200}}$$

$$= .175 \pm (1.645)(.0269)$$

$$= .175 \pm .0442$$

$$.1308 \leq p \leq .2192$$

Therefore, the production manager can estimate with 90% confidence that between 13.08% and 21.92% of the newspapers printed on that day have some type of nonconformance.

We should note that in this example, the number of successes and failures was sufficiently large that the normal distribution provides an excellent approximation for the binomial distribution. However, if the sample size is not large or the percentage of successes is either very low or very high, then the binomial distribution should be used rather than the normal distribution (References 1 and 9). The exact confidence intervals for various sample sizes and proportions of successes have been tabulated by Fisher and Yates (Reference 4).

For a given sample size, confidence intervals for proportions often seem to be wider than those for continuous variables. With continuous variables, the measurement on each respondent contributes more information than for a variable with only two categories. In other words, a categorical variable with only two possible values is a very crude measure compared with a continuous variable, so each observation contributes only a little information about the parameter we are estimating.

Problems for Section 6.5

Note: *The problems in this section can be solved using Microsoft Excel (see Section 6.10).*

6.20 The manager of a bank in a small city would like to determine the proportion of its depositors who are paid on a weekly basis. A random sample of 100 depositors is selected, and 30 state that they are paid weekly. Set up a 90% confidence interval estimate of the population proportion of the bank's depositors who are paid weekly.

6.21 An auditor for the state insurance department would like to determine the proportion of claims that are paid by a health insurance company within 2 months of receipt of the claim. A random sample of 200 claims is selected, and it is determined that 80 were paid out within 2 months of the receipt of the claim.
(a) Set up a 99% confidence interval estimate of the population proportion of the claims paid within 2 months.
(b) How can the results of (a) be used in a report to the state insurance department?

6.22 An automobile dealer would like to estimate the proportion of customers who still own the same cars they purchased 5 years earlier. A random sample of 200 customers selected from the automobile dealer's records indicates that 82 still own cars that had been purchased 5 years earlier. Set up a 95% confidence interval estimate of the population proportion of all customers who still own the same cars 5 years after they were purchased.

6.23 A stationery supply store receives a shipment of a certain brand of inexpensive ballpoint pens from the manufacturer. The owner of the store wishes to estimate the proportion of pens that are defective. A random sample of 300 pens is tested, and 30 are found to be defective.
(a) Set up a 90% confidence interval estimate of the proportion of defective pens in the shipment.
(b) The shipment can be returned if it is more than 5% defective; based on the sample results, can the owner return this shipment?
(c) Suppose that 99% confidence was used in (a). What would be the effect of this change on your answers to (a) and (b)?

6.24 The advertising director for a fast-food chain would like to estimate the proportion of high school students who are familiar with a particular commercial broadcast on radio and television in the last month. A random sample of 400 high school students indicates that 160 are familiar with the commercial. Set up a 95% confidence interval estimate of the population proportion of high school students who are familiar with the commercial.

6.25 The telephone company would like to estimate the proportion of households that would purchase an additional telephone line if it was made available at a substantially reduced installation cost. A random sample of 500 households is selected. The results indicate that 135 of the households would purchase the additional telephone line at a reduced installation cost. Set up a 99% confidence interval estimate of the population proportion of households who would purchase the additional telephone line.

6.26 The dean of a graduate school of business would like to estimate the proportion of MBA students enrolled who have access to a personal computer outside the school (either at home or at work). A sample of 150 students reveals that 105 have access to a personal computer outside the school (either at home or at work). Set up a 90% confidence interval estimate of the population proportion of students who have access to a personal computer outside the school.

6.6 SAMPLE SIZE DETERMINATION FOR THE MEAN

In each of our examples concerning confidence interval estimation, the sample size was arbitrarily determined without regard to the size of the confidence interval. In the business world, the determination of the proper sample size is a complicated procedure subject to the constraints of budget, time, and ease of selection. For example, if in the heating oil usage example, the marketing manager wants to estimate the average annual usage of single-family homes in a particular geographical area, he would try to determine in advance how good an estimate would be required. This means that he must decide how much error he is willing to allow in estimating the population average annual usage. That is, is accuracy required to be within ±10 gallons, ±25 gallons, ±50 gallons, or ±100 gallons? The marketing manager must also determine in advance how sure (confident) he wants to be of correctly estimating the true population parameter. In determining the sample size for estimating the mean, these requirements must be kept in mind along with information about the standard deviation.

To develop a formula for determining sample size, recall Equation (5.8):

$$Z = \frac{\bar{X} - \mu}{\frac{\sigma}{\sqrt{n}}}$$

where Z is the critical value corresponding to an area of $(1 - \alpha)/2$ from the center of a standardized normal distribution. Multiplying both sides of Equation (5.8) by $\sigma/\sqrt{n}$, we have

$$Z \frac{\sigma}{\sqrt{n}} = \bar{X} - \mu$$

Thus, the value of Z will be positive or negative, depending on whether $\bar{X}$ is larger or smaller than μ. The difference between the sample mean $\bar{X}$ and the population mean μ, denoted by e, is called the **sampling error**. The sampling error e can be defined as

$$e = \frac{Z\sigma}{\sqrt{n}}$$

Solving this equation for n, we have

The sample size n is equal to the product of the Z value squared and the variance σ^2, divided by the sampling error squared.

$$n = \frac{Z^2 \sigma^2}{e^2} \qquad (6.4)$$

Therefore, to determine the sample size, three factors must be known:

1. The confidence level desired, which determines the value of Z, the critical value from the normal distribution[2]
2. The sampling error permitted, e
3. The standard deviation, σ

In practice, the determination of these three quantities may not be easy. How is one to know what level of confidence to use and what sampling error is desired? Typically, these questions can be answered only by the *subject matter expert*, that is, the individual who is familiar with the variables to be analyzed. Although 95% is the most common confidence level used (in which case Z = 1.96), if one desires greater confidence, 99% might be more appropriate; if less confidence is deemed acceptable, then 90% might be utilized. For the sampling error, we should be thinking not of how much sampling error we would like to have (we really don't want any error), but of how much we can live with and still be able to provide adequate conclusions for the data.

Even when the confidence level and the sampling error are specified, an estimate of the standard deviation must be available. Unfortunately, the population standard deviation σ is rarely known. In some instances, the standard deviation can be estimated from past data. In other situations, one can develop an educated guess by taking into account the range and distribution of the variable. For example, if one assumes a normal distribution, the range is approximately equal to 6σ (i.e., ±3σ around the mean) so that σ can be estimated as range/6. If σ cannot be estimated in this manner, a *pilot* study can be conducted and the standard deviation estimated from the resulting data.

Returning to the heating oil usage example, suppose the marketing manager would like to estimate the population mean annual usage of home heating oil to within ±50 gallons of the true value, and he desires to be 95% confident of correctly estimating the true mean. Based on a study taken the previous year, he feels that the standard deviation can be estimated as 325 gallons. With this information, the sample size can be determined in the following manner for e = 50, σ = 325, and 95% confidence (Z = 1.96):

$$n = \frac{Z^2 \sigma^2}{e^2} = \frac{(1.96)^2 (325)^2}{(50)^2}$$

$$= \frac{(3.8416)(105{,}625)}{2{,}500}$$

$$= 162.31$$

Therefore, n = 163. We have chosen a sample of size 163 because the general rule used in determining sample size is to always round up to the nearest integer value in order to slightly oversatisfy the criteria desired.

We may note that, if the marketing manager utilizes these criteria, a sample of 163 should be taken—not a sample of 35. However, the standard deviation that has been used was estimated at 325 based on a previous survey. If the standard deviation obtained in the actual survey is very different from this value, the computed sampling error will be directly affected.

Problems for Section 6.6

Note: *The problems in this section can be solved using Microsoft Excel (see Section 6.10).*

6.27 A survey is planned to determine the average annual family medical expenses of employees of a large company. The management of the company wishes to be 95% confident that the sample average is correct to within ±$50 of the true average family medical expenses. A pilot study indicates that the standard deviation can be estimated as $400.

(a) How large a sample size is necessary?
(b) If management wants to be correct to within ±$25, what sample size is necessary?

6.28 If the manager of the paint supply store in Problem 6.5 on page 296 wants to estimate the average amount in a 1-gallon can to within ± 0.004 gallon with 95% confidence and also assumes that the standard deviation remains at 0.02 gallon, what sample size is needed?

6.29 If the quality control manager in Problem 6.6 on page 296 wants to estimate the average life to within ±20 hours with 95% confidence and also assumes that the process standard deviation remains at 100 hours, what sample size is needed?

6.30 If the inspection division in Problem 6.7 on page 296 wants to estimate the average amount of soft-drink fill to within ±0.01 liter with 95% confidence and also assumes that the standard deviation remains at 0.05 liter, what sample size is needed?

6.31 A consumer group would like to estimate the average monthly electric bills for the month of July for single-family homes in a large city. Based on studies conducted in other cities, the standard deviation is assumed to be $25. The group would like to estimate the average bill for July to within ±$5 of the true average with 99% confidence.
(a) What sample size is needed?
(b) If 95% confidence is desired, what sample size is necessary?

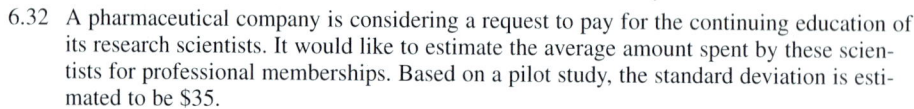

6.32 A pharmaceutical company is considering a request to pay for the continuing education of its research scientists. It would like to estimate the average amount spent by these scientists for professional memberships. Based on a pilot study, the standard deviation is estimated to be $35.
(a) What sample size is required to be 90% confident of being correct to within ±$10?
(b) If 95% confidence of being correct to within ±$20 is desired, what sample size is necessary?

6.33 An advertising agency that serves a major radio station would like to estimate the average amount of time that the station's audience spends listening to radio on a daily basis. From past studies, the standard deviation is estimated as 45 minutes.
(a) What sample size is needed if the agency wants to be 90% confident of being correct to within ±5 minutes?
(b) If 99% confidence is desired, what sample size is necessary?

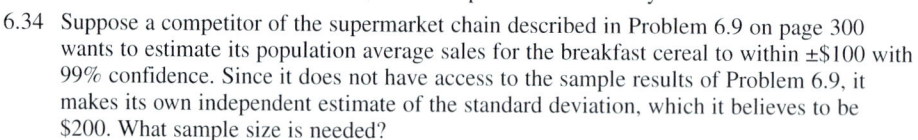

6.34 Suppose a competitor of the supermarket chain described in Problem 6.9 on page 300 wants to estimate its population average sales for the breakfast cereal to within ±$100 with 99% confidence. Since it does not have access to the sample results of Problem 6.9, it makes its own independent estimate of the standard deviation, which it believes to be $200. What sample size is needed?

6.35 Suppose that a gas utility serving a different geographical area than the one in Problem 6.13 on page 301 wishes to estimate its average waiting time to within ±5 days with 95% confidence. Since it does not have access to the sample results of Problem 6.13, it makes its own independent estimate of the standard deviation, which it believes to be 20 days. What sample size is needed?

6.7 SAMPLE SIZE DETERMINATION FOR A PROPORTION

In Section 6.6, we discussed the determination of sample size needed for the estimation of a population mean. Now suppose that the production manager in the newspaper example on page 304 wishes to determine the sample size necessary for estimating the population proportion of newspapers that have a nonconforming attribute, such as excessive ruboff, and so on. The methods of sample size determination that are utilized in estimating a population proportion are similar to those employed in estimating a mean.

To develop a formula for determining sample size, recall from Equation (5.12) that

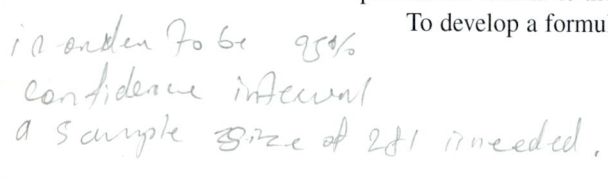

$$Z \cong \frac{p_s - p}{\sqrt{\dfrac{p(1-p)}{n}}}$$

where Z is the critical value corresponding to an area of $(1 - \alpha)/2$ from the center of a standardized normal distribution. Multiplying both sides of Equation (5.12) by $\sqrt{p(1-p)/n}$, we have

$$Z\sqrt{\frac{p(1-p)}{n}} = p_s - p$$

The sampling error, e, is equal to $(p_s - p)$, the difference between the sample proportion (p_s) and the parameter to be estimated (p). This sampling error can be defined as

$$e = Z\sqrt{\frac{p(1-p)}{n}}$$

Solving for n, we obtain

$$n = \frac{Z^2 p(1-p)}{e^2} \quad (6.5)$$

In determining the sample size for estimating a proportion, three unknowns must be defined:

1. The level of confidence desired
2. The sampling error permitted, e
3. The true proportion of "success," p

In practice, the selection of these three quantities is often difficult. Once we determine the desired level of confidence, we can obtain the appropriate Z value from the normal distribution. The sampling error e indicates the amount of error that we are willing to accept or tolerate in estimating the population proportion. The third quantity—the true proportion of success, p—is actually the population parameter that we are trying to find! Thus, how can we state a value for the very thing that we are taking a sample in order to determine?

Here there are two alternatives. First, in many situations, past information or relevant experience may be available that enables us to provide an educated estimate of p. Second, if past information or relevant experience is not available, we try to provide a value for p that would never *underestimate* the sample size needed. Referring to Equation (6.5), we observe that the quantity $p(1-p)$ appears in the numerator. Thus, we need to determine the value of p that will make $p(1-p)$ as large as possible. It can be shown that when $p = .5$, the product $p(1-p)$ achieves its maximum result. Several values of p along with the accompanying products of $p(1-p)$ are

$p = .5$, $p(1-p) = (.5)(.5) = .25$
$p = .4$, $p(1-p) = (.4)(.6) = .24$
$p = .3$, $p(1-p) = (.3)(.7) = .21$
$p = .1$, $p(1-p) = (.1)(.9) = .09$
$p = .01$, $p(1-p) = (.01)(.99) = .0099$

Therefore, when we have no prior knowledge or estimate of the true proportion p, we may want to use $p = .5$ as the most conservative way of determining the sample size. This would produce the largest sample size possible and therefore result in the highest possible cost. However, the use of $p = .5$ may result in an overestimate of the sample size since the actual sample proportion is utilized in the confidence interval. If the actual sample proportion is very different from .5, the width of the confidence interval may be substantially narrower than originally intended.

In our example, suppose the production manager wants to have 90% confidence of estimating the proportion of nonconforming newspapers to within ±.04 of its true value. In addition, since the publisher of the newspaper has not previously undertaken such a survey, no information is available from past data. Therefore, p will be set equal to .5.

With these criteria in mind, the sample size needed can be determined in the following manner with 90% confidence ($Z = 1.645$), $e = .04$, and $p = .5$:

$$n = \frac{(1.645)^2(.5)(.5)}{(.04)^2} = 422.82$$

Thus, $n = 423$. Therefore, in order to be 90% confident of estimating the proportion to within ±.04 of its true value, a sample size of 423 would be needed.

Problems for Section 6.7

Note: *The problems in this section can be solved using Microsoft Excel (see Section 6.10).*

6.36 A political pollster would like to estimate the proportion of voters who will vote for the Democratic candidate in a presidential campaign. The pollster would like 90% confidence that her prediction is correct to within ±.04 of the population proportion.
(a) What sample size is needed?
(b) If the pollster wants to have 95% confidence, what sample size is needed?
(c) If she wants to have 95% confidence and a sampling error of ±.03, what sample size is needed?

6.37 A cable television company would like to estimate the proportion of its customers who would purchase a cable television program guide. The company would like to have 95% confidence that its estimate is correct to within ±.05 of the true proportion. Past experience in other areas indicates that 30% of the customers will purchase the program guide. What sample size is needed?

6.38 A bank manager wants to be 90% confident of being correct to within ±.05 of the true population proportion of depositors who have both savings and checking accounts. What sample size is needed?

6.39 An audit test to establish the frequency of occurrence of failures to follow a specific internal control procedure is to be undertaken. The auditor decides that the maximum tolerable error rate that is permissible is 5%.
(a) What size sample is required to achieve a sample precision of ±2% with 99% confidence?
(b) What would be your answer in (a) if the maximum tolerable error rate is
   (1) 10%?
   (2) 15%?
   (3) 20%?

6.40 A large shipment of air filters is received by Joe's Auto Supply Company. The air filters are to be sampled to estimate the proportion that are unusable. From past experience, the proportion of unusable air filters is estimated to be .10.
(a) How large a random sample should be taken to estimate the true proportion of unusable air filters to within ±.07 with 99% confidence?
(b) If 95% confidence is desired with a sampling error of ±.06, what sample size should be taken?

6.41 Suppose that Milt's Motors, a competitor of the automobile dealer in Problem 6.22 on page 305, also wants to conduct a survey to determine the proportion of his customers who still own their cars 5 years after purchasing them. Suppose he wants to be 95% confident of being correct to within ±.025 of the true proportion. Note that this dealer does not have access to the sample results of Problem 6.22. What sample size is needed?

310 **Chapter 6** Estimation

6.8 ESTIMATION AND SAMPLE SIZE DETERMINATION FOR FINITE POPULATIONS

6.8.1 Estimating the Mean

In Section 5.12, we saw that when sampling without replacement from finite populations, the **finite population correction (fpc) factor** serves to reduce the standard error by a value equal to $\sqrt{(N-n)/(N-1)}$. When estimating population parameters from such samples without replacement, the fpc factor should be used for developing confidence interval estimates. Therefore, the $(1-\alpha) \times 100\%$ confidence interval estimate for the mean would become

$$\bar{X} \pm t_{n-1} \frac{S}{\sqrt{n}} \sqrt{\frac{N-n}{N-1}} \qquad (6.6)$$

In the heating oil usage example, a sample of 35 single-family homes has been selected. Suppose there is a population of 500 single-family homes served by the company. Using the finite population correction factor, we would have, with $\bar{X} = 1{,}122.7$ gallons, $S = 295.72$, $n = 35$, $N = 500$, and $t_{34} = 2.0322$ (for 95% confidence):

$$\bar{X} \pm t_{n-1} \frac{S}{\sqrt{n}} \sqrt{\frac{N-n}{N-1}} = 1{,}122.7 \pm (2.0322) \frac{295.72}{\sqrt{35}} \sqrt{\frac{500-35}{500-1}}$$

$$= 1{,}122.7 \pm (101.58)(.9653)$$

$$= 1{,}122.7 \pm 98.05$$

$$1{,}024.65 \leq \mu \leq 1{,}220.75$$

Here, since more than 5% of the population is to be sampled, the fpc factor has a moderate effect on the confidence interval estimate.

6.8.2 Estimating the Proportion

When sampling without replacement, the $(1-\alpha) \times 100\%$ confidence interval estimate of the proportion is

$$p_s \pm Z \sqrt{\frac{p_s(1-p_s)}{n}} \sqrt{\frac{N-n}{N-1}} \qquad (6.7)$$

In the production manager's study of nonconforming newspapers, a sample of 200 is selected from a population of a printing of 100,000 newspapers. The 90% confidence interval estimate is determined in the following manner when sampling without replacement. We have $p_s = 35/200 = .175$, $Z = 1.645$, $n = 200$, and $N = 100{,}000$. Thus,

$$p_s \pm Z \sqrt{\frac{p_s(1-p_s)}{n}} \sqrt{\frac{N-n}{N-1}} = .175 \pm (1.645) \sqrt{\frac{(.175)(.825)}{200}} \sqrt{\frac{100{,}000-200}{100{,}000-1}}$$

$$= .175 \pm (1.645)(.0269)\sqrt{.998}$$

$$= .175 \pm .0442(.999)$$
$$= .175 \pm .0442$$
$$.1308 \leq p \leq .2192$$

In this case, since the sample is a very small fraction of the population, the fpc factor has virtually no effect on the confidence interval estimate (as was computed on page 304).

6.8.3 Determining the Sample Size

Just as the fpc factor is used in developing confidence interval estimates, it should also be used in determining sample size when sampling without replacement. For example, when estimating the mean, the sampling error is

$$e = \frac{Z\sigma}{\sqrt{n}} \sqrt{\frac{N-n}{N-1}}$$

and when estimating the proportion, the sampling error is

$$e = Z\sqrt{\frac{p(1-p)}{n}} \sqrt{\frac{N-n}{N-1}}$$

In determining the sample size when estimating the mean, we have, from Equation (6.4),

$$n_0 = \frac{Z^2 \sigma^2}{e^2}$$

where n_0 = sample size without considering the finite population correction factor

Applying the fpc factor to this results in the actual sample size n, computed from

$$n = \frac{n_0 N}{n_0 + (N-1)} \tag{6.8}$$

In the marketing manager's survey of the annual heating oil usage, the sample size needed to be 95% confident of being correct to within ±$50 (assuming a standard deviation of 325 gallons) was 163, since n_0 was computed as 162.31. Using the fpc factor in Equation (6.8) leads to the following:

$$n = \frac{(162.31)(500)}{162.31 + (500-1)} = 122.72$$

Thus, $n = 123$. Here, since more than 30% of the population is to be sampled, the fpc factor has a substantial effect on the sample size—reducing it from 163 to 123. However, in general this may not be the case. For example, you may recall that in order to estimate the true proportion of nonconforming newspapers, the production manager needed a sample size of 423 (since n_0 was computed as 422.82). Using the fpc factor leads to

$$n = \frac{n_0 N}{n_0 + (N-1)}$$

$$= \frac{(422.82)(100{,}000)}{422.82 + (100{,}000 - 1)} = 421.04$$

Thus, $n = 422$. Here the use of the fpc factor made virtually no difference in the sample size selected.

Problems for Section 6.8

Note: *The problems in this section can be solved using Microsoft Excel (see Section 6.10).*

- 6.42 Refer to Problems 6.6 and 6.29 on pages 296 and 308. If the shipment contains a total of 2,000 light bulbs
  - (a) Set up a 95% confidence interval estimate of the true average life of light bulbs in this shipment.
  - (b) Determine the sample size needed to estimate the average life to within ±20 hours with 95% confidence.
  - (c) What would be your answers in (a) and (b) if the shipment contains 1,000 light bulbs?
- 6.43 Refer to Problem 6.27(a) on page 308. What sample size is necessary if the company has 3,000 employees?
- 6.44 Refer to Problem 6.20 on page 305. If the bank has 1,000 depositors
  - (a) Set up a 90% confidence interval estimate of the true proportion of depositors who are paid weekly.
  - (b) Determine the sample size needed to estimate the true proportion to within ±.05 with 90% confidence.
  - (c) What would be your answers in (a) and (b) if the bank has 2,000 depositors?
- 6.45 Refer to Problems 6.22 and 6.41 on pages 305 and 310. Suppose the population consists of 4,000 owners at *each* automobile dealer.
  - (a) From Problem 6.22, set up a 95% confidence interval estimate of the true proportion of customers who still own their cars 5 years after they purchased them.
  - (b) From Problem 6.41, determine what sample size is necessary to estimate the true proportion to within ±.025 with 95% confidence.
  - (c) What would be your answers in (a) and (b) if the population consists of 6,000 owners at *each* dealer?
- 6.46 Refer to Problems 6.7 and 6.30 on pages 296 and 308. If the population consists of 2,000 bottles
  - (a) Set up a 95% confidence interval estimate of the population average amount of soft drink in each bottle.
  - (b) Determine the sample size necessary to estimate the population average amount to within ±0.01 liter with 95% confidence.
  - (c) What would be your answers to (a) and (b) if the population consists of 1,000 bottles?
- 6.47 Refer to Problem 6.11 on page 300. If the number of greeting cards in the store's inventory is equal to 300
  - (a) Set up a 95% confidence interval estimate of the population average value of all greeting cards that are in its inventory.
  - (b) Compare the results obtained in (a) with those of Problem 6.11.
  - (c) What would be your answers to (a) and (b) if the store has 500 greeting cards in its inventory?

6.9 APPLICATIONS OF ESTIMATION IN AUDITING

6.9.1 Introduction to Auditing

In our discussion of estimation procedures, we have focused on estimating either the population mean or the population proportion when sampling with replacement or sampling without replacement from finite populations. One of the areas in business that makes widespread use of statistical sampling for the purposes of estimation is auditing.

Auditing may be defined as the collection and evaluation of evidence about numerical information relating to an economic entity such as a sole business proprietor, a partnership, a corporation, or a government agency to determine and report on the correspondence between the numerical information obtained and established criteria.

It is widely accepted in the field of auditing that the 100% examination of items is uneconomical and unwarranted. Among the advantages of statistical sampling are the following:

1. The sample result is objective and defensible. As the sample size is based on demonstrable statistical principles, the audit is defensible by one's superiors and in a court of law.
2. The method provides a way of estimating the sample size in advance on an objective basis.
3. The method provides an estimate of the sampling error.
4. This approach may turn out to be more accurate in drawing conclusions about the population due to the fact that the examination of large populations may be time-consuming and subject to more nonsampling error than a statistical sample.
5. Statistical samples may be combined and evaluated even though carried out by different auditors. This is due to the fact that there is a scientific basis for the sample, so the samples may be treated as if they were done by a single auditor.
6. Objective evaluation of the results of an audit is possible. The results can be projected with a known sampling error.

6.9.2 Estimating the Population Total

In auditing applications, one is usually more interested in obtaining estimates of the population total than the population mean. The **total** can be defined as follows:

The total is equal to the population size multiplied by the sample arithmetic mean.

$$\text{Total} = N\bar{X} \tag{6.9}$$

Thus, the confidence interval estimate for the total can be obtained from

$$N\bar{X} \pm N(t_{n-1})\frac{S}{\sqrt{n}}\sqrt{\frac{N-n}{N-1}} \tag{6.10}$$

To illustrate the estimated population total, we will use the following example.

● **Application** Suppose an auditor is faced with a population of 5,000 vouchers and wishes to estimate the total value of the population. A sample of 100 vouchers is selected with the following results:

Average voucher amount: $\bar{X} = \$1,076.39$

Standard deviation: $S = \$273.62$

Using Equation (6.9), the total may be computed as

$$\text{Total} = (5,000)(1,076.39) = \$5,381,950$$

From Equation (6.10), a 95% confidence interval estimate of the population total amount is obtained as follows:

$$(5{,}000)(1{,}076.39) \pm (5{,}000)(1.9842)\frac{(273.62)}{\sqrt{100}}\sqrt{\frac{5{,}000-100}{5{,}000-1}}$$

$$5{,}381{,}950 \pm 271{,}458.40(.99)$$

$$5{,}381{,}950 \pm 268{,}743.81$$

$$\$5{,}113{,}206.19 \leq \text{Population Total} \leq \$5{,}650{,}693.81$$

6.9.3 Difference Estimation

Difference estimation is used when an auditor believes that errors exist in a set of items being audited and wishes to estimate the magnitude of the errors based only on a sample. The following steps are used in difference estimation:

1. Determine the sample size required.
2. Compute the average difference in the sample ($\overline{D}$) by dividing the total difference by the sample size as shown in Equation (6.11).

> The average difference is equal to the sum of the differences divided by the sample size.
>
> $$\overline{D} = \frac{\sum_{i=1}^{n} D_i}{n} \tag{6.11}$$

3. Compute the standard deviation of the differences (S_D) as shown in Equation (6.12).

> $$S_D = \sqrt{\frac{\sum_{i=1}^{n} D_i^2 - n\overline{D}^2}{n-1}} \tag{6.12}$$

Be sure to remember that any item without a difference has a value of 0.

4. Set up a confidence interval estimate of the total difference in the population as shown in Equation (6.13).

> $$N\overline{D} \pm N(t_{n-1})\frac{S_D}{\sqrt{n}}\sqrt{\frac{N-n}{N-1}} \tag{6.13}$$

Returning to our voucher example, suppose that in the sample of 100 vouchers, there are 14 vouchers that contain errors. The values of the 14 errors (in dollars) are as follows:

| 75.41 | 38.97 | 108.54 | −37.18 | 62.75 | 118.32 | −88.84 |
| 127.74 | 55.42 | 39.03 | 29.41 | 47.99 | 28.73 | 84.05 |

For these data,

$$\bar{D} = \frac{\sum_{i=1}^{n} D_i}{n} = \frac{690.34}{100} = 6.9034$$

and

$$S_D = \sqrt{\frac{\sum_{i=1}^{n} D_i^2 - n\bar{D}^2}{n-1}}$$

$$= \sqrt{\frac{78{,}168.366 - (100)(6.9034)^2}{100 - 1}}$$

$$= \sqrt{\frac{73{,}402.673}{99}}$$

$$= \sqrt{741.441}$$

$$= 27.23$$

To obtain the confidence interval estimate for the total difference in the population, we use Equation (6.13) so that we have

$$(5{,}000)(6.9034) \pm 5{,}000(1.9842)\frac{27.23}{\sqrt{100}}\sqrt{\frac{N-n}{N-1}}$$

$$34{,}517 \pm 26{,}744.73$$

$$\$7{,}772.27 \leq \text{Total Difference} \leq \$61{,}261.73$$

Problems for Section 6.9

Note: The problems in this section can be solved using Microsoft Excel (see Section 6.10).

6.48 Referring to Problem 6.10 on page 300, suppose the savings bank branch has 2,000 depositors. Set up a 95% confidence interval estimate of the total amount held in passbook savings accounts in the branch.

6.49 Referring to Problem 6.11 on page 300 and Problem 6.47 on page 313, set up a 95% confidence interval estimate of the total value of greeting cards held in inventory.

6.50 Referring to Problem 6.12 on page 300, suppose that the corporation has 3,000 employees. Set up a 90% confidence interval estimate of the total family dental expenses for all employees in the preceding year.

6.51 An outlet of a large electronics store chain is conducting an end-of-month inventory of the merchandise in stock. It has been determined that there are 1,546 items in inventory. A sample size of 50 items is randomly selected, and an audit is conducted with the following results:

Value of merchandise: $\bar{X} = \$252.28$; $S = \$93.67$

Set up a 95% confidence interval estimate of the total estimated value of the merchandise in inventory at the end of the month.

6.52 A customer in the wholesale garment trade is often entitled to a discount for a cash payment for goods. The amount of discount varies by vendor. A sample of 150 items selected from a population of 4,000 invoices at the end of a period of time reveals that in 13 cases the customer failed to take the discount to which he was entitled. The amount (in dollars) of the 13 discounts that were not taken are as follows:

6.45	15.32	97.36	230.63	104.18	84.92	132.76
66.12	26.55	129.43	88.32	47.81	89.01	

DISCOUNT.TXT

Set up a 99% confidence interval estimate of the population total amount of discount not taken.

6.53 Econe Dresses is a small company manufacturing woman's dresses for sale to specialty stores. There are 1,200 inventory items in which the historical cost is recorded on a first-in first-out (FIFO) basis. In the past, approximately 15% of the inventory items were incorrectly priced. However, any misstatements were not thought to be significant. A sample of 120 items is selected, and the historical cost of each item is compared to the audited value. The results indicate that 15 items differ in their historical costs and audited values. These differences are as follows:

DRESSES.TXT

Sample Number	Historical Cost ($)	Audited Value ($)
5	261	240
9	87	105
17	201	276
18	121	110
28	315	298
35	411	356
43	249	211
51	216	305
60	21	210
73	140	152
86	129	112
95	340	216
96	341	402
107	135	97
119	228	220

Set up a 95% confidence interval estimate of the total population difference in the historical cost and audited value.

6.10 USING MICROSOFT EXCEL FOR CONFIDENCE INTERVALS AND SAMPLE SIZE DETERMINATION

In this chapter we have developed the confidence interval estimate for both means and proportions, we have discussed how to determine the sample size necessary to conduct a survey, and we have used the finite population correction (fpc) factor. In addition, specific applications in auditing have been introduced. In this section we will learn how the various features of Microsoft Excel can be used to compute confidence intervals and to determine sample sizes.

6-10.XLS

Table 6.1.Excel Design for Confidence sheet for computing confidence interval estimates.

	A	B	C	D	E	F
1			Confidence Intervals			
2	Mean (σ Known)		Mean (σ Unknown)		Proportion	
3	n	xxx	n	xxx	n	xxx
4	Arithmetic mean	xx.xx	Arithmetic mean	xx.xx	Successes	xxx
5	σ	xx.xx	S	xx.xx	Failures	=F3–F4
6	Confidence Level	.xx	Confidence Level	.xx	Confidence Level	.xx
7						
8	Std Error	=B5/SQRT(B3)	Std Error	=D5/SQRT(D3)	Sample Proportion	=F4/F3
9	Z	=NORMSINV(.5+(B6/2))	df	=D3–1	Z	=NORMSINV(.5+(F6/2))
10			t	=TINV(1–D6,D9)	Std Error Proportion	=SQRT((F8*(1–F8)/F3)
11	Half-width	=CONFIDENCE(1–B6,B5,B3)	Half-width	=D10*D8	Half-width	=F9*F10
12	Lower Limit	=B4–B11	Lower Limit	=D4–D11	Lower Limit	=F8–F11
13	Upper Limit	=B4+B11	Upper Limit	=D4+D11	Upper Limit	=F8+F11
14			Finite Populations			
15	N	xxxx	N	xxxx	N	xxxx
16	f.p.c.	=SQRT((B15–B3)/(B15–1))	f.p.c.	=SQRT((D15–D3)/(D15–1))	f.p.c.	=SQRT((F15–F3)/(F15–1))
17	Half-width	=B11*B16	Half-width	=D11*D16	Half-width	=F11*F16
18	Lower Limit	=B4–B17	Lower Limit	=D4–D17	Lower Limit	=F8–F17
19	Upper Limit	=B4+B17	Upper Limit	=D4+D17	Upper Limit	=F8+F17

6.10.1 Using Microsoft Excel for the Confidence Interval Estimate for the Mean (σ Known)

You may recall that in Section 3.7, the Descriptive Statistics option of the Data Analysis tool was used to obtain a set of descriptive statistics. At that time, we deferred discussion of the Confidence Interval check box in the Descriptive Statistics dialog box. Selecting this check box will allow the Data Analysis tool to compute half the width of the confidence interval for the mean when σ is known and it should be used when working with raw data.

When raw data are not available, or we want results that can dynamically change, Excel formulas can be used. Table 6.1.Excel presents the design of a sheet named Confidence that calculates all the confidence interval estimates for the means and proportions that will be developed in this worksheet. To implement this worksheet, open a new workbook and rename the active sheet Confidence. Then do the following:

① Enter the value for the sample size n in cell B3, the value for the sample arithmetic mean in cell B4, the population standard deviation σ in cell B5, and the confidence level (entered as a decimal) in cell B6. If the arithmetic mean $\overline{X}$ needs to be calculated, either the Data Analysis tool or Excel formulas can be used and the results copied to cell B4. If the value for the sample mean is available, it should be entered in cell B4. Using the computer paper production examples on page 295, we would enter 100 in cell B3, 10.998 in cell B4, .02 in cell B5, and .95 in cell B6.

❷ Now we can compute the standard error of the mean $\sigma/\sqrt{n}$ in cell B8 by entering the formula =B5/SQRT(B3).

❸ The value of Z necessary in Equation (6.1) can be obtained using the NORMSINV function (see Section 5.5). Since we wish to obtain the upper critical value, we need the cumulative percentage equal to .5 plus the confidence level in cell B6 divided by 2. This means that we enter the formula =NORMSINV(.5+(B6/2)) in cell B9.

❹ Half the width of the confidence interval can be obtained by using the CONFIDENCE function whose format is

$$\text{CONFIDENCE}(1 - \textit{level of confidence}, \sigma, n)$$

For this problem enter =CONFIDENCE(1−B6,B5,B3) in cell B11.

❺ The lower limit of the confidence interval can be obtained by subtracting the half-width in cell B11 from the arithmetic mean in cell B4, while the upper limit of the confidence interval can be obtained by adding the half-width in cell B11 to the arithmetic mean in cell B4. The formulas to perform these calculations (as shown in Table 6.1.Excel) are entered in cells B12 and B13.

The results of completing these portions of the worksheet are depicted in Figure 6.1.Excel.

	A	B
1	Confidence Intervals	
2	Mean (σ Known)	
3	n	100
4	Arithmetic mean	10.998
5	σ	0.02
6	Confidence Level	0.95
7		
8	Standard Error	0.002
9	Z	1.959961
10		
11	Half-width	0.00392
12	Lower Limit	10.99408
13	Upper Limit	11.00192

FIGURE 6.1.EXCEL Confidence interval estimate for the mean (σ known) using Microsoft Excel for the computer paper production example.

▲ WHAT IF EXAMPLE

The Confidence worksheet presented in Table 6.1.Excel allows us to study the effect of changes in $\bar{X}$, σ, n, and the level of confidence on the confidence interval estimate. For example, if we decide that we want 90% confidence instead of 95% confidence, we would change the entry in cell B6 from .95 to .90. Observe that when this is done, the Z value in cell B9 changes to 1.64485, the half-width in cell B11 changes to .00329, and the confidence limits in cells B12 and B13 change to 10.9947 and 11.0013, respectively. As before, if we are interested in seeing the effects of many different changes, we could use the Scenario Manager (see Section 1S.15) to store and use sets of alternative values of $\bar{X}$, σ, n, and the level of confidence.

6.10.2 Using Microsoft Excel for the Confidence Interval Estimate for the Mean (σ Unknown)

Although Excel does not have a Data Analysis tool to compute confidence interval estimates for the mean when σ is unknown, Excel formulas can be used instead. In a manner similar to the development of the confidence interval estimate for the mean when σ was known, we can now obtain the confidence interval estimate for the mean with σ unknown by continuing to implement the Confidence sheet design of Table 6.1.Excel.

Referring to Equation (6.2) on page 299, we see that to obtain the confidence interval estimate for the mean, we need to have the sample mean $\bar{X}$, the sample standard deviation S, and the sample size n. If the arithmetic mean $\bar{X}$ and the sample standard deviation S need to be calculated, either the Data Analysis tool or Excel formulas could be used and the results copied to cells D4 and D5, respectively. If the values for the sample mean and the sample standard deviation S are provided, they should be entered in cells D4 and D5, respectively. The steps for setting up the confidence interval are as follows:

1. Enter the value for the sample size n in cell D3, the sample arithmetic mean in cell D4, the sample standard deviation S in cell D5, and the confidence level (entered as a decimal) in cell D6. Using the heating oil usage example on page 299, we would enter 35 in cell D3, 1122.7 in cell D4, 295.72 in cell D5, and .95 in cell D6. (Since the raw data is presented in Table 6.2 on page 299, the Descriptive Statistics tool would be used to obtain the arithmetic mean and the sample standard deviation S.)

2. Now we can compute the standard error of the mean $S/\sqrt{n}$ in cell D8 by entering the formula =D5/SQRT(D3).

3. The number of degrees of freedom, equal to $n - 1$, is entered into cell D9 by using the formula =D3–1.

4. The value of t necessary in Equation (6.2) can be obtained using the TINV function whose format is

    TINV(1 – *confidence level, degrees of freedom*)

    This function provides the t value needed for the confidence level provided in cell D6, given the degrees of freedom (in cell D9). Thus, we enter =TINV(1–D6,D9) in cell D10.

5. The half-width of the confidence interval (cell D11) can be obtained by multiplying the t statistic obtained in cell D10 by the standard error of the mean obtained in cell D8.

6. The lower limit of the confidence interval (cell D12) can be obtained by subtracting the half-width (cell D11) from the arithmetic mean (cell D4), while the upper limit of the confidence interval (cell D13) can be obtained by adding the half-width to the arithmetic mean.

The results are depicted in Figure 6.2.Excel.

▲ WHAT IF EXAMPLE

This Confidence worksheet presented in Table 6.1.Excel allows us to study the effect of changes in $\bar{X}$, s, n, and the level of confidence on the confidence interval estimate. For example, if we decided that we wanted 90% confidence instead of 95% confidence, we would change the entry in cell D6 from .95 to .90. Observe that when this is done, the *t* value in cell D10 changes to 1.69092, the half-width in cell D11 changes to 84.5222, and the confidence limits in cells D12 and D13 change to 1,038.18 and 1,207.22, respectively. As before, if we are interested in seeing the effects of many different changes, we could use the Scenario Manager (see Section 1S.15) to store and use sets of alternative values of $\bar{X}$, S, n, and the level of confidence.

	C	D
	\multicolumn{2}{l	}{Confidence Intervals}
	\multicolumn{2}{l	}{Mean (σ Unknown)}
	n	35
	Arithmetic mean	1122.7
	S	295.72
	Confidence Level	0.95
	Standard Error	49.9858
	df	34
	t	2.032243
	Half-width	101.5833
	Lower Limit	1021.117
	Upper Limit	1224.283

FIGURE 6.2.EXCEL Confidence interval estimate for the mean (σ unknown) using Microsoft Excel for the heating oil usage example.

6.10.3 Using Microsoft Excel for the Confidence Interval Estimate for the Proportion

Although Excel does not have a Data Analysis tool to compute confidence interval estimates for the proportion, Excel formulas can be used instead. In a way similar to the development of the confidence interval estimate for the mean, we can now obtain the confidence interval estimate for the proportion by again continuing to implement the Confidence sheet design of Table 6.1.Excel.

Referring to Equation (6.3) on page 304, we see that to obtain the confidence interval estimate for the proportion, we need to have the sample proportion p_s and the sample size n. If the sample size and the number of successes need to be obtained from the raw data, the PivotTable Wizard (see Section 2.12) could be used and the results copied to cells F3 and F4, respectively. If the values for the sample size and the number of successes are available, they should be entered in cells F3 and F4. The steps for setting up the confidence interval for the proportion are as follows:

❶ Enter the value for the sample size *n* in cell F3, the value for the number of successes in cell F4, and the confidence level (entered as a decimal) in cell F6. The number of failures (which is the complement of the number of successes) can be computed in cell F5 by entering the formula =F3–F4. Using the nonconforming newspaper example on page 304, we would enter 200 in cell D3, 35 in cell F4, and .90 in cell F6.

❷ Now we can compute the sample proportion in cell F8 by dividing the number of successes in cell F4 by the sample size in cell F3.

❸ The value of Z necessary in Equation (6.3) can be obtained using the NORMSINV function (see Section 5.5). Since we wish to obtain the upper critical value, we need the cumulative percentage equal to .5 plus the confidence level divided by 2. This means that we enter the formula =NORMSINV(.5+(F6/2)) in cell F9.

④ Now we can compute the standard error of the proportion $\sqrt{p_s(1-p_s)/n}$ in cell F10 by entering the formula =SQRT(F8*(1–F8)/F3).

⑤ Half the width of the confidence interval can be determined by multiplying the Z statistic obtained in cell F9 by the standard error of the proportion obtained in cell F10.

⑥ The lower limit of the confidence interval can be obtained by subtracting the half-width in cell F11 from the sample proportion in cell F8, while the upper limit of the confidence interval can be obtained by adding the half-width in cell F11 to the sample proportion in cell F8.

The results are depicted in Figure 6.3.Excel.

	E	F
	Confidence Intervals	
	Proportion	
	n	200
	Successes	35
	Failures	165
	Confidence Level	0.9
	Sample Proportion	0.175
	Z	1.644853
	Std. Error Proportion	0.026868
	Half-width	0.044193
	Lower Limit	0.130807
	Upper Limit	0.219193

FIGURE 6.3.EXCEL Confidence interval estimate for the proportion using Microsoft Excel for the nonconforming newspaper example.

▲ **WHAT IF EXAMPLE**

This Confidence worksheet allows us to study the effect of changes in the number of successes, n, and the level of confidence on the confidence interval estimate. For example, if we decided that we wanted 95% confidence instead of 90% confidence, we would change the entry in cell F6 from .90 to .95. Observe that when this is done, the Z value in cell F9 changes to 1.95996, the half-width in cell F11 changes to .05266, and the confidence limits in cells F12 and F13 change to .12234 and .22766, respectively. As before, if we are interested in seeing the effects of many different changes, we can use the Scenario Manager (see Section 1S.15) to store and use sets of alternative values of the number of successes, n, and the level of confidence.

6.10.4 Using Microsoft Excel for Sample Size Determination

Although Excel does not have a Data Analysis tool to determine sample size, as was the case with confidence intervals, Excel functions can be used instead. Table 6.2.Excel contains the design for a worksheet named SampleSize that we will use for sample size determination.

Table 6.2.Excel Design for determining sample size.

	A	B	C	D
1		Sample Size Determination		
2	Mean			Proportion
3	Confidence Level	.xx	Confidence Level	.xx
4	σ	xxx	p	.xx
5	Error (e)	xxx	Error (e)	.xx
6	Z	=NORMSINV(.5+(B3/2))	Z	=NORMSINV(.5+(D3/2))
7	Sample Size	=((B6*B4)/B5)^2	Sample Size	=(D6^2*D4*(1–D4))/D5^2
8	Sample Required	=ROUNDUP(B7,0)	Sample Required	=ROUNDUP(D7,0)
9			Finite Populations	
10	N	xxxx	N	xxxx
11	n	=(B7*B10)/(B7+B10–1)	n	=(D7*D10)/(D7+D10–1)
12	Sample Required	=ROUNDUP(B11,0)	Sample Required	=ROUNDUP(D11,0)

- **Sample Size Determination for the Mean** Referring to Equation (6.4), we see that to determine the sample size for the mean, we need to specify the confidence level desired (which determines the Z value), the population standard deviation σ, and the desired sampling error e. The steps for determining the sample size for the mean are as follows:

1. Enter the confidence level (as a decimal) in cell B3, the estimate of the population standard deviation σ in cell B4, and the sampling error e in cell B5. Using the heating oil usage example on page 307, we would enter .95 in cell B3, 325 in cell B4, and 50 in cell B5.

2. The value of Z necessary in Equation (6.4) can be obtained using the NORMSINV function (see Section 5.5). Since we wish to obtain the upper critical value, we need the cumulative percentage equal to .5 plus the confidence level divided by 2. This means that we enter the formula =NORMSINV(.5+(B3/2)) in cell B6.

3. The sample size is computed in cell B7 using the formula =((B6*B4)/B5)^2.

4. Since Excel computes the result with numerous decimal places, and the convention in determining sample size is to round up our result to the next integer value, we need to use the ROUNDUP function of Microsoft Excel. The format of this function is

   =ROUNDUP(*cell, decimal places*)

   where decimal places specifies the number of decimal places to round up the value in a cell. (To round up to the next integer value, decimal places is set at zero.)

Thus, to round up the sample size obtained in cell B7 to the next integer value, we enter =ROUNDUP(B7,0) in cell B8.

The results are depicted in Figure 6.4.Excel.

▲ **WHAT IF EXAMPLE**

This Sample Size worksheet allows us to study the effect on the sample size of changes in the confidence level desired (which determines the Z value), the population standard deviation σ, and the desired sampling error, e. For example, if we decided we wanted a sampling error e of 75 instead of 50, we would enter 75 in cell B5. The resulting sample size indicated in cell B7 would be 72.1338, which is rounded up to 73 in cell B8. As before, if we are interested in seeing the effects of many different changes, we could use the Scenario Manager (see Section 1S.15) to store and use sets of alternative values of the confidence level desired (which determines the Z value), the population standard deviation σ, and the desired sampling error e.

	A	B
1	Sample Size Determination	
2	Mean	
3	Confidence Level	0.95
4	σ	325
5	Error (e)	50
6	Z	1.959961
7	Sample Size	162.3012
8	Sample Required	163

FIGURE 6.4.EXCEL Sample size determination for the mean using Microsoft Excel for the heating oil usage example.

● **Sample Size Determination for the Proportion** Referring to Equation (6.5), we see that to determine the sample size for the proportion, we need to indicate the confidence level desired (which determines the Z value), the estimated population proportion p, and the desired sampling error e. The steps for determining the sample size for the proportion are as follows:

❶ Enter the confidence level (as a decimal) in cell D3, the estimate of the population proportion p in cell D4, and the sampling error e in cell D5. Using the nonconforming newspaper example on page 310, we would enter .90 in cell D3, .5 in cell D4, and .04 in cell D5.

❷ The value of Z necessary in Equation (6.5) can be determined using the NORMSINV function (see Section 5.5). Since we wish to obtain the upper critical value, we need the cumulative percentage equal to .5 plus the confidence level divided by 2. This means that we enter the formula =NORMSINV (.5+(D3/2)) in cell D6.

❸ The sample size is computed in cell D7 using the formula =(D6^2*D4*(1–D4))/D5^2.

❹ Since Excel computes the result with numerous decimal places, and the convention in determining sample size is to round our result to the next higher integer value, we need to use the ROUNDUP function of Microsoft Excel. To do this for the sample size obtained in cell D7, we enter =ROUNDUP(D7,0) in cell D8.

The results are depicted in Figure 6.5.Excel.

▲ **WHAT IF EXAMPLE**

This Sample Size worksheet allows us to study the effect on the sample size of changes in the confidence level desired (which determines the Z value), the population proportion p, and the desired sampling error e. For example, if we decided we wanted a sampling error e of .03 instead of .04, we would enter .03 in cell D5. The resulting sample size indicated in cell D7 would be 751.539, which is rounded up to 752 in cell D8. As before, if we are interested in seeing the effects of many different changes, we could use the

Scenario Manager (see Section 1S.15) to store and use sets of alternative values of the confidence level desired (which determines the Z value), the population proportion p, and the desired sampling error e.

	C	D
	Sample Size Determination Proportion	
	Confidence Level	0.9
	p	0.5
	Error (e)	0.04
	Z	1.644853
	Sample Size	422.7408
	Sample Required	423
	Finite Populations	
	N	100,000
	n	420.9655
	Sample Required	421

FIGURE 6.5.EXCEL
Sample size determination for the proportion using Microsoft Excel for the nonconforming newspaper example.

6.10.5 Using Microsoft Excel for Bootstrapping Estimation

Although the Data Analysis tool of Microsoft Excel does not have a bootstrapping tool, the Sampling option can be used along with Excel formulas to obtain a bootstrap estimate. On a new worksheet renamed Bootstrap, we need to do the following:

① Select a random sample without replacement from a population and store the results in a worksheet renamed Data. If the sample has already been selected, copy it to this worksheet.

② Issue the command Tools | Data Analysis and then select Sampling (see Section 5.11) from the dialog box and click OK.

③ From the Sampling dialog box, select the Random option button and enter the desired sample size n in the Number of Samples edit box.

④ Compute the average of this sample.

⑤ Repeat steps 1–4 m times and sort the m averages in ascending order.

⑥ Manually determine the desired bootstrap confidence level $(1 - \alpha)$, and manually determine the values that cut off the $(\alpha/2 \times 100\%)$ smallest value and the $(\alpha/2 \times 100\%)$ largest value.

6.10.6 Using Microsoft Excel with the Finite Population Correction Factor

In Section 6.8, we used the finite population correction factor to adjust our results obtained in confidence interval estimation and sample size determination for the existence of a finite population. We may also adjust the results obtained from the Confidence and Sample Size worksheets developed in Sections 6.10.1–6.10.3 to take into account the fpc factor.

• **Using the Finite Population Correction Factor for Confidence Intervals**
To obtain the confidence interval estimate when sampling from a finite population, we can multiply the fpc factor equal to $\sqrt{(N-n)/(N-1)}$ by the previous half-width of the confidence interval [see Equations (6.6) and (6.7) on page 311]. Referring to Table 6.1.Excel on page 318, the population size can be entered in cells B15, D15, and F15, respectively, for the three confidence intervals discussed in Sections 6.10.1–6.10.3. Then the fpc factor can be defined for the confidence interval for the mean with σ known in cell B16 as =SQRT((B15–B3)/(B15–1)). Similar formulas are entered in cell D16 for the confidence interval for the mean with σ unknown and in cell F16 for the confidence interval for the proportion. The revised half-widths in row 17 are obtained by multiplying the previous half-widths in row 11 by the fpc in row 16. Then the lower and upper limits would be obtained in rows 18 and 19 by subtracting or adding the revised half-width in row 17 from the arithmetic mean in cells B4 or D4 or the sample proportion in cell F8.

Referring to the heating oil usage example on page 299, if there was a population of 500, we would enter 500 in cell D15, and the confidence limits would be computed in cells D17 and D18. Referring to the nonconforming newspaper example discussed on page 304, with a population of 100,000, we would enter 100,000 in cell F15, and the confidence limits would be computed in cells F17 and F18. Figure 6.6.Excel illustrates the results obtained for these two examples.

C	D	E	F
\multicolumn{4}{c}{Confidence Intervals}
Mean (σ Unknown)		Proportion	
n	35	n	200
Arithmetic mean	1122.7	Successes	35
S	295.72	Failures	165
Confidence Level	0.95	Confidence Level	0.9
Standard Error	49.9858	Sample Proportion	0.175
df	34	Z	1.644853
t	2.032243	Std. Error Proportion	0.026868
Half-width	101.5833	Half-width	0.044193
Lower Limit	1021.117	Lower Limit	0.130807
Upper Limit	1224.283	Upper Limit	0.219193
\multicolumn{4}{c}{Finite Populations}
N	500	N	100,000
f.p.c.	0.965331	f.p.c	0.999004
Half-width	98.0615	Half-width	0.044149
Lower Limit	1024.638	Lower Limit	0.130851
Upper Limit	1220.762	Upper Limit	0.219149

FIGURE 6.6.EXCEL Confidence interval estimates with the fpc factor using Microsoft Excel for the heating oil usage and the nonconforming newspaper examples.

• **Using the Finite Population Correction Factor for Sample Size Determination** To obtain the desired sample size when sampling from a finite population, we adjust the sample size using the fpc factor as indicated in Equation (6.8). Referring to Table 6.2.Excel, we enter the population size N in cell B10 for means, and cell D10 for proportions. Then we can enter the formula =(B7*B10)/(B7+B10–1) in cell B11 and the formula

=(D7*D10)/(D7+D10–1) in cell D11. The results obtained are rounded up to the next integer value in cells B12 and D12, respectively.

Referring to the heating oil usage example on page 312, with a population size of 500 entered in cell B10, the resulting sample size is 122.7135 in cell B11, rounded up to 123 in cell B12. Referring to the nonconforming newspaper example discussed on page 312, with a population size of 100,000 entered in cell D10, the resulting sample size is 420.9655 in cell D11, rounded up to 421 in cell D12. Figure 6.7.Excel illustrates the results obtained for these examples.

	A	B	C	D
1	Sample Size Determination			
2	Mean		Proportion	
3	Confidence Level	0.95	Confidence Level	0.9
4	σ	325	σ	0.5
5	Error (e)	50	Error (e)	0.04
6	Z	1.959961	Z	1.644853
7	Sample Size	162.3012	Sample Size	422.7408
8	Sample Required	163	Sample Required	423
9	Finite Populations			
10	N	500	N	100,000
11	n	122.7135	n	420.9655
12	Sample Required	123	Sample Required	421

FIGURE 6.7.EXCEL Sample size determination for the mean for a finite population using Microsoft Excel for the heating oil usage and nonconforming newspaper examples.

▲ **WHAT IF EXAMPLE**

The Confidence worksheet presented in Table 6.1.Excel and the SampleSize worksheet presented in Table 6.2.Excel allow us to study the effect on the confidence interval estimate and the sample size of changes in the size of the population. For example, referring to the heating oil usage example, if the population size is 1,000 instead of 500, 1,000 would be entered in cell D15 on the Confidence worksheet and cell B10 on the Sample Size worksheet. The confidence interval would then have limits between 1,022.86 and 1,222.54, and the sample size would be 140.

6.10.7 Using Microsoft Excel for Applications of Estimation in Auditing

Although the Data Analysis tool of Microsoft Excel does not have a tool for estimating the total or the differences, Excel functions can be used to obtain these estimates. Table 6.3.Excel contains the design for worksheets labeled Differences and Auditing that we will use for obtaining these estimates.

● **Estimating the Population Total** Referring to Equations (6.9) and (6.10), we see that in a manner similar to the development of the confidence interval estimate for the

Table 6.3. Excel Design for the Differences and Auditing worksheets.

DIFFERENCES WORKSHEET (FOR THE DATA ON PAGE 315)

	A	B	C	D	E	F
1	Differences			SSCalc.		
2	75.41	Sum	=SUM(A2:A15)	=(A2–$C$5)^2	SS:X with difference	=Sum(D2:D15)
3	38.97	Count	14	=(A3–$C$5)^2	SS:X without difference	=C6* – C5^2
4	108.54	n	xxx	=(A4–$C$5)^2	Sum of Squares	=F2+F3
5	–37.18	DBar	=C2/C4	.	df	=C4–1
6	62.75	Count of X w/no diff.	=C4–C3	.	Variance	=F4/F5
7	118.32			.	Std.Deviation	=SQRT(F6)
8	.			.		
9	.			.		
⋮						
15	84.05			=(A15–$C$5)^2		

THE AUDITING WORKSHEET

	A	B	C	D
1	Confidence Interval for the Total		Difference Estimation	
2				
3	n	xxx	n	=Differences!C4
4	Arithmetic Mean	xxx	Sum of Differences	=Differences!C2
5	S	xxx	Average Difference	=Differences!C5
6	Confidence Level	.xx	Confidence Level	.xx
7	N	xxxx	S	=Differences!F7
8	Total	=B7*B4	N	xxxx
9	f.p.c.	=SQRT((B7–B3)/(B7–1))	Total Difference	=D8*D5
10	Standard Error	=(B7*B5*B9)/SQRT(B3)	f.p.c.	=SQRT((D8–D3)/(D8–1))
11	df	=B3–1	Standard Error	=(D8*D7*D10)/SQRT(D3)
12	t	=TINV/(1–B6,B11)	df	=D3–1
13	Half-width	=B12*B10	t	=TINV/(1–D6,D12)
14	Lower Limit	=B8–B13	Half-width	=D13*D11
15	Upper Limit	=B8+B13	Lower Limit	=D9–D14
16			Upper limit	=D9+D14

mean when σ is unknown, we need to have the sample mean $\bar{X}$, the sample standard deviation S, the population size N, and the sample size n. If the arithmetic mean $\bar{X}$ and the sample standard deviation S need to be calculated, either the Data Analysis tool or Excel formulas can be used and the results copied to cells B4 and B5, respectively, of the Auditing worksheet. If the values for the sample mean and the sample standard deviation S are provided, they should be entered in cells B4 and B5. The steps for setting up the confidence interval with a workbook open to a new sheet renamed Auditing are as follows:

❶ Enter the value for the sample size n in cell B3, the sample arithmetic mean in cell B4, the sample standard deviation S in cell B5, the confidence level (entered as a decimal) in cell B6, and the population size N in cell B7. Using the voucher example on page 314 we would enter 100 in cell B3, 1076.39 in cell B4, 273.62 in cell B5, .95 in cell B6, and 5,000 in cell B7. (If the raw data were available, the Descriptive Statistics tool would be used to obtain the arithmetic mean and the sample standard deviation S.)

❷ The estimate of the total equal to $N\bar{X}$ is entered in cell B8 using the formula =B7*B4.

③ The fpc factor equal to $\sqrt{(N-n)/(N-1)}$ is entered in cell B9 using the formula =SQRT((B7–B3)/(B7–1)).

④ Now we can compute the standard error of the mean, $(N)(S/\sqrt{n})(\sqrt{(N-n)(N-1)})$, in cell B10 by entering the formula =(B7*B5*B9)/SQRT(B3).

⑤ The number of degrees of freedom equal to $n-1$, is entered into cell B11 by using the formula =B3–1.

⑥ The value of t necessary in Equation (6.10) can be obtained using the TINV function. Thus, we enter =TINV(1–B6,B11) in cell B12.

⑦ Half the width of the confidence interval (cell B13) can be determined by multiplying the t statistic (cell B12) by the standard error of the mean (cell B10).

⑧ The lower limit of the confidence interval can be obtained by subtracting the half-width in cell B13 from the total in cell B8, while the upper limit can be obtained by adding the half-width to the total.

The results are depicted in Figure 6.8.Excel. Any differences from those shown on page 314 are due to rounding in those calculations.

	A	B
1	Confidence Interval	
2	for the Total	
3	n	100
4	Arithmetic Mean	1076.39
5	S	273.62
6	Confidence Level	0.95
7	N	5000
8	Total	5381950
9	f.p.c.	0.990049
10	Standard Error	135448.5
11	df	99
12	t	1.984217
13	Half-width	268759.3
14	Lower Limit	5113191
15	Upper Limit	5650709

FIGURE 6.8.EXCEL Confidence interval for the population total using Microsoft Excel for the voucher example.

● **Difference Estimation** Referring to Equation (6.11) on page 315, we need to have the average difference $\bar{D}$, the sample standard deviation of the differences S_D, the population size N, and the sample size n. Continuing with the same workbook used earlier in this section, the steps for setting up the confidence interval for the total difference are as follows:

① Insert a new worksheet and rename it Differences. This worksheet is used to store the differences and compute the average difference and the standard deviation of the differences. After entering the label Differences in cell A1, enter the differences in column A beginning in row 2. For the voucher example data on page 314, since there are 14 differences, they would be entered into the cell range A2:A15.

② Compute the sum of the differences using the SUM function by entering the formula =SUM(A2:A15) in cell C2.

③ Enter 14 for the count of the number of differences in cell C3.

④ Enter 100 as the sample size *n* in cell C4.

⑤ Compute the average difference $\bar{D}$ in cell C5 by entering the formula =C2/C4.

⑥ Obtain a count of the number of observations in the sample that do not have differences by subtracting the contents of cell C3 from those of cell C4.

⑦ Next, we need to compute the standard deviation of the difference. First, for the observations that contain differences, we need to obtain the sum of the squared differences around the average difference. To compute this, we enter the formula =(A2–$C$5)^2 in cell D2 and copy it down through cell D15 (since the last difference is in cell A15). Then we sum these squared differences by entering the formula =SUM(D2:D15) in cell F2.

⑧ Now, we need to remember that there are 100–14 = 86 observations for which there is no difference (i.e., zero values for the difference). The number of these observations is contained in cell C6. To complete the computation of the variance and the standard deviation, the average must be subtracted from each of these difference values of 0 and the resulting difference is then squared. This is done in cell F3 by using the formula = C6*–C5^2.

⑨ The sum of squared differences around the average difference in cell F4 is the sum of the differences contained in cells F2 and F3.

⑩ Now we can finalize the computation of the variance and the standard deviation. The degrees of freedom, equal to *n* – 1, are entered into cell F5 using the formula =C4–1, the variance, equal to the sum of squares divided by the degrees of freedom (=F4/F5), is entered in cell F6, and the standard deviation (the square root of the variance) is computed in cell F7.

The results are depicted in Figure 6.9.Excel.

	A	B	C	D	E	F
1	Differences			SS Calc.		
2	75.41	Sum	690.34	4693.15424	SS: X with difference	69304.18
3	38.97	Count	14	1028.26684	SS: X w/no difference	4098.496
4	108.54	n	100	10329.9985	Sum of Squares	73402.68
5	-37.18	DBar	6.9034	1943.34616	df	99
6	62.75	Count of X w/no diff.	86	3118.84273	Variance	741.4412
7	118.32			12413.6588	Std. Deviation	27.22942
8	-88.84			9166.79864		
9	127.74			14601.4839		
10	55.42			2353.86048		
11	39.03			1032.11843		
12	29.41			506.547044		
13	47.99			1688.1087		
14	28.73			476.400468		
15	84.05			5951.59789		

FIGURE 6.9.EXCEL Computations in differences estimation using Microsoft Excel for the voucher example.

Now that the computation of the differences has been completed, develop the confidence interval for the total difference. With the Auditing worksheet active, we follow these steps:

❶ Copy the sample size from cell C4 of the Differences worksheet to cell D3, the sum of the differences from cell C2 of the Differences worksheet to cell D4, the average difference from cell C5 of the Differences worksheet to cell D5, and the standard deviation of the differences from cell F7 of the Differences worksheet to cell D7.

❷ Enter the confidence level (entered as a decimal) in cell D6, and the population size N in cell D8. Using the voucher example on page 315, we would enter .95 in cell D6 and 5000 in cell D8.

❸ The total difference equal to $N\bar{D}$ is entered in cell D9 using the formula =D8*D5.

❹ The remaining steps are similar to those used in steps 3–8 in computing the confidence interval estimate of the total.

The results are depicted in Figure 6.10.Excel. Any differences from those shown on page 316 are due to rounding in those calculations.

C	D
Difference Estimation	
n	100
Sum of Differences	690.34
Average Difference	6.9034
Confidence Level	0.95
S	27.22942
N	5000
Total Difference	34517
f.p.c.	0.990049
Standard Error	13479.22
df	99
t	1.984217
Half-width	26745.71
Lower Limit	7771.294
Upper Limit	61262.71

FIGURE 6.10.EXCEL Confidence interval for the total difference using Microsoft Excel for the voucher example.

6.11 ESTIMATION, SAMPLE SIZE DETERMINATION, AND ETHICAL ISSUES

Ethical issues relating to the selection of samples and the inferences that accompany them from sample surveys can arise in several ways. The major ethical issue relates to whether or not confidence interval estimates are provided along with the point estimates of the sample statistics obtained from a survey. To just indicate a point estimate of a sample statistic without also including the confidence interval limits (typically set at 95%), the sample size used, and an interpretation of the meaning of the confidence interval in terms that a layman can understand can raise ethical issues because of their omission. Failure to include a confidence interval estimate might mislead the user of the survey results into thinking that the point estimate is all that is needed to predict the population characteristics with certainty. Thus, it is

Table 6.5 Five polls of the 1992 presidential election taken between October 1 and October 4, 1992.

	Candidate		
Poll	Bush	Clinton	Perot
New York Times/CBS News	38%	46%	7%
Washington Post/ABC News	35%	48%	9%
Gallup for CNN/*USA Today*	35%	47%	10%
Harris	36%	53%	9%
Gallup for *Newsweek*	36%	44%	14%

Source: The New York Times, October 7, 1992, p. A.1.

important that the interval estimate be indicated in a prominent place in any written communication along with a simple explanation of the meaning of the confidence interval. In addition, the size of the sample should be highlighted so that the reader clearly understands the magnitude of the survey that has been undertaken.

One of the most common areas where ethical issues concerning estimation from sample surveys occurs is in the publication of the results of political polls. All too often, the results of the polls are highlighted on page one of the newspaper, and the sampling error involved along with the methodology used is printed on the page where the article is typically continued (often in the middle of the newspaper). During the 1992 Presidential campaign, *The New York Times* (October 7, 1992) presented some of these issues by publishing the results of five different polls that were taken between October 1 and October 4, 1992. Table 6.5 summarizes the results of these five polls.

Although there are many possible reasons for the differences in the results, including those discussed in Section 1.11, it is also quite possible that most of the differences are merely due to sampling error. If we assume that the sample sizes were sufficiently large to provide a sampling error of ±3.5% with 95% confidence, confidence intervals for each of the five different polls could be obtained. Confidence intervals for the percentage of voters favoring candidate Bill Clinton are presented in Table 6.6.

We may observe from Table 6.6 that all the polls resulted in confidence intervals whose major difference appears to be sampling error. Thus, in summary, to ensure an ethical interpretation of statistical results, the confidence levels, sample size, and confidence limits should be made available for all surveys.

Table 6.6 Confidence interval estimates based on five polls of the 1992 presidential election taken between October 1 and October 4, 1992.

		Confidence Interval Estimate	
Poll	Voters in Favor of Clinton	Lower Limit	Upper Limit
New York Times/CBS News	46%	42.5	49.5
Washington Post/ABC News	48%	44.5	51.5
Gallup for CNN/*USA Today*	47%	43.5	50.5
Harris	53%	49.5	56.5
Gallup for *Newsweek*	44%	40.5	47.5

6.12 ESTIMATION AND STATISTICAL INFERENCE: REVIEW AND A PREVIEW

As observed in the summary chart, in this chapter we have developed two approaches for estimating the characteristic of a population, confidence interval estimation and bootstrapping. We have investigated how we can determine the necessary sample size for a survey and also examined the finite population correction factor.

To be sure you understand what we have covered in this chapter, you should be able to answer the following conceptual questions:

1. Why is it that we can never really have 100% confidence of correctly estimating the population characteristic of interest?
2. When is the t distribution used in developing the confidence interval estimate for the mean?
3. Under what circumstances might you use the bootstrapping approach rather than the traditional confidence interval estimation?
4. Why is it true that for a given sample size n, an increase in confidence is achieved by widening (and making less precise) the confidence interval obtained?
5. How does sampling without replacement from a finite population affect the confidence interval estimate and the sample size necessary?
6. When would you want to estimate the population total instead of the population mean?
7. How does difference estimation differ from estimating the mean?

Now that we have made estimates of population characteristics such as the mean, proportion, and total using confidence intervals, in the next three chapters we turn to a hypothesis-testing approach in which we make decisions about population parameters.

Getting It All Together

Key Terms

auditing 314
bootstrapping estimation 301
confidence interval estimate 294
critical value 294
degrees of freedom 299
difference estimation 315
finite population correction factor (fpc) 311
interval estimate 292
level of confidence 294
point estimate 292
resampling distribution 302
sampling error 306
Student's t distribution 297
total 314

Chapter Review Problems

Note: The Chapter Review Problems can be solved using Microsoft Excel.

- **6.54** A market researcher for a large consumer electronics company would like to study television viewing habits of residents of a particular small city. A random sample of 40 respondents is selected, and each respondent is instructed to keep a detailed record of all television viewing in a particular week. The results are as follows:
  - Viewing time per week: $\bar{X} = 15.3$ hours, $S = 3.8$ hours
  - 27 respondents watch the evening news on at least three weeknights.

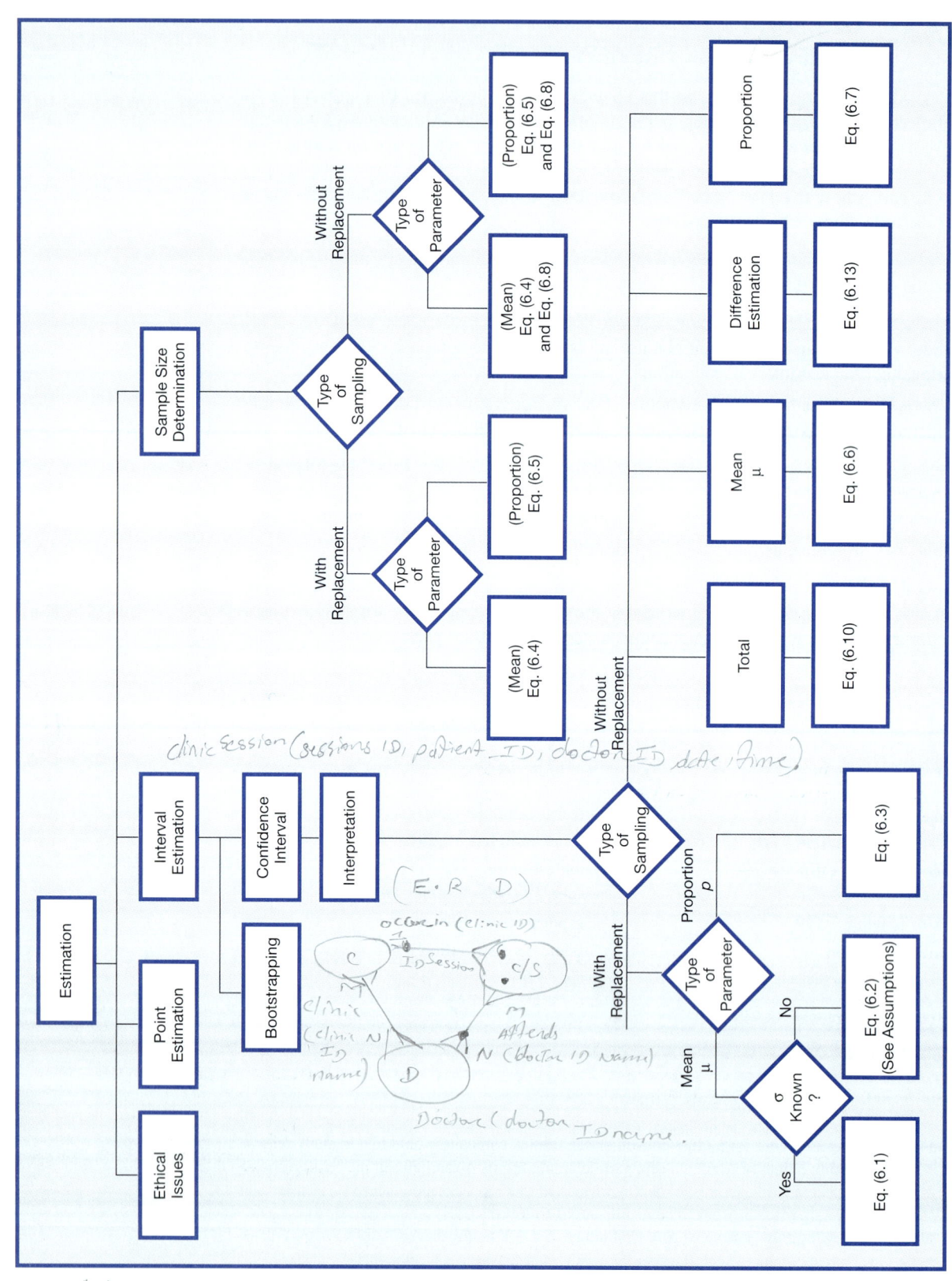

Chapter 6 summary chart.

334

(a) Set up a 95% confidence interval estimate for the average amount of television watched per week in this city.
(b) Set up a 95% confidence interval estimate for the proportion of respondents who watch the evening news on at least three nights per week.

If the market researcher wants to take another survey in a different city
(c) What sample size is required if he wishes to be 95% confident of being correct to within ±2 hours and assumes the population standard deviation is equal to 5 hours?
(d) What sample size is needed if he wishes to be 95% confident of being within ±.035 of the true proportion who watch the evening news on at least three weeknights if no previous estimate were available?

6.55 The real estate assessor for a county government wishes to study various characteristics concerning single-family houses in the county. A random sample of 70 houses reveals the following:
- Heated area of the house (in square feet): $\bar{X} = 1{,}759$, $S = 380$
- 42 houses have central air conditioning.

(a) Set up a 99% confidence interval estimate of the population average heated area of the house.
(b) Set up a 95% confidence interval estimate of the population proportion of houses that have central air conditioning.

6.56 The personnel director of a large corporation wishes to study absenteeism among clerical workers at the corporation's central office during the year. A random sample of 25 clerical workers reveals the following:
- Absenteeism: $\bar{X} = 9.7$ days, $S = 4.0$ days
- 12 clerical workers were absent more than 10 days.

(a) Set up a 95% confidence interval estimate of the average number of absences for clerical workers last year.
(b) Set up a 95% confidence interval estimate of the population proportion of clerical workers absent more than 10 days last year.

If the personnel director also wishes to take a survey in a branch office
(c) What sample size is needed if the director wishes to be 95% confident of being correct to ±1.5 days and the population standard deviation is assumed to be 4.5 days?
(d) What sample size is needed if the director wishes to be 90% confident of being correct to within ±.075 of the true proportion of workers who are absent more than 10 days if no previous estimate is available?

6.57 The market research director for Dotty's department store would like to study women's spending per year on cosmetics. A survey is to be sent to a sample of the store's credit card holders to determine:
- The average yearly amount that women spend on cosmetics.
- The population proportion of women who purchase their cosmetics primarily from Dotty's department store.

(a) If the market researcher wants to have 99% confidence of estimating the true population average to within ±$5 and the standard deviation is assumed to be $18 (based on previous surveys), what sample size is needed?
(b) If the market researcher wishes to have 90% confidence of estimating the true proportion to within ±.045, what sample size is needed?
(c) **ACTION** Based on the results in (a) and (b), how many of the store's credit card holders should be sampled? Explain.

6.58 The branch manager of an outlet of a large, nationwide bookstore chain wants to study characteristics of customers of her store, which is located near the campus of a large state university. In particular, she decides to focus on two variables: the amount of money spent by customers and whether the customers would consider purchasing educational videotapes relating to specific courses such as statistics, accounting, or calculus or graduate preparation exams such as GMAT, GRE, or LSAT. The results from a sample of 70 customers are as follows:

- Amount spent: $\bar{X} = \$28.52$, $S = \$11.39$
- 28 customers stated that they would consider purchasing educational videotapes.

(a) Set up a 95% confidence interval estimate of the population average amount spent in the bookstore.
(b) Set up a 90% confidence interval estimate of the population proportion of customers who would consider purchasing educational videotapes.

Suppose the branch manager of another store from a *different* bookstore chain wishes to conduct a similar survey in his store (which is located near another university).

(c) If he wants to have 95% confidence of estimating the true population average amount spent in his store to within ±$2 and the standard deviation is assumed to be $10, what sample size is needed?
(d) If he wants to have 90% confidence of estimating the true proportion of shoppers who would consider the purchase of videotapes to within ±.04, what sample size is needed?
(e) Based on your answers to (c) and (d), what size sample should be taken?

6.59 The branch manager of an outlet (store #1) of a large, nationwide chain of pet supply stores wants to study characteristics of customers of her store. In particular, she decides to focus on two variables: the amount of money spent by customers, and whether the customers own only one dog, only one cat, or more than one dog and/or one cat. The results from a sample of 70 customers are as follows.

- Amount of money spent: $\bar{X} = \$21.34$, $S = \$9.22$
- 37 customers own only a dog.
- 26 customers own only a cat.
- 7 customers own at least one dog and at least one cat.

(a) Set up a 95% confidence interval estimate of the population average amount spent in the pet supply store.
(b) Set up a 90% confidence interval estimate of the proportion of customers who own *only* a cat.

Suppose the branch manager of another outlet (store #2) wishes to conduct a similar survey in his store (and does not have any access to the information generated by the manager of store #1).

(c) If he wants to have 95% confidence of estimating the true population average amount spent in his store to within ±$1.50 and the standard deviation is assumed to be $10, what sample size is needed?
(d) If he wants to have 90% confidence of estimating the true proportion of customers who own only a cat to within ±.045, what sample size is needed?
(e) Based on your answers to (c) and (d), what size sample should be taken?

6.60 The owner of a restaurant serving continental food wants to study characteristics of customers of his restaurant. In particular, he decides to focus on two variables, the amount of money spent by customers and whether or not customers order dessert. The results from a sample of 60 customers are as follows:

- Amount spent: $\bar{X} = \$38.54$, $S = \$7.26$
- 18 customers purchased dessert.

(a) Set up a 95% confidence interval estimate of the population average amount spent per customer in the restaurant.
(b) Set up a 90% confidence interval estimate of the population proportion of customers who purchase dessert.

Suppose the owner of a competing restaurant wishes to conduct a similar survey in her restaurant (and does not have access to the information obtained by the owner of the first restaurant).

(c) If she wants to have 95% confidence of estimating the true population average amount spent in her restaurant to within ±$1.50 and the standard deviation is assumed to be $8, what sample size is needed?
(d) If she wants to have 90% confidence of estimating the true proportion of customers who purchase dessert to within ±.04, what sample size is needed?
(e) Based on your answers to (c) and (d), what size sample should be taken?

Chapter 6 Estimation

6.61 A representative for a large chain of hardware stores is interested in testing the product claims of a manufacturer of "Ice Melt" that is reported to melt snow and ice at temperatures as low as 0 degrees Fahrenheit. A shipment of 400 five-pound bags is purchased by the chain for distribution. The representative wants to know with 95% confidence, within ±.05, what proportion of bags of Ice Melt performs the job as claimed by the manufacturer.
   (a) How many bags does the representative need to test? What assumption should be made concerning the true proportion in the population? (This is called *destructive testing;* that is, the product being tested is destroyed by the test and is then unavailable to be sold.)

   The representative tests 50 bags, and 42 do the job as claimed.
   (b) Construct a 95% confidence interval estimate for the population proportion that will do the job as claimed.
   (c) How can the representative use the results of (b) to determine whether to sell the Ice Melt product?

6.62 An audit test to establish the frequency of occurrence of failure to follow a specific internal control procedure is to be undertaken. Suppose a sample of 50 from a population of 1,000 items is selected and it is determined that in 7 instances the internal control procedure was not followed.
   (a) Set up a 90% confidence interval estimate of the population proportion of items in which the internal control procedure was not followed.
   (b) Suppose that, in an effort to objectively determine sample size, the auditor decides that she would like to have 95% confidence of correctly estimating the population proportion to within ±.04 (4%) of the true population value. She also assumes from past experience that the maximum error rate is 10%. What sample size needs to be selected?

6.63 In a test of an attribute from a population of 15,000 items, if the maximum rate of occurrence is not expected to exceed 5% and a sample precision of ±2% is deemed satisfactory:
   (a) What sample size is required for 95% confidence?
   (b) What sample size is required for 99% confidence?
   (c) Compare the difference in your answers to (a) and (b).
   (d) If you desire a precision of 3%, what sample size is required for 95% confidence?
   (e) What sample size is required in (a) if the population size is 5,000?

6.64 An auditor for a government agency is assigned the task of evaluating reimbursement for office visits to doctors paid by Medicare. The audit is to be conducted for all Medicare payments in a particular geographical area during a certain month. A population of 25,056 visits exists in this area for this month. The auditor primarily wants to estimate the total amount paid by Medicare in this area for the month. He desires to have 95% confidence of estimating the true population average to within ±$5. Based on past experience, he estimates the standard deviation to be $30.
   (a) What sample size should be selected?

   Using the sample size selected in (a), an audit is conducted with the following results:
   - In 12 of the office visits, an incorrect amount of reimbursement was provided.
   - Amount of reimbursement: $\bar{X} = \$93.70$, $S = \$34.55$

   For the 12 office visits in which incorrect reimbursement was provided, the difference between the amount reimbursed and what should have been reimbursed (in dollars) is as follows:

   17, 25, 14, −10, 20, 40, 35, 30, 28, 22, 15, 5

   (b) Set up a 90% confidence interval estimate of the population proportion of reimbursements that contain errors.
   (c) Set up 95% confidence interval estimates of the
      (1) average reimbursement per office visit.
      (2) the total amount of reimbursements for this geographical area in this month.
   (d) Set up a 95% confidence interval estimate of the total difference between the amount reimbursed and the amount that the auditor determined should have been reimbursed.

6.65 A large computer store is conducting an end-of-month inventory of the computers in stock. It has been determined that there are 258 computers in inventory at that time. An auditor for the store would like to estimate the total value of the computers in inventory at that time. He would like to have 99% confidence that his estimate of the average value is correct to within ± $200. Based on past experience, he estimates that the standard deviation of the value of a computer is $400.
(a) What sample size should be selected?

Using the sample size selected in (a), an audit is conducted with the following results:

$$\text{Value of computers: } \bar{X} = \$3,054.13, \ S = \$384.62$$

(b) Set up a 99% confidence interval estimate of the total estimated value of the computers in inventory at the end of the month.

Team Projects

CEREAL.TXT

TP6.1 Refer to TP2.2 on page 117. Set up all appropriate estimates of population nutritional characteristics of ready-to-eat cereals. Include these estimates in any written and oral presentation to be made to the food editor of the magazine.

FRAGRANC.TXT

TP6.2 Refer to TP2.3 on page 117. Set up all appropriate estimates of population characteristics of fragrances. Include these estimates in any written and oral presentation to be made to the marketing director.

Endnotes

1. It is for this reason that the denominator of the sample variance is $n-1$ instead of n, so that S^2 will be an unbiased estimator of σ^2; that is, if

$$S^2 = \frac{\sum_{i=1}^{n}(X_i - \bar{X})^2}{n-1} \quad \text{and} \quad \sigma^2 = \frac{\sum_{i=1}^{n}(X_i - \mu)^2}{N}$$

then $E(S^2) = \sigma^2$ and therefore, S^2 is an unbiased estimator of σ^2.

2. We use Z instead of t because (a) to determine the critical value of t, we would need to know the sample size, which we don't know yet, and (b) for most studies, the sample size needed will be large enough that the normal distribution is a good approximation of the t distribution.

References

1. Cochran, W. G., *Sampling Techniques,* 3d ed. (New York: Wiley, 1977).
2. Diaconis, P., and B. Efron. "Computer-Intensive Methods in Statistics," *Scientific American,* 248, 1983, pp. 116–130.
3. Efron, B., *The Jackknife, the Bootstrap, and Other Resampling Plans* (Philadelphia: Society for Industrial and Applied Mathematics, 1982).
4. Fisher, R. A., and F. Yates, *Statistical Tables for Biological, Agricultural and Medical Research,* 5th ed. (Edinburgh: Oliver & Boyd, 1957).
5. Gunter, B., "Bootstrapping: How to Make Something from Almost Nothing and Get Statistically Valid Answers. Part I: Brave New World," *Quality Progress,* 24, December 1991, pp. 97–103.
6. Kirk, R. E., ed., *Statistical Issues: A Reader for the Behavioral Sciences* (Belmont, CA: Wadsworth, 1972).
7. Larsen, R. L., and M. L. Marx, *An Introduction to Mathematical Statistics and Its Applications,* 2d ed. (Englewood Cliffs, NJ: Prentice-Hall, 1986).
8. *Microsoft Excel Version 7* (Redmond, WA: Microsoft Corporation, 1996).
9. Snedecor, G. W., and W. G. Cochran, *Statistical Methods,* 7th ed. (Ames, IA: Iowa State University Press, 1980).

chapter 7
Fundamentals of Hypothesis Testing

CHAPTER OBJECTIVE To develop hypothesis-testing methodology as a technique for analyzing differences and making decisions.

7.1 INTRODUCTION

In Chapter 5, we began our discussion of statistical inference by developing the concept of a sampling distribution. In Chapter 6, we considered *enumerative studies* in which a statistic (such as the sample mean or sample proportion) obtained from a random sample is used to *estimate* its corresponding population parameter.

In this chapter we will begin to focus on another phase of statistical inference that is also based on sample information—hypothesis testing. We will develop a step-by-step methodology that will enable us to make inferences about a population parameter by *analyzing differences* between the results we observe (our sample statistic) and the results we would expect to obtain if some underlying hypothesis were actually true. Emphasis here is placed on the fundamental and conceptual underpinnings of **hypothesis-testing methodology**. In the two chapters that follow, numerous hypothesis-testing procedures will be presented that are frequently employed in the analysis of data obtained from studies and experiments designed under a variety of conditions.

7.2 HYPOTHESIS-TESTING METHODOLOGY

To develop hypothesis-testing methodology, we will focus on some issues based on the cereal box filling process discussed in Chapters 5 and 6. For example, the production manager is concerned with evaluating whether or not the process is working in a way that ensures that, on average, the proper amount of cereal (i.e., 368 grams) is being filled in each box. He decides to select a random sample of 25 boxes from the filling process and examine their weights to determine how close each of these boxes comes to the company's specification of an average of 368 grams per box. The production manager hopes to find that the process is working properly. However, he might find that the sampled boxes weigh too little or perhaps too much; he may then decide to halt the production process until the reason for the failure to adhere to the specified weight of 368 grams is determined. By analyzing the differences between the weights obtained from the sample and the 368-gram expectation obtained from the company's specifications, a decision based on this sample information can be made, and one of the following two conclusions will be reached:

1. The average fill in the entire process is 368 grams. No corrective action is needed.
2. The average fill is not 368 grams; either it is less than 368 grams, or it is more than 368 grams. Corrective action is needed.

7.2.1 The Null and Alternative Hypotheses

Hypothesis testing begins with some theory, claim, or assertion about a particular parameter of a population. For purposes of statistical analysis, the production manager chooses as his initial hypothesis that the process is in control; that is, the average fill is 368 grams and no corrective action is needed. The hypothesis that the population parameter is equal to the company specification is referred to as the **null hypothesis**.

A null hypothesis is always one of status quo or no difference. We commonly identify the null hypothesis by the symbol H_0. Our production manager would establish as his null hypothesis that the filling process is in control and working properly, that the mean fill per box is the 368-gram specification. This can be stated as

$$H_0: \mu = 368$$

Note that even though the production manager only has information from the sample, the null hypothesis is written in terms of the population parameter. This is because he is interested in the entire filling process, that is, (the population of) all cereal boxes being filled. The sample statistics will be used to make inferences about the entire filling process. Parallel to the American legal system, in which innocence is presumed until guilt is proven, the theoretical basis of hypothesis testing requires that the null hypothesis be considered true until evidence, such as the results observed from the sample data, indicates that it is false. If the null hypothesis is considered false, something else must be true.

Whenever we specify a null hypothesis, we must also specify an **alternative hypothesis**, or one that must be true if the null hypothesis is found to be false. The alternative hypothesis (H_1) is the opposite of the null hypothesis (H_0). For the production manager, this can be stated as

$$H_1: \mu \neq 368$$

The alternative hypothesis represents the conclusion that would be reached if there were sufficient evidence from sample information to decide that the null hypothesis is unlikely to be true and we can therefore reject it. In our example, if the weights of the sampled boxes were sufficiently above or below the expected 368-gram average specified by the company, the production manager would reject the null hypothesis in favor of the alternative hypothesis that the average amount of fill is different from 368 grams. He would therefore stop production and take whatever action is necessary to correct the problem.

Hypothesis-testing methodology is designed so that our rejection of the null hypothesis is based on evidence from the sample that our alternative hypothesis is far more likely to be true. However, failure to reject the null hypothesis is not proof that it is true. We can never prove the null hypothesis is correct because we are basing our decision only on the sample information, not on the entire population. Therefore, if we fail to reject the null hypothesis, we can only conclude that there is insufficient evidence to warrant its rejection.

To summarize some key points:

- The null hypothesis (H_0) is the hypothesis that is always tested.
- The alternative hypothesis (H_1) is set up as the opposite of the null hypothesis and represents the conclusion supported if the null hypothesis is rejected.

In what is known as *classical* hypothesis-testing methodology (see Reference 1),

- The null hypothesis always refers to a specified value of the *population parameter* (such as μ), not a *sample statistic* (such as $\bar{X}$).
- The statement of the null hypothesis *always* contains an equal sign regarding the specified value of the parameter (that is, $H_0: \mu = 368$ grams).
- The statement of the alternative hypothesis *never* contains an equal sign regarding the specified value of the parameter (that is, $H_1: \mu \neq 368$ grams).

7.2.2 The Critical Value of the Test Statistic

We can develop the logic behind the hypothesis-testing methodology by contemplating how we can determine, based only on sample information, the plausibility of the null hypothesis. Our production manager has stated as his null hypothesis that the average amount of cereal per box over the entire filling process is 368 grams (i.e., the population parameter specified by the company). He then collected a sample of boxes from the filling process, weighed each box, and computed the sample mean. We recall that a statistic from a sample is an estimate of the corresponding parameter from the population from which the sample was drawn and will likely differ from the actual parameter value because of chance or sampling error. Therefore, even if the null hypothesis were in fact true, the sample statistic would not necessarily be equal to the corresponding population parameter. Nevertheless, under such circumstances, we would expect them to be very similar to each other in value. In such a situation, there would be no evidence to reject the null hypothesis. If, for example, the sample mean were 367.6, we would be inclined to conclude that the population mean has not changed (that is, $\mu = 368$), since the sample mean is very close to the hypothesized value of 368. Intuitively, we would be thinking that it is not unlikely that we could obtain a sample mean of 367.6 from a population whose mean is 368. On the other hand, if there were a large discrepancy between the value of the statistic and its corresponding hypothesized parameter, our instinct would be to conclude that the null hypothesis is implausible or unlikely to be true. For example, if the sample average were 320, our instinct would be to conclude that the average is not 368 (that is, $\mu \neq 368$), since the sample mean is very far from the hypothesized value of 368. In such a case, we would be reasoning that it is very unlikely that the sample mean of 320 could be obtained if the population mean were really 368 and, therefore, it would be more reasonable to conclude that the population mean is not equal to 368. Here we would reject the null hypothesis. In either case, our decision would be reached because of our belief that randomly selected samples are truly representative of the underlying populations from which they are drawn.

Unfortunately, the decision-making process is not always so clear-cut and cannot be left to an individual's subjective judgment as to the meaning of "very close" or "very different." It would be arbitrary for us to determine what is very close and what is very different without using operational definitions. Hypothesis-testing methodology provides operational definitions for evaluating such differences and enables us to quantify the decision-making process so that the probability of obtaining a given sample result if the null hypothesis were true can be found. This is achieved by first determining the sampling distribution for the sample statistic (i.e., the sample mean) and then computing the particular *test statistic* based on the given sample result. Since the sampling distribution for the test statistic often follows a well-known statistical distribution, such as the normal or *t* distribution, we can use these distributions to determine the likelihood of a null hypothesis being true.

7.2.3 Regions of Rejection and Nonrejection

The sampling distribution of the test statistic is divided into two regions, a **region of rejection** (sometimes called the **critical region**) and a **region of nonrejection** (see Figure 7.1). If the test statistic falls into the region of nonrejection, the null hypothesis cannot be rejected. In our example, the production manager would conclude that the average fill amount has not changed. If the test statistic falls into the rejection region, the null hypothesis will be rejected. Here the production manager would conclude that the population mean is not 368.

The region of rejection may be thought of as consisting of the values of the test statistic that are unlikely to occur if the null hypothesis is true. On the other hand, these values are not

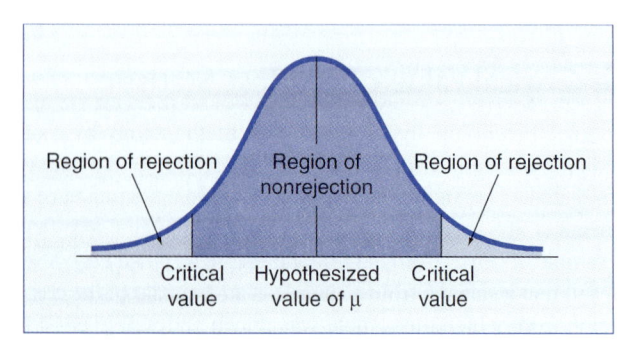

FIGURE 7.1 Regions of rejection and nonrejection in hypothesis testing.

so unlikely to occur if the null hypothesis is false. Therefore, if we observe a value of the test statistic that falls into this *critical region*, we reject the null hypothesis because that value would be unlikely if the null hypothesis were true.

To make a decision concerning the null hypothesis, we must first determine the **critical value** of the test statistic. The critical value divides the nonrejection region from the rejection region. However, the determination of this critical value depends on the size of the rejection region. As we will see in the next section, the size of the rejection region is directly related to the risks involved in using only sample evidence to make decisions about a population parameter.

7.2.4 Risks in Decision Making Using Hypothesis-Testing Methodology

When using a sample statistic to make decisions about a population parameter, there is a risk that an incorrect conclusion will be reached. Indeed, two different types of errors can occur when applying hypothesis-testing methodology:

A **Type I error** occurs if the null hypothesis H_0 is rejected when in fact it is true and should not be rejected.

A **Type II error** occurs if the null hypothesis H_0 is not rejected when in fact it is false and should be rejected.

In our cereal filling-process example, the Type I error would occur if the production manager concluded (based on sample information) that the average population amount filled was *not* 368 when in fact it was 368. On the other hand, the Type II error would occur if he concluded (based on sample information) that the average population fill amount was 368 when in fact it was not 368.

- **The Level of Significance** The probability of committing a Type I error, denoted by α (the lowercase Greek letter alpha), is referred to as the **level of significance** of the statistical test. Traditionally, the statistician controls the Type I error rate by deciding the risk level α he or she is willing to tolerate in terms of rejecting the null hypothesis when it is in fact true. Since the level of significance is specified before the hypothesis test is performed, the risk of committing a Type I error, α, is directly under the control of the individual performing the test. Researchers have traditionally selected α levels of .05 or smaller. The choice of selecting a particular risk level for making a Type I error is dependent on the cost of making a Type I error. Once the value for α is specified, the size of the rejection region is known, since α is the probability of rejection under the null hypothesis. From this fact, the critical value or values that divide the rejection and nonrejection regions can be determined.

- **The Confidence Coefficient** The complement $(1 - \alpha)$ of the probability of a Type I error is called the confidence coefficient, which, when multiplied by 100%, yields the confidence level that we studied in Section 6.2.

    The **confidence coefficient**, denoted by $1 - \alpha$, is the probability that the null hypothesis H_0 is not rejected when in fact it is true and should not be rejected.

In terms of hypothesis-testing methodology, this coefficient represents the probability of concluding that the specified value of the parameter being tested under the null hypothesis may be plausible. In our cereal filling-process example, the confidence coefficient measures the probability of concluding that the average fill per box is 368 grams when in fact it is 368 grams.

- **The β Risk** The **probability of committing a Type II error**, denoted by β (the lowercase Greek letter beta), is often referred to as the consumer's risk level. Unlike the Type I error, which statistical tests permit us to control by our selection of α, the probability of making a Type II error is dependent on the difference between the hypothesized and actual values of the population parameter. Since large differences are easier to find, if the difference between the sample statistic and the corresponding population parameter is large, β, the probability of committing a Type II error, will likely be small. For example, if the true population average (which is unknown to us) were 320 grams, there would be a small chance (β) of concluding that the average had not changed from 368. On the other hand, if the difference between the statistic and the corresponding parameter value is small, the probability of committing a Type II error will likely be large. Thus, if the true population average were really 367 grams, there would be a high probability of concluding that the population average fill amount had not changed from the specified 368 grams (and we would be making a Type II error).

- **The Power of a Test** The complement $(1 - \beta)$ of the probability of a Type II error is called the power of a statistical test.

> The **power of a** statistical **test**, denoted by $1 - \beta$, is the probability of rejecting the null hypothesis when in fact it is false and should be rejected.

In our cereal filling-process example, the power of the test is the probability of concluding that the average fill amount is not 368 grams when in fact it is actually not 368 grams.

- **Risks in Decision Making: A Delicate Balance** Table 7.1 illustrates the results of the two possible decisions (do not reject H_0 or reject H_0) that can occur in any hypothesis test. Depending on the specific decision, one of two types of errors may occur[1] or one of two types of correct conclusions may be reached.

One way in which we can control the probability of making a Type II error in a study is to increase the size of the sample. Larger sample sizes will generally permit us to detect even very small differences between the sample statistics and the population parameters. For a given level of α, increasing the sample size will decrease β and therefore increase the power of the test to detect that the null hypothesis H_0 is false. Unfortunately, however, there is always a limit to our resources. Thus, for a given sample size we must consider the trade-offs between the two possible types of errors. Since we can directly control our risk of Type I error, we can reduce our risk by selecting a lower level for α (e.g., .01 instead of .05). However, when α is decreased, β will be increased, so a reduction in risk of Type I error will result in an increased risk of Type II error. If, on the other hand, we wish to reduce β, our risk of Type II error, we could select a larger value for α (e.g., .05 instead of .01).

In our cereal filling-process example, the risk of a Type I error involves concluding that the average fill per box has changed from the hypothesized 368 grams when in fact it has not changed. The risk of a Type II error involves concluding that the average fill per box has not changed from the hypothesized 368 grams when in truth it has changed. The choice of reasonable values for α and β depends on the costs inherent in each type of error. For example,

Table 7.1 Hypothesis testing and decision making.

	Actual Situation	
Statistical Decision	H_0 True	H_0 False
Do not reject H_0	Confidence $(1 - \alpha)$	Type II error (β)
Reject H_0	Type I error (α)	Power $(1 - \beta)$

if it were very costly to change the status quo, then we would want to be very sure that a change would be beneficial, so the risk of a Type I error might be most important and would be kept very low. On the other hand, if we wanted to be very certain of detecting changes from a hypothesized mean, the risk of a Type II error would be most important, and we might choose a higher level of α.

Problems for Section 7.2

7.1 Why is it possible for the null hypothesis to be rejected when in fact it is true?

7.2 For a given sample size, if α is reduced from .05 to .01, what will happen to β?

7.3 Why is it possible that the null hypothesis will not always be rejected when it is false?

7.4 What is the relationship of α to the Type I error?

7.5 What is the relationship of β to the Type II error?

7.6 For $H_0: \mu = 100$, $H_1: \mu \neq 100$, and for a sample of size n, β will be larger if the actual value of μ is 90 than if the actual value of μ is 75. Why?

7.7 In the American legal system, a defendant is presumed innocent until proven guilty. Consider a null hypothesis H_0 that the defendant is innocent and an alternative hypothesis H_1 that the defendant is guilty. A jury has two possible decisions: convict the defendant (i.e., reject the null hypothesis) or do not convict the defendant (i.e., do not reject the null hypothesis). Explain the meaning of the risks of committing either a Type I or Type II error in this example.

7.8 Suppose the defendant in Problem 7.7 was presumed guilty until proven innocent. How would the null and alternative hypotheses differ from those in Problem 7.7? What would be the meaning of the risks of committing either a Type I or Type II error here?

7.9 How is power related to the probability of making a Type II error?

7.3　Z TEST OF HYPOTHESIS FOR THE MEAN (σ KNOWN)

Now that we have described the hypothesis-testing methodology, let us return to the question of interest to the production manager at the packaging plant. You may recall that he wants to determine whether or not the cereal filling process is in control—that the average fill per box throughout the entire packaging process remains at the specified 368 grams and no corrective action is needed. To study this, he plans to take a random sample of 25 boxes, weigh each one, and then evaluate the difference between the sample statistic and hypothesized population parameter by comparing the mean weight (in grams) from the sample to the expected mean of 368 grams specified by the company. For this filling process, the null and alternative hypotheses are

$$H_0: \mu = 368$$
$$H_1: \mu \neq 368$$

If we assume that the standard deviation σ is known, then based on the central limit theorem, the sampling distribution of the mean would follow the normal distribution and the **Z-test statistic** would be

$$Z = \frac{\bar{X} - \mu}{\frac{\sigma}{\sqrt{n}}} \tag{7.1}$$

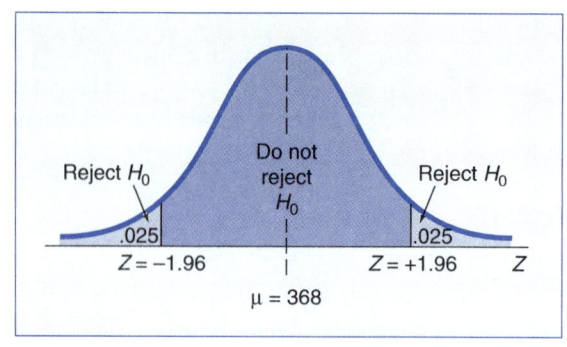

FIGURE 7.2 Testing a hypothesis about the mean (σ known) at the .05 level of significance.

In this formula, the numerator measures how far (in an absolute sense) the observed sample mean $\bar{X}$ is from the hypothesized mean μ. The denominator is the standard error of the mean, so Z represents how many standard errors $\bar{X}$ is from μ.

If the production manager decided to choose a level of significance of .05, the size of the rejection region would be .05, and the critical values of the normal distribution could be determined. These critical values can be expressed in standard-deviation units. Since the rejection region is divided into the two tails of the distribution (this is called a **two-tailed test**), the .05 is divided into two equal parts of .025 each. A rejection region of .025 in each tail of the normal distribution results in an area of .475 between the hypothesized mean and each critical value. Looking up this area in the normal distribution (Table E.2), we find that the critical values that divide the rejection and nonrejection regions are (in standard-deviation units) +1.96 and −1.96. Figure 7.2 illustrates this case; it shows that if the mean is actually 368 grams, as H_0 claims, then the values of the test statistic Z will have a standard normal distribution centered at $\mu = 368$. Observed values of Z greater than 1.96 or less than −1.96 indicate that $\bar{X}$ is so far from the hypothesized $\mu = 368$ that it is unlikely that such a value would occur if H_0 were true.

Therefore, the decision rule would be

Reject H_0 if $Z > +1.96$

or if $Z < -1.96$;

otherwise do not reject H_0.

Suppose that the sample of 25 cereal boxes indicated a sample mean ($\bar{X}$) of 372.5 grams and the population standard deviation (σ) is assumed to remain at 15 grams as specified by the company (see Section 5.9.3). Using Equation (7.1), we have

$$Z = \frac{\bar{X} - \mu}{\frac{\sigma}{\sqrt{n}}}$$

$$= \frac{372.5 - 368}{\frac{15}{\sqrt{25}}} = +1.50$$

Since $Z = +1.50$, we see that $-1.96 < +1.50 < +1.96$. Thus, our decision is not to reject H_0. We would conclude that the average fill amount is 368 grams. Alternatively, to take into account the possibility of a Type II error, we may phrase the conclusion as "there is no evidence that the average fill is different from 368 grams."

Problems for Section 7.3

Note: The problems in this section can be solved using Microsoft Excel (see Section 7.11).

7.10 Suppose the director of manufacturing at a clothing factory needs to determine whether a new machine is producing a particular type of cloth according to the manufacturer's specifications, which indicates that the cloth should have a mean breaking strength of 70 *← Null Hypothesis* pounds and a standard deviation of 3.5 pounds. A sample of 36 pieces reveals a sample mean of 69.7 pounds.
  (a) Is there evidence that the machine is not meeting the manufacturer's specifications in terms of the average breaking strength? (Use a .05 level of significance.)
  (b) What will your answer in (a) be if the standard deviation is 2.0 pounds?
  (c) What will your answer in (a) be if the sample mean is 69 pounds?

7.11 The purchase of a coin-operated laundry is being considered by a potential entrepreneur. The present owner claims that over the past 5 years, the average daily revenue was $675 with a standard deviation of $75. A sample of 30 selected days reveals a daily average revenue of $625.
  (a) Is there evidence that the claim of the present owner is not valid? (Use a .01 level of significance.)
  (b) What will your answer in (a) be if the standard deviation is $100?
  (c) What will your answer in (a) be if the sample mean is $650?

- 7.12 A manufacturer of salad dressings uses machines to dispense liquid ingredients into bottles that move along a filling line. The machine that dispenses dressings is working properly when 8 ounces are dispensed. The standard deviation of the process is 0.15 ounce. A sample of 50 bottles is selected periodically, and the filling line is stopped if there is evidence that the average amount dispensed is different from 8 ounces. Suppose that the average amount dispensed in a particular sample of 50 bottles is 7.983 ounces.
  (a) Is there evidence that the population average amount is different from 8 ounces? (Use a .05 level of significance.)
  (b) What will your answer in (a) be if the standard deviation is 0.05 ounce?
  (c) What will your answer in (a) be if the sample mean is 7.952 ounces?

7.13 Suppose that scores on an aptitude test used for determining admission to graduate study in business are known to be normally distributed with a population mean of 500 and a population standard deviation of 100.
  (a) If a random sample of 12 applicants from Stephan College has a sample mean of 537, is there any evidence that this mean score is different from the mean expected of all applicants? (Use a .01 level of significance.)
  (b) What will your answer in (a) be if the standard deviation is 50?
  (c) What will your answer in (a) be if the sample mean is 578?

7.4 SUMMARIZING THE STEPS OF HYPOTHESIS TESTING

Now that we have used hypothesis-testing methodology to draw a conclusion about the population mean in situations where the population standard deviation is known, it will be useful to summarize the steps involved.

1. State the null hypothesis, H_0.
2. State the alternative hypothesis, H_1.
3. Choose the level of significance, α.
4. Choose the sample size, n.
5. Determine the appropriate statistical technique and corresponding test statistic to use.
6. Set up the critical values that divide the rejection and nonrejection regions.
7. Collect the data and compute the sample value of the appropriate test statistic.

8. Determine whether the test statistic has fallen into the rejection or the nonrejection region.
9. Make the statistical decision.
10. Express the statistical decision in terms of the problem.

- **Steps 1 and 2** The null and alternative hypotheses must be stated in statistical terms. In testing whether the average amount filled was 368 grams, the null hypothesis was that μ equals 368, and the alternative hypothesis was that μ is not equal to 368 grams.

- **Step 3** The level of significance is specified according to the relative importance of the risks of committing Type I and Type II errors in the problem. We chose $\alpha = .05$. (This, along with the sample size, determines β.)

- **Step 4** The sample size is determined after taking into account the specified risks of committing Type I and Type II errors (i.e., selected levels of α and β) and considering budget constraints in carrying out the study. Here 25 cereal boxes were randomly selected.

- **Step 5** The statistical technique that will be used to test the null hypothesis must be chosen. Since σ was known (i.e., specified by the company to be 15 grams), a Z test was selected.

- **Step 6** Once the null and alternative hypotheses are specified and the level of significance and the sample size are determined, the critical values for the appropriate statistical distribution can be found so that the rejection and nonrejection regions can be indicated. Here the values +1.96 and −1.96 were used to define these regions since the Z-test statistic refers to the standard normal distribution.

- **Step 7** The data are collected and the value of the test statistic is computed. Here, $\bar{X} = 372.5$ grams, so $Z = +1.50$.

- **Step 8** The computed value of the test statistic is compared with the critical values for the appropriate sampling distribution to determine whether it falls into the rejection or nonrejection region. Here, $Z = +1.50$ was in the region of nonrejection, since $-1.96 < Z = +1.50 < +1.96$.

- **Step 9** The hypothesis-testing decision is made. If the test statistic falls into the nonrejection region, the null hypothesis H_0 cannot be rejected. If the test statistic falls into the rejection region, the null hypothesis is rejected. Here, H_0 was not rejected.

- **Step 10** The consequences of the hypothesis-testing decision must be expressed in terms of the actual problem involved. In our cereal filling process example, we concluded that there was no evidence that the average amount of cereal fill was different from 368 grams.

7.5 THE p-VALUE APPROACH TO HYPOTHESIS TESTING: TWO-TAILED TESTS

In recent years, with the advent of widely available statistical and spreadsheet software, an approach to hypothesis testing that has increasingly gained acceptance involves the concept of the p-value.

The ***p*-value** is the probability of obtaining a test statistic equal to or more extreme than the result obtained from the sample data, given that the null hypothesis H_0 is really true.

The *p*-value is often referred to as the *observed level of significance,* the smallest level at which H_0 can be rejected for a given set of data.

- If the *p*-value is greater than or equal to α, the null hypothesis is not rejected.
- If the *p*-value is smaller than α, the null hypothesis is rejected.

To understand the *p*-value approach, let us refer to the cereal filling-process example of Section 7.3. In that section we tested whether or not the average fill amount was equal to 368 grams (page 345). We obtained a Z value of +1.50 and did not reject the null hypothesis since +1.50 was greater than the lower critical value of −1.96 but less than the upper critical value of +1.96.

We may now use the *p*-value approach to find the probability of obtaining a test statistic Z that is *more extreme* than +1.50. When using a *two-tailed test,* this means that we need to compute the probability of obtaining a Z value greater than +1.50 along with the probability of obtaining a Z value less than −1.50. From Table E.2, the probability of obtaining a Z value above +1.50 is .5000 − .4332 = .0668. Since the standard normal distribution is symmetric, the probability of obtaining a value below −1.50 is also .0668. Thus, the *p*-value for this two-tailed test is .0668 + .0668 = .1336 (see Figure 7.3). This result may be interpreted to mean that the probability of obtaining a result equal to or more extreme than the one observed is .1336. Since this is greater than α = .05, the null hypothesis is not rejected.

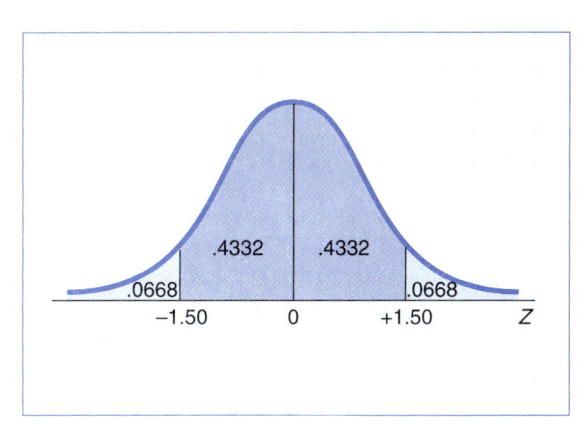

FIGURE 7.3 Finding the *p*-value for a two-tailed test.

Unless we are dealing with a test statistic that follows the normal distribution, the computation of the *p*-value can be very difficult. Thus, it is fortunate that software such as Excel (see Reference 3) routinely present the *p*-value as part of the output for many hypothesis-testing procedures.

Now that we have discussed the *p*-value approach for hypothesis testing, it will be useful to summarize the steps involved.

1. State the null hypothesis, H_0.
2. State the alternative hypothesis, H_1.
3. Choose the level of significance, α.
4. Choose the sample size, n.
5. Determine the appropriate statistical technique and corresponding test statistic to use.

6. Collect the data and compute the sample value of the appropriate test statistic.
7. Calculate the *p*-value based on the test statistic. This involves
   (a) Sketching the distribution under the null hypothesis H_0.
   (b) Placing the test statistic on the horizontal axis.
   (c) Shading in the appropriate area under the curve, based on the alternative hypothesis H_1.
8. Compare the *p*-value to α.
9. Make the statistical decision.
10. Express the statistical decision in terms of the problem.

Problems for Section 7.5

Note: *The problems in this section can be solved using Microsoft Excel (see Section 7.11).*

7.14 Compute the *p*-value in Problem 7.10(a) on page 347 and interpret its meaning.
7.15 Compute the *p*-value in Problem 7.11(a) on page 347 and interpret its meaning.
• 7.16 Compute the *p*-value in Problem 7.12(a) on page 347 and interpret its meaning.
7.17 Compute the *p*-value in Problem 7.13(a) on page 347 and interpret its meaning.

7.6 A CONNECTION BETWEEN CONFIDENCE INTERVAL ESTIMATION AND HYPOTHESIS TESTING

Both in this chapter and in Chapter 6, we have examined the two major components of statistical inference—confidence interval estimation and hypothesis testing. Although they are based on the same set of concepts, we have used them for different purposes. In Chapter 6, we used confidence intervals to estimate parameters, and in this chapter we have seen that we can use hypothesis testing for making decisions about specified values of population parameters.

In many situations we can use confidence intervals to do a test of a null hypothesis. This can be illustrated for the test of a hypothesis for a mean. Referring to the cereal filling process, we first attempted to determine whether the population average fill amount was different from 368 grams. We tested this in Section 7.3 using Equation (7.1)

$$Z = \frac{\bar{X} - \mu}{\frac{\sigma}{\sqrt{n}}}$$

Instead of testing the null hypothesis that $\mu = 368$ grams, we could also solve the problem by obtaining a confidence interval estimate of μ. If the hypothesized value of $\mu = 368$ fell into the interval, the null hypothesis would not be rejected. That is, the value 368 would not be considered unusual for the data observed. On the other hand, if the hypothesized value did not fall into the interval, the null hypothesis would be rejected, because 368 grams would then be considered an unusual value. Using Equation (6.1), the confidence interval estimate could be set up from the following data:

$n = 25,$ $\bar{X} = 372.5$ grams, $\sigma = 15$ grams (specified by the company)

For a confidence level of 95% (corresponding to a .05 level of significance—i.e., $\alpha = .05$), we have

$$\bar{X} \pm Z \frac{\sigma}{\sqrt{n}}$$

$$372.5 \pm (1.96)\frac{15}{\sqrt{25}}$$

$$372.5 \pm 5.88$$

so that

$$366.62 \leq \mu \leq 378.38$$

Since the interval includes the hypothesized value of 368 grams, we would not reject the null hypothesis, and we would conclude that there is no evidence the mean fill amount over the entire filling process is not 368 grams. This is the same decision we reached by using hypothesis-testing methodology.

Problems for Section 7.6

7.18 (a) Refer to Problem 6.5 on page 296. Is there evidence that the average amount is different from 1.0 gallon (use $\alpha = .01$)
  (b) Compare the conclusions obtained in (a) with those of Problem 6.5. Are the conclusions the same? Why?

7.19 (a) Referring to Problem 6.6 on page 296, at the .05 level of significance is there evidence that the average life is different from 375 hours?
  (b) Compare the conclusions obtained in (a) to those of Problem 6.6. Are the conclusions the same? Why?

7.20 (a) Referring to Problem 6.7 on page 296, at the .05 level of significance, is there evidence that the average amount in the bottles is different from 2.0 liters?
  (b) Compare the conclusions obtained in (a) to those of Problem 6.7. Are the conclusions the same? Why?

7.7 ONE-TAILED TESTS

In Section 7.3, we used hypothesis-testing methodology to examine the question of whether or not the average fill amount over the entire filling process (i.e., the population) was 368 grams. The alternative hypothesis (H_1: $\mu \neq 368$) contained two possibilities: Either the average could be less than 368 grams, or the average could be more than 368 grams. For this reason, the rejection region was divided into the two tails of the sampling distribution of the mean. And, as we have just observed in the previous section, since a confidence interval estimate of the mean contains a lower and upper limit respectively corresponding to the left- and right-tail critical values from the sampling distribution of the mean, we were able to use the confidence interval to do a test of the null hypothesis that the average amount of fill over the entire filling process was 368 grams.

In some situations, however, the alternative hypothesis focuses in a particular direction. For example, the chief financial officer (CFO) of the food packaging company would mainly be concerned with excess because, if more than 368 grams of cereal were actually being filled per box but the price charged to the customer was based on the 368 grams labeled on the box, the company would be losing money unnecessarily. Therefore, she would be interested in whether the average fill for the entire filling process was *above* 368 grams. To her, and strictly from a financial point of view with respect to her responsibility as CFO for the company (the ethics of which will be discussed in Section 7.12), unless the sample mean was significantly above 368 grams, the process would be considered to be working properly. For the CFO, the null and alternative hypotheses would be stated as follows:

$$H_0: \mu \leq 368 \text{ (process is working properly)}$$

$$H_1: \mu > 368 \text{ (process is not working properly)}$$

The rejection region here would be entirely contained in the upper tail of the sampling distribution of the mean since we want to reject H_0 only when the sample mean is significantly above 368 grams. When such a situation occurs where the entire rejection region is contained in one tail of the sampling distribution of the test statistic, it is called a **one-tailed** or **directional test**. If we again choose a level of significance α of .05, the critical value on the Z distribution can be determined. As seen from Table 7.2 and Figure 7.4, since the entire rejection region is in the upper tail of the standard normal distribution and contains an area of .05, the area from the mean to the critical value must be .45; thus, the critical value of the Z test statistic is +1.645, the average of +1.64 and +1.65. (We should note here that some statisticians would *round off* to two decimal places and select +1.64 as the critical value, while others would *round up* to +1.65. We prefer to interpolate between the areas .4495 and .4505 so as to select the critical value with upper-tail area as close to .05 as possible. Thus, we took the average of +1.64 and +1.65.)

The decision rule would be

Reject H_0 if $Z > +1.645$;

otherwise do not reject H_0.

Using the Z test given by Equation (7.1) on the information obtained from the sample drawn by the production manager

$$n = 25, \quad \bar{X} = 372.5 \text{ grams}, \quad \sigma = 15 \text{ grams (specified by the company)}$$

Table 7.2 Obtaining the critical value of the Z-test statistic from the standard normal distribution for a one-tailed test with $\alpha = .05$.

Z	.00	.01	.02	.03	.04	.05	.06	.07	.08	.09
0.0	.0000	.0040	.0080	.0120	.0160	.0199	.0239	.0279	.0319	.0359
0.1	.0398	.0438	.0478	.0517	.0557	.0596	.0636	.0675	.0714	.0753
0.2	.0793	.0832	.0871	.0910	.0948	.0987	.1026	.1064	.1103	.1141
0.3	.1179	.1217	.1255	.1293	.1331	.1368	.1406	.1443	.1480	.1517
0.4	.1554	.1591	.1628	.1664	.1700	.1736	.1772	.1808	.1844	.1879
⋮	⋮	⋮	⋮	⋮	⋮	⋮	⋮	⋮	⋮	⋮
1.0	.3413	.3438	.3461	.3485	.3508	.3531	.3554	.3577	.3599	.3621
1.1	.3643	.3665	.3686	.3708	.3729	.3749	.3770	.3790	.3810	.3830
1.2	.3849	.3869	.3888	.3907	.3925	.3944	.3962	.3980	.3997	.4015
1.3	.4032	.4049	.4066	.4082	.4099	.4115	.4131	.4147	.4162	.4177
1.4	.4192	.4207	.4222	.4236	.4251	.4265	.4279	.4292	.4306	.4319
1.5	.4332	.4345	.4357	.4370	.4382	.4394	.4406	.4418	.4429	.4441
1.6	.4452	.4463	.4474	.4484	.4495	.4505	.4515	.4525	.4535	.4545
1.7	.4554	.4564	.4573	.4582	.4591	.4599	.4608	.4616	.4625	.4633

Source: Extracted from Table E.2.

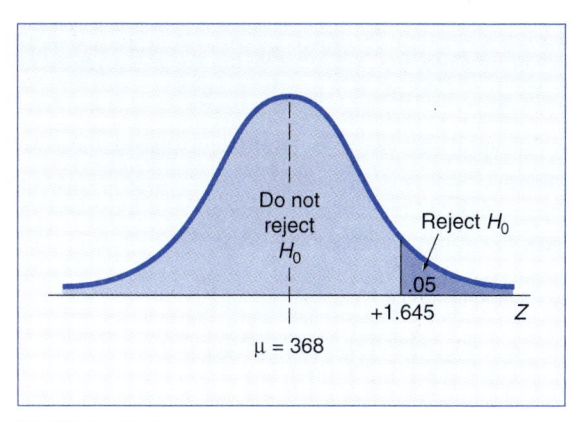

FIGURE 7.4 One-tailed test of hypothesis for a mean (σ known) at the .05 level of significance.

we have

$$Z = \frac{\bar{X} - \mu}{\frac{\sigma}{\sqrt{n}}}$$

$$= \frac{372.5 - 368}{\frac{15}{\sqrt{25}}} = +1.50$$

Since $Z = +1.50 < +1.645$, our decision would be not to reject H_0, and we would conclude that there is no evidence that the average fill per box over the entire filling process is above 368 grams. That is, even though the sample mean $\bar{X}$ exceeded 368 grams, the result from the sample is deemed due to chance or sampling error; it is not statistically significant.

Problems for Section 7.7

Note: *The problems in this section can be solved using Microsoft Excel (see Section 7.11).*

- 7.21 The Glen Valley Steel Company manufactures steel bars. If the production process is working properly, it turns out steel bars with an average length of at least 2.8 feet with a standard deviation of 0.20 foot (as determined from engineering specifications on the involved production equipment). Longer steel bars can be used or altered; shorter bars must be scrapped. A sample of 25 bars is selected from the production line. The sample indicates an average length of 2.73 feet. The company wishes to determine whether the production equipment needs any adjustment.
  (a) State the null and alternative hypotheses.
  (b) If the company wishes to test the hypothesis at the .05 level of significance, what decision would it make?

7.22 Referring to Problem 7.10 on page 347
  (a) At the .05 level of significance, is there evidence that the mean breaking strength is less than 70 pounds?
  (b) How does (a) differ from Problem 7.10? Explain.

7.23 Referring to Problem 7.12 on page 347
  (a) At the .05 level of significance, is there evidence that the average amount dispensed is less than 8 ounces?
  (b) How does (a) differ from Problem 7.12? Explain.

7.8 THE p-VALUE APPROACH TO HYPOTHESIS TESTING: ONE-TAILED TESTS

To understand the *p*-value approach for the one-tailed test, we need to realize that for such situations, we compute the probability of obtaining a value *either* greater than the computed test statistic *or* less than the computed test statistic, depending on the direction of the alternative hypothesis. To illustrate the computation of the *p*-value for the one-tailed test, we will refer to the cereal filling process example discussed in the previous section. For the chief financial officer (CFO), the null and alternative hypotheses were

$$H_0: \mu \leq 368 \text{ (process is working properly)}$$

$$H_1: \mu > 368 \text{ (process is not working properly)}$$

Since the alternative hypothesis indicates a rejection region entirely in the *upper* tail of the sampling distribution of the Z-test statistic, we need only find the probability of obtaining a Z value *above* +1.50 (see Figure 7.5). From Table E.2, the probability of obtaining a Z value above +1.50 is .5000 − .4332 = .0668. Since this *p*-value is greater than the selected level of significance (α = .05), the null hypothesis is not rejected.

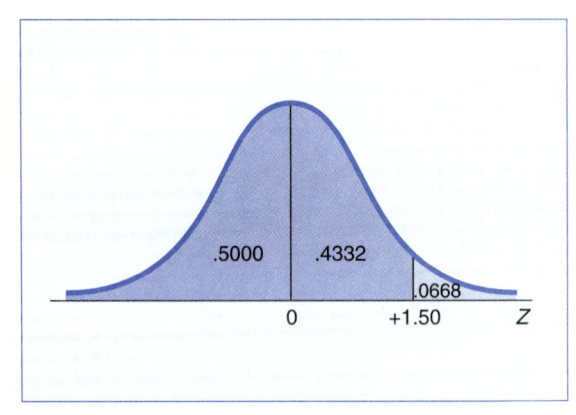

FIGURE 7.5 Determining the *p*-value for a one-tailed test.

Problems for Section 7.8

Note: *The problems in this section can be solved using Microsoft Excel (see Section 7.11).*

• 7.24 Referring to Problem 7.21 on page 353, compute the *p*-value and interpret its meaning.
  7.25 Referring to Problem 7.22 on page 353, compute the *p*-value and interpret its meaning.
  7.26 Referring to Problem 7.23 on page 353, compute the *p*-value and interpret its meaning.

7.9 t TEST OF HYPOTHESIS FOR THE MEAN (σ UNKNOWN)

7.9.1 Introduction

In most hypothesis-testing situations dealing with numerical data, the standard deviation σ of the population is unknown. However, the actual standard deviation of the population is estimated by computing S, the standard deviation of the sample. If the population is assumed to be normally distributed, you may recall from Section 6.3 that the sampling distribution of the mean will follow a *t* distribution with $n - 1$ degrees of freedom. In practice, it has been found that as long as the sample size is not very small and the population is not very skewed, the *t* distribution gives a good approximation to the sampling distribution of the mean. The test statistic for determining the difference between the sample mean $\bar{X}$ and the population mean μ when the sample standard deviation S is used is given by

$$t = \frac{\bar{X} - \mu}{\frac{S}{\sqrt{n}}} \tag{7.2}$$

where the test statistic *t* follows a *t* distribution having $n - 1$ degrees of freedom.

7.9.2 Application

To illustrate the use of the (one-sample) *t* test, suppose a manufacturer of batteries claims that the average capacity of a certain type of battery that the company produces is at least 140 ampere-hours. An independent consumer protection agency wishes to test the credibility of the manufacturer's claim and measures the capacity of a random sample of 20 batteries from a recently produced batch. The results, in ampere-hours, are as follows:

| 137.4 | 140.0 | 138.8 | 139.1 | 144.4 | 139.2 | 141.8 | 137.3 | 133.5 | 138.2 |
| 141.1 | 139.7 | 136.7 | 136.3 | 135.6 | 138.0 | 140.9 | 140.6 | 136.7 | 134.1 |

AMPHRS.TXT

Since the consumer protection agency is interested in whether or not the manufacturer's claim is being overstated, the test is one-tailed and the following null and alternative hypotheses are established:

$$H_0: \mu \geq 140 \text{ ampere-hours}$$

$$H_1: \mu < 140 \text{ ampere-hours}$$

If a level of significance of $\alpha = .05$ is selected, the critical value of the *t* distribution with $20 - 1 = 19$ degrees of freedom can be obtained from Table E.3, as illustrated in Figure 7.6 and Table 7.3. Since the alternative hypothesis H_1 that $\mu < 140$ ampere-hours is directional, the entire rejection region of .05 is contained in the left (lower) tail of the *t* distribution.

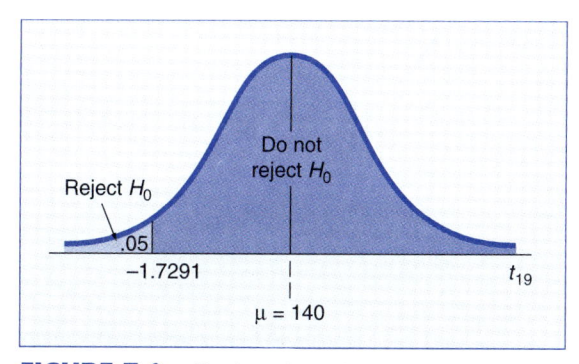

FIGURE 7.6 Testing a hypothesis about the mean (σ unknown) at the .05 level of significance with 19 degrees of freedom.

Table 7.3 Determining the critical value from the t table for an area of .05 in one tail with 19 degrees of freedom.

Degrees of Freedom	Upper-Tail Areas					
	.25	.10	.05	.025	.01	.005
1	1.0000	3.0777	6.3138	12.7062	31.8207	63.6574
2	0.8165	1.8856	2.9200	4.3027	6.9646	9.9248
3	0.7649	1.6377	2.3534	3.1824	4.5407	5.8409
4	0.7407	1.5332	2.1318	2.7764	3.7469	4.6041
5	0.7267	1.4759	2.0150	2.5706	3.3649	4.0322
⋮	⋮	⋮	⋮	⋮	⋮	⋮
16	0.6901	1.3368	1.7459	2.1199	2.5835	2.9208
17	0.6892	1.3334	1.7396	2.1098	2.5669	2.8982
18	0.6884	1.3304	1.7341	2.1009	2.5524	2.8784
19	0.6876	1.3277	1.7291	2.0930	2.5395	2.8609
20	0.6870	1.3253	1.7247	2.0860	2.5280	2.8453

Source: Extracted from Table E.3.

From the *t* table as given in Table E.3, a replica of which is shown in Table 7.3, the critical value is −1.7291. The decision rule is

$$\text{Reject } H_0 \text{ if } t < t_{19} = -1.7291;$$

otherwise do not reject H_0.

For this data set,

$$\sum_{i=1}^{n} X_i = 2{,}769.4 \qquad \sum_{i=1}^{n} X_i^2 = 383{,}613.16 \qquad n = 20$$

Thus,

$$\bar{X} = \frac{\sum_{i=1}^{n} X_i}{n} = \frac{2{,}769.4}{20} = 138.47$$

and

$$S^2 = \frac{\sum_{i=1}^{n} X_i^2 - n\bar{X}^2}{n-1} = \frac{383{,}613.16 - (20)(138.47)^2}{20-1} = 7.0706$$

so that

$$S = 2.66$$

Using Equation (7.2), we have

$$t = \frac{\bar{X} - \mu}{\frac{S}{\sqrt{n}}} = \frac{138.47 - 140}{\frac{2.66}{\sqrt{20}}} = -2.57$$

Since $t = -2.57 < t_{19} = -1.7291$ or using Microsoft Excel (see Figure 7.2.Excel on page 369), the one-tailed p-value is equal to .0093 < .05, the decision is to reject H_0. There is evidence to believe that the manufacturer's claim is overstated, and the consumer protection agency should initiate some corrective measure against the company.

7.9.3 Assumptions of the One-Sample t Test

For a given sample size n, the test statistic t follows a t distribution with n − 1 degrees of freedom. As we observe in Table E.3, based on available degrees of freedom, each of the rows corresponds to a particular t distribution.

The one-sample t test is considered a classical parametric procedure. As such, it makes a variety of stringent assumptions that must hold if we are to be assured that the results we obtain from employing the test are valid. In particular, to use the one-sample t test, it is assumed that the obtained numerical data are independently drawn and represent a random sample from a population that is normally distributed.

As we learned in Section 5.6, the normality assumption can be checked in several ways. A determination of how closely the actual data match the normal distribution's theoretical properties can be made by a descriptive analysis of the obtained statistics along with a graphical analysis to provide a visual interpretation. Thus, by exploring the sample data through a study of its descriptive summary measures along with a graphical analysis (a stem-and-leaf display, a box-and-whisker plot, and a normal probability plot), we may draw our own conclusions as to the likelihood that the underlying population is at least approximately normally distributed. Using the battery capacity ampere-hours data, Figure 7.7 depicts Excel output displaying the descriptive summary measures, box-and-whisker plot, and a normal probability plot. From these, there is no reason to believe that the assumption of underlying population normality is violated to any great degree, and we may conclude that the results obtained by the consumer protection agency are valid.

	A	B	C	D	E	F
1				Stem and Leaf Display		
2				for Battery Capacity Data		
3				leaf unit:	0.1	
4						
5				133	5	
6	n	20		134	1	
7	mean	138.47		135	6	
8	median	138.50		136	3 7 7	
9	std. dev.	2.66		137	3 4	
10	minimum	133.50		138	0 2 8	
11	maximum	144.40		139	1 2 7	
12				140	0 6 9	
13				141	1 8	
14				142		
15				143		
16				144	4	
17						

FIGURE 7.7 Excel output for studying the assumptions necessary to employ the t test. Panel A.

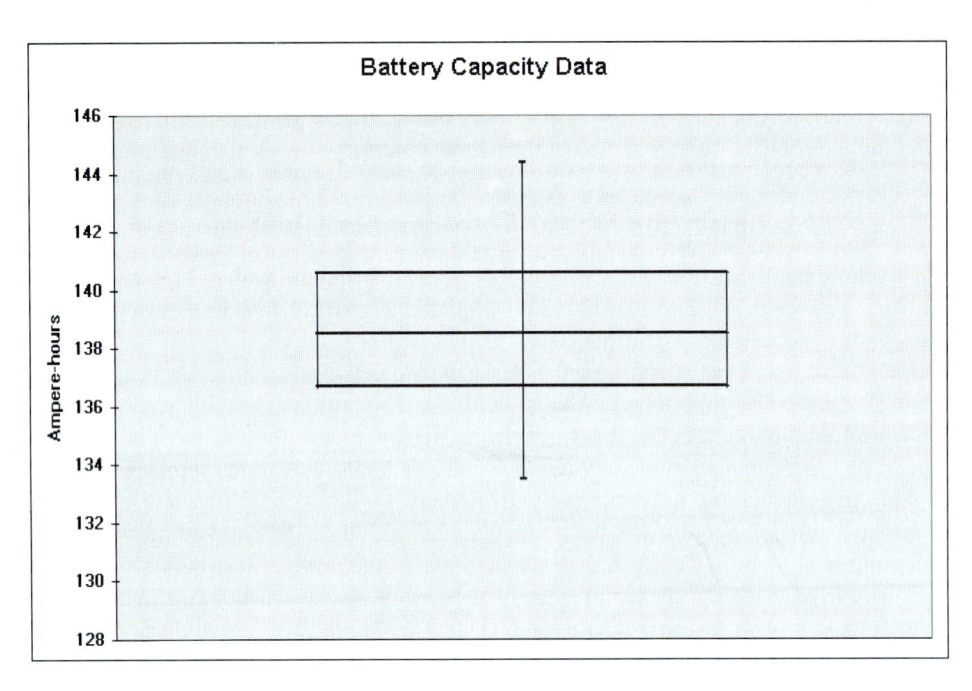

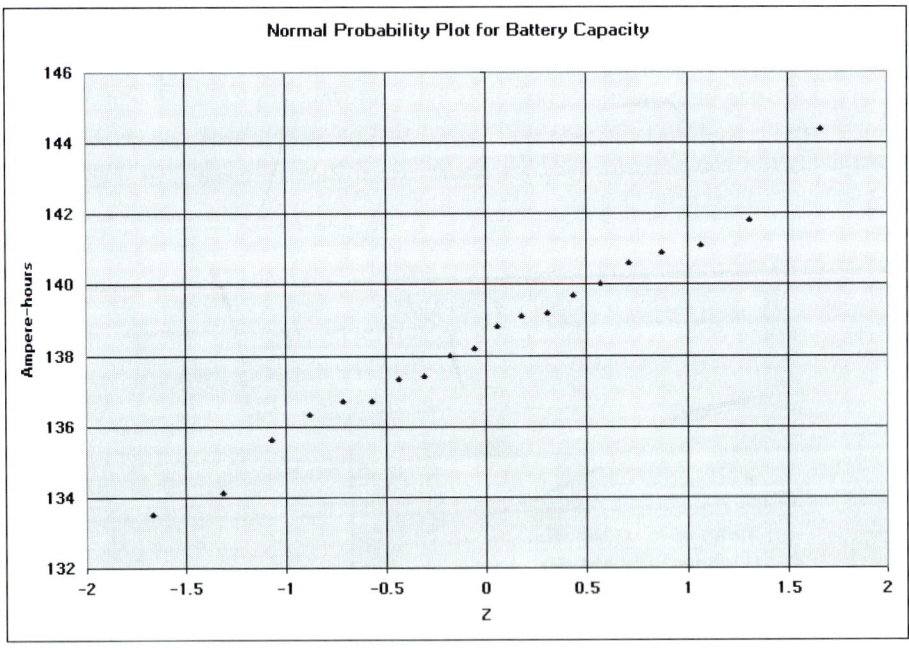

FIGURE 7.7 Panels B and C.

Although the t test is robust if the shape of the population from which the sample is drawn departs somewhat from a normal distribution, particularly when the sample size is large enough to enable the test statistic t to be influenced by the central limit theorem (see Section 5.9), erroneous conclusions may be drawn and statistical power can be lost if the t test is incorrectly used. Thus, if the sample size n is small (i.e., less than 30), and we cannot easily make the assumption that the underlying population from which the sample was drawn is normally distributed, other, distribution-free testing procedures are likely to be more powerful (see Reference 1).

Problems for Section 7.9

Note: *The problems in this section can be solved using Microsoft Excel (see Section 7.11).*

- 7.27 A consumers' advocate group would like to evaluate the average energy efficiency rating (EER) of window-mounted, large-capacity (i.e., in excess of 7,000 Btu) air-conditioning units. A random sample of 36 such air-conditioning units is selected and tested for a fixed period of time with their EER recorded as follows:

  EER.TXT

8.9	9.1	9.2	9.1	8.4	9.5	9.0	9.6	9.3
9.3	8.9	9.7	8.7	9.4	8.5	8.9	8.4	9.5
9.3	9.3	8.8	9.4	8.9	9.3	9.0	9.2	9.1
9.8	9.6	9.3	9.2	9.1	9.6	9.8	9.5	10.0

  (a) Using the .05 level of significance, is there evidence that the average EER is different from 9.0?
  (b) What assumptions are being made in order to perform this test?
  (c) Find the *p*-value and interpret its meaning.
  (d) What will your answer in (a) be if the last data value is 8.0 instead of 10.0?

7.28 A manufacturer of plastics wants to evaluate the durability of rectangularly molded plastic blocks that are to be used in furniture. A random sample of 50 such plastic blocks is examined and the hardness measurements (in Brinell units) are recorded as follows:

  PLASTIC.TXT

283.6	273.3	278.8	238.7	334.9	302.6	239.9	254.6	281.9	270.4
269.1	250.1	301.6	289.2	240.8	267.5	279.3	228.4	265.2	285.9
279.3	252.3	271.7	235.0	313.2	277.8	243.8	295.5	249.3	228.7
255.3	267.2	255.3	281.0	302.1	256.3	233.0	194.4	291.9	263.7
273.6	267.7	283.1	260.9	274.8	277.4	276.9	259.5	262.0	263.5

  (a) Using the .05 level of significance, is there evidence that the average hardness of the plastic blocks exceeds 260 (in Brinell units)?
  (b) What assumptions are being made in order to perform this test?
  (c) Find the *p*-value and interpret its meaning.
  (d) What will your answer in (a) be if the first data value is 233.6 instead of 283.6?

- 7.29 The manager of the credit department for an oil company would like to determine whether the average monthly balance of credit card holders is equal to $75. An auditor selects a random sample of 100 accounts and finds that the average owed is $83.40 with a sample standard deviation of $23.65.
  (a) Using the .05 level of significance, should the auditor conclude that there is evidence that the average balance is different from $75?
  (b) Find the *p*-value and interpret its meaning.
  (c) What will your answer in (a) be if the standard deviation is $37.26?
  (d) What will your answer in (a) be if the sample mean is $78.81?

7.30 A manufacturer of detergent claims that the mean weight of a particular box of detergent is 3.25 pounds. A random sample of 64 boxes revealed a sample average of 3.238 pounds and a sample standard deviation of 0.117 pound.
  (a) Using the .01 level of significance, is there evidence that the average weight of the boxes is different from 3.25 pounds?
  (b) Find the *p*-value and interpret its meaning.
  (c) What will your answer in (a) be if the standard deviation is 0.05 pound?
  (d) What will your answer in (a) be if the sample mean is 3.211 pounds?

7.31 The director of admissions at a large university would like to advise parents of incoming students concerning the cost of textbooks during a typical semester. A sample of 100 students enrolled in the university indicates a sample average cost of $315.40 with a sample standard deviation of $43.20.
  (a) Using the .10 level of significance, is there evidence that the population average is above $300?

(b) Find the *p*-value and interpret its meaning.
(c) What will your answer in (a) be if the standard deviation is $75 and the .05 level of significance is used?
(d) What will your answer in (a) be if the sample average is $305.11?

7.32 A machine being used for packaging seedless golden raisins has been set so that, on average, 15 ounces of raisins will be packaged per box. The quality control engineer wishes to test the machine setting and selects a sample of 30 consecutive raisin packages filled during the production process. Their weights are recorded as follows:

15.2	15.3	15.1	15.7	15.3	15.0	15.1	14.3	14.6	14.5
15.0	15.2	15.4	15.6	15.7	15.4	15.3	14.9	14.8	14.6
14.3	14.4	15.5	15.4	15.2	15.5	15.6	15.1	15.3	15.1

(a) Is there evidence that the mean weight per box is different from 15 ounces? (Use $\alpha = .05$.)
(b) To perform the test in part (a), we must assume that the observed sequence in which the data were collected is random. What other assumptions must be made to perform the test? Discuss.
(c) What will your answers in (a) be if the weights for the last two packages are 16.3 and 16.1 instead of 15.3 and 15.1?

7.33 The director of admissions for a well-known business school claims that GMAT scores among applicants to the MBA program have increased substantially over the past year. The average score for all applicants last year was 520. The following data represent the GMAT scores for a random sample of 20 applicants during this year:

| 560 | 500 | 670 | 460 | 590 | 490 | 540 | 550 | 750 | 620 |
| 510 | 520 | 380 | 580 | 600 | 550 | 570 | 640 | 490 | 600 |

(a) Is there evidence that the average score has increased? (Use $\alpha = .05$.)
(b) What assumption must hold in order to perform the test in part (a)?
(c) Evaluate this assumption through a graphical approach. Discuss.
(d) What will your answer in (a) be if the last value is 500 instead of 600?

7.34 A manufacturer of automobile batteries claims that his product will last on average at least four years (i.e., 48 months). A consumers' advocate group would like to evaluate this longevity claim and selects a random sample of 28 such batteries to test. The data below indicate the length of time (in months) that each of these batteries lasted (i.e., performed properly before failure).

42.3	39.6	25.0	56.2	37.2	47.4	57.5
39.3	39.2	47.0	47.4	39.7	57.3	51.8
31.6	45.1	40.8	42.4	38.9	42.9	34.1
49.0	41.5	60.1	34.6	50.4	30.7	44.1

(a) Is there evidence that the average battery life is less than 48 months? (Use $\alpha = .05$.)
(b) What assumption must hold in order to perform the test in part (a)?
(c) Evaluate this assumption through a graphical approach. Discuss.
(d) What would be your answer in (a) if the last two values were 50.7 and 54.1 instead of 30.7 and 44.1?

7.10 ONE-SAMPLE Z TEST FOR THE PROPORTION

In some situations, we want to test a hypothesis pertaining to the population proportion p of values that are in a particular category rather than the population mean value. A random sample can be selected from the population, and the **sample proportion**, $p_s = X/n$, would be computed. The value of this statistic would then be compared to the hypothesized value of the parameter, p, so that a decision pertaining to the hypothesis can be made.

To carry out this hypothesis test if certain assumptions can be met, the sampling distribution of a proportion will follow a standardized normal distribution (see Section 5.10). To evaluate the magnitude of the difference between the sample proportion p_s and the hypothesized population proportion p, the test statistic Z given in Equation (7.3) can be used:

$$Z \cong \frac{p_s - p}{\sqrt{\frac{p(1-p)}{n}}} \qquad (7.3)$$

where $p_s = \dfrac{X}{n} = \dfrac{\text{number of successes in sample}}{\text{sample size}}$ = observed proportion of successes

p = proportion of successes from the null hypothesis

The test statistic Z is approximately normally distributed.

Alternatively, instead of examining the *proportion* of successes in a sample, as in Equation (7.3), we could study the *number* of successes in a sample. The test statistic Z for determining the magnitude of the difference between the number of successes in a sample and the hypothesized or expected number of successes in the population is presented in Equation (7.4):

$$Z \cong \frac{X - np}{\sqrt{np(1-p)}} \qquad (7.4)$$

Again, this test statistic Z is approximately normally distributed. You may recall from Section 5.10 that although the random variable X (the number of successes in the sample) follows a binomial distribution, if the sample size is large enough [i.e., both $np \geq 5$ and $n(1-p) \geq 5$], the normal distribution provides a good approximation to the binomial distribution.

Aside from possible rounding errors, the test statistic Z given by Equations (7.3) and (7.4) will provide exactly the same results. The two alternative forms of the test statistic are equivalent because the numerator of Equation (7.4) is n times the numerator of Equation (7.3), and the denominator of Equation (7.4) is also n times the denominator of Equation (7.3). The choice of which of these two formulas to use is up to the user.

To illustrate the (one-sample) Z test for a hypothesized proportion, let us return to the cereal filling-process example discussed earlier in this chapter. Suppose that the production manager is also concerned with the sealing process for filled boxes. Once the package inside the box is filled, it is supposed to be sealed so that it is airtight. Based on past experience, however, it is known that 1 out of 10 packages (i.e., 10% or .10) initially do not meet standards for sealing and must be "reworked" in order to pass inspection. To alter this situation, suppose the production manager implements a newly developed sealing system on a trial basis. After a 1-day "break-in" period, he takes a random sample of 200 boxes that represent daily output at the plant and, through inspection, finds that 11 need rework. The production manager would want to determine whether there is evidence that, under the new sealing system, the proportion of defective packages has improved (i.e., has decreased below .10).

In terms of proportions (rather than percentages), the null and alternative hypotheses can be stated as follows:

$$H_0: p \geq .10$$
$$H_1: p < .10$$

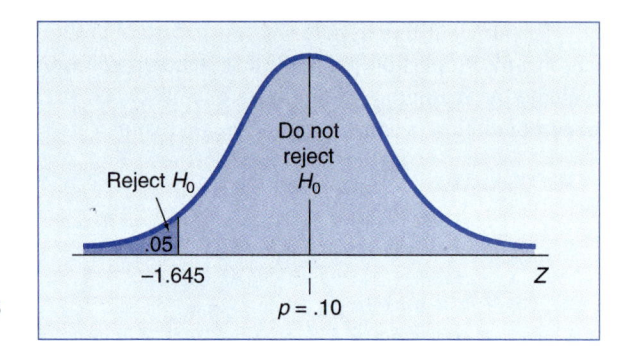

FIGURE 7.8 One-tailed test of hypothesis for a proportion at the .05 level of significance.

Since the production manager is interested in whether or not there has been a significant reduction in the proportion of defective packages owing to the new process, the test is one-tailed. If a level of significance α of .05 is selected, the rejection and nonrejection regions would be set up as in Figure 7.8, and the decision rule would be

$$\text{Reject } H_0 \text{ if } Z < -1.645;$$

otherwise do not reject H_0.

From our data,

$$p_S = \frac{11}{200} = .055$$

Using Equation (7.3), we have

$$Z \cong \frac{p_S - p}{\sqrt{\dfrac{p(1-p)}{n}}} = \frac{.055 - .10}{\sqrt{\dfrac{(.10)(.90)}{200}}} = \frac{-.045}{\sqrt{.00045}} = \frac{-.045}{.0212} = -2.12$$

or using Equation (7.4), we have

$$Z \cong \frac{X - np}{\sqrt{np(1-p)}} = \frac{11 - 200(.10)}{\sqrt{(200)(.10)(.90)}} = \frac{11 - 20}{\sqrt{18}} = \frac{-9}{4.243} = -2.12$$

Since $-2.12 < -1.645$, we reject H_0. Thus, the manager may conclude that there is evidence that the proportion of defectives with the new system is less than .10.

● **Finding the *p*-value**   As an alternative approach toward making a hypothesis-testing decision, we may also compute the *p*-value for this situation (see Sections 7.5 and 7.8). Since a one-tailed test is involved in which the rejection region is located only in the lower tail (see Figure 7.9), we need to find the area below a Z value of −2.12. From Table E.2, this

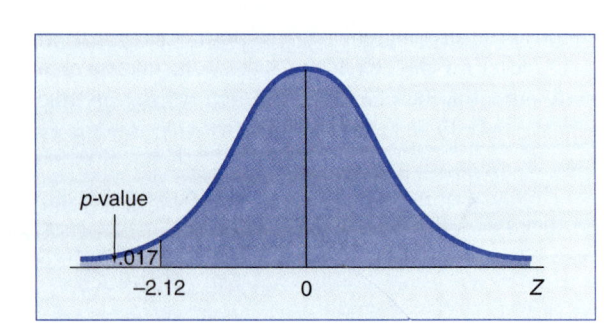

FIGURE 7.9 Determining the *p*-value for a one-tailed test.

362   Chapter 7   Fundamentals of Hypothesis Testing

probability will be .5000 − .4830 = .0170 (see Figure 7.3.Excel on page 371). Since this value is less than $\alpha = .05$, the null hypothesis can be rejected.

Problems for Section 7.10

Note: *The problems in this section can be solved using Microsoft Excel (see Section 7.11).*

7.35 Prove that the formula on the right-hand side of Equation (7.3) on page 361 is equivalent to the formula on the right-hand side of Equation (7.4).

7.36 A television manufacturer had claimed in its warranty that in the past not more than 10% of its television sets needed any repair during their first 2 years of operation. To test the validity of this claim, a government testing agency selects a sample of 100 sets and finds that 14 sets required some repair within their first 2 years of operation. Using the .01 level of significance,
(a) Is the manufacturer's claim valid or is there evidence that the claim is not valid?
(b) Compute the *p*-value and interpret its meaning.
(c) What would be your answer in (a) if 18 sets required some repair?

7.37 The Giansante Company, provider of extermination services, claims that no more than 15% of its customers need repeated treatment after a 90-day warranty period. To determine the validity of this claim, a consumer organization selects a sample of 100 customers and finds that 22 needed repeated treatment after the 90-day warranty period.
(a) Is there evidence at the .05 level of significance that the claim is not valid (i.e., that the proportion needing treatment is greater than .15)?
(b) Compute the *p*-value and interpret its meaning.
(c) What would be your answer in (a) if the .01 level of significance was used?
(d) What would be your answer in (a) if 18 homes needed treatment?

• 7.38 The personnel director of a large insurance company is interested in reducing the turnover rate of data processing clerks in the first year of employment. Past records indicate that 25% of all new hires in this area are no longer employed at the end of 1 year. Extensive new training approaches are implemented for a sample of 150 new data processing clerks. At the end of a 1-year period, of these 150 individuals, 29 are no longer employed.
(a) At the .01 level of significance, is there evidence that the proportion of data processing clerks who have gone through the new training and are no longer employed is less than .25?
(b) Compute the *p*-value and interpret its meaning.
(c) What would be your answer in (a) if 22 of the individuals are no longer employed?

• 7.39 The marketing manager for an automobile manufacturer is interested in determining the proportion of new compact-car owners who would have purchased a passenger-side inflatable auto bag if it had been available for an additional cost of $300. The manager believes from previous information that the proportion is .30. Suppose that a survey of 200 new compact-car owners is selected and 79 indicate that they would have purchased the inflatable air bags.
(a) At the .10 level of significance, is there evidence that the population proportion is different from .30?
(b) Compute the *p*-value and interpret its meaning.
(c) What would be your answer in (a) if 70 indicated that they would have purchased inflatable air bags?

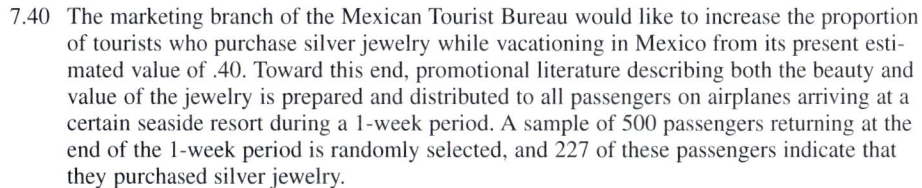

7.40 The marketing branch of the Mexican Tourist Bureau would like to increase the proportion of tourists who purchase silver jewelry while vacationing in Mexico from its present estimated value of .40. Toward this end, promotional literature describing both the beauty and value of the jewelry is prepared and distributed to all passengers on airplanes arriving at a certain seaside resort during a 1-week period. A sample of 500 passengers returning at the end of the 1-week period is randomly selected, and 227 of these passengers indicate that they purchased silver jewelry.
(a) At the .05 level of significance, is there evidence that the proportion has increased above the previous value of .40?
(b) Compute the *p*-value and interpret its meaning.
(c) What would be your answer in (a) if 213 passengers indicated that they purchased silver jewelry?

7.11 USING MICROSOFT EXCEL FOR ONE-SAMPLE TESTS

In this chapter we have developed the one-sample test of hypothesis for the mean and the proportion. In this section we will learn how the various features of Microsoft Excel can be used for these tests of hypotheses.

7.11.1 Using Microsoft Excel for the One-Sample Z Test for the Mean (σ Known)

Although Excel does not have a Data Analysis tool to test a hypothesis for the mean when σ is known, Excel formulas can be used instead. The design of a Calculations worksheet for this test, is shown in Table 7.1.Excel.

Referring to Equation (7.1) on page 345, we see that to test the hypothesis for the mean, we need to have the sample mean $\bar{X}$, population standard deviation σ, the value of μ in the null hypothesis, and the sample size n. If the arithmetic mean $\bar{X}$ needs to be calculated, the Excel AVERAGE function can be used and the results copied to cell B5. Open a new workbook and rename a sheet Calculations. The steps for setting up the test of hypothesis would then be as follows:

1 Enter the value for the sample size n in cell B4, the sample arithmetic mean in cell B5, and the population standard deviation σ in cell B6. Using the cereal filling-process example on page 346, we enter 25 in cell B4, 372.5 in cell B5, and 15 in cell B6.

Table 7.1.Excel Design for testing the hypothesis for a mean (σ Known).

	A	B
1	One Sample Tests of Hypothesis	
2		
3	Z Test for the Mean	
4	n	xxx
5	Arithmetic Mean	xxx
6	Standard Deviation	xxx
7	Standard Error	=B6/SQRT(B4)
8	Null Hypothesis $\mu=$	xxx
9	α	.xx
10	Z Test Statistic	=(B5–B8)/B7
11	Two-Tailed Test	
12	Lower Critical Value	=NORMSINV(B9/2)
13	Upper Critical Value	=NORMSINV(1–B9/2)
14	p-value	=2*(1–NORMSDIST(ABS(B10)))
15	Decision	=IF(B14<B9,"Reject","Do not Reject")
16	One-Tailed Test (Lower)	
17	Lower Critical Value	=NORMSINV(B9)
18	p-value	=NORMSDIST(B10)
19	Decision	=IF(B18<B9,"Reject","Do not Reject")
20	One-Tailed Test (Upper)	
21	Upper Critical Value	=NORMSINV(1–B9)
22	p-value	=1–NORMSDIST(B10)
23	Decision	=IF(B22<B9,"Reject","Do not Reject")

❷ Now we can compute the standard error of the mean $\sigma/\sqrt{n}$ in cell B7 by entering the formula =B6/SQRT(B4).

❸ We now enter the value of μ stated in the null hypothesis H_0 in cell B8 and the value of the level of significance α in cell B9. Using our example, we enter 368 in cell B8 and .05 in cell B9.

❹ The Z-test statistic indicated in Equation (7.1) can now be computed in cell B10 by entering the formula =(B5–B8)/B7 in cell B10.

❺ If a two-tailed test is desired, the lower critical value is the Z value corresponding to a cumulative area of $\alpha/2$, and the upper critical value corresponds to a cumulative area of $1-\alpha/2$. These critical values can be obtained by using the NORMSINV function (see Section 5.5). Thus, we enter =NORMSINV(B9/2) in cell B12 and =NORMSINV(1–B9/2) in cell B13.

❻ Since a two-tailed test is being performed, we need to obtain the probability of obtaining a Z value in both the upper and lower tails. This is done by using the NORMSDIST function to find the probability of exceeding the absolute value of the Z-test statistic in cell B10 and multiplying by two, =2*(1–NORMSDIST(ABS(B10))).

❼ The decision involves a comparison between the p-value and the level of significance α. As discussed in Section 7.4, we reject the null hypothesis H_0 if the p-value is less than α; otherwise we don't reject the null hypothesis H_0. This decision can be made in Excel by using the IF function, whose format is

        IF(comparison, action if comparison holds, action if comparison fails)

where the actions are either values to display or formulas to calculate.

For this example we enter =IF(B14<B9,"Reject","Do not Reject") in cell B15. Excel will compare the contents of cell B14 (the p-value) with the contents of cell B9 (the α value). If the contents of cell B14 are less than the contents of cell B9, then the value "Reject" will be displayed in the cell; otherwise the value "Do not Reject" will be displayed. In our example, since .1336 > .05, the null hypothesis is not rejected and "Do not reject" is displayed in the cell.

❽ If a one-tailed test in which the rejection region is in the lower tail is desired, the entire rejection region of size α is in the lower tail. Thus, the lower critical value represents the Z value corresponding to a cumulative area of α and is obtained in cell B17 using the formula =NORMSINV(B9). The p-value represents the cumulative area up to the value of the test statistic and is obtained in cell B18 using the formula =NORMSDIST(B10). The decision made follows a logic similar to the two-tailed test in that the p-value is compared to α, and the null hypothesis is rejected if the p-value is less than α.

❾ If a one-tailed test in which the rejection region is in the upper tail is desired, the entire rejection region of size α is in the upper tail. Thus, the upper critical value represents the Z value corresponding to a cumulative area of $1-\alpha$ and is obtained in cell B21 using the formula =NORMSINV(1–B9). The p-value represents the cumulative area greater than the value of the test statistic and is obtained in cell B22 using the formula =1–NORMSDIST(B10). The decision then follows a logic similar to the other one- and two-tailed tests in that the p-value is compared to α and the null hypothesis is rejected if the p-value is less than α.

The results are illustrated in Figure 7.1.Excel.

	A	B
1	One-Sample Tests of Hypothesis	
2		
3	Z Test for the Mean	
4	n	25
5	Arithmetic Mean	372.5
6	Standard Deviation	15
7	Standard Error	3
8	Null Hypothesis μ=	368
9	α	0.05
10	Z Test Statistic	1.5
11	Two-Tailed Test	
12	Lower Critical Value	-1.95996108
13	Upper Critical value	1.959961082
14	p-value	0.133614458
15	Decision	Do not reject
16	One-tailed Test (lower)	
17	Lower Critical Value	-1.644853
18	p-value	0.933192771
19	Decision	Do not reject
20	One-Tailed test (Upper)	
21	Upper Critical Value	1.644853
22	p-value	0.066807229
23	Decision	Do not reject

FIGURE 7.1.EXCEL One-sample Z test for the mean using Microsoft Excel for the cereal filling-process example.

▲ **WHAT IF EXAMPLE**

The Calculations worksheet presented in Table 7.1.Excel allows us to study the effect of changes in $\bar{X}$, σ, n, and the level of confidence on the test of hypothesis for the mean. For example, if we wanted to determine the effect of a change in the population standard deviation σ on the test statistic, we could change the value of σ in cell B6 from 15 to 10. We observe that the Z-test statistic changes to 2.25, the two-tailed p-value becomes .024, and the null hypothesis is now rejected. As before, if we are interested in seeing the effects of many different changes, we could use the Scenario Manager (see Section 1S.15) to store and use sets of alternative values of $\bar{X}$, σ, n, and the level of confidence.

7.11.2 Using Microsoft Excel for the One-Sample t Test for the Mean (σ Unknown)

Although Excel does not have a Data Analysis tool to test a hypothesis for the mean when σ is unknown, Excel formulas can be used instead. The design for a Calculations worksheet for this test is shown in Table 7.2.Excel.

7-11-2.XLS

Referring to Equation (7.2) on page 355, we see that to test the hypothesis for the mean, we need to have the sample mean $\bar{X}$, sample standard deviation S, the value of μ in the null hypothesis, and the sample size n. If the arithmetic mean $\bar{X}$ and the sample standard deviation S need to be calculated, the AVERAGE and STDEV functions could be used in cells B5 and

Table 7.2. Excel Design for testing the hypothesis for a mean (σ unknown)

	A	B	C	D
1	One-Sample Tests of Hypothesis			
2				
3	t Test for the Mean			
4	n	xxx		
5	Arithmetic Mean	=AVERAGE(Data!A2:A21)		
6	Standard Deviation	=STDEV(Data!A2:A21)		
7	Standard Error	=B6/SQRT(B4)		
8	Null Hypothesis $\mu =$	xxx		
9	α	.xx		
10	df	=B4–1		
11	T Test Statistic	=(B5–B8)/B7		
12	Two-Tailed Test			
13	Lower Critical Value	= –(TINV(B9,B10))		
14	Upper Critical Value	=TINV(B9,B10)		
15	p-value	=TDIST(ABS(B11),B10,2)		
16	Decision	=IF(B15<B9,"Reject","Do not Reject")		
17	One-Tailed Test (Lower)			
18	Lower Critical Value	= –(TINV(2*B9,B10))		
19	p-value	=IF(B11<0,D20,D21)	IF Calculations	
20	Decision	=IF(B19<B9,"Reject","Do not Reject")	TDIST Calculation	=TDIST(ABS(B11),B10,1)
21	One-Tailed Test (Upper)		1–TDIST Calculation	=1–D20
22	Upper Critical Value	=(TINV(2*B9, 10))		
23	p-value	=IF(B11<0,D21,D20)		
24	Decision	=IF(B23<B9,"Reject","Do not Reject")		

B6. Open a new workbook, rename the active sheet Data, and enter the battery capacity data (page 355) in the range A2:A21. Insert a worksheet named Calculations and do the following to set up the test of hypothesis:

❶ Enter the value for the sample size *n* in cell B4, the sample arithmetic mean in cell B5, and the sample standard deviation *S* in cell B6. For the battery capacity example on page 355, enter the sample size of 20 in cell B4, enter the formula =AVERAGE(Data!A2:A21) in cell B5, and enter the formula =STDEV(Data!A2:A21) in cell B6.

❷ Now we can compute the standard error of the mean $S/\sqrt{n}$ in cell B7 by entering the formula =B6/SQRT(B4).

❸ We now enter the value of μ stated in the null hypothesis H_0 in cell B8 and the value of the level of significance α in cell B9. In our example, we enter 140 in cell B8 and .05 in cell B9.

❹ The degrees of freedom equal to $n - 1$, is now entered in cell B10 using the formula =B4–1.

❺ The *t*-test statistic indicated in Equation (7.2) can now be computed in cell B11 by entering the formula =(B5–B8)/B7.

❻ If a two-tailed test is desired, the lower critical value is the *t* value corresponding to a cumulative area of $\alpha/2$, and the upper critical value corresponds to a cumulative area of $1 - \alpha/2$. These critical values can be obtained by using the TINV function (see

Section 6.10.2). This function provides the upper critical value, splitting the level of significance α into the two tails of the distribution. Thus, to obtain the lower critical value, we enter =–(TINV(B9,B10)) in cell B13 and =TINV(B9,B10) in cell B14.

⑦ Since a two-tailed test is being performed, we need to find the probability of obtaining a *t* value in both the upper and lower tails. This is done by using the TDIST function whose format is

$$\text{TDIST}(\text{ABS}(X), df, \textit{tails})$$

where ABS(X) = the absolute value of the *t*-test statistic

df = the degrees of freedom

tails = 1 for a one-tailed test and 2 for a two-tailed test

Thus, we enter the formula =TDIST(ABS(B11),B10,2) in cell B15. The result for our data, .0186, represents the probability of obtaining a *t*-test statistic either greater than 2.5734 or less than –2.5734.

⑧ The decision about rejecting the null hypothesis involves a comparison between the *p*-value and the level of significance α. As in the case of the Z test for the mean discussed in Section 7.11.1, we reject the null hypothesis H_0 if the *p*-value is less than α; otherwise we do not reject the null hypothesis H_0. This comparison can be made by entering the formula =IF(B15<B9,"Reject","Do not Reject") in cell B16. Excel will compare the contents of cell B15 (the *p*-value) with the contents of cell B9 (the α value). If the contents of B15 are less than B9, then the value "Reject" will be displayed in the cell; otherwise the value "Do not reject" will be displayed. In our example, since the value in B15, .0186004, is less than the value in B9, .05, the null hypothesis is rejected.

⑨ If a one-tailed test in which the rejection region is in the lower tail is desired, the entire rejection region of size α is in the lower tail. Thus, the lower critical value represents the *t* value corresponding to a cumulative area of α. Since the TINV function provides the upper-tail *t* value for a two-tailed test, we need to provide twice the α level indicated in cell B9 and also enter a minus sign to obtain the lower-tail value. Thus, we use the formula =–(TINV(2*B9, B10)) in cell B18.

⑩ The *p*-value represents the cumulative area up to the value of the test statistic. However, the TDIST function provides the upper-tail area corresponding to a positive-valued *t* statistic. In a one-tailed test with a lower-tail rejection region, if the test statistic is negative, the TDIST function can be used directly, but if the test statistic is positive, the *p*-value is the cumulative area up to the test statistic. To express this dichotomy, we enter the formula = IF(B11<0,TDIST(ABS(B11),B10,1),1–TDIST(ABS(B11),B10, 1)) in cell B19. Instead, to simplify this formula (and the one to be entered into cell B23) we can enter the formulas =TDIST(ABS(B11),B10,1) in cell D20 and 1–D20 in cell D21. This leads to the more easily read =IF(B11<0,D20,D21) formula as the entry for cell B19. In cell B20, in a similar manner as in the two-tailed test, the *p* value in cell B19 is compared to the α level in cell B9, and a decision to reject or not reject the null hypothesis is made.

⑪ If a one-tailed test in which the rejection region is in the upper tail is desired, the entire rejection region of size α is in the upper tail. The upper critical value is determined as in step 9, except the minus sign is not placed before the TINV function.

⑫ The *p*-value is entirely contained in the upper tail of the *t* distribution. In a one-tailed test with an upper-tail rejection region, if the test statistic is positive, the TDIST function can be used directly, but if the test statistic is negative, the p-value is the cumulative area up to the test statistic. Using the simplification explained in Step 10, the formula =IF(B11<0,D21,D20) can be used to determine the p-value in cell B23. As in the case of the other one- and two-tailed tests, the p-value is compared to α, and the null hypothesis is rejected if the p-value is less than α.

The results are illustrated in Figure 7.2.Excel.

	A	B	C	D
1	One-Sample Tests of Hypothesis			
2				
3	T Test for the Mean			
4	n	20		
5	Arithmetic Mean	138.47		
6	Standard Deviation	2.658867982		
7	Standard Error	0.594540955		
8	Null Hypothesis $\mu=$	140		
9	α	0.05		
10	df	19		
11	T Test Statistic	-2.57341397		
12	Two-Tailed Test			
13	Lower Critical Value	-2.0930247		
14	Upper Critical value	2.093024705		
15	p-value	0.01861004		
16	Decision	Reject		
17	One-tailed Test (Lower)			
18	Lower Critical Value	-1.72913133		
19	p-value	0.00930502	IF calculations	
20	Decision	Reject	TDIST calculation	0.009305
21	One-Tailed test (Upper)		1-TDIST calculation	0.990695
22	Upper Critical Value	1.729131327		
23	p-value	0.99069498		
24	Decision	Do not reject		

FIGURE 7.2.EXCEL One-sample t test for the mean using Microsoft Excel for the battery capacity example.

▲ WHAT IF EXAMPLE

The Calculations worksheet presented in Table 7.2.Excel allows us to study the effect of changes in $\bar{X}$, σ, n, and the level of confidence on the test of hypothesis for the mean. For example, if we wanted to determine the effect of a change in one of the data values on the test statistic, we could change the smallest observation from 133.5 to 153.5. We observe that $\bar{X}$ changes to 139.47, S becomes 4.075, the t-test statistic changes to $-.582$, the one-tailed p-value becomes .2838, and the null hypothesis is not rejected. As before, if we are interested in seeing the effects of many different changes, we could use the Scenario Manager (see Section 1S.15) to store and use sets of alternative values of $\bar{X}$, σ, n, and the level of confidence.

7.11.3 Using Microsoft Excel for the One-Sample Z Test for the Proportion

Although Excel does not have a Data Analysis tool to test a hypothesis for the proportion, Excel formulas can be used instead. The design for a Calculations worksheet for this test is shown in Table 7.3.Excel.

Referring to Equation (7.3) on page 361, we see that to obtain the test of hypothesis for the proportion, we need to have the sample proportion p_S and the sample size n. If the sample size and the number of successes need to be obtained from the raw data, the PivotTable Wizard (see Section 2.12) could be used and the results copied to cells B4 and B5, respectively. If the values for the sample size and the number of successes are available, they should be entered in cells B4 and B5. Open a new workbook and rename the active sheet Calculations. The steps for setting up the test of hypothesis for the proportion would be the same as setting up the Z test for the mean, except that the number of successes (instead of the average) would be entered in cell B5, the sample proportion (instead of the standard deviation) would be entered in cell B6, the hypothesized value of p would be entered in cell B7, and the standard error would be computed using the formula =SQRT((B7*(1–B7))/B4) in cell B8. Using the cereal filling-process example on page 361, we enter 200 in cell B4, 11 in cell B5, .10 in cell B7, and .05 in cell B9. The results are illustrated in Figure 7.3.Excel.

Table 7.3.Excel Design for testing the hypothesis for a proportion.

	A	B	C	D
1	One-Sample Tests of Hypothesis			
2				
3	Z Test for the Proportion			
4	n	xxx		
5	Number of Successes	xxx		
6	Sample Proportion	=B5/B4		
7	Null Hypothesis p=	.xxx		
8	Standard Error	=SQRT((B7*(1–B7))/B4)		
9	α	.xx		
10	Z Test Statistic	=(B6–B7)/B8		
11	Two-Tailed Test			
12	Lower Critical Value	=NORMSINV(B9/2)		
13	Upper Critical Value	=NORMSINV(1–B9/2)		
14	p-value	=2*(1–NORMSDIST(ABS(B10)))		
15	Decision	=IF(B14<B9,"Reject","Do not Reject")		
16	One-Tailed Test (Lower)			
17	Lower Critical Value	=NORMSINV(B9)		
18	p-value	=NORMSDIST(B10)		
19	Decision	=IF(B18<B9,"Reject","Do not Reject")		
20	One-Tailed Test (Upper)			
21	Upper Critical Value	=NORMSINV(1–B9)		
22	p-value	=1–NORMSDIST(B10)		
23	Decision	=IF(B22<B9,"Reject","Do not Reject")		

	A	B
1	One-Sample Tests of Hypothesis	
2		
3	Z Test for the Proportion	
4	n	200
5	Number of Successes	11
6	Sample Proportion	0.055
7	Null Hypothesis p=	0.1
8	Standard Error	0.021213203
9	α	0.05
10	Z Test Statistic	-2.12132034
11	Two-Tailed Test	
12	Lower Critical Value	-1.95996108
13	Upper Critical value	1.959961082
14	p-value	0.033894732
15	Decision	Reject
16	One-tailed Test (lower)	
17	Lower Critical Value	-1.644853
18	p-value	0.016947366
19	Decision	Reject
20	One-Tailed test (Upper)	
21	Upper Critical Value	1.644853
22	p-value	0.983052634
23	Decision	Do not reject

FIGURE 7.3.EXCEL One-sample Z test for the proportion using Microsoft Excel for the cereal filling-process example.

▲ WHAT IF EXAMPLE

The Calculations worksheet presented in Table 7.3.Excel allows us to study the effect of changes in the number of successes, n, and the level of confidence on the test of hypothesis for the proportion. For example, if we wanted to determine the effect of a change in the number of successes on the Z-test statistic, we could change the number of successes in cell B5 from 11 to 15. We observe that the sample proportion in cell B6 changes to .075, the Z-test statistic changes to -1.1785, the one-tailed p-value becomes .1193, and the null hypothesis is not rejected. As before, if we are interested in seeing the effects of many different changes, we could use the Scenario Manager (see Section 1S.15) to store and use sets of alternative values of the number of successes, n, and the level of confidence.

7.12 POTENTIAL HYPOTHESIS-TESTING PITFALLS AND ETHICAL ISSUES

To this point we have studied the fundamental concepts of hypothesis-testing methodology. We have learned how it is used for analyzing differences between sample estimates (i.e., statistics) of hypothesized population characteristics (i.e., parameters) in order to make decisions about the underlying characteristics. We have also learned how to evaluate the risks involved in making these decisions.

When planning to carry out a test of hypothesis based on some designed experiment or research study under investigation, several questions need to be raised to ensure that proper methodology is used:

1. What is the goal of the experiment or research? Can it be translated into a null and alternative hypothesis?
2. Is the hypothesis test going to be two-tailed or one-tailed?
3. Can a random sample be drawn from the underlying population of interest?
4. What kind of *measurements* will be obtained from the sample? Are the sampled outcomes of the random variable going to be numerical or categorical?
5. At what significance level, or risk of committing a Type I error, should the hypothesis test be conducted?
6. Is the intended sample size large enough to achieve the desired power of the test for the level of significance chosen?
7. What statistical test procedure is to be used on the sampled data and why?
8. What kind of conclusions and interpretations can be drawn from the results of the hypothesis test?

Questions like these need to be raised and answered in the planning stage of a survey or designed experiment, so a person with substantial statistical training should be consulted and involved early in the process. All too often such an individual is consulted far too late in the process, after the data have been collected. Typically in such a situation, all that can be done at such a late stage is to choose the statistical test procedure that would be best for the obtained data. We are forced to assume that certain biases that have been built into the study (because of poor planning) are negligible. But this is a large assumption. Good research involves good planning. To avoid biases, adequate controls must be built in from the beginning.

We need to distinguish between what is poor research methodology and what is unethical behavior. Ethical considerations arise when a researcher is manipulative of the hypothesis-testing process. The following are some of the ethical issues that arise when dealing with hypothesis-testing methodology:

- Data collection method—randomization
- Informed consent from human subjects being "treated"
- Type of test—two-tailed or one-tailed
- Choice of level of significance α
- Data snooping
- Cleansing and discarding of data
- Reporting of findings

- **Data Collection Method—Randomization** To eliminate the possibility of potential biases in the results, we must use proper data collection methods. To be able to draw meaningful conclusions, the data we obtain must be the outcomes of a random sample from some underlying population or the outcomes from some experiment in which a **randomization** process was employed. Potential subjects should not be permitted to self-select for a study. In a similar manner, a researcher should not be permitted to purposely select the subjects for the study. Aside from the potential ethical issues that may be raised, such a lack of randomization can result in serious coverage errors or selection biases and destroy the value of any study.

- **Informed Consent from Human Subjects Being "Treated"** Ethical considerations require that any individual who is to be subjected to some "treatment" in an experiment be apprised of the research endeavor and any potential behavioral or physical side effects and provide informed consent with respect to participation. A researcher is not permitted to dupe or manipulate the subjects in a study.

- **Type of Test—Two-tailed or One-tailed** If we have prior information that leads us to test the null hypothesis against a specifically directed alternative, then a one-tailed test will be more powerful than a two-tailed test. On the other hand, we should realize that if we are interested only in *differences* from the null hypothesis, not in the *direction* of the difference, the two-tailed test is the appropriate procedure to utilize. This is an important point. For example, if previous research and statistical testing have already established the difference in a particular direction, or if an established scientific theory states that it is only possible for results to occur in one direction, then a one-tailed or directional test may be employed. However, these conditions are not often satisfied in practice, and it is recommended that one-tailed tests be used cautiously. Using arguments based on ethical principles, Fleiss (see Reference 2) and other statisticians have stated that, in the overwhelming majority of research studies, a two-tailed test should be employed, particularly if the intention is to report the results to professional colleagues at meetings or in published journal articles. A major reason for this more conservative approach to testing is to enable us to draw more appropriate conclusions on data that may yield unexpected, counterintuitive results.

- **Choice of Level of Significance α** In a well-designed experiment or study, the **level of significance α** is selected in advance of data collection. One cannot be permitted to alter the level of significance, after the fact, to achieve a specific result. This would be **data snooping**. One answer to this issue of level of significance is to always report the *p*-value, not just the results of the test.

- **Data Snooping** Data snooping is never permissible. It would be unethical to perform a hypothesis test on a set of data, look at the results, and then select whether it should be two-tailed or one-tailed and/or choose the level of significance. These steps must be done first, as part of the planned experiment or study, before the data are collected, for the conclusions drawn to have meaning. In those situations in which a statistician is consulted by a researcher late in the process, with data already available, it is imperative that the null and alternative hypotheses be established and the level of significance chosen prior to carrying out the hypothesis test.

- **Cleansing and Discarding of Data** Data cleansing is not data snooping. Data cleansing is an important part of an overall analysis—remember GIGO (garbage in, garbage out). In the data preparation stage of editing, coding, and transcribing, one has an opportunity to review the data for any observation whose measurement seems to be extreme or unusual. After this has been done, the outcomes of the numerical variables in the data set should be organized into stem-and-leaf displays and box-and-whisker plots in preparation for further data presentation and *confirmatory analysis*. This *exploratory data analysis* stage gives us another opportunity to cleanse the data set by flagging outlier observations that need to be checked against the original data. In addition, the exploratory data analysis enables us to examine the data graphically with respect to the assumptions underlying a particular hypothesis test procedure needed for confirmatory analysis.

    The process of data cleansing raises a major ethical question. Should an observation be removed from a study? The answer is a qualified "yes." If it can be determined that a measurement is incomplete or grossly in error owing to some equipment problem or unusual behavioral occurrence unrelated to the study, a decision to discard the observation may be made. Sometimes there is no choice—an individual may decide to quit a particular study he has been participating in before a final measurement can be made. In a well-designed experiment or study, the researcher would plan, in advance, decision rules regarding the possible discarding of data.

- **Reporting of Findings**   When conducting research, it is vitally important to document both good and bad results so that individuals who follow up on such research do not have to "reinvent the wheel." It would be inappropriate to report the results of hypothesis tests that show statistical significance but not those for which there was insufficient evidence in the findings.

- **Ethical Considerations: A Summary**   Again, when discussing ethical issues concerning the hypothesis-testing methodology, the key is *intent*. We must distinguish between poor confirmatory data analysis and unethical practice. Unethical behavior occurs when a researcher willfully causes a selection bias in data collection, manipulates the treatment of human subjects without informed consent, uses data snooping to select the type of test (two-tailed or one-tailed) and/or level of significance to his or her advantage, hides the facts by discarding observations that do not support a stated hypothesis, or fails to report pertinent findings.

7.13 HYPOTHESIS-TESTING METHODOLOGY: A REVIEW AND A PREVIEW

As observed in the chapter summary chart, this chapter presented the fundamental underpinnings of hypothesis-testing methodology. To be sure you understand the material covered in this chapter, you should be able to answer the following conceptual questions:

1. What is the difference between a null hypothesis (H_0) and an alternative hypothesis (H_1)?
2. What is the difference between a Type I and Type II error?
3. What is meant by the power of a test?
4. What is the difference between a one-tailed and a two-tailed test?
5. What is meant by a *p*-value?
6. How can a confidence interval estimate for the population mean provide conclusions to the corresponding hypothesis test for the population mean?
7. What is the step-by-step hypothesis testing methodology?
8. What are some of the ethical issues to be concerned with when performing a hypothesis test?

Check over the list of questions to see whether you know the answers and could (1) explain your answers to someone who did not read this chapter and (2) give reference to specific readings or examples that support your answer. Also, reread any of the sections that may have seemed unclear to see whether they make sense now.

In the two chapters that follow, we shall be building on the foundations of hypothesis testing that we have discussed here. We will present a set of procedures that may be employed to verify or confirm statistically the results of studies and experiments designed under a variety of conditions.

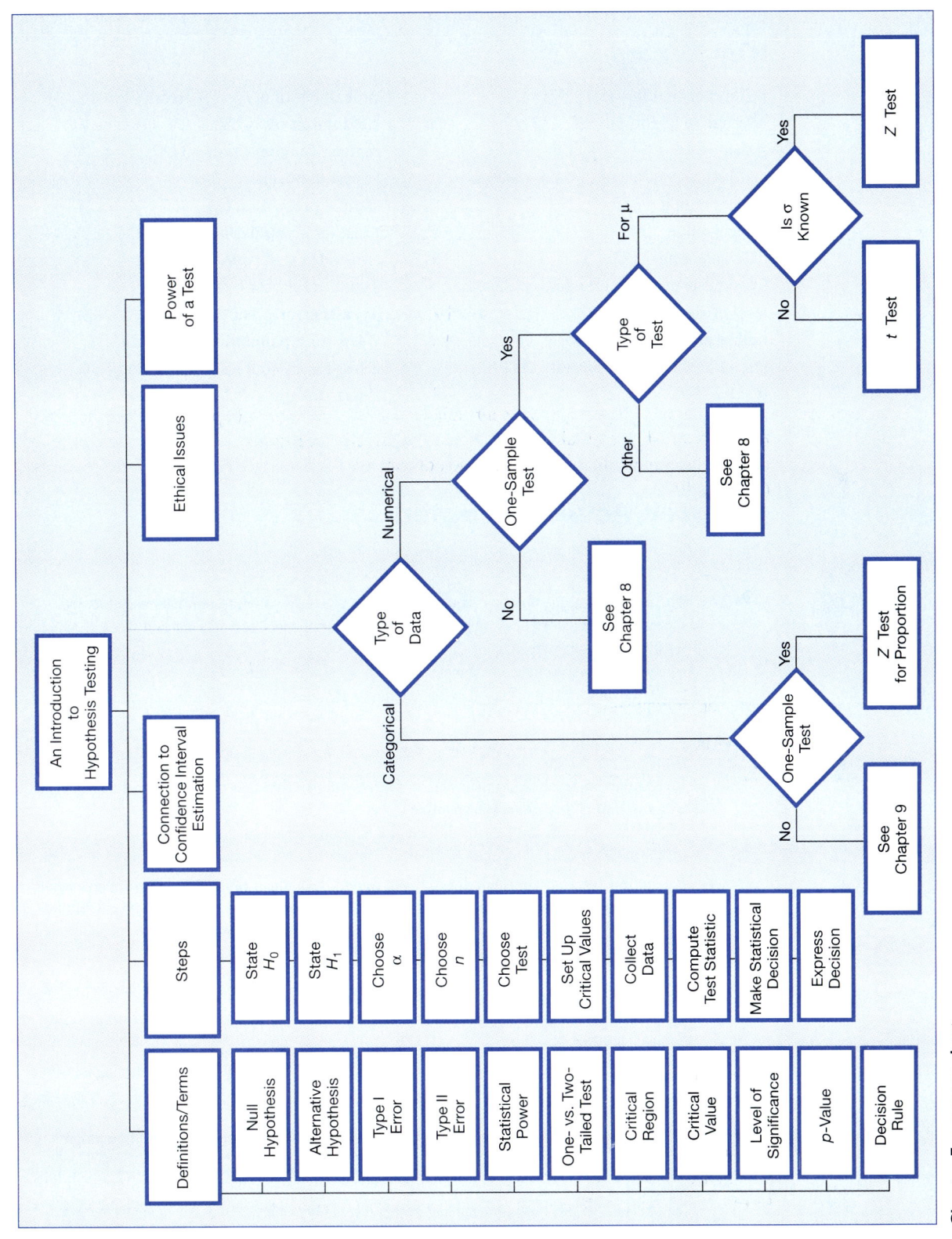

Chapter 7 summary chart.

Getting It All Together

Key Terms

α (level of significance)   373
alternative hypothesis (H_1)   341
β risk   343
confidence coefficient $(1 - \alpha)$ 343
critical region   342
critical value   343
data snooping   373
hypothesis-testing methodology   340
level of significance   373
null hypothesis (H_0)   340
one-tailed or directional test   352
p-value   349
power of a test $(1 - \beta)$ 344

probability of a Type II error (β)   343
randomization   372
region of nonrejection   342
region of rejection   342
sample proportion (p)   360
t test for a population mean   354
two-tailed test   346
Type I error   343
Type II error   343
Z test for a population mean   345
Z test for a population proportion   360
Z-test statistic   345

Chapter Review Problems

Note: *The Chapter Review Problems may be solved using Microsoft Excel.*

7.41 The owner of a gasoline service station wants to study gasoline purchasing habits by motorists at his station. A random sample of 60 motorists during a certain week is selected with the following results:
- Amount purchased: $\bar{X} = 11.3$ gallons, $S = 3.1$ gallons
- 11 motorists purchased super unleaded gasoline.

(a) At the .05 level of significance, is there evidence that the average purchase is different from 10 gallons?
(b) Find the p-value in (a).
(c) At the .05 level of significance, is there evidence that less than 20% of the motorists purchase super unleaded gasoline?
(d) What will your answer in (a) be if the sample average is 10.3 gallons?
(e) What would be your answer in (b) if 7 motorists purchased super unleaded gasoline?

7.42 An auditor for a government agency is assigned to the task of evaluating reimbursement for office visits to doctors paid by Medicare. The audit is to be conducted for all Medicare payments in a particular geographical area during a certain month. An audit is conducted on a sample of 75 of the reimbursements with the following results:
- In 12 of the office visits, an incorrect amount of reimbursement was provided.
- Amount of reimbursement: $\bar{X} = \$93.70$, $S = \$34.55$

(a) At the .05 level of significance, is there evidence that the average reimbursement is less than $100?
(b) At the .05 level of significance, is there evidence that the proportion of incorrect reimbursements in the population is greater than .10?
(c) What will your answer in (a) be if the sample average is $90?
(d) What would be your answer in (b) if 15 office visits had incorrect reimbursements?

7.43 Referring to the sample of houses selected by the real estate assessor in Problem 6.55 on page 335
(a) At the .01 level of significance, is there evidence that the average heated area of the house is less than 2,000 square feet?

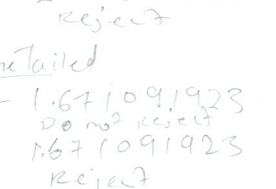

(b) At the .05 level of significance, is there evidence that the proportion of houses that have central air-conditioning is different from .50?
(c) Compare the results obtained in (a) and (b) to those of Problem 6.55.
(d) What will your answer in (a) be if the sample average is 1,959 square feet?
(e) What will your answer in (b) be if the level of significance is .10?

7.44 Referring to the sample of customers from the bookstore in Problem 6.58 on page 335
(a) At the .05 level of significance, is there evidence that the average amount spent is different from $30?
(b) At the .10 level of significance, is there evidence that the proportion of customers interested in purchasing the videotapes is greater than .20?
(c) Compare the results obtained in (a) and (b) to those of Problem 6.58.
(d) What would be your answer in (a) if you wanted to test whether the average amount spent was different from $25?
(e) What would be your answer in (b) if 18 customers said they would consider purchasing educational videotapes?

7.45 Referring to the sample of customers of the pet supply store in Problem 6.59 on page 336
(a) At the .05 level of significance, is there evidence that the average amount spent is different from $20?
(b) At the .10 level of significance, is there evidence that the proportion of customers who own only a cat is different from .15?
(c) Compare the results obtained in (a) and (b) to those of Problem 6.59.
(d) What will your answer in (a) be if the sample standard deviation is $6.22?
(e) What will your answer in (b) be if you want to test whether the proportion who own only a cat is different from .25 and you use the .05 level of significance?

7.46 Referring to the sample of customers of the restaurant in Problem 6.60 on page 336
(a) At the .05 level of significance, is there evidence that the average amount spent is different from $35?
(b) At the .10 level of significance, is there evidence that the proportion of customers who purchase dessert is different from .25?
(c) Compare the results obtained in (a) and (b) to those of Problem 6.60.
(d) What will your answer in (a) be if the sample average is $36.27?
(e) What will your answer in (b) be if the number of customers who purchase dessert is 23?

Endnotes

1. An easy way to remember which probability goes with which type of error is to note that α is the first letter of the Greek alphabet and it is used to represent the probability of a Type I error. The letter β is the second letter of the Greek alphabet and is used to represent the probability of a Type II error. (If you have trouble remembering the Greek alphabet, note that the word *alphabet* tells you its first two letters.)

References

1. Berenson, M. L., and D. M. Levine, *Basic Business Statistics Concepts and Applications,* 6th ed. (Englewood Cliffs, NJ: Prentice Hall, 1996).
2. Fleiss, J. L., *Statistical Methods for Rates and Proportions,* 2d ed. (New York: Wiley, 1981).
3. *Microsoft Excel Version 7* (Redmond, WA: Microsoft Corporation, 1996).

chapter 8

Two-Sample and c-Sample Tests with Numerical Data

CHAPTER OBJECTIVE To extend the basic principles of hypothesis-testing methodology to two-sample and c-sample tests involving numerical data.

8.1 INTRODUCTION

In Chapter 7, we introduced the fundamental concepts of hypothesis-testing methodology. When dealing with a single sample of numerical data, we used a Z test or a t test to determine whether the population mean μ is equal to some specified (i.e., hypothesized) value and also used the Z test to determine whether the population proportion p is equal to some specified (i.e., hypothesized) value.

In this chapter we extend the basic principles of hypothesis-testing methodology to the commonly used two-sample and c-sample tests involving numerical data.

8.2 CHOOSING THE APPROPRIATE TEST PROCEDURE

Recall from Section 7.4 that when summarizing the steps involved in hypothesis-testing methodology (see pages 347–348), a major consideration is the selection of the appropriate statistical technique and its corresponding test statistic. Part of a good data analysis is to understand the assumptions underlying each of the hypothesis-testing procedures we encounter and to select the one most appropriate for a given set of conditions. As we shall see in this chapter, all hypothesis-testing procedures can be broadly described as either *parametric* or *nonparametric*.

8.2.1 Parametric Procedures

In Chapter 7, we used parametric procedures, the Z test [given by Equation (7.1)] and the t test [given by Equation (7.2)], to test a hypothesis about a population mean. In this chapter we will examine three additional parametric procedures, the t test for the difference in the means of two independent populations, the F test for the difference between two variances, and the F test for the difference in the means of c groups. All parametric procedures have three distinguishing characteristics:

> **Parametric test procedures** may be defined as those that (1) require the level of measurement attained on the collected data to be in the form of an *interval* scale or *ratio* scale (see Section 1.8), (2) involve hypothesis testing of specified parameter values (such as $\mu = 368$ grams), and (3) require a stringent set of assumptions.

8.2.2 Nonparametric Procedures

We must choose other testing procedures when

1. The measurements attained on the data are only categorical (i.e., *nominally* scaled) or in ranks (i.e., *ordinally* scaled).
2. The assumptions underlying the use of the parametric methods cannot be met.

When parametric methods of hypothesis testing are not applicable, as in such circumstances as these, nonparametric methods of hypothesis testing are an alternative. (References 1 and 3).

> **Nonparametric test procedures** may be broadly defined as either (1) those whose test statistic does not depend on the form of the underlying population distribution from which the sample data were drawn, or (2) those for which the data are of insufficient strength (i.e., *nominally* scaled or *ordinally* scaled) to warrant meaningful arithmetic operations.

There are five major advantages to using nonparametric procedures:

1. They may be used on all types of data—categorical data (nominal scaling), data in rank form (ordinal scaling), as well as data that have been measured more precisely (interval or ratio scaling).
2. They are generally easy to apply and quick to compute when the sample sizes are small. Sometimes they are as simple as counting how often some feature appears in the data.
3. They make fewer, less stringent assumptions (which are more easily met) than do the parametric procedures. Hence, they enjoy wider applicability and yield a more general, broad-based set of conclusions.
4. Nonparametric methods permit the solution of problems that do not involve the testing of population parameters.
5. Depending on the particular procedure selected, nonparametric methods may be equally (or almost) as powerful as the corresponding parametric procedure when the assumptions of the latter are met and, when they are not met, may be more powerful.

Although nonparametric methods may be advantageously employed in a variety of situations, they have three major shortcomings:

1. It is disadvantageous to use nonparametric methods when all the assumptions of the parametric procedures can be met.
2. As the sample size gets larger, data manipulations required for nonparametric procedures are sometimes laborious unless appropriate computer software is available.
3. Often special tables of critical values are needed for the test statistics obtained by using nonparametric procedures, and these are not as readily available as are the tables of critical values needed for the test statistics obtained by using parametric procedures (Z, t, F and χ^2).

8.2.3 Importance of Assumptions in Test Selection

The sensitivity of parametric procedures to violations in the assumptions has been considered in the statistical literature (References 2 and 3). Some parametric test procedures are said to be **robust** because they are relatively insensitive to slight violations in the assumptions. However, with gross violations in the assumptions, both the true level of significance (α) and the power of a test ($1 - \beta$) may differ sharply from what otherwise would be expected. In such cases, a parametric test would be invalid, and a nonparametric procedure should be selected instead.

On the other hand, it is disadvantageous to use a nonparametric procedure when all the assumptions of the corresponding parametric test can be met. Unless a parametric procedure is employed in these instances, we would not be taking full advantage of the data. Information is lost when we convert such collected data (from an interval or ratio scale) to either ranks (ordinal scale) or categories (nominal scale). In particular, in such circumstances, some very quick and simple nonparametric tests have much less power than their corresponding parametric procedures and should usually be avoided.

As we investigate several hypothesis-testing procedures in this chapter, we will see how part of a good data analysis is to understand the assumptions underlying each of the procedures we shall be encountering and to select the one most appropriate for a given set of conditions.

8.2.4 Choosing the Appropriate Test Procedure When Comparing Two Independent Samples

Over the years, many statistical test procedures have been developed that enable us to make comparisons and examine differences between two groups based on independent samples containing numerical data. Thus, an important issue faced by anyone involved in hypothesis testing is the criteria to be used for selection of a particular statistical procedure from among the many that are available. Part of a good data analysis is to understand the assumptions underlying each of the hypothesis-testing techniques and to select the one most appropriate for a given set of conditions. Other criteria for test selection involve the simplicity of the procedure, the generalizability of the conclusions to be drawn, the accessibility of tables of critical values for the test statistic, the availability of computer software packages that contain the test procedure, and last, but certainly not least, the statistical power of the procedure.

In each of the following two sections, we shall describe a hypothesis-testing procedure that examines differences between two independent groups based on samples containing numerical data.

8.3 POOLED-VARIANCE t TEST FOR DIFFERENCES IN TWO MEANS

8.3.1 Introduction

Let us first extend the hypothesis-testing concepts developed in Chapter 7 to situations in which we would like to determine whether there is any difference between the means of two independent populations. Suppose we consider two independent populations, each having a mean and standard deviation, symbolically represented as follows:

Population 1	Population 2
μ_1, σ_1	μ_2, σ_2

Let us also suppose that a random sample of size n_1 is taken from the first population and a random sample of size n_2 is drawn from the second population and, further, that the data collected in each sample pertain to some numerical random variable of interest.

The test statistic used to determine the difference between the population means is based on the difference between the sample means $(\bar{X}_1 - \bar{X}_2)$. Because of the central limit theorem discussed in Section 5.9, this test statistic will follow the standard normal distribution for large enough sample sizes. The **Z-test** statistic is

$$Z = \frac{(\bar{X}_1 - \bar{X}_2) - (\mu_1 - \mu_2)}{\sqrt{\dfrac{\sigma_1^2}{n_1} + \dfrac{\sigma_2^2}{n_2}}} \qquad (8.1)$$

where
$\bar{X}_1$ = mean from the sample taken from population 1
μ_1 = mean of population 1
σ_1^2 = variance of population 1
n_1 = size of the sample taken from population 1
$\bar{X}_2$ = mean from the sample taken from population 2
μ_2 = mean of population 2

σ_2^2 = variance of population 2

n_2 = size of the sample taken from population 2

8.3.2 Developing the Pooled-Variance *t* Test

However, as we mentioned previously, in most cases we do not know the actual standard deviation of either of the two populations. The only information usually obtainable is the sample means ($\bar{X}_1$ and $\bar{X}_2$) and sample standard deviations (S_1 and S_2). If the assumptions made are that the samples are randomly and independently drawn from normally distributed populations and, further, that the *population variances are equal* (i.e., $\sigma_1^2 = \sigma_2^2$), a *pooled-variance t test* can be used to determine whether there is a significant difference between the means of the two populations.

The test to be performed can be either two-tailed or one-tailed, depending on whether we are testing if the two population means are merely *different* or if one mean is *greater than* the other mean.

Two-Tailed Test	One-Tailed Test	One-Tailed Test
$H_0: \mu_1 = \mu_2$ or $\mu_1 - \mu_2 = 0$	$H_0: \mu_1 \geq \mu_2$ or $\mu_1 - \mu_2 \geq 0$	$H_0: \mu_1 \leq \mu_2$ or $\mu_1 - \mu_2 \leq 0$
$H_1: \mu_1 \neq \mu_2$ or $\mu_1 - \mu_2 \neq 0$	$H_1: \mu_1 < \mu_2$ or $\mu_1 - \mu_2 < 0$	$H_1: \mu_1 > \mu_2$ or $\mu_1 - \mu_2 > 0$

where μ_1 = mean of population 1
μ_2 = mean of population 2

To test the null hypothesis of no difference in the means of two independent populations

$$H_0: \mu_1 = \mu_2$$

against the alternative that the means are not the same

$$H_1: \mu_1 \neq \mu_2$$

the following **pooled-variance *t*-test** statistic can be computed:

$$t = \frac{(\bar{X}_1 - \bar{X}_2) - (\mu_1 - \mu_2)}{\sqrt{S_p^2 \left(\frac{1}{n_1} + \frac{1}{n_2}\right)}} \qquad (8.2)$$

where

$$S_p^2 = \frac{(n_1 - 1)S_1^2 + (n_2 - 1)S_2^2}{(n_1 - 1) + (n_2 - 1)}$$

and

S_p^2 = pooled variance

$\bar{X}_1$ = mean of the sample taken from population 1

S_1^2 = variance of the sample taken from population 1

n_1 = size of the sample taken from population 1

$\bar{X}_2$ = mean of the sample taken from population 2

S_2^2 = variance of the sample taken from population 2

n_2 = size of the sample taken from population 2

From Equation (8.2), we may observe that the pooled-variance t test gets its name because the test statistic requires the pooling or combining of the two sample variances S_1^2 and S_2^2 to obtain S_p^2, the best estimate of the variance common to both populations under the assumption that the two population variances are equal.

The pooled-variance t-test statistic follows a t distribution with $n_1 + n_2 - 2$ degrees of freedom. For a given level of significance, α, we may reject the null hypothesis if the computed t-test statistic exceeds the upper-tailed critical value $t_{n_1+n_2-2}$ from the t distribution or if the computed test statistic falls below the lower-tailed critical value $-t_{n_1+n_2-2}$ from the t distribution. This means that the decision rule is

$$\text{Reject } H_0 \text{ if } t > t_{n_1+n_2-2}$$

$$\text{or if } t < -t_{n_1+n_2-2};$$

otherwise do not reject H_0.

The decision rule and regions of rejection are displayed in Figure 8.1.

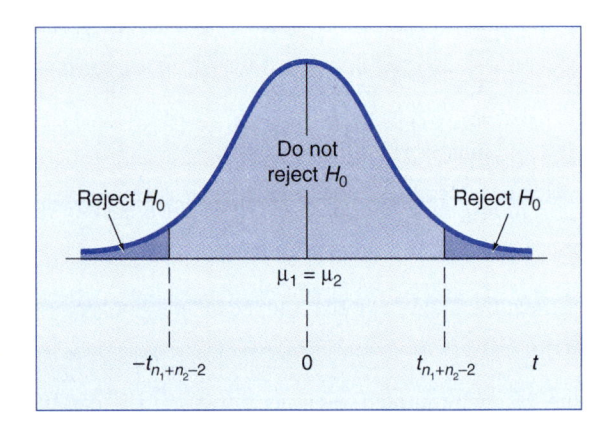

FIGURE 8.1 Rejection regions for a two-tailed test for the difference between two means.

8.3.3 Application

To demonstrate the use of the pooled-variance t test, suppose a financial analyst wishes to compare the average dividend yields of stocks traded on the New York Stock Exchange (NYSE) with those traded "over the counter" on the NASDAQ national market listing. Random samples of 21 companies from the NYSE and 25 stocks from the NASDAQ national market listing are selected, and the results are presented in Table 8.1.

If the analyst wishes to determine whether there is evidence of a difference in average dividend yield between the two populations of stock listings, the null and alternative hypotheses would be

$$H_0: \mu_1 = \mu_2 \text{ or } \mu_1 - \mu_2 = 0$$
$$H_1: \mu_1 \neq \mu_2 \text{ or } \mu_1 - \mu_2 \neq 0$$

Assuming that the samples are taken from underlying normal populations with equal variances, the pooled-variance t test can be used. If the test were conducted at the $\alpha = .05$ level of significance, the t-test statistic would follow a t distribution with $21 + 25 - 2 = 44$ degrees of freedom. From Table E.3 of Appendix E, the critical values for this two-tailed test are $+2.0154$ and -2.0154 and, as depicted in Figure 8.2, the decision rule is

Table 8.1 Comparing dividend yields* of selected issues from the NYSE and the NASDAQ national market listing.

DIVYIELD.TXT

NYSE Listing ($n_1 = 21$)		NASDAQ Listing ($n_2 = 25$)	
Company	Dividend Yield	Company	Dividend Yield
American Express	3.4	Atlantic SE Airlines	1.2
Anheuser-Busch	2.7	Boral Ltd.	5.1
Bristol-Myers-Squibb	5.4	Cathay Bancorp	4.3
Dayton-Hudson	2.1	Cit Fed Bancorp	0.8
Dresser Industries	3.0	CPB	3.2
Ford Motor	3.1	First Essex Bancorp	3.0
General Electric	3.0	Goulds Pumps	3.8
General Mills	3.5	Harper Group	1.3
IBM	1.6	Innovex	2.2
Kellogg Co.	2.6	Intel Corp.	0.4
Merck & Co.	3.6	Lindberg Corp.	2.7
NYNEX	6.4	Nature's Sunshine Prod.	1.5
Occidental Petroleum	5.3	Newcor	2.1
Pfizer Inc.	3.0	PCA International	3.3
PPG Inc.	3.0	T Rowe Price Assoc.	1.8
Sara Lee Corp.	2.9	PSB Holdings Corp.	2.4
Texaco Inc.	5.0	Research Inc.	4.6
Texas Instruments	0.9	Seacoast Banking Corp.	2.8
Whirlpool Corp.	2.2	Span-America Med. Sys.	1.8
Winn-Dixie	3.1	Sumitomo Bank of Cal.	3.6
Xerox Corp.	2.9	TCA Cable TV	2.2
		United Fire & Casualty	2.8
		West Coast Bancorp	1.7
		Whitney Holding Corp.	2.6
		Worthington Industries	2.1

*Dividend yield is the ratio of the annual dividend per share to the closing price per share, expressed as a percentage.

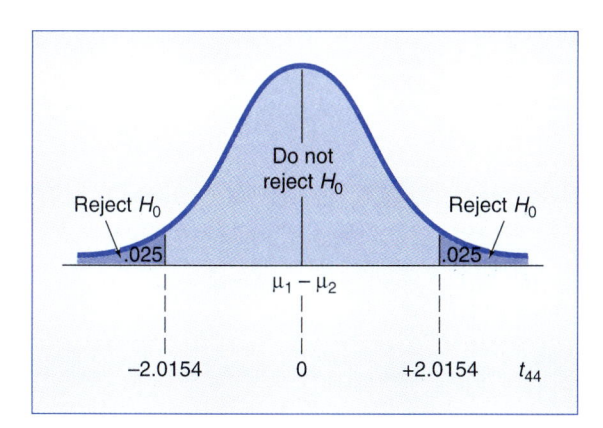

FIGURE 8.2 Two-tailed test of hypothesis for the difference between the means at the .05 level of significance.

Reject H_0 if $t > t_{44} = +2.0154$

or if $t < -t_{44} = -2.0154$;

otherwise do not reject H_0.

Using the data in Table 8.1, a set of summary statistics are computed and displayed in Table 8.2.

Table 8.2 Some summary statistics on dividend yields.

NYSE Listing	NASDAQ Listing
$n_1 = 21$	$n_2 = 25$
$\overline{X}_1 = 3.27$	$\overline{X}_2 = 2.53$
$S_1^2 = 1.698$	$S_2^2 = 1.353$
$S_1 = 1.30$	$S_2 = 1.16$
$X_{smallest_1} = 0.9$	$X_{smallest_2} = 0.4$
$Q_{1_1} = 2.65$	$Q_{1_2} = 1.75$
$Median_1 = 3.0$	$Median_2 = 2.4$
$Q_{3_1} = 3.55$	$Q_{3_2} = 3.25$
$X_{largest_1} = 6.4$	$X_{largest_2} = 5.1$

For our data we have

$$t = \frac{(\overline{X}_1 - \overline{X}_2) - (\mu_1 - \mu_2)}{\sqrt{S_p^2\left(\frac{1}{n_1} + \frac{1}{n_2}\right)}}$$

where

$$S_p^2 = \frac{(n_1 - 1)S_1^2 + (n_2 - 1)S_2^2}{(n_1 - 1) + (n_2 - 1)}$$

$$= \frac{20(1.30)^2 + 24(1.16)^2}{21 + 25 - 2}$$

$$= \frac{66.432}{44}$$

$$= 1.510$$

and therefore

$$t = \frac{3.27 - 2.53}{\sqrt{1.510\left(\frac{1}{21} + \frac{1}{25}\right)}}$$

$$= \frac{0.74}{\sqrt{0.132}}$$

$$= \frac{0.74}{0.364}$$

$$= 2.03$$

Using a .05 level of significance, the null hypothesis (H_0) is rejected because $t = +2.03 > t_{44} = +2.0154$. If the null hypothesis were true, there would be an $\alpha = .05$ probability of obtaining a t-test statistic either larger than $+2.0154$ standard deviations or smaller than -2.0154 standard deviations from the center of the t distribution. The p-value, or probability of obtaining a difference between the two sample means even larger than the 0.74 observed here, which translates to a test statistic t with a distance even farther from the center of the t distribution than ±2.03 standard deviations, would be slightly less than .05 if the null hypothesis of no difference in the means were true. [Using Excel, (see Section 8.3.5), the p-value is actually computed as .048.] Since the p-value is less than α, we have sufficient evidence that the null hypothesis is not true and we reject it.

The null hypothesis is rejected because the test statistic t has fallen into the region of rejection. The financial analyst may conclude that there is evidence of a difference in the mean dividend yields for the two groups. Companies traded on the NYSE appear to have significantly higher dividend yields than do companies trading over the counter on the NASDAQ national market listing.

We should note that in our analyst's study, the two groups have unequal sample sizes. When the two sample sizes are equal (i.e., $n_1 = n_2$), the formula for the pooled variance can be simplified to

$$S_p^2 = \frac{S_1^2 + S_2^2}{2}$$

8.3.4 Summary

In testing for the difference between the means, we have assumed that we are sampling from normally distributed populations with equal variances. We must examine the consequences for the pooled-variance t test of departures from each of these assumptions. For situations in which we cannot or do not wish to make the assumption that the two populations with equal variances are actually normally distributed, the pooled-variance t test is robust (i.e., not sensitive) to moderate departures from the assumption of normality, provided the sample sizes are large. In such situations, the pooled-variance t test may be used without serious effect on its power. On the other hand, if sample sizes are small and we cannot or do not wish to make the assumption of normally distributed populations, two choices exist. Either some *normalizing transformation* (see References 3 and 9) on each of the outcomes can be made and the pooled-variance t test used, or some nonparametric procedure such as the *Wilcoxon rank sum test* (to be covered in Section 8.4), which does not depend on the assumption of normality for the two populations, can be performed.

For situations in which we cannot or do not wish to make the assumption that the two normally distributed populations have equal population variances, the *Behrens-Fisher problem* is said to exist (see Reference 1), and a *separate-variance t´ test*, developed by Satterthwaite (see Reference 10), may be used.

8.3.5 Using Microsoft Excel for the Pooled-Variance *t* Test

In this section we developed the pooled-variance t test to determine whether any differences exist between the means of two groups. We will now explain how Microsoft Excel can be used for this procedure. If raw data are available, the *t* Test: Two-Sample Assuming Equal Variances option of the Data Analysis tool is used. If raw data are not available, Excel formulas can be used to obtain the t-test statistic and to determine whether there are significant differences between the means of two groups.

- **Using the Data Analysis Tool for the Two-Sample *t* Test Assuming Equal Variances** If raw data are available, the *t* Test: Two-Sample Assuming Equal Variances option of the Data Analysis tool may be used to test for the difference in the means of two groups. To illustrate the use of this tool for the *t* test, we will refer to the dividend yield data of Table 8.1 on page 385. With the YIELD-1.XLS workbook open,

8-3-5.XLS

❶ Click on the Data sheet tab. You will note that the data are set up in columns A and B with the column headings NYSE and NASDAQ.

❷ Select Tools | Data Analysis. Select t-Test: Two-Sample Assuming Equal Variances from the Analysis tools list box. Click the OK button.

❸ (a) In the Variable 1 Range edit box, enter A1:A22.
   (b) In the Variable 2 Range edit box, enter B1:B26.
   (c) In the Hypothesized Mean Difference edit box, enter 0.
   (d) Select the labels check box.
   (e) In the Alpha edit box, enter the level of significance (for these data enter .05).

❹ Select the New Worksheet Ply option button and enter the name T-Test for Two Means. The dialog box should now resemble the one illustrated in Figure 8.1.Excel. Click the OK button.

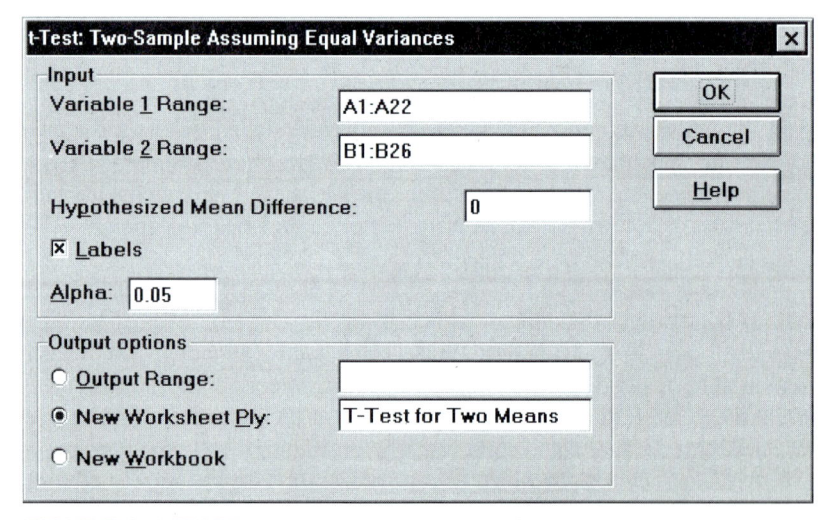

FIGURE 8.1.EXCEL Dialog box for Microsoft Excel t Test: Two-Sample Assuming Equal Variances Data Analysis option.

Figure 8.2.Excel represents the output obtained from Excel for the dividend yield data of page 385.

From Figure 8.2.Excel, we observe that Excel has computed the means, variances, degrees of freedom, the *t*-test statistic, the critical values, and the *p*-values. The *t*-test statistic equal to 2.0335 is greater than the two-tailed critical value of 2.015367, so the null hypothesis may be rejected. Alternatively, since the *p*-value for the two-tailed test equals .048, which is less than .05, the null hypothesis may be rejected.

- **Working with Stacked Data** To use the Data Analysis tool for the *t* test for the difference between two means (and also for some other procedures), the data for each of the two groups must be placed in separate columns. This arrangement is often referred to as an

	A	B	C
1	t-Test: Two-Sample Assuming Equal Variances		
2			
3		NYSE	NASDAQ
4	Mean	3.271429	2.532
5	Variance	1.697143	1.352267
6	Observations	21	25
7	Pooled Variance	1.509029	
8	Hypothesized Mean Difference	0	
9	df	44	
10	t Stat	2.033519	
11	P(T<=t) one-tail	0.024029	
12	t Critical one-tail	1.68023	
13	P(T<=t) two-tail	0.048058	
14	t Critical two-tail	2.015367	

FIGURE 8.2.EXCEL Output obtained from t Test: Two Sample Assuming Equal Variances for the dividend yield data.

unstacked data format. Sometimes data are organized in a *stacked* format, in which the response variable (e.g., dividend yield) for both groups is in a single column, and the grouping variable (e.g., type of exchange) is in a different column. Such a format is shown in Table 8.1.Excel.

If the data were in stacked format the NASDAQ dividend yields would need to be moved to a new column and the first column, containing the exchange names, deleted.

Table 8.1.Excel Dividend yield data in stacked format.

Exchange	Dividend Yield
NYSE	3.4
NYSE	2.7
.	.
.	.
NYSE	2.9
NASDAQ	1.2
NASDAQ	5.1
.	.
.	.
NASDAQ	2.1

- **Using Microsoft Excel Formulas for the Two Sample t Test Assuming Equal Variances When Raw Data Are Not Available** It is sometimes the case that only summary data such as the mean and standard deviation for each of two groups are available. For such situations, Excel formulas can be used to compute test statistics, critical values, and *p*-values. The design for a worksheet named TwoGroups for this test is shown in Table 8.2.Excel.

Referring to Equation (8.2) on page 383, we see that to test the hypothesis for the mean, we need to have the sample mean $\bar{X}$, the sample standard deviation S, and the sample size n

Table 8.2.Excel Design for testing the differences between two means.

	A	B	C	D
1	t Test of Differences Between Two Means			
2				
3	Mean Group 1	xxx		
4	S Group 1	xxx		
5	n Group 1	xxx		
6	df Group 1	=B5 – 1		
7	Mean Group 2	xxx		
8	S Group 2	xxx		
9	n Group 2	xxx		
10	df Group 2	=B9 – 1		
11	Hypothesized Difference	xxx		
12	α	.xx		
13	Total df	=B6+B10		
14	Pooled Variance	=((B6*B4^2)+(B10*B8^2))/(B13)		
15	Difference in Sample Means	=B3–B7		
16	t-Test Statistic	=(B15–B11)/SQRT(B14*(1/B5+1/B9))		
17	Two-Tailed Test			
18	Lower Critical Value	=–(TINV(B12,B13))		
19	Upper Critical Value	=TINV(B12,B13)		
20	p-Value	=TDIST(ABS(B16),B13,2)		
21	Decision	=IF(B20<B12,"Reject","Do Not Reject")		
22	One-Tailed Test (Lower)			
23	Lower Critical Value	=–(TINV(2*B12,B13))		
24	p-Value	=IF(B16<0,D25,D26)	IF calculations	
25	Decision	=IF(B24<B12,"Reject","Do Not Reject")	TDIST calculation	=TDIST(ABS(B16),B13,1)
26	One-Tailed Test (Upper)		1–TDIST calculation	=1–D25
27	Upper Critical Value	=(TINV(2*B12,B13))		
28	p-Value	=IF(B16<0,D26,D25)		
29	Decision	=IF(B28<B13,"Reject","Do Not Reject")		

for each of the two groups. If the arithmetic means and the sample standard deviations need to be computed, the Excel AVERAGE and STDEV functions could be used and the results copied to the ranges B3:B4 and B7:B8 respectively. With the YIELD-1.XLS workbook still open, insert a new worksheet and rename it TwoGroups. The steps for setting up the test of hypothesis for the difference between two means are as follows:

❶ For group 1, enter the value for the sample arithmetic mean in cell B3, the sample standard deviation S in cell B4, and the sample size n in cell B5. For group 2, enter the value for the sample arithmetic mean in cell B7, the sample standard deviation S in cell B8, and the sample size n in cell B9. Using the dividend yield example on page 388, if we were going to use Excel formulas instead of the Data Analysis tool, we would enter 3.27143 in cell B3, 1.30274 in cell B4, the sample size of 21 in cell B5, 2.532 in cell B7, 1.16287 in cell B8, and 25 in cell B9.

❷ The degrees of freedom for each of the two groups are computed in cells B6 (=B5–1) and B10 (=B9–1).

❸ The hypothesized difference between the population means (equal to 0 for the dividend yield example) is entered in cell B11.

④ The level of significance α is entered in cell B12 (.05 for the dividend yield problem).

⑤ The total degrees of freedom is computed in cell B13 by adding the contents of cells B6 and B10.

⑥ The pooled variance is computed in cell B14 using the formula
=((B6*B4^2)+(B10*B8^2))/B13.

⑦ The difference between the two sample means is computed in cell B15, and the test statistic is computed in cell B16 by entering the formula
=(B15−B11)/SQRT(B14*(1/B5+1/B9)).

⑧ The critical values, p-value, and decisions for one- and two-tailed tests are presented in the range B17:B29 and the range C24:D26. These follow the design used for the one-sample test for the mean developed in Section 7.11.2.

The results are illustrated in Figure 8.3.Excel.

	A	B	C	D
1	T-Test of Differences Between Two Means			
2				
3	Mean Group 1	3.27143		
4	S Group 1	1.30274		
5	n Group 1	21		
6	df Group 1	20		
7	Mean Group 2	2.532		
8	S Group 2	1.16287		
9	n Group 2	25		
10	df Group 2	24		
11	Hypothesized Difference	0		
12	α	0.05		
13	Total df	44		
14	Pooled Variance	1.509023396		
15	Difference in Sample Means	0.73943		
16	T-Test Statistic	2.033526405		
17	Two-Tailed Test			
18	Lower Critical Value	-2.0153675		
19	Upper Critical value	2.0153675		
20	p-value	0.048057653		
21	Decision	Reject		
22	One-tailed Test (Lower)			
23	Lower Critical Value	-1.68023007		
24	p-value	0.975971174	IF calculations	
25	Decision	Do not reject	TDIST calculation	0.024029
26	One-Tailed test (Upper)		1-TDIST calculation	0.975971
27	Upper Critical Value	1.680230071		
28	p-value	0.024028826		
29	Decision	Reject		

FIGURE 8.3.EXCEL t test for the differences between Two Means obtained from Microsoft Excel for the dividend yield data.

> ▲ **WHAT IF EXAMPLE**
>
> The TwoGroups worksheet presented in Table 8.2.Excel allows us to study the effect of changes in $\bar{X}$, S, n, and the level of confidence on the test of hypothesis for the differences between two means. For example, if we wanted to determine the effect of a change in one of the sample standard deviations on the test statistic, we could change the standard deviation in the first group from 1.30274 to 2.30274. We observe that the pooled variance changes to 3.1479, the t-test statistic changes to 1.408, the two-tailed p-value becomes .166, and the null hypothesis is not rejected. We also see that since there is a large difference in the two sample standard deviations, it is possible that the validity of the t test is seriously in doubt in this situation. As before, if we are interested in seeing the effects of many different changes, we could use the Scenario Manager (see Section 1S.15) to store and use sets of alternative values of $\bar{X}$, σ, n, and the level of confidence.

Problems for Section 8.3

Note: *The problems in this section can be solved using Microsoft Excel.*

- 8.1 The quality control manager at a light bulb factory would like to determine whether there is any difference in the average life of bulbs manufactured on two different types of machines. The process standard deviation is 110 hours for machine I and 125 hours for machine II. A random sample of 25 light bulbs obtained from machine I indicates a sample mean of 375 hours, and a similar sample of 25 from machine II indicates a sample mean of 362 hours. Using the .05 level of significance
    (a) Is there any evidence of a difference in the average life of bulbs produced by the two types of machines?
    (b) Compute the p-value in part (a) and interpret its meaning.

  8.2 The purchasing director for an industrial parts factory is investigating the possibility of purchasing a new type of milling machine. She has determined that the new machine will be bought if there is evidence that the parts produced have a higher average breaking strength than those from the old machine. The process standard deviation of the breaking strength is 10 kilograms for the old machine and 9 kilograms for the new machine. A sample of 100 parts taken from the old machine indicates a sample mean of 65 kilograms, whereas a similar sample of 100 from the new machine indicates a sample mean of 72 kilograms. Using the .01 level of significance
    (a) Is there evidence that the purchasing director should buy the new machine?
    (b) Compute the p-value in part (a) and interpret its meaning.

- 8.3 Management of the Sycamore Steel Company wishes to determine whether there is any difference in performance between the day shift of workers and the evening shift of workers. A sample of 100 day-shift workers reveals an average output of 74.3 parts per hour with a sample standard deviation of 16 parts per hour. A sample of 100 evening-shift workers reveals an average output of 69.7 parts per hour with a sample standard deviation of 18 parts per hour. At the .10 level of significance
    (a) Is there evidence of a difference in average output between the day shift and the evening shift?
    (b) Find the p-value in part (a) and interpret its meaning.

  8.4 An independent testing agency has been contracted to determine whether there is any difference in the gasoline mileage output of two different gasolines on the same model automobile. Gasoline A is tested on 200 cars and produces a sample average of 18.5 miles per gallon with a sample standard deviation of 4.6 miles per gallon. Gasoline B is tested on a sample of 100 cars and produces a sample average of 19.34 miles per gallon with a sample standard deviation of 5.2 miles per gallon. At the .05 level of significance

(a) Is there evidence of a difference in the average performance of the two gasolines?
(b) Find the *p*-value in part (a) and interpret its meaning.

8.5 A carpet manufacturer is studying differences between two of its major outlet stores. The company is particularly interested in the time it takes for customers to receive carpeting that has been ordered from the plant. Data concerning a sample of delivery times for the most popular type of carpet are summarized as follows:

	Store A	Store B
$\bar{X}$	34.3 days	43.7 days
S	2.4 days	3.1 days
n	41	31

(a) At the .01 level of significance, is there evidence of a difference in the average delivery time for the two outlet stores?
(b) Find the *p*-value in part (a) and interpret its meaning.

8.6 Suppose the manager of a pet supply store wants to determine whether there is a difference in the amount of money spent by owners of dogs and owners of cats. (Customers who own both were disregarded in this analysis.) The results for a sample of 37 dog owners and 26 cat owners are summarized as follows:

	Purchases for Dogs	Purchases for Cats
$\bar{X}$	$26.47	$19.16
S	$ 9.45	$ 8.52
n	37	26

(a) At the .05 level of significance, is there evidence of a difference in the average amount of money spent in the pet supply store between dog owners and cat owners?
(b) What assumptions must be made in order to do part (a) of this problem?
(c) Find the *p*-value in part (a) and interpret its meaning.

8.7 An industrial psychologist wishes to study the effects of motivation on sales in a particular firm. Of 24 new salespersons being trained, 12 are to be paid at an hourly rate and 12 on a commission basis. The 24 individuals are randomly assigned to the two groups. The following data represent the sales volume (in thousands of dollars) achieved during the first month on the job.

SALESVOL.TXT

Hourly Rate		Commission	
256	212	224	261
239	216	254	228
222	236	273	234
207	219	285	225
228	225	237	232
241	230	277	245

(a) Is there evidence that wage incentives (through commission) yield greater average sales volume? (Use $\alpha = .01$.)
(b) What assumptions must be made in order to do part (a) of this problem?
(c) Find the *p*-value in part (a) and interpret its meaning.

8.8 A manufacturer is developing a nickel-metal hydride battery to be used in cellular telephones instead of nickel-cadmium batteries. The director of quality control decides to evaluate the newly developed battery against the widely used nickel-cadmium battery with

NICKELBAT.TXT

respect to performance. A random sample of 25 nickel-cadmium batteries and a random sample of 25 of the newly developed nickel-metal hydride batteries are placed in cellular telephones of the same brand and model. The performance measure of interest is the talking time (in minutes) prior to recharging. The results are as follows:

Nickel-Cadmium Battery		Nickel-Metal Hydride Battery	
54.5	71.0	78.3	103.0
67.0	67.8	79.8	95.4
41.7	56.7	81.3	91.1
64.5	69.7	69.4	46.4
86.8	70.4	82.8	87.3
40.8	74.9	82.3	71.8
72.5	75.4	62.5	83.2
76.9	64.9	77.5	85.0
81.0	104.4	85.3	74.3
83.3	90.4	85.3	85.5
82.0	72.8	86.1	72.1
71.8	58.7	41.1	74.1
68.8		112.3	

(a) Is there evidence of a difference in the two types of batteries with respect to average talking time (in minutes) prior to recharging? (Use $\alpha = .05$.)
(b) What assumptions must be made in order to do part (a) of this problem?
(c) Find the *p*-value in part (a) and interpret its meaning.

8.9 The following data represent the annual effective yields, in percent, on money market accounts from a sample of 10 New York area commercial banks and from a sample of 10 New York area savings banks on a recent day:

NYBANKS.TXT

Commercial Banks	Yield	Savings Banks	Yield
Banco Popular	2.25	Anchor Savings	2.43
Bank of N.Y.	2.32	Apple Bank Savings	2.53
Chase Manhattan	2.02	Carteret Savings (N.J.)	2.38
Chemical	1.92	Crossland Savings	2.50
Citibank	2.02	Dime Savings Bank	3.00
EAB	1.82	Emigrant Savings	2.50
First Fidelity (N.J.)	2.10	First Fed (Rochester)	2.55
Marine Midland	2.38	Green Point Savings	3.20
Midlantic Bank (N.J.)	2.30	Home Savings Amer	2.50
Republic Nat'l	2.28	People's Bank (Conn.)	2.02

(a) Is there evidence of a difference in the average effective yields on money market accounts in the two types of banks in the New York area? (Use $\alpha = .05$.)
(b) What assumptions must be made in order to do part (a) of this problem?
(c) Find the *p*-value in part (a) and interpret its meaning.

8.4 WILCOXON RANK SUM TEST FOR DIFFERENCES IN TWO MEDIANS

8.4.1 Introduction

If sample sizes are small and we cannot or do not wish to make the assumption that the data in each group are taken from normally distributed populations, two choices exist. Either the pooled-variance t test or separate-variance t' test (see Reference 1) may be used, whichever is more appropriate, following some *normalizing transformation* on the data (see References 1 and 9), or some nonparametric procedure that does not depend on the assumption of normality for the two populations can be employed. In this section we introduce the **Wilcoxon rank sum test**, a widely used, very simple, and powerful nonparametric procedure for testing for differences between the medians of two populations. The Wilcoxon rank sum test has proven to be almost as powerful as its parametric counterpart (the t test) under conditions appropriate to the latter and is likely to be more powerful when the stringent assumptions of the t test are not met.

In addition, the Wilcoxon rank sum test is an excellent procedure to choose when only ordinal-type data can be obtained, as is often the case when dealing with studies in consumer behavior, marketing research, and experimental psychology. The parametric t test would not be used in such situations since these procedures require that the obtained data are measured on at least an interval scale.

8.4.2 Procedure

To perform the Wilcoxon rank sum test, we must replace the observations in the two samples of size n_1 and n_2 with their combined ranks (unless the obtained data contained the ranks initially). The ranks are assigned in such a manner that rank 1 is given to the smallest of the $n = n_1 + n_2$ combined observations, rank 2 is given to the second smallest, and so on until rank n is given to the largest. If several values are tied, we assign each the average of the ranks that would otherwise have been assigned.

For convenience, whenever the two sample sizes are unequal, we will let n_1 represent the *smaller*-sized sample and n_2 the *larger*-sized sample. The Wilcoxon rank sum test statistic T_1 is the sum of the ranks assigned to the n_1 observations in the smaller sample. (For equal-sized samples, either group may be selected for determining T_1.)

For any integer value n, the sum of the first n consecutive integers is equal to $n(n + 1)/2$. For example, $1 + 2 + 3 + 4 = 10 = 4(4 + 1)/2$. The test statistic T_1 plus the sum of the ranks assigned to the n_2 items in the second sample, T_2, must therefore be equal to this value; that is,

$$T_1 + T_2 = \frac{n(n + 1)}{2} \tag{8.3}$$

so that Equation (8.3) can serve as a check on the ranking procedure.

The test of the null hypothesis can be either two-tailed or one-tailed, depending on whether we are testing if the two population medians are merely *different* or if one median is *greater than* the other median.

	Two-Tailed Test	One-Tailed Test	One-Tailed Test
	$H_0: M_1 = M_2$	$H_0: M_1 \geq M_2$	$H_0: M_1 \leq M_2$
	$H_1: M_1 \neq M_2$	$H_1: M_1 < M_2$	$H_1: M_1 > M_2$

where M_1 = median of population 1 with n_1 sample observations
       M_2 = median of population 2 with n_2 sample observations

When the sizes of both samples n_1 and n_2 are less than or equal to 10, Table E.6 may be used to obtain the critical values of the test statistic T_1 for both one- and two-tailed tests at various levels of significance. For a two-tailed test and for a particular level of significance, α, if the computed value of T_1 equals or exceeds the upper critical value or is less than or equal to the lower critical value, the null hypothesis may be rejected. For one-tailed tests having the alternative $H_1: M_1 < M_2$, the decision rule is to reject the null hypothesis if the observed value of T_1 is less than or equal to the lower critical value. For one-tailed tests having the alternative $H_1: M_1 > M_2$ the decision rule is to reject the null hypothesis if the observed value of T_1 equals or exceeds the upper critical value.

To demonstrate how to use Table E.6 to obtain the critical values for the T_1 test statistic, let us suppose that the sample sizes for our two groups are 8 and 10 and we wish to choose a level of significance α of .05. From Table 8.3, which is a replica of Table E.6, if $n_1 = 8$, $n_2 = 10$, and $\alpha = .05$, we observe that the lower and upper critical values for a two-tailed test are, respectively, 53 and 99. If the computed value of the test statistic T_1 falls between these critical values, the null hypothesis could not be rejected. If, however, the computed value of the test statistic equals or exceeds 99 or is less than or equal to 53, the null hypothesis would be rejected.

For large sample sizes, the test statistic T_1 is approximately normally distributed. The following large-sample approximation formula may be used for testing the null hypothesis when sample sizes are outside the range of Table E.6:

$$Z = \frac{T_1 - \mu_{T_1}}{\sigma_{T_1}} \qquad (8.4)$$

Table 8.3 Obtaining the lower- and upper-tail critical values T_1 for the Wilcoxon rank sum test where $n_1 = 8$, $n_2 = 10$, and $\alpha = .05$.

		α				n_1			
			4	5	6	7	8	9	10
n_1	One-Tailed	Two-Tailed				(Lower, Upper)			
9	.025	.05	14,42	22,53	31,65	40,79	51,93	62,109	
	.01	.02	13,43	20,55	28,68	37,82	47,97	59,112	
	.005	.01	11,45	18,57	26,70	35,84	45,99	56,115	
	.05	.10	17,43	26,54	35,67	45,81	56,96	69,111	82,128
10	.025	.05	15,45	23,57	32,70	42,84	53,99	65,115	78,132
	.01	.02	13,47	21,59	29,73	39,87	49,103	61,119	74,136
	.005	.01	12,48	19,61	27,75	37,89	47,105	58,122	71,139

Source: Extracted from Table E.6.

where $n_1 \leq n_2$

T_1 = sum of the ranks assigned to the n_1 observations in sample 1

μ_{T_1} = mean value of T_1

σ_{T_1} = standard deviation of T_1

μ_{T_1}, the mean value of the test statistic T_1, can be computed from

$$\mu_{T_1} = \frac{n_1(n+1)}{2}$$

and σ_{T_1}, the standard deviation of T_1, is calculated from

$$\sigma_{T_1} = \sqrt{\frac{n_1 n_2 (n+1)}{12}}$$

so that Equation (8.4) may be rewritten as

$$Z = \frac{T_1 - \dfrac{n_1(n+1)}{2}}{\sqrt{\dfrac{n_1 n_2 (n+1)}{12}}} \qquad (8.5)$$

Based on α, the level of significance selected, the null hypothesis may be rejected if the computed Z value falls in the appropriate region of rejection, depending on whether a two-tailed or a one-tailed test is used (see Figure 8.3).

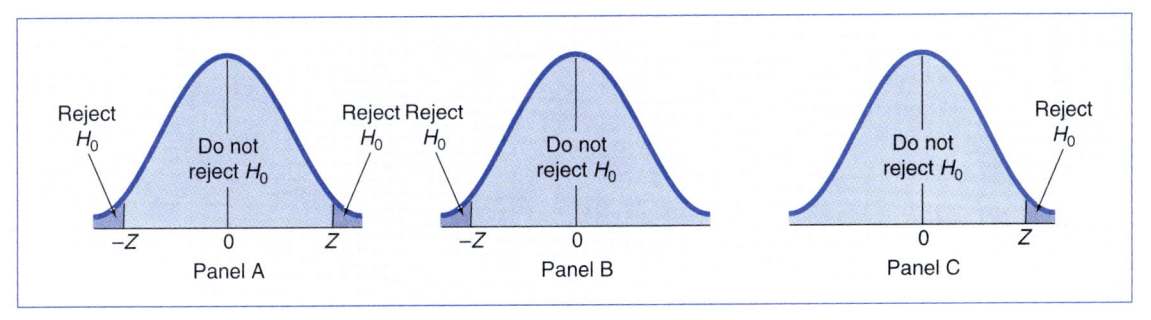

FIGURE 8.3 Determining the rejection region: Panel A, two-tailed test ($M_1 \neq M_2$); Panel B, one-tailed test ($M_1 < M_2$); Panel C, one-tailed test ($M_1 > M_2$).

8.4.3 Application

To demonstrate the use of the Wilcoxon rank sum test, let us refer to the problem faced by our financial analyst (see page 384) who wants to determine whether there is any difference in the average dividend yield of stocks traded on the NYSE versus those traded over the counter on the NASDAQ national market listing. Dividend yields for random samples of 21 issues from the NYSE and 25 issues from the NASDAQ national market listing are displayed in Table 8.1 (page 385), and summary statistics are presented in Table 8.2 (page 386).

If, as a result of an exploratory data analysis, the financial analyst does not wish to make the stringent assumption that the samples were taken from normally distributed populations, the Wilcoxon rank sum test may be used for evaluating possible differences in the median dividend yields.[1] Since the financial analyst is not specifying which of the two groups is likely to possess a greater median dividend yield, the test is two-tailed, and the following null and alternative hypotheses are established:

$$H_0: M_1 = M_2 \text{ (the median dividend yields are equal)}$$

$$H_1: M_1 \neq M_2 \text{ (the median dividend yields are different)}$$

To perform the Wilcoxon rank sum test, we form the combined ranking of the dividend yields obtained from the $n_1 = 21$ companies on the NYSE and the $n_2 = 25$ companies on the NASDAQ national market listing. The combined ranking of the dividend yields is displayed in Table 8.4. (We note that a rank of 1 is given to Intel, the company with the lowest dividend yield, and a rank of 46 is given to NYNEX, the company with the highest dividend yield in the combined ranking.)

We then obtain the test statistic T_1, the sum of the ranks assigned to the smaller sample:

$$T_1 = 35 + 20.5 + \cdots + 24.5 = 585.5$$

Table 8.4 Forming the combined ranks.

NYSE Listing ($n_1 = 21$)		NASDAQ Listing ($n_2 = 25$)	
Company	Dividend Yield Combined Rank	Company	Dividend Yield Combined Rank
American Express	35	Atlantic SE Airlines	4
Anheuser-Busch	20.5	Boral Ltd.	43
Bristol-Myers-Squibb	45	Cathay Bancorp	40
Dayton-Hudson	12	Cit Fed Bancorp	2
Dresser Industries	28	CPB	33
Ford Motor	31.5	First Essex Bancorp	28
General Electric	28	Goulds Pumps	39
General Mills	36	Harper Group	5
IBM	7	Innovex	15
Kellogg Co.	18.5	Intel Corp	1
Merck & Co.	37.5	Lindberg Corp	20.5
NYNEX	46	Nature's Sunshine Prod.	6
Occidental Petroleum	44	Newcor	12
Pfizer Inc.	28	PCA International	34
PPG Inc.	28	T Rowe Price Assoc.	9.5
Sara Lee Corp.	24.5	PSB Holdings Corp.	17
Texaco Inc.	42	Research Inc.	41
Texas Instruments	3	Seacoast Banking Corp.	22.5
Whirlpool Corp.	15	Span-America Med. Sys.	9.5
Winn-Dixie	31.5	Sumitomo Bank of Cal.	37.5
Xerox Corp.	24.5	TCA Cable TV	15
		United Fire & Casualty	22.5
		West Coast Bancorp	8
		Whitney Holdings	18.5
		Worthington Industries	12

Source: Data are taken from Table 8.1.

As a check on the ranking procedure we also obtain $T_2 = 495.5$ and use Equation (8.3) to show that the sum of the first $n = 46$ integers in the combined ranking is equal to $T_1 + T_2$:

$$T_1 + T_2 = \frac{n(n+1)}{2}$$

$$585.5 + 495.5 = \frac{46(47)}{2} = 1{,}081$$

To test the null hypothesis of no difference in the median dividend yields in the two populations, we use the large-sample approximation formula [Equation (8.4)]. Choosing the .05 level of significance, the critical values from the standard normal distribution (Table E.2) are ±1.96 (see Figure 8.4). The decision rule would be

Reject H_0 if $Z > +1.96$

or if $Z < -1.96$;

otherwise do not reject H_0.

Using Equation (8.5), we have

$$Z = \frac{T_1 - \dfrac{n_1(n+1)}{2}}{\sqrt{\dfrac{n_1 n_2 (n+1)}{12}}}$$

$$= \frac{585.5 - \dfrac{21(47)}{2}}{\sqrt{\dfrac{21(25)(47)}{12}}}$$

$$= \frac{585.5 - 493.5}{45.35}$$

$$= 2.03$$

Since $Z = +2.03 > +1.96$, the decision is to reject H_0. The p-value, or probability of obtaining a test statistic T_1 even larger than the 585.5 observed here, which translates to a test statistic Z with a distance even farther from the center of the standard normal distribution than ±2.03 standard deviations, is .0424 if the null hypothesis of no difference in the medians is true. Since the p-value is less than $\alpha = .05$, we have sufficient reason to reject the null hypothesis.

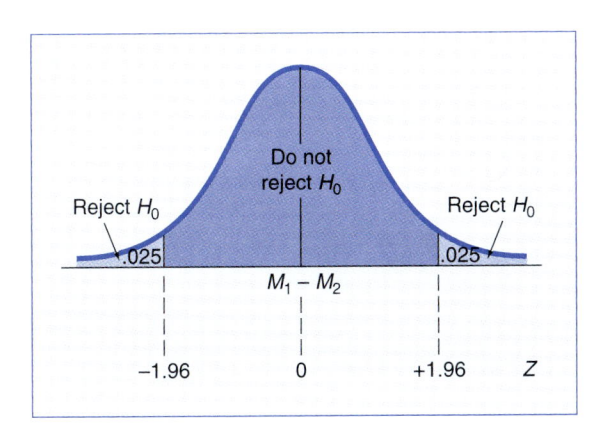

FIGURE 8.4 A two-tailed test of hypothesis for the difference in medians at the .05 level of significance.

8.4 Wilcoxon Rank Sum Test for Differences in Two Medians

The null hypothesis is rejected because the test statistic Z has fallen into the region of rejection. Thus, without having to make the stringent assumption of normality in the original populations, the financial analyst may conclude that there is evidence of a difference in the median dividend yields for the two groups. Companies traded on the NYSE appear to have significantly higher dividend yields than do the companies on the NASDAQ national market listing.

8.4.4 Using Microsoft Excel for the Wilcoxon Rank Sum Test

In this section we developed the Wilcoxon rank sum test as a nonparametric alternative to the pooled-variance t test for the dividend yield data. Although Excel does not have a Data Analysis tool for the Wilcoxon test, Excel formulas can be used instead. The design for the Calculations sheet for this test is shown in Table 8.3.Excel.

To implement this sheet, the ranks of each dividend yield must first be determined and added to the Data sheet. To do this, begin by opening the YIELD-2.XLS workbook file and then select the Data sheet. Note that the data are arranged in stacked order, a requirement for the steps that follow. (Had the data been unstacked as it is in the YIELD-1.XLS workbook

Table 8.3.Excel Design for obtaining the Wilcoxon test statistic.

8-4-4.XLS

	A	B
1	Wilcoxon Rank Sum Test	
2		
3	Sum of Ranks Group 1	=SUM(Data!E:E)
4	Sample Size Group 1	=COUNT(Data!E:E)
5	Sum of Ranks Group 2	=SUM(Data!F:F)
6	Sample Size Group 2	=COUNT(Data!F:F)
7	Total Sample Size n	=B4+B6
8	Total Sum of Ranks	=(B7*(B7+1))/2
9	Total Sums Check	=IF(B8=B3+B5,"Passed","Failed--review manually assigned ranks.")
10	Large-Sample Approximation	=IF(AND(B4<=10,B6<=10),"Invalid--Use Table E6. Do not use this sheet.","Valid--Use calculations on this sheet.")
11	α	0.05
12	T_1	=IF(B4<=B6,B3,B5)
13	Mean of T_1	=IF(B4<=B6,B4*(B7+1)/2,B6*(B7+1)/2)
14	Standard Error	=SQRT(B4*B6*(B7+1)/12)
15	Z-Test Statistic	=(B12–B13)/B14
16	Two-Tailed Test	
17	Lower Critical Value	=NORMSINV(B11/2)
18	Upper Critical Value	=NORMSINV(1–B11/2)
19	p-Value	=2*(1–NORMSDIST(ABS(B15)))
20	Decision	=IF(B19<B11,"Reject","Do Not Reject")
21	One-Tailed Test (Lower)	
22	Lower Critical Value	=NORMSINV(B11)
23	p-Value	=NORMSDIST(B15)
24	Decision	=IF(B23<B11,"Reject","Do not Reject")
25	One-Tailed Test (Upper)	
26	Upper Critical Value	=NORMSINV(1–B11)
27	p-Value	=1–NORMSDIST(B15)
28	Decision	=IF(B27<B11,"Reject","Do not Reject")

used in Section 8.3.5, the data would first need to be stacked by moving all the data values into one column and by adding a column to identify the groups.) To determine the ranks for the two groups, do the following:

1. Sort the data in ascending order by dividend yield by selecting any cell in the dividend yield column (B) and then selecting the command Data | Sort. As was first described in Section 2.2.3, select the Ascending and Header Row option buttons. Click the OK button. The data are now sorted by dividend yield.

2. Create a ranks column in column C by entering the column heading Rank in cell C1 and the value 1 in cell C2. With cell C2 highlighted, select Edit | Fill | Series to use the shortcut for entering ranks first described in Section 5.6.5. Select the Columns and Linear option buttons as before, but this time enter the stop value 46. Click the OK button. The ranks 2 to 46 are entered into the range C3:C47.

3. Now we must manually check for tied values and give each of the tied observations the average of the ranks for which they are tied. For example, there are two dividend yields of 1.8, which are tied for ranks 9 and 10. Change the ranks for these values to 9.5 instead of 9 and 10. Follow the same procedure for the other tied values.

4. Having assigned ranks, we now need to separate these ranks by group. One way to do this is by using the **advanced data filter feature** of Microsoft Excel. To use this feature, first copy the column headings in cells A1 and C1 (Exchange and Rank) to cells D1 and E1, respectively. Then to select the NASDAQ ranks, enter the value NASDAQ in cell D2. At this point, the top of the Data sheet should look like Figure 8.4.Excel.

	A	B	C	D	E
1	Exchange	Div. Yield	Rank	Exchange	Rank
2	NASDAQ	0.4	1	NASDAQ	
3	NASDAQ	0.8	2		
4	NYSE	0.9	3		
5	NASDAQ	1.2	4		
6	NASDAQ	1.3	5		
7	NASDAQ	1.5	6		
8	NYSE	1.6	7		

FIGURE 8.4.EXCEL Data sheet before using Data Filter command.

5. Now, with the Data sheet still active, select Data | Filter | Advanced Filter, and in the Advanced Filter dialog box, select the Copy to Another Location option button, and enter A1:C47 in the List Range edit box, D1:D2 in the Criteria Range edit box, and E1 in the Copy to edit box (see Figure 8.5.Excel). Click the OK button. All of the NASDAQ ranks now appear in Column E.

6. To copy the NYSE ranks to column F, first copy the column heading Rank in cell C1 to cell F1. Change the value in cell D2 to NYSE. Then repeat the previous step, this time entering F1 in the Copy to edit box (all other choices remain the same). After clicking the OK button, the NYSE ranks will appear in column F. To avoid confusion, change the column heading in E1 to NASDAQ and the column heading in cell F1 to NYSE. The top of the Data sheet should now look similar to Figure 8.6.Excel.

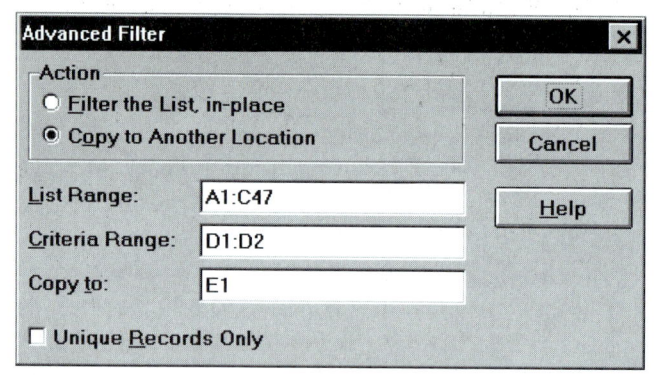

FIGURE 8.5.EXCEL Advanced Filter dialog box.

	A	B	C	D	E	F
1	Exchange	Div. Yield	Rank	Exchange	NASDAQ	NYSE
2	NASDAQ	0.4	1	NYSE	1	3
3	NASDAQ	0.8	2		2	7
4	NYSE	0.9	3		4	12
5	NASDAQ	1.2	4		5	15
6	NASDAQ	1.3	5		6	18.5
7	NASDAQ	1.5	6		8	20.5
8	NYSE	1.6	7		9.5	24.5

FIGURE 8.6.EXCEL Top of Data sheet after Step 6.

With the two columns of ranks now determined, the Calculations sheet can now be implemented. Begin by entering the formula =SUM(Data!E:E) in cell B3, =COUNT(Data!E:E) in cell B4, =SUM(Data!F:F) in cell B5, and =COUNT(Data!F:F) in cell B6 to calculate the sum of the ranks and the sample size of the two groups. Note that these formulas use the special column range notation and therefore do not require the manual determination of how many rows of data there are for the two groups. (Alternatively, if the sum of the ranks and sample sizes were previously known, they could be directly entered into these cells.) Implement the rest of the Calculations sheet as follows:

❶ Enter the total sample size in cell B7 using the formula =B4+B6.

❷ Using Equation (8.3) on page 395, compute the sum of the ranks using the formula =(B7*(B7+1))/2 in cell B8. This will serve to double-check the manual checking for tied values done earlier.

❸ Enter the formula =IF(B8=B3+B5,"Passed","Failed--review manually assigned ranks.") in cell B9 to compare the double-check value in cell B8 with the sum of the two sums in cells B3 and B5. Should the comparison fail, cell B9 will display the message "Failed--review manually assigned ranks." warning the user to go back and review the assigned ranks before continuing.

❹ In cell B10, we determine whether we should use Table E.6 (if the sample sizes of both groups *are* less than or equal to 10), or whether we can use the large-sample normal approximation as is assumed by the calculations on this sheet. To compare both sample sizes in one formula, we can combine the IF and AND functions using the format

=IF(AND(comparison$_1$,comparison$_2$) action to be taken if comparison holds, action to be taken if both comparisons do not hold).

In this example, we would enter =IF(AND(B4<=10,B6<=10,"Invalid--Use Table E.6. Do not use this sheet.","Valid--Use calculations on this sheet.") in cell B12. This will display an appropriate message stating whether or not the two sample sizes are less than or equal to 10.

5 Enter the level of significance α in cell B11 (.05 for this problem).

6 Compute the T_1 test statistic in cell B12 based on whether group 1 or group 2 has the smaller sample size by entering the formula =IF(B4<=B6,B3,B5).

7 Cells B13 and B14 are used to compute the average and standard error of T_1 so that the test statistic shown in Equation (8.4) can be computed in cell B15. In cell B13, enter the formula =IF(B4<=B6,B4*(B7+1)/2,B6*(B7+1)/2), and in cell B14, enter the formula =SQRT(B4*B6*(B7+1)/12). Compute the Z-test statistic in cell B15 by using the formula =(B12–B13)/B14.

8 The critical values, p-values, and decisions in the cell range B16:B28 follow the format used in the one-sample Z test for the mean developed in Section 7.11.1.

The results are illustrated in Figure 8.7.Excel.

	A	B
1	Wilcoxon Rank Sum Test	
2		
3	Sum of Ranks Group 1	495.5
4	Sample Size group 1	25
5	Sum of Ranks Group 2	585.5
6	Sample Size group 2	21
7	Total Sample Size n	46
8	Total Sum of Ranks	1081
9	Total Sums Check	Passed
10	Large-sample Approximation	Valid—Use calculations on this sheet.
11	α	0.05
12	T1	585.5
13	Mean of T1	493.5
14	Standard Error	45.34589287
15	Z Test Statistic	2.028849675
16	Two-Tailed Test	
17	Lower Critical Value	−1.95996108
18	Upper Critical value	1.959961082
19	p-value	0.042473466
20	Decision	Reject
21	One-tailed Test (Lower)	
22	Lower Critical Value	−1.644853
23	p-value	0.978763267
24	Decision	Do not Reject
25	One-Tailed test (Upper)	
26	Upper Critical Value	1.644853
27	p-value	0.021236733
28	Decision	Reject

FIGURE 8.7.EXCEL Excel Wilcoxon Test for the Difference between Two Medians obtained from Microsoft Excel for the dividend yield data.

As may be seen in Figure 8.7.Excel, the sum of the ranks for each of the two groups has been computed, and a check has been performed to determine whether the sum of the ranks in the two groups adds up to the correct value. The value of T_1 has been computed as 585.5, and the critical value may be obtained from the normal distribution. The Z-test statistic is 2.0288, and using a two-tailed test, this is greater than the critical value of approximately 1.96, so the null hypothesis of equality between the medians of the two groups can be rejected. Alternatively, using the *p*-value approach, the null hypothesis is rejected since the *p*-value of .042 is less than the level of significance of .05.

Problems for Section 8.4

Note: The problems in this section may be solved using Microsoft Excel.

TESTRANK.TXT

8.10 A statistics professor taught two special sections of a basic course in which the 10 students in each section were considered outstanding. She used a "traditional" method of instruction (T) in one section and an "experimental" method (E) in the other. At the end of the semester she ranked the students on the basis of their performance from 1 (worst) to 20 (best).

T	1	2	3	5	9	10	12	13	14	15
E	4	6	7	8	11	16	17	18	19	20

For this instructor, was there any evidence of a difference in performance based on the two methods? (Use $\alpha = .05$.)

INTRANK.TXT

8.11 The Director of Human Resources at a 1,200-bed New York City hospital is evaluating candidates for the position of Administrator of the Billings and Payments Department. Among the applicants, 22 are invited for interviews. Following the interviews, the rankings (1 = most preferred) of the candidates (based on interview, academic record, and prior experience) are as presented here, broken down by "type" of master's degree obtained—MBA versus MPH.

MBA Candidates		MPH Candidates	
1	2	3	6
4	5	7	10
8	9	13	14
11	12	16	18
15	17	19	20
		21	22

Is there evidence that the MBA candidates are more preferred than the MPH candidates? (Use $\alpha = .05$.)

TRAIN2.TXT

8.12 A New York television station decides to do a story comparing two commuter railroads in the area—the Long Island Rail Road (LIRR) and New Jersey Transit (NJT). The researchers at the station sample the performance of several scheduled train runs of each railroad, 10 for the LIRR and 12 for the NJT. The data on how many minutes early (negative numbers) or late (positive numbers) each train was are presented here:

LIRR: 5 −1 39 9 12 21 15 52 18 23

NJT: 8 24 10 4 12 5 4 9 15 33 14 7

(a) Is there evidence that the railroads differ in their median tendencies to be late? (Use $\alpha = .01$.)
(b) What conclusions about the lateness of the two railroads can be made?

8.13 Refer to the data for Problem 8.7 on page 393. Using a .01 level of significance, is there evidence that wage incentives (through commission) yield greater median sales volume?

📁 SALESVOL.TXT

• **8.14** Refer to the data for Problem 8.8 on page 393–394.
   (a) Using a .05 level of significance, is there evidence of a difference in the two types of batteries with respect to median talking time (in minutes) prior to recharging?
   (b) What assumptions must be made in order to do part (a) of this problem?
   (c) Compare the results obtained in (a) with those of Problem 8.8. Discuss.

📁 NICKBAT.TXT

8.15 Refer to Problem 8.9 on page 394.
   (a) Using a .05 level of significance, is there evidence of a difference in the median effective yields on money market accounts in the two types of banks in the New York area?
   (b) What assumptions must be made in order to do part (a) of this problem?
   (c) Compare the results obtained in (a) with those of Problem 8.9. Discuss.

📁 NYBANKS.TXT

8.5 F TEST FOR DIFFERENCES IN TWO VARIANCES

8.5.1 Introduction

In the previous two sections we examined two procedures for testing for differences in central tendency (i.e., differences in the means or the medians) between two independent populations. In many situations, however, we may also be interested in testing whether two independent populations have the same variability. Either we may be interested in studying the variances of two populations as a "means to an end," that is, testing the assumption of equal variances to determine whether the pooled-variance t test is appropriate for comparing the two population means (Section 8.3), or we may be truly interested in studying the variances of two populations as an "end in itself."

8.5.2 Development

To test for the equality of the variances of two independent populations, a statistical procedure has been devised that is based on the ratio of the two sample variances. If the data from each population are assumed to be normally distributed, then the ratio S_1^2/S_2^2 follows a distribution called the F distribution (see Table E.5), which was named after the famous statistician R. A. Fisher. From Table E.5 (a replica of which appears as Table 8.5), we can see that the critical values of the F distribution depend on *two* sets of degrees of freedom. The degrees of freedom in the numerator of the ratio pertain to the first sample, and the degrees of freedom in the denominator of the ratio pertain to the second sample. The **F-test** statistic for testing the equality between two variances is defined as follows:

> The F-test statistic is equal to the variance in group 1 divided by the variance in group 2.
>
> $$F = \frac{S_1^2}{S_2^2} \tag{8.6}$$

where
n_1 = size of sample taken from population 1
n_2 = size of sample taken from population 2
$n_1 - 1$ = degrees of freedom from sample 1 (i.e., the numerator degrees of freedom)
$n_2 - 1$ = degrees of freedom from sample 2 (i.e., the denominator degrees of freedom)

S_1^2 = variance of sample 1

S_2^2 = variance of sample 2

In testing the equality of two variances, either two-tailed or one-tailed tests can be employed, depending on whether we are testing if the two population variances are *different* or if one variance is *greater than* the other variance. These situations are depicted in Figure 8.5.

For a given level of significance, α, to test the null hypothesis of equality of variances

$$H_0: \sigma_1^2 = \sigma_2^2$$

against the alternative hypothesis that the two population variances are not equal

$$H_1: \sigma_1^2 \neq \sigma_2^2$$

we may reject the null hypothesis if the computed F-test statistic exceeds the upper-tailed critical value $F_{U(n_1-1),(n_2-1)}$ from the F distribution or if the computed test statistic falls below the lower-tailed critical value $F_{L(n_1-1),(n_2-1)}$ from the F distribution. That is, the decision rule is

$$\text{Reject } H_0 \text{ if } F > F_{U(n_1-1),(n_2-1)}$$

$$\text{or if } F < F_{L(n_1-1),(n_2-1)};$$

otherwise do not reject H_0.

This decision rule and rejection region are displayed in Panel A of Figure 8.5.

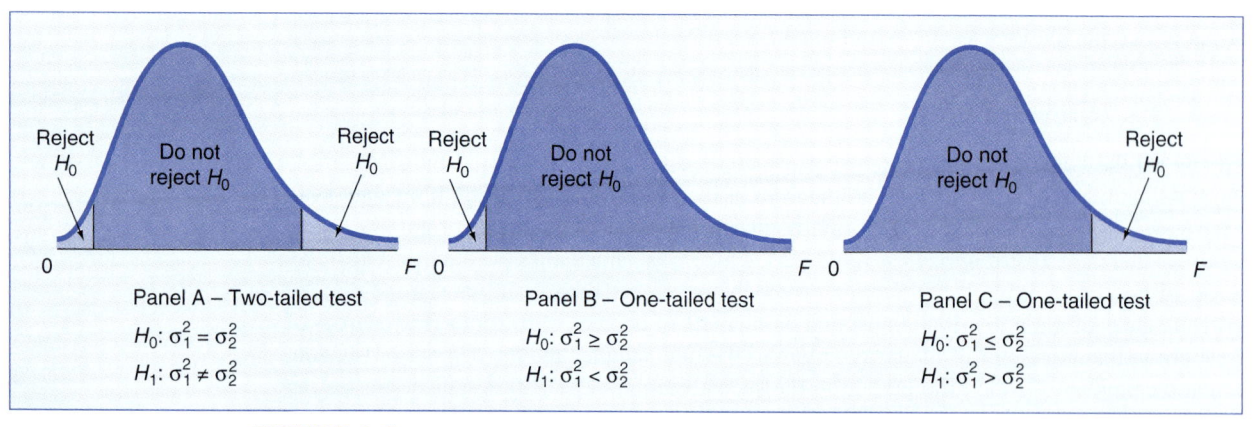

FIGURE 8.5 Determining the rejection region for testing a hypothesis about the equality of two population variances: Panel A, two-tailed test; Panel B, one-tailed test; Panel C, one-tailed test.

8.5.3 Application

To demonstrate how we may test for the equality of two variances, we can return to the financial analyst's study of dividend yields from two groups of stocks. The data for this are displayed in Table 8.1 on page 385, and the summary measures for the two samples are presented in Table 8.2 on page 386.

To test for the equality of the two population variances, we have the following null and alternative hypotheses:

$$H_0: \sigma_1^2 = \sigma_2^2$$

$$H_1: \sigma_1^2 \neq \sigma_2^2$$

Because this is a two-tailed test, the rejection region is split into the lower and upper tails of the F distribution. If a level of significance of $\alpha = .05$ is selected, each rejection region would contain .025 of the distribution.

The upper-tail critical value of the F distribution with 20 and 24 degrees of freedom can be obtained directly from Table E.5, a replica of which is presented in Table 8.5. Since there are 20 degrees of freedom in the numerator and 24 degrees of freedom in the denominator, the upper-tail critical value can be found by looking in the column labeled "20" and the row labeled "24," which pertain to an upper-tail area of .025. Thus, the upper-tail critical value of this F distribution is 2.33.

Table 8.5 Obtaining the critical value of F with 20 and 24 degrees of freedom for an upper-tail area of .025.

	Numerator df$_1$					
Denominator df$_2$	1	2	...	15	20	24
1	647.8	799.5	...	984.9	993.1	997.2
2	38.51	39.00	...	39.43	39.45	39.46
3	17.44	16.04	...	14.25	14.17	14.12
4	12.22	10.65	...	8.66	8.56	8.51
⋮	⋮	⋮	⋮	⋮	⋮	⋮
21	5.83	4.42	...	2.53	2.42	2.37
22	5.79	4.38	...	2.50	2.39	2.33
23	5.75	4.35	...	2.47	2.36	2.30
24	5.72	4.32	...	2.44	2.33	2.27

Source: Extracted from Table E.5.

● **Obtaining Lower-Tail Critical Values** Any lower-tail critical value on the F distribution can be obtained from

$$F_{L(n_1-1),(n_2-1)} = \frac{1}{F_{U(n_2-1),(n_1-1)}} \quad (8.7)$$

where $F_{L(n_1-1),(n_2-1)}$ = lower-tail critical value of the F distribution with $n_1 - 1$ and $n_2 - 1$ degrees of freedom

$F_{U(n_2-1),(n_1-1)}$ = upper-tail critical value of the F distribution with $n_2 - 1$ and $n_1 - 1$ degrees of freedom

$n_1 - 1$ = degrees of freedom from sample 1

$n_2 - 1$ = degrees of freedom from sample 2

Therefore, in this example we have

$$F_{L(20,24)} = \frac{1}{F_{U(24,20)}}$$

To compute our desired lower-tail critical value, we need to obtain the upper-tail .025 value of F with 24 degrees of freedom in the numerator and 20 degrees of freedom in the denominator and take its reciprocal. From Table E.5, this upper-tail value is 2.41. Therefore, from Equation (8.7),

$$F_{L(20,24)} = \frac{1}{F_{U(24,20)}} = \frac{1}{2.41} = 0.415$$

As depicted in Figure 8.6, the decision rule is

$$\text{Reject } H_0 \text{ if } F > F_{U(20,24)} = 2.33$$
$$\text{or if } F < F_{L(20,24)} = 0.415;$$
$$\text{otherwise do not reject } H_0.$$

Using Equation (8.6) for the financial analyst's data (see Table 8.2 on page 386), we compute the following F-test statistic:

$$F = \frac{S_1^2}{S_2^2}$$
$$= \frac{1.698}{1.353} = 1.25$$

Therefore, since $F_{L(20,24)} = 0.415 < F = 1.25 < F_{U(20,24)} = 2.33$, we do not reject H_0. The analyst would conclude that there is no evidence of a difference in the variability of dividend yields for the two populations. Thus, if we could assume that the two populations are normally distributed, the pooled-variance t test would be appropriate for comparing differences in the average dividend yields because there is no evidence that the population variances are not equal. On the other hand, if we do not feel that the normality assumption is viable, we should use the Wilcoxon rank sum test to determine whether there are differences in the median dividend yields in the two populations.

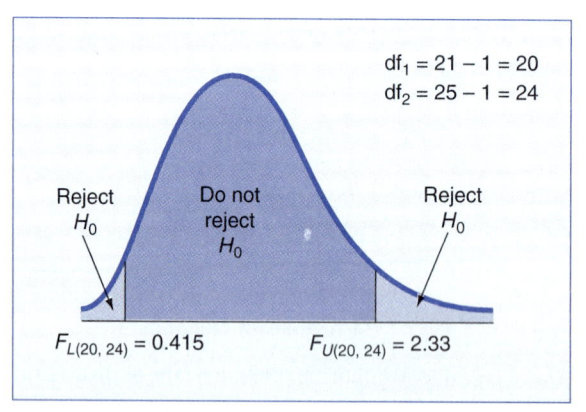

FIGURE 8.6 Regions of rejection and nonrejection for a two-tailed test for the equality of two variances at the .05 level of significance with 20 and 24 degrees of freedom.

8.5.4 Caution

In testing for the equality of two population variances, we should be aware that the test assumes that each of the two populations is normally distributed. This means that if the assumption of normality for each population is met, the F-test statistic follows an F distribution with $n_1 - 1$ and $n_2 - 1$ degrees of freedom. Unfortunately, this F-test statistic is not *robust* to departures from this assumption (Reference 2), particularly when the sample sizes in the two groups are not equal. Therefore, if the populations are not at least approximately normally distributed, the accuracy of the procedure can be seriously affected (References 2 and 3 present other procedures for testing the equality of two variances).

8.5.5 Using Microsoft Excel for the *F* Test for Differences in Two Variances

In this section we developed the *F* test to determine whether any differences exist between the variances of two groups. We will now explain how Microsoft Excel can be used for this procedure. If raw data are available, we use the *F* Test: Two Sample for Variances option of the Data Analysis tool. If raw data are not available, Excel formulas can be used to obtain the *F*-test statistic and to determine whether there are significant differences between the variances of two groups.

● **Using the Data Analysis Tool for the Two-Sample *F* Test for the Difference Between Variances** If we have raw data, we can use the *F* Test: Two Sample for Variances option of the Data Analysis tool. To illustrate the use of this tool, we will again refer to the dividend yield data of Table 8.1 on page 385. With this YIELD-1.XLS workbook containing these data open,

8–5–5.XLS

① Click the Data sheet tab. You will note that the data are set up in columns A and B with heads indicating the names of the two groups.

② Select Tools | Data Analysis, select F-Test: Two Sample for Variances from the Analysis tools list box, and click the OK button. As illustrated in Figure 8.8.Excel, the *F* Test: Two Sample for Variances dialog box appears.

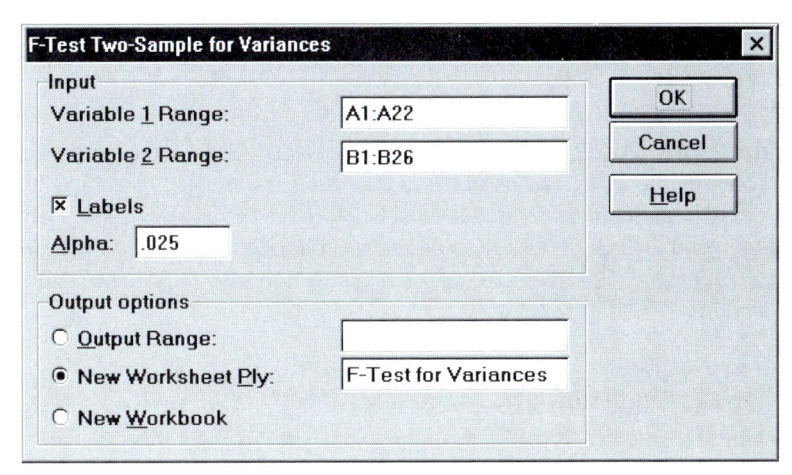

FIGURE 8.8.EXCEL Dialog box for Microsoft Excel *F* Test: Two Sample for Variances data analysis option.

③ (a) In the Variable 1 Range edit box, enter A1:A22.
   (b) In the Variable 2 Range edit box, enter B1:B26.
   (c) Select the Labels check box.
   (d) In the Alpha edit box, enter the *one-tailed* level of significance (for these data, enter .025).

> In Version 5 the *F*-test option of the Data Analysis tool provides only a one-tailed test. In Version 5 if the variance for the data listed in the first data range is greater than the variance of the data listed in the second data range, the *p*-value will represent the upper tailed area. In Version 7, the *F*-statistic represents the larger variance divided by the smaller variance, and the *p*-value represents the upper tailed area.

❹ Select the New Worksheet Ply option button and enter the name F-Test for Variances. Click the OK button.

Figure 8.9.Excel represents the output obtained from Excel.

From Figure 8.9.Excel, we observe that Excel has computed the means, variances, degrees of freedom, the *F*-test statistic, the upper critical value, and the *p*-value. The *F*-test statistic equal to 1.255 is less than the upper critical value of 2.3273, so the null hypothesis cannot be rejected. Alternatively, since the *p*-value for the upper-tail test is .295, which is greater than .025, the null hypothesis cannot be rejected.

	A	B	C
1	F-Test Two-Sample for Variances		
2			
3		NYSE	NASDAQ
4	Mean	3.271429	2.532
5	Variance	1.697143	1.352267
6	Observations	21	25
7	df	20	24
8	F	1.255036	
9	P(F<=f) one-tail	0.295002	
10	F Critical one-tail	2.327269	

FIGURE 8.9.EXCEL Output obtained from *F* Test: Two Sample for Variances for the dividend yield data.

● **Working with Stacked Data** To use the Data Analysis tool for the *F* test for the difference between two variances, as in the case of the *t* test for the difference between two means, the data for each of the two groups must be placed in separate columns. If the response variable of interest (such as dividend yield) had been in one column and the grouping variable (such as exchange) stacked in another, such data would have to be unstacked first as discussed in Section 8.3.5.

● **Using Microsoft Excel Formulas for the Two-Sample *F* Test for the Difference Between Two Variances When Raw Data Are Not Available** Sometimes only summary data such as the mean and standard deviations are available. For such situations, you can use Excel formulas to compute test statistics, critical values, and *p*-values. The design for this worksheet, named TwoGroups, is shown in Table 8.4.Excel on page 411.

Referring to Equation (8.6) on page 405, we see that to test the hypothesis for the variance, we need to have the sample standard deviation *S* and the sample size *n* for each of the two groups. (If the sample standard deviations need to be computed, the STDEV function could be used and the results copied to cells B3 and B6, respectively.) To implement the TwoGroups sheet, open a new workbook and rename the active sheet TwoGroups (or insert a new worksheet named TwoGroups in the workbook containing your data). Then do the following:

Table 8.4.Excel Design for Testing the Difference between Two Variances.

	A	B	C	D
1	F-Test for Variances			
2				
3	S Group 1	xxxx		
4	n Group 1	xx		
5	df Group 1	=B4–1		
6	S Group 2	xxxx		
7	n Group 2	xx		
8	df Group 2	=B7–1		
9	F Ratio of Variances	=B3^2/B6^2		
10	α	.xx		
11	Two-Tailed Test			
12	Lower Critical Value	=FINV(1–B10/2,B5,B8)		
13	Upper Critical value	=FINV(B10/2,B5,B8)		
14	p-value	=IF(B9>1,2*D15,2*D16)	FDIST calculations	
15	Decision	=IF(OR(B14<B10,B14>1–B10),"Reject","Do not reject")	FDIST calculation	=FDIST(B9,B5,B8)
16	One-tailed Test (Lower)		1–FDIST calculation	=1–D15
17	Lower Critical Value	=FINV(1-B10,B5,B8)		
18	p-value	=D16		
19	Decision	=IF(B18<B10,"Reject","Do not reject")		
20	One-Tailed test (Upper)			
21	Upper Critical Value	=FINV(B10,B5,B8)		
22	p-value	=D15		
23	Decision	=IF(B22<B10,"Reject","Do not reject")		

❶ For group 1, enter the value for the sample standard deviation S in cell B3, and the sample size n in cell B4. For group 2, enter the value for the sample standard deviation S in cell B6, and the sample size n in cell B7. Again using the dividend yield example, if we were going to use the Excel formulas shown in Table 8.4.Excel instead of the Data Analysis tool, we would enter 1.30274 in cell B3, the sample size of 21 in cell B4, 1.16287 in cell B6, and 25 in cell B7.

❷ The degrees of freedom for each of the two groups are computed in cell B5 (=B4–1) and B8 (=B7–1).

❸ The ratio of the two variances is computed in cell B9 using the formula =B3^2/B6^2.

❹ The level of significance α is entered in cell B10 (.05 for this example).

❺ The critical values for the F test are obtained using the FINV function whose format is

$$\text{FINV}(p, df_1, df_2)$$

where p = upper-tailed p-value (the probability that F will be greater than the specific value indicated)

df_1 = degrees of freedom in the numerator

df_2 = degrees of freedom in the denominator

For the two-tailed test, the lower critical value in cell B12 is obtained using the formula =FINV(1–B10/2,B5,B8) while the upper critical value in cell B13 is obtained using the formula =FINV(B10/2,B5,B8).

⑥ The *p*-value is obtained by using the FDIST function whose format is

$$\text{FDIST}(F, \text{df}_1, \text{df}_2)$$

where
F = value for the F statistic
df_1 = degrees of freedom in the numerator
df_2 = degrees of freedom in the denominator

This function computes the probability of exceeding a value of F for the degrees of freedom indicated in the formula. The formula for the *p*-value will differ depending on whether or not the variance in group 1 is larger than the variance in group 2. If the F ratio is greater than 1, the *p*-value represents the probability in the upper tail of the F distribution, while if the F value is less than 1, the *p*-value represents the area in the lower tail of the distribution. In either case, since a two-tailed test is used, the probability obtained for the *p*-value in a single tail of the distribution needs to be doubled. Thus, in cell B14, we use the IF function and the FDIST function to provide the formula

=IF(B9>1,2*(D15,2*D16))

Note that the FDIST calculation =FDIST(B9,B5,B8) has been made in cell D15 and the 1–FDIST calculation =1–D15 has been made in cell D16.

⑦ The decision for this two-tailed test depends on whether the test statistic is in the upper or lower tail of the F distribution. To determine this we can use an IF function and OR function in the format.

=IF(OR(comparison₁,comparison₂), action to be taken if either comparison holds, action to be taken if neither comparison holds).

For this example we would enter the formula

=IF(OR(B14<B10,B14>1–B10),"Reject","Do Not Reject")

in cell B15. Thus, if the *p*-value in cell B14 was less than α or greater than $1-\alpha$, the null hypothesis would be rejected; otherwise the null hypothesis would not be rejected.

⑧ For the one-tailed test in the lower tail, the lower critical value would be computed in cell B17 using the FINV function with a probability of $1-\alpha$ of exceeding the critical value. The *p*-value in cell B18, equal to the 1–FDIST calculation in cell D16, would represent the probability of obtaining a result that was less than the F statistic obtained, and the decision rule in cell B19 would reject the null hypothesis if the *p*-value was less than α.

⑨ For the one-tailed test in the upper tail, the upper critical value would be computed in cell B21 using the FINV function with a probability α of exceeding the critical value. The *p*-value in cell B22, equal to the FDIST calculation in cell D15, would represent the probability of obtaining a result that was greater than the F statistic obtained, and the decision rule in cell B23 would reject the null hypothesis if the *p*-value was less than α.

The results are illustrated in Figure 8.10.Excel.

▲ WHAT IF EXAMPLE

The TwoGroups worksheet allows us to study the effect of changes in the sample standard deviations, the sample size, and level of significance on the test of hypothesis for the difference between two variances. Changing the standard deviation of the first group from

1.30274 to 2.30274 changes the results of the F test for the difference between the variances. The F ratio becomes 3.92, the p-value for the two-tailed test becomes .0018, and the null hypothesis is rejected. As before, if we are interested in seeing the effects of many different changes, we could use the Scenario Manager (see Section 1S.15) to store and use sets of alternative values of S, n, and the level of confidence.

	A	B	C	D
1	F-Test for Variances			
2				
3	S Group 1	1.30274		
4	n Group 1	21		
5	df Group 1	20		
6	S Group 2	1.16287		
7	n Group 2	25		
8	df Group 2	24		
9	F Ratio of Variances	1.255027271		
10	α	0.05		
11	Two-Tailed Test			
12	Lower Critical Value	0.415358414		
13	Upper Critical value	2.327269044		
14	p-value	0.590015074	FDIST calculations	
15	Decision	Do not reject	FDIST calculation	0.295008
16	One-tailed Test (Lower)		1-FDIST calculation	0.704992
17	Lower Critical Value	0.480202544		
18	p-value	0.704992463		
19	Decision	Do not reject		
20	One-Tailed test (Upper)			
21	Upper Critical Value	2.026663282		
22	p-value	0.295007537		
23	Decision	Do not reject		

FIGURE 8.10.EXCEL F Test for the difference between two variances obtained from Microsoft Excel for the dividend yield data.

Problems for Section 8.5

Note: *The problems in this section may be solved using Microsoft Excel.*

- 8.16 Suppose that the following information is available for two groups:

$$n_1 = 10 \quad S_1^2 = 13.7 \quad n_2 = 10 \quad S_2^2 = 16.9$$

(a) At the .05 level of significance, is there evidence of a difference between σ_1^2 and σ_2^2?
(b) What is the relationship in (a) between the lower critical value and the upper critical value? Under what conditions will this relationship hold? Explain.
(c) Suppose that we had wanted to perform a one-tailed test. At the .05 level of significance, what is the upper-tailed critical value of the F-test statistic to determine if there is evidence that $\sigma_1^2 > \sigma_2^2$?
(d) Suppose that we had wanted to perform a one-tailed test. At the .05 level of significance, what is the lower-tailed critical value of the F-test statistic to determine if there is evidence that $\sigma_1^2 < \sigma_2^2$?

8.17 Suppose that the following information is available for two groups:

$$n_1 = 16 \quad S_1^2 = 47.3 \quad n_2 = 13 \quad S_2^2 = 36.4$$

(a) At the .05 level of significance, is there evidence of a difference between σ_1^2 and σ_2^2?
(b) Suppose that we had wanted to perform a one-tailed test. At the .05 level of significance, what is the upper-tailed critical value of the F-test statistic to determine if there is evidence that $\sigma_1^2 > \sigma_2^2$?
(c) Suppose that we had wanted to perform a one-tailed test. At the .05 level of significance, what is the lower-tailed critical value of the F-test statistic to determine if there is evidence that $\sigma_1^2 < \sigma_2^2$?

8.18 A professor in the Accountancy Department of a business school claims that there is much more variability in the final exam scores of students taking the introductory accounting course as a requirement than for students taking the course as part of their major. Random samples of 13 nonaccounting majors and 10 accounting majors are taken from the professor's class roster in his large lecture and the following results are computed based on the final exam scores:

$$n_{NA} = 13 \quad S_{NA}^2 = 210.2 \quad n_A = 10 \quad S_A^2 = 36.5$$

(a) At the .05 level of significance, is there evidence to support the professor's claim?
(b) Find the p-value.

8.19 Referring to Problem 8.5 on page 393
(a) At the .01 level of significance, is there evidence of a difference in the variances of the shipping time between the two outlets?
(b) Find the p-value.

8.20 Referring to Problem 8.6 on page 393
(a) At the .05 level of significance, is there evidence of a difference in the variances of the amount spent between dog and cat owners?
(b) Find the p-value.

8.21 Referring to the data for Problem 8.8 on page 393–394 *(NICKBAT.TXT)*
(a) Using a .05 level of significance, is there evidence of a difference in the variances in talking time (in minutes) prior to recharging between the two types of batteries?
(b) Based on the results obtained in part (a), which test should have been chosen, the t test of Problem 8.8 or the Wilcoxon rank sum test of Problem 8.14 (page 405)? Discuss.

8.22 Referring to the data for Problem 8.9 on page 394 *(NYBANKS.TXT)*
(a) Using a .05 level of significance, is there evidence of a difference in the variances in effective yields on money market accounts between the two types of banks in the New York area?
(b) Based on the results obtained in part (a), which test should have been chosen, the t test of Problem 8.9 or the Wilcoxon rank sum test of Problem 8.15 (page 405)? Discuss.

8.6 ONE-WAY ANOVA F TEST FOR DIFFERENCES IN c MEANS

It is frequently of interest to compare differences in results among several groups. Many industrial applications involve experiments in which the groups or levels pertaining to only one **factor** of interest (e.g., baking temperature or flavor preference) are considered. A factor such as baking temperature may have several *numerical* levels (e.g., 300°, 350°, 400°, 450°) or a factor such as flavor preference may have several *categorical* levels (vanilla, chocolate, mocha, strawberry, pistachio). Such designed one-factor experiments in which subjects or experimental units are randomly assigned to groups or levels of a single factor are called **one-way** or **completely randomized design models**.

8.6.1 Introduction

When the outcome measurements across the c groups are continuous and certain assumptions are met, a methodology known as **analysis of variance** (or **ANOVA**) may be employed to compare the means of the groups. In a sense, the term "analysis of variance" appears to be a misnomer since the objective is to analyze differences among the group means. However, through an analysis of the variation in the data, both among and within the c groups, we will be able to draw conclusions about possible differences in group means. In ANOVA we subdivide the total variation in the outcome measurements into that which is attributable to differences *among* the c groups and that which is due to chance or attributable to inherent variation *within* the c groups (see Figure 8.7). "Within-group" variation is considered **experimental error,** while "among-group" variation is attributable to treatment effects.

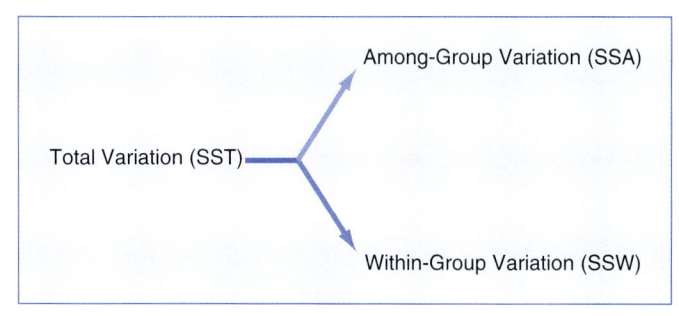

FIGURE 8.7 Partitioning the total variation in a completely randomized model.

8.6.2 Development

Under the assumptions that the c groups or levels of the factor being studied represent populations whose outcome measurements are randomly and independently drawn, follow a normal distribution, and have equal variances, the null hypothesis of no differences in the population means

$$H_0: \mu_1 = \mu_2 = \cdots = \mu_c$$

may be tested against the alternative that not all the c population means are equal

$$H_1: \text{Not all } \mu_j \text{ are equal (where } j = 1, 2, \ldots, c)$$

Figure 8.8 presents a picture of what a true null hypothesis would look like when five groups are compared and the assumptions of normality and equality of variances hold. The five populations representing the different levels of the factor are identical and therefore superimpose on one another. The properties of central tendency, variation, and shape are identical for each.

On the other hand, suppose that the null hypothesis were false with level 4 having the largest mean, level 1 having the second largest mean, and no differences in the other population means. Figure 8.9 presents a pictorial representation of this.

We note that, except for differences in central tendency (i.e., $\mu_4 > \mu_1 > \mu_2 = \mu_3 = \mu_5$), the five populations are the same in appearance.

To perform an ANOVA test of equality of population means, we subdivide the total variation in the outcome measurements into two parts, that which is attributable to differences

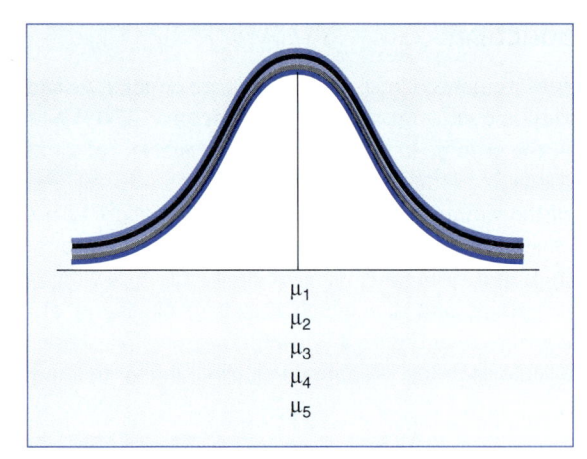

FIGURE 8.8 All five populations have the same mean: $\mu_1 = \mu_2 = \mu_3 = \mu_4 = \mu_5$.

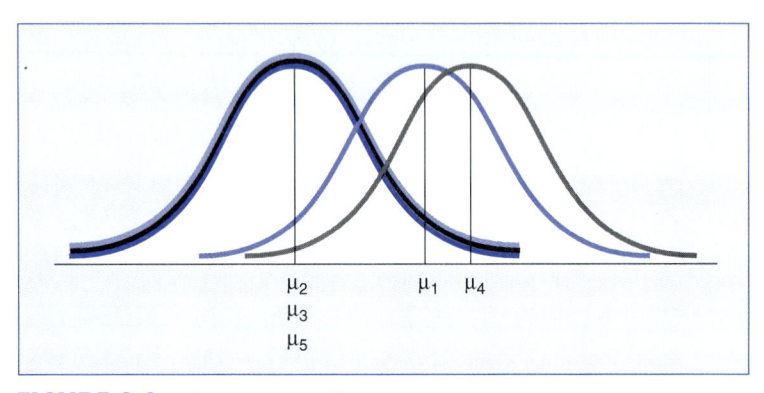

FIGURE 8.9 A treatment effect is present: $\mu_4 > \mu_1 > \mu_2 = \mu_3 = \mu_5$.

among the groups and that which is due to inherent variation within the groups. The **total variation** is usually represented by the **Sum of Squares Total** (or **SST**). Since under the null hypothesis the population means of the c groups are presumed equal, a measure of the total variation among all the observations can be obtained by summing the squared differences between each individual observation and the **overall** or **grand mean** $\overline{\overline{X}}$ that is based on all the observations in all the groups combined. The total variation is defined as follows:

The total variation (SST) is equal to the sum of the squared differences between each value and the overall mean.

$$\text{Total variation (SST)} = \sum_{j=1}^{c} \sum_{i=1}^{n_j} \left(X_{ij} - \overline{\overline{X}} \right)^2 \tag{8.8}$$

where

$$\overline{\overline{X}} = \frac{\sum_{j=1}^{c} \sum_{i=1}^{n_j} X_{ij}}{n} = \text{overall or grand mean}$$

X_{ij} = ith observation in group or level j

n_j = number of observations in group j

n = total number of observations in all groups combined
$(n = n_1 + n_2 + \cdots + n_c)$

c = number of groups or levels of the factor of interest

The **among-group variation**, usually called the **Sum of Squares Among** groups (or **SSA**), is measured by the sum of the squared differences between the sample mean of each group $\overline{X}_j$ and the overall or grand mean $\overline{\overline{X}}$, weighted by the sample size n_j in each group.[2] The among-group variation is computed from

$$\text{Among-group variation (SSA)} = \sum_{j=1}^{c} n_j \left(\overline{X}_j - \overline{\overline{X}} \right)^2 \tag{8.9}$$

where
c = number of groups or levels being compared
n_j = number of observations in group or level j
$\overline{X}_j$ = sample mean of group j
$\overline{\overline{X}}$ = overall or grand mean

The **within-group variation**, usually called the **Sum of Squares Within** groups (or **SSW**), measures the difference between each observation and the mean of its own group and cumulates the squares of these differences over all groups. The within-group variation may be computed as

$$\text{Within-group variation (SSW)} = \sum_{j=1}^{c} \sum_{i=1}^{n_j} \left(X_{ij} - \overline{X}_j \right)^2 \tag{8.10}$$

where
X_{ij} = ith observation in group or level j
$\overline{X}_j$ = sample mean of group j

Since c levels of the factor are being compared, there are $c - 1$ degrees of freedom associated with the sum of squares among groups. Since each of the c levels contributes $n_j - 1$ degrees of freedom and

$$\sum_{j=1}^{c} (n_j - 1) = n - c$$

there are $n - c$ degrees of freedom associated with the sum of squares within groups. In addition, there are $n - 1$ degrees of freedom associated with the sum of squares total because each observation X_{ij} is being compared to the overall or grand mean $\overline{\overline{X}}$ based on all n observations.

If each of these sums of squares is divided by its associated degrees of freedom, we obtain three *variances* or **mean square** terms—**MSA**, **MSW**, and **MST**:

$$\text{MSA} = \frac{\text{SSA}}{c - 1} \tag{8.11a}$$

$$\text{MSW} = \frac{\text{SSW}}{n-c} \qquad (8.11\text{b})$$

$$\text{MST} = \frac{\text{SST}}{n-1} \qquad (8.11\text{c})$$

Since a variance is computed by dividing the sum of squared differences by its appropriate degrees of freedom, the mean square terms are all variances.

Although the primary interest is in comparing the means of the c groups or levels of a factor to determine whether a treatment effect exists among the c groups, the ANOVA procedure derives its name from the fact that this is achieved by analyzing variances. If the null hypothesis is true and there are no real differences in the c group means, all three mean square terms—MSA, MSW, and MST—provide estimates of the variance σ^2 inherent in the data. Thus, to test the null hypothesis

$$H_0: \mu_1 = \mu_2 = \cdots = \mu_c$$

against the alternative

$$H_1: \text{Not all } \mu_j \text{ are equal (where } j = 1, 2, \ldots, c)$$

we compute the test statistic F, the ratio of MSA to MSW, as follows:

The F-test statistic is equal to the Mean Square Among groups divided by the Mean Square Within groups.

$$F = \frac{\text{MSA}}{\text{MSW}} \qquad (8.12)$$

The F statistic follows an **F distribution** with $c-1$ and $n-c$ degrees of freedom. For a given level of significance, α, we may reject the null hypothesis if the computed F-test statistic exceeds the upper-tailed critical value $F_{U(c-1, n-c)}$ from the F distribution (see Table E.5). That is, as shown in Figure 8.10, our decision rule is

$$\text{Reject } H_0 \text{ if } F > F_{U(c-1, n-c)};$$

otherwise do not reject H_0.

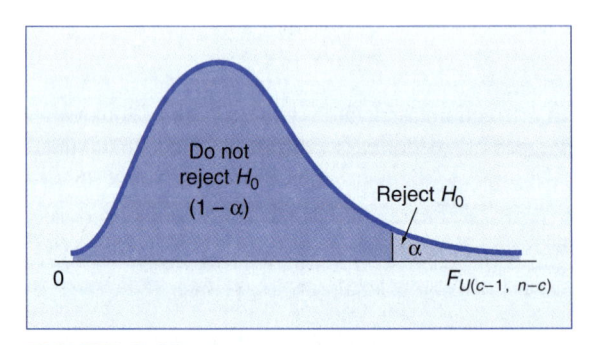

FIGURE 8.10 Regions of rejection and nonrejection when using ANOVA to test H_0.

If the null hypothesis were true, we should expect the computed F statistic to be approximately equal to 1, since both the numerator and denominator mean square terms are estimating the true variance σ^2 inherent in the data. On the other hand, if H_0 is false (and there are real differences in the means), we should expect the computed F statistic to be substantially *larger* than 1 because the numerator, MSA, would be estimating the treatment effect or differences among groups in addition to the inherent variability in the data, while the denominator, MSW, would be measuring only the inherent variability. Hence, the ANOVA procedure yields an F test in which the null hypothesis can be rejected at a selected α level of significance *only* if the computed F statistic is large enough to exceed $F_{U(c-1, n-c)}$, the upper-tail critical value of the F distribution having $c-1$ and $n-c$ degrees of freedom, as illustrated in Figure 8.10.

The results of an analysis of variance procedure are usually displayed in an **ANOVA summary table**, the format for which is presented in Table 8.6. The entries in this table include the sources of variation (i.e., among group, within group, and total), the degrees of freedom, the sums of squares, the mean squares (i.e., the variances), and the calculated F statistic. In addition, the *p*-value (i.e., the probability of obtaining an F statistic as large as or larger than the one obtained, given that the null hypothesis is true) is included in the ANOVA table of most statistical and spreadsheet packages. This enables us to make direct conclusions about the null hypothesis without referring to a table of critical values of the F distribution. If the *p*-value is less than the chosen level of significance α, the null hypothesis is rejected.

Table 8.6 ANOVA summary table.

Source	Degrees of Freedom	Sums of Squares	Mean Square (Variance)	F
Among groups	$c-1$	$SSA = \sum_{j=1}^{c} n_j (\overline{X}_j - \overline{\overline{X}})^2$	$MSA = \dfrac{SSA}{c-1}$	$F = \dfrac{MSA}{MSW}$
Within groups	$n-c$	$SSW = \sum_{j=1}^{c} \sum_{i=1}^{n_j} (X_{ij} - \overline{X}_j)^2$	$MSW = \dfrac{SSW}{n-c}$	
Total	$n-1$	$SST = \sum_{j=1}^{c} \sum_{i=1}^{n_j} (X_{ij} - \overline{\overline{X}})^2$		

8.6.3 Application

To illustrate the one-way ANOVA F test, suppose the production manager at the plant in which cereal is being filled in 368-gram-sized boxes is considering the replacement of an old machine that directly affects output in the production process. Three competing suppliers have permitted the production manager to use their particular equipment on a trial basis. The purchasing prices and servicing contracts for these three brands of machines are essentially the same. To make a purchasing decision, the production manager decides to conduct an experiment to determine whether there are any significant differences among the three brands of machines in the average time (in seconds) it takes factory workers using them to complete the filling process. Fifteen factory workers of similar experience, ability, and age are randomly assigned to receive training on one of the three brands of machines in such a manner that there are five factory workers for each machine. After an appropriate amount of training and practice, the production manager measures the time (in seconds) it takes the factory workers to complete the filling process using their respective equipment. The results of this experiment are displayed in Table 8.7 along with some summary computations. A *scatter plot* is depicted in Figure 8.11 so that we can visually inspect the data and see how the measurements (in seconds) distribute around their own group means as well as around the overall group mean $\bar{\bar{X}}$. We also get a sense for how each group mean compares to the overall mean. Thus, by examining Figure 8.11, we have the opportunity to observe possible trends or relationships across the groups as well as patterns within groups and, most importantly, this enables us to consider any potential violations in assumptions required by a particular testing procedure. Had the sample sizes in each group been larger, stem-and-leaf displays and a box-and-whisker plot, which provide additional visual information, would also have been obtained.[3]

We note from Table 8.7 and Figure 8.11 that there are differences in the *sample* means for the three machines. It takes, on average, 24.93 seconds to complete the filling process using machine I; 22.61 seconds using machine II; and 20.59 seconds using machine III. The question that must be answered is whether these sample results are sufficiently different for the production manager to decide that the *population* averages are not all equal.

The null hypothesis states that there is no difference among the groups in the average time to complete the filling process. Thus, substituting 1, 2, 3 for I, II, III, we have

Table 8.7 Time (in seconds) to complete a cereal filling process using three different machines.

	Machine		
	I	II	III
	25.40	23.40	20.00
	26.31	21.80	22.20
	24.10	23.50	19.75
	23.74	22.75	20.60
	25.10	21.60	20.40
Mean	$\bar{X}_1 = 24.93$	$\bar{X}_2 = 22.61$	$\bar{X}_3 = 20.59$

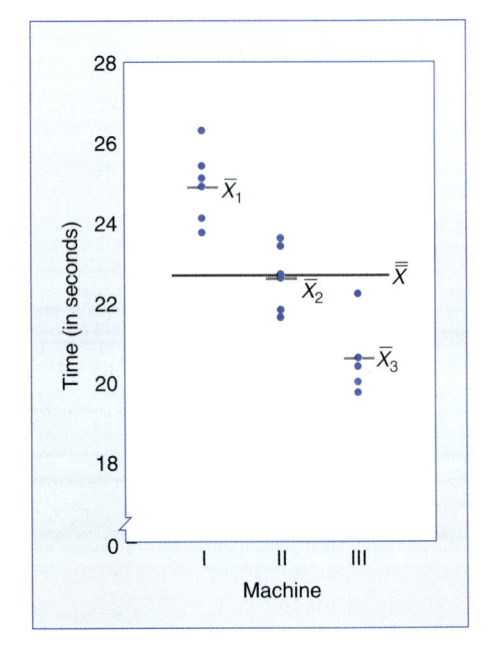

FIGURE 8.11 Scatter plot of time (in seconds) to complete a task using three different machines. *Source:* Table 8.7.

$$H_0: \mu_1 = \mu_2 = \mu_3$$

The alternative hypothesis states that there is a *treatment effect;* that is, at least one of the machines differs with respect to the average time required for completing the filling process.

$$H_1: \text{Not all the means are equal}$$

To construct the ANOVA summary table, we first compute the sample means in each group (see Table 8.7 on page 420); then we compute the overall or grand mean

$$\overline{\overline{X}} = \frac{\sum_{j=1}^{c}\sum_{i=1}^{n_j} X_{ij}}{n} = \frac{25.40 + 26.31 + \cdots + 23.40 + \cdots + 20.40}{15}$$

$$= \frac{340.65}{15} = 22.71$$

followed by the sums of squares:

$$\text{SSA} = \sum_{j=1}^{c} n_j \left(\overline{X}_j - \overline{\overline{X}}\right)^2 = (5)(24.93 - 22.71)^2 + (5)(22.61 - 22.71)^2 + (5)(20.59 - 22.71)^2$$
$$= (5)(2.22)^2 + (5)(-.10)^2 + (5)(-2.12)^2$$
$$= 24.642 + .05 + 22.472$$
$$= 47.164$$

$$\text{SSW} = \sum_{j=1}^{c}\sum_{i=1}^{n_j}\left(X_{ij} - \overline{X}_j\right)^2 = \begin{pmatrix}(25.40 - 24.93)^2 \\ +(26.31 - 24.93)^2 \\ +(24.10 - 24.93)^2 \\ +(23.74 - 24.93)^2 \\ +(25.10 - 24.93)^2\end{pmatrix} + \begin{pmatrix}(23.40 - 22.61)^2 \\ +(21.80 - 22.61)^2 \\ +(23.50 - 22.61)^2 \\ +(22.75 - 22.61)^2 \\ +(21.60 - 22.61)^2\end{pmatrix} + \begin{pmatrix}(20.00 - 20.59)^2 \\ +(22.20 - 20.59)^2 \\ +(19.75 - 20.59)^2 \\ +(20.60 - 20.59)^2 \\ +(20.40 - 20.59)^2\end{pmatrix}$$

$$= \begin{pmatrix}.2209 \\ + 1.9044 \\ + .6889 \\ + 1.4161 \\ + .0289\end{pmatrix} + \begin{pmatrix}.6241 \\ + .6561 \\ + .7921 \\ + .0196 \\ + 1.0201\end{pmatrix} + \begin{pmatrix}.3481 \\ + 2.5921 \\ + .7056 \\ + .0001 \\ + .0361\end{pmatrix}$$

$$= 11.0532$$

$$\text{SST} = \sum_{j=1}^{c}\sum_{i=1}^{n_j}\left(X_{ij} - \overline{\overline{X}}\right)^2 = \begin{pmatrix}(25.40 - 22.71)^2 \\ +(26.31 - 22.71)^2 \\ +(24.10 - 22.71)^2 \\ +(23.74 - 22.71)^2 \\ +(25.10 - 22.71)^2\end{pmatrix} + \begin{pmatrix}(23.40 - 22.71)^2 \\ +(21.80 - 22.71)^2 \\ +(23.50 - 22.71)^2 \\ +(22.75 - 22.71)^2 \\ +(21.60 - 22.71)^2\end{pmatrix} + \begin{pmatrix}(20.00 - 22.71)^2 \\ +(22.20 - 22.71)^2 \\ +(19.75 - 22.71)^2 \\ +(20.60 - 22.71)^2 \\ +(20.40 - 22.71)^2\end{pmatrix}$$

$$= \begin{pmatrix} 7.2361 \\ +12.9600 \\ +1.9321 \\ +1.0609 \\ +5.7121 \end{pmatrix} + \begin{pmatrix} .4761 \\ +.8281 \\ +.6241 \\ +.0016 \\ +1.2321 \end{pmatrix} + \begin{pmatrix} 7.3441 \\ +.2601 \\ +8.7616 \\ +4.4521 \\ +5.3361 \end{pmatrix}$$

$$= 58.2172$$

The respective mean square terms are obtained by dividing these sums of squares by their corresponding degrees of freedom. Since $c = 3$ and $n = 15$, we have

$$\text{MSA} = \frac{\text{SSA}}{c-1} = \frac{47.164}{3-1} = \frac{47.164}{2} = 23.582$$

$$\text{MSW} = \frac{\text{SSW}}{n-c} = \frac{11.0532}{15-3} = \frac{11.0532}{12} = 0.9211$$

so that using Equation (8.12) to test H_0, we obtain

$$F = \frac{\text{MSA}}{\text{MSW}} = \frac{23.582}{0.9211} = 25.60$$

If a .05 level of significance is chosen, the critical value of the F statistic would be obtained from Table E.5, a replica of which is presented as Table 8.8. The values in the body of this table refer to selected upper-tailed percentage points of the F distribution. In our productivity study, since there are 2 degrees of freedom in the numerator of the F ratio and 12 degrees of freedom in the denominator, the critical value of F at the .05 level of significance is 3.89. Because our computed test statistic $F = 25.60$ exceeds this critical F value, the null hypothesis may be rejected (see Figure 8.12). The production manager may conclude that there is a significant difference in the average time required to complete the filling process on the three machines.

The corresponding ANOVA summary table is presented in Table 8.9 and contains the exact p-value for the calculated value of F obtained from the Excel spreadsheet package (see

Table 8.8 Obtaining the critical value of F with 2 and 12 degrees of freedom at the .05 level.

Denominator, df_2	Numerator, df_1					
	1	2	3	4	5	6
⋮	⋮	⋮	⋮	⋮	⋮	⋮
5	6.61	5.79	5.41	5.19	5.05	4.95
6	5.99	5.14	4.76	4.53	4.39	4.28
7	5.59	4.74	4.35	4.12	3.97	3.87
8	5.32	4.46	4.07	3.84	3.69	3.58
9	5.12	4.26	3.86	3.63	3.48	3.37
10	4.96	4.10	3.71	3.48	3.33	3.22
11	4.84	3.98	3.59	3.36	3.20	3.09
12	4.75	3.89	3.49	3.26	3.11	3.00
13	4.67	3.81	3.41	3.18	3.03	2.92
14	4.60	3.74	3.34	3.11	2.96	2.85

Source: Extracted from Table E.5.

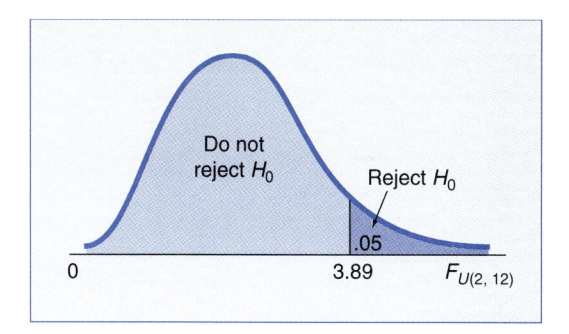

FIGURE 8.12 Regions of rejection and nonrejection for the analysis of variance at the .05 level of significance with 2 and 12 degrees of freedom.

Table 8.9 Analysis-of-variance table for the productivity study.

Source	Degrees of Freedom	Sums of Squares	Mean Square (Variance)	F	p-value
Among groups (machines)	3 − 1 = 2	47.1640	23.5820	25.60	.000
Within groups (machines)	15 − 3 = 12	11.0532	.9211		
Total	15 − 1 = 14	58.2172			

Reference 8). Note that the p-value or probability of obtaining an F statistic of 25.60 or larger when the null hypothesis is true is 0.000. Since this p-value is less than the specified α of .05, the null hypothesis is rejected.

8.6.4 Reflection

Let us review what we have just developed. From Table 8.7 on page 420 and Figure 8.11 on page 420, we noted that there were differences among the three sample means. Under the null hypothesis that the population means of the three groups were presumed equal, a measure of the total variation (or SST) among all the workers was obtained by summing up the squared differences between each observation and the overall mean, 22.71, based on all the observations. The total variation was then subdivided into two separate components (see Figure 8.7 on page 415), one portion consisting of variation among the groups and the other consisting of variation within the groups.

Why is there variation among the values; that is, why are the observations not all the same? One reason is that by treating people differently (in this case, giving them different machines to use) we affect their productivity. This would explain part of the reason why the groups have different means: the bigger the effect of the treatment, the more variation in group means we will find. But there is another reason for variability in the results, which is that people are naturally variable whether we treat them alike or not. So even within a particular group where everyone gets the same treatment (i.e., machine) there is variability. Because it occurs within each group, it is called within-group variation (or SSW).

The differences among the group means are called the among-group variation (or SSA). Part of the among-group variation, as we noted, is due to the effect of being in different groups. But even if there is no real effect of being in different groups (i.e., the null hypothesis is true), there will likely be differences among the group means. This is because variability among workers will make the sample means different just because we have different samples. Therefore, if the null hypothesis is true, then the among-group variation will estimate the population variability just as well as the within-group variation. But if the null hypothesis is false, then the among-group variation will be larger. This fact forms the basis for the one-way ANOVA F test of differences in group means.

Now let's look at what we have just accomplished. Again, from Table 8.7 on page 420 and Figure 8.11 on page 420, we noted that there were differences among the three sample means. Using the one-way ANOVA F test, the production manager has found sufficient evidence to conclude that there is a significant *treatment effect* across the levels (or groups) of the factor of interest, machine brands. That is, there is evidence that the population means differ with respect to the time required to complete the filling process.

What we don't yet know, however, is which machine or machines differ from the other(s). All we know is that there is sufficient evidence to state that the population means are not all the same; that is, at least one or some combination of them is significantly different. To determine exactly which machine or machines differ, we shall make all possible pairwise comparisons between machines and use a procedure developed by John Tukey (and later modified, independently by Tukey and by C. Y. Kramer, for situations in which the sample sizes differ) to draw our conclusions. (See References 6, 7, and 11.)

8.6.5 Multiple Comparisons: The Tukey-Kramer Procedure

In the cereal filling machines study, the analysis of variance was used to determine whether there is a difference among several groups in the average time to complete a task. Once differences in the means of the groups are found, it is important that we determine which particular groups are different.

Although many procedures are available (see References 6 and 9), we will focus on the **Tukey-Kramer procedure** to determine which of the c means are significantly different from the others. This method is an example of a *post hoc* (or **a posteriori**) comparison procedure, since the hypotheses of interest are formulated *after* the data have been inspected.

The Tukey-Kramer procedure enables us to simultaneously examine comparisons between all pairs of groups. The first step involved is to compute the differences $\bar{X}_j - \bar{X}_{j'}$ (where $j \neq j'$) among all $c(c-1)/2$ pairs of means. The **critical range** for the Tukey-Kramer procedure is then obtained as follows:

$$\text{Critical range} = Q_{U(c, n-c)} \sqrt{\frac{\text{MSW}}{2}\left(\frac{1}{n_j} + \frac{1}{n_{j'}}\right)} \tag{8.13}$$

where

$Q_{U(c, n-c)}$ = upper-tailed critical value of the Studentized range (see Table E.7) with c and $(n-c)$ degrees of freedom

MSW = Mean Square Within groups

n_j = sample size for group j

$n_{j'}$ = sample size for group j'

If the sample sizes differ, a critical range would be computed for each pairwise comparison of sample means. The final step is to compare each of the $c(c-1)/2$ pairs of means against its corresponding critical range. A specific pair would be declared significantly different if the absolute difference in the sample means $|\bar{X}_j - \bar{X}_{j'}|$ exceeds the critical range.

To apply the Tukey-Kramer procedure, we return to the productivity study. Using the ANOVA procedure, we concluded that there is a difference in the average time to complete a task by using the three machines. Since there are three groups, there are $(3)(3-1)/2 = 3$ possible pairwise comparisons to be made. From Table 8.7 on page 420, the absolute mean differences are

1. $|\bar{X}_1 - \bar{X}_2| = |24.93 - 22.61| = 2.32$
2. $|\bar{X}_1 - \bar{X}_3| = |24.93 - 20.59| = 4.34$
3. $|\bar{X}_2 - \bar{X}_3| = |22.61 - 20.59| = 2.02$

Only one critical range needs to be obtained here because the three groups have equal-sized samples. To determine the critical range, from Table 8.9 on page 423, we have MSW = .9211 and $n_j = 5$. From Table E.7, for $\alpha = .05$, $c = 3$, and $n - c = 15 - 3 = 12$, the upper-tailed critical value of $Q_{U(3,12)}$ is 3.77 (see Table 8.10). From Equation (8.13), we have

$$\text{Critical range} = 3.77 \sqrt{\left(\frac{.9211}{2}\right)\left(\frac{1}{5} + \frac{1}{5}\right)} = 1.618$$

Since 2.32 > 1.618, 4.34 > 1.618, and 2.02 > 1.618, it would be concluded that there is a significant difference between each pair of means. Hence, the production manager would purchase machine III because the average time for completing the task on it is fastest.

Table 8.10 Obtaining the Studentized range Q statistic for $\alpha = .05$ with 3 and 12 degrees of freedom.

Denominator Degrees of Freedom	\multicolumn{7}{c	}{Numerator Degrees of Freedom}
	2	3	4	5	6	7	8
1	18.0	27.0	32.8	37.1	40.4	43.1	45.4
2	6.09	8.3	9.8	10.9	11.7	12.4	13.0
3	4.50	5.91	6.82	7.50	8.04	8.48	8.85
⋮	⋮	⋮	⋮	⋮	⋮	⋮	⋮
11	3.11	3.82	4.26	4.57	4.82	5.03	5.20
12	3.08	3.77	4.20	4.51	4.75	4.95	5.12
13	3.06	3.73	4.15	4.45	4.69	4.88	5.05
14	3.03	3.70	4.11	4.41	4.64	4.83	4.99

Source: Extracted from Table E.7.

8.6.6 ANOVA Assumptions

In our cereal filling machines study, it seems that the analysis is now complete. But is it? Aside from our exploratory investigations in Figure 8.11 on page 420, we have not yet thoroughly evaluated the assumptions underlying the one-way F test. How can the production manager know whether the one-way F test is an appropriate procedure for analyzing his experimental data?

In Chapter 7 and in this chapter we mentioned the assumptions made in the application of each hypothesis-testing procedure and the consequences of departures from these assumptions. To employ the one-way ANOVA F test, we must also make certain assumptions about the data being investigated. There are three major assumptions in the analysis of variance:

1. Randomness and independence of errors
2. Normality
3. Homogeneity of variance

The first assumption, **randomness** and **independence of errors**, must be met for all procedures discussed in this chapter, not only those dealing with ANOVA, since the validity of any experiment depends on random sampling and/or the randomization process. To avoid biases in the outcomes, it is essential that either the obtained data be considered as randomly and independently drawn from the c populations or that the items or subjects be randomly assigned to the c levels of the factor of interest (i.e., the treatment groups). Therefore, the assumption of randomness and independence refers not to haphazard mistakes, but to the difference of each observed value from its own group mean. The assumption is that these differences should be independent for each observed value. This means that the difference (or *error*) for one observation should not be related to the difference (or *error*) for any other observation. This assumption might be violated, for example, in our cereal filling machines study if one worker aided another in completing the filling process. Most often, however, this assumption is violated when data are collected over a period of time, because observations made at adjacent time points may be more alike than those made at very different times. Consider, for example, temperature recorded every day for a month. The temperature on a given day is likely to be near what it was the day before, but less likely to be close to the temperature several weeks later.

Again, departures from this assumption can seriously affect inferences from the analysis of variance. These problems are discussed more thoroughly in References 2, 3, and 9.

The second assumption, **normality**, states that the values in each group are normally distributed. Just as in the case of the t test, the one-way ANOVA F test is fairly robust against departures from the normal distribution. As long as the distributions are not extremely different from a normal distribution, the level of significance of the analysis-of-variance test is usually not greatly affected by lack of normality, particularly for large samples. When only the normality assumption is seriously violated, nonparametric alternatives to the one-way ANOVA F test are available (see Section 8.7).

The third assumption, **homogeneity of variance**, states that the variance within each population should be equal for all populations (i.e., $\sigma_1^2 = \sigma_2^2 = \cdots = \sigma_c^2$). This assumption is needed in order to combine or pool the variances within the groups into a single within-group source of variation, SSW. If there are equal sample sizes in each group, inferences based on the F distribution may not be seriously affected by unequal variances. If, however, there are unequal sample sizes in different groups, unequal variances from group to group can have serious effects on drawing inferences from the analysis of variance. Thus, from the perspective of computational simplicity, robustness, and power, there should be equal sample sizes in all groups whenever possible.

When only the homogeneity-of-variance assumption is violated, alternative procedures are available (see References 2 and 3). However, if both the normality and homogeneity-of-variance assumptions have been violated, an appropriate *data transformation* may be used that will both normalize the data and reduce the differences in variances (see References 1 and 9) or, alternatively, a more general nonparametric procedure may be employed (see References 2 and 3).

8.6.7 Using Microsoft Excel for the One-Way ANOVA *F* Test

In this section we developed the one-way analysis of variance F test to determine whether any differences exist between the means of two or more groups. We will now explain how Microsoft Excel can be used for this procedure. If raw data are available, the ANOVA: Single Factor option of the Data Analysis tool can be used.

To illustrate the use of this tool for the One-Way ANOVA F test, open the FILL-1.XLS workbook which contains the cereal filling machines data of Table 8.7 on page 420. Then do the following:

8–6.XLS

① Click on the Data sheet tab. Note that the data have been set up in columns A, B, and C with column headings indicating the names of the three machines.

② Select Tools | Data Analysis, select Anova: Single Factor from the Analysis tools list box and click the OK button. In the Anova: Single Factor dialog box which appears, do the following:

③ **(a)** In the Input Range edit box, enter A1:C6.
  (b) Select the Grouped by Columns option button.
  (c) Select the Labels in First Row check box.
  (d) In the Alpha edit box, enter the level of significance (for these data, enter .05).

④ Select the New Worksheet Ply option button and enter the name OneWay ANOVA. The dialog box should now appear similar to the one illustrated in Figure 8.11.Excel. Click the OK button.

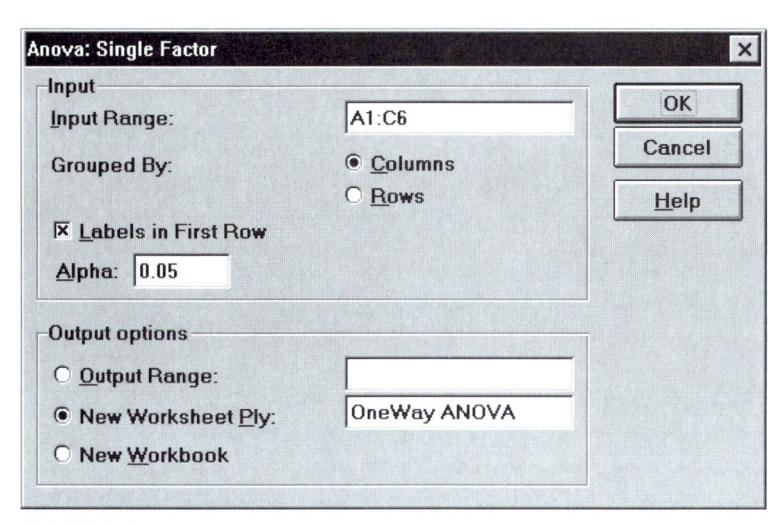

FIGURE 8.11.EXCEL Dialog box for Microsoft Excel ANOVA: Single Factor data analysis option.

Microsoft Excel performs the ANOVA and produces the results shown in Figure 8.12.Excel.
From Figure 8.12.Excel, we observe that in the first table, labeled Summary, Excel has computed the sample size, sum, mean, and variance for each group. The second table, labeled ANOVA, consists of the ANOVA summary table for these data, in which the Sum of Squares (SS), degrees of freedom, Mean Squares (MS), the F-test statistic, the p-value, and the critical value are displayed. For these data, the F-test statistic, equal to 25.602, is greater than the critical value of 3.885, so the null hypothesis may be rejected. Alternatively, since the p-value for the two-tailed test equals .0000468 (4.68E-05 in scientific notation), which is less than .05, the null hypothesis may be rejected.

● **Working with Stacked Data** As in the case of the t test for the difference between two means, and the F test for the difference between two variances, to use the Data Analysis tool for the OneWay ANOVA, the data for each of the c groups must be in separate columns. If the data were in a stacked format, they would have to be unstacked first as discussed in Section 8.3.5.

	A	B	C	D	E	F	G
1	Anova: Single Factor						
2							
3	SUMMARY						
4	Groups	Count	Sum	Average	Variance		
5	Machine I	5	124.65	24.93	1.0648		
6	Machine II	5	113.05	22.61	0.778		
7	Machine III	5	102.95	20.59	0.9205		
8							
9							
10	ANOVA						
11	Source of Variation	SS	df	MS	F	P-value	F crit
12	Between Groups	47.164	2	23.582	25.602	4.68E-05	3.88529
13	Within Groups	11.0532	12	0.9211			
14							
15	Total	58.2172	14				

FIGURE 8.12.EXCEL Results obtained from **ANOVA: Single Factor** for the cereal filling machines data.

▲ WHAT IF EXAMPLE

We can use the Data sheet to explore the effect of changing data values on the Analysis of Variance results. For example, suppose that the third value for Machine III was 29.75 instead of 19.75. If we change this value on the Data sheet and then use the Data Analysis ANOVA: Single Factor tool for the revised data, we would find the mean for the third machine would become 22.59, the F statistic for the test for the difference between the means would become 1.462, and the p-value would be .27. For these results, at the .05 level of significance, since 1.462 < 3.885 or the p-value = .27 > .05, the null hypothesis cannot be rejected, and we would conclude that there is no evidence of a difference in the average processing time among the three machines.

8.6.8 Using Microsoft Excel for the Tukey-Kramer Multiple Comparisons Procedure

In addition to the one-way ANOVA, in Section 8.6.5, we also developed the Tukey-Kramer procedure for making multiple comparisons between groups of means. Although Excel does not have a Data Analysis tool for the Tukey-Kramer multiple comparisons procedure, Excel formulas can be used instead. The design for the worksheet for this test, named TukeyKramer for this comparison, is shown in Table 8.5.Excel.

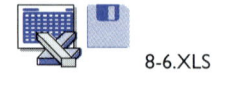

Referring to Equation (8.13) on page 424, we see that in order to use the Tukey-Kramer procedure, we need to have the sample mean $\bar{X}$ and the sample size in each group. In addition, we also need to know the Mean Squares Within groups, and the value of the Studentized range Q statistic. With the exception of the Studentized range Q statistic, each of these can be copied from the ANOVA table obtained by using the Data Analysis ANOVA: One Factor option. To illustrate how we develop multiple comparisons using the Tukey-Kramer procedure, refer to the cereal filling machines data of Table 8.7 on page 420. With the workbook

Table 8.5. Excel Design for obtaining Tukey-Kramer multiple comparisons.

	A	B
1	Tukey-Kramer Multiple Comparisons	
2		
3	Mean Group 1	xxx
4	n Group 1	xx
5	Mean Group 2	xxx
6	n Group 2	xx
7	Mean Group 3	xxx
8	n Group 3	xx
9	MSW	xxx
10	Q Statistic	xxx
11	Comparison of Group 1 to Group 2	
12	Absolute Difference	=ABS(B3–B5)
13	Standard Error of Difference	=SQRT((B9/2)*((1/B4)+(1/B6)))
14	Critical Range	=B10*B13
15	Means of Groups 1 and 2 are	=IF(B12>B14,"Different","Not Different")
16	Comparison of Group 1 to Group 3	
17	Absolute Difference	=ABS(B3–B7)
18	Standard Error of Difference	=SQRT((B9/2)*((1/B4)+(1/B8)))
19	Critical Range	=B10*B18
20	Means of Groups 1 and 3 are	=IF(B17>B19,"Different","Not Different")
21	Comparison of Group 2 to Group 3	
22	Absolute Difference	=ABS(B5–B7)
23	Standard Error of Difference	=SQRT((B9/2)*((1/B6)+(1/B8)))
24	Critical Range	=B10*B23
25	Means of Groups 2 and 3 are	=IF(B22>B24,"Different","Not Different")

that was developed in Section 8.6.7 open, insert a worksheet named TukeyKramer and do the following:

① Transfer the mean and sample size for each of the three groups and the Mean Squares Within Groups (MSW) from the OneWay ANOVA sheet to the cell range B3:B9. (Because the results on the OneWay ANOVA sheet cannot dynamically change, using Edit | Copy and Edit | Paste to transfer values would be a better choice than using formulas to transfer values, as is done in many other examples in this text.)

② Enter the Studentized range Q statistic into cell B10. For the data of Table 8.7, with 3 and 20 degrees of freedom, at the .05 level of significance, the value of this cell would be 3.77 (see Table 8.10 on page 425).

③ The absolute difference in the means of groups 1 and 2 is computed in cell B12 using the formula =ABS(B3–B5).

④ The standard error of the difference is computed in cell B13 using the formula =SQRT((B9/2)*((1/B4)+(1/B6))).

⑤ The critical range is computed in cell B14 by multiplying the Q statistic in cell B10 by the standard error of the difference computed in cell B13.

⑥ In cell B15 the absolute difference in the means of the two groups is compared to the critical range. If the absolute difference is greater than the critical range, the two group means are considered to be significantly different. This comparison is done using the formula =IF(B12>B14,"Different," "Not Different").

⑦ Similar comparisons for groups 1 and 3 and groups 2 and 3 are developed in cells B16:B20 and B21:B25, respectively.

The results are presented in Figure 8.13.Excel.

	A	B
3	Mean Group 1	24.93
4	n Group 1	5
5	Mean Group 2	22.61
6	n Group 2	5
7	Mean Group 3	20.59
8	n Group 3	5
9	MSW	0.9211
10	Q Statistic	3.77
11	Comparison of Group 1 to Group 2	
12	Absolute Difference	2.32
13	Standard Error of Difference	0.429208574
14	Critical Range	1.618116324
15	Means of Groups 1 and 2 are	Different
16	Comparison of Group 1 to Group 3	
17	Absolute Difference	4.34
18	Standard Error of Difference	0.429208574
19	Critical Range	1.618116324
20	Means of Groups 1 and 3 are	Different
21	Comparison of Group 2 to Group 3	
22	Absolute Difference	2.02
23	Standard Error of Difference	0.429208574
24	Critical Range	1.618116324
25	Means of Groups 2 and 3 are	Different

FIGURE 8.13.EXCEL Tukey-Kramer multiple comparisons obtained from Microsoft Excel for the cereal filling machines data.

Problems for Section 8.6

Note: *The problems in this section may be solved using Microsoft Excel.*

8.23 How does the one-way ANOVA F test differ from the test for the differences between two variances of Section 8.5?

8.24 Compare and contrast the assumptions of the one-way ANOVA F test and the assumptions of the t test for the difference between the means of two populations. Discuss fully.

INSPGM.TXT

8.25 The personnel manager of a large insurance company wishes to evaluate the effectiveness of four different sales-training programs designed for new employees. A group of 32 recently hired college graduates are randomly assigned to the four programs so that there are eight subjects in each program. At the end of the month-long training period, a standard exam is administered to the 32 subjects; the scores are shown on page 431:
(a) Construct an appropriate graph, plot, or chart of the data.
(b) Describe any trends or relationships that might be apparent within or among the groups.
(c) Does the variation within the groups seem to be similar for all groups? Discuss.

Programs			
A	B	C	D
66	72	61	63
74	51	60	61
82	59	57	76
75	62	60	84
73	74	81	58
97	64	55	65
87	78	70	69
78	63	71	80

(d) If conditions are appropriate, at an $\alpha = .05$ level of significance, use the one-way F test to determine whether there is evidence of a difference in the four sales-training programs.
(e) Based on your results in (d), if appropriate, use the Tukey-Kramer procedure to make all pairwise comparisons of the training programs. (Use an overall level of significance of .05.)
(f) **ACTION** Prepare an executive summary that the personnel manager might send to the vice-president for operations in this company.

8.26 A metallurgist tests five different alloys for tensile strength. He tests several samples of each alloy; the tensile strengths of each sample are given here.

TENSILE.TXT

Alloy 1:	12.4	19.8	15.2	14.8	18.5
Alloy 2:	8.9	11.6	10.0	10.3	
Alloy 3:	10.5	13.8	12.1	11.9	12.6
Alloy 4:	12.8	14.2	15.9	14.1	
Alloy 5:	16.4	15.9	17.8	20.3	

The metallurgist would like to know if there is evidence of a difference in the tensile strengths of the various alloys and, if so, which ones are significantly stronger than the others.
(a) Completely analyze the data. (Use $\alpha = .05$.)
(b) **ACTION** Describe the results to the metallurgist in a memo.

8.27 A statistics professor wants to study four different strategies of playing the game of Blackjack (Twenty-One). The four strategies are
    (1) Dealer's strategy
    (2) Five-count strategy
    (3) Basic ten-count strategy
    (4) Advanced ten-count strategy

A calculator that could play Blackjack is utilized and data from five sessions of each strategy are collected. The profits (or losses) from each session are as follows:

BLACKJ.TXT

		Strategy	
Dealer's	Five Count	Basic Ten Count	Advanced Ten Count
−$56	−$26	+$16	+$60
−$78	−$12	+$20	+$40
−$20	+$18	−$14	−$16
−$46	−$8	+$ 6	+$12
−$60	−$16	−$25	+$ 4

The professor wants to know whether there is evidence of a difference among the four strategies and, if so, which strategies are superior with respect to potential profitability.
(a) Completely analyze the data. (Use $\alpha = .01$.)
(b) **ACTION** Write a letter to the professor explaining your findings.

LOCATE.TXT

8.28 The retailing manager of a food chain wishes to determine whether product location has any effect on the sale of pet toys. Three different aisle locations are to be considered: front, middle, and rear. A random sample of 18 stores is selected with 6 stores randomly assigned to each aisle location. The size of the display area and price of the product are constant for all stores. At the end of a 1-week trial period, the sales volume (in thousands of dollars) of the product in each store was as follows:

\multicolumn{3}{c	}{Aisle Location}
Front	Middle	Rear
8.6	2.0	4.6
7.2	3.2	2.8
5.4	2.4	6.0
4.0	1.8	2.2
5.0	1.4	2.8
6.2	1.6	4.0

(a) At the .05 level of significance, is there evidence of a difference in average sales between the various aisle locations?
(b) If appropriate, use the Tukey-Kramer procedure to determine which aisle locations are different in average sales. (Use $\alpha = .05$.)
(c) What should the retailing manager conclude? Discuss fully the retailing manager's options with respect to aisle locations.

WORKATT.TXT

8.29 To examine effects of the work environment on attitude toward work, an industrial psychologist randomly assigns a group of 18 recently hired sales trainees to three "home rooms"—six trainees per room. Each room is identical except for wall color. One is light green, another is light blue, and the third is deep red.

During the week-long training program, the trainees stay mainly in their respective home rooms. At the end of the program, an attitude scale is used to measure each trainee's attitude toward work (a low score indicates a poor attitude, a high score a good attitude). The following data are obtained:

\multicolumn{3}{c	}{Room Color}
Light Green	Light Blue	Deep Red
46	59	34
51	54	29
48	47	43
42	55	40
58	49	45
50	44	34

Based on these data, the industrial psychologist wants to determine whether there is evidence that work environment (i.e., color of room) has an effect on attitude toward work and, if so, which room color(s) significantly enhances attitude.
(a) Completely analyze the data. (Use $\alpha = .05$.)
(b) **ACTION** Based on this thorough analysis, draft a report discussing the implications of the findings for office design in large firms, knowing that the industrial psycholo-

gist might use this when meeting with the vice-president for human resources at the company.

8.30 A senior partner in a brokerage firm wishes to determine whether there is really any difference between long-run performance of different categories of people hired as customers' representatives. The junior members of the firm are classified into four groups: professionals who have changed careers, recent business school graduates, former salesmen, and brokers hired from competing firms. A random sample of six individuals in each of these categories is selected, and a detailed performance score is obtained.

BROKER.TXT

Customer Representative Backgrounds

Professionals	Business School Grads	Salesmen	Brokers
88	65	61	83
85	73	67	87
95	54	74	90
96	72	65	84
91	81	68	92
88	69	77	94

(a) Is there evidence of a difference in the average performance score for the various categories? (Use $\alpha = .05$.)
(b) **ACTION** Write a memo to the senior partner explaining your findings.

8.7 KRUSKAL-WALLIS RANK TEST FOR DIFFERENCES IN c MEDIANS

8.7.1 Introduction

The **Kruskal-Wallis rank test** for differences in c medians (where $c > 2$) may be considered an extension of the Wilcoxon rank sum test for two independent samples discussed in Section 8.4. Thus, the Kruskal-Wallis test enjoys the same power properties relative to the one-way ANOVA F test as does the Wilcoxon rank sum test relative to the t test for two independent samples (Section 8.3). That is, the Kruskal-Wallis procedure has proven to be almost as powerful as the F test under conditions appropriate to the latter and even more powerful than the classical procedure when its assumptions (see Section 8.6.6) are violated.

8.7.2 Development

The Kruskal-Wallis rank test is most often used to test whether c independent sample groups have been drawn from populations possessing equal medians. That is, we may test

$$H_0: M_1 = M_2 = \cdots = M_c$$

against the alternative

$$H_1: \text{Not all } M_j \text{ are equal (where } j = 1, 2, \ldots, c\text{)}.$$

For such situations, it is necessary to assume that

1. The c samples are randomly and independently drawn from their respective populations.
2. The underlying random phenomenon of interest is continuous (to avoid ties).

3. The observed data constitute at least an ordinal scale of measurement, both within and among the c samples.
4. The c populations have the same *variability*.
5. The c populations have the same *shape*.

Interestingly, the Kruskal-Wallis procedure still makes less stringent assumptions than does the F test. To employ the Kruskal-Wallis procedure, the measurements need only be ordinal over all sample groups, and the common population distributions need only be continuous—their common shapes are irrelevant. On the other hand, to utilize the classical F test, the level of measurement must be more sophisticated, and we must assume that the c samples are coming from underlying normal populations having equal variances.

To perform the Kruskal-Wallis rank test, we must first (if necessary) replace the observations in the c samples with their combined ranks such that rank 1 is given to the smallest of the combined observations and rank n to the largest of the combined observations (where $n = n_1 + n_2 + \cdots + n_c$). If any values are tied, they are assigned the average of the ranks they would otherwise have been assigned if ties had not been present in the data.

The Kruskal-Wallis test statistic H may be computed from

$$H = \left[\frac{12}{n(n+1)} \sum_{j=1}^{c} \frac{T_j^2}{n_j} \right] - 3(n+1) \tag{8.14}$$

where
n = total number of observations over the combined samples, (i.e., $n = n_1 + n_2 + \cdots + n_c$)

n_j = number of observations in the jth sample; $j = 1, 2, \ldots, c$

T_j = sum of the ranks assigned to the jth sample

T_j^2 = square of the sum of the ranks assigned to the jth sample

As the sample sizes in each group get large (greater than 5), the test statistic H may be approximated by the chi-square distribution with $c - 1$ degrees of freedom. A chi-square distribution is a skewed distribution whose shape depends solely on the number of degrees of freedom. As the number of degrees of freedom increases, a chi-square distribution becomes more symmetrical. Table E.4 contains various upper-tail areas for chi-square distributions pertaining to different degrees of freedom. A portion of this table is displayed in Table 8.12 on page 436.

Thus, for any selected level of significance α, the decision rule would be to reject the null hypothesis if the computed value of H exceeds the upper-tail critical χ^2 value and not to reject the null hypothesis if H is less than or equal to the critical χ^2 value (see Figure 8.13). That is,

Reject H_0 if $H > \chi^2_{c-1}$;

otherwise do not reject H_0.

8.7.3 Application

To illustrate the Kruskal-Wallis rank test for differences in c medians, let us return to our cereal filling machines study of the previous section. You may recall that the production manager at the plant in which cereal is being filled in 368-gram sized boxes is considering replacing an old machine that directly affects output in the production process. To this end, he has conducted an

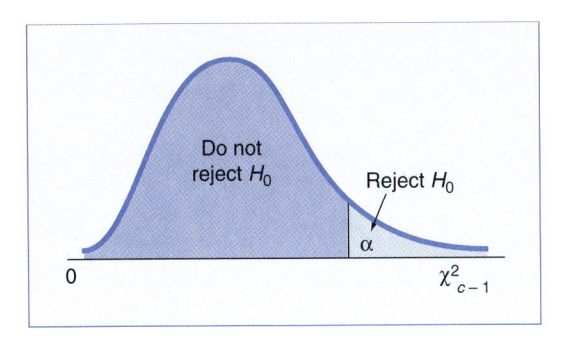

FIGURE 8.13 Determining the rejection region.

experiment to determine whether there are any significant differences among the three new machines in the average time (in seconds) it takes factory workers using them to complete the filling process. Fifteen factory workers of similar experience, ability, and age were randomly assigned to receive training on one of the three brands of machines in such a manner that there are five factory workers for each machine. After an appropriate amount of training and practice, the production manager measures the time (in seconds) it takes the factory workers to complete the filling process using their respective equipment. The results of this experiment are displayed in Table 8.7 on page 420 along with some summary computations, and a scatter plot is depicted in Figure 8.11 on page 420 so that a visual, exploratory evaluation of potential trends, relationships, and violations in assumptions of particular testing procedures can be made. If the production manager does not wish to make the assumption that the time measurements (in seconds) are normally distributed across the underlying populations, the nonparametric Kruskal-Wallis rank test for differences in the three population medians can be used.

The null hypothesis to be tested is that the median times to complete the filling process on the three machines are equal; the alternative is that at least one of the machines differs from the others. Thus, substituting 1, 2, 3 for I, II, III, we have

$$H_0: M_1 = M_2 = M_3$$

H_1: Not all the medians are equal.

Converting the 15 time measurements in Table 8.7 to ranks, we obtain Table 8.11.

Table 8.11 Converting data to ranks.

	Machine	
I	II	III
14	9	2
15	6	7
12	10	1
11	8	4
13	5	3

Source: Data are taken from Table 8.7 on page 420.

We note that in the combined ranking, the third employee assigned to machine III completed the filling process fastest and received a rank of 1. The first employee assigned to machine III was second fastest and received a rank of 2. The second employee assigned to machine I completed the filling process in the slowest time and received a rank of 15.

8.7 Kruskal-Wallis Rank Test for Differences in c Medians

After all the ranks are assigned, we then obtain the sum of the ranks for each group:

Rank sums: $T_1 = 65 \quad T_2 = 38 \quad T_3 = 17$

As a check on the rankings, we have

$$T_1 + T_2 + T_3 = \frac{n(n+1)}{2}$$

$$65 + 38 + 17 = \frac{(15)(16)}{2}$$

$$120 = 120$$

Choosing a .05 level of significance, to test the null hypothesis of equal population medians, we use Equation (8.14):

$$H = \left[\frac{12}{n(n+1)} \sum_{j=1}^{c} \frac{T_j^2}{n_j}\right] - 3(n+1)$$

$$= \left\{\frac{12}{(15)(16)}\left[\frac{(65)^2}{5} + \frac{(38)^2}{5} + \frac{(17)^2}{5}\right]\right\} - 3(16)$$

$$= \left(\frac{12}{240}\right)[1,191.6] - 48$$

$$= 59.58 - 48 = 11.58$$

Using Table E.4, the upper-tail critical χ^2 value with $c - 1 = 2$ degrees of freedom and corresponding to a .05 level of significance is 5.991 (see Table 8.12, which is a replica of Table E.4). Since the computed value of the test statistic H exceeds the critical value, we may reject the null hypothesis and conclude that not all the machines are the same with respect to median time required for a worker to complete the filling process. (That is, if the null hypothesis were really true, the probability of obtaining such a result or one even more extreme is less than .05.) You may note that these are the same results obtained using the one-way ANOVA F test in Section 8.6.

Since we rejected the null hypothesis and concluded that there is evidence of a significant difference among the machines in the median time it takes to complete the filling process (as we did with the one-way ANOVA procedure), the next step would be a simultaneous comparison of all possible pairs of machines to determine which one or ones differ from the oth-

Table 8.12 Obtaining the approximate χ^2 critical value for the Kruskal-Wallis test at the .05 level of significance with 2 degrees of freedom.

Degrees of Freedom	.995	.99	.975	.95	.90	...	.10	.05	.025
					Upper-Tail Area				
1	—	—	0.001	0.004	0.016	...	2.706	3.841	5.024
2	0.010	0.020	0.051	0.103	0.211	...	4.605	5.991	7.378
3	0.072	0.115	0.216	0.352	0.584	...	6.251	7.815	9.348
4	0.207	0.297	0.484	0.711	1.064	...	7.779	9.488	11.143
5	0.412	0.554	0.831	1.145	1.610	...	9.236	11.071	12.833

Source: Extracted from Table E.4.

ers. As a follow-up to the Kruskal-Wallis rank test, a *post hoc* multiple comparison procedure proposed by O. J. Dunn (see References 1 and 4) could be undertaken.

8.7.4 Using Microsoft Excel for the Kruskal-Wallis Rank Test for the Difference in c Medians

In this section we developed the Kruskal-Wallis rank sum test as a nonparametric alternative to the one-way analysis of variance. Although Excel does not have a Data Analysis tool for the Kruskal-Wallis test, Excel formulas can be used instead in a manner similar to the development of the Wilcoxon test in Section 8.4.4. The design for the Calculations sheet for this test is shown in Table 8.6.Excel.

Table 8.6.Excel Design for Calculations sheet obtaining the Kruskal-Wallis test statistic.

	A	B
1	Kruskal-Wallis Test	
2		
3	Sum of Ranks Group 1	=SUM(Data! E:E)
4	Sample Size Group 1	=COUNT(Data! E:E)
5	Sum of Ranks Group 2	=SUM(Data!F:F)
6	Sample Size Group 2	=COUNT(Data!F:F)
7	Sum of Ranks Group 3	=SUM(Data!G:G)
8	Sample Size Group 3	=COUNT(Data!G:G)
9	Total Sample Size n	=B4+B6+B8
10	Total Sum of Ranks	=(B9*(B9+1))/2
11	Total Sums Check	=IF(B10=B3+B5+B7,"Passed","Failed--Review manually assigned ranks")
12	α	.xx
13	Sum of Squared Ranks/Sample Size	=(B3^2/B4)+(B5^2/B6)+(B7^2/B8)
14	H	=(12/(B9*(B9+1)))*B13–(3*(B9+1))
15	Number of Groups	xx
16	Critical Value	=CHIINV(B12,B15–1)
17	p-Value	=CHIDIST(B14,B15–1)
18	Decision	=IF(B17<B12,"Reject","Do not Reject")

To implement this sheet, the ranks of each cereal filling machine must first be determined and added to the Data sheet. To do this, begin by opening the FILL-2.XLS workbook and select the Data sheet. Note that the data is arranged in stacked order, a requirement for this procedure. (Had the data been unstacked as they are in the FILL-1.XLS workbook used in Section 8.6.7, the data would first need to be stacked by moving all the data values into one column and by adding a column to identify the groups.) To determine the ranks for the three groups, do the following:

8-7-4.XLS

① Sort the data in ascending order by processing time. To do this, select any cell in the processing time column (B) and then select the command Data | Sort. As was first described in Section 2.2.3, select the Ascending and Header Row option buttons, and then click the OK button. The data is now sorted by processing time.

② Create a ranks column in column C by entering the column heading Rank in cell C1 and the value 1 in cell C2. With cell C2 highlighted, select Edit | Fill | Series to use the shortcut for entering ranks first described in Section 5.6.5. Select the Columns and

Linear option buttons as before, but this time enter the stop value 15. Click the OK button. The ranks 2 to 15 are entered into the range C3:C16.

❸ Manually check for tied values and give each of the tied observations the average of the ranks for which they are tied. (For the processing time data, there are no ties.)

❹ Separate these ranks by group. One way to do this is by using the Advanced Data Filter feature of Microsoft Excel first discussed in Section 8.4.4. To use this feature, first copy the column headings in cells A1 and C1 (Machine and Rank) to cells D1 and E1, respectively. Then to select the machine M1 ranks, enter the value M1 in cell D2.

❺ With the Data sheet still active, select the command Data | Filter | Advanced Filter and in the Advanced Filter dialog box select the Copy to Another Location option button, and enter A1:C16 in the List Range edit box, D1:D2 in the Criteria Range edit box, and E1 in the Copy to edit box. Click the OK button. All of the machine M1 ranks now appear in Column E.

❻ To copy the machine M2 ranks to column F, copy the column heading Rank in cell C1 to cell F1 and change the value in cell D2 to M2. Then repeat the previous step, this time entering F1 in the Copy to edit box (all other choices remain the same). Click the OK button. The machine M2 ranks appear in column F.

❼ To copy the machine M3 ranks to column G, copy the column heading Rank in cell C1 to cell G1 and change the value in cell D2 to M3. Then repeat step 5, this time entering G1 in the Copy to edit box (all other choices remain the same). After clicking the OK button, the machine M3 ranks will appear in column G.

To avoid confusion, change the column heading in cell E1 to Machine 1, the column heading in cell F1 to Machine 2, and the column heading in cell G1 to Machine 3.

With the three columns of ranks now generated, the Calculations sheet can be implemented. Formulas using the SUM and COUNT functions and containing the special column range notation can be used to obtain the sums of the ranks and the sample sizes for the two groups on that sheet. To implement the calculations sheet, continue with the same workbook and insert a worksheet, renaming it Calculations. Enter the formulas =SUM(Data!E:E) and =COUNT(Data!E:E) in the cell range B3 and B4, and similar formulas in B5:B8 as shown in Table 8.6.Excel. Implement the rest of the Calculations sheet as follows:

❶ Enter the total sample size in cell B9 using the formula =B4+B6+B8.

❷ Using Equation (8.3) on page 395, compute the sum of the ranks using the formula =(B9*(B9+1))/2 in cell B10.

❸ Enter the formula =IF(B10=B3+B5+B7,"Passed","Failed—Review manually assigned ranks.") in cell B11 to compare the double-check value with the sum of the three sums in cells B3, B5, and B7. Should the comparison fail, cell B11 will display the message "Failed—Review manually assigned ranks." warning the user to go back and review the assigned ranks before continuing.

❹ Enter the level of significance α in cell B12.

❺ Compute the sum of the squared ranks divided by the sample size for each of the three groups using the formula =(B3^2/B4)+(B5^2/B6)+(B7^2/B8).

❻ Compute the H statistic using the formula =(12/(B9*(B9+1)))*B13–(3*(B9+1)).

❼ Enter the number of groups, 3, in cell B15.

❽ The critical values of χ^2 for the Kruskal-Wallis test are obtained using the CHIINV function whose format is

$$\text{CHIINV}(p, df)$$

where p = upper-tailed p-value (the probability that χ^2 will be greater than the specific value indicated)

   df = degrees of freedom

Thus, for this test we enter the formula =CHIINV(B12,B15–1) in cell B16.

⑨ The p-value is obtained by using the CHIDIST function whose format is

$$\text{CHIDIST}(\chi^2, df)$$

where χ^2 = value for the χ^2 statistic

   df = degrees of freedom

This function computes the probability of exceeding the value of χ^2 provided for the number of degrees of freedom indicated in the formula. For this Kruskal-Wallis test, we enter the formula =CHIDIST(B14,B15–1) in cell B17.

⑩ The decision to reject the null hypothesis is made if the p-value in cell B17 is less than the significance level in cell B12 using the formula =IF(B17<B12,"Reject," "Do not reject").

The results are presented in Figure 8.14.Excel.

	A	B
1	Kruskal-Wallis Test	
2		
3	Sum of Ranks Group 1	65
4	Sample Size Group 1	5
5	Sum of Ranks Group 2	38
6	Sample Size Group 2	5
7	Sum of Ranks Group 3	17
8	Sample Size Group 3	5
9	Total Sample Size n	15
10	Total Sum of Ranks	120
11	Total Sums Check	Passed
12	α	0.05
13	Sum of Squared Ranks/Sample Size	1191.6
14	H	11.58
15	Number of groups	3
16	Critical Value	5.991476357
17	p-value	0.003057982
18	Decision	Reject

FIGURE 8.14.EXCEL   Kruskal-Wallis Test for the Difference between Medians for the cereal filling machines data.

Problems for Section 8.7

Note: The problems in this section may be solved using Microsoft Excel.

- **8.31** An industrial psychologist desires to test whether the reaction times of assembly-line workers are equivalent under three different learning methods. From a group of 25 new employees, nine

INDPSYCH.TXT

are randomly assigned to method A, eight to method B, and eight to method C. After the learning period, the workers are given a task to complete, and their reaction times are measured. The following data present the rankings of the reaction times from 1 (fastest) to 25 (slowest):

	Method	
A	B	C
2	1	5
3	6	7
4	8	11
9	15	12
10	16	13
14	17	18
19	21	24
20	22	25
23		

Is there evidence of a difference in the median reaction times for these learning methods? (Use $\alpha = .01$.)

8.32 A quality engineer in a company manufacturing electronic audio equipment is inspecting a new type of battery. A batch of 20 batteries are randomly assigned to four groups (so that there are five batteries per group). Each group of batteries is then subjected to a particular pressure level—low, normal, high, very high. The batteries are simultaneously tested under these pressure levels and the times to failure (in hours) are recorded:

BATFAIL.TXT

	Pressure		
Low	Normal	High	Very High
8.0	7.6	6.0	5.1
8.1	8.2	6.3	5.6
9.2	9.8	7.1	5.9
9.4	10.9	7.7	6.7
11.7	12.3	8.9	7.8

The engineer, by experience, knows such data are coming from populations that are not normally distributed, and he wants to use a nonparametric procedure for purposes of data analysis.
(a) At the .05 level of significance, analyze the data to determine whether there is evidence of a difference in the four pressure levels with respect to median battery life.
(b) **ACTION** Write a memo to the quality engineer expressing your findings.
(c) Recommend a warranty policy with respect to battery life.

8.33 The quality engineer in a plant manufacturing stereo equipment wants to study the effect of temperature on the failure time of a particular electronic component. She designs an experiment wherein 24 of these components, all from the same batch, are randomly assigned to one of three levels of temperature and then simultaneously activated. The rank order of their times to failure (i.e., a rank of 1 is given to the first component to burn out) are as follows:

COMPFAIL.TXT

Temperature		
150°F	200°F	250°F
4	2	1
7	8	3
10	11	5
13	12	6
18	17	9
21	19	14
22	20	15
24	23	16

(a) At the .05 level of significance, is there evidence of a temperature effect on the median life of this type of electronic component?

(b) **ACTION** Draft a report to the quality engineer based on your findings.

8.34 Use the Kruskal-Wallis rank test to answer part (d) of Problem 8.25 (performance scores) on page 430. (Use $\alpha = .05$.) Are there any differences in your present results from those previously obtained? Discuss.

8.35 Use the Kruskal-Wallis rank test to answer part (a) of Problem 8.26 (tensile strength) on page 431. (Use $\alpha = .05$.) Are there any differences in your present results from those previously obtained? Discuss.

8.36 Use the Kruskal-Wallis rank test to answer part (a) of Problem 8.27 (profitability) on page 431. (Use $\alpha = .01$.) Are there any differences in your present results from those previously obtained? Discuss.

8.8 HYPOTHESIS TESTING BASED ON TWO AND c SAMPLES OF NUMERICAL DATA: A REVIEW

As shown in the chapter summary chart, this chapter presented several widely used hypothesis-testing procedures that enable us to compare statistics computed from two and c samples of numerical data in order to make inferences about possible differences in the parameters of the populations. To be sure you understand what has been covered in this chapter, you should be able to answer the following conceptual questions:

1. What are some of the criteria used in the selection of a particular hypothesis-testing procedure?

2. Under what conditions should the pooled-variance t test be selected to examine possible differences in the means of two independent populations?

3. Under what conditions should the Wilcoxon rank sum test be selected to examine possible differences in the medians of two independent populations?

4. Under what conditions should the F test be selected to examine possible differences in the variances of two independent populations?

5. What are the major assumptions of ANOVA?

6. Under what conditions should the one-way ANOVA F test be selected to examine possible differences in the means of c independent populations?

7. Under what conditions should the Kruskal-Wallis rank test be selected to examine possible differences in the medians of c independent populations?

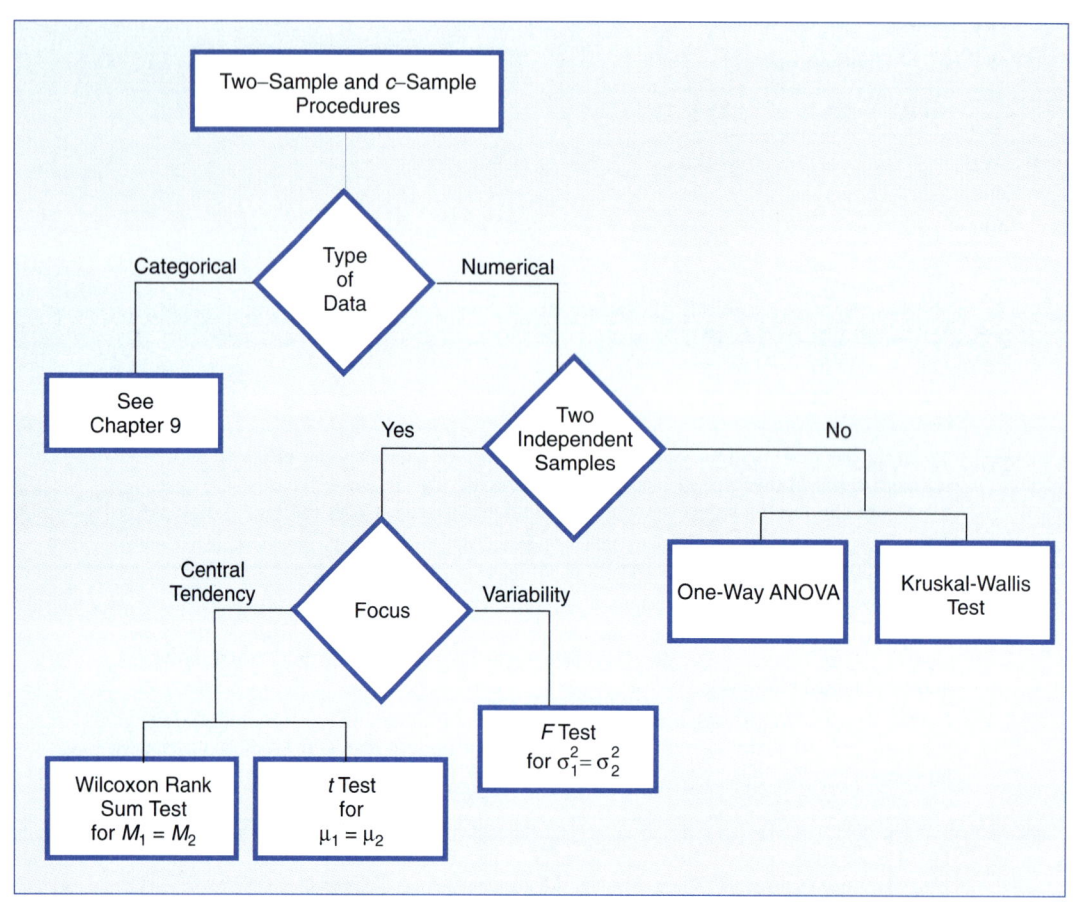

Chapter 8 summary chart.

8. When and how should multiple comparison procedures for evaluating pairwise combinations of the group means be used?

Check over the list of questions to see whether you know the answers and could (1) explain your answers to someone who did not read this chapter and (2) give reference to specific readings or examples that support your answer. Also, reread any of the sections that may have seemed unclear to see whether they make sense now.

Getting It All Together

Key Terms

a posteriori *424*
advanced data filter feature *401*
among-group variation (SSA) *417*
analysis of variance (ANOVA) *415*
ANOVA summary table *419*
completely randomized design model *414*
critical range *424*
experimental error *415*

F distribution *418*
F test for differences in two variances *405*
factor *414*
homogeneity of variance *426*
independence of errors *425*
Kruskal-Wallis rank test *433*
mean squares (MSA, MSW, MST) *417*
nonparametric test procedures *380*

normality   426
one-way ANOVA F test   414
overall or grand mean   416
parametric test procedures   380
pooled-variance t test for differences in two means   383
randomness   425
robust   381

Sum of Squares Among (SSA)   417
Sum of Squares Total (SST)   416
Sum of Squares Within (SSW)   417
total variation (SST)   416
Tukey-Kramer procedure   424
Wilcoxon rank sum test for differences in two medians   395
within-group variation (SSW)   417

Chapter Review Problems

Note: *The Chapter Review Problems can be solved using Microsoft Excel.*

- 8.37 The R & M department store has two charge plans available for its credit account customers. The management of the store wishes to collect information about each plan and study the differences between the two plans. It is interested in the average monthly balance. A random sample of 25 accounts of plan A and 50 accounts of plan B is selected with the following results:

Plan A	Plan B
$n_A = 25$	$n_B = 50$
$\overline{X}_A = \$75$	$\overline{X}_B = \$110$
$S_A = \$15$	$S_B = \$14.14$

  Use statistical inference (confidence intervals or tests of hypothesis) to draw conclusions about *each* of the following:

  Note: Use a level of significance of .01 (99% confidence) throughout.
  (a) Average monthly balance of all plan B accounts.
  (b) Is there evidence that the average monthly balance of plan A accounts is different from $105?
  (c) Is there evidence of a difference in the variances (in the monthly balances) between plan A and plan B?
  (d) Is there evidence of a difference in the average monthly balance between plan A and plan B?
  (e) Compute the p-values in parts (b)–(d) and interpret their meaning.
  (f) Based on the results of parts (a)–(e), what would you tell the management about the two plans?

8.38 A large public utility wishes to compare the consumption of electricity during the summer season for single-family homes in two counties that it services. For each household sampled, the monthly electric bill is recorded with the following results:

	County I	County II
$\overline{X}$	$115	$98
S	$30	$18
n	25	21

  Use statistical inference (confidence intervals or tests of hypothesis) to draw conclusions about *each* of the following:

  Note: Use a level of significance of .05 (95% confidence) throughout.
  (a) Population average monthly electric bill for County I.

(b) Is there evidence that the average bill in County II is above $80?
(c) Is there evidence of a difference in the variances between bills in County I and County II?
(d) Is there evidence that the average monthly bill is higher in County I than County II?
(e) Compute the *p*-values in parts (b)–(d) and interpret their meaning.
(f) Based on the results of parts (a)–(e), what would you tell the utility about the consumption of electricity in the two counties?

8.39 The manager of computer operations of a large company wishes to study computer usage of two departments within the company, the Accounting Department and the Research Department. A random sample of 5 jobs from the Accounting Department in the last week and 6 jobs from the Research Department in the last week is selected, and the processing time (in seconds) for each job is recorded with the following results:

Department	Processing Time (in Seconds)
Accounting	9 3 8 7 12
Research	4 13 10 9 9 6

Use statistical inference (confidence intervals or tests of hypothesis) to draw conclusions about *each* of the following:

Note: Use a level of significance of .05 (95% confidence) throughout.
(a) Average processing time for all jobs in the Accounting Department.
(b) Is there evidence that the average processing time in the Research Department is greater than 6 seconds?
(c) Is there evidence of a difference in the variances in the processing time between the two departments?
(d) What assumption must be made in order to answer (c)?
(e) Is there evidence of a difference in the mean processing time between the Accounting Department and the Research Department?
(f) What assumption(s) is (are) needed to answer (e)?
(g) Compute the *p*-values in (b), (c), and (e) and interpret their meaning.
(h) Is there evidence of a difference in the median processing time between the Accounting Department and the Research Department?
(i) Based on the results of (a)–(h), what should the manager write in his report to the Director of Information Systems concerning the two departments?

8.40 A computer professor is interested in studying the amount of time it would take students enrolled in the Introduction to Computers course to write and run a program in Visual Basic. The professor hires you to analyze the following results (in minutes) from a random sample of nine students:

10,   13,   9,   15,   12,   13,   11,   13,   12

(a) At the .05 level of significance, is there evidence that the population average amount is greater than 10 minutes? What would you tell the professor?
(b) Suppose that when checking her results, the computer professor realizes that the fourth student needed 51 minutes rather than the recorded 15 minutes to write and run the Visual Basic program. At the .05 level of significance, reanalyze the revised data in part (a). What would you tell the professor now?
(c) The professor is perplexed by these paradoxical results and requests an explanation from you regarding the justification for the difference in your findings in parts (a) and (b). Discuss.
(d) A few days later, the professor calls to tell you that the dilemma is completely resolved. The original number 15 is correct, and therefore your findings in part (a) are being used in the article she is writing for a computer magazine. Now she wants to hire you to compare the results from that group of introductory students against those from a sample of 11 computer majors in order to determine whether there is evidence that computer majors can write a Visual Basic program (on average) in less time than introductory students. The sample mean for the computer majors is 8.5 minutes, and the sample standard deviation is 2.0 minutes. At the .05 level of significance, completely analyze these data. What would you tell the professor?

(e) A few days later the professor calls again to tell you that a reviewer of her article wants her to include the *p*-value for the "correct" result in part (a). Discuss the concept of *p*-value and give the *p*-value in part (a).

8.41 The director of training for a company manufacturing electronic equipment is interested in determining whether different training methods have an effect on the productivity of assembly-line employees. She randomly assigns 42 recently hired employees into two groups of 21, of which the first receive a computer-assisted, individual-based training program and the other receive a team-based training program. Upon completion of the training, the employees are evaluated on the time (in seconds) it took to assemble a part. The results are as follows:

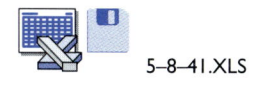

5-8-41.XLS

Computer-Assisted, Individual-Based Program		Team-Based Program	
19.4	16.7	22.4	13.8
20.7	19.3	18.7	18.0
21.8	16.8	19.3	20.8
14.1	17.7	15.6	17.1
16.1	19.8	18.0	28.2
16.8	19.3	21.7	20.8
14.7	16.0	30.7	24.7
16.5	17.7	23.7	17.4
16.2	17.4	23.2	20.1
16.4	16.8	12.3	15.2
18.5		16.0	

(a) Using a .05 level of significance, is there evidence of a difference in the average time to assemble a part between the two programs?
(b) Using the .05 level of significance, is there evidence of a difference in the variances between the two programs?
(c) Using a .05 level of significance, is there evidence of a difference in the median time to assemble a part between the two programs?
(d) Compare the results of (c) with those of (a).

Case Study C—Test-Marketing and Promoting a Ball-Point Pen

PEN.TXT

EPC Advertising has been hired by a well-established manufacturer of pens to develop a series of advertisements and national promotions for the upcoming holiday season. To prepare for this project, Nat Berry, Research Director at EPC Advertising, decides to initiate a study of the effect of advertising on product perception. After conferring with Kate Hansen, the chief of the statistics group, an experiment is designed in which five different advertisements are to be compared in the marketing of a ball-point pen. Advertisement A tends to greatly undersell the pen's characteristics. Advertisement B tends to slightly undersell the pen's characteristics. Advertisement C tends to slightly oversell the pen's characteristics. Advertisement D tends to greatly oversell the pen's characteristics. Advertisement E attempts to correctly state the pen's characteristics. A sample of 30 adult respondents, taken from a larger focus group, is randomly assigned to the five advertisements (so that there are six respondents to each). After reading the advertisement and developing a sense of "product expectation," all respondents unknowingly received the same pen to evaluate. The respondents were permitted to test their pen and the plausibility of the advertising copy. The respondents are then asked to rate the pen from 1 to 7 on the following three product characteristic scales:

	Extremely Poor			Neutral			Extremely Good
Appearance	1	2	3	4	5	6	7
Durability	1	2	3	4	5	6	7
Writing performance	1	2	3	4	5	6	7

The *combined* scores of three ratings (appearance, durability, and writing performance) for the 30 respondents are as follows:

Product ratings for five advertisements

A	B	C	D	E
15	16	8	5	12
18	17	7	6	19
17	21	10	13	18
19	16	15	11	12
19	19	14	9	17
20	17	14	10	14

As a research assistant to Kate Hansen, you have been assigned to work on this project. You have an appointment to discuss this project with her at her office and when you arrive she states: "Hi, come on in and sit down. I'd offer you some coffee but I've just been called to a meeting so we'll make this brief. You know, Nat Berry feels he is really on to something here—test-marketing the advertising copy—and I hope he's right. I'd like you to thoroughly analyze the data that were obtained from the experiment. I'd like to have a detailed report on my desk in four days that summarizes your findings and includes, as an appendix, a discussion of the statistical analysis utilized. Then let's get together for lunch, go over the details, and prepare for the presentations to Nat and the CEO. Do you have any questions before you get started? No? Well, good luck—and don't hesitate to give me a call if something comes up."

———ONE WEEK LATER

PENEAD.TXT

Following your presentation at the meeting with the research team from the agency's public relations group, their chief statistician John Mack suggests that an appropriate nonparametric method be considered for analyzing the data. "Many researchers would argue," John Mack observes, "that the 'product characteristic scales' used do not truly satisfy the criteria of *interval* or *ratio* scaling and, therefore, that nonparametric methods are more appropriate." "John, that might make for an interesting statistics argument," Nat Berry exclaims, "we just have to be as sure as possible that we're practicing good data analysis." John Mack follows: "Fine, but would you also take a look at the following data that I've collected in a similar manner to your experiment? In this new data set the sampled audience was composed only of high-school students who are part of a focus group, not adults. And the data are the ratings or responses only for students who were exposed to advertising copy E, which, as you may recall, attempted to correctly state the pen's characteristics. Please take a look at the corresponding group of adults in your study and analyze the differences in their responses."

The aforementioned combined ratings data for a sample of 8 *high-school student* respondents are as follows:

$$14, \quad 13, \quad 15, \quad 9, \quad 11, \quad 13, \quad 12, \quad 16$$

You leave the room thinking about John Mack's last remarks. You decide to answer John's requests by evaluating whether there is evidence of a difference in the combined ratings of adult versus high-school student respondents subjected to advertisement E (which attempts to correctly state the pen's characteristics). Your answer will identify possible differences in perceptions of adults and that of students with respect to the product. You will be preparing a detailed report on this matter for Kate Hansen.

Team Projects

TP8.1 Refer to TP 2.2 on page 117 and TP 3.1 on page 165. Your group, the _____ Corporation, has been hired by the food editor of a popular family magazine to study the cost and nutritional characteristics of ready-to-eat cereals. Armed with Special Data Set 2 of Appendix D on pages D6–D7, the _____ Corporation is ready to

CEREAL.TXT

(a) Determine if there is evidence of a difference in the average cost of cereals based on whether or not the level of calories per serving is below 155 or at or above 155.
(b) Determine if there is evidence of a difference in the mean amount of sugar in high-fiber cereals versus that in low- and moderate-fiber cereals combined.
(c) Determine if there is a difference in the average cost per serving of ready-to-eat cereals based on their classification as high fiber, moderate fiber, and low fiber.
(d) Write and submit an executive summary describing the results in parts (a) and (b), clearly specifying all hypotheses, selected levels of significance, and the assumptions of the chosen test procedures.
(e) Prepare and deliver a 5-minute oral presentation to the food editor of the magazine.

TP8.2 Refer to TP 2.3 on page 117 and TP 3.2 on page 166. Your group, the _____ Corporation, has been hired by the marketing director of a manufacturer of well-known men's and women's fragrances to study the characteristics of currently available fragrances. Armed with Special Data Set 3 of Appendix D on pages D8–D9, the _____ Corporation is ready to

FRAGRANC.TXT

(a) Determine if there is evidence of a difference in the average cost of men's fragrances versus women's fragrances.
(b) Separately determine if there is evidence of a difference in the mean cost of men's and women's perfumes based on intensity (very strong, strong, medium, or mild.)
(c) Write and submit an executive summary describing the results in parts (a) and (b), clearly specifying all hypotheses, selected levels of significance, and the assumptions of the chosen test procedures.
(d) Prepare and deliver a 5-minute oral presentation to the marketing director.

Endnotes

1. To test for differences in the median dividend yields, it must be assumed that the distributions of dividend yields in both populations from which the random samples were drawn are identical, except possibly for differences in location (i.e., the medians).

2. Among-group variation is sometimes referred to as between-group variation. In these situations the sum of squares term is referred to as **Sum of Squares Between** groups or **SSB**.

3. In addition to this exploratory data analysis, a more confirmatory approach to examining the assumptions of a particular testing procedure would be taken before deciding if the procedure was viable for a given set of data. For the one-way ANOVA F test, the major assumptions are that the sample data in each group are randomly and independently drawn from an underlying normal population and that these populations have equal variability (see Figures 8.8 and 8.9 on page 416). To test for normality, see Reference 2. To test for equality of population variances, see References 1 and 3.

References

1. Berenson, M. L., and D. M. Levine *Basic Business Statistics: Concepts and Applications*, 6th ed. (Englewood Cliffs, NJ: Prentice Hall, 1996).

2. Conover, W. J., *Practical Nonparametric Statistics,* 2d ed. (New York: Wiley, 1980).

3. Daniel, W. W., *Applied Nonparametric Statistics,* 2d ed. (Boston, MA: PWS Kent, 1990).

4. Dunn, O. J., "Multiple Comparisons Using Rank Sums," *Technometrics,* 1964, Vol. 6, pp. 241–252.

5. Hicks, C. R., *Fundamental Concepts in the Design of Experiments,* 4th ed. (New York: Holt, Rinehart and Winston, 1995).

6. Kirk, R. E., *Experimental Design,* 2d ed. (Belmont, CA: Brooks-Cole, 1982).

7. Kramer, C. Y., "Extension of Multiple Range Tests to Group Means with Unequal Numbers of Replications," *Biometrics,* 1956, Vol. 12, pp. 307–310.

8. *Microsoft Excel Version 7* (Redmond, WA: Microsoft Press, 1996).

9. Neter, J., M. H. Kutner, C.J. Nachtsheim, and W. Wasserman, *Applied Linear Statistical Models,* 4th ed. (Homewood, IL: Richard D. Irwin, 1996).

10. Satterthwaite, F. E., "An approximate distribution of estimates of variance components," *Biometrics Bulletin,* 1946, Vol. 2, pp. 110–114.

11. Tukey, J. W., "Comparing Individual Means in the Analysis of Variance," *Biometrics,* 1949, Vol. 5, pp. 99–114.

chapter 9

Two-Sample and c-Sample Tests with Categorical Data

CHAPTER OBJECTIVE To extend the basic principles of hypothesis-testing methodology to situations involving differences in the proportions in two or more groups and tests of independence between two categorical variables.

9.1 INTRODUCTION

In the preceding two chapters we have been concerned with hypothesis-testing procedures that are used to analyze numerical data and to test the proportion in a single population. In Chapter 7, a variety of one-sample tests were presented, and in Chapter 8, several two-sample and c-sample tests were developed for numerical data. In this chapter we shall extend our discussion of hypothesis-testing methodology to consider procedures that are used to analyze differences in population proportions based on two independent samples and c independent samples. In addition, we shall extend our earlier discussions on the theory of probability in Sections 4.7 and 4.8 by presenting a more confirmatory analysis of the hypothesis of independence in the joint responses to two categorical variables. Once again, in this chapter, emphasis will be given to the assumptions behind the use of the various tests.

9.2 Z TEST FOR DIFFERENCES IN TWO PROPORTIONS (INDEPENDENT SAMPLES)

9.2.1 Introduction

Often we are concerned with making comparisons and analyzing differences between two populations in terms of some categorical characteristic. A test for the difference between two proportions based on independent samples can be performed using two different methods. In this section we present a procedure whose test statistic Z is approximated by a standard normal distribution. In Section 9.3, we will develop a procedure whose test statistic χ^2 is approximated by a chi-square distribution with 1 degree of freedom. We will find that the results will be equivalent.

9.2.2 Development

When evaluating differences between two proportions based on independent samples, a Z test can be employed. The test statistic Z used to determine the difference between the two population proportions is based on the difference between the two sample proportions $(p_{S_1} - p_{S_2})$. As we discussed in Section 5.10, this test statistic may be approximated by a standard normal distribution for large enough sample sizes. As shown in Equation (9.1), the Z-test statistic is

$$Z \cong \frac{(p_{S_1} - p_{S_2}) - (p_1 - p_2)}{\sqrt{\bar{p}(1-\bar{p})\left(\frac{1}{n_1} + \frac{1}{n_2}\right)}} \tag{9.1}$$

with

$$\bar{p} = \frac{X_1 + X_2}{n_1 + n_2} \qquad p_{S_1} = \frac{X_1}{n_1} \qquad p_{S_2} = \frac{X_2}{n_2}$$

where
p_{S_1} = sample proportion obtained from population 1
X_1 = number of successes in sample 1
n_1 = size of the sample taken from population 1

p_1 = proportion of successes in population 1
p_{S_2} = sample proportion obtained from population 2
X_2 = number of successes in sample 2
n_2 = size of the sample taken from population 2
p_2 = proportion of successes in population 2
$\bar{p}$ = pooled estimate of the population proportion

Under the null hypothesis, it is assumed that the two population proportions are equal. We should note that $\bar{p}$, **the pooled estimate for the population proportion**, is based on the null hypothesis. Therefore, when testing for equality in the two population proportions, we obtain an overall estimate of the common population proportion by combining or pooling the two sample proportions. This estimate, $\bar{p}$, is the number of successes in the two samples combined ($X_1 + X_2$) divided by the total sample size from the two sample groups ($n_1 + n_2$).

A distinguishing feature of employing this Z test to evaluate differences in population proportions is that we may be interested in determining either whether there is *any difference* in the proportion of successes in the two groups (two-tailed test) or whether one group has a *higher* proportion of successes than the other group (one-tailed test).[1]

Two-Tailed Test	One-Tailed Test	One-Tailed Test
$H_0: p_1 = p_2$	$H_0: p_1 \geq p_2$	$H_0: p_1 \leq p_2$
$H_1: p_1 \neq p_2$	$H_1: p_1 < p_2$	$H_1: p_1 > p_2$

where p_1 = proportion of successes in population 1
            p_2 = proportion of successes in population 2

To test the null hypothesis of no difference in the proportions of two independent populations

$$H_0: p_1 = p_2$$

against the alternative that the two population proportions are not the same

$$H_1: p_1 \neq p_2$$

we may use the test statistic Z, given by Equation (9.1), and, for a given level of significance, α, we would reject the null hypothesis if the computed Z-test statistic exceeds the upper-tailed critical value from the standard normal distribution or if the computed test statistic falls below the lower-tailed critical value from the standard normal distribution.

9.2.3   Application

To illustrate the use of the Z test for the homogeneity of two proportions, suppose a human resources director is investigating employee perception of the fairness of two performance evaluation methods. To test for differences between the two methods, 160 employees are randomly assigned to be evaluated by one of the methods: 78 employees to method 1, where individuals provide feedback to supervisory queries as part of their evaluation process, and 82 employees to method 2, where individuals provide self-assessments of their work performance. Following the evaluations, the employees were asked whether they considered the performance evaluation process to be fair or unfair. In the first sample (method 1), there were 63 "fair" ratings, and in the second sample (method 2), there were 49 "fair" ratings. These results are displayed in Table 9.1.

Table 9.1 A comparison of the performance evaluation data.

	Evaluation Method	
	1	2
Sample sizes:	$n_1 = 78$	$n_2 = 82$
Fair ratings:	$X_1 = 63$	$X_2 = 49$

The null and alternative hypotheses are

$$H_0: p_1 = p_2 \text{ or } p_1 - p_2 = 0$$
$$H_1: p_1 \neq p_2 \text{ or } p_1 - p_2 \neq 0$$

The test is to be carried out at the .01 level of significance, so the critical values are –2.58 and +2.58 (see Figure 9.1) and our decision rule is

Reject H_0 if $Z > +2.58$

or if $Z < -2.58$;

otherwise do not reject H_0.

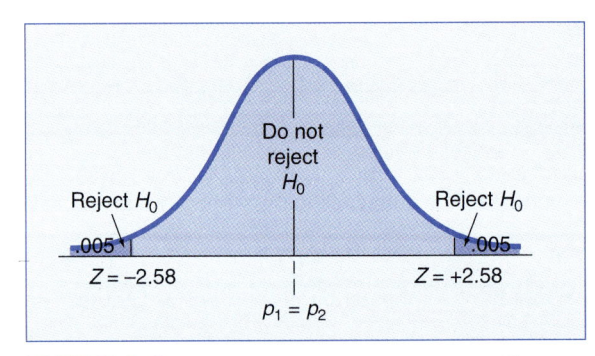

FIGURE 9.1 Testing a hypothesis about the difference between two proportions at the .01 level of significance.

For our data, we have

$$Z \cong \frac{(p_{S_1} - p_{S_2}) - (p_1 - p_2)}{\sqrt{\bar{p}(1 - \bar{p})\left(\frac{1}{n_1} + \frac{1}{n_2}\right)}}$$

where

$$p_{S_1} = \frac{X_1}{n_1} = \frac{63}{78} = .808 \qquad p_{S_2} = \frac{X_2}{n_2} = \frac{49}{82} = .598$$

and

$$\bar{p} = \frac{X_1 + X_2}{n_1 + n_2} = \frac{63 + 49}{78 + 82} = \frac{112}{160} = .70$$

so that

$$Z \cong \frac{.808 - .598}{\sqrt{(.70)(.30)\left(\frac{1}{78} + \frac{1}{82}\right)}}$$

$$= \frac{.210}{\sqrt{(.2100)\left(\frac{160}{6,396}\right)}}$$

$$= \frac{.210}{\sqrt{.005253}}$$

$$= \frac{.210}{.0725} = +2.90$$

Using a .01 level of significance, the null hypothesis (H_0) is rejected because $Z = +2.90 > +2.58$. If the null hypothesis were true, there would be an $\alpha = .01$ probability of obtaining a Z-test statistic either larger than +2.58 standard deviations or smaller than –2.58 standard deviations from the center of the Z distribution. The p-value, or probability of obtaining a difference in the two sample proportions even larger than the one observed here (which translates into a test statistic Z even farther from the center than ±2.90 standard deviations) is .0038 (obtained from Table E.2). This means that if the null hypothesis were true, the probability of obtaining a Z-test statistic below –2.90 is .5000 – .4981 = .0019 and, similarly, the probability of obtaining a Z-test statistic above +2.90 is .5000 – .4981 = .0019. Thus, for this two-tailed test, the p-value is .0019 + .0019 = .0038. Since .0038 < α = .01, the null hypothesis is rejected. There is evidence to conclude that the two evaluation methods are significantly different with respect to employee perceptions of fairness; that is, a greater proportion of employees found method 1 (employee feedback) fairer than method 2 (self-assessment).

9.2.4 Using Microsoft Excel for the Z Test for Differences in Two Proportions

In this section we have developed the Z test for differences between two proportions using the normal distribution. Although Excel does not have a Data Analysis tool to test the hypothesis for differences in two proportions, we can use Excel formulas. The design for the Calculations sheet for this test is shown in Table 9.1.Excel.

To implement this design, open a new workbook and rename the active sheet Calculations. Referring to Equation (9.1) on page 450, we see that to perform the Z test for differences in two proportions using the normal distribution, we need the sample proportion and sample size for both groups. If the number of successes and the sample size for each group need to be obtained from the raw data, the PivotTable Wizard (see Section 2.12) can be used and the results copied to the cell ranges B3:B4 and B6:B7, respectively. For the performance evaluation problem these values are known, so the values 63, 78, 49, and 82 can be entered into these cells. To complete the implementation of the Calculations sheet:

❶ Compute the sample proportion for group 1 in cell B5 using the formula =B3/B4 and the sample proportion for group 2 in cell B8 using the formula =B6/B7.

❷ Compute the average proportion in cell B9 using the formula =(B3+B6)/(B4+B7).

❸ Compute the difference between the two sample proportions in cell B10 using the formula =B5–B8.

Table 9.1.Excel Design for Calculations sheet testing the hypothesis for the differences between two proportions.

	A	B
1	Z Test for Two Proportions	
2		
3	Successes in Group 1	xxx
4	Sample Size Group 1	xxx
5	Proportion Group 1	=B3/B4
6	Successes in Group 2	xxx
7	Sample Size Group 2	xxx
8	Proportion Group 2	=B6/B7
9	Average Proportion	=(B3+B6)/(B4+B7)
10	Difference in Two Proportions	=B5–B8
11	Hypothesized Difference	.xx
12	α	.xx
13	Z	=(B10–B11)/SQRT(B9*(1–B9)*(1/B4+1/B7))
14	Two-Tailed Test	
15	Lower Critical Value	=NORMSINV(B12/2)
16	Upper Critical Value	=NORMSINV(1–B12/2)
17	p-Value	=2*(1–NORMSDIST(ABS(B13)))
18	Decision	=IF(B17<B12,"Reject","Do not Reject")
19	One-Tailed Test (Lower)	
20	Lower Critical Value	=NORMSINV(B12)
21	p-Value	=NORMSDIST(B13)
22	Decision	=IF(B21<B12,"Reject","Do not Reject")
23	One-Tailed Test (Upper)	
24	Upper Critical Value	=NORMSINV(1–B12)
25	p-Value	=1–NORMSDIST(B13)
26	Decision	=IF(B25<B12,"Reject","Do not Reject")

❹ Enter the hypothesized difference in cell B11 and the level of significance α in cell B12.

❺ Compute the value of the Z-test statistic in cell B13 using the formula

=(B10–B11)/SQRT(B9*(1–B9)*(1/B4+1/B7))

❻ Use an approach similar to the one used in Section 7.11 (the tests of hypothesis for the mean and proportion) to obtain the critical value, p-value, and decision.

Using the performance evaluation data example from Table 9.1 on page 452, we enter 0 in cell B11, and .01 in cell B12. The results are illustrated in Figure 9.1.Excel.

From Figure 9.1.Excel, we observe that the Z-test statistic is computed as 2.899, the two-tailed p-value is .003742, and the null hypothesis is rejected since Z = 2.899 > 2.5758 or the p-value = .003742 < .01.

> ▲ **WHAT IF EXAMPLE**
>
> This Calculations sheet allows us to explore the effect of changes in the number of successes and sample size in a group, and in the level of significance on the results of the Z test. For example, if we wanted to determine the effect of a change in the number of

successes in group 1 from 63 to 53, we would change cell B3 to 53. The proportion of successes in group 1 would change to .6795 in cell B5, the average proportion in cell B9 would change to .6375, the Z-test statistic would be 1.0775, and the two-tailed *p*-value would be .281. With these results, at the .01 level of significance since Z = 1.0775 < 2.58 or the *p*-value = .281 > .01, the null hypothesis could not be rejected. We would conclude that there is no evidence of a difference in the proportion of successes in the two groups.

	A	B
1	Z Test for Two Proportions	
2		
3	Successes in Group 1	63
4	Sample Size Group 1	78
5	Proportion Group 1	0.807692308
6	Successes in Group 2	49
7	Sample Size Group 2	82
8	Proportion Group 2	0.597560976
9	Average Proportion	0.7
10	Difference in Two Proportions	0.210131332
11	Hypothesized Difference	0
12	α	0.01
13	Z	2.899181485
14	Two-Tailed Test	
15	Lower Critical Value	-2.57583451
16	Upper Critical value	2.575834515
17	p-value	0.003741516
18	Decision	Reject
19	One-tailed Test (Lower)	
20	Lower Critical Value	-2.32634193
21	p-value	0.998129242
22	Decision	Do not reject
23	One-Tailed test (Upper)	
24	Upper Critical Value	2.326341928
25	p-value	0.001870758
26	Decision	Reject

FIGURE 9.1.EXCEL Using Microsoft Excel for the Z Test for differences in two proportions for the performance evaluation data.

Problems for Section 9.2

Note: *The problems in this section can be solved using Microsoft Excel.*

9.1 We wish to determine whether there is any difference in the popularity of football between college-educated males and non-college-educated males. A sample of 100 college-educated males reveals 55 football fans, and a sample of 200 non-college-educated males reveals 125 football fans.
(a) Is there any evidence of a difference in the popularity of football between college-educated and non-college-educated males at the .01 level of significance?
(b) Compute the *p*-value in (a) and interpret its meaning.
(c) What would be your answers to (a) and (b) if 135 non-college-educated males considered themselves football fans?

9.2 A marketing study conducted in a large city shows that of 100 married women who work full time, 43 eat dinner at a restaurant at least one night during a typical workweek, while 27 out of a sample of 100 married women who do not work full time eat dinner out at least once a week.
  (a) Using a .05 level of significance, is there evidence of a difference between the two groups of married women in the proportion who eat dinner out at least once a week?
  (b) Compute the p-value in (a) and interpret its meaning.

9.3 A professor of accountancy was studying the readability of the annual reports of two major companies. A random sample of 100 certified public accountants was selected. Fifty were randomly assigned to read the annual report of Company A, and the other 50 were to read the annual report of Company B. Based on a standard measure of readability, 17 found Company A's annual report "understandable" and 23 found Company B's annual report "understandable."
  (a) At the .10 level of significance, is there any evidence of a difference between the two companies in the proportion of CPAs who find the annual reports understandable?
  (b) Compute the p-value in (a) and interpret its meaning.
  (c) What would be your answers to (a) and (b) if 33 found Company B's annual report "understandable"?

9.4 The director of marketing for a company manufacturing laundry detergent conducts an experiment to compare customer satisfaction with the laundry detergent product based on level of temperature used for the wash. A random sample of 500 individuals who agree to participate in the experiment are asked to use the product on a standard-sized load under a low-temperature setting. A second random sample of 500 participants are asked to use the same product on a standard-sized load under a high-temperature setting. Of the 500 participants who use the low-temperature setting, 280 are happy with the cleansing outcome. Of the 500 participants who use the high-temperature setting, 320 are happy with the results.
  (a) At the .05 level of significance, is there evidence that the detergent is preferred when used with the high-temperature setting than with the low-temperature setting?
  (b) Compute the p-value in (a) and interpret its meaning.

9.5 The manager of a campus bookstore conducts a survey to investigate whether there are any differences between males and females with respect to the consideration of the purchase of educational videotapes. Depending on the answer, her goal is to develop pertinent promotional material that will lead to increased sales in educational videotapes over the coming semester. Of the 40 males in the survey, 13 state they would consider purchasing educational videotapes. Of the 30 females in the survey, 15 say they would consider purchasing educational videotapes.
  (a) At the .01 level of significance, is there evidence of a difference in the proportion of males and females who would consider purchasing educational videotapes?
  (b) Compute the p-value in (a) and interpret its meaning.
  (c) Given the results in (a) and (b), what should the bookstore manager do with respect to a future promotional campaign? Discuss your recommendations.
  (d) What would be your answers to (a)–(c) if 25 females said they would consider purchasing educational videotapes?

9.3 χ^2 Test for Differences in Two Proportions (Independent Samples)

9.3.1 Introduction

In the previous section we described the Z test for the difference between two proportions based on independent samples. In this section, rather than directly comparing *proportions* of success, we will view the data in terms of the *frequency* of success in two groups. We will develop a procedure whose test statistic χ^2 is approximated by a chi-square distribution with 1 degree of freedom. The results obtained with the χ^2 test are equivalent to those obtained by using the Z test of Section 9.2.

9.3.2 Development

If it is of interest to compare the tallies or counts of categorical responses between two independent groups, a two-way **table of cross-classifications** can be developed (see Section 2.11) to display the frequency of occurrence of successes and failures for each group. Such a table is also called a **contingency table**, which, as you may recall, was used in Chapter 4 to define and study probability from an objective empirical approach. In this section, however, we develop methodology for a more confirmatory analysis of data presented in such contingency tables.

To illustrate the use of this technique, let us return to the performance evaluation study presented in the previous section. Table 9.2 is a schematic layout of a cross-classification table resulting from the study, and Table 9.3 is the contingency table displaying the actual data from the study.

where
X_1 = number of employees who believe method 1 is fair
X_2 = number of employees who believe method 2 is fair
$n_1 - X_1$ = number of employees who believe method 1 is unfair
$n_2 - X_2$ = number of employees who believe method 2 is unfair
$X = X_1 + X_2$ = the number of fair ratings
$n - X = (n_1 - X_1) + (n_2 - X_2)$ = the number of unfair ratings
n_1 = number of employees in the sample evaluated under method 1
n_2 = number of employees in the sample evaluated under method 2
$n = n_1 + n_2$ = total number of employees evaluated in the study

The contingency table displayed in Table 9.3 has two rows, indicating whether the employees perceived their evaluations to be fair (i.e., success) or unfair (i.e., failure), and two columns, one for each method of performance evaluation. Such a table is called a **2 × 2 table**. The cells in the table indicate the frequency of successes and failures for each row and column combination. The row totals indicate the number of fair and unfair ratings; the column totals are the sample sizes evaluated by each method. The proportion of employees who perceive their evaluations to be fair is obtained by dividing the number of fair ratings for a particular method by the number of employees evaluated by that method. A methodology known as the χ^2 **test for homogeneity of proportions** may then be employed to compare the proportions for the two methods.

To test the null hypothesis of no differences in the two population proportions

$$H_0: p_1 = p_2$$

Table 9.2 Design of a 2 × 2 contingency table for the performance evaluation study.

Employee Perception	Evaluation Method 1	Evaluation Method 2	Totals
Fair	X_1	X_2	X
Unfair	$n_1 - X_1$	$n_2 - X_2$	$n - X$
Totals	n_1	n_2	n

Table 9.3 2 × 2 contingency table for comparing observed employee perceptions of fairness in the performance evaluation study.

Employee Perception	Evaluation Method 1	Evaluation Method 2	Totals
Fair	63	49	112
Unfair	15	33	48
Totals	78	82	160

against the alternative that the two population proportions are different

$$H_1: p_1 \neq p_2$$

we obtain the χ^2-test statistic, which is given by the following:

> The χ^2-test statistic is equal to the squared difference between the observed and expected frequencies, divided by the expected frequency in each cell, summed over all cells.
>
> $$\chi^2 = \sum_{\text{all cells}} \frac{(f_o - f_e)^2}{f_e} \quad (9.2)$$

where f_o = **observed frequency** or actual tally in a particular cell of a 2×2 contingency table

f_e = **theoretical or expected frequency** we would expect to find in a particular cell if the null hypothesis is true

To compute the expected frequency (f_e) in any cell requires an understanding of its conceptual foundation. If the null hypothesis is true and the proportion of fair ratings is equal for each population, then the sample proportions computed from the two groups should differ from each other only by chance, since they would each be providing an estimate of the common population proportion p. In such a situation, a statistic that pools or combines these two separate estimates into one overall or average estimate of the population proportion p would provide more information than any one of the two separate estimates. This statistic, given by the symbol $\bar{p}$, therefore represents the overall or average proportion of fair ratings for the two groups combined (i.e., the total number of fair ratings divided by the total number of employees rated). Using the notation for proportions given in Section 9.2, this can be stated as follows:

> The average proportion $\bar{p}$ is equal to the sum of the "successes" in the two groups, divided by the sum of the sample sizes of the two groups.
>
> $$\bar{p} = \frac{(X_1 + X_2)}{(n_1 + n_2)} = \frac{X}{n} \quad (9.3)$$

Its complement, $1 - \bar{p}$, represents the overall or average proportion of "failures" over the two groups.

To obtain the expected frequency (f_e) for each cell of fair ratings (i.e., the first row in the contingency table), we multiply the sample size (or column total) for an evaluation method by $\bar{p}$. To obtain the expected frequency (f_e) for each cell of unfair ratings (i.e., the second row in the contingency table), we multiply the sample size (or column total) by $(1 - \bar{p})$.

The test statistic shown in Equation (9.2) approximately follows a chi-square distribution with **degrees of freedom** equal to the number of rows in the contingency table minus 1 times the number of columns in the table minus 1:

$$\text{Degrees of freedom} = (r - 1)(c - 1)$$

where r = number of rows in the table

$$c = \text{number of columns in the table}$$

For our 2 × 2 contingency table, there is 1 degree of freedom; that is,

$$\text{Degrees of freedom} = (2 - 1)(2 - 1) = 1$$

Using a level of significance α, the null hypothesis may be rejected in favor of the alternative if the computed χ^2-test statistic exceeds χ^2, the upper-tailed critical value from the **chi-square distribution** with 1 degree of freedom. That is, the decision rule is to reject H_0 if

$$\chi^2 > \chi_1^2$$

as illustrated in Figure 9.2.

Referring to Equation (9.2), if the null hypothesis is true, the computed χ^2-test statistic should be close to 0 since the squared difference between what we actually observe in each cell (f_0) and what we theoretically expect (f_e) would be very small. On the other hand, if H_0 is false and there are real differences in the population proportions, we should expect the computed χ^2-test statistic to be large since the discrepancy between what we actually observe in each cell and what we theoretically expect will be magnified when we square the differences. However, what constitutes a large difference in a cell is relative. The same actual difference between f_0 and f_e would contribute more to the χ^2-test statistic from a cell in which only a few observations are expected (f_e) than from a cell where many observations are expected. This is because a standardizing adjustment is made for the size of the cell—the squared difference between f_0 and f_e is divided by the expected frequency (f_e) in the cell. The χ^2-test statistic given in Equation (9.2) is then obtained by summing each standardized value $(f_0 - f_e)^2/f_e$ over all the cells of the contingency table.

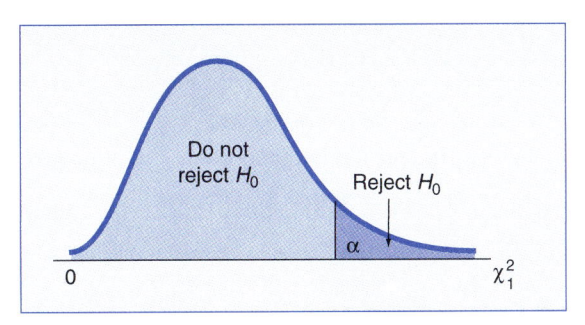

FIGURE 9.2 Testing a hypothesis for the difference between two proportions using the χ^2 test.

9.3.3 Application

To illustrate the use of the χ^2 test for homogeneity of two proportions, we again turn our attention to our human resource director's performance evaluation study, the results of which are displayed in Table 9.3 on page 457.

The null hypothesis (H_0: $p_1 = p_2$) states that, when comparing the two evaluation methods, there is no difference in the proportion of employees with respect to the perception of fairness. Using Equation (9.3) on page 458, we can estimate the common parameter p, the true proportion of employees who believe these methods to be fair; using $\bar{p}$, the overall or average proportion, is computed as

$$\bar{p} = \frac{(X_1 + X_2)}{(n_1 + n_2)} = \frac{X}{n}$$

$$= \frac{(63 + 49)}{(78 + 82)} = \frac{112}{160}$$

$$= .70$$

The estimated proportion who do not feel that these evaluation methods are fair is the complement, $(1 - \bar{p})$, or .30. Multiplying these two proportions by method 1's sample size gives the number of employees expected to perceive their evaluations as fair and the number expected to perceive their evaluations as unfair. In a similar manner, multiplying the two respective proportions by method 2's sample size yields the corresponding expected frequencies for that group. All these expected frequencies are presented in Table 9.4, next to the corresponding observed frequencies taken from Table 9.3.

Table 9.4 2 × 2 contingency table for comparing *observed* (f_0) and *expected* (f_e) employee perceptions of fairness for the performance evaluation data.

	Evaluation Method				
	1		2		
Employee Perception	Observed	Expected	Observed	Expected	Totals
Fair	63	54.6	49	57.4	112
Unfair	15	23.4	33	24.6	48
Totals	78	78	82	82	160

To test the null hypothesis of homogeneity of proportions

$$H_0: p_1 = p_2$$

against the alternative that the true population proportions are not equal

$$H_1: p_1 \neq p_2$$

we use the actual and expected data from Table 9.4 to compute the χ^2-test statistic given by Equation (9.2). The calculations are presented in Table 9.5.

If a .01 level of significance is chosen, the critical value of the χ^2-test statistic could be obtained from Table E.4, a replica of which is presented as Table 9.6. The chi-square distribution is a skewed distribution whose shape depends solely on the number of degrees of freedom. As the number of degrees of freedom increases, the chi-square distribution becomes more symmetrical.

Table 9.5 Computation of the χ^2-test statistic for the performance evaluation data.

f_0	f_e	$(f_0 - f_e)$	$(f_0 - f_e)^2$	$(f_0 - f_e)^2/f_e$
63	54.6	+8.4	70.56	1.293
49	57.4	−8.4	70.56	1.229
15	23.4	−8.4	70.56	3.015
33	24.6	+8.4	70.56	2.868
				8.405

Table 9.6 Obtaining the χ^2 critical value from the chi-square distribution with 1 degree of freedom using a .01 level of significance.

Degrees of Freedom	Upper-Tail Areas (α)						
	.995	.99	...	.05	.025	.01	.005
1			...	3.841	5.024	**6.635**	7.879
2	0.010	0.020	...	5.991	7.378	9.210	10.597
3	0.072	0.115	...	7.815	9.348	11.345	12.838
4	0.207	0.297	...	9.488	11.143	13.277	14.860
5	0.412	0.554	...	11.071	12.833	15.086	16.750

Source: Extracted from Table E.4.

The values in the body of Table 9.6 refer to selected upper-tailed areas of the chi-square distribution. Since a χ^2-test statistic for a 2 × 2 table has 1 degree of freedom, and we are testing at the $\alpha = .01$ level of significance, the critical value of the χ^2-test statistic is 6.635 (see Figure 9.3). Since our computed χ^2-test statistic of 8.405 exceeds this critical value, the null hypothesis may be rejected. There is evidence to conclude that the two evaluation methods are significantly different with respect to employee perceptions of fairness. An examination of Table 9.1 on page 452 indicates that a greater proportion of employees found method 1 (employee feedback) fairer than method 2 (self-assessment).

As we shall observe in Section 9.6, output obtained by using statistical software or spreadsheet packages such as Microsoft Excel for contingency table analysis typically contains the cross-classification table (with both the observed and expected frequencies), the computed χ^2-test statistic, and the *p*-value. Whenever the *p*-value is provided, we do not need the critical value of the test statistic to make our decision. We can simply compare the obtained *p*-value to our selected level of significance, α. If the *p*-value is less than α, the null hypothesis is rejected; if the *p*-value is greater than or equal to α, then H_0 is not rejected. In our performance evaluation study, since the *p*-value = .0038, which is less than $\alpha = .01$, the null hypothesis is rejected. There is evidence of a difference in the two proportions; that is, employees found method 1 (employee feedback) to be fairer than method 2 (self-assessment).

● **Caution** For the test to give accurate results, the χ^2 test for 2 × 2 tables assumes that each expected frequency is at least 5. If this assumption is not satisfied, other procedures, such as *Fisher's exact test* (see Reference 2), can be used.

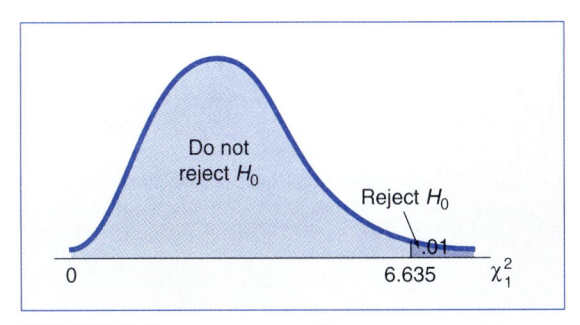

FIGURE 9.3 Finding the χ^2 critical value with 1 degree of freedom at the .01 level of significance.

9.3.4 Testing for the Equality of Two Proportions by Z and by χ^2: A Comparison of Results

We have seen in our human resource director's performance evaluation study that both the Z test based on the standard normal distribution and the χ^2 test based on the chi-square distribution with 1 degree of freedom have led to the same conclusion. This can be explained by the interrelationship between the standard normal distribution and a chi-square distribution with 1 degree of freedom. The χ^2-test statistic will always be the square of the Z-test statistic.[2] For instance, in this study, the computed Z-test statistic is +2.90 and the computed χ^2-test statistic is 8.405. Except for rounding error, we note that this latter value is the square of +2.90 [i.e., $(+2.90)^2 \equiv 8.405$]. Also, if we compare the critical values of the test statistics from the two distributions, we can see that at the .01 level of significance the χ_1^2 value of 6.635 is (except for rounding) the square of the Z values of ±2.58 (i.e., $\chi_1^2 = Z^2$).

- **Advantage of the Z Test Over the χ^2 Test**   From this discussion it should be clear that when testing the null hypothesis of homogeneity of proportions

$$H_0: p_1 = p_2$$

against the alternative that the true population proportions are not equal

$$H_1: p_1 \neq p_2$$

the Z test and the χ^2 test are equivalent methods. However, if we are specifically interested in determining whether there is evidence of a *directional difference*, such as $p_1 > p_2$, then the Z test must be used with the entire rejection region located in one tail of the standard normal distribution.

- **Advantage of the χ^2 Test Over the Z Test**   On the other hand, if we wish to make comparisons and evaluate differences in the proportions among c groups or levels of some factor, we will be able to extend the χ^2 test for such purposes. The Z test, however, cannot be employed if there are more than two groups.

Problems for Section 9.3

Note: *The problems in this section can be solved using Microsoft Excel (see Section 9.6).*

9.6 Why shouldn't the χ^2 test be used for differences in proportions when the expected frequencies in some of the cells are too small?

9.7 Do Problem 9.1 on page 455 using the χ^2 test.

9.8 Do Problem 9.2 on page 456 using the χ^2 test.

9.9 Do Problem 9.3 on page 456 using the χ^2 test.

9.10 Do Problem 9.4 on page 456 using the χ^2 test.

9.11 In an effort to compare the efficacy of two medical approaches to removing plaque that clogs arteries, Dr. Eric J. Topol conducted a study in which he randomly assigned 1,012 heart patients to have either directional coronary atherectomy or balloon angioplasty. (See E. Topol et al., "A Comparison of Directional Atherectomy with Coronary Angioplasty in Patients with Coronary Artery Disease," *The New England Journal of Medicine*, July 22, 1993, Vol. 329, pp. 221–227.) Of the 512 patients given the atherectomy, 44 died or suffered heart attacks within 6 months of treatment. Of the 500 patients given angioplasty, 23 died or suffered heart attacks within 6 months of treatment.
(a) At the .01 level of significance, is there evidence of a difference in the two medical approaches with respect to the proportion of deaths or heart attacks within 6 months of treatment?
(b) Compute the *p*-value in (a) and interpret its meaning.

(c) Given the results in (a) and (b), what should the physicians conclude with respect to the two approaches?

(d) How should a congressional policymaker react to the results in (a) and (b)?

(e) How should a hospital CEO react to the results in (a) and (b) if the medical facility is given a fixed sum of money per treatment?

(f) How should an insurance company CEO react to the results in (a) and (b) if the company has to pay for each treatment?

(g) What would be your answers in (a) and (b) if 34 of the 512 patients given the atherectomy died or suffered heart attacks?

9.4 χ^2 TEST FOR DIFFERENCES IN c PROPORTIONS (INDEPENDENT SAMPLES)

The χ^2 test can be extended to the general case in which there are c independent populations to be compared. Thus, if there is interest in evaluating differences in the proportions among c groups or levels of some factor, the χ^2 test can be used for such purposes. The contingency table would have two rows and c columns. To test the null hypothesis of no differences in the proportions among the c populations

$$H_0: p_1 = p_2 = \cdots = p_c$$

against the alternative that not all the c population proportions are equal

$$H_1: \text{Not all } p_j \text{ are equal (where } j = 1, 2, \ldots, c\text{).}$$

we use Equation (9.2) and compute the test statistic

$$\chi^2 = \sum_{\text{all cells}} \frac{(f_o - f_e)^2}{f_e}$$

where f_o = observed frequency in a particular cell of a $2 \times c$ contingency table

f_e = theoretical or expected frequency in a particular cell if the null hypothesis is true

To compute the expected frequency (f_e) in any cell, we must realize that if the null hypothesis were true and the proportions were equal across all c populations, then the c sample proportions should be differing from each other only by chance. This is because they would each be providing estimates of the common population proportion p. In such a situation, a statistic that would pool or combine these c separate estimates into one overall or average estimate of the population proportion p would provide more information than any one of the c separate estimates alone. Expanding on Equation (9.3) on page 458, the statistic $\bar{p}$ represents the overall or average proportion over all c groups combined:

$$\bar{p} = \frac{(X_1 + X_2 + \cdots + X_c)}{(n_1 + n_2 + \cdots + n_c)} = \frac{X}{n} \qquad (9.4)$$

To obtain the expected frequency (f_e) for each cell in the first row in the contingency table, we multiply each respective sample size (or column total) by $\bar{p}$. To obtain the expected frequency (f_e) for each cell in the second row in the contingency table, we multiply each respective sample size (or column total) by $(1 - \bar{p})$. The test statistic shown in Equation (9.2) approximately follows a chi-square distribution with degrees of freedom equal to the number of rows in the contingency table minus 1 times the number of columns in the table minus 1. For a **2 × c contingency table,** there are $c - 1$ degrees of freedom; that is,

$$\text{Degrees of freedom} = (2 - 1)(c - 1) = c - 1$$

Using a level of significance α, the null hypothesis may be rejected in favor of the alternative if the computed χ^2-test statistic exceeds the upper-tailed critical value from a χ^2 distribution with $c - 1$ degrees of freedom. That is, the decision rule is

$$\text{Reject } H_0 \text{ if } \chi^2 > \chi^2_{c-1};$$

otherwise do not reject H_0.

This is depicted in Figure 9.4.

For the χ^2 test to give accurate results when dealing with $2 \times c$ contingency tables, all expected frequencies must be large. For such situations, there has been much debate among statisticians as to the definition of "large." Some statistical researchers (see Reference 4) have found that the test gives accurate results as long as all expected frequencies equal or exceed 0.5. Other statisticians, more conservative in their approach, require that no more than 20% of the cells contain expected frequencies less than 5 and no cells have expected frequencies less than 1 (see Reference 3). We suggest that a reasonable compromise between these points of view is to make sure that all expected frequencies are at least 1. To accomplish this, it may be necessary to collapse two or more low-frequency categories into one category in the contingency table prior to performing the test. Such merging of categories usually results in expected frequencies sufficiently large to conduct the χ^2 test accurately. If the combining or pooling of categories is undesirable, alternative procedures are available (see References 1 and 6).

To illustrate the χ^2 test for equality or homogeneity of proportions when there are more than two groups, suppose a real estate developer has just received approval from the city council to develop a parcel of land that is to contain 4,000 apartment dwellings. Among the numerous items needed for each apartment is a circuit-breaker box to be installed in a kitchen cabinet. Several manufacturers make such circuit-breaker boxes, and the real estate developer wishes to contract with only one supplier. From the architectural and engineering design specifications approved for this development, it is necessary that the circuit-breaker box be able to tolerate a stipulated current level without malfunctioning in order to be considered for use. Five of the vendors who have placed a bid claim that their products will meet the stipulated current requirement; these vendors have passed the first phase of the contract competition. However, since the circuit-breaker box is a relatively inexpensive item and the prices offered by the five vendors in bidding for the contract are very similar, the real estate director decides to design an experiment to evaluate the capability of each of the competing boxes. Random samples of 400 boxes are obtained from each vendor and subjected to a peak-current test (i.e.,

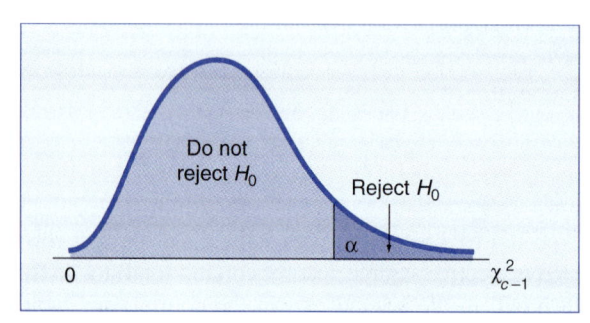

FIGURE 9.4 Testing for differences among c proportions using the χ^2 test.

Table 9.7 Cross-classification of *observed* frequencies from peak-current experiment on five brands of circuit-breaker boxes.

Result of Peak-Current Experiment	Brands of Circuit-Breaker Boxes					Totals
	1	2	3	4	5	
Malfunctioning boxes	92	66	94	144	104	500
Properly working boxes	308	334	306	256	296	1,500
Totals	400	400	400	400	400	2,000

one in excess of the stipulated level of current). Table 9.7 presents, for each of the five vendors' products, the number of boxes that malfunction (i.e., at least one of the current breakers in the box fails to flip appropriately) during the test and the number of boxes that continue to work properly under the peak-current condition.

Under the null hypothesis of no differences among the five vendors' products with respect to the proportion of malfunctioning or nonconforming boxes, we can use Equation (9.4) to calculate an estimate of p, the population proportion of malfunctioning circuit-breaker boxes. That is, $\bar{p}$, the overall or average proportion of malfunctioning boxes from all five competing vendors, is computed as

$$\bar{p} = \frac{(X_1 + X_2 + \cdots + X_c)}{(n_1 + n_2 + \cdots + n_c)} = \frac{X}{n}$$

$$= \frac{(92 + 66 + 94 + 144 + 104)}{(400 + 400 + 400 + 400 + 400)} = \frac{500}{2,000}$$

$$= .25$$

The estimated proportion of properly working circuit-breaker boxes in the population is the complement, $(1 - \bar{p})$, or .75. Multiplying these two proportions by the sample size used for each vendor's product results in the expected frequencies of malfunctioning and properly working boxes. These are presented in Table 9.8.

To test the null hypothesis of homogeneity or equality of proportions

$$H_0: p_1 = p_2 = p_3 = p_4 = p_5$$

against the alternative that not all the five proportions are equal

$$H_1: \text{Not all } p_j \text{ are equal (where } j = 1, 2, \ldots, 5).$$

we use the observed and expected data from Tables 9.7 and 9.8 to compute the χ^2-test statistic given by Equation (9.2). The calculations are presented in Table 9.9.

Table 9.8 Cross-classification of *expected* frequencies from the peak-current experiment data.

Result	Brands of Circuit-Breaker Boxes					Totals
	1	2	3	4	5	
Malfunctioning boxes	100	100	100	100	100	500
Properly working boxes	300	300	300	300	300	1,500
Totals	400	400	400	400	400	2,000

9.4 χ^2 Test for Differences in *c* Proportions (Independent Samples)

Table 9.9 Computation of the χ^2-test statistic for the peak-current experiment data.

f_0	f_e	(f_0-f_e)	$(f_0-f_e)^2$	$(f_0-f_e)^2/f_e$
92	100	−8	64	0.640
66	100	−34	1,156	11.560
94	100	−6	36	0.360
144	100	44	1,936	19.360
104	100	4	16	0.160
308	300	8	64	0.213
334	300	34	1,156	3.853
306	300	6	36	0.120
256	300	−44	1,936	6.453
296	300	−4	16	0.053
				42.772

If a .01 level of significance is chosen, the critical value of the χ^2-test statistic could be obtained from Table E.4. In our peak-current experiment, since five vendor products are being evaluated, there are $(2 - 1)(5 - 1) = 4$ degrees of freedom. The critical value of χ^2 with 4 degrees of freedom at the $\alpha = .01$ level of significance is 13.277. Since our computed test statistic $\chi^2 = 42.772$ exceeds this critical value, or alternatively since the p-value obtained from Microsoft Excel = .0000 < .01, the null hypothesis may be rejected (see Figure 9.5). It may be stated that, at the .01 level of significance, there is sufficient evidence to conclude that the five brands of circuit-breaker boxes are different with respect to the proportion that malfunction during a peak-current test.

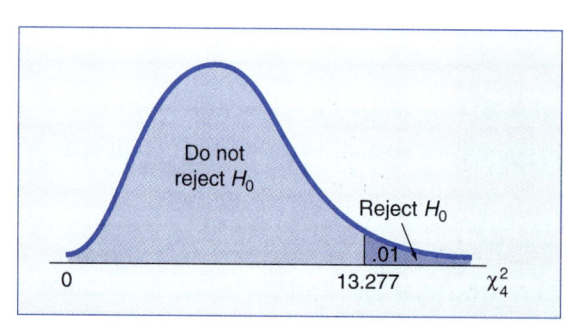

FIGURE 9.5 Testing for the equality of five proportions at the .01 level of significance with 4 degrees of freedom.

Rejecting the null hypothesis in a χ^2 test of equality of proportions in a $2 \times c$ table only allows us to conclude that not all the brands of circuit-breaker boxes are equal with respect to the proportion that malfunction. Interest would then focus on which brand or brands are different from the others with respect to performance. Since the result of the χ^2 test for homogeneity of proportions does not specifically answer these questions, other approaches are needed. One such confirmatory approach that may be used following rejection of the null hypothesis of equal proportions is the Marascuilo procedure (see References 1, 5, and 6).

Problems for Section 9.4

Note: *The problems in this section can be solved using Microsoft Excel (see Section 9.6).*

9.12 Why can't the normal approximation be used for determining differences in the proportion of successes in more than two groups?

9.13 The faculty council of a large university would like to determine the opinion of various groups on a proposed trimester academic calendar. A random sample of 100 undergraduate students, 50 graduate students, and 50 faculty members is selected with the following results:

Opinion	Undergraduate	Graduate	Faculty
Favor trimester	63	27	30
Oppose trimester	37	23	20
Totals	100	50	50

(a) At the .01 level of significance, is there evidence of a difference in attitude toward the trimester between the various groups?
(b) **ACTION** What should the faculty council report to the president of the university concerning the attitude toward the trimester academic calendar?
(c) What would be your answers to (a) and (b) if 53 of the 100 undergraduates favored the trimester?

9.14 Dr. Lawrence K. Altman reported the results of a clinical trial (*The New York Times*, May 1, 1993, p. 7) comparing the effectiveness of four drug regimens randomly assigned for treatment of patients following the onset of a heart attack. A total of 40,845 patients were studied. Each was given one of the four drug regimens. The outcome measure compared was the proportion of severe adverse events (i.e., deaths or disabling stroke) reported within 30 days of treatment. His data are presented here:

	Drug Regimen				
Result	A	B	C	D	Totals
Severe	714	785	754	820	3,073
Not severe	9,630	9,543	9,042	9,557	37,772
Totals	10,344	10,328	9,796	10,377	40,845

where: A = accelerated TPA with intravenous heparin
B = combined TPA and streptokinase, with intravenous heparin
C = streptokinase with subcutaneous heparin
D = streptokinase with intravenous heparin

(a) Describe the methodology you are using to analyze the data (at $\alpha = .05$) to determine whether there is evidence of a significant difference among the four drug regimens with respect to the proportion of patients suffering severe adverse events (i.e., death or disabling stroke) within 30 days following treatment for heart attack.
(b) State your findings.
(c) Discuss the impact that your findings may have on the community of health care administrators and policymakers if a dose of TPA costs $2,400 per patient while a dose of streptokinase costs $240 per patient.

9.15 The quality control manager of an automobile parts factory would like to know whether there is a difference in the proportion of defective parts produced on different days of the work week. Random samples of 100 parts produced on each day of the week were selected with the results shown on page 468.

Result	Mon.	Tues.	Wed.	Thurs.	Fri.
Number of defective parts	12	7	7	10	14
Number of acceptable parts	88	93	93	90	86
Totals	100	100	100	100	100

(a) At the .05 level of significance, is there evidence of a difference in the proportion of defective parts produced on various days of the week?

(b) What would be your answer to (a) if 24 of the 100 parts produced on Friday were defective?

9.16 A manufacturer of automobile batteries wishes to determine whether there are any differences in three media (magazine, TV, radio) in terms of consumer recall of an ad. The results of an advertising study are as follows:

Recall Ability	Magazine	TV	Radio	Totals
Number of persons remembering ad	25	10	7	42
Number of persons not remembering ad	73	93	108	274
Totals	98	103	115	316

Media

(a) At the .10 level of significance, determine whether there is evidence of a media effect with respect to the proportion of individuals who can recall the ad.

(b) **ACTION** Write an executive summary for the manufacturer.

(c) What would be your answer in (a) if 17 of the 115 individuals who heard the radio ad could recall it?

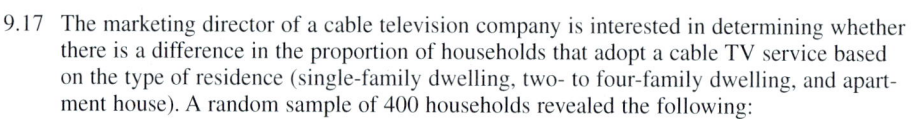

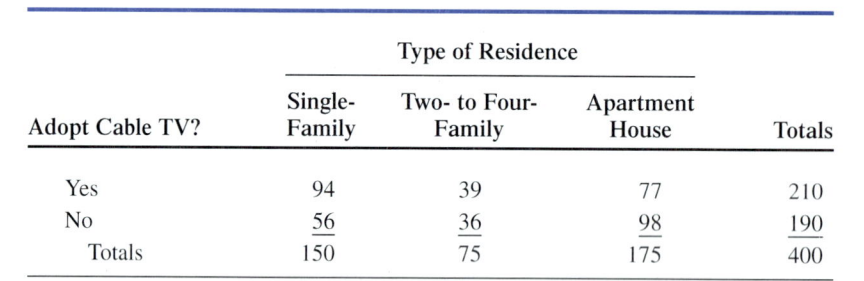

9.17 The marketing director of a cable television company is interested in determining whether there is a difference in the proportion of households that adopt a cable TV service based on the type of residence (single-family dwelling, two- to four-family dwelling, and apartment house). A random sample of 400 households revealed the following:

Adopt Cable TV?	Single-Family	Two- to Four-Family	Apartment House	Totals
Yes	94	39	77	210
No	56	36	98	190
Totals	150	75	175	400

Type of Residence

(a) At the .01 level of significance, is there evidence of a difference among the types of residence with respect to the proportion of households that adopt the cable TV service?

(b) **ACTION** Write an executive summary for the marketing director.

9.5 χ^2 TEST OF INDEPENDENCE

We have just seen how the χ^2 test can be used to evaluate potential differences among the proportion of successes in any number of populations. For a contingency table that has r rows and c columns, the χ^2 test can be generalized as a test of independence. In these situations, we shall be able to extend our earlier discussions on the rules of probability in Sections 4.7 and 4.8 by presenting a confirmatory analysis based on a hypothesis of independence in the joint responses to two categorical variables.

As a test of independence, the null and alternative hypotheses would be

H_0: The two categorical variables are independent (i.e., there is no relationship between them).

H_1: The two categorical variables are related (i.e., dependent).

And we once more use Equation (9.2) and compute the test statistic

$$\chi^2 = \sum_{\text{all cells}} \frac{(f_0 - f_e)^2}{f_e}$$

The decision rule is to reject the null hypothesis at an α level of significance if the computed value of the test statistic exceeds the upper-tailed critical value from a chi-square distribution with $(r-1)(c-1)$ degrees of freedom (see Table E.4). That is,

Reject H_0 if $\chi^2 > \chi^2_{(r-1)(c-1)}$;

otherwise, do not reject H_0.

Many researchers consider the χ^2 test for independence as an alternative approach to viewing the χ^2 test for equality of proportions. The test statistics are the same and the decision rules are the same, but the stated hypotheses and the conclusion to be drawn are different. Thus, in the performance evaluation study of Section 9.3, we concluded that there is evidence of a difference in the two methods with respect to the proportion of employees who perceive the evaluations as fair. From a different viewpoint, we could conclude that there is a significant relationship between the evaluation method used and the perception of fairness. Similarly, in the peak-current experiment of Section 9.4, we concluded that there is evidence of a difference in the brands with respect to the proportion that malfunction under a peak-current test. Taking a different perspective, we could conclude that there is a significant relationship between brand of circuit-breaker box and performance under a specified peak current. Nevertheless, there is a fundamental distinction between the two types of tests. The major difference is in the sampling scheme used.

In a test for equality of proportions, we have one factor of interest with two or more levels. These levels represent samples drawn from independent populations. The categorical responses in each sample group or level are usually dichotomized into two levels—*success* and *failure*. The objective is to make comparisons and evaluate differences in the proportions of success among the various levels.

On the other hand, in a test for independence, we have two factors of interest, each with two or more levels. One sample is drawn, and the joint responses to the two categorical variables are tallied into the cells of the contingency table that represent particular levels of each variable.

To illustrate the χ^2 test for independence, let us suppose that a survey is undertaken by a central Nassau County (New York) branch of a large nationwide chain of real estate bro-

Table 9.10 *Observed frequency of responses to a survey cross-classifying architectural style and geographical location for 233 single-family houses.*

Style	East Meadow	Farmingdale	Levittown	Totals
Cape	31	14	52	97
Expanded ranch	2	1	12	15
Colonial	6	8	9	23
Ranch	16	20	24	60
Split-level	19	17	2	38
Totals	74	60	99	233

(Geographical Location spans East Meadow, Farmingdale, Levittown)

kerage offices to profile single-family homes in some neighboring communities. One question of interest to the branch manager is whether there is a relationship between architectural style (cape, expanded ranch, colonial, ranch, and split-level) and geographical location (East Meadow, Farmingdale, and Levittown). Suppose a random sample of $n = 233$ single-family homes is selected and a two-way tally is obtained for each combination of architectural style and geographic location. The resulting 5×3 contingency table is presented in Table 9.10.

From the totals tallied in Table 9.10, we observe that with respect to architectural style, 97 of the sampled homes are considered capes, 15 are expanded ranches, 23 are colonials, 60 are ranches, and 38 are split-levels. With respect to geographic location, 74 of the sampled homes are located in East Meadow, 60 are in Farmingdale, and 99 are in Levittown. The observed frequencies in the cells of the 5×3 contingency table represent the joint tallies of the sampled homes with respect to architectural style and geographic location.

The null and alternative hypotheses are

H_0: There is no relationship between architectural style and geographic location.

H_1: There is a relationship between architectural style and geographic location.

To test this null hypothesis of independence against the alternative that there is a relationship between the two categorical variables, we use Equation (9.2) and compute the test statistic

$$\chi^2 = \sum_{\text{all cells}} \frac{(f_0 - f_e)^2}{f_e}$$

where f_0 = *observed frequency* or actual tally in a particular cell of the $r \times c$ contingency table

f_e = theoretical frequency we would *expect* to find in a particular cell if the null hypothesis of independence was true

To compute the *expected frequency* (f_e) in any cell, we may utilize the probability rules developed in Section 4.8. That is, if the null hypothesis of independence is true, then we could use the multiplication rule for independent events [which we discussed on page 184; see Equation (4.8)] to determine the joint probability or proportion of responses expected for any cell combination. For example, under the null hypothesis of independence, the probability or proportion of responses expected in the upper-left corner cell representing cape houses in East Meadow would be the product of the two separate probabilities, that is,

$$P(\text{Cape and East Meadow}) = P(\text{Cape}) \times P(\text{East Meadow}).$$

Here, the proportion of cape houses, $P(\text{Cape})$, is 97/233 or .416, while the proportion of houses located in East Meadow, $P(\text{East Meadow})$, is 74/233 or .318. If the null hypothesis were true and architectural style and geographic location were independent, the expected proportion or probability $P(\text{Cape and East Meadow})$ would equal the product of the separate probabilities, .416 × .318, or .132. The expected frequency (f_e) for that particular cell combination would then be the product of the survey sample size n and this probability, that is, 233 × .132, or 30.8.

As a second example, to compute the expected frequency (f_e) for the lower-right corner cell representing split-level houses in Levittown under the null hypothesis of independence, we have

$$P(\text{Split-level and Levittown}) = P(\text{Split-level}) \times P(\text{Levittown}).$$

Here, the proportion of split-level houses, $P(\text{Split-level})$, is 38/233 or .163, while the proportion of houses located in Levittown, $P(\text{Levittown})$, is 99/233 or .425. If the null hypothesis were true and architectural style and geographic location were independent, the expected proportion or probability $P(\text{Split-level and Levittown})$ would equal the product of the separate probabilities, .163 × .425, or .069. The expected frequency (f_e) for that particular cell combination would then be the product of the survey sample size n and this probability, that is, 233 × .069, or 16.1. The f_e values for the remainder of the 5 × 3 contingency table would be obtained in a similar manner (see Table 9.11).

An easier way to compute expected frequencies, which does not require calculation of probabilities, is

$$f_e = \frac{\text{row sum} \times \text{column sum}}{n} \quad (9.5)$$

where

row sum = sum of all the frequencies in the row

column sum = sum of all the frequencies in the column

n = sample size

Table 9.11 *Expected frequency of responses to the architectural style survey.*

	Geographical Location			
Style	East Meadow	Farmingdale	Levittown	Totals
Cape	30.8	25.0	41.2	97
Expanded ranch	4.8	3.9	6.4	15
Colonial	7.3	5.9	9.8	23
Ranch	19.1	15.5	25.5	60
Split-level	12.1	9.8	16.1	38
Totals	74	60	99	233

For example, using Equation (9.5) for the upper-left corner cell (cape houses in East Meadow), we have

$$f_e = \frac{\text{row sum} \times \text{column sum}}{n} = \frac{(97)(74)}{233} = 30.8$$

while for the lower-right corner cell (split-level houses in Levittown), we have

$$f_e = \frac{\text{row sum} \times \text{column sum}}{n} = \frac{(38)(99)}{233} = 16.1$$

All other f_e values could be obtained in a similar manner (see Table 9.11).

The test statistic shown in Equation (9.2) approximately follows a chi-square distribution with degrees of freedom equal to the number of rows in the contingency table minus 1 times the number of columns in the table minus 1. For an $r \times c$ contingency table, there are $(r-1)(c-1)$ degrees of freedom; that is,

$$\text{Degrees of freedom} = (r-1)(c-1)$$

Using a level of significance α, the null hypothesis of independence may be rejected in favor of the alternative that the two categorical variables are related if the computed χ^2-test statistic exceeds the upper-tailed critical value from a chi-square distribution with $(r-1)(c-1)$ degrees of freedom. That is, the decision rule is to

$$\text{Reject } H_0 \text{ if } \chi^2 > \chi^2_{(r-1)(c-1)};$$

otherwise do not reject H_0.

This is depicted in Figure 9.6.

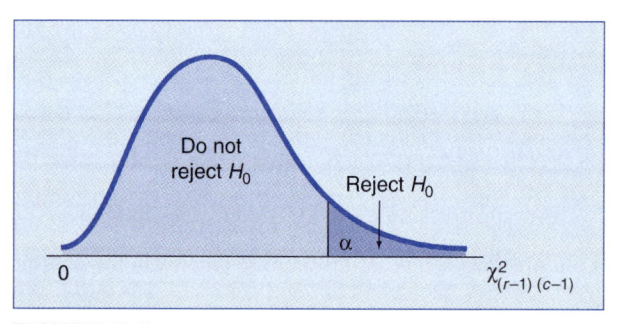

FIGURE 9.6 Testing for independence in an $r \times c$ contingency table using the χ^2 test.

The χ^2-test statistic for these data would then be computed as indicated in Table 9.12. Using a level of significance, α, of .05, $\chi^2 = 42.9607$ exceeds 15.507, the upper-tail critical value from the χ^2 distribution (see Table E.4) with $(5-1)(3-1) = 8$ degrees of freedom, and the null hypothesis of independence is rejected (see Figure 9.7). That is, there is evidence of a relationship between architectural style and geographical location. Examination of the table of observed frequencies (see Table 9.10 on page 470) shows that capes are found in greater abundance than expected in Levittown but are underrepresented in Farmingdale. Expanded ranches are overrepresented in Levittown. Split-levels are overrepresented in East Meadow and in Farmingdale but are vastly underrepresented in Levittown. The p-value or probability of obtaining a χ^2 value of 42.9607 or one even more extreme if the null hypothesis of independence were true is equal to .0000, which is less than .05.

Table 9.12 Computation of the χ^2-test statistic for the architectural style contingency table.

f_0	f_e	(f_0-f_e)	$(f_0-f_e)^2$	$(f_0-f_e)^2/f_e$
31	30.8	+0.2	0.04	0.0013
14	25.0	−11.0	121.00	4.8400
52	41.2	+10.8	116.64	2.8311
2	4.8	−2.8	7.84	1.6333
1	3.9	−2.9	8.41	2.1564
12	6.4	+5.6	31.36	4.9000
6	7.3	−1.3	1.69	2.3151
8	5.9	+2.1	4.41	0.7475
9	9.8	−0.8	0.64	0.0653
16	19.1	−3.1	9.61	0.5031
20	15.5	+4.5	20.25	1.3065
24	25.5	−1.5	2.25	0.0882
19	12.1	+6.9	47.61	3.9347
17	9.8	+7.2	51.84	5.2898
2	16.1	−14.1	198.81	12.3484
				42.9607

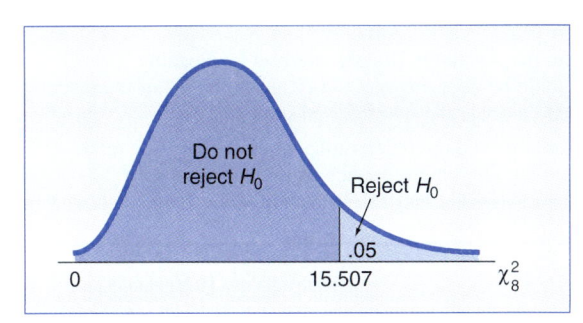

FIGURE 9.7 Testing for independence in the architectural style example at the .05 level of significance with 8 degrees of freedom.

● **Caution** As in the case of the $2 \times c$ contingency tables, to assure accurate results, use of the χ^2 test when dealing with $r \times c$ contingency tables requires that all expected frequencies be "large." The same rules suggested for employing the χ^2 test in the case of the $2 \times c$ contingency tables on page 464 may be used. Again, we suggest that all expected frequencies are at least 1. For cases in which one or more expected frequency is less than 1, the test may be performed after collapsing two or more low-frequency row categories or two or more low-frequency column categories into one category. Such merging of row or column categories will usually result in expected frequencies sufficiently large to conduct the χ^2 test accurately.

Problems for Section 9.5

Note: The problems in this section may be solved using Microsoft Excel (see Section 9.6).

9.18 If a contingency table has four rows and three columns, how many degrees of freedom would there be for the χ^2 test for independence?

9.19 When performing a χ^2 test for independence in a contingency table with r rows and c columns, determine the upper-tailed critical value of the χ^2-test statistic in each of the following circumstances:
  (a) $\alpha = .05$, $r = 4$ rows, $c = 5$ columns
  (b) $\alpha = .01$, $r = 4$ rows, $c = 5$ columns
  (c) $\alpha = .01$, $r = 4$ rows, $c = 6$ columns
  (d) $\alpha = .01$, $r = 3$ rows, $c = 6$ columns
  (e) $\alpha = .01$, $r = 6$ rows, $c = 3$ columns

9.20 A nationwide market research study is undertaken to determine the preferences of various age groups of males for different sports. A random sample of 1,000 men is selected, and each individual is asked to indicate his favorite sport. The results are as follows:

		Sport			
Age Group	Baseball	Football	Basketball	Hockey	Totals
Under 20	26	47	41	36	150
20–29	38	84	80	48	250
30–39	72	68	38	22	200
40–49	96	48	30	26	200
50 and over	134	44	18	4	200
Totals	366	291	207	136	1,000

  (a) At the .01 level of significance, is there evidence of a relationship between men's ages and their preference in sports?
  (b) Compute the *p*-value in (a) and interpret its meaning.
  (c) What would be your answers to (a) and (b) if 114 of the respondents aged 50 and over preferred baseball and 64 preferred football?

9.21 Suppose a survey is taken to determine whether there is a relationship between place of residence and automobile preference. A random sample of 200 car owners from large cities, 150 from suburbs, and 150 from rural areas is selected with the following results:

		Automobile Preference				
Residence	GM	Ford	Chrysler	European	Asian	Totals
Large city	64	40	26	8	62	200
Suburb	53	35	24	6	32	150
Rural	53	45	30	6	16	150
Totals	170	120	80	20	110	500

  (a) At the .05 level of significance, is there evidence of a relationship between place of residence and automobile preference?
  (b) Compute the *p*-value in (a) and interpret its meaning.
  (c) What would be your answers to (a) and (b) if 16 of the respondents from a large city preferred Chrysler and 18 preferred a European automobile?

9.22 During the Vietnam War, a lottery system was instituted to choose males to be drafted into the military. Numbers representing days of the year were "randomly" selected; men born on days of the year with low numbers were drafted first, while those with high numbers were not drafted. The following shows how many low (1–122), medium (123–244), and high (245–366) numbers were drawn for birthdates in each quarter of the year:

| | Quarter of Year | | | | |
Number Set	Jan.–Mar.	Apr.–Jun.	Jul.–Sep.	Oct.–Dec.	Totals
Low	21	28	35	38	122
Medium	34	22	29	37	122
High	36	41	28	17	122
Totals	91	91	92	92	366

(a) Is there evidence that the numbers drawn were related to the time of year? (Use $\alpha = .05$.)
(b) Would you conclude that the lottery drawing appears to have been random?
(c) Compute the *p*-value in (a) and interpret its meaning.
(d) What would be your answers to (a)–(c) if the frequencies were

23	30	32	37
27	30	34	31
41	31	26	24

9.23 A large corporation is interested in determining whether an association exists between the commuting time of their employees and the level of stress-related problems observed on the job. A study of 116 assembly-line workers reveals the following:

| | | Stress | | |
Commuting Time	High	Moderate	Low	Totals
Under 15 min.	9	5	18	32
15–45 min.	17	8	28	53
Over 45 min.	18	6	7	31
Totals	44	19	53	116

(a) At the .01 level of significance, is there evidence of a relationship between commuting time and stress?
(b) Compute the *p*-value in (a) and interpret its meaning.
(c) What would be your answers to (a) and (b) if the .05 level of significance was used?

9.6 USING MICROSOFT EXCEL FOR χ^2 TESTS

In Sections 9.3–9.5, we developed the χ^2 test to determine whether there was any difference in the proportion of successes in two or more populations and to test for the independence of two categorical variables. Although Excel does not have a Data Analysis tool to completely perform the χ^2 test, Excel formulas can be used instead. The design for the Calculations sheet for this test is shown in Table 9.2.Excel for the test for a 2 × 2 contingency table. These calculations can be extended to the 2 × c and r × c tables.

To implement this design, open a new workbook and rename the active sheet Calculations. Referring to Equation (9.2) on page 458, we see that in order to compute the χ^2-test statistic, we need to obtain the observed and expected frequencies for each cell of the contingency table. If the observed frequencies need to be obtained from the raw data, the PivotTable Wizard (see Section 2.12) can be used and the results copied to the appropriate cells of the contingency

Table 9.2.Excel   Design for Calculations Sheet for the χ^2 Test.

	A	B	C	D
1		Chi-Square Test		
2				
3		Observed		
4	Employee Perception	Method 1	Method 2	Total
5	Fair	xxx	xxx	=SUM(B5:C5)
6	Unfair	xxx	xxx	=SUM(B6:C6)
7	Total	=SUM(B5:B6)	=SUM(C5:C6)	=SUM(B7:C7)
8				
9		Expected		
10	Employee Perception	Method 1	Method 2	Total
11	Fair	=D5*$B$7/$D$7	=D5*$C$7/$D$7	=SUM(B11:C11)
12	Unfair	=D6*B7/$D$7	=D6*$C$7/$D$7	=SUM(B12:C12)
13	Total	=SUM(B11:B12)	=SUM(C11:C12)	=SUM(B13:C13)
14				
15	p-Value	=CHITEST(B5:C6,B11:C12)		
16	Number of Rows (R)	xx		
17	Number of Columns (C)	xx		
18	Degrees of Freedom	=(B16–1)*(B17–1)		
19	α	.xx		
20	Critical Value	=CHIINV(B19,B18)		
21	Chi-Square Statistic	=CHIINV(B15,B18)		
22	Decision	=IF(B15<B19,"Reject","Do not Reject")		

table (for the 2 × 2 table illustrated in Table 9.2.Excel, the observed frequencies are located in cells B5, C5, B6, and C6). For the performance evaluation problem, the observed frequencies are known so the values 63, 15, 49, and 33 can be entered into these cells.

Microsoft Excel does not currently have a Data Analysis tool that will directly calculate the χ^2-test statistic from a Pivot Table or from a table of observed frequencies. However, it does include the CHITEST function that will compute the p-value for the χ^2-test statistic, given that a set of observed and expected frequencies are provided. To complete the implementation of the Calculations sheet, do the following:

① Enter the formulas that use the SUM function in cells D5, D6, B7, C7, and D7 to obtain the observed row totals, column totals, and the overall total.

② Equation (9.5) on page 471 can now be used to obtain the set of expected frequencies. For cell B11, the row total in cell D5 is multiplied by the column total in cell B7 and divided by the overall total from cell D7. Similar formulas are used to obtain the other expected frequencies in cells C11, B12, and C12.

③ The expected row, column, and overall totals are obtained by using the SUM function as discussed in step 1.

④ Now that the observed and expected frequencies have been obtained, the CHITEST function can be used to obtain the p-value of the χ^2-test statistic. The format of this function is

CHITEST(*range of observed frequencies, range of expected frequencies*)

For our 2 × 2 table, the observed frequencies are located in cells B5:C6, and the expected frequencies are located in cells B11:C12. Thus, we enter the formula =CHITEST(B5:C6,B11:C12) in cell B15.

⑤ The number of rows and columns are entered in cells B16 and B17, respectively. The degrees of freedom are then computed in cell B18 using the formula =(B16–1)*(B17–1) and the level of significance α is entered in cell B19.

⑥ The critical value is obtained by using the CHIINV function. The format of this function is

$$\text{CHIINV}(p, \text{df})$$

where p = upper-tailed p-value (the probability that χ^2 will be greater than the specific value indicated)

  df = degrees of freedom

Thus, for this test we enter the formula =CHIINV(B19,B18) in cell B20. With α = .01 and 1 degree of freedom, the critical value is 6.6348913.

⑦ The CHIINV function is also used to convert the p-value obtained in cell B15 into the χ^2-test statistic computed in cell B21.

⑧ The p-value in cell B15 is then compared to the level of significance in cell B19. If the p-value is less than the level of significance, the null hypothesis is rejected; if it is equal to or greater than the level of significance, the null hypothesis is not rejected.

The results are illustrated in Figure 9.2.Excel where we observe that the χ^2-test statistic is computed as 8.405, the p-value is .00374, and the null hypothesis is rejected since χ^2 = 8.405 > 6.63489 or the p-value = .00374 < .01.

	A	B	C	D
1		Chi-Square Test		
2				
3		Observed		
4	Employee Perception	Method 1	Method 2	Total
5	Fair	63	49	112
6	Unfair	15	33	48
7	Total	78	82	160
8				
9		Expected		
10	Employee Perception	Method 1	Method 2	Total
11	Fair	54.6	57.4	112
12	Unfair	23.4	24.6	48
13	Total	78	82	160
14				
15	p-value	0.003741383		
16	Number of Rows (R)	2		
17	Number of Columns (C)	2		
18	Degrees of Freedom	1		
19	α	0.01		
20	Critical Value	6.634891297		
21	Chi-Square Statistic	8.405176777		
22	Decision	Reject		

FIGURE 9.2.EXCEL Using Microsoft Excel to perform the Chi-Square Test for the performance evaluation data.

> ▲ **WHAT IF EXAMPLE**
>
> The Calculations sheet allows us to explore the effect of changes in the observed frequencies and in the level of significance on the results of the χ^2 test for the difference between proportions. For example, if we wanted to determine the effect of a change in the number of employees who consider method 1 fair from 63 to 53 and the number of employees who consider method 1 unfair from 15 to 25, we would change cell B5 to 53 and cell B6 to 25. The row totals would change to 102 in cell D5 and 58 in cell D6, and the expected frequencies would change to 49.725 in cell B11, 52.275 in cell C11, 28.275 in cell B12, and 29.275 in cell C12. In addition, the χ^2-test statistic would be 1.161, and the two-tailed p-value would be .281. With these results, at the .01 level of significance since $\chi^2 = 1.161 < 6.635$ or the p-value $= .281 > .01$, the null hypothesis could not be rejected. We would conclude that there is no evidence of a difference between the two methods in the proportion who believe the evaluation is fair.

9.7 HYPOTHESIS TESTING BASED ON CATEGORICAL DATA: A REVIEW

As observed in the chapter summary chart, there are different approaches to categorical data analysis. Hypothesis-testing methodology was developed separately for analyzing categorical response data obtained from two independent samples, and c independent samples. In addition, we extended our earlier discussions on the rules of probability in Sections 4.7 and 4.8 by presenting a more confirmatory analysis of the hypothesis of independence in the joint responses to two categorical variables. Once again, the assumptions and conditions behind the use of the various tests were emphasized. To be sure you understand the material we have covered, you should be able to answer the following conceptual questions:

1. Under what conditions should the Z test be used to examine possible differences in the proportions of two independent populations?
2. Under what conditions should the χ^2 test be used to examine possible differences in the proportions of two independent populations?
3. What are the similarities and distinctions between the Z and χ^2 tests for differences in population proportions?
4. Under what conditions should the χ^2 test be used to examine possible differences in the proportions of c independent populations?
5. Under what conditions should the χ^2 test of independence be used?

Check over the list of questions to see whether you know the answers and could (1) explain your answers to someone who did not read this chapter and (2) give reference to specific readings or examples that support your answer. Also, reread any of the sections that may have seemed unclear to see whether they make sense now.

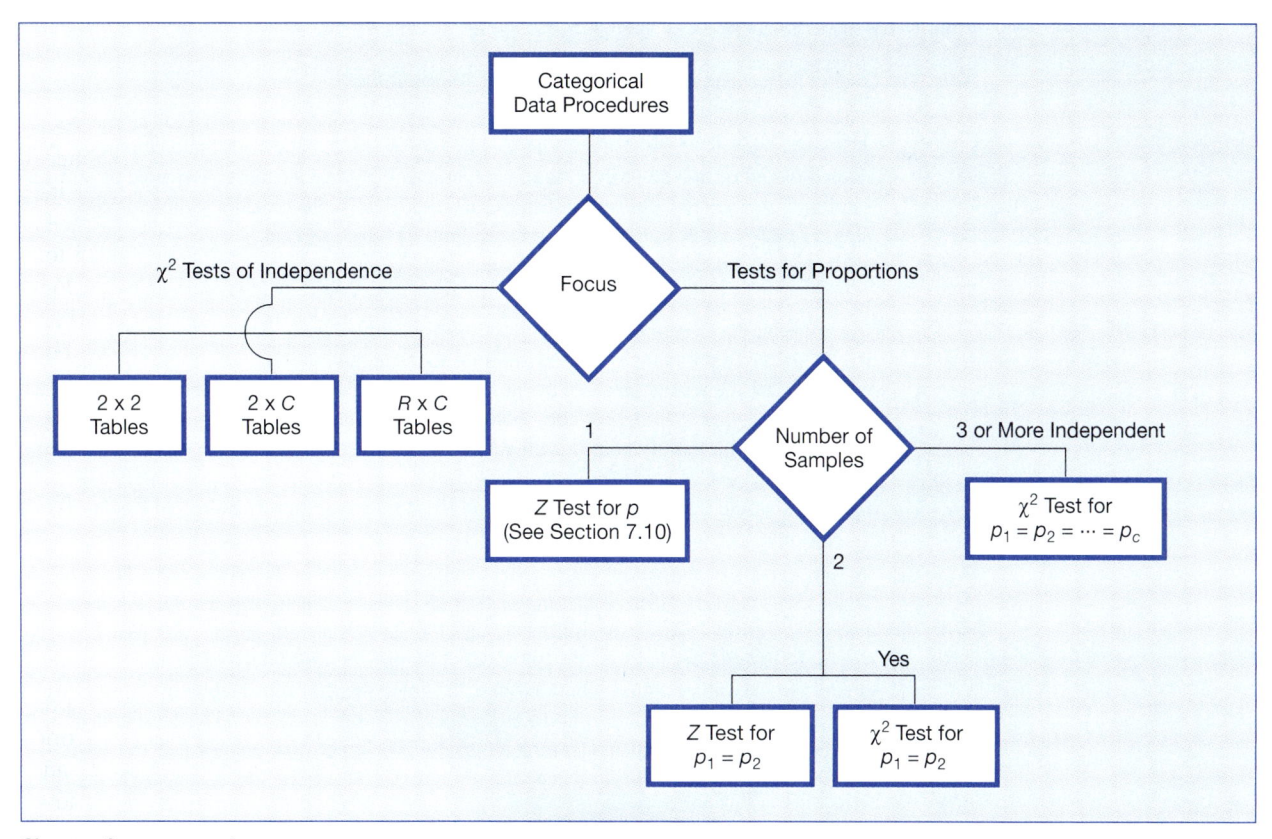

Chapter 9 summary chart.

Getting It All Together

Key Terms

2 × 2 table   457
2 × c contingency table   464
χ^2 test for differences in c proportions   463
χ^2 test for difference in two proportions   456
χ^2 test for homogeneity of proportions   457
χ^2 test of independence   469
chi-square distribution   459
contingency table   457

degrees of freedom   458
expected frequencies (f_e)   458
observed frequencies (f_o)   458
pooled estimate ($\bar{p}$) for the population
   proportion   451
table of cross-classifications   457
theoretical or expected frequencies   458
Z test for differences in two proportions   450

Chapter Review Problems

Note: *The Chapter Review Problems can be solved using Microsoft Excel.*

9.24 A housing survey of single-family homes in two suburban New York State counties is conducted to determine the proportion of such homes that have gas heat. A sample of 300 single-family homes in County A indicates that 185 have gas heat, and a sample of 200 single-family homes in County B indicates that 75 have gas heat.

Note: Use a level of significance of .01 throughout the problem.
(a) Use *two* different statistical tests to determine whether there is evidence of a difference between the two counties in the proportion of single-family houses that have gas heat.
(b) Compute the *p*-value in (a) and interpret its meaning.
(c) Compare the results obtained by the two methods in (a). Are your conclusions the same?
(d) If you wanted to know whether there is evidence that County A has a higher proportion of single-family houses with gas heat, what method would you use to perform the statistical test?
(e) What would be your answers to (a) and (b) if 95 single-family homes in County B had gas heat?

9.25 A "physician's health study" of the effectiveness of aspirin in the reduction of heart attacks was begun in 1982 and completed in 1987 (see C. Hennekens et al., "Findings from the Aspirin Component of the Ongoing Physician's Health Study," *The New England Journal of Medicine*, January 28, 1988, Vol. 318, pp. 262–264). Of 11,037 male medical doctors in the United States who took one 325-mg buffered aspirin tablet every other day, 104 suffered heart attacks during the 5-year period of the study. Of 11,034 male medical doctors in the United States who took a placebo (i.e., a pill that, unknown to the participants in the study, contained no active ingredients), 189 suffered heart attacks during the 5-year period of the study.
(a) At the .01 level of significance, is there evidence that the proportion having heart attacks is lower for the male medical doctors in the United States who received the buffered aspirin every other day than for those who received the placebo?
(b) Compute the *p*-value in (a). Does this lead you to believe that taking one buffered aspirin pill every other day was effective in reducing the incidence of heart attacks? Explain.
(c) Why is it not appropriate to use the χ^2 test in (a)?
(d) What would be your answers in (a) and (b) if 149 of the male medical doctors taking the placebo suffered heart attacks?

9.26 A statistician wishes to study the distribution of three types of cars (subcompact, compact, and full size) sold in the four geographical regions of the United States (Northeast, South, Midwest, West). A random sample of 200 cars is selected with the following results:
- Of 60 cars sold in the Northeast, 25 were subcompacts, 20 were compacts, and 15 were full size.
- Of 40 cars sold in the South, 10 were subcompacts, 10 were compacts, and 20 were full size.
- Of 50 cars sold in the Midwest, 15 were subcompacts, 15 were compacts, and 20 were full size.
- Of 50 cars sold in the West, 20 were subcompacts, 15 were compacts, and 15 were full size.

(a) At the .05 level of significance, is there evidence of a relationship between type of car and geographical region?
(b) Compute the *p*-value in (a) and interpret its meaning.
(c) Estimate with 95% confidence the true proportion of full-size cars sold in the Northeast.
(d) From the given data, can you obtain a reasonable 95% confidence interval estimate of the proportion of full-size cars sold throughout the United States? Discuss.
(e) What would be your answer to (a) and (b) if, of 40 cars sold in the South, 5 were subcompacts, 10 were compacts, and 25 were full-size?

Case Study D—Airline Satisfaction Survey

As Chief Research Associate, you are attending this week's meeting of the board of directors at the invitation of Mike Drucker, the Senior Vice-President for Marketing and Promotion at Peller Airlines. Mike is at the podium. "Ladies and gentlemen, I'd like to inform you that the data from our quarterly survey have now been edited and entered into our computer system. This data set constitutes a random sample of 1,600 adult passengers who flew with us on mainland routes over the 2-week period ending last Friday. It is imperative that we continue to monitor the airline services we provide through these quarterly surveys. By keeping tabs on our customers and taking the market pulse, we can continue to make those improvements that will guarantee that our passengers remain loyal to Peller and, if they are satisfied, they will encourage their friends and families to fly with us." "Thank you Mike," interrupted Lorena Martinez who recently took over as CEO of Peller Airlines, "will you kindly remind us of the central theme in this particular quarterly survey and give us your time table for presentation and discussion." "Certainly, Ms. Martinez. The theme, ma'am, for this survey deals with passenger satisfaction and its potential relationships with reasons for travel and disposition of luggage. Gender differences will also be explored. As for a time table, Dr. Tom Foley, the Director of Central Processing, assures me that the initial printouts will be ready by noon today. Since it takes two working days for data cleansing and preliminary analyses, I will be ready to make our presentations at next week's board meeting. I'll need 15 minutes for the presentation and I request an additional 15 minutes for questions, answers, and board discussion." "Very good, Mike, it sounds like your Marketing and Promotion Department has got a handle on this one. On behalf of the board, I would like to grant your requests and state that we look forward to your presentation next week. Please keep me informed if you need anything to expedite the analysis." "Thank you, ma'am," Mike replied, and he took his seat.

Two Days Later in Mike Drucker's Office You are sitting in the office of the senior vice-president for Marketing and Promotion, Mike Drucker, awaiting his entrance. He is on the telephone in the vestibule discussing the computer printouts that he is holding. Suddenly, the conversation ends and Mike enters his office smiling. "Well, here it is," Mike Drucker exclaims triumphantly as he takes his seat at his desk. "Dr. Foley says that the data appear to be clean, all error checks worked. You did a great job!" You nod and acknowledge the praise, aware that this is just the beginning. Mike continues, "Let me turn this over to you again. Take a good look at the important theme questions, along with the responses and the cross-tabulations. I'd like some confirmatory answers to the following:

1. Our well-known jingle says '*Peller is the airline of first choice for one out of two, so if you're not the one, it's time you pick us too!*' I really hate that jingle—but people like it, it's catchy. But is it accurate? Is there evidence that the proportion of passengers who state that Peller is the airline of choice differs from .50?

2. For advertising and promotional purposes, it is important to know if a gender effect is present with respect to customer satisfaction. Is there evidence of a difference between males and females in terms of the proportion of these passengers who claim Peller is the airline of choice?

3. Many of us have argued over whether the primary reason for a trip affects customer satisfaction. Is there evidence of a difference among the various primary reasons for flying with respect to the proportion of the passengers who claim Peller is the airline of choice?

4. For advertising and promotional purposes, it is important to study potential associations between such factors as primary reasons for flying and luggage disposition. If handling the luggage is perceived as an important service feature, we will have to make sure Peller provides better and quicker service for those who check their luggage in as well as more

room on the aircraft for those who carry their luggage on board. Is there evidence of a relationship between primary reasons for flying and luggage disposition?

I know we're under the gun on this one, but I'd like you to have this analysis on my desk first thing Monday morning. Please prepare an executive summary and attach all tables and charts, list all hypotheses that you are testing, the levels of significance you chose for the tests, and the conclusions to be drawn. Also, please prepare to give me and my staff an informal 10-minute presentation on your findings. Do you have any questions?" Taking a deep breath, you say "no, not at this time. I'm ready to work." "Thanks," Mike continues, as he ushers you out of his office, "I'll see you first thing Monday, but if any questions come up don't hesitate to call me."

Responses to Theme Portion of Peller Airlines Quarterly Satisfaction Survey

1. What is your gender?
   Male…960   Female…640
2. *Before* this trip, would you have considered Peller to be your airline of choice?
   Yes…816   No…784
3. Now that you've taken this trip, do you consider Peller as your airline of choice?
   Yes…832   No…768
4. What was the primary reason for taking this trip?
   Business…880
   Emergency…64
   Moving/Transit…96
   Pleasure…560
5. What did you do with your baggage for this trip?
   Carried it all on board…768
   Checked it all in…592
   Carried some and checked some…192
   Had no luggage…48

Theme Question 2

	Gender		
Choice *After*	Males	Females	Totals
Yes	512	320	832
No	448	320	768
Totals	960	640	1,600

Theme Question 3

	Primary Reason				
Choice *After*	Business	Emergency	Moving/In Transit	Pleasure	Totals
Yes	455	20	42	315	832
No	425	44	54	245	768
Totals	880	64	96	560	1,600

Theme Question 4

Primary Reason	Baggage Disposition				Totals
	Carry All	Check All	Do Both	No Luggage	
Business	653	83	103	41	880
Emergency	47	14	1	2	64
Moving/In Transit	6	78	9	3	96
Pleasure	62	417	79	2	560
Totals	768	592	192	48	1,600

Endnotes

1. If the hypothesized difference is 0 (i.e., $p_1 - p_2 = 0$ or $p_1 = p_2$), the numerator in Equation (9.1) becomes $p_{s_1} - p_{s_2}$.
2. Examine Figures 9.1 and 9.3. Observe that in a two-tailed test, the two critical values, $+Z$ and $-Z$, denote rejection regions in the tails of the standard normal distribution. Each contains an area of $\alpha/2$. Also observe that the one critical value, χ^2_1, denotes a rejection region in the upper tail of a chi-square distribution having 1 degree of freedom. The upper tail of this chi-square distribution contains an area of α. Since the Z-test statistic from the standard normal distribution ranges from $-\infty$ to $+\infty$ and the χ^2-test statistic from the chi-square distribution ranges from 0 to $+\infty$, we see that by squaring the Z value we obtain the χ^2_1 value.

References

1. Berenson, M. L., and D. M. Levine, *Basic Business Statistics: Concepts and Applications*, 6th ed. (Englewood Cliffs, NJ: Prentice-Hall, 1996).
2. Daniel, W. W., *Applied Nonparametric Statistics*, 2d ed. (Boston, MA: PWS Kent, 1990).
3. Dixon, W. J., and F. J. Massey, Jr., *Introduction to Statistical Analysis*, 4th ed. (New York: McGraw-Hill, 1983).
4. Lewontin, R. C., and J. Felsenstein, "Robustness of Homogeneity Tests in 2 × n Tables," *Biometrics*, March 1965, Vol. 21, pp. 19–33.
5. Marascuilo, L. A., "Large-Sample Multiple Comparisons," *Psychological Bulletin*, 1966, Vol. 65, pp. 280–290.
6. Marascuilo, L. A., and M. McSweeney, *Nonparametric and Distribution-Free Methods for the Social Sciences* (Monterey, CA: Brooks/Cole, 1977).
7. *Microsoft Excel Version 7* (Redmond, WA: Microsoft Corp., 1996).

chapter 10

Statistical Applications in Quality and Productivity Management

CHAPTER OBJECTIVE To provide an introduction to the history of quality and Deming's 14 points of management, and to demonstrate the use of a variety of control charts.

10.1 INTRODUCTION

In this chapter we will focus on statistical applications in quality and productivity management. The pioneer of this methodology, W. A. Shewhart, said more than a half-century ago (see Reference 22) that

> The long range contribution of statistics depends not so much upon getting a lot of highly trained statisticians into industry as it does in creating a statistically minded generation of physicists, chemists, engineers and others who will in any way have a hand in developing and directing the production processes of tomorrow.

We begin with a historical perspective on quality and productivity and discuss the evolution of management styles. We will then develop the underlying theory behind the topic of control charts. We will also consider two management planning tools useful for process improvement, the process flow diagram and the fishbone diagram. The subsequent discussion of Deming's 14 points of management sets the stage for the development of a variety of control charts used for different types of data. In addition, an intriguing experiment known as "the parable of the red beads" will be examined to highlight the different types of variation inherent in a set of data and to reinforce the importance of management's responsibility to improve systems.

10.2 QUALITY AND PRODUCTIVITY: A HISTORICAL PERSPECTIVE

By the mid-1980s, it had become clear that a global economy had developed in which companies located in an individual country had to compete not only with local and national competitors, but also with competitors from all parts of the world. This global economy developed because of many factors, including the rapid expansion in worldwide communications and the exponential increase in the availability and power of computer systems. In such an environment, it is vitally important that business organizations be able to respond rapidly to changes in market conditions by incorporating the most effective managerial approaches available.

The development of this global economy has also led to a reemergence of an interest in the area of quality improvement in the United States. Evidence of the renewed interest can be seen in the increasing importance being placed on the competition for the Malcolm Baldrige Award (see Reference 10), given annually to companies making the greatest strides in improving quality and customer satisfaction with their products and services. Among the companies who have won this award are Motorola, Xerox, Federal Express, Cadillac Motor Company, Ritz-Carlton Hotels, AT&T Universal Card Services, and Eastman Chemical Company.

We will present the background to this reemergence of interest in quality and productivity by briefly examining the historical development of management in four phases (see References 4, 12, and 13). We may think of first-generation management as **management by doing,** the type of management practiced by primitive hunter-gatherer societies in which individuals produced something for themselves or their tribal unit whenever the product was required.

Beginning in the Middle Ages, the rise of craft guilds in Europe led to a second generation of management, **management by directing.** Craft guilds managed the training of apprentices and journeymen and determined standards of quality and workmanship for the products made by the guild.

The creation of the assembly line, in conjunction with the Industrial Revolution (see References 6 and 9), brought about a third generation of management, **management by control,** in which workers were divided into those who actually did the work (i.e., the laborers) and those who planned and supervised the work (i.e., the managers). This approach took responsibility for quality out of the hands of the individual worker and put it under the inspector, foreman, and other managers. The American engineer Frederick W. Taylor attempted to develop methods for overcoming this division by developing a scientific management approach that required a detailed study of each job. The management by control style also contained a hierarchical structure that emphasized individual accountability to a set of predetermined goals. This approach has been commonly practiced in the United States since the development of the factory environment at the beginning of the twentieth century.

The last several decades have seen the competitive advantage enjoyed by the United States in the post–World War II period gradually eliminated. The United States emerged from World War II with its industrial base intact, unlike most European countries and Japan. Thus, the United States was in a monopolistic position, and the rest of the world eagerly awaited whatever consumer products it was able to produce. In such a supplier's economy, both labor and management had little incentive to critically examine the ways in which they were operating with the goal of making production more efficient.

In addition, the redevelopment of Japanese industry, beginning in 1950, with the assistance of individuals such as W. Edwards Deming, Joseph Juran, Kaoru Ishikawa, and others was based on an emphasis on quality and continuous improvement of products and services. The approach pioneered by these individuals has led to a fourth generation of management that has been called **management by process.** It is often referred to as **total quality management** or **TQM**. One of the principal features of this approach is a focus on the continuous improvement of processes. This managerial style is characterized by an emphasis on teamwork, a focus on the customer (broadened to include everyone who is a part of a process), and a quick reaction to change. Management by process has a strong statistical foundation based on a thorough knowledge of variability, a systems perspective, and belief in continuous improvement. Such statistical tools as Pareto diagrams, histograms, and control charts, and management planning tools such as process flow and fishbone diagrams, are an integral part of this approach. Control charts, as a tool for studying the variability of a system, are useful for helping managers determine how to improve a process. A general introduction to the theory underlying control charts will be the subject of our next section.

10.3 THE THEORY OF CONTROL CHARTS

If we are examining data that have been collected sequentially over a period of time, it is imperative that the variable of interest be plotted at successive time periods. One such graph, originally developed by Shewhart (see References 22, 23, and 24), is the control chart.

The **control chart** is a means of monitoring variation in the characteristic of a product or service by (1) focusing on the time dimension in which the system produces products or services and (2) studying the nature of the variability in the system. The control chart may be used to study past performance and/or to evaluate present conditions. Data collected from a control chart may form the basis for process improvement. Control charts may be used for different types of variables—for categorical variables such as the proportion of airplane flights of a particular company that are more than 15 minutes late on a given day, for discrete variables such as the count of the number of paint blemishes in a panel of a car door, and for continuous variables such as the amount of apple juice contained in 1-liter bottles.

In addition to providing a visual display of data representing a process, the principal focus of the control chart is the attempt to separate special or assignable causes of variation from chance or common causes of variation.

> **Special or assignable causes of variation** represent large fluctuations or patterns in the data that are not inherent to a process. These fluctuations are often caused by

changes in a system that represent either problems to be fixed or opportunities to exploit.

Chance or **common causes of variation** represent the inherent variability that exists in a system. These consist of the numerous small causes of variability that operate randomly or by chance.

The distinction between the two causes of variation is crucial because special causes of variation are considered to be those that are not part of a process and are correctable or exploitable without changing the system, while common causes of variation may be reduced only by changing the system. Such systemic changes are the responsibility of management.

Control charts allow us to monitor the process and determine the presence of special causes. There are two types of errors that control charts help prevent. The first type of error involves the belief that an observed value represents special cause variation when in fact it is due to the common cause variation of the system. Treating such a common cause of variation as special cause variation can result in *tampering* with or overadjustment of a process with an accompanying increase in variation. The second type of error involves treating special cause variation as if it is common cause variation and thus not taking immediate corrective action when it is necessary. Although these errors can still occur when a control chart is used, they are far less likely.

The most typical form of control chart will set control limits that are within ±3 standard deviations[1] of the statistical measure of interest (be it the average, the proportion, the range, etc.). In general, this may be stated as

$$\text{Process average} \pm 3 \text{ standard deviations} \qquad (10.1)$$

so that

Upper control limit (UCL) = process average + 3 standard deviations

Lower control limit (LCL) = process average − 3 standard deviations

Once these control limits are set, the control chart is evaluated from the perspective of (1) discerning any pattern that might exist in the values over time, and (2) determining whether any points fall outside the control limits. Figure 10.1 illustrates three different situations.

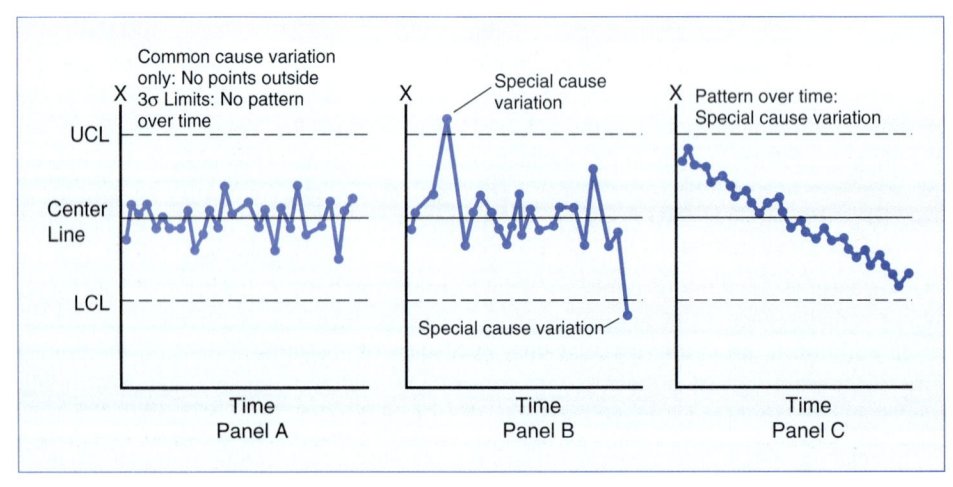

FIGURE 10.1 Three control chart patterns.

In Panel A of Figure 10.1, we observe a process that is stable and contains only common cause variation, one in which there does not appear to be any pattern in the ordering of values over time, and one in which there are no points that fall outside the 3 standard deviation control limits. Panel B, on the contrary, contains two points that fall outside the 3 standard deviation control limits. Each of these points would have to be investigated to determine the special causes that led to their occurrence. Although Panel C does not have any points outside the control limits, it has a series of consecutive points above the average value (the center line) as well as a series of consecutive points below the average value. In addition, a long-term overall downward trend in the value of the variable is clearly visible. Such a situation would call for corrective action to determine what might account for this pattern prior to initiating any changes in the system.

The detection of a trend is not always so obvious. Two easy rules (see Reference 8) for indicating the presence of a trend are (1) having eight consecutive points above the center line (or eight consecutive points below the center line), or (2) having eight consecutive points that are increasing (or eight consecutive points that are decreasing).

Once all special causes of variation have been explained and eliminated, the process involved can be examined on a continuing basis until there are no patterns over time or points outside the 3 standard deviation limits. When the process contains only common cause variation, its performance is predictable (at least in the near future).

To reduce common cause variation, it is necessary to alter the system that is producing the product or service. In our next section, we will discuss some management planning tools that are extremely valuable in helping to understand a process so that a system can be improved.

10.4 SOME TOOLS FOR STUDYING A PROCESS: FISHBONE (ISHIKAWA) AND PROCESS FLOW DIAGRAMS

10.4.1 Introduction

Before we can determine the appropriate control charts to be used for a set of data, we need to define further what we mean by a process.

A **process** is a sequence of steps that describe an activity from beginning to completion.

The concept of a process may be viewed schematically in Figure 10.2. Using this approach, all work is viewed as a set of processes. These processes need to be analyzed in order to develop process knowledge so that variation can be reduced. Variation in the process may be reduced by first removing special cause variation. Then, common cause variation can be reduced by changing the process. This will lead to improved quality and more satisfied customers. Thus, the analysis of process variation and tools for developing process knowledge are the subjects of this section, while the 14 points of Deming's management approach for improving processes will be the subject of the next section.

To understand any process of interest, it is useful to become familiar with two management planning tools, the *fishbone* (or *Ishikawa*) *diagram* and the *process flow diagram*.

10.4.2 The Fishbone (or Ishikawa) Diagram

The **fishbone diagram** was originally developed by Kaoru Ishikawa (see References 13, 16, and 28) to represent the relationship between some effect that could be measured and the set of possible causes that produce the effect. For this reason it is also known as a *cause-and-effect diagram*.

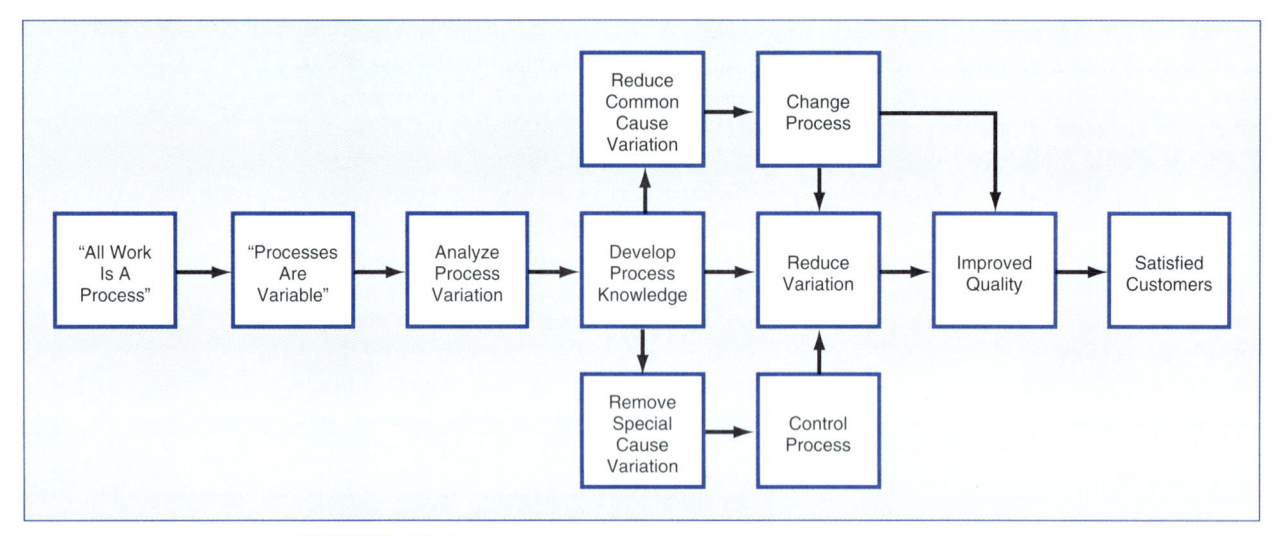

FIGURE 10.2 **The concept of a process.** *Source:* Reprinted from R. Snee, "Statistical Thinking and Its Contributions to Total Quality," *American Statistician,* 1990, Vol. 44, pp. 116–121.

The name *fishbone diagram* (see Figure 10.3) comes from the way various causes are arranged on the diagram. Typically, the effect or problem is shown on the right side, and the major causes are listed on the left side of the diagram. These causes are often subdivided into four categories (usually manpower, methods, material, and machinery in a manufacturing environment and equipment, policies, procedures, and people in a service environment). Within each major category specific causes are listed as branches and subbranches of the major category tree. Thus, the impression of the cause-and-effect diagram is that of a "fishbone."

The fishbone diagram can be useful in understanding processes in a variety of applications. Figure 10.4 represents a fishbone diagram for a problem many have encountered—being late for work. Before a fishbone diagram can be constructed, a group of individuals should collectively "brainstorm" a list of causes for the effect of interest. In Figure 10.4 (which relates to Marilyn Levine, an elementary school teacher and the spouse of one of this text's authors), we observe that each of the four major categories has been subdivided into several causes, with subcauses appearing on each branch. For example, under materials, five causes are listed: gas, information, student lesson assignments, breakfast, and other meals.

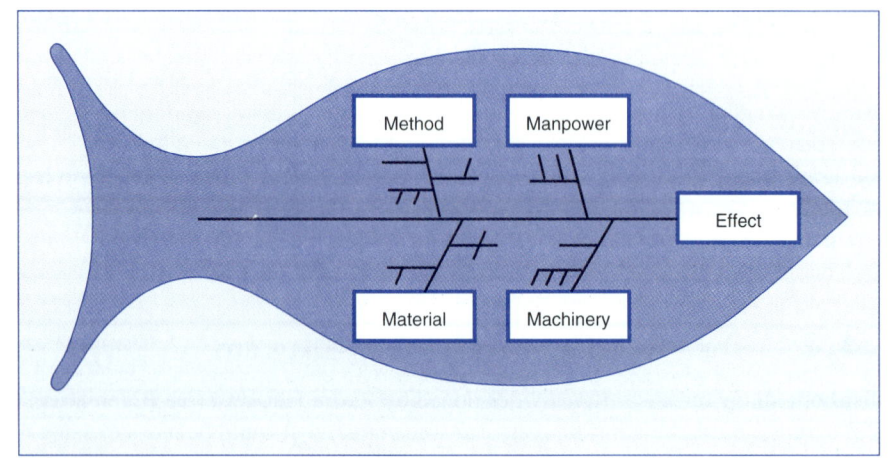

FIGURE 10.3 The fishbone diagram.

490 **Chapter 10** Statistical Applications in Quality and Productivity Management

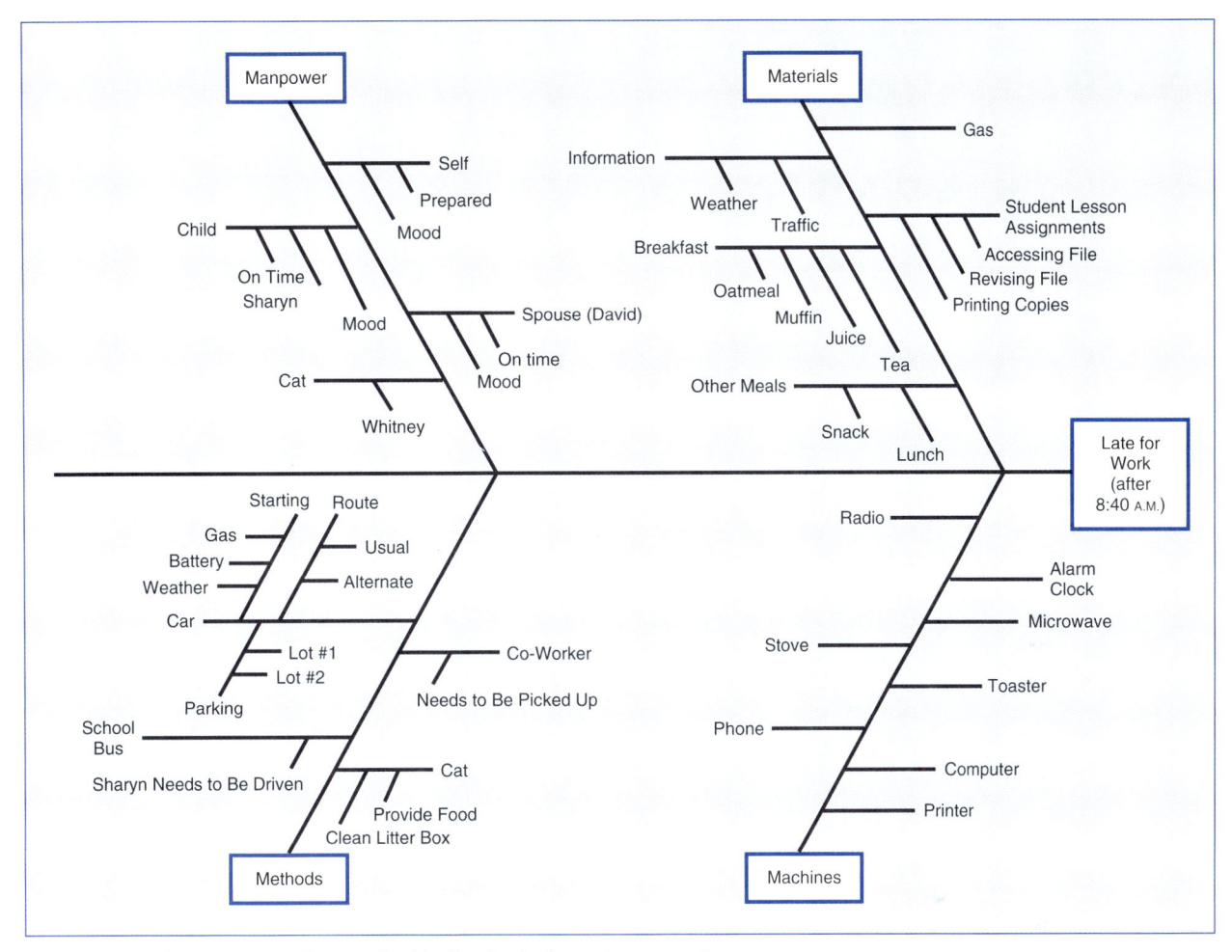

FIGURE 10.4 Fishbone diagram for Marilyn Levine's getting-to-work process.

For student lesson assignments, such subcauses as accessing existing file, revising existing file, and printing copies are shown. Similar sets of causes are illustrated for methods, manpower, and machines.

As a second example, Figure 10.5 represents a fishbone diagram that might be constructed to represent the process of studying the utilization (or actually the underutilization) of hospital operating rooms on Saturday. Under the category of people, among the possible causes are registered nurse (RN) staff, medical doctor staff, and housekeeping staff, each with subbranches. Under the category of policies, among the causes are informed consent, personnel, supplies, and marketing plans. Similar detailed sets of causes are provided for the procedures and equipment categories.

10.4.3 Process Flow Diagrams

A second useful management planning tool for understanding a process is a **process flow diagram**. This diagram allows us to see a flow of steps in a process from its beginning to its termination. Such a diagram is invaluable in understanding a process. Three commonly used process flow symbols are depicted in Figure 10.6. The oval termination symbol is used at the beginning and end of a process as a start/stop symbol. The rectangular process symbol is used

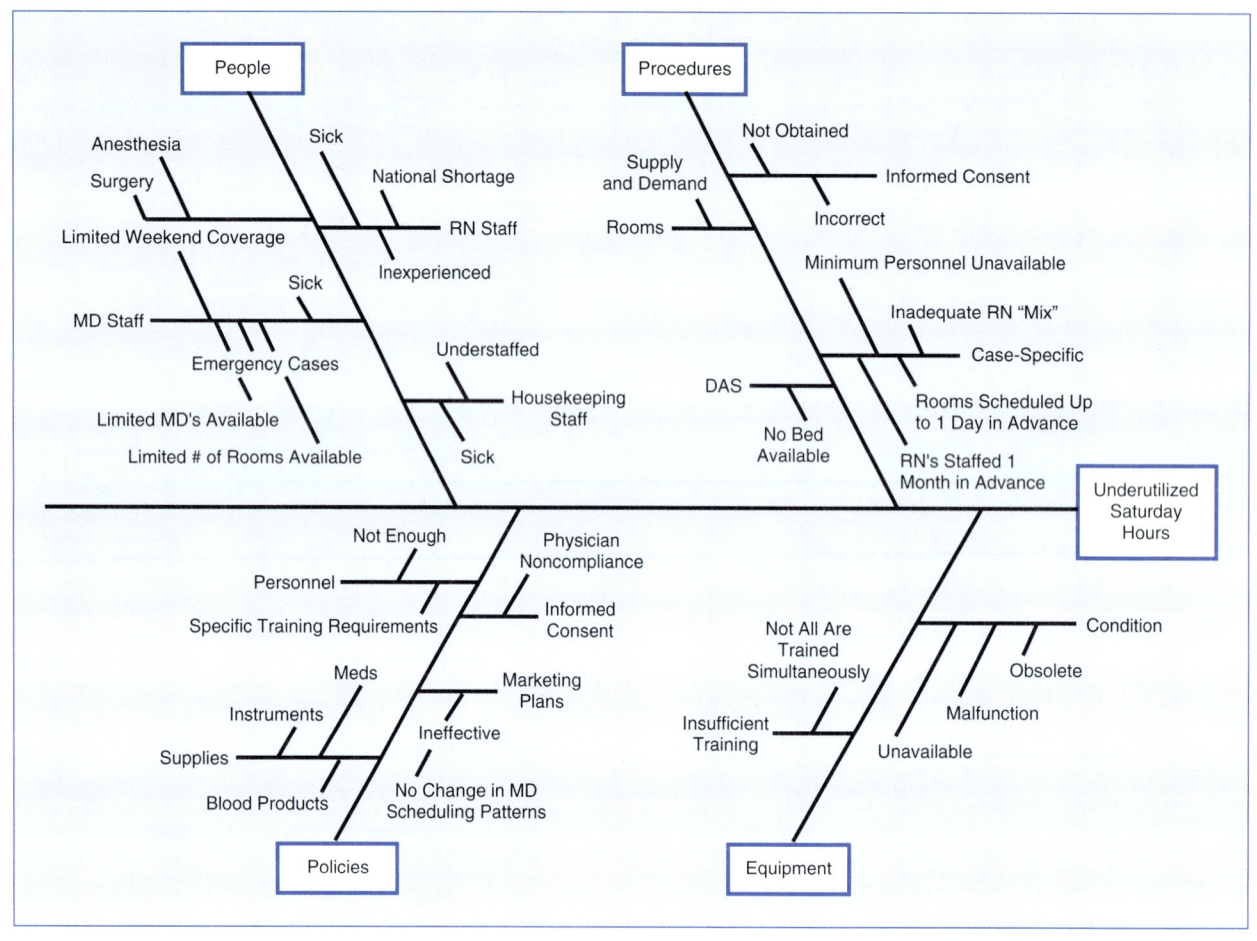

FIGURE 10.5 Fishbone diagram for the utilization of hospital operating rooms on Saturday.

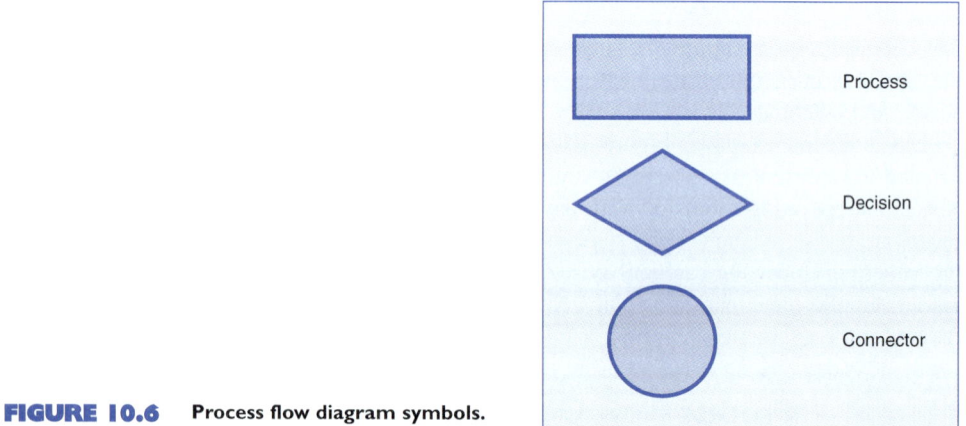

FIGURE 10.6 Process flow diagram symbols.

to indicate that a step of the process is to be performed. The decision diamond symbol provides one way in, but at least two ways out, so that alternative paths in the process can be described.

Now that we have defined some of the symbols used, we may illustrate the process flow diagram with two examples.

492 **Chapter 10** Statistical Applications in Quality and Productivity Management

The first situation refers to the problem of getting to work on time that was depicted in the fishbone diagram of Figure 10.4 on page 491. Figure 10.7 represents the process flow diagram for this circumstance. We observe that this process flow diagram begins with a concern about whether the alarm has gone off and proceeds to whether the bathroom is currently available, whether the student assignments are ready, whether her daughter has to be driven to school, and whether a co-worker has to be picked up on the way to work. In summary, the process flow diagram, by forcing us to document the process involved, provides us with a more thorough understanding of the various aspects of the process and gives insights into where delays may be likely to develop.

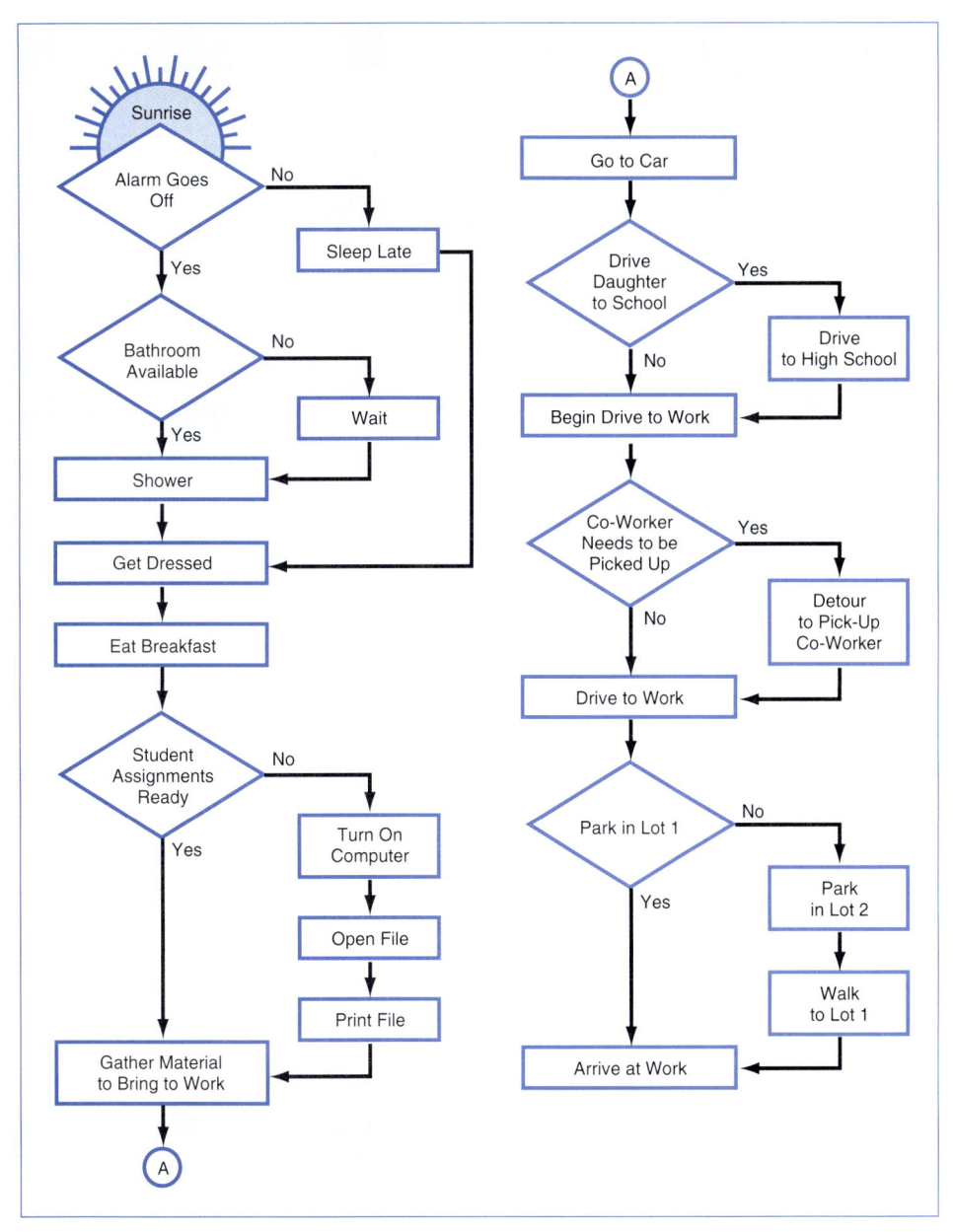

FIGURE 10.7 Process flow diagram for Marilyn Levine's getting-to-work process.

10.4 Some Tools for Studying a Process: Fishbone (Ishikawa) and Process Flow Diagrams

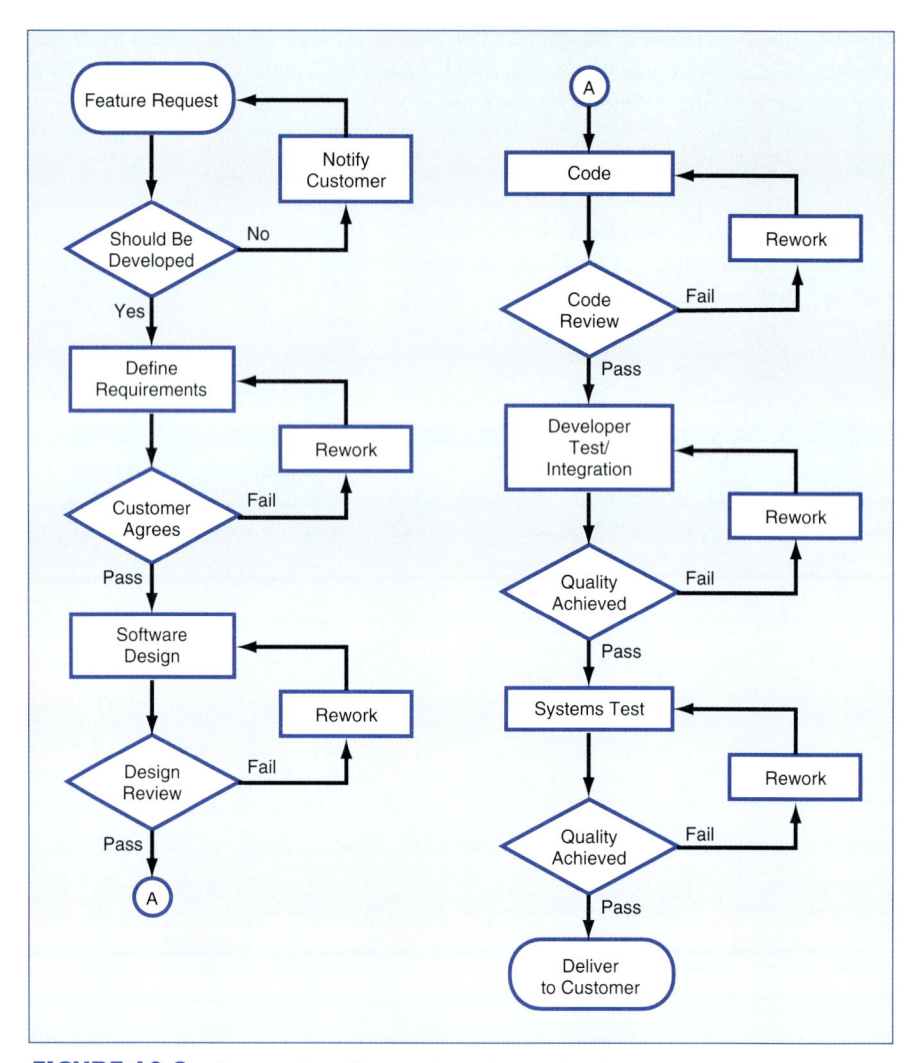

FIGURE 10.8 Process flow diagram for software development.

A second example of a process flow diagram, involving the development of software, is depicted in Figure 10.8. In this instance, the software to be developed passes through a series of steps, each of which contains an inspection aspect (with rework if necessary) before being delivered to a customer.

Now that we have discussed these management planning tools that help us understand a process, in the next section we will consider a managerial approach that may be used when processes need to be changed.

Problems for Section 10.4

10.1 Compare and contrast the fishbone diagram and the process flow diagram.

10.2 Set up fishbone and process flow diagrams for your own personal process of getting to school or work in the morning.

10.3 (a) Set up fishbone and process flow diagrams for the registration process at your school.
(b) On the basis of the diagrams developed in (a), what improvements can you suggest in the registration process?

10.4 (a) Set up fishbone and process flow diagrams for your own personal process of studying for a statistics exam.
(b) On the basis of the diagrams developed in (a), what improvements can you make in the way that you go about studying for your statistics exam?

10.5 You are planning to have a dinner party for eight people at your house. The party will consist of cocktails and hors d'oeuvres, soup, salad, entree, and dessert.
(a) Set up fishbone and process flow diagrams for the process of preparing and serving the food and drinks for the dinner party.
(b) On the basis of the diagrams developed in (a), what improvements can you make in the way that you were planning to prepare for the dinner party?

10.5 DEMING'S 14 POINTS: A THEORY OF MANAGEMENT BY PROCESS

The high quality of Japanese products and the economic miracle of Japanese development after World War II are well-known facts. What is not as readily known, particularly by young people today, is the fact that, prior to the 1950s, Japan had acquired the unenviable reputation of producing shoddy consumer products of poor quality. Thus, the question must surely be asked, what happened to change this reputation? Part of the answer lies in the fact that by 1950, top management of Japanese companies, in alliance with the Union of Japanese Science and Engineering, realized that quality was a vital factor for being able to export consumer products successfully. Some Japanese engineers had been exposed to the contribution that the Shewhart control charts had made toward the American war effort during World War II (see References 9 and 27). Thus, several American experts, including W. Edwards Deming, were invited to Japan during the early 1950s. The rigorous application of a total quality management approach led to improved productivity in Japanese industry. Owing primarily to his experiences in Japan, Deming developed his approach to management based on the following **14 points**:

1. Create constancy of purpose for improvement of product and service.
2. Adopt the new philosophy.
3. Cease dependence on inspection to achieve quality.
4. End the practice of awarding business on the basis of price tag alone. Instead, minimize total cost by working with a single supplier.
5. Improve constantly and forever every process for planning, production, and service.
6. Institute training on the job.
7. Adopt and institute leadership.
8. Drive out fear.
9. Break down barriers between staff areas.
10. Eliminate slogans, exhortations, and targets for the workforce.
11. Eliminate numerical quotas for the workforce and numerical goals for management.
12. Remove barriers that rob people of pride of workmanship. Eliminate the annual rating or merit system.
13. Institute a vigorous program of education and self-improvement for everyone.
14. Put everyone in the company to work to accomplish the transformation.

Point 1, create constancy of purpose, refers to how an organization deals with problems that arise both at present and in the future. The focus is on the constant improvement of a product or service. This improvement process is illustrated by the **Shewhart cycle** of Figure 10.9.

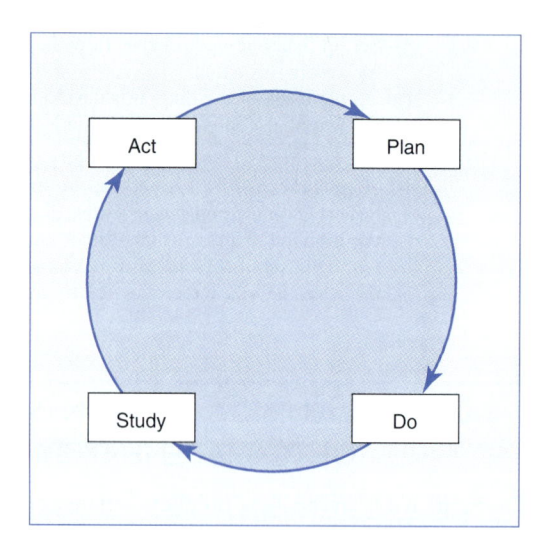

FIGURE 10.9 The Shewhart-Deming PDSA cycle.

Unlike the traditional manufacturing approach of "design it, make it, try to sell it," the Shewhart cycle represents a continuous cycle of "plan, do, study, and act." The first step, *planning*, represents the initial design phase for planning a change in a manufacturing or service process. The second step, *doing*, involves carrying out the change, preferably on a small scale. In doing this, planned experiments can be a particularly valuable approach. The third step, *studying* or *checking*, involves an analysis of the results using statistical tools to determine what was learned. The fourth step, *acting*, involves the acceptance of the change, its abandonment, or further study of the change under different conditions. With this approach, the process starts with the customer as the most important element of the production or service process.

The key aspect of this approach is the overwhelming concern for future problems. Improvement of processes must go hand in hand with innovation and the development of new products. The importance of innovation can be illustrated by an analogy to the "Fosbury Flop" technique[2] of high jumping. Prior to the development of this technique of high jumping, the accepted approach was to use a technique called the "Western Roll." If someone had just worked at improving his ability using the Western Roll, he still would not have been able to compete with others who had adopted what was at that time the newer and better Fosbury Flop technique. Thus, innovation must go hand in hand with process improvement.

Point 2, adopt the new philosophy, refers to the urgency with which American companies need to realize that we are in a new economic age that differs drastically from the post–World War II period of American dominance (see Reference 9). It is a commonly accepted fact that, as part of human nature, people will not act until a crisis is at hand, because they prefer to continue doing things in ways that they believe have produced successful results in the past. However, in this new economic age, American management is often afflicted with what Deming called a set of "deadly diseases"—including lack of constancy of purpose, emphasis on short-term profits, fear of unfriendly takeover, evaluation of performance and merit rating systems, and excessive mobility of management. Finally, management philosophy needs to accept the notion that higher quality costs less, *not* more. However, it does require an up-front investment to get improved quality. This investment pays off with tremendous dividends.

Point 3, cease dependence on mass inspection to achieve quality, implies that any inspection whose purpose is to improve quality is too late because the quality is already built into the product. It would be better to focus on making it right the first time. Among the difficulties involved in mass inspection (besides high costs) are the failure of inspectors to agree on nonconforming items and the problem of separating good and bad items. Such difficulties can be illustrated with an example taken from Scherkenbach (see Reference 21) and depicted in

> **FINISHED FILES ARE THE RESULT OF YEARS OF SCIENTIFIC STUDY COMBINED WITH THE EXPERIENCE OF MANY YEARS**

FIGURE 10.10   **An example of the proofreading process.**   Source: W. W. Scherkenbach, *The Deming Route to Quality and Productivity: Road Maps and Roadblocks* (Washington, D.C.: CEEP/Press, 1987).

Figure 10.10. Suppose your job involves proofreading this sentence with the objective of counting the number of occurrences of the letter "F." Perform this task and note the number of occurrences of the letter F that you discover.

People usually see either three Fs or six Fs. The correct number is six Fs. The number seen is dependent on the method used in reading the sentence. One is likely to find three Fs if the paragraph is read phonetically and six Fs if one counts the number of Fs carefully. The point of the exercise is to show that, if we have such a simple process as counting Fs leading to inconsistency of "inspector's" results, what will happen when a process fails to contain a clear operational definition of nonconforming? Certainly, in such situations much more variability from inspector to inspector will occur.

Point 4, ending the practice of awarding business on the basis of price tag alone, represents the antithesis of lowest bidder awards. It focuses on the fact that there can be no real long-term meaning to price without a knowledge of the quality of the product. A lowest bidder approach ignores the advantages in reduced variation of a single supplier and fails to consider the advantages of the development of a long-term relationship between purchaser and supplier. Such a relationship would allow the supplier to be innovative and would tend to make the supplier and purchaser partners in achieving success.

Point 5, improve constantly and forever the system of production and services, reinforces the importance of the continuous focus of the Shewhart cycle and the belief that quality needs to be built in at the design stage. Attaining quality is viewed as a never-ending process in which smaller variation translates into a reduction in the economic losses that occur in the manufacturing of a product whose characteristics are variable. Such an approach stands in contrast to one whose only concern is in meeting specifications. This latter approach does not associate any economic loss with products whose characteristics are within specification limits.

Point 6, institute training, reflects the needs of all employees including production workers, engineers, and managers. It is critically important for management to understand the differences between special causes and common causes of variation so that proper action can be taken in each circumstance. In particular, training needs to focus on developing standards for acceptable work that do not change on a daily basis. Also, management needs to realize that people learn in different ways; some learn better with written instructions, some learn better with verbal instructions. In addition, management must decide who should be trained and in what. Often, corporate training in areas such as statistics is wasted because of (1) limited screening of who gets trained, (2) no emphasis on how to manage what has been learned, and (3) little insistence on the use of what has been learned.

Point 7, adopt and institute leadership, relates to the distinction between leadership and supervision. The aim of leadership should be to improve the system and achieve greater consistency of performance.

Points 8–12 [drive out fear, break down barriers between staff areas, eliminate slogans, eliminate numerical quotas for the workforce and numerical goals for management, and remove barriers to pride of workmanship including the annual rating and merit system (this may be the most controversial point)] are all related to how the performance of an employee is to be evaluated.

The quota system for the production worker is viewed as detrimental for several reasons. First, it has a negative effect on the quality of the product, since supervisors are more inclined to pass inferior products through the system when they need to meet work quotas. Such flexible standards of work reduce the pride of workmanship of the individual and perpetuate a system in which peer pressure holds the upper half of the workers to no more than the quota rate.

In addition, the emphasis on targets and exhortations may place an improper burden on the worker, since it is management's job to improve the system, not to expect workers to produce beyond the system (this will be clearly illustrated in Section 10.7).

Third, the annual performance rating system can rob the manager of his or her pride of workmanship because this system of evaluation more often than not fails to provide a meaningful measure of performance. For many managers, the only customer is his or her supervisor. In too many cases (see Reference 25), efforts are focused on either distorting the data or distorting the system to produce the desired set of results, rather than on efforts to improve the system. Such an approach stifles teamwork because there is often a reduced tangible reward for working together across functional areas. Finally, it rewards people who work successfully within the system, rather than people who work to improve the system.

Point 13, encourage education and self-improvement for everyone, reflects the notion that the most important resource of any organization is its people. Efforts that improve the knowledge of people in the organization also serve to increase the assets of the organization.

Point 14, take action to accomplish the transformation, again reflects the approach of management as a process in which one continuously strives toward improvement in a never-ending cycle.

Now that we have provided a brief introduction to the Deming philosophy and linked management by process to fundamental control chart ideas, we shall develop several control charts that are used in industry.

10.6 CONTROL CHARTS FOR THE PROPORTION AND NUMBER OF NONCONFORMING ITEMS— THE p AND np CHARTS

10.6.1 Introduction

Let us turn our attention to various types of control charts that are used to monitor processes and determine whether special cause variation and common cause variation are present in a process. One type of commonly used control chart are the **attribute charts**, which are used when sampled items are classified according to whether they conform or do not conform to operationally defined requirements. The p and np charts that will be discussed in this section are based on either the proportion of nonconforming items (***p* chart**) or the number of nonconforming items (***np* chart**) in a sample.

10.6.2 The p Chart

You may recall that we studied proportions in Chapters 4 and 5. We discussed the binomial distribution in Section 4.13. In Equation (5.11) of Section 5.10, we defined the proportion as X/n, and in Equation (5.11), we defined the standard deviation of the proportion as

$$\sigma_{p_s} = \sqrt{\frac{p(1-p)}{n}}$$

Using Equation (10.1) on page 488, we may establish control limits for the proportion of nonconforming[3] items from the sample or subgroup data as

$$\bar{p} \pm 3\sqrt{\frac{\bar{p}(1-\bar{p})}{\bar{n}}} \qquad (10.2)$$

so that

$$\text{LCL} = \bar{p} - 3\sqrt{\frac{\bar{p}(1-\bar{p})}{\bar{n}}} \qquad (10.3a)$$

$$\text{UCL} = \bar{p} + 3\sqrt{\frac{\bar{p}(1-\bar{p})}{\bar{n}}} \qquad (10.3b)$$

where
X_i = number of nonconforming items in subgroup i
n_i = sample or subgroup size for subgroup i
$p_i = X_i/n_i$
k = number of subgroups taken
$\bar{n}$ = average subgroup size
$\bar{p}$ = average proportion of nonconforming items

For equal n_i,

$$\bar{n} = n_i \quad \text{and} \quad \bar{p} = \frac{\sum_{i=1}^{k} p_i}{k}$$

or in general,

$$\bar{n} = \frac{\sum_{i=1}^{k} n_i}{k} \quad \text{and} \quad \bar{p} = \frac{\sum_{i=1}^{k} X_i}{\sum_{i=1}^{k} n_i}$$

Any negative value for the lower control limit will mean that the lower control limit does not exist.

We shall illustrate the use of the p chart by referring to the plan of a large hotel located in a resort city to improve the quality of its services. One aspect of its services to hotel guests is represented by the readiness of a hotel room when the guest first enters his or her assigned room. From the viewpoint of initial impressions, it is particularly important that all amenities that are supposed to be available (soap, towels, complimentary guest basket, etc.) are actually available in the room and equally important that all appliances such as the radio, television, and telephone are in proper working order. Hotel management decided to study this process for a 4-week period

Table 10.1 Nonconforming hotel rooms at check-in over a 4-week period.

Day	Rooms Studied	Rooms Not Ready	Proportion	Day	Rooms Studied	Rooms Not Ready	Proportion
1	200	16	.080	15	200	18	.090
2	200	7	.035	16	200	13	.065
3	200	21	.105	17	200	15	.075
4	200	17	.085	18	200	10	.050
5	200	25	.125	19	200	14	.070
6	200	19	.095	20	200	25	.125
7	200	16	.080	21	200	19	.095
8	200	15	.075	22	200	12	.060
9	200	11	.055	23	200	6	.030
10	200	12	.060	24	200	12	.060
11	200	22	.110	25	200	18	.090
12	200	20	.100	26	200	15	.075
13	200	17	.085	27	200	20	.100
14	200	26	.130	28	200	22	.110

by taking a daily sample of 200 rooms for which guests are holding reservations. Thus, before each arrival it determined whether the room in the sample contained any nonconformances in terms of the availability of amenities and the working order of all appliances. Table 10.1 lists the number and proportion of nonconforming rooms for each day in the 4-week period.

For these data, $k = 28$, $\sum_{i=1}^{k} p_i = 2.315$, and $n_i = 200$

Thus,

$$\bar{p} = \frac{2.315}{28} = .0827$$

so that using Equation (10.2) we have

$$.0827 \pm 3\sqrt{\frac{(.0827)(.9173)}{200}}$$

$$.0827 \pm .0584$$

Thus,

$$\text{UCL} = .0827 + .0584 = .1411$$

and

$$\text{LCL} = .0827 - .0584 = .0243$$

The control chart for the data of Table 10.1 is displayed in Figure 10.11. An examination of Figure 10.11 seems to indicate a process in a state of statistical control, with the individual points distributed around $\bar{p}$ without any pattern. Thus, any improvement in this system of making rooms ready for guests must come from the reduction of common-cause variation. As we have stated previously, such system alterations are the responsibility of management.

Now that we have examined a situation in which the sample or subgroup size does not vary, we need to turn to a more general situation in which the subgroup size varies over time. As a rule, as long as none of the subgroup sizes n_i differs from the average subgroup size $\bar{n}$

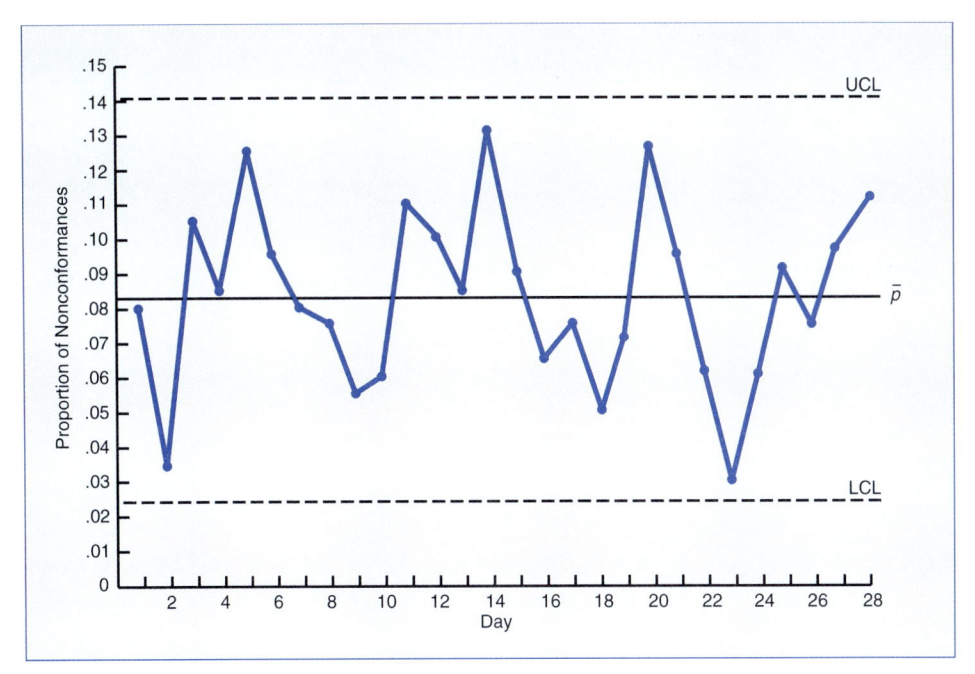

FIGURE 10.11 The p chart for the proportion of nonconforming rooms upon guest arrival. *Source:* Table 10.1.

by more than ±25% of $\bar{n}$ (see Reference 8), Equation (10.2) may be used to obtain the control limits for the p chart.

To illustrate the use of the p chart when the subgroup sizes are unequal, we will investigate the production of gauze sponges at a factory. The number of nonconforming sponges and the number of sponges produced daily for a period of 32 days are displayed in Table 10.2. For these data,

Table 10.2 Nonconforming sponges produced daily for a 32-day period.

Day	Sponges Produced	Nonconforming Sponges	Proportion	Day	Sponges Produced	Nonconforming Sponges	Proportion
1	690	21	.030	17	575	20	.035
2	580	22	.038	18	610	16	.026
3	685	20	.029	19	596	15	.025
4	595	21	.035	20	630	24	.038
5	665	23	.035	21	625	25	.040
6	596	19	.032	22	615	21	.034
7	600	18	.030	23	575	23	.040
8	620	24	.039	24	572	20	.035
9	610	20	.033	25	645	24	.037
10	595	22	.037	26	651	25	.038
11	645	19	.029	27	660	21	.032
12	675	23	.034	28	685	19	.028
13	670	22	.033	29	671	17	.025
14	590	26	.044	30	660	22	.033
15	585	17	.029	31	595	24	.040
16	560	16	.029	32	600	16	.027

$$k = 32, \quad \sum_{i=1}^{k} n_i = 19,926, \quad \text{and} \quad \sum_{i=1}^{k} X_i = 665$$

Thus,

$$\bar{n} = \frac{19,926}{32} = 622.69 \quad \text{and} \quad \bar{p} = \frac{665}{19,926} = .033$$

so that we have

$$.033 \pm 3\sqrt{\frac{(.033)(1-.033)}{622.69}}$$

$$.033 \pm .021$$

Thus,

$$\text{LCL} = .033 - .021 = .012$$

and

$$\text{UCL} = .033 + .021 = .054$$

The control chart for the data of Table 10.2 is displayed in Figure 10.12. An examination of Figure 10.12 seems to indicate a process in a state of statistical control, without any points beyond the control limits and without a pattern. Thus, any improvements in the system of producing gauze sponges must come from the reduction of common cause variation.

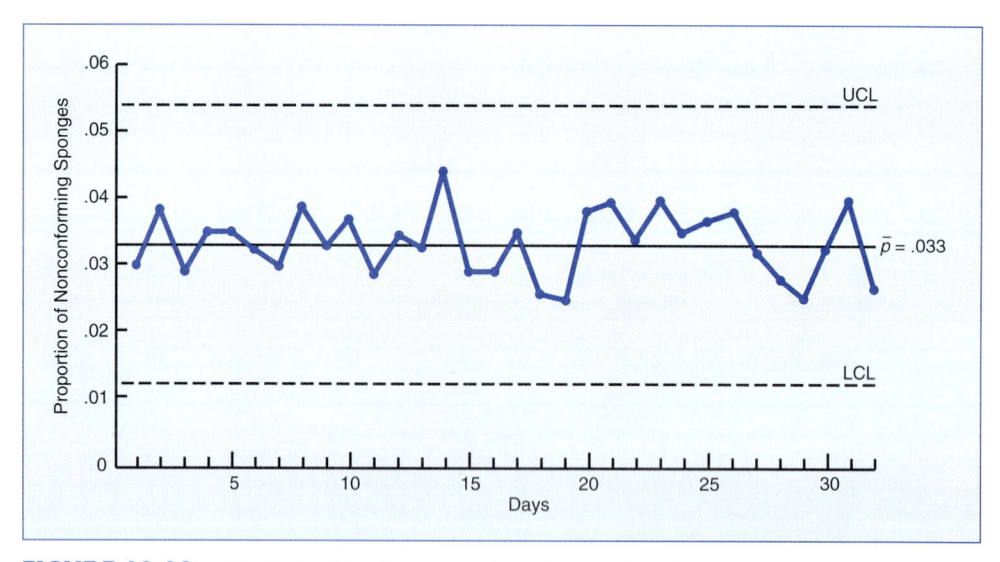

FIGURE 10.12 The p chart for the proportion of nonconforming sponges.

10.6.3 The np Chart

When the subgroups are all of the same size, a desirable alternative to the p chart is the np chart. You may recall that in Section 4.13.2 we defined the standard error of the number of "successes" or nonconforming items as

$$\sigma = \sqrt{np(1-p)}$$

Thus, using the normal approximation to the binomial distribution and Equation (10.1), we may set up control limits for the number of nonconforming items as follows:

$$\bar{X} \pm 3\sqrt{\bar{X}(1 - \bar{p})} \qquad (10.4)$$

so that

$$\text{LCL} = \bar{X} - 3\sqrt{\bar{X}(1 - \bar{p})} \qquad (10.5a)$$

$$\text{UCL} = \bar{X} + 3\sqrt{\bar{X}(1 - \bar{p})} \qquad (10.5b)$$

where

$$\bar{X} = \frac{\sum_{i=1}^{k} X_i}{k}$$

$$\bar{p} = \frac{\sum_{i=1}^{k} X_i}{nk}$$

n = subgroup size

k = number of subgroups

To illustrate the np chart, we return to the nonconforming hotel room data of Table 10.1 on page 500 that we used for the p chart. For these data,

$$k = 28, \quad n = 200, \quad \text{and} \quad \sum_{i=1}^{k} X_i = 463$$

Thus,

$$\bar{X} = \frac{463}{28} = 16.536 \quad \text{and} \quad \bar{p} = \frac{463}{(200)28} = .0827$$

so that we have

$$16.536 \pm 3\sqrt{16.536(1 - .0827)}$$

$$16.536 \pm 11.684$$

Thus,

$$\text{UCL} = 16.536 + 11.684 = 28.22$$

and

$$\text{LCL} = 16.536 - 11.684 = 4.852$$

The np chart for these data is displayed in Figure 10.13. Note that this figure provides precisely the same results as the p chart of Figure 10.11 on page 501, but indicates the number of nonconforming items rather than the proportion. The choice between the two approaches (available only for equal subgroups) is a matter of personal preference.

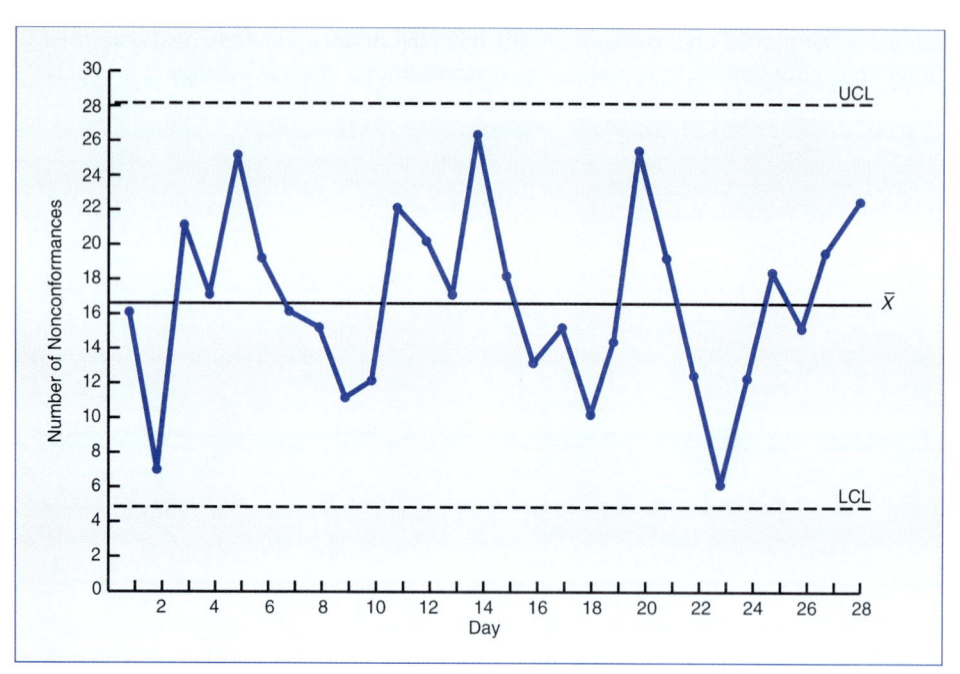

FIGURE 10.13 The *np* chart for the nonconforming hotel rooms. *Source:* Data are taken from Table 10.1.

10.6.4 Using Microsoft Excel for the *p* and *np* Charts

In this section we have studied two control charts, the *p* chart for the proportion and the *np* chart for the number of nonconforming items. Although Microsoft Excel does not have a Data Analysis tool to compute the control limits and center line of a control chart, we can use Excel formulas to perform the computations. Once the control limits and center line have been computed, the Chart Wizard can be used to plot the data along with the control limits and center line. The design for the Data and Calculations sheets for this purpose are shown in Table 10.1.Excel.

10-6-4.XLS

Referring to Equation (10.2) on page 499, we see that to compute the control limits for the *p* chart, we need to obtain $\bar{n}$, the average sample size, and $\bar{p}$, the average proportion. Using the nonconforming hotel room data from Table 10.1 on page 500, open a new workbook and name a sheet Data. The Data sheet may be set up with days in column A, Rooms Studied in column B, Rooms Not Ready in column C, and the proportion for each day entered in column D. Begin by opening a new workbook and rename the active sheet Data. The steps for creating and entering the data for the Data sheet are as follows:

❶ Since the days are to be entered in rows 2–29 of column A, enter 1 in cell A2 and then select Edit | Fill | Series as a short cut as was done in Section 5.6.5. Select the Columns and Linear options buttons as before, but this time enter 28 in the Stop Value edit box. Click the OK button, and the day values 2 through 28 are entered in cells A2 through A29, respectively.

❷ Since the sample size on each day is constant at 200, enter 200 in cell B2 and copy this value down the column to cell B29.

❸ Enter the Rooms Not Ready data for each day in the cell range C2:C29.

❹ The proportion of rooms not ready each day can be obtained by dividing the entry in column C by the entry in column B. Enter =C2/B2 in cell D2 and then copy this formula down the column to cell D29.

Table 10.1.Excel Design for data and calculations sheets for the p and np charts.

Data Sheet

	A	B	C	D
1	Day	Rooms Studied	Rooms Not Ready	p
2	1	xxx	xxx	=C2/B2
3	.	.	.	=C3/B3
4	.	.	.	=C4/B4
5	.	.	.	=C5/B5
⋮	⋮	⋮	⋮	⋮
29	28	xxx	xxx	=C29/B29

Calculations Sheet

	A	B
1	Calculations for p and np Charts	
2		
3	nbar	=SUM(Data!B2:B29)/COUNT(Data!B2:B29)
4	pbar	=SUM(Data!C2:C29)/SUM(Data!B2:B29)
5	p Chart	
6	LCL	=B4–3*SQRT(B4*(1–B4)/B3)
7	Center Line	=B4
8	UCL	=B4+3*SQRT(B4*(1–B4)/B3)
9	=IF(STDEV(Data!B:B)=0,"","Subgroup sizes are different, np chart should not be done.")	
10	np Chart	
11	XBAR	=AVERAGE(Data!C2:C29)
12	LCL	=B11–3*SQRT(B11*(1–B4))
13	Center Line	=B11
14	UCL	=B11+3*SQRT(B11*(1–B4))

Once these four columns have been entered on the Data sheet, we can obtain the control limits and the center line by inserting a new worksheet and renaming it Calculations and doing the following with the new Calculations sheet active:

❶ Compute the average sample size $\bar{n}$ in cell B3 by dividing the sum of the sample sizes in the range Data!B2:B29 by the number of days using the formula =SUM(Data!B2:B29)/COUNT(Data!B2:B29).

❷ Obtain the average proportion $\bar{p}$ in cell B4 by dividing the sum of the nonconforming rooms on all the days by the sum of the sample sizes using the formula =SUM(Data!C2:C29)/SUM(Data!B2:B29).

❸ Determine the lower control limit in cell B6 using the formula =B4–3*SQRT(B4*(1–B4)/B3). Use the formula =B4 to copy the $\bar{p}$ value for use as the center line value. Determine the upper control limit in cell B8 using the formula =B4+3*SQRT(B4*(1–B4)/B3).

The top part of Figure 10.2.Excel on page 507 illustrates the calculations for the p chart.

Now that the control limits and center line have been computed, we need to copy their values to columns E, F, and G of the Data sheet. Click the Data sheet tab to make the Data sheet the active sheet. After entering the column headings LCL, Center, and UCL in the cell range E1:G1, enter the formulas =Calculations!$B$6 in cell E2, =Calculations!$B$7 in cell F2, and =Calculations!$B$8 in cell G2 that will duplicate the lower control limit, center line value, and the upper control limit from the Calculations sheet. Then, to plot a line for each of these variables across the days, copy these three formulas down their columns through row 29.

We are now ready to use the Chart Wizard (see Section 2.7) to obtain the control chart for the proportion. With the Data sheet still active, select Insert | Chart | As New Sheet. The selections in the five dialog boxes for the Chart Wizard are as follows:

Dialog box 1: Enter the range A1:A29, D1:G29 for the five variables of Days (on the X-axis), p, LCL, Center, and UCL. Be sure to include the *comma* in your entry. Click the Next button.

Dialog box 2: Select the XY Scatter chart and click the Next button.

Dialog box 3: Select format 2, which shows points plotted as parts of a connecting line, and click the Next button.

Dialog box 4: Select the Data Series in Columns options button and enter 1 in the First Columns edit box and 1 in the First Rows edit box. Click the Next button.

Dialog box 5: Select No for Add a Legend, enter the Chart title p Chart for Rooms Not Ready, Days in the Category (X) edit box, and Proportion in the Value (Y) edit box. Click the Finish button.

Microsoft Excel produces a chart on a new sheet. To refine this chart, select the UCL line. Then select Format | Selected Data Series. In the Format Data Series dialog chart box, select the dashed line from the Style drop-down list under Line and select the None option button under Marker. (You can also make a selection from the Color drop-down list under Line if you want to change the color of the line.) Click the OK button. The UCL line changes to a dashed line.

Change the LCL line to a dashed line by first selecting that line and then repeating the procedure used to alter the UCL line. Then, select the center line and select Format | Selected Data Series. This time, however, leave the Line Style unchanged while selecting the None option button under Marker (you can also change the Line Color, if desired). Click the OK button. The p chart will now be similar to the one illustrated in Figure 10.1.Excel. Rename the chart sheet p Chart.

Now that we have obtained the p chart, we can implement the np chart portion of the Calculations sheet. Recall from Section 10.6.3, that np charts can only be used when the subgroups are of equal size. To check this condition, enter the formula =IF(STDEV(Data!B:B)=0,"","Subgroup sizes are different, np chart should not be done.") in cell A9. Should the subgroup sizes (the Rooms Studied values in column B of the Data sheet) differ, then from Equation 3.7 on page 138, the standard deviation cannot equal zero and the warning message "Subgroup sizes are different, np chart should not be done." will be displayed in row 9. (Otherwise, that row will appear blank.)

Continue by entering the formula =AVERAGE(Data!C2:C29) in cell B11 to compute Xbar, the average number of rooms not ready, and enter in cells B12, B13, and B14 the formulas shown in Table 10.1.Excel for the lower control limit, the center line, and the upper control limit, respectively. Figure 10.2.Excel illustrates the completed Calculations sheet.

Now that the control limits and center line for the np chart have been computed, we need to copy their values to columns H, I, and J of the Data sheet. After entering the column headings for npLCL, npCenter, and npUCL in cells H1:J1, enter the formulas =Calculations!B$12 in cell H2, =Calculations!B$13 in cell I2, and =Calculations!B$14 in cell J2. That will duplicate the lower control limit, center line value, and upper control limit from the Calculations sheet. Then, to plot a line for each of these variables across the days, copy these three formulas down their columns through row 29.

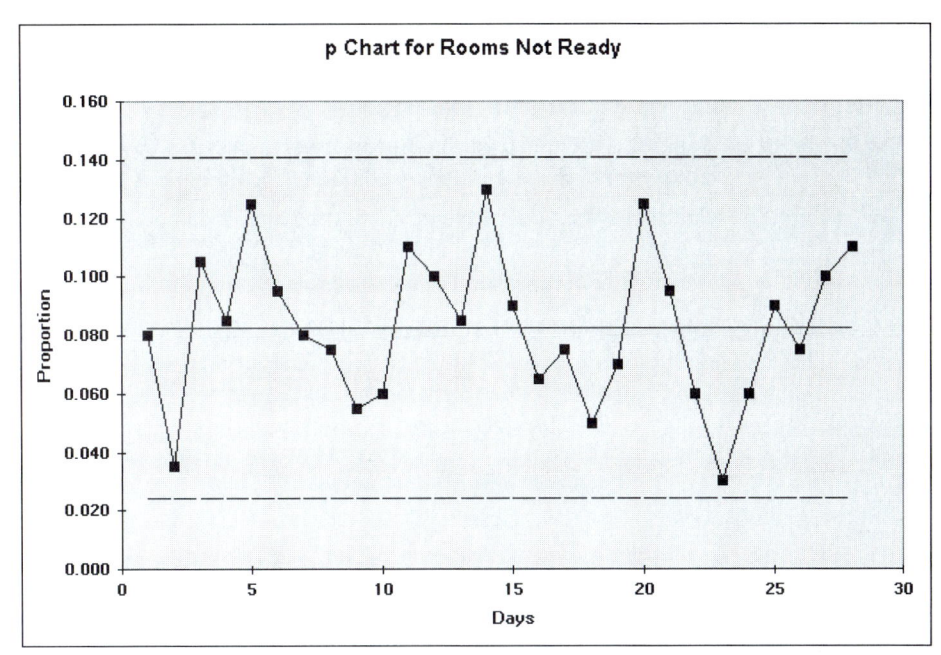

FIGURE 10.1.EXCEL The *p* chart obtained from Microsoft Excel for the nonconforming hotel room data.

FIGURE 10.2.EXCEL Calculations sheet for the *p* chart and the *np* chart obtained from Microsoft Excel for the nonconforming hotel room data.

We are now ready to use the Chart Wizard (see Section 2.7) to obtain the control chart for the number of successes. With the Data sheet active, select Insert | Chart | As New Sheet. The entries and choices in the 5 dialog boxes for the Chart Wizard are as follows:

Dialog box 1: Enter the range A1:A29, C1:C29, H1:J29 for the five variables of Days (on the X-axis), Rooms Not Ready (on the Y-axis), LCL, Center, and UCL. Be sure to include the two commas as part of this entry. Click the Next button.

10.6 Control Charts for the Proportion and Number of Nonconforming Items—The *p* and *np* Charts

Dialog box 2: Select the *XY* Scatter chart and click the Next button.

Dialog box 3: Select format 2, which shows points plotted as parts of a connecting line, and click the Next button.

Dialog box 4: Select the Columns option button, enter 1 in the First Columns edit box, enter 1 in the First rows edit box. Click the Next button.

Dialog box 5: Select No for Add a Legend, enter the Chart title np Chart for Rooms Not Ready, Days in the Category (*X*) edit box, and Rooms Not Ready in the value (*Y*) edit box. Click the Finish button.

Finish the chart process by modifying the formatting of the UCL, LCL and Center lines as discussed under the *p* chart procedure earlier in this section. The *np* chart will now be similar to the one illustrated in Figure 10.3.Excel. Rename the chart sheet np chart.

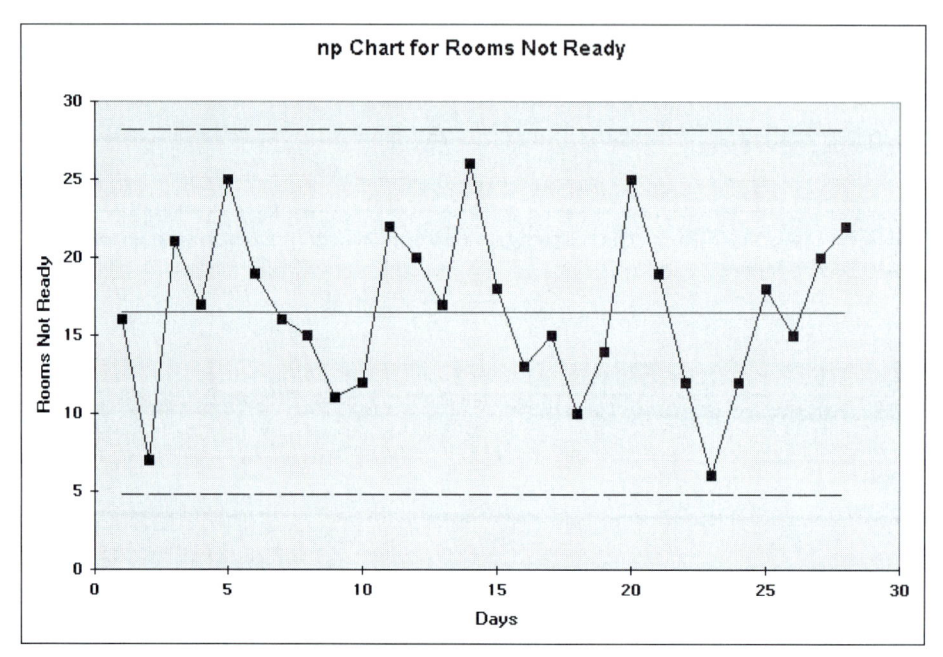

FIGURE 10.3.EXCEL The *np* chart obtained from Microsoft Excel for the nonconforming hotel room data.

▲ WHAT IF EXAMPLE

The design of the Data and Calculations sheets allows us to explore the effect of changes in the data on either the *p* or *np* chart. For example, we could change the value in cell C16 of the Data sheet for the number of nonconforming rooms on day 14 from 26 to 30. Once this was done, we would see that the LCL would be .0247, the center line $\bar{p}$ would be .0834, and the UCL would be .142. The *p* chart obtained from the Chart Wizard would change to reflect these changes and would now indicate that the result for day 14 was above the UCL. As before, if we are interested in seeing the effects of many different changes, we could use the Scenario Manager (see Section 1S.15) to store and use sets of alternative data values.

Problems for Section 10.6

Note: The problems in this section may be solved using Microsoft Excel.

- **10.6** The Commuters Watchdog Council of a railroad that serves a large metropolitan area wishes to monitor the on-time performance of the railroad during the morning rush hour. Suppose that a train is defined as being late if it arrives more than 5 minutes after the scheduled arrival time. A total of 235 trains are scheduled during the rush hour each morning. The results for a 4-week period (based on a 5-day work week) are as follows:

    RRSPC.TXT

Day	Late Arrivals	Day	Late Arrivals	Day	Late Arrivals
1	17	8	24	15	35
2	25	9	20	16	18
3	22	10	36	17	23
4	27	11	21	18	24
5	32	12	23	19	26
6	23	13	67	20	35
7	16	14	24		

   (a) Set up a p chart for the proportion of late arrivals and indicate whether the arrival process is in statistical control during this period.
   (b) Set up an np chart for the number of late arrivals and indicate whether the arrival process is in statistical control during this period.
   (c) Compare the results of the p chart obtained in (a) with those of the np chart obtained in (b).
   (d) What effect would it have on your conclusion in (a) or (b) if you knew that there had been a 4-inch snowstorm on the morning of day 13?
   (e) What effect will there be on the results obtained in (a) or (b) if the number of late arrivals on day 13 is 47?

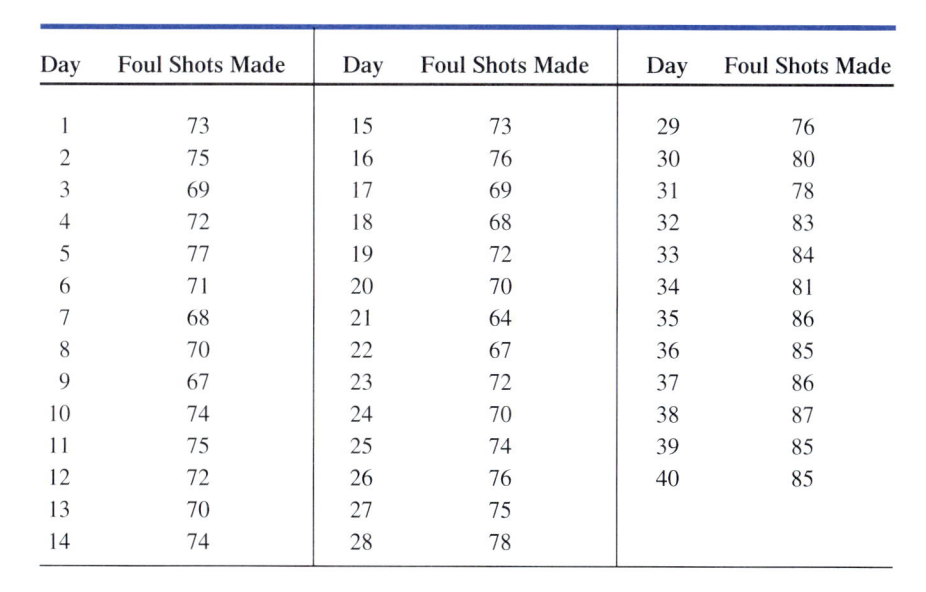

- **10.7** A professional basketball player has embarked on a program to study his ability to shoot foul shots. On each day in which a game is not scheduled, he intends to shoot 100 foul shots. He maintains records over a period of 40 days of practice, with the following results:

    FOULSPC.TXT

Day	Foul Shots Made	Day	Foul Shots Made	Day	Foul Shots Made
1	73	15	73	29	76
2	75	16	76	30	80
3	69	17	69	31	78
4	72	18	68	32	83
5	77	19	72	33	84
6	71	20	70	34	81
7	68	21	64	35	86
8	70	22	67	36	85
9	67	23	72	37	86
10	74	24	70	38	87
11	75	25	74	39	85
12	72	26	76	40	85
13	70	27	75		
14	74	28	78		

(a) Set up a *p* chart for the proportion of successful foul shots. Do you think that the player's foul-shooting process is in statistical control? If not, why not?
(b) Set up an *np* chart for the number of foul shots made and indicate whether the foul-shooting process is in statistical control during this period.
(c) What if you were told that after the first 20 days, the player changed his method of shooting foul shots? How might this information change the conclusions you drew in (a) or (b)?
(d) If you knew this information prior to doing (a) or (b), how might you have done the *p* or *np* chart differently?

10.8 A private mail delivery service has a policy of guaranteeing delivery by 10:30 A.M. of the morning after a package is picked up. Suppose that management of the service wishes to study delivery performance in a particular geographic area over a 4-week time period based on a 5-day work week. The total number of packages delivered daily and the number of packages that were not delivered by 10:30 A.M. are recorded, and the results are as follows:

MAILSPC.TXT

Day	Packages Delivered	Packages Not Arriving Before 10:30 A.M.	Day	Packages Delivered	Packages Not Arriving Before 10:30 A.M.
1	136	4	11	157	6
2	153	6	12	150	9
3	127	2	13	142	8
4	157	7	14	137	10
5	144	5	15	147	8
6	122	5	16	132	7
7	154	6	17	136	6
8	132	3	18	137	7
9	160	8	19	153	11
10	142	7	20	141	7

(a) Set up a *p* chart for the proportion of packages that are not delivered before 10:30 A.M.
(b) Does the process give an out-of-control signal?

10.9 The superintendent of a school district is interested in studying student absenteeism at a particular elementary school during December and January. The school had 537 students registered during this time period. The results are as follows:

ABSSPC.TXT

Day	Students Absent	Day	Students Absent	Day	Students Absent
1	39	13	44	25	49
2	46	14	39	26	39
3	38	15	53	27	72
4	46	16	68	28	55
5	53	17	101	29	50
6	52	18	70	30	42
7	56	19	54	31	48
8	61	20	52	32	46
9	51	21	46	33	45
10	55	22	45	34	49
11	52	23	42	35	41
12	49	24	44	36	47

Note: The first 17 days are from December, and the last 19 days are from January.

(a) Set up a *p* chart for the proportion of students who were absent during December and January. Does the process give an out-of-control signal?
(b) Set up an *np* chart for the number of students absent and indicate whether the absenteeism process is in statistical control during this period.
(c) Compare the results of the *p* chart obtained in (a) with those of the *np* chart obtained in (b).
(d) If the superintendent wants to develop a process to reduce absenteeism, how should she proceed?
(e) Suppose that day 17 was not included in the analysis. Do (a) or (b) without this day and compare the results.

● 10.10 A bottling company of Sweet Suzy's Sugarless Cola maintains daily records of the occurrences of unacceptable cans flowing from the filling and sealing machine. Nonconformities such as improper filling amount, dented cans, and cans that are improperly sealed are noted. Data for 1 month's production (based on a 5-day work week) are as follows:

COLASPC.TXT

Day	Cans Filled	Unacceptable Cans	Day	Cans Filled	Unacceptable Cans
1	5,043	47	12	5,314	70
2	4,852	51	13	5,097	64
3	4,908	43	14	4,932	59
4	4,756	37	15	5,023	75
5	4,901	78	16	5,117	71
6	4,892	66	17	5,099	68
7	5,354	51	18	5,345	78
8	5,321	66	19	5,456	88
9	5,045	61	20	5,554	83
10	5,113	72	21	5,421	82
11	5,247	63	22	5,555	87

(a) Set up a *p* chart for the proportion of unacceptable cans for the month. Does the process give an out-of-control signal?
(b) If management wants to develop a process for reducing the proportion of unacceptable cans, how should it proceed?

10.11 The manager of the accounting office of a large hospital is interested in studying the problem of errors in the entry of account numbers into the computer system. A group of 200 account numbers are selected from each day's output, and each is inspected to determine whether it is a conforming item. The results for a period of 39 days are as follows:

ERRORSPC.TXT

Day	Nonconforming Items	Day	Nonconforming Items	Day	Nonconforming Items	Day	Nonconforming Items
1	3	11	0	21	13	31	21
2	5	12	6	22	5	32	2
3	2	13	9	23	2	33	4
4	11	14	2	24	0	34	2
5	6	15	8	25	14	35	8
6	15	16	28	26	10	36	30
7	8	17	16	27	9	37	0
8	1	18	5	28	7	38	0
9	25	19	10	29	6	39	1
10	4	20	30	30	1		

10.6 Control Charts for the Proportion and Number of Nonconforming Items—The *p* and *np* Charts

(a) Set up a *p* chart for the proportion of nonconforming items. Does the process give an out-of-control signal?
(b) Set up an *np* chart for the number of nonconforming items and indicate whether the process is in statistical control during this period.
(c) Compare the results of the *p* chart obtained in (a) with those of the *np* chart obtained in (b).
(d) On the basis of the results of (a) or (b), what would you now do as a manager to improve the process of account number entry?

TELESPC.TXT

10.12 A manager of a regional office of a local telephone company has as one of her job responsibilities the task of processing requests for additions, changes, or deletions of telephone service. She forms a service improvement team to look at the corrections in terms of central office equipment and facilities required to process the orders that are issued to service requests. Data collected over a period of 30 days reveals the following:

Day	Orders	Corrections	Day	Orders	Corrections
1	690	80	16	831	91
2	676	88	17	816	80
3	896	74	18	701	96
4	707	94	19	761	78
5	694	70	20	851	85
6	765	95	21	678	65
7	788	73	22	915	74
8	794	103	23	698	68
9	694	100	24	821	72
10	784	103	25	750	101
11	812	70	26	600	91
12	759	83	27	744	64
13	781	64	28	698	67
14	682	64	29	820	105
15	802	72	30	732	112

(a) Set up a *p* chart for the proportion of corrections. Does the process give an out-of-control signal?
(b) What would you now do as a manager to improve the processing of requests for changes in telephone service?

10.7 THE RED BEAD EXPERIMENT: UNDERSTANDING PROCESS VARIABILITY

We began this chapter with a review of the history of quality and productivity and then developed the concept of common-cause variation and special-cause variation that led us to a discussion of Deming's 14 points. We discussed the important management planning tools of fishbone and process flow diagrams for understanding a process, and thus far we have covered the *p* and *np* control chart procedures. In this section, to enhance our understanding of these two types of variation, we will discuss what has become a famous parable, the **red bead experiment**.

The experiment involves the selection of beads from a box that typically contains 4,000 beads.[4] Several different scenarios can be used for conducting the experiment. The one that we will use here begins with the following:

A facilitator (who will play the role of company foreman) asks members of the audience to volunteer for the jobs of workers (at least four are needed), inspectors (two are needed), chief inspector (one is needed), and recorder (one is needed). A worker's job consists of using a paddle that has five rows of 10 bead-size holes to select 50 beads from the box of beads.

Once the participants have been selected, the foreman explains the jobs to the particular participants. The job of the workers is to produce white beads, since red beads are unacceptable to the customers. Strict procedures are to be followed. Work standards call for the production of 50 beads by each worker (a strict quota system), no more and no less than 50. Management has established a standard that no more than two red beads per worker are to be produced on any given day. The paddle is dipped into the box of beads so that when it is removed, each of the 50 holes contains a bead. Once this is done, the paddle is carried to each of the two inspectors, who independently record the count of red beads. The chief inspector compares their counts and announces the results to the audience. The recorder writes down the number of red beads next to the name of the worker.

Once all the people know their jobs, "production" can begin. Suppose that on the first "day," the number of red beads "produced" by the four workers (call them Alyson, David, Peter, and Sharyn) was 9, 12, 13, and 7, respectively. How should management react to the day's production when the standard says that no more than two red beads per worker should be produced? Should all the workers be reprimanded, or should only David and Peter be given a stern warning that they will be fired if they don't improve?

Suppose that production continues for an additional 2 days; Table 10.3 summarizes the results for all three days.

Table 10.3 Red bead experiment results for four workers over 3 days.

	Day			
Name	1	2	3	All 3 Days
Alyson	9	11	6	26
David	12	12	8	32
Peter	13	6	12	31
Sharyn	7	9	8	24
All four workers	41	38	34	113
Average ($\bar{X}$)	10.25	9.5	8.5	9.42

From Table 10.3, we may observe several phenomena. On each day, some of the workers were above the average and some below the average. On day 1, Sharyn did best, but on day 2, Peter (who had the worst record on day 1) was best, and on day 3, Alyson was best.

How then can we explain all this variation? An answer can be provided by using Equation (10.4) to develop an *np* chart. For these data, we have

$$k = 4 \text{ workers} \times 3 \text{ days} = 12, \quad n = 50, \quad \text{and} \quad \sum_{i=1}^{k} X_i = 113$$

Thus,

$$\bar{X} = \frac{113}{12} = 9.42 \quad \text{and} \quad \bar{p} = \frac{113}{(50)(12)} = .1883$$

so that we have

$$\bar{X} \pm 3\sqrt{\bar{X}(1-\bar{p})}$$

$$9.42 \pm 3\sqrt{(9.42)(1-.1883)}$$

$$9.42 \pm 8.30$$

Thus,

$$UCL = 9.42 + 8.30 = 17.72$$

and

$$LCL = 9.42 - 8.30 = 1.12$$

Figure 10.14 represents the *np* control chart for the data of Table 10.3. We observe from Figure 10.14 that all of the points are within the control limits and there are no patterns in the results. The differences between the workers *merely represent common cause variation inherent in a stable system.*

In conclusion, there are four morals to the parable of the red beads:

1. Variation is an inherent part of any process.
2. Workers work within a system over which they have little control. It is the system that primarily determines their performance.
3. Only management can change the system.
4. Some workers will always be above the average, and some workers will always be below the average.

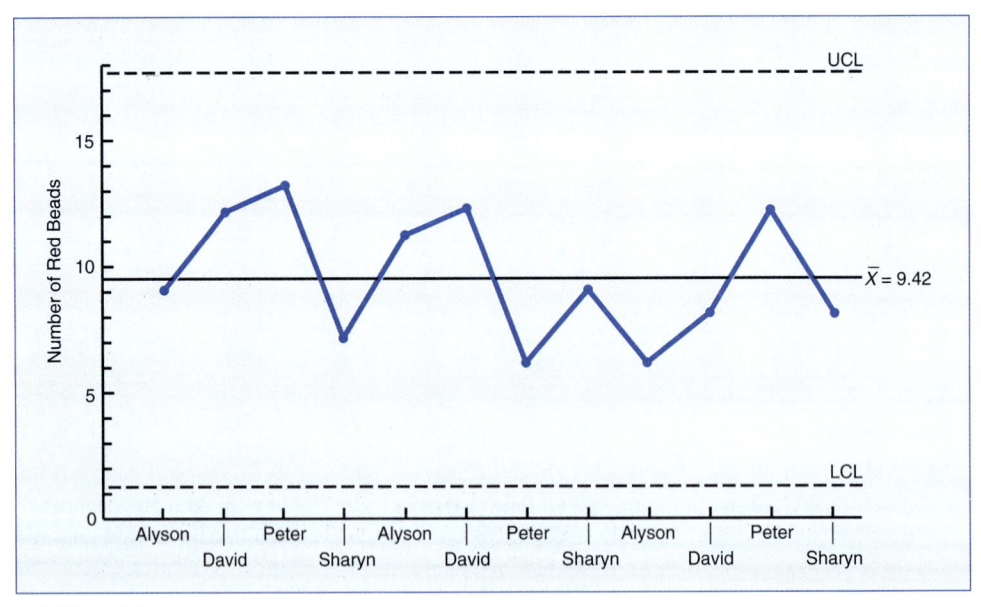

FIGURE 10.14 The *np* chart for the red bead experiment.

Problems for Section 10.7

10.13 How do you think many managers would have reacted after day 1? Day 2? Day 3?

10.14 **(Class Project)** Obtain a version of the red bead experiment for your class.
(a) Conduct the experiment in the same way as described in Section 10.7.

(b) Remove 400 red beads from the bead box before beginning the experiment. How do your results differ from those obtained in (a)? What does this tell you about the effect of the "system" on the workers?

10.8 CONTROL CHARTS FOR THE RANGE (R) AND THE MEAN ($\bar{X}$)

10.8.1 Introduction

Whenever a characteristic of interest is measured on an interval or ratio scale, **variables control charts** can be used to monitor a process. Because measures from these more powerful scales provide more information than the proportion or number of nonconforming items, these charts are more sensitive in detecting special cause variation than the p or np charts. Variables charts are typically used in pairs. One chart monitors the variation in a process, while the other monitors the process average. The chart that monitors variability must be examined first because if it indicates the presence of out-of-control conditions, the interpretation of the chart for the average will be misleading. Although several alternative pairs of charts can be considered (see References 8, 13, 18, and 20), in this text we will study the control charts for the range and average.

10.8.2 The R Chart: A Control Chart for Dispersion

Before obtaining control limits for the mean, we need to develop a control chart for the range, the **R chart**. This will enable us to determine whether the variability in a process is in control or whether shifts are occurring over time. If the process range is in control, then it can be used to develop the control limits for the average.

From Equation (10.1), we observe that to develop control limits for the range, we need to obtain an estimate of the average range and the standard deviation of the range. As can be seen in Equation (10.6), these control limits are a function not only of the d_2 **factor**, which represents the relationship between the standard deviation and the range for varying sample sizes, but also of the d_3 **factor**, which represents the relationship between the standard deviation and the standard deviation of the range for varying sample sizes. Values for these factors are presented in Table E.8. Thus, we may set up the following control limits for the range over k consecutive sequences or periods of time.

$$\bar{R} \pm 3\bar{R}\frac{d_3}{d_2} \tag{10.6}$$

so that

$$\text{LCL} = \bar{R} - 3\bar{R}\frac{d_3}{d_2} \tag{10.7a}$$

$$\text{UCL} = \bar{R} + 3\bar{R}\frac{d_3}{d_2} \tag{10.7b}$$

where

$$\overline{R} = \frac{\sum_{i=1}^{k} R_i}{k}$$

Referring to Equations (10.7a) and (10.7b), we may simplify the calculations by utilizing the **D_3 factor**, equal to $1 - 3(d_3/d_2)$, and the **D_4 factor**, equal to $1 + 3(d_3/d_2)$, to obtain the control limits as shown in Equations (10.8a) and (10.8b).

$$\text{LCL} = D_3 \overline{R} \qquad (10.8a)$$

$$\text{UCL} = D_4 \overline{R} \qquad (10.8b)$$

To illustrate the use of the R chart, let us refer to the following example. Suppose that the management of the hotel discussed in Section 10.6 also wants to analyze the check-in process. In particular, they want to study the amount of time it takes to deliver luggage (as measured from the time the guest completes check-in procedures to the time the luggage arrives in the guest's room). Data are recorded over a 4-week (Sunday–Saturday) period, and subgroups of 5 deliveries are selected (on a certain shift) on each day for analysis. The summary results (in minutes) are recorded in Table 10.4.

For these data,

$$k = 28 \quad \text{and} \quad \sum_{i=1}^{k} R_i = 104.41$$

Thus,

$$\overline{R} = \frac{104.41}{28} = 3.729$$

Table 10.4 Subgroup average and range for luggage delivery times over a 4-week period.

Day	Subgroup Average $\overline{X}_i$ (in minutes)	Subgroup Range R_i (in minutes)	Day	Subgroup Average $\overline{X}_i$ (in minutes)	Subgroup Range R_i (in minutes)
1	5.32	3.85	15	5.21	3.26
2	6.59	4.27	16	4.68	2.92
3	4.88	3.28	17	5.32	3.37
4	5.70	2.99	18	4.90	3.55
5	4.07	3.61	19	4.44	3.73
6	7.34	5.04	20	5.80	3.86
7	6.79	4.22	21	5.61	3.65
8	4.93	3.69	22	4.77	3.38
9	5.01	3.33	23	4.37	3.02
10	3.92	2.96	24	4.79	3.80
11	5.66	3.77	25	5.03	4.11
12	4.98	3.09	26	5.11	3.75
13	6.83	5.21	27	6.94	4.57
14	5.27	3.84	28	5.71	4.29

From Table E.8 for $n = 5$, we obtain $d_2 = 2.326$ and $d_3 = .864$. Using Equations (10.6) and (10.7), we have

$$3.729 \pm 3 \frac{(.864)(3.729)}{2.326}$$

$$3.729 \pm 4.155$$

so that

$$\text{UCL} = 3.729 + 4.155 = 7.884$$

and

$$\text{LCL} = 3.729 - 4.155 \quad \text{so that LCL does not exist.}$$

Alternatively, using Equation (10.8), from Table E.8, $D_3 = 0$ and $D_4 = 2.114$. Thus,

$$\text{UCL} = (2.114)(3.729) = 7.883$$

and

$$\text{LCL does not exist.}$$

We note that the lower control limit for R does not exist because a negative range is impossible to attain. The R chart is displayed in Figure 10.15. An examination of Figure 10.15 does not indicate any individual ranges outside the control limits.

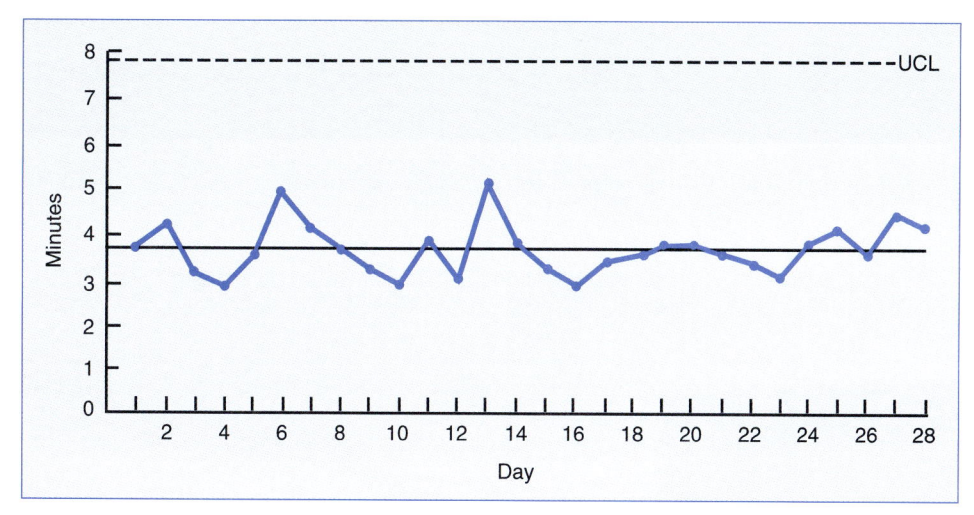

FIGURE 10.15 The R chart for luggage delivery times. *Source:* Table 10.4.

10.8.3 The $\bar{X}$ Chart

Now that we have determined that the control chart for the range is in control, we may continue by examining the control chart for the process average, the $\bar{X}$ **chart**.

The control chart for $\bar{X}$ uses subgroups of size n that are obtained over k consecutive sequences or periods of time. From Equation (10.1), we observe that to compute control limits for the average, we need to obtain an estimate of the average of the subgroup averages (which we shall call $\bar{\bar{X}}$) and the standard deviation of the average (which we called the standard error of the mean $\sigma_{\bar{X}}$ in Chapter 5). These control limits are a function of the d_2 factor, which represents the relationship between the standard deviation and the range for varying sample sizes. The range may be used to estimate the standard deviation as long as the subgroup size is no more than 10 (see References 13, 18, 20). Thus, we may set up the following control limits:

$$\bar{\bar{X}} \pm 3 \frac{\bar{R}}{d_2 \sqrt{n}} \qquad (10.9)$$

where

$$\bar{\bar{X}} = \frac{\sum_{i=1}^{k} \bar{X}_i}{k} \quad \text{and} \quad \bar{R} = \frac{\sum_{i=1}^{k} R_i}{k}$$

$\bar{X}_i$ = the sample mean of n observations at time i

R_i = the range of n observations at time i

k = number of subgroups

so that

$$\text{LCL} = \bar{\bar{X}} - 3 \frac{\bar{R}}{d_2 \sqrt{n}} \qquad (10.10a)$$

$$\text{UCL} = \bar{\bar{X}} + 3 \frac{\bar{R}}{d_2 \sqrt{n}} \qquad (10.10b)$$

Referring to Equations (10.10a) and (10.10b), we may simplify the calculations by utilizing the **A_2 factor**, equal to $3/(d_2 \sqrt{n})$, to obtain the control limits as displayed in Equations (10.11a) and (10.11b).

$$\text{LCL} = \bar{\bar{X}} - A_2 \bar{R} \qquad (10.11a)$$

$$\text{UCL} = \bar{\bar{X}} + A_2 \bar{R} \qquad (10.11b)$$

Thus, returning to our luggage delivery time example from Table 10.4, we may compute

$$k = 28, \quad \sum_{i=1}^{k} \bar{X}_i = 149.97, \quad \text{and} \quad \sum_{i=1}^{k} R_i = 104.41$$

so that

$$\bar{\bar{X}} = \frac{149.97}{28} = 5.356 \quad \text{and} \quad \bar{R} = \frac{104.41}{28} = 3.729$$

From Table E.8 for $n = 5$, we obtain $d_2 = 2.326$. Thus, using Equation (10.10), we have

$$5.356 \pm 3 \frac{3.729}{(2.326)\sqrt{5}}$$

$$5.356 \pm 2.151$$

Therefore,
$$LCL = 5.356 - 2.151 = 3.205$$
and
$$UCL = 5.356 + 2.151 = 7.507$$

Alternatively, using Equation (10.11), from Table E.9, we have $A_2 = .577$ and
$$LCL = 5.356 - (.577)(3.729) = 5.356 - 2.152 = 3.204$$
$$UCL = 5.356 + (.577)(3.729) = 5.356 + 2.152 = 7.508$$

These results are the same, except for rounding error.

The control chart for the luggage delivery time data of Table 10.4 is displayed in Figure 10.16. An examination of Figure 10.16 does not reveal any points outside the control limits, although there is a large amount of variability among the 28 subgroup means.

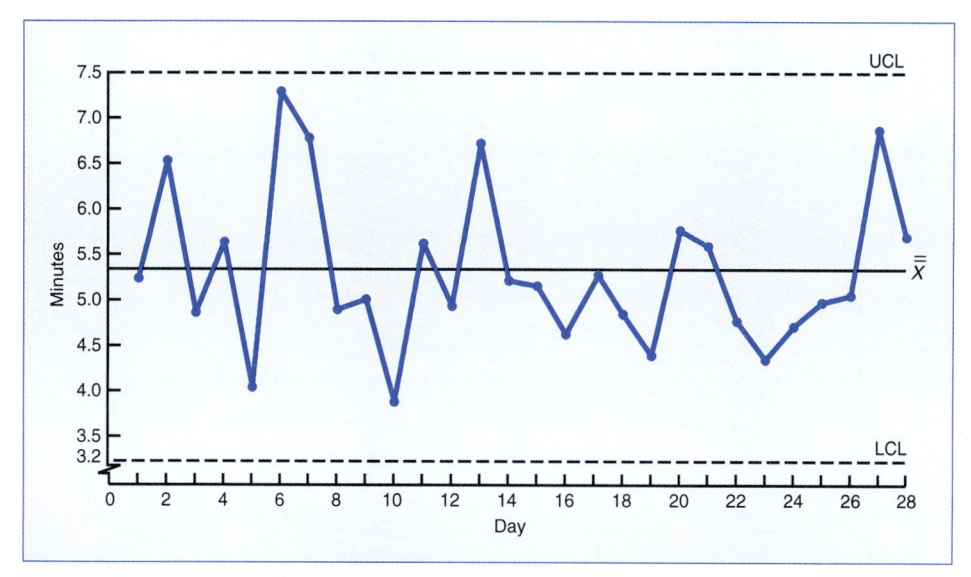

FIGURE 10.16 The $\bar{X}$ chart for the average luggage delivery time data. *Source:* Table 10.4.

10.8.4 Using Microsoft Excel for the R and $\bar{X}$ Charts

In this section we have studied two control charts, the R chart for the range and the $\bar{X}$ chart for the mean. Although Microsoft Excel does not have a Data Analysis tool to compute the control limits and center line of a control chart, we can use Excel formulas instead. Once the control limits and center line have been computed, the Chart Wizard can be used to plot the data along with the control limits and center line. The design for the Data and Calculations sheets for this purpose is shown in Table 10.2.Excel.

To implement the Calculations sheet, begin by opening the LUGGAGE.XLS workbook. This workbook contains a Data sheet in which the day, Xbar, and range values from Table 10.4 on page 516 have been entered previously. (If only the raw data were available as is the case for problems 10.16 and 10.17 on pages 524–526, the raw data could be stored on a separate sheet, and the arithmetic means and ranges could be computed on the Data sheet for each day using the MAXIMUM, MINIMUM, and AVERAGE functions.) The LUGGAGE workbook

10–8–4.XLS

Table 10.2.Excel Design for the Data and Calculations sheets for the R and $\bar{X}$ charts.

Data Sheet

	A	B	C
1	Day	XBar	Range
2	1	xxx	xxx
3	.	.	.
4	.	.	.
5	.	.	.
⋮	⋮	⋮	⋮
29	28	xxx	xxx

Calculations Sheet

	A	B
1	Calculations for R and XBAR Charts	
2		
3	R Chart	
4	RBar	=AVERAGE(Data!C2:C29)
5	Number of Obs. in sample	xx
6	D3 Factor	=VLOOKUP($B$5,CCFactors!$A$4:$D$28,2)
7	D4 Factor	=VLOOKUP($B$5,CCFactors!$A$4:$D$28,3)
8	LCL	=B6*B4
9	Center Line	=B4
10	UCL	=B7*B4
11		
12	XBar Chart	
13	Average XBar	=AVERAGE(Data!B2:B29)
14	A2 Factor	=VLOOKUP($B$5,CCFactors!$A$4:$D$28,4)
15	LCL	=B13–B14*B4
16	Center Line	=B13
17	UCL	=B13+B14*B4

also includes a CCFactors sheet that contains a table of control chart factors, adapted from Table E.8 on page E–15. The use of this sheet is explained later in this section.

To begin insert a new sheet and rename it Calculations. To obtain the D_3 and D_4 factors and the average range necessary to obtain the control limits for the R chart, do the following:

❶ Enter the formula =AVERAGE(Data!C2:C29), in cell B4 to compute the average range.

❷ Enter the number of observations in each sample in cell B5 (5 for these data).

❸ Obtain in the D_3 factor based on the number of observations in each sample. This can be accomplished by using the VLOOKUP function to lookup the proper D_3 factor value in the CCFactors sheet table. The format of the VLOOKUP function is:

VLOOKUP (*Lookup value, lookup table range, lookup column number*)

where   *lookup value* is the value to be matched to a value in the first column of the lookup table,

lookup table range is the cell range containing the body of the lookup table, (excluding column headings), and

lookup column number is the number of columns to move over to find the value to be returned.

For cell B6, enter the formula =VLOOKUP($B$5,CCFactors!$A$4:$D$28,2) which will take the value in cell B5 and match it to a value in the first column of Table (CCFactors!A4:A28) and then return a value found in the same row in the second column of the table (CCFactors!B$:B28). For the luggage delivery data, the value in cell B5 is 5 so the value of B6 will be 0. [Using the table formed by the range CCFactors!A4:D28, the 5 is matched to the 5 that appears in cell A7 and the value returned is the value found in cell B7 (in column 2) which is zero.]

④ Copy the formula in cell B6 to B7 and change the value of the lookup column number to 3 (to look up the D4 factor). For the luggage delivery data this will produce a value of 2.114.

⑤ Enter the formulas for the lower control limit, the center line, and the upper control limit as shown in Table 10.2.Excel.

The Calculations sheet should now look similar to the top part of Figure 10.5.Excel on page 522.

Now that the control limits and center line have been computed, we need to copy their values to columns D, E, and F of the Data sheet. Click the Data sheet tab to make the Data sheet the active sheet. After entering the column headings for LCL, Center, and UCL in the cell range D1:F1, enter the formulas =Calculations!$B$8 in cell D2, =Calculations!$B$9 in cell E2, and =Calculations!$B$10 in cell F2 that duplicate the lower control limit, center line value, and upper control limit from the Calculations sheet. Then, to plot a line for each of these variables across the days, copy these three formulas down their columns through row 29.

We are now ready to use the Chart Wizard (see Section 2.7) to obtain the control chart for the range. With the Data sheet still active, select Insert | Chart | As New Sheet. The entries and choices in the five dialog boxes for the Chart Wizard are as follows:

Dialog box 1: Enter the input range A1:A29, C1:F29 for the five variables of Days (on the X-axis), range, LCL, Center, and UCL. (Be sure to include the comma as part of your entry.) Click the Next button.

Dialog box 2: Select the XY Scatter chart and click the Next button.

Dialog box 3: Select format 2, which shows points plotted as part of a connecting line, and click the Next button.

Dialog box 4: Select the Columns option button. Enter 1 in the First Column edit box and 1 in the First Rows edit box. Click the Next button.

Dialog box 5: Select No for Add a Legend and enter the Chart title R Chart for Luggage Delivery, Days in the Category (X) edit box, and Minutes in the Value (Y) edit box. Click the Finish button.

Finish the chart process by modifying the lines as described in Section 10.6.4. The finished R chart will be similar to the one illustrated in Figure 10.4.Excel.

Now that we have completed the R chart (and renamed the sheet R Chart), we are ready to obtain the $\bar{X}$ chart. Using Equations (10.11a) and (10.11b) on page 518, we need to compute $\bar{\bar{X}}$, the average of all the sample averages. With the Calculations sheet active, we need to do the following:

① Compute $\bar{\bar{X}}$ in cell B13 by using the formula =AVERAGE(Data!B2:B29).

② Copy the formula in cell B6 to cell B12 and change the value of the lookup column number to 4 (to look up the A2 factor). For the luggage delivery data, the A2 factor value returned is .577.

③ Enter the formulas for the lower control limit, the center line and the upper control limit in cells B15, B16, and B17, respectively, as shown in Table 10.2.Excel. Figure 10.5.Excel illustrates the completed Calculations sheet.

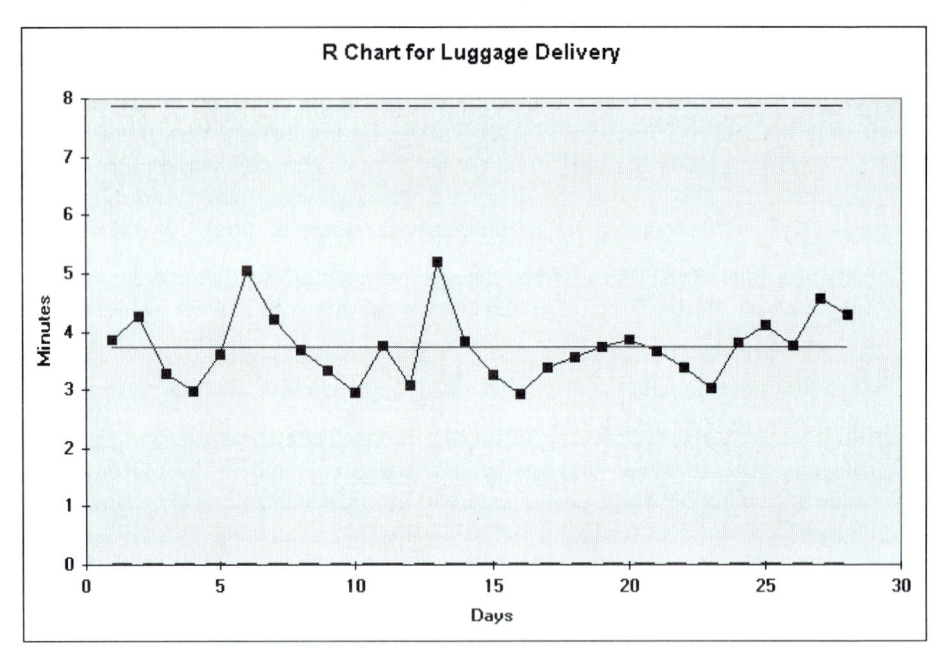

FIGURE 10.4.EXCEL The R chart obtained from Microsoft Excel for the luggage delivery time data.

	A	B
1	Calculations for R and XBar Charts	
2		
3	R Chart	
4	RBar	3.72892857
5	No. of obs. in sample	5
6	D3 factor	0
7	D4 factor	2.114
8	LCL	0
9	Center	3.72892857
10	UCL	7.882955
11		
12	XBar Chart	
13	Average XBar	5.35607143
14	A2 factor	0.577
15	LCL	3.20447964
16	Center	5.35607143
17	UCL	7.50766321

FIGURE 10.5.EXCEL Calculations sheet for the R chart and the $\bar{X}$ chart obtained from Microsoft Excel for the luggage delivery time data.

Now that the control limits and center line for the $\bar{X}$ chart have been computed, we need to copy their values to columns G, H, and I of the Data sheet. With the Data sheet active, enter the column headings XLCL, XCenter, and XUCL in the cell range G1:I1. Then enter the formulas =Calculation!$B$15 in cell G2, =Calculations!$B$16 in cell H2, and =Calculations!$B$17 in cell I2 that duplicate the lower control limit, the center line value, and the upper control limit from the Calculations sheet. Copy these three formulas down the columns through row 29.

We are now ready to use the Chart Wizard (see Section 2.7) to obtain the control chart for the number of successes. With the Data sheet active, issue the command Insert | Chart | As New Sheet. The selections in the five dialog boxes for the Chart Wizard are as follows:

Dialog box 1: Enter the input range A1:A29, B1:B29, G1:I29 for the five variables of Days (on the X-axis), arithmetic mean, LCL, Center, and UCL. (Be sure to include the two commas in your entry.) Click the Next button.

Dialog box 2: Select the XY Scatter chart and click the Next button.

Dialog box 3: Select format 2, which shows points plotted as parts of a connecting line, and click the Next button.

Dialog box 4: Select the Columns option button and enter 1 in the First Columns edit box and 1 in the First Rows edit box. Click the Next button.

Dialog box 5: Select No for Add a Legend and enter the Chart title XBar Chart for Luggage Delivery, Days in the Category (X) edit box, and Minutes in the Value (Y) edit box. Click the Finish button.

Finish the chart process by modifying the lines as described in Section 10.6.4. The finished $\bar{X}$ chart will be similar to the one illustrated in Figure 10.6.Excel.

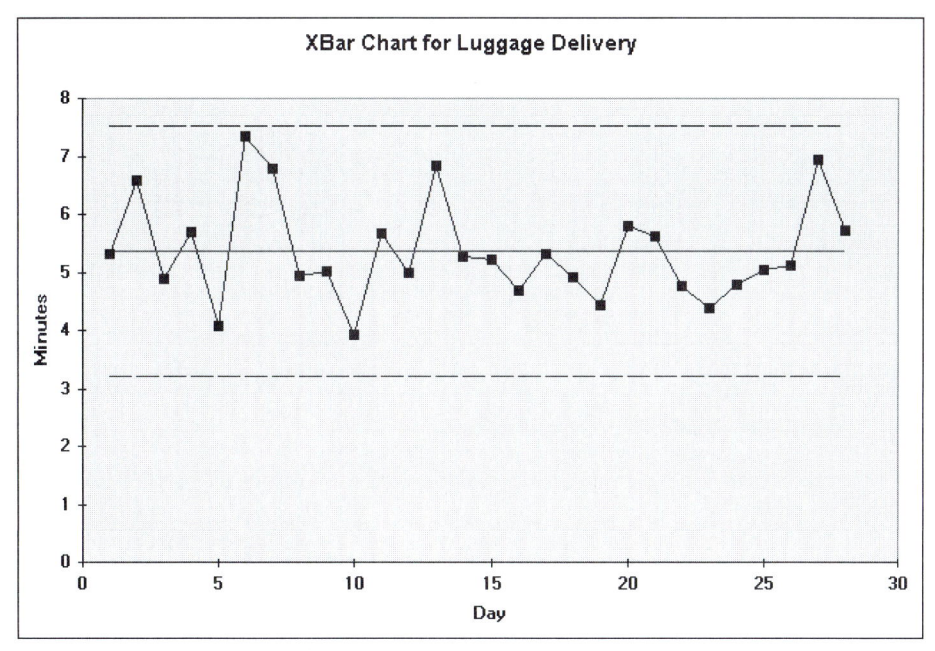

FIGURE 10.6.EXCEL The $\bar{X}$ chart obtained from Microsoft Excel for the luggage delivery time data.

▲ WHAT IF EXAMPLE

The design of the Data and Calculations sheets allows us to explore the effect of changes in the data on either the R or $\bar{X}$ chart. For example, if $\bar{X}$ on day 6 is 9.34 instead of 7.34, we could change the value in cell B7 on the Data sheet. Once this was done, we would observe that the LCL would be 3.2759, the center line $\bar{X}$ would be 5.4275, and the UCL would be 7.579. The $\bar{X}$ chart obtained from the Chart Wizard would change to reflect these changes and would now indicate that the result for day 6 was above the UCL. As before, if we are interested in seeing the effects of many different changes, we could use the Scenario Manager (see Section 1S.15) to store and use sets of alternative data values.

Problems for Section 10.8

Note: The problems in this section may be solved using Microsoft Excel.

BULBLIFE.TXT

10.15 The following data pertaining to incandescent light bulbs represent the average life and range for 30 subgroups of 5 light bulbs each.

Subgroup Number	Subgroup Mean $\bar{X}_i$	Subgroup Range R_i	Subgroup Number	Subgroup Mean $\bar{X}_i$	Subgroup Range R_i
1	790	52	16	845	42
2	845	56	17	891	38
3	857	116	18	859	65
4	846	89	19	826	70
5	843	65	20	828	37
6	877	73	21	854	52
7	861	38	22	847	49
8	891	84	23	868	40
9	866	76	24	851	43
10	816	72	25	870	64
11	806	61	26	857	53
12	835	55	27	851	59
13	797	59	28	834	68
14	803	47	29	842	57
15	818	69	30	825	74

(a) Set up a control chart for the range.
(b) Set up a control chart for the average light bulb life.
(c) On the basis of the results of (a) and (b), what conclusions can you draw about the process?

BANKTIME.TXT

10.16 The manager of a branch of a local bank wants to study waiting times of customers for teller service during the peak 12 noon to 1 P.M. lunch hour. A subgroup of four customers is selected (one at each 15-minute interval during the hour), and the time in minutes is measured from the point each customer enters the line to when he or she begins to be served. The results over a 4-week period are as follows:

Day	Time in Minutes			
1	7.2	8.4	7.9	4.9
2	5.6	8.7	3.3	4.2
3	5.5	7.3	3.2	6.0
4	4.4	8.0	5.4	7.4
5	9.7	4.6	4.8	5.8
6	8.3	8.9	9.1	6.2

(continued)

(continued)

Day	Time in Minutes			
7	4.7	6.6	5.3	5.8
8	8.8	5.5	8.4	6.9
9	5.7	4.7	4.1	4.6
10	1.7	4.0	3.0	5.2
11	2.6	3.9	5.2	4.8
12	4.6	2.7	6.3	3.4
13	4.9	6.2	7.8	8.7
14	7.1	6.3	8.2	5.5
15	7.1	5.8	6.9	7.0
16	6.7	6.9	7.0	9.4
17	5.5	6.3	3.2	4.9
18	4.9	5.1	3.2	7.6
19	7.2	8.0	4.1	5.9
20	6.1	3.4	7.2	5.9

(a) Set up control charts for the arithmetic mean and the range.
(b) On the basis of the results in (a), indicate whether the process is in control in terms of these charts.

• **10.17** The manager of a warehouse for a local telephone company is involved in an important process that receives expensive circuit boards and returns them to central stock so they may be used at a later date when a circuit or new telephone service is needed. The timely return and processing of these units are critical in providing good service to field customers and reducing capital expenditures of the corporation. The following data represent the number of units handled by each of a subgroup of five employees over a 30-day period.

WAREHSE.TXT

	Employee				
Day	1	2	3	4	5
1	114	499	106	342	55
2	219	319	162	44	87
3	64	302	38	83	93
4	258	110	98	78	154
5	127	140	298	518	275
6	151	176	188	268	77
7	24	183	202	81	104
8	41	249	342	338	69
9	93	189	209	444	151
10	111	207	143	318	129
11	205	281	250	468	79
12	121	261	183	606	287
13	225	83	198	223	180
14	235	439	102	330	190
15	91	32	190	70	150
16	181	191	182	444	124
17	52	190	310	245	156
18	90	538	277	308	171

(continued)

(continued)

Day	Employee 1	2	3	4	5
19	78	587	147	172	299
20	45	265	126	137	151
21	410	227	179	298	342
22	68	375	195	67	72
23	140	266	157	92	140
24	145	170	231	60	191
25	129	74	148	119	139
26	143	384	263	147	131
27	86	229	474	181	40
28	164	313	295	297	280
29	257	310	217	152	351
30	106	134	175	153	69

(a) Set up control charts for the arithmetic mean and the range.
(b) On the basis of the results in (a), indicate whether the process is in control in terms of these charts.

AUTOREP.TXT

10.18 The service manager of a large automobile dealership wants to study the length of time required for a particular type of repair in his shop. A subgroup of 10 cars needing this repair is selected on each day for a period of 4 weeks. The results (service time in hours) are recorded as follows:

Day	Subgroup Average $\bar{X}_i$	Subgroup Range R_i	Day	Subgroup Average $\bar{X}_i$	Subgroup Range R_i
1	3.73	5.23	11	3.64	5.37
2	3.16	4.82	12	3.27	4.42
3	3.56	4.98	13	3.16	4.85
4	3.01	4.28	14	3.39	4.44
5	3.87	5.74	15	3.85	5.06
6	3.90	5.42	16	3.90	4.99
7	3.54	4.08	17	3.72	4.67
8	3.32	4.55	18	3.51	4.37
9	3.29	4.48	19	3.34	4.53
10	3.83	5.09	20	3.99	5.28

(a) Set up all appropriate control charts and determine whether the service time process is in a state of statistical control.
(b) If the service manager wants to develop a process to reduce service time, how should he proceed?

PHLEVEL.TXT

10.19 The manager of a private swimming pool facility monitors the pH (alkalinity-acidity) level of the swimming pool by taking hourly readings from 8 A.M. to 6 P.M. daily. The results for a 3-week period summarized on a daily basis are presented as follows:

Day	Average $\bar{X}_i$	Range R_i	Day	Average $\bar{X}_i$	Range R_i
1	7.34	0.16	12	7.39	0.16
2	7.41	0.12	13	7.40	0.18
3	7.30	0.11	14	7.35	0.17
4	7.28	0.19	15	7.39	0.22
5	7.23	0.17	16	7.42	0.20
6	7.30	0.20	17	7.40	0.18
7	7.35	0.15	18	7.37	0.18
8	7.38	0.19	19	7.41	0.22
9	7.32	0.14	20	7.36	0.15
10	7.38	0.19	21	7.40	0.12
11	7.43	0.23			

(a) Set up a control chart for the range.
(b) Set up a control chart for the average daily pH level.
(c) On the basis of the results of (a) and (b), what conclusions can you draw about the process?

10.9 SUMMARY AND OVERVIEW

As can be observed in the chapter summary chart, in this chapter we have introduced the topic of quality and productivity by discussing the Deming approach to management and by developing several different types of control charts. Readers interested in the Deming approach are encouraged to examine References 1, 4, 5, 7, 12, 13, 14, 15, 19, 25, 28, and 29. Readers interested in additional control chart procedures should access References 8, 13, 18, and 20. To be sure you understand the material we have covered, you should be able to answer the following conceptual questions:

1. What are the differences in approach between management by control and management by process?
2. What is the difference between common causes of variation and special causes of variation?
3. What should be done to improve a process when special causes of variation are present?
4. What should be done to improve a process when only common causes of variation are present?
5. How can process flow and fishbone diagrams be used to improve processes?
6. Under what circumstances can the *np* chart be used?
7. What is the difference between attribute control charts and variables control charts?
8. Why are the $\bar{X}$ and range charts used together?
9. What principles did you learn from the red bead experiment?

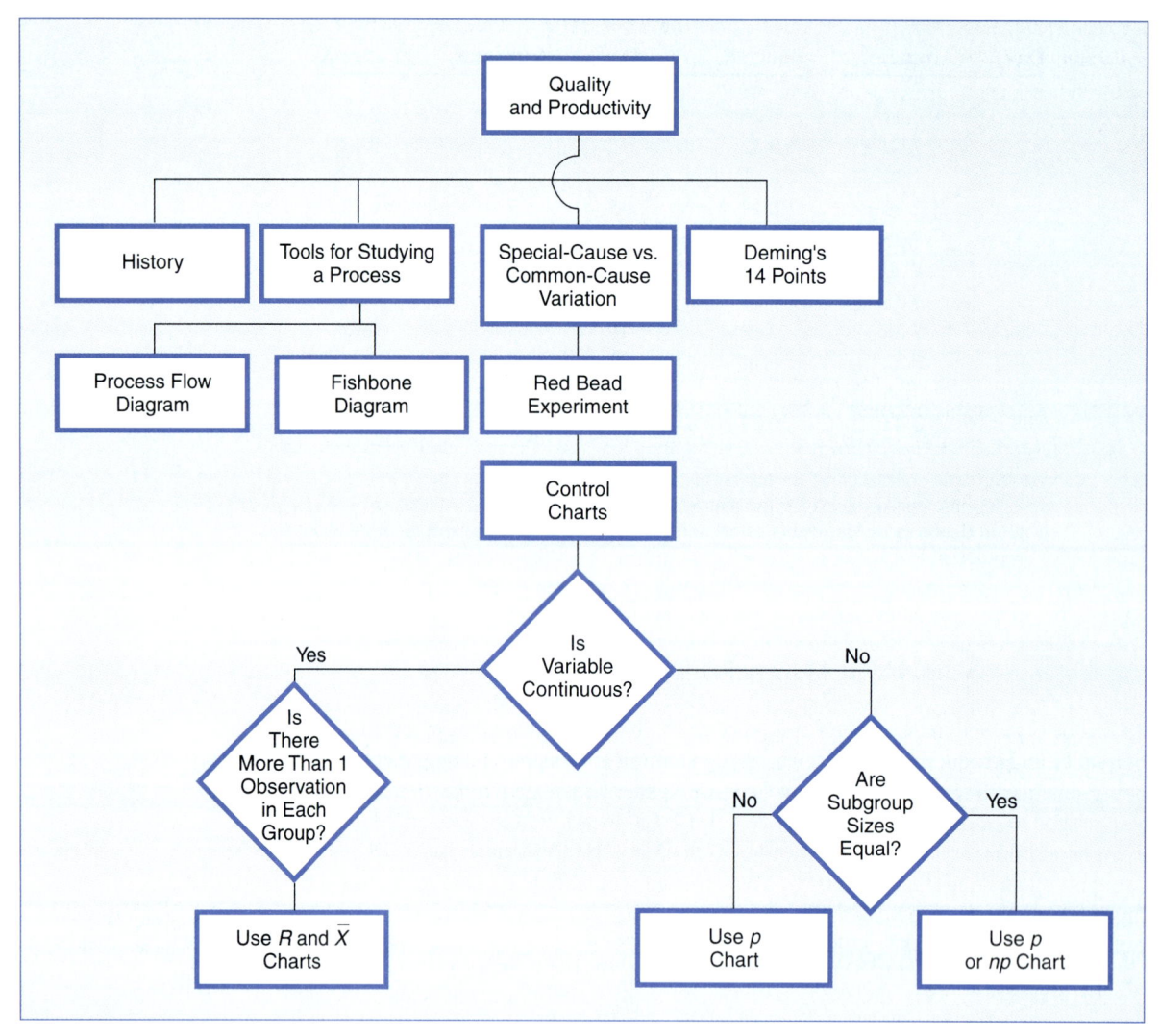

Chapter 10 summary chart.

Getting It All Together

Key Terms

A_2 factor 518
attribute charts 498
common causes of variation 488
control chart 487
d_2 factor 515
d_3 factor 515
D_3 factor 516
D_4 factor 516
Deming's 14 points 495
fishbone diagram 489

lower control limit (LCL) 488
management by control 486
management by directing 486
management by doing 486
management by process 487
np chart 498
p chart 498
process 489
process flow diagram 491
R chart 515

red bead experiment   512
Shewhart cycle   495
special or assignable causes of variation   487
total quality management (TQM)   487

upper control limit (UCL)   488
variables control charts   515
$\bar{X}$ chart   517

Chapter Review Problems

Note: *The Chapter Review problems may be solved using Microsoft Excel.*

10.20 On each morning for a period of 4 weeks, record your pulse rate (in beats per minute) just after you get out of bed and also just before you go to sleep at night. Set up $\bar{X}$ and R charts for pulse rate and determine whether it is in a state of statistical control. Explain.

10.21 **(Class Project)** The table of random numbers (Table E.1) can be used to simulate the selecting of different colored balls from an urn as follows:
  (1) Start in the row corresponding to the day of the month you were born plus the year in which you were born. For example, if you were born October 15, 1971, you would start in row 15 + 71 = 86. If your total exceeds 100, subtract 100 from the total.
  (2) Two-digit random numbers are to be selected.
  (3) If the random number selected is between 00 and 94, consider the ball to be white; if the random number is 95–99, consider the ball to be red.

Each student is to select 100 such two-digit random numbers and report the number of "red balls" in the sample. A control chart is to be set up for the number of (or the proportion of) red balls. What conclusions can you draw about the system of selecting red balls? Are all the students part of the system? Is anyone outside the system? If so, what explanation can you give for someone who has too many red balls? If a bonus is paid to the top 10% of the students (those 10% with the fewest red balls), what effect would that have on the rest of the students? Discuss.

10.22 As chief operating officer of a local community hospital, you have just returned from a 3-day seminar on quality and productivity. It is your intention to implement many of the ideas that you learned at the seminar. You have decided to maintain control charts for the upcoming month for the following variables: proportion of rework in the laboratory (based on 1,000 daily samples), and time (in hours) between receipt of a specimen at the laboratory and completion of the work (based on a subgroup of 10 specimens per day). The data collected are summarized as follows:

S-10-22.XLS

Hospital summary data.

Day	Processing Time $\bar{X}_i$	R_i	Proportion of Rework in Laboratory
1	1.72	3.57	0.048
2	2.03	3.98	0.052
3	2.18	3.54	0.047
4	1.90	3.49	0.046
5	2.53	3.99	0.039
6	2.26	3.34	0.086
7	2.11	3.36	0.051
8	2.35	3.52	0.043

(continued)

(continued)

	Processing Time		
Day	$\bar{X}_i$	R_i	Proportion of Rework in Laboratory
9	2.06	3.39	0.046
10	2.01	3.24	0.040
11	2.13	3.62	0.045
12	2.18	3.37	0.036
13	2.31	3.97	0.048
14	2.37	4.06	0.057
15	2.78	4.27	0.052
16	2.12	3.21	0.046
17	2.27	3.48	0.041
18	2.49	3.62	0.032
19	2.32	3.19	0.042
20	2.43	3.67	0.053
21	2.25	3.10	0.041
22	2.31	3.58	0.037
23	2.07	3.26	0.039
24	2.33	3.40	0.050
25	2.36	3.52	0.048
26	2.47	3.82	0.054
27	2.28	3.97	0.046
28	2.17	3.60	0.035
29	2.54	3.92	0.075
30	2.63	3.86	0.046

You are to make a presentation to the chief executive officer of the hospital and the board of directors. You need to prepare a report that summarizes the conclusions obtained from analyzing control charts for these variables. In addition, it is expected that you will recommend additional variables for which control charts are to be maintained. Finally, it is your intention to explain how the Deming philosophy of management by process can be implemented in the context of your hospital's environment.

Endnotes

1. Recall from Section 5.3 that in the normal distribution, $\mu \pm 3\sigma$ includes almost all (99.73%) of the observations in the population. Although the calculations used in control charts are based on the normal distribution, we should stress that in analytical studies, the concept of population has no applicability. The subject of interest is a process, not a population from which a sample is taken.

2. This example was related to the authors by Brian Joiner of Joiner Associates, who credits it to Ed Pindy of the Philadelphia Electric Company.

3. In this chapter we use the term *nonconforming items*, while in Chapters 4–6 when we discussed proportions we used the term *success*.

4. Unknown to the participants in the experiment, 3,200 of the beads are white and 800 are red.

References

1. Aguayo, R., *Dr. Deming The American Who Taught the Japanese about Quality* (New York: Lyle Stuart, 1990).
2. Brassard, M., *The Memory Jogger Plus* (Methuen, MA: GOAL/QPC, 1989).
3. Cryer, J. D., and T. P. Ryan, "The estimation of sigma for an X chart: MR/d_2 or S/c_4?" *Journal of Quality Technology*, 1990, Vol. 22, pp. 187–192.
4. Deming, W. E., *Out of the Crisis* (Cambridge, MA: MIT Center for Advanced Engineering Study, 1986).
5. Deming, W. E., *The New Economics for Business, Industry, and Government* (Cambridge, MA: MIT Center for Advanced Engineering Study, 1993).
6. Dobson, J. M., *A History of American Enterprise* (Englewood Cliffs, NJ: Prentice-Hall, 1988).
7. Gabor, A., *The Man Who Discovered Quality* (New York: Time Books, 1990).
8. Gitlow, H., A. Oppenheim, and R. Oppenheim, *Tools and Methods for the Improvement of Quality*, 2d ed. (Homewood, IL: Richard D. Irwin, 1994).
9. Halberstam, D., *The Reckoning* (New York: William Morrow, 1986).
10. Holusha, J., "The Baldrige badge of courage—and quality," *New York Times,* October 21, 1990, p. F12.
11. Joiner, B. J., "The key role of statisticians in the transformation of North American industry," *American Statistician,* 1985, Vol. 39, pp. 224–234.
12. Joiner, B. L., *Fourth Generation Management* (New York: McGraw-Hill, 1994).
13. Levine, D. M., P. P. Ramsey, and M. L. Berenson, *Business Statistics for Quality and Productivity* (Englewood Cliffs, NJ: Prentice-Hall, 1995).
14. Main, J., "The curmudgeon who talks tough on quality," *Fortune,* June 25, 1984, pp. 118–122.
15. Mann, N. R., *The Keys to Excellence: The Story of the Deming Philosophy* (Los Angeles: Prestwick Books, 1987).
16. *The Memory Jogger II: A Pocket Guide of Tools for Continuous Improvement and Effective Planning* (Methuen, MA: GOAL/QPC, 1994).
17. *Microsoft Excel Version 7,* (Redmond, WA: Microsoft Corp., 1996).
18. Montgomery, D. C., *Introduction to Statistical Quality Control,* 3d ed. (New York: John Wiley, 1996).
19. Port, O., "The push for quality," *Business Week,* June 8, 1987, pp. 130–135.
20. Ryan, T. P., *Statistical Methods for Quality Improvement* (New York: John Wiley, 1989).
21. Scherkenbach, W. W., *The Deming Route to Quality and Productivity: Road Maps and Roadblocks* (Washington, D.C.: CEEP Press, 1986).
22. Shewhart, W. A., "The applications of statistics as an aid in maintaining quality of manufactured products," *Journal of the American Statistical Association,* 1925, Vol. 20, pp. 546–548.
23. Shewhart, W. A., *Economic Control of Quality of Manufactured Products* (New York: Van Nostrand-Reinhard, 1931, reprinted by the American Society for Quality Control, Milwaukee, 1980).
24. Shewhart, W. A., and W. E. Deming, *Statistical Methods from the Viewpoint of Quality Control* (Washington, D.C.: Graduate School, Dept. of Agriculture, 1939, Dover Press, 1986).
25. Sholtes, P. R., *An Elaboration on Deming's Teaching on Performance Appraisal* (Madison, WI: Joiner Associates, 1987).
26. Skrebec, Q. R., "Ancient process control and its modern implications," *Quality Progress,* 1990, Vol. 23, pp. 49–52.
27. Wallis, W. A., "The statistical research group 1942–1945," *Journal of the American Statistical Association*, 1980, Vol. 75, pp. 320–335.
28. Walton, M., *The Deming Management Method* (New York: Perigee Books, Putnam Publishing Group, 1986).
29. Walton, M., *Deming Management at Work* (New York: G. P. Putnam, 1990).

chapter 11

Simple Linear Regression and Correlation

CHAPTER OBJECTIVE

To develop the simple linear regression model as a means of using one variable to predict another variable and to study correlation as a measure of the strength of the association between two variables.

11.1 INTRODUCTION

In previous chapters we focused primarily on a single numerical response variable, such as the amount of out-of-state tuition. We studied various measures of statistical description (see Chapter 3) and applied different techniques of statistical inference to make estimates and draw conclusions about our numerical response variable (see Chapters 6–8). In this and the following chapter we will concern ourselves with problems involving two or more numerical variables as a means of viewing the relationships that exist between them. Two techniques will be discussed: regression and correlation.

Regression analysis is used primarily for the purpose of prediction. Our goal in regression analysis is the development of a statistical model that can be used to predict the values of a **dependent** or **response variable** based on the values of at least one **explanatory** or **independent variable**. In this chapter we shall focus on a simple regression model—one that would utilize a single numerical independent variable X to predict the numerical dependent variable Y. In Chapter 12, we shall develop a multiple regression model—one that could utilize several explanatory variables $(X_1, X_2, ..., X_p)$ to predict a numerical dependent variable Y.[1]

Correlation analysis, in contrast to regression, is used to measure the strength of the association between numerical variables. For example, in Section 11.8 we will determine the correlation between the price of a six-pack of soft drinks and the price of chicken in different cities. In this instance, the objective is not to use one variable to predict another, but rather to measure the strength of the association or covariation that exists between two numerical variables.

11.2 THE SCATTER DIAGRAM

Methods of regression and correlation analysis will be applied to two examples in this chapter. In the first, suppose that the management of a chain of package delivery stores would like to develop a model for predicting the weekly sales (in thousands of dollars) for individual stores. A random sample of 20 stores was selected from among all the stores in the chain. In developing such a model, many explanatory variables could be considered. However, we will begin our discussion with a simple regression model in which only one variable is used to predict the values of a dependent variable. Thus, we will develop a model to predict weekly sales (the dependent variable Y) based on the number of customers (the explanatory or independent variable X). The results for a sample of 20 stores are summarized in Table 11.1. Such data, however, can be presented in a form that is more visually interpretable.

In Chapter 2, when information concerning the out-of-state tuition rates in Texas was studied, various graphs (e.g., histograms, polygons, and ogives) were developed for data presentation. In a regression analysis involving one independent and one dependent variable, the individual values are plotted on a two-dimensional graph called a **scatter diagram**. Each value is plotted at its particular X- and Y-coordinates. The scatter diagram for the data in Table 11.1 is shown in Figure 11.1.

An examination of Figure 11.1 indicates a clearly increasing relationship between number of customers (X) and weekly sales (Y). As the number of customers increases, weekly sales increase. The exact mathematical form of the model expressing the relationship as well as methods for estimating the weekly sales for a given number of customers will be examined in subsequent sections of this chapter.

Table 11.1 Number of customers and weekly sales for a sample of 20 package delivery stores.

Stores	Customers	Sales ($000)
1	907	11.20
2	926	11.05
3	506	6.84
4	741	9.21
5	789	9.42
6	889	10.08
7	874	9.45
8	510	6.73
9	529	7.24
10	420	6.12
11	679	7.63
12	872	9.43
13	924	9.46
14	607	7.64
15	452	6.92
16	729	8.95
17	794	9.33
18	844	10.23
19	1,010	11.77
20	621	7.41

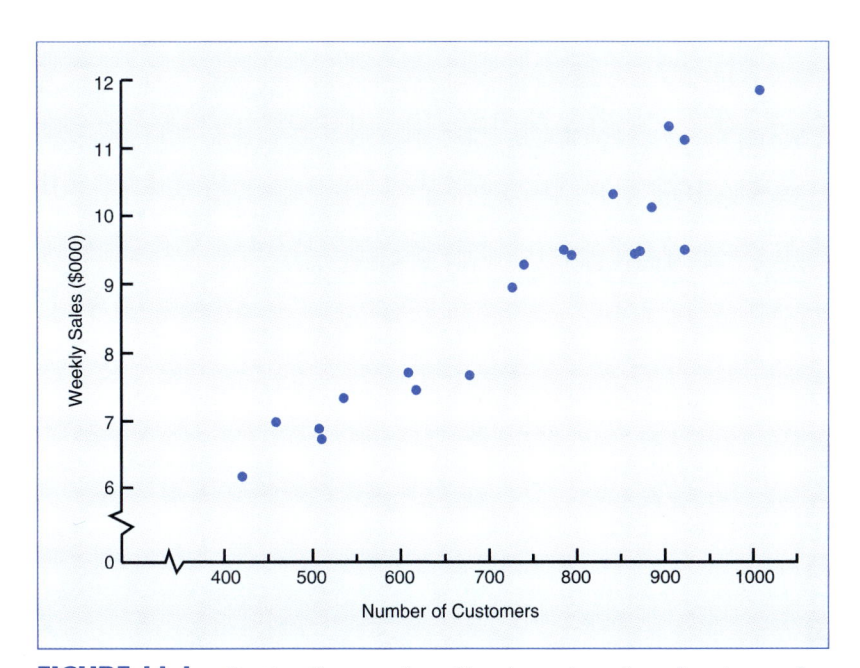

PACKAGE.TXT

FIGURE 11.1 Scatter diagram of weekly sales and number of customers for the package delivery stores. *Source:* Data are taken from Table 11.1.

11.2 The Scatter Diagram

Problems for Section 11.2

Note: *The problems in this section can be solved using Microsoft Excel (see Sections 11.6.3 and 11.7).*

PETFOOD.TXT

- **11.1** The marketing manager of a large supermarket chain would like to determine the effect of shelf space on the sales of pet food. A random sample of 12 equal-sized stores is selected with the following results:

  Pet food sales.

Store	Shelf Space, X (feet)	Weekly Sales, Y (hundreds of dollars)
1	5	1.6
2	5	2.2
3	5	1.4
4	10	1.9
5	10	2.4
6	10	2.6
7	15	2.3
8	15	2.7
9	15	2.8
10	20	2.6
11	20	2.9
12	20	3.1

  Set up a scatter diagram.

S-SITE.XLS

- **11.2** Over the past 25 years, a chain of discount women's clothing stores has increased market share by increasing the number of locations in the chain. A systematic approach to site selection was never utilized. Site selection was primarily based on what was considered to be a great location or a great lease. This year, with a strategic plan for opening several new stores, the director of special projects and planning is being asked to develop an approach to forecasting annual sales for all new stores. The following represents the square footage and the annual sales (in thousands of dollars) for a sample of 14 stores in the chain:

  Site selection.

Store	Square Feet	Annual Sales ($000)
1	1,726	3,681
2	1,642	3,895
3	2,816	6,653
4	5,555	9,543
5	1,292	3,418
6	2,208	5,563
7	1,313	3,660
8	1,102	2,694
9	3,151	5,468
10	1,516	2,898
11	5,161	10,674
12	4,567	7,585
13	5,841	11,760
14	3,008	4,085

  Set up a scatter diagram.

- **11.3** A company manufacturing parts would like to develop a model to estimate the number of worker-hours required for production runs of varying lot size. A random sample of 14 production runs (2 each for lot sizes 20, 30, 40, 50, 60, 70, and 80) is selected with the following results:

WORKHRS.TXT

Production worker-hours.

Lot Size	Worker-Hours
20	50
20	55
30	73
30	67
40	87
40	95
50	108
50	112
60	128
60	135
70	148
70	160
80	170
80	162

Set up a scatter diagram.

11.4 An agronomist would like to determine the effect of a natural organic fertilizer on the yield of tomatoes. In this experiment five differing amounts of fertilizer are to be used on 10 equivalent plots of land: 0, 10, 20, 30, and 40 pounds per 100 square feet. The levels of fertilizer are randomly assigned to the plots of land with the following results:

TOMYIELD.TXT

Tomato yield.

Plot	Amount of Fertilizer, X (in pounds per 100 square feet)	Yield, Y (in pounds)
1	0	6
2	0	8
3	10	11
4	10	14
5	20	18
6	20	23
7	30	25
8	30	28
9	40	30
10	40	34

Set up a scatter diagram.

11.5 A limousine service operating from a suburban county wants to determine the length of time it would take to transport passengers from various locations to a major metropolitan airport during nonpeak times. A sample of 12 trips on a particular day during nonpeak times indicates the following as shown on page 538:

LIMO.TXT

11.2 The Scatter Diagram **537**

Airport travel.

Distance (miles)	Time (minutes)
10.3	19.71
11.6	18.15
12.1	21.88
14.3	24.21
15.7	27.08
16.1	22.96
18.4	29.38
20.2	37.24
21.8	36.84
24.3	40.59
25.4	41.21
26.7	38.19

Set up a scatter diagram.

11.3 TYPES OF REGRESSION MODELS

In the scatter diagram plotted in Figure 11.1 on page 535, a rough idea of the type of relationship that exists between the variables can be observed. The nature of the relationship can take many forms, ranging from simple ones to extremely complicated mathematical functions. The simplest relationship consists of a straight-line or **linear relationship**. An example of this relationship is shown in Figure 11.2.

The straight-line (linear) model can be represented as

$$Y_i = \beta_0 + \beta_1 X_i + \epsilon_i \tag{11.1}$$

where
β_0 = Y intercept for the population
β_1 = slope for the population
ϵ_i = random error in Y for observation i

In this model, the **slope** of the line β_1 represents the expected change in Y per unit change in X. It represents the amount that Y changes (either positively or negatively) for a particular unit

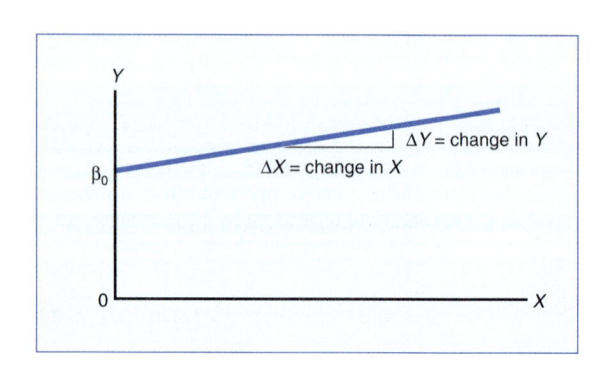

FIGURE 11.2 A positive straight-line relationship.

change in X. The **Y intercept** β_0 represents the average value of Y when X equals 0. The last component of the model, ϵ_i, represents the random error in Y for each observation i that occurs.

The selection of the proper mathematical model is influenced by the distribution of the X and Y values on the scatter diagram. This can be seen readily from an examination of Panels A–F in Figure 11.3. Thus, from Panel A, we note that the values of Y are generally increasing linearly as X increases. This panel is similar to Figure 11.1, which illustrates the relationship between number of customers and sales. Panel B is an example of a negative linear relationship. As X increases, we note that the values of Y are decreasing. An example of this type of relationship might be the price of a particular product and the amount of sales. Panel C shows a set of data in which there is very little or no relationship between X and Y. High and low values of Y appear at each value of X. The data in Panel D show a positive curvilinear relationship between X and Y. The values of Y are increasing as X increases, but this increase tapers off beyond certain values of X. An example of this positive curvilinear relationship might be the age and maintenance cost of a machine. As a machine gets older, the maintenance cost may rise rapidly at first but then level off beyond a certain number of years. Panel E shows a parabolic or U-shaped relationship between X and Y. As X increases, at first Y decreases; but as X continues to increase, Y not only stops decreasing but actually increases above its minimum value. An example of this type of relationship could be the number of

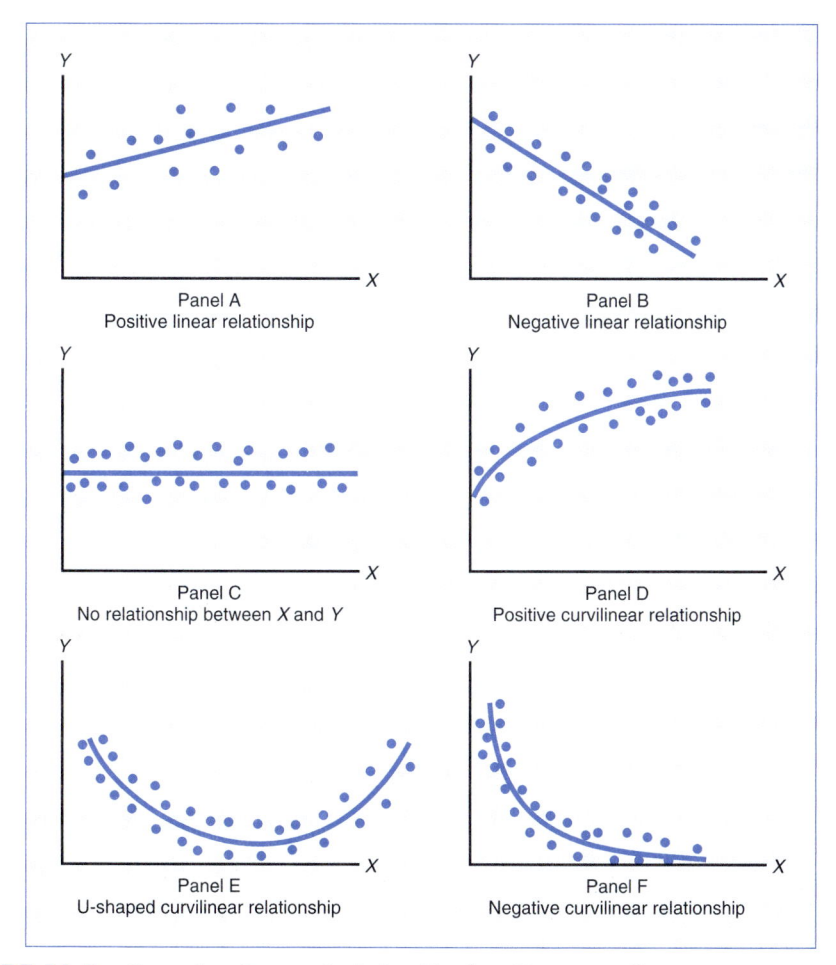

FIGURE 11.3 **Examples of types of relationships found in scatter diagrams.**

errors per hour at a task and the number of hours worked. The number of errors per hour would decrease as the individual becomes more proficient at the task, but then would increase beyond a certain point because of factors such as fatigue and boredom. Finally, Panel F indicates an exponential or negative curvilinear relationship between X and Y. In this case, Y decreases very rapidly as X first increases, but then decreases much less rapidly as X increases further. An example of this exponential relationship could be the resale value of a particular type of automobile and its age. In the first year, the resale value drops drastically from its original price; however, the resale value then decreases much less rapidly in subsequent years.

In this section we have briefly examined a variety of different models that could be used to represent the relationship between two variables. Although scatter diagrams can be extremely helpful in determining the mathematical form of the relationship, more sophisticated statistical procedures are available to determine the most appropriate model for a set of variables. In subsequent sections of this chapter, we shall primarily focus on building statistical models for fitting *linear* relationships between variables.

11.4 DETERMINING THE SIMPLE LINEAR REGRESSION EQUATION

If we refer to the scatter diagram in Figure 11.1 on page 535, we notice that sales appear to increase linearly as a function of the number of customers. The question that must be addressed in regression analysis involves the determination of the particular straight-line model that is the best fit to these data.

11.4.1 The Least Squares Method

In the preceding section we hypothesized a statistical model to represent the relationship between two variables in a population for a chain of package delivery stores. However, as shown in Table 11.1 on page 535, we have obtained data from only a random sample of the population. If certain assumptions are valid (see Section 11.9), the sample Y intercept (b_0) and the sample slope (b_1) can be used as estimates of the respective population parameters (β_0 and β_1). Thus, the sample regression equation representing the straight-line regression model would be

> The predicted value of Y equals the Y intercept plus the slope times the X value.
>
> $$\hat{Y}_i = b_0 + b_1 X_i \qquad (11.1a)$$

where
$\hat{Y}_i$ = predicted value of Y for observation i

X_i = value of X for observation i

This equation requires the determination of two **regression coefficients**—b_0 (the Y intercept) and b_1 (the slope) in order to predict values of Y. Once b_0 and b_1 are obtained, the straight line is known and can be plotted on the scatter diagram. Then we can make a visual comparison of how well our particular statistical model (a straight line) fits the original data. That is, we can see whether the original data lie close to the fitted line or deviate greatly from the fitted line.

Simple linear regression analysis is concerned with finding the straight line that fits the data best. The best fit means that we wish to find the straight line for which the differences between the actual values (Y_i) and the values that would be predicted from the fitted line of regression ($\hat{Y}_i$) are as small as possible. Because these differences will be both positive and negative for different observations, mathematically we minimize

$$\sum_{i=1}^{n}(Y_i - \hat{Y}_i)^2$$

where
Y_i = actual value of Y for observation i
$\hat{Y}_i$ = predicted value of Y for observation i

Since $\hat{Y}_i = b_0 + b_1 X_i$, we are minimizing

$$\sum_{i=1}^{n}[Y_i - (b_0 + b_1 X_i)]^2$$

which has two unknowns, b_0 and b_1.

A mathematical technique that determines the values of b_0 and b_1 that minimizes this difference is known as the **least squares method**. Any values for b_0 and b_1 other than those determined by the least squares method would result in a greater sum of squared differences between the actual value of Y and the predicted value of Y.

In using the least squares method, we obtain the following two equations, called the normal equations:

$$\sum_{i=1}^{n} Y_i = nb_0 + b_1 \sum_{i=1}^{n} X_i \qquad (11.2a)$$

$$\sum_{i=1}^{n} X_i Y_i = b_0 \sum_{i=1}^{n} X_i + b_1 \sum_{i=1}^{n} X_i^2 \qquad (11.2b)$$

From these two equations, we must solve for b_1 and b_0. However, in this text, we shall take the position that in solving regression equations, Excel spreadsheet software will be used to perform the (often tedious) calculations. However, to understand how the results displayed in the output of this software have been computed for the case of **simple linear regression**, we will also illustrate many of the computations involved.

Referring to Equations (11.2a) and (11.2b), since there are two equations with two unknowns, we can solve these equations simultaneously for b_1 and b_0 as follows:

$$b_1 = \frac{\sum_{i=1}^{n} X_i Y_i - n\overline{X}\overline{Y}}{\sum_{i=1}^{n} X_i^2 - n\overline{X}^2} \qquad (11.3)$$

and

$$b_0 = \overline{Y} - b_1 \overline{X} \qquad (11.4)$$

where

$$\bar{Y} = \frac{\sum_{i=1}^{n} Y_i}{n} \quad \text{and} \quad \bar{X} = \frac{\sum_{i=1}^{n} X_i}{n}$$

Examining Equations (11.3) and (11.4), we see that there are five quantities that must be calculated to determine b_1 and b_0. These are n, the sample size; $\sum_{i=1}^{n} X_i$, the sum of the X values; $\sum_{i=1}^{n} Y_i$, the sum of the Y values; $\sum_{i=1}^{n} X_i^2$, the sum of the squared X values, and $\sum_{i=1}^{n} X_i Y_i$, the sum of the cross product of X and Y. For our data in Table 11.1, the number of customers is used to predict the weekly sales in a store. The computation of the various sums needed (including $\sum_{i=1}^{n} Y_i^2$, the sum of the squared Y values that will be used in Section 11.5) are presented in Table 11.2.

Using Equations (11.3) and (11.4), we can compute the values of b_0 and b_1:

$$b_1 = \frac{\sum_{i=1}^{n} X_i Y_i - n\bar{X}\bar{Y}}{\sum_{i=1}^{n} X_i^2 - n\bar{X}^2}$$

Table 11.2 Computations for the package delivery sales data.

Store	Customers X	Sales Y	X²	Y²	XY
1	907	11.20	822,649	125.4400	10,158.40
2	926	11.05	857,476	122.1025	10,232.30
3	506	6.84	256,036	46.7856	3,461.04
4	741	9.21	549,081	84.8241	6,824.61
5	789	9.42	622,521	88.7364	7,432.38
6	889	10.08	790,321	101.6064	8,961.12
7	874	9.45	763,876	89.3025	8,259.30
8	510	6.73	260,100	45.2929	3,432.30
9	529	7.24	279,841	52.4176	3,829.96
10	420	6.12	176,400	37.4544	2,570.40
11	679	7.63	461,041	58.2169	5,180.77
12	872	9.43	760,384	88.9249	8,222.96
13	924	9.46	853,776	89.4916	8,741.04
14	607	7.64	368,449	58.3696	4,637.48
15	452	6.92	204,304	47.8864	3,127.84
16	729	8.95	531,441	80.1025	6,524.55
17	794	9.33	630,436	87.0489	7,408.02
18	844	10.23	712,336	104.6529	8,634.12
19	1,010	11.77	1,020,100	138.5329	11,887.70
20	621	7.41	385,641	54.9081	4,601.61
Totals	14,623	176.11	11,306,209	1,602.0971	134,127.90

where

$$\bar{Y} = \frac{\sum_{i=1}^{n} Y_i}{n} = \frac{176.11}{20} = 8.8055$$

$$\bar{X} = \frac{\sum_{i=1}^{n} X_i}{n} = \frac{14{,}623}{20} = 731.15$$

so that

$$b_1 = \frac{134{,}127.90 - (20)(731.15)(8.8055)}{11{,}306{,}209 - 20(731.15)^2}$$

$$= \frac{5{,}365.08}{614{,}603} = +.00873$$

and

$$b_0 = \bar{Y} - b_1 \bar{X}$$
$$= 8.8055 - (.00873)(731.15) = +2.423$$

Thus, the equation for the *best* straight line for these data is

$$\hat{Y}_i = 2.423 + .00873 X_i$$

The slope b_1 was computed as $+.00873$. This means that for each increase of one unit in X, the value of Y is estimated to increase by an average of .00873 unit. This means that for each increase of one customer, the fitted model predicts that the expected weekly sales are estimated to increase by .00873 thousands of dollars or $8.73 (or we can say that for each increase of 100 customers, weekly sales are expected to increase by $873). Thus, the slope can be viewed as representing the portion of the weekly sales that are estimated to vary according to the number of customers.

The Y intercept b_0 was computed to be $+2.423$ (thousands of dollars). The Y intercept represents the average value of Y when X equals 0. Since the number of customers is unlikely to be 0, this Y intercept can be viewed as expressing the portion of the weekly sales that varies with factors other than the number of customers.

The regression model that has been fit to the data can now be used to predict the weekly sales. For example, let us say that we would like to use the fitted model to predict the average weekly sales for a store with 600 customers. We can determine the predicted value by substituting $X = 600$ into our regression equation,

$$\hat{Y}_i = 2.423 + .00873(600) = 7.661$$

Thus, the predicted average weekly sales for a store with 600 customers is 7.661 thousands of dollars or $7,661.

11.4.2 Predictions in Regression Analysis: Interpolation Versus Extrapolation

When using a regression model for prediction purposes, it is important that we consider only the relevant range of the independent variable in making our predictions. This **relevant range** encompasses all values from the smallest to the largest X used in developing the regression model. Hence, when predicting Y for a given value of X, we may *interpolate* within this relevant range of the X values, but we may not *extrapolate* beyond the range of X values. For example, when we use the number of customers to predict weekly sales, we note from Table 11.1 that the number of customers varies from 420 to 1,010. Therefore, predictions of weekly sales should be made only for stores that have between 420 and 1,010 customers. Any prediction of weekly sales outside this range of number of customers presumes that the fitted relationship holds outside the 420–1,010 range.

Problems for Section 11.4

Note: *Excel may be used to obtain the results for these problems (see Sections 11.6.3 and 11.7).*

PETFOOD.TXT

- 11.6 Referring to Problem 11.1 (pet food sales) on page 536
  - (a) Assuming a linear relationship, use the least squares method to find the regression coefficients b_0 and b_1.
  - (b) Interpret the meaning of the slope b_1 in this problem.
  - (c) Predict the average weekly sales (in hundreds of dollars) of pet food for stores with 8 feet of shelf space for pet food.
  - (d) Suppose that sales in store 12 are 2.6. Do parts (a)–(c) with this value and compare the results.

S-SITE.XLS

11.7 Referring to Problem 11.2 (site selection) on page 536
  - (a) Assuming a linear relationship, use the least squares method to find the regression coefficients b_0 and b_1.
  - (b) Interpret the meaning of the slope b_1 in this problem.
  - (c) Predict the average annual sales for a store that contains 4,000 square feet.
  - (d) Suppose that sales in store 7 were 2,660. Do parts (a)–(c) with this value and compare the results.

WORKHRS.TXT

- 11.8 Referring to Problem 11.3 (production worker-hours) on page 537
  - (a) Assuming a linear relationship, use the least squares method to find the regression coefficients b_0 and b_1.
  - (b) Interpret the meaning of the Y intercept b_0 and the slope b_1 in this problem.
  - (c) Predict the average number of worker-hours required for a production run with a lot size of 45.
  - (d) Why would it not be appropriate to predict the average number of worker-hours required for a production run with lot size of 100? Explain.
  - (e) Suppose that the worker-hours for the lot size of 60 were 117 and 119. Do parts (a)–(c) with these values and compare the results.

TOMYIELD.TXT

11.9 Referring to Problem 11.4 (tomato yield) on page 537
  - (a) Assuming a linear relationship, use the least squares method to find the regression coefficients of b_0 and b_1.
  - (b) Interpret the meaning of the Y intercept b_0 and the slope b_1 in this problem.
  - (c) Predict the average yield of tomatoes for a plot that has been given 15 pounds per 100 square feet of natural organic fertilizer.
  - (d) Why would it not be appropriate to predict the average yield for a plot that has been fertilized with 100 pounds per 100 square feet? Explain.
  - (e) Suppose the yield with 40 pounds of fertilizer was 30 and 32 pounds. Do parts (a)–(c) with these values and compare the results.
  - (f) What would the results of (e) lead you to think about the relationship between amount of fertilizer and yield for amounts of 40 pounds or more?

LIMO.DAT

11.10 Referring to Problem 11.5 (airport travel) on pages 537–538
  - (a) Assuming a linear relationship, use the least squares method to find the regression coefficients of b_0 and b_1.
  - (b) Interpret the meaning of the Y intercept b_0 and the slope b_1 in this problem.
  - (c) Use the regression model developed in (a) to predict the number of minutes to transport someone from a location that is 21 miles from the airport.
  - (d) Suppose the distance for the last trip was 36.7 miles and the time was 65 minutes. Do parts (a)–(c) with these values and compare the results.
  - (e) What would the results of (d) lead you to think about the usefulness of the regression model?

11.5 STANDARD ERROR OF THE ESTIMATE

In the preceding section we used the least squares method to develop an equation to predict the weekly sales based on the number of customers. Although the least squares method results in the line that fits the data with the minimum amount of variation, the regression equation is not a perfect predictor, unless all the observed data points fall on the regression line. Just as we do not expect all data values to be exactly equal to their arithmetic *mean*, neither can we expect all data points to fall exactly on the regression line. The regression line serves only as an approximate predictor of a Y value for a given value of X. Therefore, we need to develop a statistic that measures the variability of the actual Y values, from the predicted Y values, in the same way that we developed (see Chapter 3) a measure of the variability of each observation around its mean. The measure of variability around the line of regression (its standard deviation) is called the **standard error of the estimate**.

The variability around the line of regression is illustrated in Figure 11.4 for the package delivery sales data. We can see from Figure 11.4 that, although the predicted line of regression falls near many of the actual values of Y, there are several values above the line of regression as well as below the line of regression, so that

$$\sum_{i=1}^{n}(Y_i - \hat{Y}_i) = 0$$

The standard error of the estimate, given by the symbol S_{YX} is defined as

$$S_{YX} = \sqrt{\frac{\sum_{i=1}^{n}(Y_i - \hat{Y}_i)^2}{n-2}} \qquad (11.5)$$

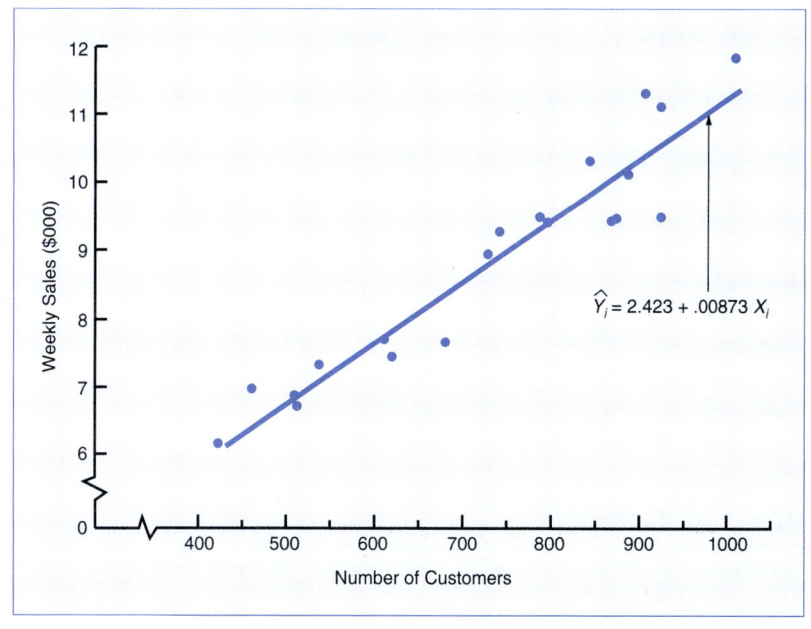

FIGURE 11.4 Scatter diagram and line of regression for the package delivery sales data.

where Y_i = actual value of Y for a given X_i
$\hat{Y}_i$ = predicted value of Y for a given X_i

The computation of the standard error of the estimate using Equation (11.5) would first require determining the predicted value of Y for each X value in the sample. The computation can be simplified because of the following identity:

$$\sum_{i=1}^{n}(Y_i - \hat{Y}_i)^2 = \sum_{i=1}^{n} Y_i^2 - b_0 \sum_{i=1}^{n} Y_i - b_1 \sum_{i=1}^{n} X_i Y_i$$

The standard error of the estimate S_{YX} can thus be obtained using the following formula:

$$S_{YX} = \sqrt{\frac{\sum_{i=1}^{n} Y_i^2 - b_0 \sum_{i=1}^{n} Y_i - b_1 \sum_{i=1}^{n} X_i Y_i}{n-2}} \qquad (11.6)$$

For the package delivery sales example, from Table 11.2 on page 542 we have determined that

$$\sum_{i=1}^{n} Y_i^2 = 1{,}602.0971 \qquad \sum_{i=1}^{n} Y_i = 176.11 \qquad \sum_{i=1}^{n} X_i Y_i = 134{,}127.90$$

$$b_0 = 2.423 \qquad b_1 = +.00873$$

Therefore, using Equation (11.6), the standard error of the estimate S_{YX} can be computed as

$$S_{YX} = \sqrt{\frac{\sum_{i=1}^{n} Y_i^2 - b_0 \sum_{i=1}^{n} Y_i - b_1 \sum_{i=1}^{n} X_i Y_i}{n-2}}$$

$$= \sqrt{\frac{1{,}602.0971 - (2.423)(176.11) - (.00873)(134{,}127.90)}{20-2}}$$

$$= \sqrt{\frac{4.446}{18}} = \sqrt{.247}$$

$$= .497$$

This standard error of the estimate, equal to .497 (i.e., $497) represents a measure of the variation around the fitted line of regression. It is measured in units of the dependent variable Y. The interpretation of the standard error of the estimate, then, is analogous to that of the standard deviation. Just as the standard deviation measures variability around the arithmetic mean, the standard error of the estimate measures variability around the fitted line of regression. As we shall see in Sections 11.12–11.14, the standard error of the estimate can be used to make inferences about a predicted value of Y and to determine whether a statistically significant relationship exists between the two variables.

Problems for Section 11.5

Note: *The problems in this section can be solved using Microsoft Excel (see Sections 11.6.3 and 11.7).*

- **11.11** Referring to the pet food sales problem (pages 536 and 544), compute the standard error of the estimate.

PETFOOD.TXT

11.12 Referring to the site selection problem (pages 536 and 544), compute the standard error of the estimate.

- **11.13** Referring to the production worker-hours problem (pages 537 and 544), compute the standard error of the estimate.

11.14 Referring to the tomato yield problem (pages 537 and 544), compute the standard error of the estimate.

11.15 Referring to the airport travel problem (pages 537 and 544), compute the standard error of the estimate.

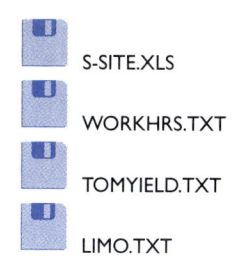

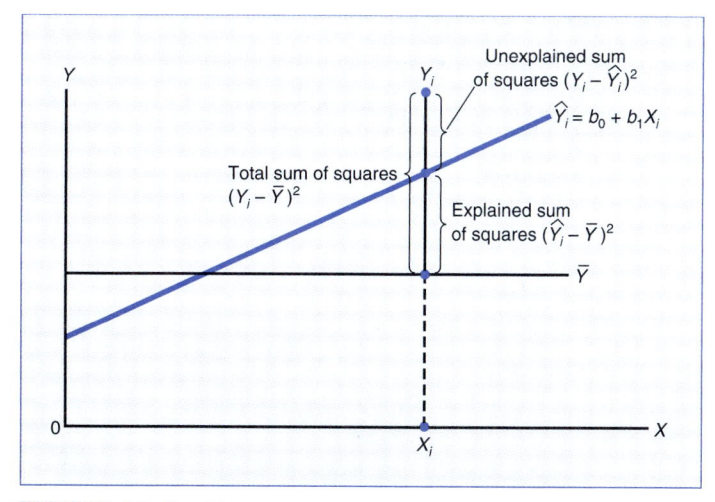

S-SITE.XLS

WORKHRS.TXT

TOMYIELD.TXT

LIMO.TXT

11.6 MEASURES OF VARIATION IN REGRESSION AND CORRELATION

11.6.1 Obtaining the Sum of Squares

To examine how well the independent variable predicts the dependent variable in our statistical model, we need to develop several measures of variation. The first measure, the **total sum of squares** (SST), is a measure of variation of the Y_i values around their mean $\bar{Y}$. In a regression analysis the total sum of squares can be subdivided into **explained variation** or **regression sum of squares** (SSR), that which is attributable to the relationship between X and Y, and **unexplained variation** or **error sum of squares** (SSE), that which is attributable to factors other than the relationship between X and Y. These different measures of variation can be seen in Figure 11.5.

The regression sum of squares (SSR) represents the difference between $\bar{Y}$ (the average value of Y) and $\hat{Y}_i$ (the value of Y that would be predicted from the regression relationship). The error sum of squares (SSE) represents that part of the variation in Y that is not explained by the regression. It is based on the difference between Y_i and $\hat{Y}_i$.

These measures of variation can be represented as follows:

$$\text{Total sum of squares} = \text{regression sum of squares} + \text{error sum of squares}$$
$$\text{SST} = \text{SSR} + \text{SSE} \tag{11.7}$$

FIGURE 11.5 Measures of variation in regression.

where

$$\text{SST} = \text{total sum of squares} = \sum_{i=1}^{n}(Y_i - \overline{Y})^2 = \sum_{i=1}^{n} Y_i^2 - n\overline{Y}^2 \qquad (11.8)$$

$$\begin{aligned}\text{SSE} &= \text{unexplained variation or error sum of squares} \\ &= \sum_{i=1}^{n}(Y_i - \hat{Y}_i)^2 \\ &= \sum_{i=1}^{n} Y_i^2 - b_0 \sum_{i=1}^{n} Y_i - b_1 \sum_{i=1}^{n} X_i Y_i \end{aligned} \qquad (11.9)$$

$$\begin{aligned}\text{SSR} &= \text{explained variation or regression sum of squares} \\ &= \sum_{i=1}^{n}(\hat{Y}_i - \overline{Y})^2 \\ &= \text{SST} - \text{SSE} \\ &= b_0 \sum_{i=1}^{n} Y_i + b_1 \sum_{i=1}^{n} X_i Y_i - n\overline{Y}^2 \end{aligned} \qquad (11.10)$$

Examining the unexplained variation or error sum of squares [Equation (11.9)], we may recall that $\sum_{i=1}^{n}(Y_i - \hat{Y}_i)^2$ was the numerator under the square root in the computation of the standard error of the estimate [see Equation (11.5)]. Therefore, in the process of computing the standard error of the estimate, we have already computed the following error sum of squares:

$$\begin{aligned}\text{SSE} &= \sum_{i=1}^{n} Y_i^2 - b_0 \sum_{i=1}^{n} Y_i - b_1 \sum_{i=1}^{n} X_i Y_i \\ &= 1{,}602.0971 - (2.423)(176.11) - (.00873)(134{,}127.90) \\ &= 4.446\end{aligned}$$

In addition,

$$\begin{aligned}\text{SST} &= \text{total sum of squares} \\ &= \sum_{i=1}^{n} Y_i^2 - n\overline{Y}^2 \\ &= 1{,}602.0971 - 20(8.8055)^2 \\ &= 1{,}602.0971 - 1{,}550.7366 \\ &= 51.3605\end{aligned}$$

and

SSR = explained variation or regression sum of squares

$$= b_0 \sum_{i=1}^{n} Y_i + b_1 \sum_{i=1}^{n} X_i Y_i - n\overline{Y}^2$$

$$= (2.423)(176.11) + (.00873)(134,127.90) - 20(8.8055)^2$$

$$= 46.9145$$

We note also, from Equation (11.7), that

$$\text{SST} = \text{SSR} + \text{SSE}$$
$$51.3605 = 46.9145 + 4.4460$$

11.6.2 The Coefficient of Determination

Now that SSR, SSE, and SST have been defined, the **coefficient of determination** r^2 can be defined as

> The coefficient of determination is equal to the regression sum of squares divided by the total sum of squares.
>
> $$r^2 = \frac{\text{regression sum of squares}}{\text{total sum of squares}} = \frac{\text{SSR}}{\text{SST}} \quad (11.11a)$$

Thus, the coefficient of determination measures the proportion of variation that is explained by the independent variable in the regression model. For the package delivery sales example,

$$r^2 = \frac{46.9145}{51.3605} = .913$$

Therefore, 91.3% of the variation in weekly sales can be explained by the variability in the number of customers from store to store. This is an example where there is a strong linear relationship between two variables, since the use of a regression model has reduced the variability in predicting weekly sales by 91.3%. Only 8.7% of the sample variability in weekly sales can be explained by factors other than what is accounted for by the linear regression model.

To interpret the coefficient of determination—particularly when dealing with multiple regression models—some researchers suggest that an *adjusted* r^2 be computed to reflect both the number of explanatory variables in the model and the sample size. In simple linear regresssion, we define the **adjusted** r^2 as

$$r^2_{adj} = 1 - \left[(1 - r^2)\frac{n-1}{n-2}\right] \quad (11.11b)$$

Thus for our package delivery sales data, since $r^2 = .913$ and $n = 20$,

$$r^2_{adj} = 1 - \left[(1 - r^2)\frac{20-1}{20-2}\right]$$

$$= 1 - \left[(1 - .913)\frac{19}{18}\right]$$

$$= 1 - .092$$

$$= .908$$

This result is similar to the one obtained without adjustment for degrees of freedom.

11.6.3 Using the Microsoft Excel Chart Wizard and TREND Function for Regression Analysis

In this chapter we have developed the least squares method to compute the regression coefficients and used the regression model obtained to predict the sales for a given number of customers for the package delivery stores. We can use the Chart Wizard that we previously discussed in Chapters 2 and 3 to obtain a scatter diagram and the line of regression for these data. With the PACKAGE.XLS workbook open, click on the Data sheet tab. Note that this sheet contains the data of Table 11.1 on page 535. Then, with the Data sheet still active, select Insert | Chart | As New Sheet. For the five dialog boxes of the Chart Wizard that appear, do the following:

11-6-3.XLS

Dialog box 1: Enter the range B1:C21 and click the Next button.
Dialog box 2: Select the XY (scatter) chart and click the Next button.
Dialog box 3: Select format 1 that shows unconnected points and click the Next button.
Dialog box 4: Select the Columns option button and enter 1 in the First Columns edit box and 1 in the First Rows edit box. Click the Next button.
Dialog box 5: Select No for Add a Legend and enter the Chart title Regression Analysis, Customers in the Category (X) edit box, and Sales in the Value (Y) edit box. Click the Finish button.

Figure 11.1.Excel displays the scatter diagram we have obtained. Rename the sheet as Trend.

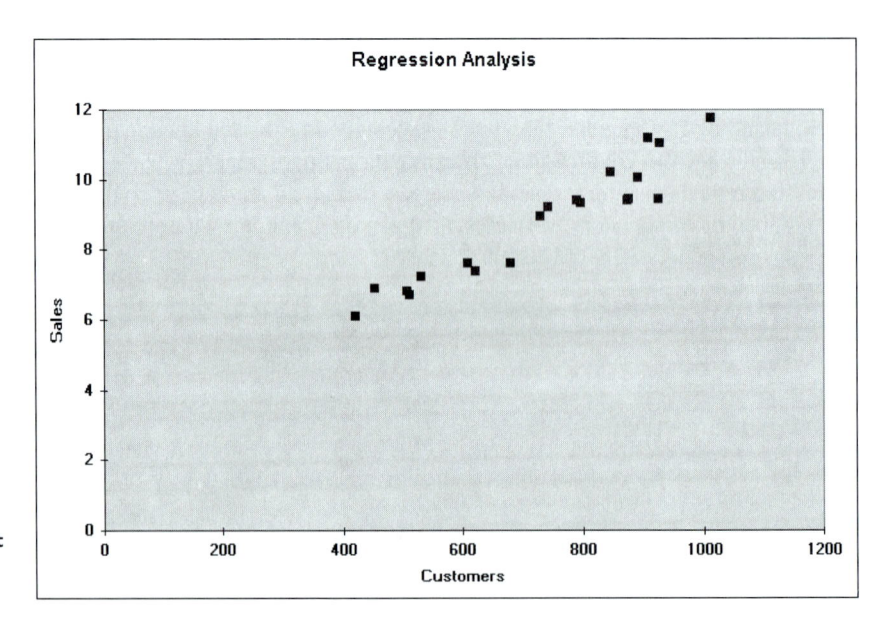

FIGURE 11.1.EXCEL Scatter diagram obtained from the Excel Chart Wizard for the package delivery sales data.

550   Chapter 11   Simple Linear Regression and Correlation

To superimpose a line of regression on this scatter diagram, select the plotted points by clicking on any single point. (Clicking causes many of the points to be highlighted.) Next, select the command Insert | Trendline. When the Trendline dialog box appears, select the Type Tab and then select Linear as the Trend/Regression Type as is illustrated in Figure 11.2.Excel. Then click the Options Tab. In the Options Tab, select the Automatic Trendline Name option button and select the Display Equation on Chart and the Display R-Squared Value on Chart check boxes (see Figure 11.3.Excel). Click the OK button. Excel adds a regression line, the regression equation, and the value of r^2 to the chart. (You can reposition the equation and r^2 by dragging them with the mouse pointer.) Figure 11.4.Excel illustrates the scatter diagram after these changes have been made.

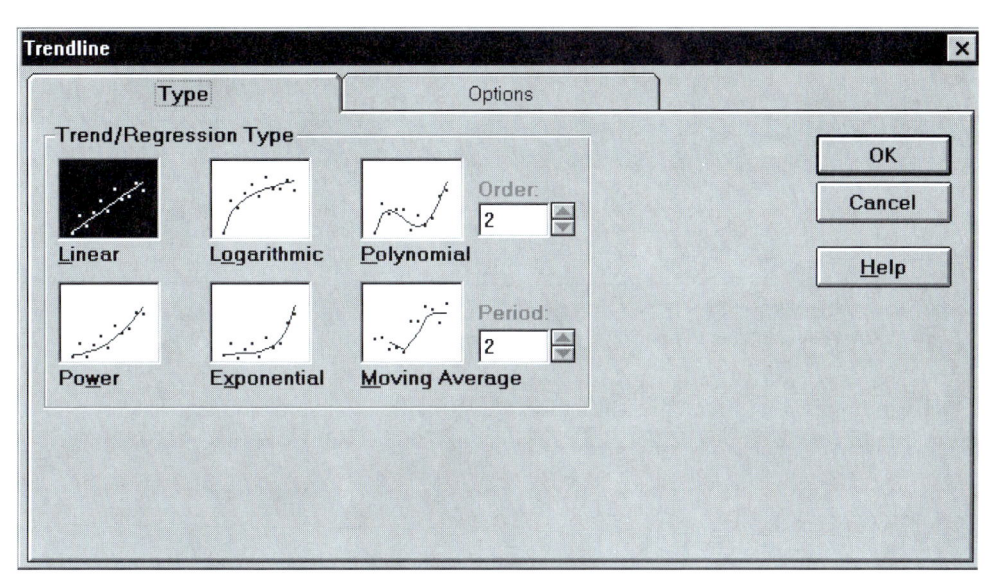

FIGURE 11.2.EXCEL Trendline Type Tab dialog box.

FIGURE 11.3.EXCEL Trendline Options Tab dialog box.

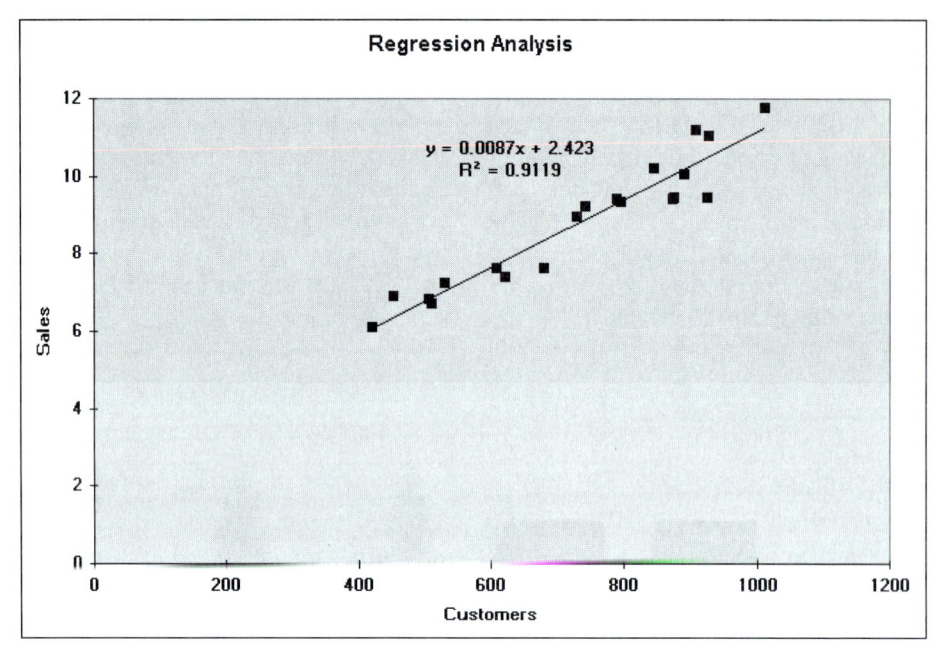

FIGURE 11.4.EXCEL Scatter diagram with regression line and r^2 obtained from Excel's Chart Wizard for the package delivery sales data.

From Figure 11.4.Excel, we can observe that the line of regression that has been fit to these data has a Y intercept of 2.423 and a slope of 0.0087, with an R^2 of 0.9119.

> ▲ **WHAT IF EXAMPLE**
>
> One of the advantages of using the Chart Wizard's TRENDLINE feature instead of the Data Analysis tool that will be discussed in Section 11.7 is that changes in the data will be immediately reflected in the fitted model. This gives us an opportunity to study the effect of changes in individual points on the fit of the regression model. Begin by removing the text of the regression equation and R^2 value by selecting them and issuing the command Edit | Clear | All. Then, for example, change the sales for the nineteenth observation to 14.77 from 11.77 on the Data sheet. The scatter diagram and fitted trend line on the Trend sheet will change as shown in Figure 11.5.Excel.
>
> We can find the new values for the Y intercept and the regression equation by selecting the plotted points and repeating the Insert | Trendline Command explained earlier in this section. We then can observe that what seemed to be a relatively small change in a single Y value has had a major impact on the fit of the regression model. The Y intercept changed from 2.423 to 1.5779, the slope changed from .0087 to .0101, and r^2 changed from .9119 to .8054.
>
> As before, if we were interested in seeing the effects of many different changes, we could use the Scenario Manager (see Section 1S.15) to store and use sets of alternative data values.

Now that we have fit the regression model to a set of data, we need to be able to predict the average value of Y for a given value of X as was explained in Section 11.4.1 on page 540.

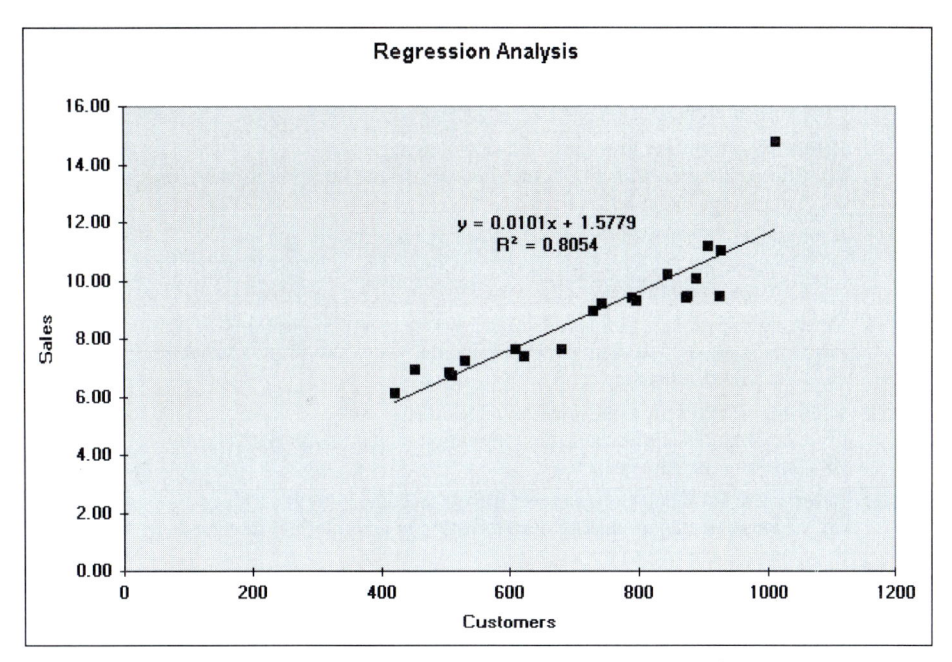

FIGURE 11.5.EXCEL Scatter diagram with regression line and r^2 obtained from Excel's Chart Wizard for the revised package delivery sales data.

This may be accomplished through the use of the Excel TREND function. The general format of the **TREND function** is

=TREND(*range of* Y *variable, range of* X *variable, value of* X)

Suppose we want to use Excel to obtain a prediction of the average sales when there are 600 customers, as we did in Section 11.4.1. With the workbook developed earlier in this section still open (or with the PACKAGE.XLS workbook open), we can obtain the prediction by using the simple Calculations sheet shown in Table 11.1.Excel. To implement, begin by inserting a new sheet and renaming it Calculations. Enter the number of customers in cell B3 and the formula =TREND(Data!C2:C21,Data!B2:B21,B3) in cell B4.

Figure 11.6.Excel presents the results for the data of Table 11.1.

Table 11.1.Excel Design of the Calculations sheet for the prediction of average sales.

	A	B
1	Prediction of the Average Sales	
2		
3	No. of Customers	xxx
4	Predicted Sales	=TREND(Data!C2:C21,Data!B2:B21,B3)

	A	B
1	Prediction of the Average Sales	
2		
3	No. of Customers	600
4	Predicted Sales	7.6606473

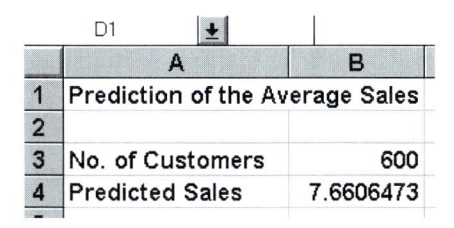

FIGURE 11.6.EXCEL Predicted value of Y using the TREND function for the package delivery data.

Problems for Section 11.6

Note: The problems in this section can be solved using Microsoft Excel (see Sections 11.6.3 and 11.7).

PETFOOD.TXT

- **11.16** Referring to Problem 11.6 (pet food sales) on page 544
  (a) Compute the coefficient of determination r^2 and interpret its meaning.
  (b) Compute the adjusted r^2.

S-SITE.XLS

11.17 Referring to Problem 11.7 (site selection) on page 544
  (a) Compute the coefficient of determination r^2 and interpret its meaning.
  (b) Compute the adjusted r^2.

WORKHRS.TXT

- **11.18** Referring to Problem 11.8 (production worker-hours) on page 544
  (a) Compute the coefficient of determination r^2 and interpret its meaning.
  (b) Compute the adjusted r^2.

TOMYIELD.TXT

11.19 Referring to Problem 11.9 (tomato yield) on page 544
  (a) Compute the coefficient of determination r^2 and interpret its meaning.
  (b) Compute the adjusted r^2.

LIMO.TXT

11.20 Referring to Problem 11.10 (airport travel) on page 544
  (a) Compute the coefficient of determination r^2 and interpret its meaning.
  (b) Compute the adjusted r^2.

11.21 When will the unexplained variation or error sum of squares be equal to 0?

11.22 When will the explained variation or sum of squares due to regression be equal to 0?

11.7 USING THE DATA ANALYSIS TOOL FOR REGRESSION

11-7.XLS

The Data Analysis Tool of Microsoft Excel can be used instead of the TREND function to obtain a more complete regression analysis. To access the **Regression option** of Data Analysis, open the PACKAGE.XLS workbook, containing the package delivery data of Table 11.1 on page 535 and click the Data sheet tab. Select Tools | Data Analysis, select Regression in the Analysis Tools list box, and click the OK button. In the Regression dialog box that appears, do the following:

(a) Enter C1:C21 in the Input *Y* Range edit box.

(b) Enter B1:B21 in the Input *X* Range edit box.

(c) Select the Labels check box.

(d) Select the Confidence Level check box and set the level to 95%. (This will provide a 95% confidence interval estimate for the regression coefficients, which will be studied in Section 11.14.)

(e) Select the New Worksheet Ply option button and enter Regression as the name.

(f) Select the Residuals, Standardized Residuals, and Residual Plots check boxes. This information will be studied in Section 11.10. Select the Line Fit Plots check box if you wish to obtain a scatter diagram with a fitted regression line, as was illustrated in Figure 11.4.Excel on page 552. The dialog box should now appear similar to the one illustrated in Figure 11.7.Excel. Click the OK button to have Excel perform the regression analysis.

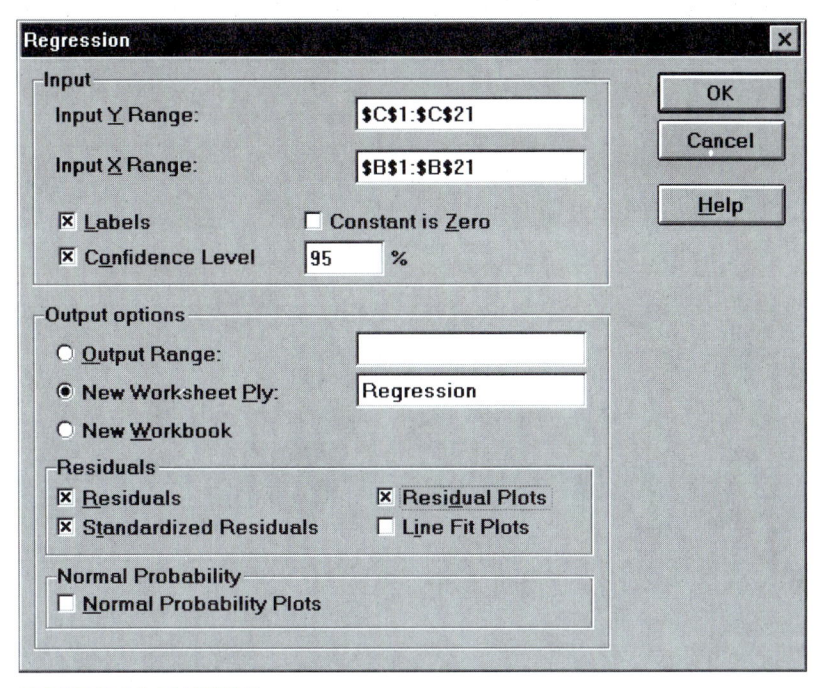

FIGURE 11.7.EXCEL Regression dialog box from Excel's Data Analysis tool.

Figure 11.8.Excel presents the results. Note that the Residuals plot will appear on the right-hand side of the sheet. (You may need to use the horizontal scroll bars in order to see the plot on the screen.)

The text portion of the Panel A output is divided into four areas. The first area, called regression statistics, provides the values of the coefficient of correlation r, the coefficient of determination r^2, the adjusted r^2, the standard error of the estimate (labeled Standard Error), and the sample size (labeled Observations). The second area presents output in a table called ANOVA, which will be discussed in detail in Chapter 12 when we cover multiple regression. However, if we examine the column labeled SS (for sum of squares), we can find the regression sum of squares (SSR) equal to 46.8335409, the residual or error sum of squares (SSE) equal to 4.5269541, and the total sum of squares (SST) equal to 51.360495. Any minor differences compared to those obtained in Section 11.6 are due to the fact that when we use calculator formulas, rounding errors can occur because the number of significant digits that can be used is limited.

The third area of the output provides information concerning the regression coefficients b_0 and b_1. The column labeled coefficients provides the values of the Y intercept b_0 and the slope b_1. The Y intercept is equal to 2.4230444, and the slope is equal to 0.00872934. Once again, any minor differences from those obtained in Section 11.4 are due to rounding errors. The remainder of the information in this portion of the output will be discussed in Section 11.14.

The fourth area of Panel A includes the predicted sales for each of the 20 data points along with values for the residuals and standardized residuals that will be discussed in Section 11.10. Panel B, containing a plot of the residuals and the X values, will also be discussed in Section 11.10.

	A	B	C	D	E	F
1	Summary Output					
2						
3	Regression Statistics					
4	Multiple R	0.9549132				
5	R Square	0.91185922				
6	Adjusted R Square	0.90696251				
7	Standard Error SYX	0.501495215				
8	Observations	20				
9						
10	ANOVA					
11		df	SS		MS	F	Significance F
12	Regression	1	SSR	46.8335409	46.8335409	186.22	6.20621E-11
13	Residual	18	SSE	4.526954104	0.25149745		
14	Total	19	SST	51.360495			
15							
16		Coefficients	Standard Error	t Stat	P-value		
17	Intercept b_0	2.423044396	0.480964609	5.037885009	8.5539E-05		
18	Customers b_1	0.008729338	0.00063969	13.64619912	6.2062E-11		
19							
20							
21							
22	Residual Output						
23							
24	Observation	Predicted Sales	Residuals	Standard Residuals			
25	1	10.34055412	0.859445883	1.713766867			
26	2	10.50641154	0.543588457	1.083935482			
27	3	6.840089511	-8.95107E-05	-0.000178488			
28	4	8.891483981	0.318516019	0.635132719			
29	5	9.310492213	0.109507787	0.218362576			
30	6	10.18342603	-0.10342603	-0.206235329			
31	7	10.05248596	-0.602485958	-1.201379276			
32	8	6.875006863	-0.145006863	-0.289149047			
33	9	7.040864289	0.199135711	0.397083971			
34	10	6.089366428	0.030633572	0.061084475			
35	11	8.350265014	-0.720265014	-1.43623507			
36	12	10.03502728	-0.605027281	-1.206446769			
37	13	10.48895287	-1.028952866	-2.05177006			
38	14	7.721752666	-0.081752666	-0.163017839			
39	15	6.368705249	0.551294751	1.099302116			
40	16	8.786731923	0.163268077	0.325562582			
41	17	9.354138904	-0.024138904	-0.048133867			
42	18	9.790605813	0.439394187	0.876168256			
43	19	11.23967595	0.530324051	1.057485766			
44	20	7.8439634	-0.4339634	-0.865339066			

FIGURE 11.8.EXCEL Regression analysis output obtained from Excel's Data Analysis tool for the package delivery sales data. Panel A.

Customers Residual Plot

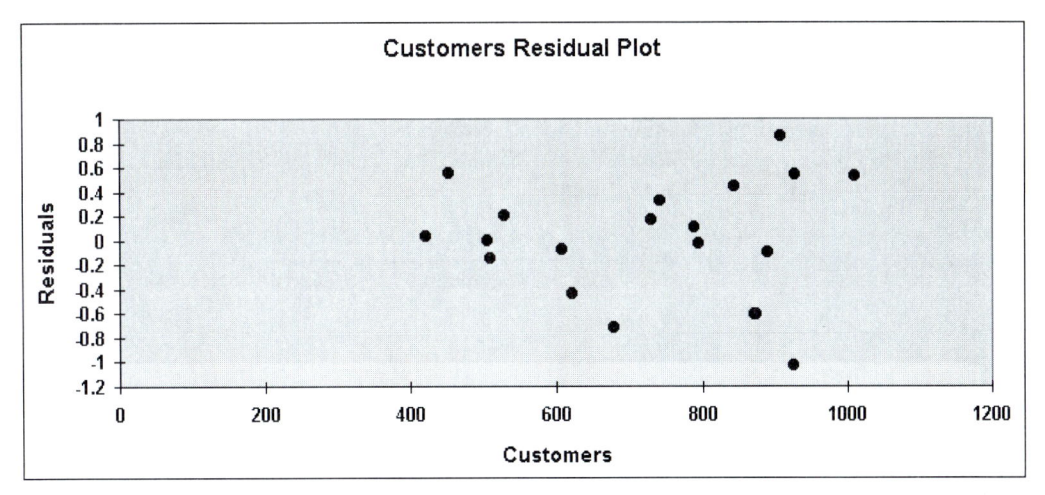

FIGURE 11.8.EXCEL Panel B.

11.8 CORRELATION—MEASURING THE STRENGTH OF THE ASSOCIATION

11.8.1 The Correlation Coefficient

In our discussion of the relationship between two variables thus far, we have been concerned with the prediction of the dependent variable Y based on the independent variable X. In contrast to a regression analysis, in a correlation analysis we are only interested in measuring the degree of association between two variables.

The strength of a relationship between two variables in a population is usually measured by the **coefficient of correlation** ρ, whose values range from -1 for perfect negative correlation up to $+1$ for perfect positive correlation. Figure 11.6 illustrates these three different types of association between variables. In Panel A of Figure 11.6, there is a perfect negative linear relationship between X and Y so that Y will decrease in a perfectly predictable manner as X

Panel A
Perfect negative correlation ($\rho = -1$)

Panel B
No correlation ($\rho = 0$)

Panel C
Perfect positive correlation ($\rho = +1$)

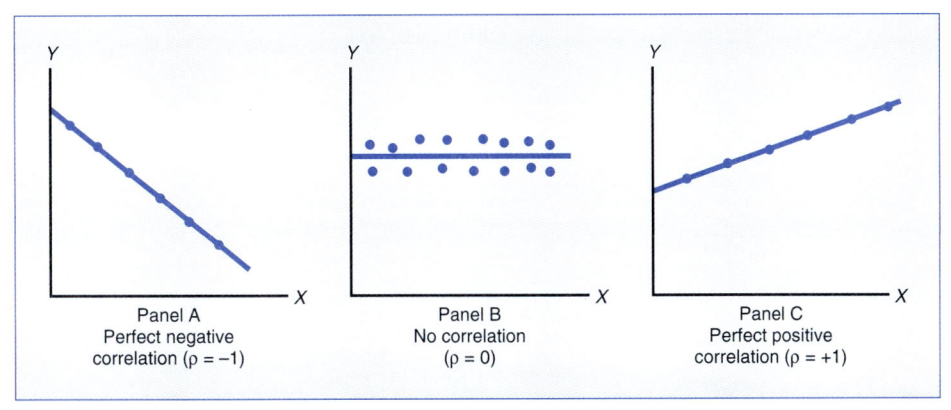

FIGURE 11.6 Types of association between variables.

increases. Panel B is an example in which there is no relationship between X and Y. As X increases, there is no change in Y, so there is no association between the values of X and the values of Y. Panel C depicts a perfect positive correlation between X and Y. In this case, Y increases in a perfectly predictable manner as X increases.

For regression-oriented problems, the sample coefficient of correlation (r) may be obtained from Equation (11.11a) as follows:

$$r^2 = \frac{\text{regression sum of squares}}{\text{total sum of squares}} = \frac{\text{SSR}}{\text{SST}}$$

so that

$$r = \sqrt{r^2} \qquad (11.12)$$

In simple linear regression, r takes the sign of b_1. If b_1 is positive, r is positive. If b_1 is negative, r is negative. If b_1 is 0, r is 0.

In the package delivery sales example, since $r^2 = .913$ and the slope b_1 is positive, the coefficient of correlation is computed as +.956. The closeness of the correlation coefficient to +1.0 implies a strong association between number of customers and weekly sales.

We have now computed and interpreted the correlation coefficient in terms of its regression viewpoint. As we mentioned at the beginning of this chapter, however, regression and correlation are two separate techniques, with regression being concerned with prediction and correlation with association. In many applications, we are concerned only with measuring association between variables, not with using one variable to predict another.

If only a correlation analysis is being performed on a set of data, the sample correlation coefficient r can be computed directly using the following formula,

$$r = \frac{\sum_{i=1}^{n}(X_i - \overline{X})(Y_i - \overline{Y})}{\sqrt{\sum_{i=1}^{n}(X_i - \overline{X})^2}\sqrt{\sum_{i=1}^{n}(Y_i - \overline{Y})^2}} \qquad (11.13a)$$

or, alternatively, using the "calculator" formula:

$$r = \frac{\sum_{i=1}^{n}X_iY_i - n\overline{X}\overline{Y}}{\sqrt{\sum_{i=1}^{n}X_i^2 - n\overline{X}^2}\sqrt{\sum_{i=1}^{n}Y_i^2 - n\overline{Y}^2}} \qquad (11.13b)$$

To illustrate such an example, suppose we want to measure the strength of the association in the price of two different grocery items in various cities throughout the world. The price of a six-pack of a brand-name cola soft drink and of 1 pound of chicken is determined at a supermarket located in a sample of nine different cities. The results are summarized in Table 11.3.

Table 11.3 Price (in $) of a six-pack of a brand-name cola soft drink and of 1 pound of chicken in a sample of nine cities.

City	Brand-Name Cola (X) (1 six-pack)	Chicken (Y) (1 pound)
Frankfurt	3.27	3.06
Hong Kong	2.22	2.34
London	2.28	2.27
Manila	3.04	1.51
Mexico City	2.33	1.87
New York	2.69	1.65
Paris	4.07	3.09
Sydney	2.78	2.36
Tokyo	5.97	4.85

For the data of Table 11.3, we compute the following values:

$$\sum_{i=1}^{n} X_i = 28.65 \quad \sum_{i=1}^{n} X_i^2 = 102.66 \quad \sum_{i=1}^{n} Y_i = 23.00$$

$$n = 9 \quad \sum_{i=1}^{n} Y_i^2 = 67.132 \quad \sum_{i=1}^{n} X_i Y_i = 81.854$$

From this, we obtain

$$\bar{X} = \frac{28.65}{9} = 3.183$$

$$\bar{Y} = \frac{23.00}{9} = 2.5556$$

so that from Equation (11.13b)

$$r = \frac{\sum_{i=1}^{n} X_i Y_i - n\bar{X}\bar{Y}}{\sqrt{\sum_{i=1}^{n} X_i^2 - n\bar{X}^2} \sqrt{\sum_{i=1}^{n} Y_i^2 - n\bar{Y}^2}}$$

$$= \frac{81.854 - 9(3.183)(2.5556)}{\sqrt{102.66 - 9(3.183)^2} \sqrt{67.132 - 9(2.5556)^2}}$$

$$= \frac{81.8540 - 73.2172}{\sqrt{11.4594} \sqrt{8.3522}}$$

$$= +.883$$

The coefficient of correlation, $r = +.883$, between the price of a brand-name cola soft drink and of chicken indicates a very strong association. A higher price of the cola is strongly associated with a higher price of chicken. In Section 11.14, we will use these sample results to determine whether there is any evidence of a significant association between these variables in the population.

11.8.2 Using the Microsoft Excel CORREL Function for Correlation Analysis

The most direct way to obtain the correlation coefficient between two variables with Excel is to use the CORREL function. The general format of the CORREL function is

=CORREL(*range of Y variable, range of X variable*)

The 11-8-2.XLS workbook illustrates the use of this function to obtain the correlation coefficient for the international grocery price data of Table 11.3 on page 559. Open this workbook and click the Calculations sheet tab. Note for this small example the data of Table 11.3 have been placed on the Calculations sheet and not on a separate Data sheet as they could have been. Cell B14 computes the correlation coefficient using the formula =CORREL(B4:B12,C4:C12) because the range of the Y variable is B4:B12 and the range of the X variable is C4:C12. Figure 11.9.Excel presents the correlation coefficient obtained.

	A	B	C
1	Correlation of Cola and Chicken		
2			
3	City	Cola	Chicken
4	Frankfurt	3.27	3.06
5	Hong Kong	2.22	2.34
6	London	2.28	2.27
7	Manila	3.04	1.51
8	Mexico City	2.33	1.87
9	New York	2.69	1.65
10	Paris	4.07	3.09
11	Sydney	2.78	2.36
12	Tokyo	5.97	4.85
13			
14	Correlation	0.8828512	

FIGURE 11.9.EXCEL Correlation coefficient obtained from Excel's CORREL function for the international grocery price data.

Problems for Section 11.8

Note: *The problems in this section can be solved using Microsoft Excel.*

11.23 Under what circumstances will the coefficient of correlation be negative?

• 11.24 Referring to Problem 11.16 (pet food sales) on page 554, compute the coefficient of correlation.

11.25 Referring to Problem 11.17 (site selection) on page 554, compute the coefficient of correlation.

• 11.26 Referring to Problem 11.18 (production worker-hours) on page 554, compute the coefficient of correlation.

11.27 Referring to Problem 11.19 (tomato yield) on page 554, compute the coefficient of correlation.

11.28 Referring to Problem 11.20 (airport travel) on page 554, compute the coefficient of correlation.

11.29 Suppose we also want to measure the strength of the association in the price (in dollars) of a six-pack of a brand-name cola soft drink and 100 tablets of a brand-name pain reliever worldwide. We again determine these prices in a nine-city sample of local supermarkets. The results are as follows:

City	Brand-Name Cola (six-pack)	Pain Reliever (100 tablets)
Frankfurt	3.27	17.22
Hong Kong	2.22	6.21
London	2.28	9.17
Manila	3.04	14.61
Mexico City	2.33	4.85
New York	2.69	6.09
Paris	4.07	13.08
Sydney	2.78	8.04
Tokyo	5.97	8.39

SUPERMKT.TXT

(a) Compute the coefficient of correlation r between the price of the cola soft drink and the pain reliever.
(b) Is the price of the cola soft drink more correlated with the price of chicken or with that of the pain reliever? Explain.

11.30 Suppose we also want to measure the strength of the association in the price (in $) of a women's haircut and a men's brand-name dress shirt in a sample of nine different international cities. The results are as follows:

CITY.TXT

City	Women's Haircut	Men's Dress Shirt
Frankfurt	29.85	49.41
Hong Kong	22.56	29.32
London	33.79	42.12
Manila	12.04	35.22
Mexico City	15.49	25.04
New York	34.87	37.85
Paris	27.73	55.28
Sydney	25.64	38.58
Tokyo	27.45	38.69

Compute the coefficient of correlation r between the women's haircut and the men's dress shirt.

11.9 ASSUMPTIONS OF REGRESSION AND CORRELATION

In our investigations into hypothesis testing and the analysis of variance, we have noted that the appropriate application of a particular statistical procedure is dependent on how well a set of assumptions for that procedure are met. The assumptions necessary for regression and correlation analysis are analogous to those of the analysis of variance, since they fall under the general heading of linear models (Reference 12). Although there are some differences in the assumptions made by the regression model and by correlation (see Reference 12), this topic is beyond the scope of this text and we will consider only the former.

The four major **assumptions of regression** are

1. Normality
2. Homoscedasticity

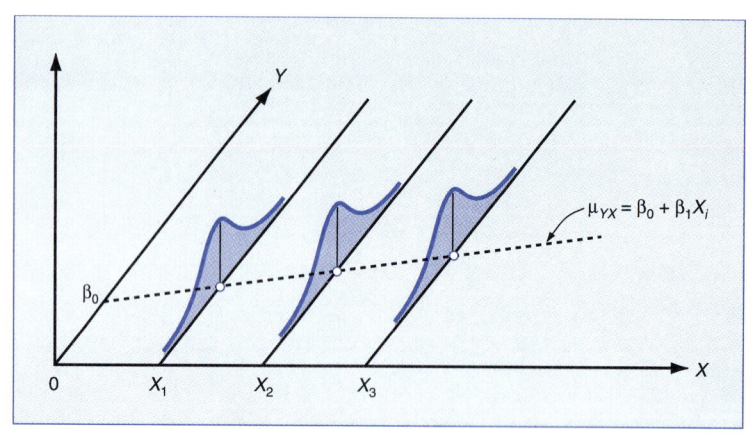

FIGURE 11.7 Assumptions of regression.

3. Independence of errors
4. Linearity

The first assumption, **normality**, requires that the values of Y be normally distributed at each value of X (see Figure 11.7). Like the t test and the ANOVA F test, regression analysis is fairly robust against departures from the normality assumption. As long as the distribution of Y_i values around each level of X is not extremely different from a normal distribution, inferences about the line of regression and the regression coefficients will not be seriously affected.

The second assumption, **homoscedasticity**, requires that the variation around the line of regression be constant for all values of X. This means that Y varies the same amount when X is a low value as when X is a high value (see Figure 11.7). The homoscedasticity assumption is important for using the least squares method of determining the regression coefficients. If there are serious departures from this assumption, either data transformations or weighted least squares methods (Reference 12) can be applied.

The third assumption, **independence of errors**, requires that the error (residual difference between observed and predicted values of Y) should be independent for each value of X. This assumption often refers to data that are collected over a period of time. When data are collected in this manner, the residuals for a particular time period are often correlated with those of the previous time period.

The fourth assumption, **linearity**, states that the relationship among variables is linear. Two variables could be perfectly related in a nonlinear fashion, and the linear correlation coefficient would be 0, indicating no relationship. Such nonlinear models will be discussed in Sections 12.10 and 12.12.

11.10 RESIDUAL ANALYSIS

11.10.1 Introduction

In the preceding discussion of the package delivery sales data, we have relied on a simple regression model in which the dependent variable is predicted based on a straight-line relationship with a single independent variable. In this section we shall use a graphical approach called **residual analysis** to evaluate the appropriateness of the regression model that has been fitted to the data. In addition, this approach will also allow us to study potential violations in the assumptions of our regression model (see Section 11.9).

11.10.2 Evaluating the Aptness of the Fitted Model

The **residual** or estimated error values (e_i) are defined as the difference between the observed (Y_i) and predicted ($\hat{Y}_i$) values of the dependent variable for given values X_i. Thus, the following definition applies:

> The residual equals the observed value of Y minus the predicted value of Y.
>
> $$e_i = Y_i - \hat{Y}_i \qquad (11.14)$$

We may evaluate the aptness of the fitted regression model by plotting the residuals on the vertical axis against the corresponding X_i values of the independent variable on the horizontal axis. If the fitted model is appropriate for the data, *there will be no apparent pattern in this plot of the residuals versus X_i*. However, if the fitted model is not appropriate, there will be a relationship between the X_i values and the residuals e_i. Such a pattern can be observed in Figure 11.8. Figure 11.8(a) depicts a situation in which there is a significant simple linear relationship between X and Y. However, a curvilinear model between the two variables seems more appropriate. This effect is highlighted in Figure 11.8(b), the residual plot of e_i versus X_i. In (b) there is a clear curvilinear effect between X_i and e_i. By plotting the residuals, we have essentially filtered out or removed the *linear* trend of X with Y, thereby exposing the lack of fit in the simple linear model. Thus, from (a) and (b), we can conclude that the curvilinear model is a better fit and should be evaluated in place of the simple linear model (see Section 12.10 for further discussion of fitting curvilinear models).

Having considered Figure 11.8, let us return to the evaluation of the package delivery sales data. Table 11.4 on page 564 lists the observed, predicted, and residual values of the response variable (weekly sales) in the simple linear model we have fitted. In addition to the residuals, we can also compute the standardized residuals and the Studentized residuals. The **standardized residuals** represent each residual divided by its standard error. The **Studentized residuals**, expressed as Equation (11.15), are the standardized residuals adjusted for the distance from the average X value.

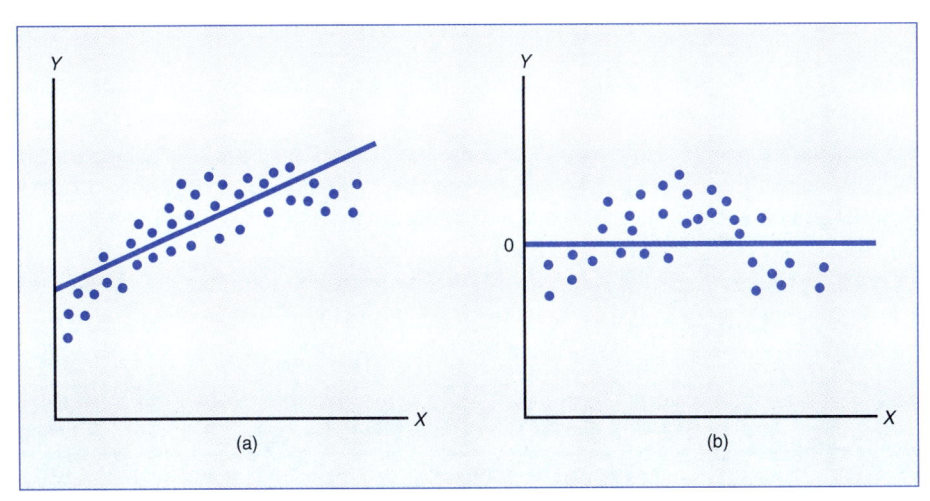

FIGURE 11.8 Studying the appropriateness of the simple linear regression model.

Table 11.4 Observed, predicted, and residual values for the package delivery sales data.

		Weekly Sales			
Observation	Customers X_i	Observed	Predicted	Residual	Studentized Residual, SR_i
1	907	11.200	10.341	0.859	1.81
2	926	11.050	10.506	0.544	1.15
3	506	6.840	6.840	−0.000	−0.00
4	741	9.210	8.891	0.319	0.65
5	789	9.420	9.310	0.110	0.22
6	889	10.080	10.183	−0.103	−0.22
7	874	9.450	10.052	−0.602	−1.25
8	510	6.730	6.875	−0.145	−0.31
9	529	7.240	7.041	0.199	0.42
10	420	6.120	6.089	0.031	0.07
11	679	7.630	8.350	−0.720	−1.48
12	872	9.430	10.035	−0.605	−1.26
13	924	9.460	10.489	−1.029	−2.18
14	607	7.640	7.722	−0.082	−0.17
15	452	6.920	6.369	0.551	1.21
16	729	8.950	8.787	0.163	0.33
17	794	9.330	9.354	−0.024	−0.05
18	844	10.230	9.791	0.439	0.91
19	1,010	11.770	11.240	0.530	1.17
20	621	7.410	7.844	−0.434	−0.90

$$\text{Studentized Residual} = SR_i = \frac{e_i}{S_{YX}\sqrt{1-h_i}} \quad (11.15)$$

where

$$h_i = \frac{1}{n} + \frac{(X_i - \overline{X})^2}{\sum_{i=1}^{n}(X_i - \overline{X})^2}$$

These Studentized residuals allow us to consider the magnitude of the residuals in units that reflect the standardized variation around the line of regression. The Studentized residuals have been plotted against the independent variable (number of customers) in Figure 11.9. From this we may observe that although there is widespread scatter in the residual plot, there is no apparent pattern or relationship between the Studentized residuals and X_i. The residuals appear to be evenly spread above and below 0 for the differing values of X. Thus, we may conclude for the package delivery sales data that the fitted model appears to be appropriate.

11.10.3 Evaluating the Assumptions

● **Homoscedasticity** The assumption of homoscedasticity (see Section 11.9) can also be evaluated from a plot of SR_i with X_i. For the package delivery sales data, there do not appear to be major differences in the variability of SR_i for different X_i values as is the case in

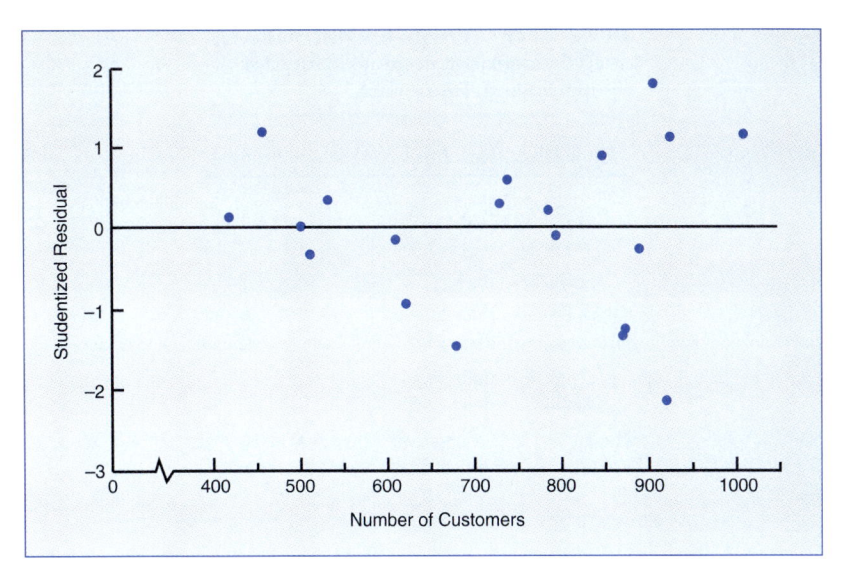

FIGURE 11.9 Plotting the Studentized residuals versus the number of customers in the package delivery sales example.

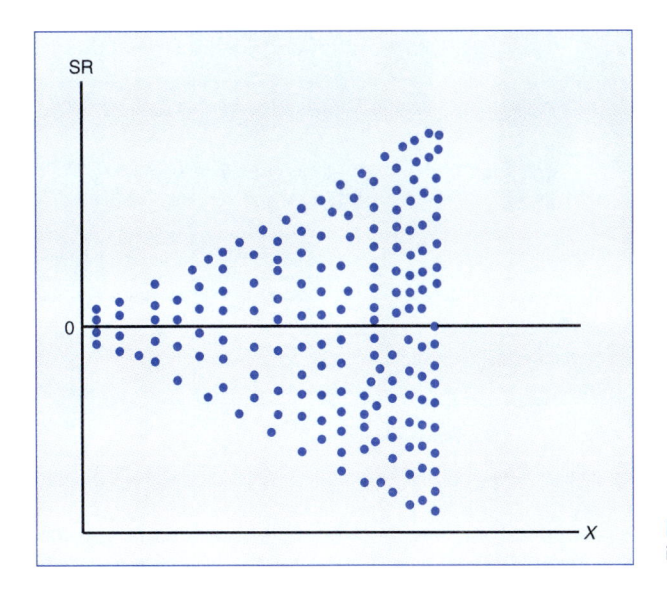

FIGURE 11.10 Violations in homoscedasticity.

Figure 11.9. Thus, we may conclude that for our fitted model there is no apparent violation in the assumption of equal variance at each level of X.

If we wish to observe a case in which the homoscedasticity assumption is violated, we should examine the *hypothetical* plot of SR_i with X_i in Figure 11.10. In this hypothetical plot, there appears to be a *fanning effect* in which the variability of the residuals increases as X increases, demonstrating the lack of homogeneity in the variances of Y_i at each level of X.

• **Normality** The normality assumption of regression (see Section 11.9) can also be evaluated from a residual analysis by tallying the Studentized residuals into a frequency distribution and displaying the results in a histogram (see Chapter 2).

For the package delivery sales data, the Studentized residuals have been tallied into a frequency distribution as indicated in Table 11.5 with the results displayed in Figure 11.11. It is

Table 11.5 Frequency distribution of 20 Studentized residual values for the package delivery data.

Studentized Residuals	Number
−2.8 but less than −2.0	1
−2.0 but less than −1.2	3
−1.2 but less than −0.4	2
−0.4 but less than +0.4	8
+0.4 but less than +1.2	4
+1.2 but less than +2.0	2
+2.0 but less than +2.8	0
Totals	20

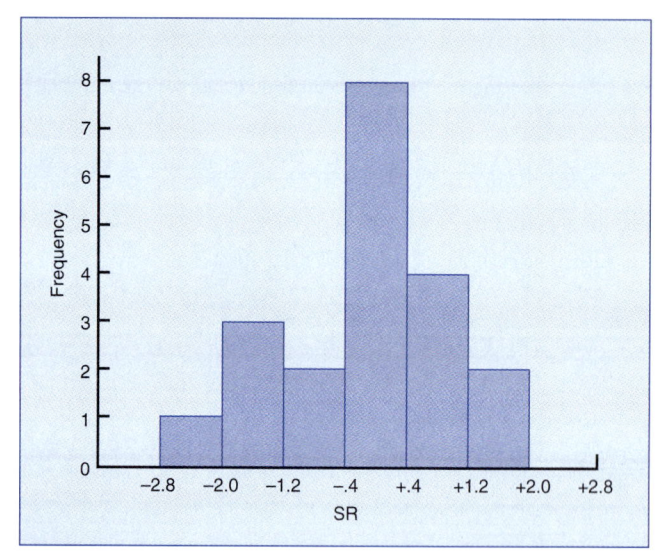

FIGURE 11.11 Plotting the Studentized residuals for the package delivery data.

difficult to evaluate the normality assumption for a sample of only 20 observations, and formal test procedures are beyond the scope of this text (see Reference 13). Although we could have also developed a normal probability plot (see Section 5.6), we can see from Figure 11.11 that the data appear to be approximately bell shaped. Thus, it seems reasonable to conclude that there is no overwhelming evidence of a violation of the normality assumption.

● **Independence** The independence assumption discussed in Section 11.9 can be evaluated by plotting the residuals in the order or sequence in which the observed data were obtained. Data collected over periods of time often exhibit an *autocorrelation* effect among successive observations. That is, there exists a correlation between a particular observation and the values that precede and succeed it. Such patterns, which violate the assumption of independence, are readily apparent in the plot of the residuals versus the time in which they were collected. This effect may be measured by the Durbin-Watson statistic, which will be the subject of Section 11.11.

11.10.4 Using Microsoft Excel for Residual Analysis

Now that we have used residual analysis to evaluate the aptness of the regression model and to test the assumptions of the model, we can determine how Excel can be accessed to obtain the pertinent charts. In Section 11.7, we used the regression option of the Data Analysis tool to obtain the output illustrated in Figure 11.8.Excel on page 556.

The residual plot provided by Excel is illustrated in Panel B of Figure 11.8.Excel. The residuals are plotted on the vertical axis, and the X variable, number of customers, is plotted on the X-axis. Figure 11.8.Excel is similar to Figure 11.9 on page 565 except that the residuals have been used in place of the Studentized residuals. We can also observe from Panel A that the residuals and standardized residuals for each observation are provided in the output.

Problems for Section 11.10

Note: *The problems in this section can be solved using Microsoft Excel.*

- 11.31 Referring to the pet food sales problem (pages 536, 544, and 546), perform a residual analysis on your results and determine the adequacy of the fit of the model.
- 11.32 Referring to the site selection problem (pages 536, 544, and 547), perform a residual analysis on your results and determine the adequacy of the fit of the model.
- 11.33 Referring to the production worker-hours problem (pages 537, 544, and 547), perform a residual analysis on your results and determine the adequacy of the fit of the model.
- 11.34 Referring to the tomato yield problem (pages 537, 544, and 547), perform a residual analysis on your results and determine the adequacy of the fit of the model.
- 11.35 Referring to the airport travel problem (pages 537, 544, and 547), perform a residual analysis on your results and determine the adequacy of the fit of the model.

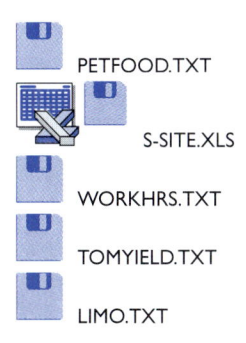

PETFOOD.TXT
S-SITE.XLS
WORKHRS.TXT
TOMYIELD.TXT
LIMO.TXT

11.11 MEASURING AUTOCORRELATION: THE DURBIN-WATSON STATISTIC

11.11.1 Introduction

One of the assumptions of the basic regression model we have been considering is the independence of the residuals. This assumption is often violated when data are collected over sequential periods of time, because a residual at any one point in time may tend to be similar to residuals at adjacent points in time. Thus, positive residuals would be more likely followed by positive residuals, and negative residuals would be more likely followed by negative residuals. Such a pattern in the residuals is called **autocorrelation**. When substantial autocorrelation is present in a set of data, the validity of a fitted regression model may be in serious doubt.

11.11.2 Residual Plots to Detect Autocorrelation

As mentioned in Section 11.10, the easiest way to detect autocorrelation in a set of data is to plot the residuals or standardized residuals in time order. If a positive autocorrelation effect is present, clusters of residuals with the same sign will be present, and an apparent pattern will be readily detected. To illustrate the autocorrelation effect, we shall consider the following example.

Recall that in Sections 11.2–11.4, we developed a regression model to predict weekly sales based on the number of customers for a sample of 20 package delivery stores. Suppose that the manager of the seventeenth package delivery store listed in Table 11.1 on page 535 wants to predict weekly sales based on the number of customers for a period of 15 weeks. In this situation, since data are collected over a period of 15 *consecutive* weeks at the *same* store, we would need to be concerned with the autocorrelation effect of the residuals. The data for this store are summarized in Table 11.6. Figure 11.12 represents partial Excel output.

	A	B	C	D	E	F
1	Summary Output					
2						
3	Regression Statistics					
4	Multiple R	0.810829997				
5	R Square	0.657445284				
6	Adjusted R Square	0.631094922				
7	Standard Error	0.936036681				
8	Observations	15				
9						
10	ANOVA					
11		df	SS	MS	F	Significance F
12	Regression	1	21.86043264	21.86043264	24.95014	0.000245105
13	Residual	13	11.39014069	0.876164669		
14	Total	14	33.25057333			
15						
16		Coefficients	Standard Error	t Stat	P-value	
17	Intercept	-16.0321936	5.310167093	-3.019150493	0.009869	
18	Customers	0.030760228	0.006158189	4.995011683	0.000245	
19						
20						
21						
22	Residual Output					
23						
24	Observation	Predicted Sales	Residuals	Standard Residuals		
25	1	8.39142731	0.93857269	1.002709305		
26	2	8.545228449	-0.285228449	-0.304719307		
27	3	9.714117107	-2.234117107	-2.386783715		
28	4	10.26780121	-1.187801208	-1.268968655		
29	5	9.96019893	-0.13019893	-0.13909597		
30	6	9.929438702	0.160561298	0.171533126		
31	7	10.51388303	0.496116969	0.530018725		
32	8	10.88300577	0.606994235	0.648472701		
33	9	11.0368069	1.033193095	1.10379552		
34	10	11.8058126	0.744187399	0.795040851		
35	11	11.22136827	0.698631728	0.746372169		
36	12	9.898678474	0.371321526	0.396695485		
37	13	11.77505237	0.024947627	0.026652403		
38	14	13.19002285	-1.040022854	-1.111091984		
39	15	9.837158019	-0.197158019	-0.210630654		

FIGURE 11.12 Excel output for single package delivery store.

568 Chapter 11 Simple Linear Regression and Correlation

Table 11.6 Customers and sales for a period of 15 consecutive weeks.

Week	Customers	Sales ($000)
1	794	9.33
2	799	8.26
3	837	7.48
4	855	9.08
5	845	9.83
6	844	10.09
7	863	11.01
8	875	11.49
9	880	12.07
10	905	12.55
11	886	11.92
12	843	10.27
13	904	11.80
14	950	12.15
15	841	9.64

We note from Figure 11.12 that r^2 is .657, indicating that 65.7% of the variation in sales can be explained by variation in the number of customers. In addition, the Y intercept, b_0, is −16.032, while the slope, b_1, is .03076. However, before we can accept the validity of this model, we must undertake proper analyses of the residuals. Since the data have been collected over a consecutive period of 15 weeks, the residuals should be plotted over time to see whether a pattern exists. Figure 11.13 represents such a plot for the 15-week sales data.

From Figure 11.13, we observe that the points tend to fluctuate up and down in a cyclical pattern. This cyclical pattern would give us strong cause for concern about the autocor-

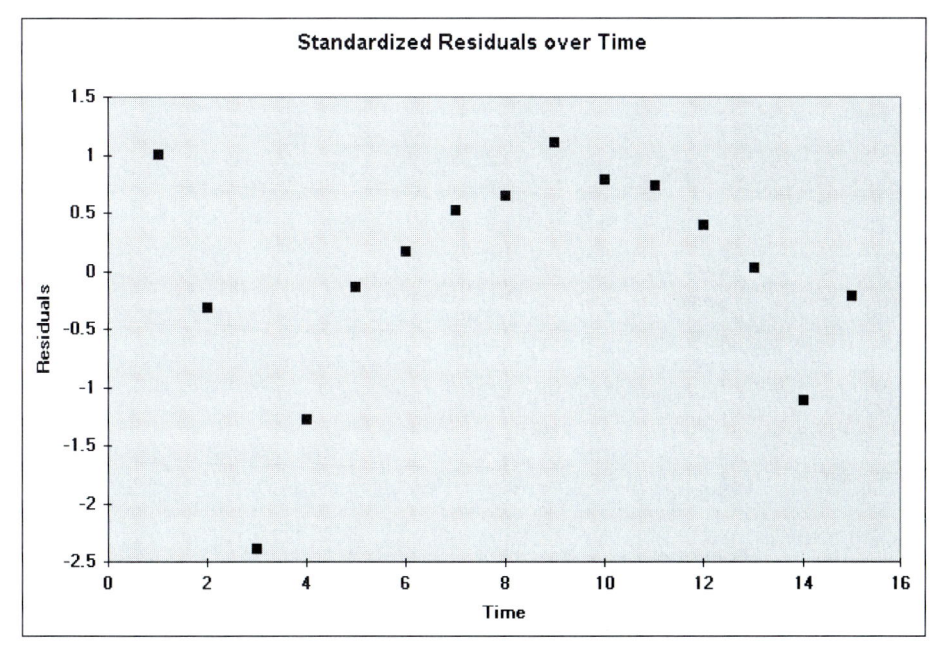

FIGURE 11.13 Excel plot of residuals over time for the 15-week sales data.

relation of the residuals and, hence, a violation in the assumption of independence of the residuals.

11.11.3 The Durbin-Watson Procedure

In addition to residual plots, autocorrelation can also be detected and measured by using the **Durbin-Watson statistic**. This statistic measures the correlation of each residual and the residual for the time period immediately preceding the one of interest. The Durbin-Watson statistic (D) is defined as follows:

$$D = \frac{\sum_{i=2}^{n}(e_i - e_{i-1})^2}{\sum_{i=1}^{n} e_i^2} \qquad (11.16)$$

where e_i = residual at the time period i

Although the computation of the Durbin-Watson statistic can be obtained using a Microsoft Excel formula, (see Figure 11.10.Excel on page 573), for illustrative purposes, the computations for the 15-week sales data are summarized in Table 11.7.

To better understand what the Durbin-Watson statistic is measuring, we need to examine the composition of the D statistic presented in Equation (11.16). The numerator $\sum_{i=2}^{n}(e_i - e_{i-1})^2$ represents the squared difference in two successive residuals, summed from the second observation to the nth observation. The denominator represents the sum of the

Table 11.7 Computation of the Durbin-Watson statistic for the regression analysis of the single package delivery store.

Week	Sales (Y_i)	$\hat{Y}_i$	$e_i = Y_i - \hat{Y}_i$	e_{i-1}	$(e_i - e_{i-1})$	$(e_i - e_{i-1})^2$	e_i^2
1	9.33	8.3914	0.93857	*	*	*	0.88092
2	8.26	8.5452	−0.28523	0.93857	−1.22380	1.49769	0.08136
3	7.48	9.7141	−2.23412	−0.28523	−1.94889	3.79817	4.99128
4	9.08	10.2678	−1.18780	−2.23412	1.04632	1.09478	1.41087
5	9.83	9.9602	−0.13020	−1.18780	1.05760	1.11852	0.01695
6	10.09	9.9294	0.16056	−0.13020	0.29076	0.08454	0.02578
7	11.01	10.5139	0.49612	0.16056	0.33556	0.11260	0.24613
8	11.49	10.8830	0.60699	0.49612	0.11088	0.01229	0.36844
9	12.07	11.0368	1.03319	0.60699	0.42620	0.18165	1.06749
10	12.55	11.8058	0.74419	1.03319	−0.28901	0.08352	0.55381
11	11.92	11.2214	0.69863	0.74419	−0.04556	0.00208	0.48809
12	10.27	9.8987	0.37132	0.69863	−0.32731	0.10713	0.13788
13	11.80	11.7751	0.02495	0.37132	−0.34637	0.11997	0.00062
14	12.15	13.1900	−1.04002	0.02495	−1.06497	1.13416	1.08165
15	9.64	9.8372	−0.19716	−1.04002	0.84286	0.71042	0.03887

$$\sum_{i=2}^{n}(e_i - e_{i-1})^2 = 10.058 \qquad \sum_{i=1}^{n} e_i^2 = 11.39$$

squared residuals, $\sum_{i=1}^{n} e_i^2$. When successive residuals are positively autocorrelated, the value of D will approach 0. If the residuals are not correlated, the value of D will be close to 2. (If there is negative autocorrelation, which rarely happens, D will be greater than 2 and could even approach its maximum value of 4.)

For the data in Table 11.7, we use Equation (11.16) and obtain

$$D = \frac{10.058}{11.39} = .883$$

The crux of the issue in using the Durbin-Watson statistic is the determination of when the autocorrelation is large enough to make the D statistic fall sufficiently below 2 to cause concern about the validity of the model. The answer to this question is dependent on the number of observations being analyzed and the number of independent variables in the model (in simple linear regression, $p = 1$). Table 11.8 has been extracted from Appendix E, Table E.9, the table of the Durbin-Watson statistic.

From Table 11.8, we observe that two values are shown in the table for each combination of level of significance (α), n (sample size), and p (the number of independent variables in the model). The first value, d_L, represents the lower critical value when there is no autocorrelation in the data. If D is below d_L, we may conclude that there is evidence of autocorrelation among the residuals. Under such a circumstance, the least squares methods that we have considered in this chapter are inappropriate and alternative methods need to be used (see References 5 and 12). The second value, d_U, represents the upper critical value of D above which we would conclude that there is no evidence of autocorrelation among the residuals. If D is between d_L and d_U, we are unable to make a definite conclusion.

Thus, as illustrated in Table 11.8, for our data concerning the single package delivery store, with one independent variable ($p = 1$), and 15 observations ($n = 15$), $d_L = 1.08$ and $d_U = 1.36$. Since $D = 0.883 < 1.08$, we may conclude that there is autocorrelation among the residuals. Our analysis of the data of Figure 11.12, using the least squares method was inappropriate due to the presence of serious autocorrelation among the residuals. We need to consider the alternative approaches discussed in Reference 12.

Table 11.8 Finding critical values of the Durbin-Watson statistic.

	\multicolumn{10}{c}{$\alpha = .05$}
	$p = 1$		$p = 2$		$p = 3$		$p = 4$		$p = 5$	
n	d_L	d_U	d_L	d_U	d_L	d_U	d_L	d_U	d_L	d_U
15	1.08	1.36	.95	1.54	.82	1.75	.69	1.97	.56	2.21
16	1.10	1.37	.98	1.54	.86	1.73	.74	1.93	.62	2.15
17	1.13	1.38	1.02	1.54	.90	1.71	.78	1.90	.67	2.10
18	1.16	1.39	1.05	1.53	.93	1.69	.82.	1.87	.71	2.06
19	1.18	1.40	1.08	1.53	.97	1.68	.86	1.85	.75	2.02
⋮	⋮	⋮	⋮	⋮	⋮	⋮	⋮	⋮	⋮	⋮
90	1.63	1.68	1.61	1.70	1.59	1.73	1.57	1.75	1.54	1.78
95	1.64	1.69	1.62	1.71	1.60	1.73	1.58	1.75	1.56	1.78
100	1.65	1.69	1.63	1.72	1.61	1.74	1.59	1.76	1.57	1.78

Note: n = number of observations; p = number of independent variables.
Source: Table E.9.

11.11.4 Using Microsoft Excel to Study Autocorrelation

11-11-4.XLS

We have used two approaches in this section to evaluate the autocorrelation of data collected over time—residual plots and the Durbin-Watson statistic. To use Microsoft Excel for these approaches, we would first use the Data Analysis regression tool, discussed in Section 11.7, to produce the residuals required for these analyses.

To generate a residual plot, the Chart Wizard could be used to produce a *XY* Plot of the residuals over time (the observation column), following a procedure similar to the one described in Section 11.6.3. This will produce a plot like the one illustrated in Figure 11.3.

To generate the Durbin-Watson statistic requires designing a sheet such as the Calculations sheet in Table 11.3.Excel that contains formulas that use the SUMXMY2 and the SUMSQ Excel functions.

The SUMXMY2 function computes the sum of the squared differences between a set of *X* and *Y* values $\sum_{i=1}^{n}(X_i - Y_i)^2$. The format of the function is

$$=\text{SUMXMY2}(range\ of\ X\ variable,\ range\ of\ Y\ variable)$$

The SUMSQ function computes the sum of the squared values of a variable, $\sum_{i=1}^{n} X_i^2$. The format of the function is

$$=\text{SUMSQ}(range\ of\ the\ X\ variable)$$

Examining Equation (11.16) on page 570 for the Durbin-Watson statistic, we can note that the numerator is the sum of the squared differences between the successive residuals e_i and e_{i-1}, while the denominator is the sum of the squared residuals e_i^2.

To implement this sheet, open the 11-11-4.XLS workbook, that contains the data of Table 11.6 on page 569 and a Regression sheet that contains the output produced by using the Data Analysis regression tool. Insert a new sheet and rename it Calculations. For this example, copy the residuals from the range C25:C39 on the Regression sheet to the range A4:A18 on the Calculations sheet.

Then enter the formula =SUMXMY2(A5:A18,A4:A17) in cell D3 to compute the sum of the squared differences between the successive residuals e_i and e_{i-1}. Next enter the formula =SUMSQ(A4:A18) in cell D4 to compute the sum of the squared residuals e_i^2. Finally, enter the formula =D3/D4 in cell D5 to compute the Durbin-Watson statistic. The results obtained for the data of Table 11.6 are displayed in Figure 11.10.Excel on page 573.

Table 11.3.Excel Design of the Calculations sheet for computing the Durbin-Watson statistic.

	A	B	C	D
1	Calculations for the Durbin-Watson Statistic			
2				
3	Residuals		Squared Difference of Residuals	=SUMXMY2(A5:A18,A4:A17)
4	x.xxx		Squared Residuals	=SUMSQ(A4:A18)
5	x.xxx		Durbin-Watson Statistic	=D3/D4
	.			
	.			
18	x.xxx			

	A	B	C	D
1	Calculations for the Durbin-Watson Statistic			
2				
3	Residuals		Squared Difference of Residuals	10.05752
4	0.938573		Squared Residuals	11.39014
5	-0.28523		Durbin-Watson Statistic	0.883003
6	-2.23412			
7	-1.1878			
8	-0.1302			
9	0.160561			
10	0.496117			
11	0.606994			
12	1.033193			
13	0.744187			
14	0.698632			
15	0.371322			
16	0.024948			
17	-1.04002			
18	-0.19716			

FIGURE 11.10.EXCEL Using Microsoft Excel to calculate the Durbin-Watson statistic for the sales data.

Problems for Section 11.10

11.36 Under what circumstances would it be important to compute the Durbin-Watson statistic? Explain.

11.37 Referring to Problem 11.1 (pet food sales) on page 536, is it necessary to compute the Durbin-Watson statistic? Explain.

11.38 Referring to Problem 11.1, under what circumstances would it be necessary to obtain the Durbin-Watson statistic before proceeding with the least squares method of regression analysis?

PETFOOD.TXT

11.39 Suppose the residuals for a set of data collected over 10 consecutive time periods are as follows:

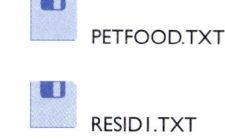

RESID1.TXT

Time Period	Residual
1	−5
2	−4
3	−3
4	−2
5	−1
6	+1
7	+2
8	+3
9	+4
10	+5

(a) Plot the residuals over time. What conclusions can you reach about the pattern of the residuals over time?
(b) Compute the Durbin-Watson statistic.
(c) Based on (a) and (b), what conclusion can you reach about the autocorrelation of the residuals?

RESID2.TXT

11.40 Suppose that the residuals for a set of data collected over 15 consecutive time periods are as follows:

Time Period	Residual
1	+4
2	−6
3	−1
4	−5
5	+2
6	+5
7	−2
8	+7
9	+6
10	−3
11	+1
12	+3
13	0
14	−4
15	−7

(a) Plot the residuals over time. What conclusions can you reach about the pattern of the residuals over time?
(b) Compute the Durbin-Watson statistic. At the .05 level of significance, is there evidence of positive autocorrelation among the residuals?
(c) Based on (a) and (b), what conclusion can you reach about the autocorrelation of the residuals?

11.12 CONFIDENCE INTERVAL ESTIMATE FOR PREDICTING μ_{YX}

11.12.1 Obtaining the Confidence Interval Estimate

In Sections 11.1–11.7, we were concerned with the use of regression and correlation solely for the purpose of description. The least squares method has been utilized to determine the regression coefficients and to predict the value of Y from a given value of X. In addition, the standard error of the estimate has been discussed along with the coefficients of correlation and determination.

Now that we have used residual analysis in Section 11.10 to assure ourselves that the assumptions of the least squares regression model have not been violated and that the straight-line model is appropriate, we may concern ourselves with making inferences about the relationship between the variables in a population based on our sample results. In this section we will discuss methods of making predictive inferences about the mean of Y, and in the following section we will predict an individual response value Y_I.

You may recall that in Section 11.4 the fitted regression equation was used to make predictions about the value of Y for a given X. In the package delivery sales example, we predicted that the average weekly sales for stores with 600 customers would be 7.661 (thousands of dollars). This estimate, however, is merely a point estimate of the population average value. In Chapter 6, we developed the concept of the confidence interval as an estimate of the population average. In a similar fashion, a **confidence interval estimate for the mean response** can now be developed to make inferences about the average predicted value of Y:

$$\hat{Y}_i \pm t_{n-2} S_{YX} \sqrt{h_i} \tag{11.17}$$

where

$$h_i = \frac{1}{n} + \frac{(X_i - \bar{X})^2}{\sum_{i=1}^{n}(X_i - \bar{X})^2}$$

$\hat{Y}_i$ = predicted value of Y; $\hat{Y}_i = b_0 + b_1 X_i$

S_{YX} = standard error of the estimate

n = sample size

X_i = given value of X

An examination of Equation (11.17) indicates that the width of the confidence interval is dependent on several factors. For a given level of confidence, increased variation around the line of regression, as measured by the standard error of the estimate, results in a wider interval. However, as would be expected, increased sample size reduces the width of the interval. In addition, the width of the interval also varies at different values of X. When predicting Y for values of X close to $\bar{X}$, the interval is much narrower than for predictions for X values more distant from the mean. This effect can be seen from the square root portion of Equation (11.17) and from Figure 11.14.

As displayed in Figure 11.14, the interval estimate of the true mean of Y varies *hyperbolically* as a function of the closeness of the given X to $\bar{X}$. When predictions are to be made for X values that are distant from the average value of X, the much wider interval is the trade-off for predicting at such values of X. Thus, as depicted in Figure 11.14, we observe a *confidence band effect* for the predictions.

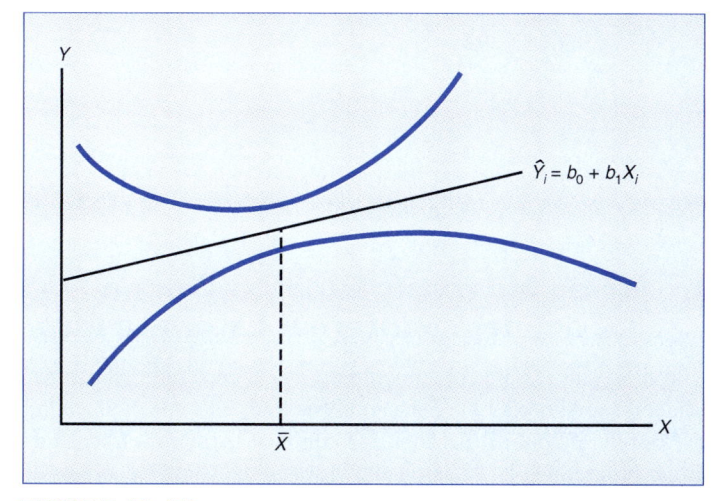

FIGURE 11.14 Interval estimates of μ_{YX} for different values of X.

Let us now use Equation (11.17) for our package delivery sales example. Suppose we desire a 95% confidence interval estimate of the true average weekly sales for all stores with 600 customers. We compute the following:

$$\hat{Y}_i = 2.423 + .00873\, X_i$$

and for $X_i = 600$, we obtain $\hat{Y}_i = 7.661$.

Also,

$$\overline{X} = 731.15 \qquad S_{YX} = .497$$

$$\sum_{i=1}^{n}(X_i - \overline{X})^2 = \sum_{i=1}^{n} X_i^2 - n\overline{X}^2 = 11{,}306{,}209 - 20(731.15)^2 = 614{,}603$$

From Table E.3, $t_{18} = 2.1009$. Thus,

$$\hat{Y}_i \pm t_{n-2}\, S_{YX}\sqrt{h_i}$$

where

$$h_i = \frac{1}{n} + \frac{(X_i - \overline{X})^2}{\sum_{i=1}^{n}(X_i - \overline{X})^2}$$

so that we have

$$\hat{Y}_i \pm t_{n-2}\, S_{YX}\sqrt{\frac{1}{n} + \frac{(X_i - \overline{X})^2}{\sum_{i=1}^{n}(X_i - \overline{X})^2}}$$

and

$$7.661 \pm (2.1009)(.497)\sqrt{\frac{1}{20} + \frac{(600 - 731.15)^2}{614{,}603}}$$

$$= 7.661 \pm (1.044)\sqrt{\frac{1}{20} + \frac{(-131.15)^2}{614{,}603}}$$

$$= 7.661 \pm (1.044)\sqrt{.078}$$

$$= 7.661 \pm .292$$

so

$$7.369 \le \mu_{YX} \le 7.953$$

Therefore, our estimate is that the average weekly sales is between 7.369 and 7.953 (thousand dollars) for stores with 600 customers.

11.12.2 Using Microsoft Excel to Obtain a Confidence Interval Estimate for μ_{YX}

11-12-2.XLS

In this section we have learned how to compute the confidence interval estimate for the average predicted value μ_{YX}. Although the Data Analysis tool does not produce this estimate, we can use results from the output of that tool and combine them with Excel formulas to compute it. Examining Equation (11.17) on page 575, recall that four quantities are required for this calculation, the predicted Y value $\hat{Y}_i$, the critical value of t, the standard error of the estimate S_{YX}, and the h_i statistic. Table 11.4.Excel presents the design for a Calculations sheet that

Table 11.4.Excel Design for the Calculations sheet to obtain confidence and prediction limits.

	A	B
1	Calculating the Confidence Interval Estimate	
2		
3	No. of Customers	xxx
4	No. of Observations	xx
5	Degrees of Freedom	=B4–2
6	t value	=TINV(0.05,B5)
7	Mean Number of Customers	=AVERAGE(Data!$B$2:$B$21)
8	Sum of Squared Differences	=SUM(Data!D2:D21)
9	Standard Error of the Estimate	xxx
10	h_i statistic	=1/B4+(B3–B7)^2/B8
11		
12	Predicted Sales	=TREND(Data!C2:C21,Data!B2:B21,B3)
13	Half-width	=B6*B9*SQRT(B10)
14	Lower Confidence Limit	=B12–B13
15	Upper Confidence Limit	=B12+B13
16	Half-width	=B6*B9*SQRT(1+B10)
17	Lower Prediction Limit	=B12–B16
18	Upper Prediction Limit	=B12+B16

computes confidence limits and prediction limits (discussed in Section 11.13.2) using the four quantities.

To implement rows 1 through 15 of this sheet, open the PACKAGE.XLS workbook, insert a new sheet and rename it Calculations. Enter the number of customers in cell B3, the number of observations in cell B4, and the standard error of the estimate in cell B9. The latter two values can be copied from the Regression Statistics table of the SUMMARY OUTPUT portion of the output produced by the Data Analysis regression tool. For the package delivery data, these values are 600, 20, and .5014952 respectively. Then do the following:

❶ Enter the formula =B4–2 to calculate the degrees of freedom in cell B5.

❷ Enter the formula =TINV(0.05,B5) to calculate the critical t statistic, as was done previously in Chapter 6, in cell B6.

❸ Calculate the average number of customers ($\bar{X}$) by entering the formula =AVERAGE(Data!$B$2:$B$21) in cell B7.

❹ Click on the Data sheet tab to make that sheet active. Calculate $(X_i - \bar{X})^2$ for each observation by entering the formula =(B2–Calculations!$B$7)^2 in cell D2 of that sheet and then copying the formula down to row 21.

❺ Click on the Calculations sheet tab to make this sheet active again. Enter the formula =SUM(Data!D2:D21) in cell B8 to sum the squared differences $\sum_{i=1}^{n}(X_i - \bar{X})^2$ entered in the Data sheet in the previous step.

❻ Enter the formula =1/B4+(B3–B7)^2 in cell B10 to calculate the h_i statistic.

❼ Enter the formula =TREND(Data!C2:C21,Data!B2:B21,B3) in cell B12 to compute the predicted sales as discussed in Section 11.6.3.

❽ Enter the formula =B6*B9*SQRT(B10) to compute the half-width of the confidence interval, equal to the $t\, S_{YX} \sqrt{h_i}$ term.

⑨ Enter the formula for the confidence limits, =B12–B13 and =B12+B13 in cells B14 and B15, respectively.

Figure 11.11.Excel shows the implemented Calculations sheet. (Section 11.13.2 discusses the implementation of rows 16-18 of this Calculations sheet.) This sheet computes the lower confidence limit to be 7.3664182 and the upper confidence limit to be 7.9548764. Differences between these results and the results obtained in Section 11.12.1 are due to rounding errors introduced into those calculations.

	A	B
1	Calculating the Confidence Interval Estimate	
2		
3	No. of Customers	600
4	No. of Observations	20
5	Degrees of Freedom	18
6	t value	2.1009237
7	Mean No. of Customers	731.15
8	Sum of Squared Difference	614602.55
9	Standard Error of the Estimate	0.5014952
10	Hi Statistic	0.0779861
11		
12	Predicted Sales	7.6606473
13	Half-width	0.2942291
14	Lower Confidence Limit	7.3664182
15	Upper Confidence Limit	7.9548764
16	Half-width	1.0939152
17	Lower Prediction Limit	6.5667321
18	Upper Prediction Limit	8.7545625

FIGURE 11.11.EXCEL
Confidence and prediction limits obtained from Excel for the package delivery sales data.

Problems for Section 11.12

Note: *The problems in this section can be solved using Microsoft Excel.*

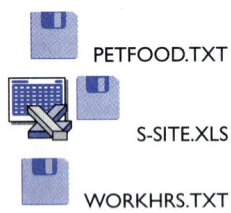

PETFOOD.TXT

S-SITE.XLS

WORKHRS.TXT

TOMYIELD.TXT

LIMO.TXT

• 11.41 Referring to the pet food sales problem (pages 536, 544, and 546), set up a 90% confidence interval estimate of the average weekly sales for all stores that have 8 feet of shelf space for pet food.

11.42 Referring to the site selection problem (pages 536, 544, and 547), set up a 95% confidence interval estimate of the average sales for stores with 4,000 square feet.

• 11.43 Referring to the production worker-hours problem (pages 537, 544, and 547), set up a 90% confidence interval estimate of the average worker hours for all production runs with a lot size of 45.

11.44 Referring to the tomato yield problem (pages 537, 544, and 547), set up a 90% confidence interval estimate of the average yield for all tomatoes that have been fertilized with 15 pounds per 100 square feet of natural organic fertilizer.

11.45 Referring to the airport travel problem (pages 537, 544, and 547), set up a 95% confidence interval estimate of the average travel time for all distances of 21 miles.

11.13 PREDICTION INTERVAL ESTIMATE FOR AN INDIVIDUAL RESPONSE Y_I

11.13.1 Obtaining the Prediction Interval Estimate

In addition to the need to obtain a confidence interval estimate for the average value, it is often important to be able to predict the response that would be obtained for an individual value. Although the form of the prediction interval estimate is similar to the confidence interval estimate of Equation (11.17), the prediction interval is estimating an individual value, not a parameter. Thus, the **prediction interval for an individual response** Y_I at a particular value X_i is provided in Equation (11.18).

$$\hat{Y}_i \pm t_{n-2}\, S_{YX}\sqrt{1 + h_i} \qquad (11.18)$$

where h_i, $\hat{Y}_i$, S_{YX}, n, and X_i are defined as in Equation (11.17) on page 575.

Suppose we desire a 95% prediction interval estimate of the weekly sales for an individual store with 600 customers. We compute the following:

$$\hat{Y}_i = 2.423 + .00873 X_i$$

and for $X_i = 600$, $\hat{Y}_i = 7.661$.

Also, from Section 11.12.1,

$$\bar{X} = 731.15, \qquad S_{YX} = .497, \qquad \text{and } h_i = .078$$

From Table E.3, $t_{18} = 2.1009$. Thus, from Equation (11.18)

$$\hat{Y}_i \pm t_{n-2}\, S_{YX}\sqrt{1 + h_i}$$

so that

$$7.661 \pm (2.1009)(.497)\sqrt{1 + .078}$$
$$= 7.661 \pm (1.044)\sqrt{1.078}$$
$$= 7.661 \pm 1.084$$

so

$$6.577 \leq Y_I \leq 8.745$$

Therefore, with 95% confidence, our estimate is that the weekly sales for an individual store that has 600 customers is between 6.577 and 8.745 (thousand dollars). We note that this prediction interval is much wider than the confidence interval estimate obtained in Section 11.12 for the average value.

11.13.2 Using Microsoft Excel to Obtain a Prediction Interval Estimate for Y_I

Now that we have developed the prediction interval estimate for the average predicted value Y_I, we can continue to implement the Calculations sheet presented in Table 11.4.Excel on page 577. Examining Equation (11.18) on page 577, we may observe that the only difference between this equation and Equation (11.17) for the confidence limits for μ_{YX} is that there is an additional value of 1 under the square root. Thus, to obtain the prediction limits, we can enter the formula =B6*B9*SQRT(1+B10) in cell B16 and the formulas =B12–B16 and =B12+B16 in cells B17 and B18, respectively.

The results are illustrated in Figure 11.11.Excel on page 578. Once again, any minor differences from the results obtained in Section 11.13.1 are due to rounding errors.

Problems for Section 11.13

Note: *The problems in this section can be solved using Microsoft Excel.*

- **11.46** Referring to Problem 11.41 (pet food sales) on page 578
  - (a) Set up a 90% prediction interval of the weekly sales of an individual store that has 8 feet of shelf space for pet food.
  - (b) Explain the difference in the results obtained in (a) and those in Problem 11.41.

PETFOOD.TXT

11.47 Referring to Problem 11.42 (site selection) on page 578
  - (a) Set up a 95% prediction interval of the sales of an individual store that has 4,000 square feet.
  - (b) Explain the difference in the results obtained in (a) and those in Problem 11.42.

S-SITE.XLS

- **11.48** Referring to Problem 11.43 (production worker-hours) on page 578, set up a 90% prediction interval of the number of worker-hours for a single lot size of 45.

WRKHRS.TXT

11.49 Referring to Problem 11.44 (tomato yield) on page 578, set up a 90% prediction interval of the yield of tomatoes for an individual plot that has been fertilized with 25 pounds per 100 square feet of natural organic fertilizer.

TOMYIELD.TXT

11.50 Referring to Problem 11.45 (airport travel) on page 578, set up a 95% prediction interval of the travel time for an individual trip of 21 miles.

LIMO.TXT

11.14 INFERENCES ABOUT THE POPULATION PARAMETERS IN REGRESSION AND CORRELATION

In the preceding two sections we used statistical inference to develop a confidence interval estimate for μ_{YX}, the true mean value of Y, and a prediction interval for Y_I, an individual observation. In this section, statistical inference will be used to draw conclusions about the population slope β_1 and the population correlation coefficient ρ.

We can determine whether a significant relationship between the variables X and Y exists by testing whether β_1 (the true slope) is equal to 0. If this hypothesis is rejected, one could conclude that there is evidence of a linear relationship. The null and alternative hypotheses could be stated as follows:

$$H_0: \beta_1 = 0 \quad \text{(There is no relationship.)}$$
$$H_1: \beta_1 \neq 0 \quad \text{(There is a relationship.)}$$

and the test statistic for this is given by

> The t statistic equals the difference between the sample slope and population slope divided by the standard error of the slope.
>
> $$t = \frac{b_1 - \beta_1}{S_{b_1}} \qquad (11.19)$$

where
$$S_{b_1} = \frac{S_{YX}}{\sqrt{\sum_{i=1}^{n} X_i^2 - n\overline{X}^2}}$$

and the test statistic t follows a t distribution with $n - 2$ degrees of freedom.

Returning to our package delivery sales example, let us now test whether the sample results enable us to conclude that a significant relationship between the number of customers and the weekly sales exists at the .05 level of significance. The results from Sections 11.4 and 11.5 gave the following information:

$$b_1 = +.00873 \quad n = 20 \quad S_{YX} = .497$$

$$\overline{X} = 731.15 \quad \sum_{i=1}^{n} X_i^2 = 11,306,209$$

Therefore, to test the existence of a relationship at the .05 level of significance, we have

$$S_{b_1} = \frac{S_{YX}}{\sqrt{\sum_{i=1}^{n} X_i^2 - n\overline{X}^2}}$$

$$= \frac{.497}{\sqrt{11,306,209 - 20(731.15)^2}} = \frac{.497}{\sqrt{614,603}} = .000634$$

and, under the null hypothesis, $\beta_1 = 0$ so that

$$t = \frac{b_1}{S_{b_1}}$$

$$= \frac{.00873}{.000634} = 13.77$$

Since $t = 13.77 > t_{18} = 2.1009$, we reject H_0. Hence, we can conclude that there is a significant linear relationship between average weekly sales and the number of customers (see Figure 11.15).

A second, equivalent method for testing the existence of a linear relationship between the variables is to set up a confidence interval estimate of β_1 and to determine whether the hypothesized value ($\beta_1 = 0$) is included in the interval. The confidence interval estimate of β_1 would be obtained by using the following formula:

$$b_1 \pm t_{n-2} S_{b_1} \tag{11.20}$$

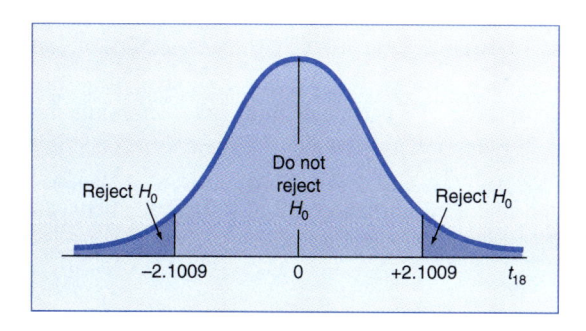

FIGURE 11.15 Testing a hypothesis about the population slope at the .05 level of significance with 18 degrees of freedom.

11.14 Inferences About the Population Parameters in Regression and Correlation

If there were a 95% confidence interval estimate desired here, we would have $b_1 = +.00873$, $t_{18} = 2.1009$, and $S_{b_1} = .000634$. Thus,

$$b_1 \pm t_{n-2} S_{b_1} = +.00873 \pm (2.1009)(.000634)$$
$$= +.00873 \pm .00133$$
$$+.0074 \le \beta_1 \le +.01006$$

From Equation (11.20), the true slope is estimated with 95% confidence to be between +.0074 and +.01006 (i.e., $7.40 to $10.06). Since these values are above 0, we can conclude that there is a significant linear relationship between weekly sales and number of customers. On the other hand, had the interval included 0, no relationship would have been determined.

A third method for examining the existence of a linear relationship between two variables involves the sample correlation coefficient r. The existence of a relationship between X and Y, which was tested using Equation (11.19), could be tested in terms of the correlation coefficient with equivalent results. Testing for the existence of a linear relationship between two variables is the same as determining whether there is any significant correlation between them. The population correlation coefficient ρ is hypothesized as equal to 0. Thus, the null and alternative hypotheses would be

$H_0: \rho = 0$ (There is no correlation.)

$H_1: \rho \ne 0$ (There is correlation.)

The test statistic for determining the existence of a significant correlation is given by

$$t = \frac{r - \rho}{\sqrt{\dfrac{1 - r^2}{n - 2}}} \qquad (11.21)$$

where the test statistic t follows a t distribution with $n - 2$ degrees of freedom.

To demonstrate that this statistic produces the same result as the test for the existence of a slope [Equation (11.19)], we will use the package delivery sales data. For these data, $r = +.956$, $r^2 = .913$, and $n = 20$, so testing the null hypothesis, we have

$$t = \frac{r}{\sqrt{\dfrac{1 - r^2}{n - 2}}}$$

$$= \frac{.956}{\sqrt{\dfrac{1 - .913}{20 - 2}}} = 13.75$$

We may note that this t value is, except for possible rounding error, the same as that obtained by using Equation (11.19). Therefore, in a linear regression analysis Equations (11.19) and (11.21) give equivalent alternative ways of determining the existence of a relationship between two variables. However, if the sole purpose of a particular study is to determine the existence of correlation, then Equation (11.21) is more appropriate. For instance, in Section 11.8, we studied the association of the price of a six-pack of a brand-name cola soft drink and the price of chicken. Had we wanted to determine the significance of the correlation between these two variables, we could have used Equation (11.21) as follows:

$$H_0: \rho = 0 \quad \text{(There is no correlation.)}$$
$$H_1: \rho \neq 0 \quad \text{(There is correlation.)}$$

If a level of significance of .05 was selected, we would have (see Figure 11.16)

$$t = \frac{r}{\sqrt{\dfrac{1-r^2}{n-2}}}$$

$$= \frac{.883}{\sqrt{\dfrac{1-(.883)^2}{9-2}}} = \frac{.883}{.1774} = +4.98$$

Since $t = 4.98 > t_7 = 2.3646$, we reject H_0. Since the null hypothesis has been rejected, we would conclude that there is evidence of an association between the price of the brand-name cola soft drink and the price of chicken.

When inferences concerning the population slope were discussed, confidence intervals and tests of hypothesis were used interchangeably. However, when examining the correlation coefficient, the development of a confidence interval becomes more complicated because the shape of the sampling distribution of the statistic r varies for different values of the true correlation coefficient. Methods for developing a confidence interval estimate for the correlation coefficient are presented in Reference 12.

Now that we have discussed the test of hypothesis and the confidence interval estimate for the slope, we can return to the Excel output of Figure 11.8.Excel on page 556. In the third section of the text output in Panel A, next to the Y intercept and the slope, is a column labeled Standard Error that provides the standard error of each of the regression coefficients. As indicated, the standard error of the slope is 0.00063969. In the next column, the t statistic for testing the slope equal to 13.6461991 is provided followed in the next column by the p-value of .000000000062062 (expressed in scientific notation as 6.2062E-11). In the last two columns, Microsoft Excel provides the lower and upper confidence interval estimates for the slope. Once again, any minor differences from the results obtained in Section 11.14 are due to rounding errors.

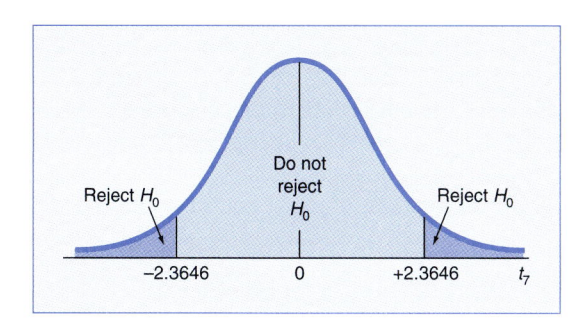

FIGURE 11.16 Testing for the existence of correlation at the .05 level of significance with 7 degrees of freedom.

Problems for Section 11.14

Note: *The problems in this section can be solved using Microsoft Excel.*

- **11.51** Referring to the pet food sales problem (pages 536, 544, and 546), at the .05 level of significance, is there evidence of a linear relationship between shelf space and sales?

  PETFOOD.TXT

- **11.52** Referring to the site selection problem (pages 536, 544, and 547), at the .05 level of significance, is there evidence of a linear relationship between annual sales and square footage?

  S-SITE.XLS

WORKHRS.TXT

TOMYIELD.TXT

LIMO.TXT

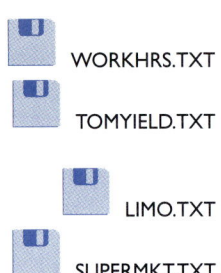

SUPERMKT.TXT

CITY.TXT

- **11.53** Referring to the production worker-hours problem (pages 537, 544, and 547), at the .01 level of significance, is there evidence of a linear relationship between lot size and worker-hours?
- **11.54** Referring to the tomato yield problem (pages 537, 544, and 547), at the .01 level of significance, is there evidence of a linear relationship between the amount of fertilizer used and the yield of tomatoes?
- **11.55** Referring to the airport travel problem (pages 537, 544, and 547), at the .05 level of significance, is there evidence of a relationship between distance and travel time?
- **11.56** Referring to Problem 11.29 on page 560, at the .01 level of significance, is there evidence of a linear relationship between the price of a six-pack of brand-name cola soft drink and 100 tablets of a pain reliever?
- **11.57** Referring to Problem 11.30 on page 561, at the .05 level of significance, is there evidence of a linear relationship between the cost of a women's haircut and a men's dress shirt?

11.15 PITFALLS IN REGRESSION AND ETHICAL ISSUES

11.15.1 Introduction

Regression and correlation analysis are perhaps the most widely used and, unfortunately, the most widely misused statistical techniques that are applied to business and economics. The difficulties frequently come from the following sources:

1. Lacking an awareness of the assumptions of least squares regression
2. Knowing how to evaluate the assumptions of least squares regression
3. Knowing what the alternatives to least squares regression are if a particular assumption is violated
4. Thinking that correlation implies causation
5. Using a regression model without knowledge of the subject matter

11.15.2 The Pitfalls of Regression

The widespread availability of spreadsheet and statistical software has removed the computational block that prevented many users from applying regression analysis to situations that required forecasting. With this positive development of availability comes the realization that, for many users, the access to powerful techniques has not been accompanied by an understanding of how to use regression analysis properly. How can a user be expected to know what the alternatives to least squares regression are if a particular assumption is violated, when he or she in many instances is not even aware of the assumptions of regression, let alone how the assumptions can be evaluated?

The necessity of going beyond the basic number crunching—of computing the Y intercept, the slope, and r^2—can be illustrated by referring to Table 11.9, a classical pedagogical piece of statistical literature that deals with the importance of observation through scatter plots and residual analysis.

Anscombe (Reference 2) showed that for the four data sets given in Table 11.9, the following results would be obtained:

$$\hat{Y}_i = 3.0 + .5X_i$$

$$S_{YX} = 1.236$$

$$S_{b_1} = .118$$

$$r^2 = .667$$

Table 11.9 Four sets of artificial data.

Data Set A		Data Set B		Data Set C		Data Set D	
X_i	Y_i	X_i	Y_i	X_i	Y_i	X_i	Y_i
10	8.04	10	9.14	10	7.46	8	6.58
14	9.96	14	8.10	14	8.84	8	5.76
5	5.68	5	4.74	5	5.73	8	7.71
8	6.95	8	8.14	8	6.77	8	8.84
9	8.81	9	8.77	9	7.11	8	8.47
12	10.84	12	9.13	12	8.15	8	7.04
4	4.26	4	3.10	4	5.39	8	5.25
7	4.82	7	7.26	7	6.42	19	12.50
11	8.33	11	9.26	1	7.81	8	5.56
13	7.58	13	8.74	13	12.74	8	7.91
6	7.24	6	6.13	6	6.08	8	6.89

Source: F. J. Anscombe, "Graphs in Statistical Analysis," *American Statistician*, Vol. 27 (1973), pp. 17–21.

$$\text{SSR} = \text{explained variation} = \sum_{i=1}^{n} (\hat{Y}_i - \overline{Y})^2 = 27.50$$

$$\text{SSE} = \text{unexplained variation} = \sum_{i=1}^{n} (Y_i - \hat{Y}_i)^2 = 13.75$$

$$\text{SST} = \text{total variation} = \sum_{i=1}^{n} (Y_i - \overline{Y})^2 = 41.25$$

Thus, with respect to the pertinent statistics associated with a simple linear regression, the four data sets are identical. Had we stopped our analysis at this point, valuable information in the data would be lost. Table 11.10 gives the standardized residuals e_i/S_{YX} for each data set.

Table 11.10 Standardized Residuals.

	Data Set A	Data Set B	Data Set C	Data Set D	
X_i	e_i/S_{YX}	e_i/S_{YX}	e_i/S_{YX}	X_i	e_i/S_{YX}
4	−.599	−1.536	.314	8	−.340
5	.145	−.614	.185	8	−1.003
6	1.002	.105	.064	8	.574
7	−1.359	.614	−.065	8	1.489
8	−.041	.922	−.186	8	1.189
9	1.059	1.027	−.315	8	.032
10	.032	.922	−.437	8	−1.416
11	−.138	.614	−.558	19	.000
12	1.487	.105	−.687	8	−1.165
13	−1.554	−.614	2.622	8	.736
14	−.033	−1.536	−.937	8	−.089

Source: F. J. Anscombe, "Graphs in Statistical Analysis," *American Statistician*, Vol. 27 (1973), pp. 17–21.

When the standardized residuals are plotted[2] against $\hat{Y}$, we see how different the data sets are. Panels A, B, C, and D of Figure 11.17 graphically depict, for each data set, a plot of the standardized residuals against the fitted values $\hat{Y}$. While the plot for data set A does not show any obvious anomalies, this is not the case for data sets B, C, and D. The parabolic form of the residual plot for B probably indicates that the basic simple linear regression model should also include a curvilinear term, as will be developed in Section 12.10. The plot for data set C clearly depicts what may very well be an *outlying* observation. If this is the case, we may deem it appropriate to remove the outlier and reestimate the basic model.[3] The result of this exercise would probably be a relationship much different from what was originally uncovered. Similarly, the plot for data set D would be evaluated cautiously because the fitted model is so dependent on the outcome of a single response ($X_8 = 19$ and $Y_8 = 12.50$).

In summary, residual plots are of vital importance to a complete regression analysis. The information they provide is so basic to a credible analysis that such plots should *always* be included as part of a regression analysis. Thus, a strategy that might be employed to avoid the first three pitfalls of regression listed would involve the following approach:

1. Always start with a scatter plot to observe the possible relationship between X and Y.
2. Check the assumptions of regression after the regression model has been fit, *before* moving on to using the results of the model.
3. Plot the residuals (or standardized or Studentized residuals) versus the independent variable. This will enable you to determine whether the model fit to the data is an appropriate one and will allow you to check visually for violations of the homoscedasticity assumption.

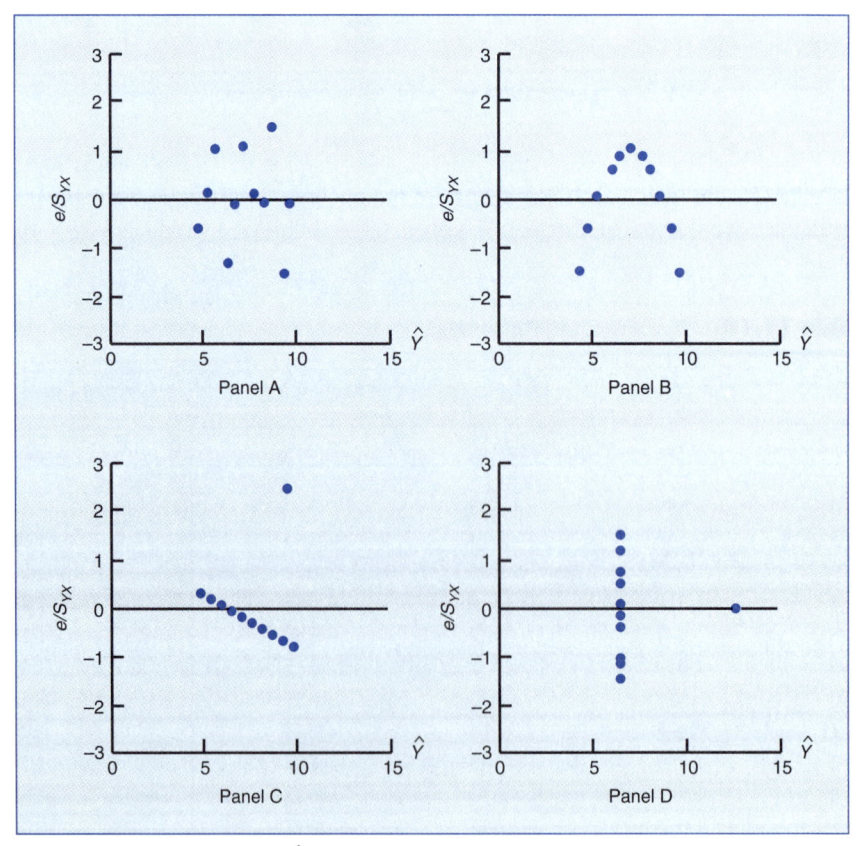

FIGURE 11.17 Plot of $\hat{Y}_i$ versus standardized residuals. *Source:* F. J. Anscombe, "Graphs in Statistical Analysis," *American Statistician*, Vol. 27 (1973), pp. 17–21.

4. Use a histogram, stem-and-leaf display, box-and-whisker plot, or normal probability plot of the residuals to evaluate graphically whether the normality assumption has been seriously violated.

5. If the data have been collected in sequential order, plot the residuals in time order and compute the Durbin-Watson statistic.

6. If the evaluation done in 3–5 indicates violations in the assumptions, use alternative methods to least squares regression or alternative least squares models (curvilinear or multiple regression), depending on what the evaluation has indicated.

7. If the evaluation done in 3–5 does not indicate violations in the assumptions, then the inferential aspects of the regression analysis can be undertaken. Confidence and prediction intervals can be developed, and tests for the significance of the regression coefficients can be done.

● **Cautions** In addition to the first three pitfalls considered above, two other pitfalls need to be mentioned. One involves the mistaken belief that correlation implies causation. In many instances, the covariation between variables is spurious in that the relationship is actually caused by a third factor that has not been or cannot be measured.

Another pitfall involves the fact that a good-fitting model does not necessarily mean that the model can be used for prediction. An individual with knowledge of the subject matter would have to be convinced that the process that produced the data will remain stable in the future in order to use the model for predictive purposes.

11.15.3 Ethical Considerations

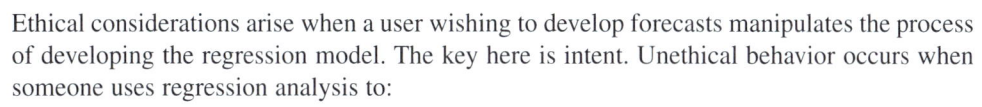

Ethical considerations arise when a user wishing to develop forecasts manipulates the process of developing the regression model. The key here is intent. Unethical behavior occurs when someone uses regression analysis to:

1. Forecast a response variable of interest with the willful intent of possibly excluding certain variables from consideration in the model.

2. Delete observations from the model to obtain a better model without giving reasons for deleting these observations.

3. Make forecasts without providing an evaluation of the assumptions when he or she knows that the assumptions of least squares regression have been violated.

All of these situations should make us realize even more the importance of following the steps given in Section 11.15.2 and knowing the assumptions of regression, how to evaluate them, and what to do if any of them have been violated.

11.16 SUMMARY AND OVERVIEW

As seen in the chapter summary chart, we developed the simple linear regression model, discussed the assumptions of the model, and showed how these assumptions could be evaluated. To be sure you understand what has been covered, you should be able to answer the following conceptual questions:

1. What is the interpretation of the Y intercept and the slope in a regression model?
2. What is the interpretation of the coefficient of determination?
3. Why should a residual analysis always be done as part of the development of a regression model?

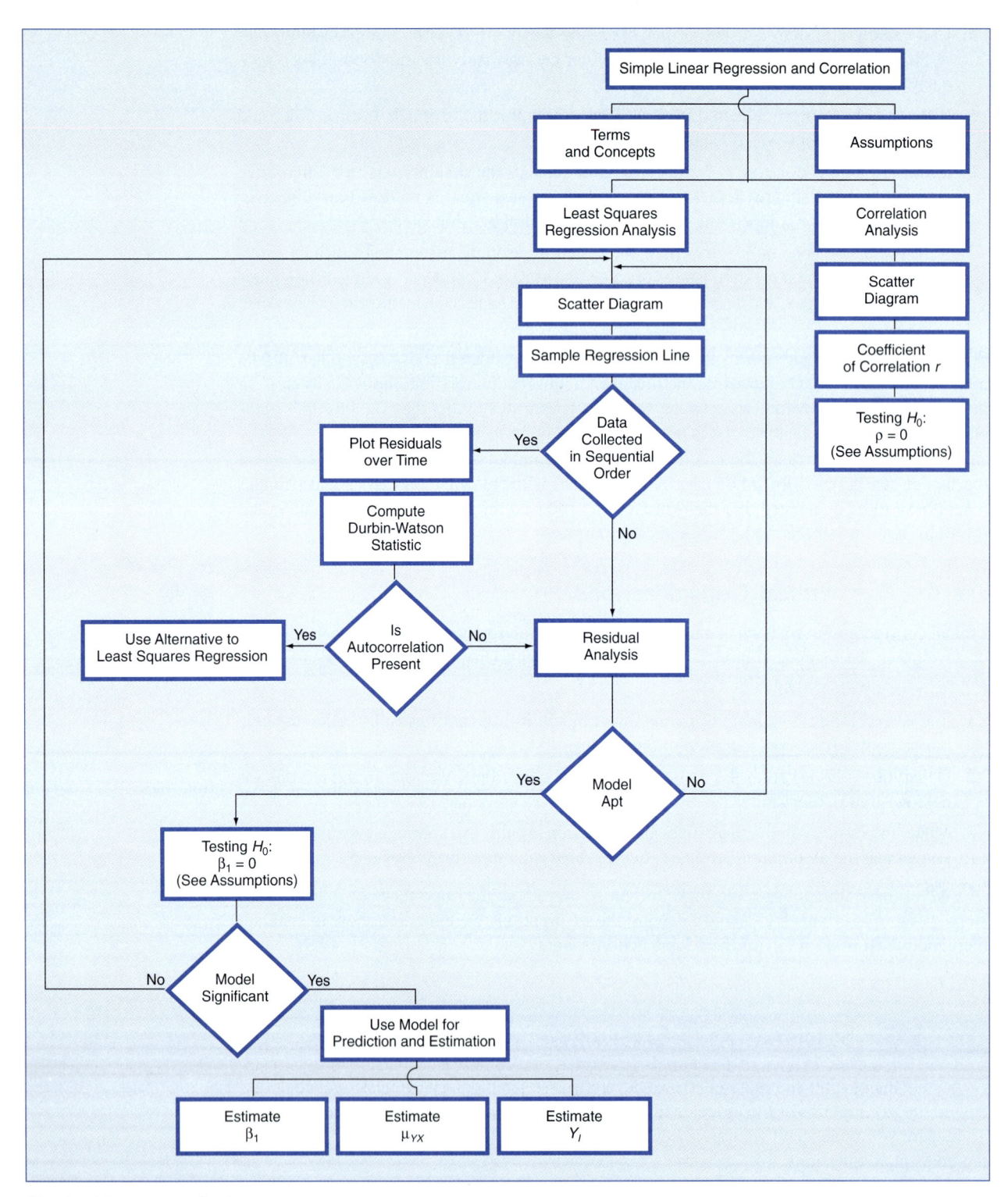

Chapter 11 summary chart.

588 **Chapter 11** Simple Linear Regression and Correlation

4. What are the assumptions of regression analysis and how can they be evaluated?
5. What is the Durbin-Watson statistic and when and how should it be used in regression analysis?
6. What is the difference between a confidence interval estimate of the mean response μ_{YX} and a prediction interval estimate of Y_I?

In Chapter 12, we will continue our discussion of regression analysis by considering a variety of multiple regression models.

Getting It All Together

Key Terms

adjusted r^2 549
assumptions of regression   561
autocorrelation   567
coefficient of correlation   557
coefficient of determination   549
confidence interval estimate for the mean response   575
correlation analysis   534
dependent variable   534
Durbin-Watson statistic   570
error sum of squares (SSE)   547
explained variation   547
explanatory variable   534
homoscedasticity   562
independence of error   562
independent variable   534
least squares method   541
linearity   562
linear relationship   534
normality   562

prediction interval for an individual response   579
regression analysis   534
regression coefficient   540
Regression option   554
regression sum of squares (SSR)   547
relevant range   543
residuals   563
residual analysis   562
response variable   534
scatter diagram   534
simple linear regression   541
slope   538
standard error of the estimate   545
standardized residuals   563
Studentized residuals   563
total sum of squares (SST)   547
TREND function   553
unexplained variation   547
Y intercept   539

Chapter Review Problems

Note: *The Chapter Review Problems can be solved using Microsoft Excel.*

- 11.58 A statistician for an American automobile manufacturer would like to develop a statistical model for predicting delivery time (the days between the ordering of the car and the actual delivery of the car) of custom-ordered new automobiles. The statistician believes there is a linear relationship between the number of options ordered on the car and delivery time. A random sample of 16 cars is selected with the results given at the top of page 590:

DELIVERY.TXT

Relating delivery time with options ordered

Car	Options Ordered, X	Delivery Time, Y (in days)	Car	Options Ordered, X	Delivery Time, Y (in days)
1	3	25	9	12	44
2	4	32	10	12	51
3	4	26	11	14	53
4	7	38	12	16	58
5	7	34	13	17	61
6	8	41	14	20	64
7	9	39	15	23	66
8	11	46	16	25	70

(a) Set up a scatter diagram.
(b) Use the least squares method to find the regression coefficients b_0 and b_1.
(c) Interpret the meaning of the Y intercept b_0 and the slope b_1 in this problem.
(d) If a car with 16 options is ordered, how many days would you predict it would take to be delivered?
(e) Compute the standard error of the estimate.
(f) Compute the coefficient of determination r^2 and interpret its meaning in this problem.
(g) Compute the adjusted r^2 and compare it with the coefficient of determination r^2.
(h) Compute the coefficient of correlation r.
(i) Set up a 95% confidence interval estimate of the average delivery time for all cars ordered with 16 options.
(j) Set up a 95% prediction interval estimate of the delivery time for an individual car that was ordered with 16 options.
(k) At the .05 level of significance, is there evidence of a linear relationship between number of options and delivery time?
(l) Set up a 95% confidence interval estimate of the population slope.
(m) Perform a residual analysis on your results and determine the adequacy of the fit of the model.
(n) What assumptions about the relationship between the number of options and delivery time would the statistician need to make to use this regression model for predictive purposes in the future?

BET.TXT

11.59 An official of a local racetrack would like to develop a model to forecast the amount of money bet (in millions of dollars) based on attendance. A random sample of 15 days is selected with the following results:

Relating betting with attendance.

Day	Attendance (000)	Amount Bet ($000,000)	Day	Attendance (000)	Amount Bet ($000,000)
1	14.5	0.70	9	16.3	0.71
2	21.2	0.83	10	32.1	1.04
3	11.6	0.62	11	27.6	0.97
4	31.7	1.10	12	34.8	1.13
5	46.8	1.27	13	29.3	0.91
6	31.4	1.02	14	19.2	0.68
7	40.0	1.15	15	16.3	0.63
8	21.0	0.80			

Hint: Determine which are the independent and dependent variables.
(a) Set up a scatter diagram.
(b) Assuming a linear relationship, use the least squares method to find the regression coefficients b_0 and b_1.
(c) Interpret the meaning of the slope b_1 in this problem.
(d) Predict the amount bet for a day on which attendance is 20,000.
(e) Compute the standard error of the estimate.
(f) Compute the coefficient of determination r^2 and interpret its meaning in this problem.
(g) Compute the coefficient of correlation r.
(h) Compute the Durbin-Watson statistic and, at the .05 level of significance, determine whether there is any autocorrelation in the residuals.
(i) Based on the results of (h), what conclusions can you reach concerning the validity of the model fit in (b)?
(j) Set up a 95% confidence interval estimate of the average amount of money bet when attendance is 20,000.
(k) Set up a 95% prediction interval for the amount of money bet on a day in which attendance is 20,000.
(l) At the .05 level of significance, is there evidence of a linear relationship between the amount of money bet and attendance?
(m) Set up a 95% confidence interval estimate of the true slope.
(n) Discuss why you should not predict the amount bet on a day in which the attendance exceeded 46,800 or was below 11,600.
(o) Perform a residual analysis on your results and determine the adequacy of the fit of the model.
(p) Suppose that the amount bet on day 5 was 1.5 million dollars. Do (a)–(o) with these data and compare the difference in the results.

11.60 The owner of a large chain of ice-cream stores would like to study the effect of atmospheric temperature on sales during the summer season. A random sample of 21 days is selected with the results given as follows:

ICECREAM.TXT

Relating sales to temperature.

Day	Daily High Temperature (°F)	Sales per Store ($000)	Day	Daily High Temperature (°F)	Sales per Store ($000)
1	63	1.52	12	75	1.92
2	70	1.68	13	98	3.40
3	73	1.80	14	100	3.28
4	75	2.05	15	92	3.17
5	80	2.36	16	87	2.83
6	82	2.25	17	84	2.58
7	85	2.68	18	88	2.86
8	88	2.90	19	80	2.26
9	90	3.14	20	82	2.14
10	91	3.06	21	76	1.98
11	92	3.24			

Hint: Determine which are the independent and dependent variables.
(a) Set up a scatter diagram.
(b) Assuming a linear relationship, use the least squares method to find the regression coefficients b_0 and b_1.
(c) Interpret the meaning of the slope b_1 in this problem.

(d) Predict the sales per store for a day in which the temperature is 83°F.
(e) Compute the standard error of the estimate.
(f) Compute the coefficient of determination r^2 and interpret its meaning in this problem.
(g) Compute the coefficient of correlation r.
(h) Compute the adjusted r^2 and compare it with the coefficient of determination r^2.
(i) Compute the Durbin-Watson statistic and, at the .05 level of significance, determine whether there is any autocorrelation in the residuals.
(j) Based on the results of (i), what conclusions can you reach concerning the validity of the model fit in (b)?
(k) Set up a 95% confidence interval estimate of the average sales per store for all days in which the temperature is 83°F.
(l) Set up a 95% prediction interval for the sales per store on a day in which the temperature is 83°F.
(m) At the .05 level of significance, is there evidence of a linear relationship between temperature and sales?
(n) Set up a 95% confidence interval estimate of the true slope.
(o) Discuss how different your results might be if the model had been based upon temperature measured according to the Celsius (°C) scale.
(p) Perform a residual analysis on your results and determine the adequacy of the fit of the model.
(q) Suppose that the amount of sales on day 21 was 1.75. Do (a)–(p) and compare the difference in the results.

11.61 Suppose we want to develop a model to predict selling price based on assessed value. A sample of 30 recently sold single-family houses in a small western city is selected to study the relationship between selling price and assessed value (the houses in the city had been reassessed at full value 1 year prior to the study). The results are as follows:

HOUSE1.TXT

Relating selling price to assessed value.

Observation	Assessed Value ($000)	Selling Price ($000)	Observation	Assessed Value ($000)	Selling Price ($000)
1	78.17	94.10	16	84.36	106.70
2	80.24	101.90	17	72.94	81.50
3	74.03	88.65	18	76.50	94.50
4	86.31	115.50	19	66.28	69.00
5	75.22	87.50	20	79.74	96.90
6	65.54	72.00	21	72.78	86.50
7	72.43	91.50	22	77.90	97.90
8	85.61	113.90	23	74.31	83.00
9	60.80	69.34	24	79.85	97.30
10	81.88	96.90	25	84.78	100.80
11	79.11	96.00	26	81.61	97.90
12	59.93	61.90	27	74.92	90.50
13	75.27	93.00	28	79.98	97.00
14	85.88	109.50	29	77.96	92.00
15	76.64	93.75	30	79.07	95.90

Hint: First determine which are the independent and dependent variables.
(a) Plot a scatter diagram and, assuming a linear relationship, use the least squares method to find the regression coefficients b_0 and b_1.
(b) Interpret the meaning of the Y intercept b_0 and the slope b_1 in this problem.

(c) Use the regression model developed in (a) to predict the selling price for a house whose assessed value is $70,000.
(d) Compute the standard error of the estimate.
(e) Compute the coefficient of determination r^2 and interpret its meaning in this problem.
(f) Compute the coefficient of correlation r.
(g) Compute the adjusted r^2 and compare it with the coefficient of determination r^2.
(h) At the .10 level of significance, is there evidence of a linear relationship between selling price and assessed value?
(i) Set up a 90% confidence interval estimate of the average selling price for houses with an assessed value of $70,000.
(j) Set up a 90% prediction interval estimate of the selling price of an individual house with an assessed value of $70,000.
(k) Set up a 90% confidence interval estimate of the population slope.
(l) Perform a residual analysis on your results and determine the adequacy of the fit of the model.

11.62 Suppose we want to develop a model to predict assessed value based on heating area. A sample of 15 single-family houses is selected in a different city. The assessed value (in thousands of dollars) and the heating area of the houses (in thousands of square feet) are recorded with the following results:

HOUSE2.TXT

Relating assessed value to heating area.

House	Assessed Value ($000)	Heating Area of Dwelling (thousands of square feet)
1	84.4	2.00
2	77.4	1.71
3	75.7	1.45
4	85.9	1.76
5	79.1	1.93
6	70.4	1.20
7	75.8	1.55
8	85.9	1.93
9	78.5	1.59
10	79.2	1.50
11	86.7	1.90
12	79.3	1.39
13	74.5	1.54
14	83.8	1.89
15	76.8	1.59

Hint: First determine which are the independent and dependent variables.
(a) Plot a scatter diagram and, assuming a linear relationship, use the least squares method to find the regression coefficients b_0 and b_1.
(b) Interpret the meaning of the Y intercept b_0 and the slope b_1 in this problem.
(c) Use the regression model developed in (a) to predict the assessed value for a house whose heating area is 1,750 square feet.
(d) Compute the standard error of the estimate.
(e) Compute the coefficient of determination r^2 and interpret its meaning in this problem.
(f) Compute the coefficient of correlation r.
(g) Compute the adjusted r^2 and compare it with the coefficient of determination r^2.
(h) At the .10 level of significance, is there evidence of a linear relationship between assessed value and heating area?

(i) Set up a 90% confidence interval estimate of the average assessed value for houses with a heating area of 1,750 square feet.
(j) Set up a 90% prediction interval estimate of the assessed value of an individual house with a heating area of 1,750 square feet.
(k) Set up a 90% confidence interval estimate of the population slope.
(l) Perform a residual analysis on your results and determine the adequacy of the fit of the model.
(m) Suppose that the assessed value for the fourth house was 79.7. Do (a)–(l) and compare the results.

GPIGMAT.TXT

• 11.63 The Director of Graduate Studies at a large college of business would like to be able to predict the grade point index (GPI) of students in an MBA program based on Graduate Management Aptitude Test (GMAT) score. A sample of 20 students who had completed 2 years in the program is selected; the results are as follows:

Relating GPI to GMAT score.

Observation	GMAT Score	GPI	Observation	GMAT Score	GPI
1	688	3.72	11	567	3.07
2	647	3.44	12	542	2.86
3	652	3.21	13	551	2.91
4	608	3.29	14	573	2.79
5	680	3.91	15	536	3.00
6	617	3.28	16	639	3.55
7	557	3.02	17	619	3.47
8	599	3.13	18	694	3.60
9	616	3.45	19	718	3.88
10	594	3.33	20	759	3.76

Hint: First determine which are the independent and dependent variables.
(a) Plot a scatter diagram and assuming a linear relationship, use the least squares method to find the regression coefficients b_0 and b_1.
(b) Interpret the meaning of the Y intercept b_0 and the slope b_1 in this problem.
(c) Use the regression model developed in (a) to predict the GPI for a student with a GMAT score of 600.
(d) Compute the standard error of the estimate.
(e) Compute the coefficient of determination r^2 and interpret its meaning in this problem.
(f) Compute the coefficient of correlation r.
(g) Compute the adjusted r^2 and compare it with the coefficient of determination r^2.
(h) At the .05 level of significance, is there evidence of a linear relationship between GMAT score and GPI?
(i) Set up a 95% confidence interval estimate for the average GPI of students with a GMAT score of 600.
(j) Set up a 95% prediction interval estimate of the GPI for a particular student with a GMAT score of 600.
(k) Set up a 95% confidence interval estimate of the population slope.
(l) Perform a residual analysis on your results and determine the adequacy of the fit of the model.
(m) Suppose the GPIs of the 19th and 20th students were incorrectly entered. The GPI for student 19 should be 3.76, and the GPI for student 20 should be 3.88. Do (a)–(l) and compare the results.

594 **Chapter 11** Simple Linear Regression and Correlation

11.64 The manager of the purchasing department of a large banking organization would like to develop a model to predict the amount of time it takes to process invoices. Data are collected from a sample of 30 days with the following results:

5-11-64.XLS

Relating time to invoices processed.

Day	Invoices Processed	Completion Time (hours)	Day	Invoices Processed	Completion Time (hours)
1	149	2.1	16	169	2.5
2	60	1.8	17	190	2.9
3	188	2.3	18	233	3.4
4	19	0.3	19	289	4.1
5	201	2.7	20	45	1.2
6	58	1.0	21	193	2.5
7	77	1.7	22	70	1.8
8	222	3.1	23	241	3.8
9	181	2.8	24	103	1.5
10	30	1.0	25	163	2.8
11	110	1.5	26	120	2.5
12	83	1.2	27	201	3.3
13	60	0.8	28	135	2.0
14	25	0.4	29	80	1.7
15	173	2.0	30	29	0.5

Hint: Determine which are the independent and dependent variables.
(a) Set up a scatter diagram.
(b) Assuming a linear relationship, use the least squares method to find the regression coefficients b_0 and b_1.
(c) Interpret the meaning of the Y intercept b_0 and the slope b_1 in this problem.
(d) Use the regression model developed in (b) to predict the amount of time it would take to process 150 invoices.
(e) Compute the standard error of the estimate.
(f) Compute the coefficient of determination r^2 and interpret its meaning.
(g) Compute the coefficient of correlation.
(h) Compute the Durbin-Watson statistic and, at the .05 level of significance, determine whether there is any autocorrelation in the residuals.
(i) Based on the results of (h), what conclusions can you reach concerning the validity of the model fit in (b)?
(j) At the .05 level of significance, is there evidence of a linear relationship between the amount of time and the number of invoices processed?
(k) Set up a 95% confidence interval estimate of the average amount of time taken to process 150 invoices.
(l) Set up a 95% prediction interval estimate of the amount of time it would take to process 150 invoices on a particular day.
(m) Perform a residual analysis on your results and determine the adequacy of the fit of the model.

11.65 Crazy Dave, a well-known baseball analyst, would like to study various team statistics for the 1995 baseball season to determine which variables might be useful in predicting the number of wins achieved by teams during the season. He has decided to begin by using the team earned run average (E.R.A.) to predict the number of wins. The data for the 28 major league teams are shown on page 596.

BB95.TXT

Relating wins to E.R.A.

American League			National League		
Team	Wins	E.R.A.	Team	Wins	E.R.A.
Boston	86	4.39	Florida	67	4.27
Cleveland	100	3.83	Cincinnati	85	4.03
Kansas City	70	4.49	Chicago Cubs	73	4.13
Minnesota	56	5.76	San Francisco	67	4.86
Toronto	56	4.88	Los Angeles	78	3.66
California	78	4.49	Pittsburgh	58	4.70
Seattle	78	4.52	San Diego	70	4.13
Texas	74	4.66	New York Mets	69	3.88
Detroit	60	5.49	St. Louis	62	4.09
Chicago White Sox	68	4.85	Philadelphia	69	4.21
Milwaukee	65	4.82	Atlanta	90	3.44
Oakland	67	4.93	Montreal	66	4.11
Baltimore	71	4.31	Houston	76	4.06
New York Yankees	79	4.56	Colorado	77	4.97

Note: Each team played 144 games in the 1995 season.

Hint: Determine which are the independent and dependent variables.

(a) Set up a scatter diagram.
(b) Assuming a linear relationship, use the least squares method to find the regression coefficients b_0 and b_1.
(c) Interpret the meaning of the Y intercept b_0 and the slope b_1 in this problem.
(d) Use the regression model developed in (b) to predict the number of wins for a team with an E.R.A. of 4.00.
(e) Compute the standard error of the estimate.
(f) Compute the coefficient of determination r^2 and interpret its meaning.
(g) Compute the coefficient of correlation.
(h) At the .05 level of significance, is there evidence of a linear relationship between the number of wins and the E.R.A.?
(i) Set up a 95% confidence interval estimate of the average number of wins for a team with an E.R.A. of 4.00.
(j) Set up a 95% prediction interval estimate of the number of wins for an individual team that has an E.R.A. of 4.00.
(k) Set up a 95% confidence interval estimate of the slope.
(l) Perform a residual analysis on your results and determine the adequacy of the fit of the model.
(m) The 28 teams constitute a population. In order to use statistical inference [as in (h)–(k)], the data must be assumed to represent a random sample. What "population" would this sample be drawing conclusions about?

Case Study E—Predicting Sunday Newspaper Circulation

You are employed in the marketing department of a large nationwide newspaper chain. The parent company is interested in investigating the feasibility of beginning a Sunday edition for some of its newspapers. However, before proceeding with a final decision, it needs to estimate the amount of Sunday circulation that would be expected. In particular, it wishes to predict the Sunday circulation that would be obtained by newspapers (in three different cities) that have daily circulations of 200,000, 400,000, and 600,000, respectively.

You have been asked to develop a model that would enable you to make a prediction of the expected Sunday circulation and to write a report that presents your results and summarizes your findings. Toward this end, data collected from a sample of 35 newspapers are as follows:

NEWSCIRC.TXT

	Circulation (in 000)	
Paper	Sunday	Daily
Des Moines Register	344.522	206.204
Philadelphia Inquirer	982.663	515.523
Tampa Tribune	408.343	321.626
New York Times	1,762.015	1,209.225
New York News	983.240	781.796
Sacramento Bee	338.355	273.844
Los Angeles Times	1,531.527	1,164.388
Boston Globe	798.298	516.981
Cincinnati Enquirer	348.744	198.832
Orange Co. Register	407.760	354.843
Miami Herald	553.479	444.581
Chicago Tribune	1,133.249	733.775
Detroit News	1,215.149	481.766
Houston Chronicle	620.752	449.755
Kansas City Star	423.305	288.571
Omaha World Herald	284.611	223.748
Denver Post	417.779	252.624
St. Louis Post-Dispatch	585.681	391.286
Portland Oregonian	440.923	337.672
Washington Post	1,165.567	838.902
Long Island Newsday	960.308	825.512
San Francisco Chronicle	704.322	570.364
Chicago Sun Times	559.093	537.780
Minneapolis Star Tribune	685.975	412.871
Baltimore Sun	488.506	391.952
Pittsburgh Press	557.000	220.465
Rocky Mountain News	432.502	374.009
Boston Herald	235.084	355.628
New Orleans Times-Picayune	324.241	272.280
Charlotte Observer	299.451	238.555
Hartford Courant	323.084	231.177
Rochester Democrat and Chronicle	262.048	133.239
St. Paul Pioneer Press	267.781	201.860
Providence Journal-Bulletin	268.060	197.120
L.A. Daily News	202.614	185.736

Source: From *Gale Directory of Publications*: 1994, 126th edition. Edited by Donald P. Boyden and John Krol. Gale Research, 1994 Copyright © 1994 by Gale Research, Inc. Reprinted by permission of the publisher.

Endnotes

1. In Section 12.12, we shall investigate multiple regression models in which at least one of the independent variables is categorical (see dummy variable models). Regression models in which the dependent variable is categorical involve the use of logistic regression (see References 3 and 10).

2. It is interesting and instructive to note that had we constructed the residual plots using the independent variable as the X-axis (instead of the estimated values $\hat{Y}$), the same conclusions would prevail.

3. Methods that analyze the impact of individual observations on the regression model are part of *influence analysis* (see References 1, 3, 4, 5, 8, 9, and 15).

References

1. Andrews, D. F., and D. Pregibon, "Finding the Outliers that Matter," *Journal of the Royal Statistical Society*, Ser. B., 1978, Vol. 40, pp. 85–93.
2. Anscombe, F. J., "Graphs in Statistical Analysis," *American Statistician*, 1973, Vol. 27, pp. 17–21.
3. Berenson, M. L. and D. M. Levine, *Basic Business Statistics: Concepts and Applications*, 6th ed. (Englewood Cliffs, NJ: Prentice-Hall, 1996).
4. Belsley, D. A., E. Kuh, and R. Welsch, *Regression Diagnostics: Identifying Influential Data and Sources of Collinearity* (New York: John Wiley, 1980).
5. Cook, R. D., and S. Weisberg, *Residuals and Influence in Regression* (New York: Chapman and Hall, 1982).
6. Conover, W. J., *Practical Nonparametric Statistics*, 2d ed. (New York: John Wiley, 1980).
7. Draper, N. R., and H. Smith, *Applied Regression Analysis*, 2d ed. (New York: John Wiley, 1981).
8. Hoaglin, D.C., and R. Welsch, "The Hat Matrix in Regression and ANOVA," *The American Statistician*, 1978, Vol. 32, pp. 17–22.
9. Hocking, R. R., "Developments in Linear Regression Methodology: 1959–1982," *Technometrics*, 1983, Vol. 25, pp. 219–250.
10. Hosmer, D., and S. Lemeshow, *Applied Logistic Regression* (New York: John Wiley, 1989).
11. *Microsoft Excel Version 7* (Redmond, WA: Microsoft Corp., 1996).
12. Neter, J., M. H. Kutner, C. J. Nachtsheim, and W. Wasserman, *Applied Linear Statistical Models*, 4th ed (Homewood, IL: Richard D. Irwin, 1996).
13. Ramsey, P. P., and P. H. Ramsey, "Simple Tests of Normality in Small Samples," *Journal of Quality Technology*, 1990, Vol. 22, pp. 299–309.
14. Tukey, J.W., "Data Analysis, Computation and Mathematics," *Quarterly Journal of Applied Mathematics*, 1972, Vol. 30, pp. 51–65.
15. Velleman, P. F., and R. Welsch, "Efficient Computing of Regression Diagnostics," *The American Statistician*, 1981, Vol. 35, pp. 234–242.

chapter 12

Multiple Regression Models

CHAPTER OBJECTIVE

To develop the multiple regression model as an extension of the simple linear regression model and to evaluate the contribution of each independent variable to the regression model. In addition, to measure the coefficient of partial determination; to develop and test the curvilinear regression model; and to introduce dummy variables in regression analysis.

12.1 INTRODUCTION

In our discussion of the simple regression model in the preceding chapter, we focused on a model in which one independent or explanatory variable X was used to predict the value of a dependent or response variable Y. You may recall that a simple regression model was developed to predict the weekly sales based on the number of customers for a chain of package delivery stores. It is often the case that a better-fitting model can be developed if more than one explanatory variable is considered. Thus, in this chapter we shall extend our discussion to consider **multiple regression** models in which several explanatory variables can be used to predict the value of a dependent variable.[1]

12.2 DEVELOPING THE MULTIPLE REGRESSION MODEL

Suppose we want to develop a regression model to predict heating oil consumption of single-family homes during the month of January. A sample of 15 homes with similar square footage built by a particular housing developer throughout the United States is selected for analysis. Although many variables could be considered, for simplicity only two explanatory variables are to be evaluated here—the average daily atmospheric temperature, as measured in degrees Fahrenheit, outside the house during that month (X_1) and the amount of insulation, as measured in inches, in the attic of the house (X_2). The results are presented in Table 12.1. With two explanatory variables in the multiple regression model, a scatter diagram of the points can be plotted on a three-dimensional graph as shown in Figure 12.1.

For a particular investigation, when there are several explanatory variables present, the simple linear regression model of Chapter 11 [Equation (11.1)] can be extended by assuming a linear relationship between each explanatory variable and the dependent variable. For example, with P explanatory variables, the multiple linear regression model is expressed as

HTNGOIL.TXT

Table 12.1 Monthly heating oil consumption, atmospheric temperature, and amount of attic insulation for a random sample of 15 single-family homes.

Observation	Monthly Consumption of Heating Oil (Gallons)	Average Daily Atmospheric Temperature (°F)	Amount of Attic Insulation (Inches)
1	275.3	40	3
2	363.8	27	3
3	164.3	40	10
4	40.8	73	6
5	94.3	64	6
6	230.9	34	6
7	366.7	9	6
8	300.6	8	10
9	237.8	23	10
10	121.4	63	3
11	31.4	65	10
12	203.5	41	6
13	441.1	21	3
14	323.0	38	3
15	52.5	58	10

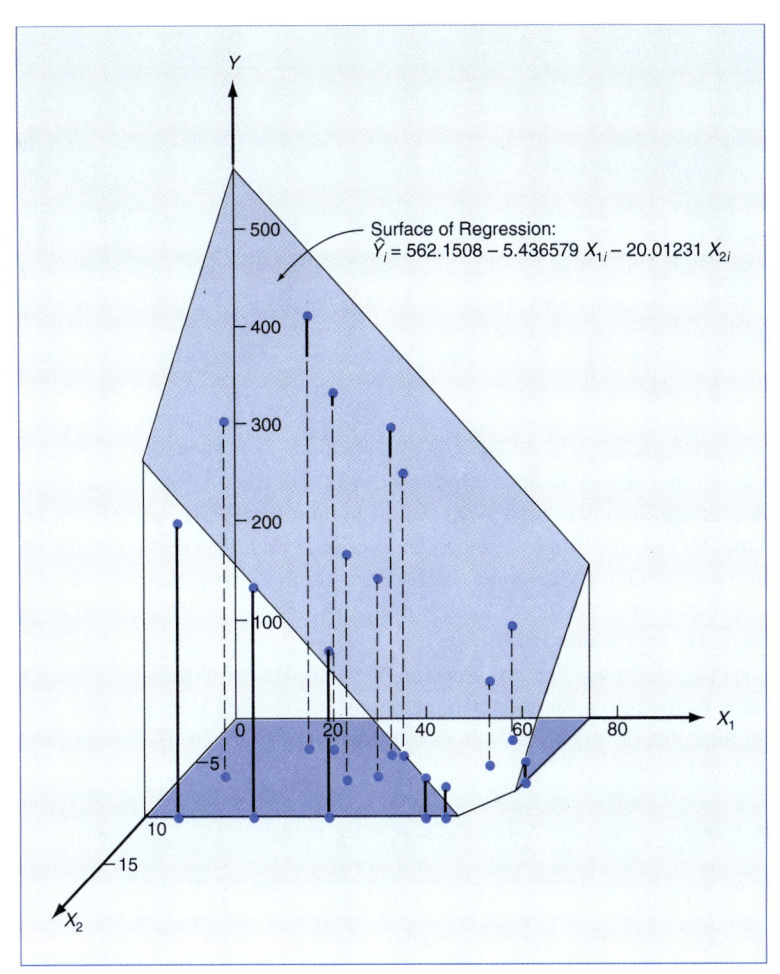

FIGURE 12.1 Scatter diagram of average daily atmospheric temperature X_1, amount of attic insulation X_2, and monthly heating oil consumption Y with indicated regression plane fitted by least squares method.

$$Y_i = \beta_0 + \beta_1 X_{1i} + \beta_2 X_{2i} + \beta_3 X_{3i} + \cdots + \beta_P X_{Pi} + \epsilon_i \quad (12.1a)$$

where $\beta_0 = Y$ intercept

β_1 = slope of Y with variable X_1 holding variables $X_2, X_3, \ldots, X_P$ constant

β_2 = slope of Y with variable X_2 holding variables $X_1, X_3, \ldots, X_P$ constant

β_3 = slope of Y with variable X_3 holding variables $X_1, X_2, X_4, \ldots, X_P$ constant

$\vdots$

β_p = slope of Y with variable X_p holding variables $X_1, X_2, X_3, \ldots, X_{P-1}$ constant

ϵ_i = random error in Y for observation i

For data with two explanatory variables, the multiple linear regression model is expressed as

12.2 Developing the Multiple Regression Model

$$Y_i = \beta_0 + \beta_1 X_{1i} + \beta_2 X_{2i} + \epsilon_i \qquad (12.1\text{b})$$

where $\beta_0 = Y$ intercept

β_1 = slope of Y with variable X_1 holding variable X_2 constant

β_2 = slope of Y with variable X_2 holding variable X_1 constant

ϵ_i = random error in Y for observation i

This multiple linear regression model can be compared to the simple linear regression model [Equation (11.1)] expressed as

$$Y_i = \beta_0 + \beta_1 X_i + \epsilon_i$$

In the case of the simple linear regression model, we should note that the slope β_1 represents the unit change in the mean of Y per unit change in X and does not take into account any other variables besides the single independent variable included in the model. On the other hand, in the multiple linear regression model [Equation (12.1b)], the slope β_1 represents the change in the mean of Y per unit change in X_1, taking into account the effect of X_2. It is referred to as a **net regression coefficient**.

As in the case of simple linear regression, the sample regression coefficients (b_0, b_1, and b_2) are used as estimates of the true parameters (β_0, β_1, and β_2). Thus, the regression equation for a multiple linear regression model with two explanatory variables would be

$$\hat{Y}_i = b_0 + b_1 X_{1i} + b_2 X_{2i} \qquad (12.2)$$

Using the least squares method, the values of the three sample regression coefficients may be obtained by accessing Excel (see Reference 12). Figure 12.2 presents partial output for the monthly heating oil consumption data.

From Figure 12.2, we observe that the computed values of the regression coefficients in this problem are

$$b_0 = 562.151 \qquad b_1 = -5.43658 \qquad b_2 = -20.0123$$

Therefore, the multiple regression equation can be expressed as

$$\hat{Y}_i = 562.151 - 5.43658 X_{1i} - 20.0123 X_{2i}$$

where $\hat{Y}_i$ = predicted average amount of heating oil consumed (gallons) during January for observation i

X_{1i} = average daily atmospheric temperature (°F) during January for observation i

X_{2i} = amount of attic insulation (inches) for observation i

The interpretation of the regression coefficients is analogous to that of the simple linear regression model. The Y intercept b_0, computed as 562.151, estimates the expected number of gallons of heating oil that would be consumed in January when the average daily atmospheric temperature is 0°F for a home that is not insulated (a house with 0 inches of attic insulation). The slope of average daily atmospheric temperature with heating oil consumption (b_1, com-

	A	B	C	D	E	F
1	Summary Output					
2						
3	Regression Statistics					
4	Multiple R	0.982654757				
5	R Square	0.965610371				
6	Adjusted R Square	0.959878766				
7	Standard Error	26.01378323				
8	Observations	15				
9						
10	ANOVA					
11		df	SS	MS	F	Significance F
12	Regression	2	228014.6263	114007.3132	168.4712028	1.65411E-09
13	Residual	12	8120.603016	676.716918		
14	Total	14	236135.2293			
15						
16		Coefficients	Standard Error	t Stat	P-value	
17	Intercept b_0	562.1510092	21.09310433	26.65093769	4.77868E-12	
18	Temperature b_1	-5.436580588	0.336216167	-16.16989642	1.64178E-09	
19	Insulation b_2	-20.01232067	2.342505227	-8.543127434	1.90731E-06	
20						
21						
22						
23	Residual Output					
24						
25	Observation	Pred. Heating Oil	Residuals	Std. Residuals		
26	1	284.6508237	-9.350823711	-0.359456509		
27	2	355.3263714	8.473628646	0.325736113		
28	3	144.564579	19.73542095	0.758652472		
29	4	45.20670231	-4.406702309	-0.169398748		
30	5	94.1359276	0.164072399	0.006307133		
31	6	257.2333452	-26.33334524	-1.012284334		
32	7	393.1478599	-26.44785994	-1.016686412		
33	8	318.5351579	-17.93515786	-0.689448271		
34	9	236.986449	0.813550957	0.031273842		
35	10	159.6094702	-38.20947019	-1.468816352		
36	11	8.650064348	22.74993565	0.874533913		
37	12	219.1772811	-15.67728112	-0.60265287		
38	13	387.9458549	53.15414512	2.04330699		
39	14	295.5239849	27.47601511	1.05620989		
40	15	46.70612846	5.793871536	0.222723142		

FIGURE 12.2 Partial output obtained from Excel for the monthly heating oil consumption data.

puted as –5.43658) means that, for a home with a *given* number of inches of attic insulation, the expected heating oil consumption is estimated to *decrease* by 5.43658 gallons per month for each 1°F increase in average daily atmospheric temperature. The slope of amount of attic insulation with heating oil consumption (b_2, computed as –20.0123) means that, for a month with a *given* average daily atmospheric temperature, the expected heating oil consumption is estimated to *decrease* by 20.0123 gallons for each additional inch of attic insulation.

Problems for Section 12.2

Note: *The problems in this section should be solved using Microsoft Excel (see Sections 11.7 and 12.14).*

12.1 Explain the difference in the interpretation of the regression coefficients in simple linear regression and multiple linear regression.

• 12.2 A marketing analyst for a major shoe manufacturer is considering the development of a new brand of running shoes. In particular, the marketing analyst wishes to determine which variables can be used in predicting durability (or the effect of long-term impact). The following two independent variables are to be considered:

X_1 (FOREIMP), a measurement of the forefoot shock-absorbing capability

X_2 (MIDSOLE), a measurement of the change in impact properties over time

along with the dependent variable Y (LTIMP), which is a measure of the long=term ability to absorb shock after a repeated impact test. A random sample of 15 types of currently manufactured running shoes was selected for testing. Using Excel, the following (partial) output is provided:

(a) Assuming that each independent variable is linearly related to long-term impact, state the multiple regression equation.
(b) Interpret the meaning of the slopes in this problem.

ANOVA	df	SS	MS	F	Significance F
Regression	2	12.61020	6.30510	97.69	0.0001
Residual	12	0.77453	0.06454		
Total	14	13.38473			

Variable	Coefficients	Standard Error	t Stat	p-value
Intercept	-0.02686	.06905	-0.39	
Foreimp	0.79116	.06295	12.57	.0000
Midsole	0.60484	.07174	8.43	.0000

Shoe durability data.

12.3 A mail-order catalog business selling personal computer supplies, software, and hardware maintains a centralized warehouse for the distribution of products ordered. Management is currently examining the process of distribution from the warehouse and is interested in studying the factors that affect warehouse distribution costs. Currently, a small handling fee is added to the order, regardless of the amount of the order. Data have been collected over the past 24 months indicating the warehouse distribution costs, the sales, and the number of orders received. The results are as follows:

Distribution cost data.

WARECOST.TXT

Month	Distribution Cost ($000)	Sales ($000)	Orders	Month	Distribution Cost ($000)	Sales ($000)	Orders
1	52.95	386	4,015	13	62.98	372	3,977
2	71.66	446	3,806	14	72.30	328	4,428
3	85.58	512	5,309	15	58.99	408	3,964
4	63.69	401	4,262	16	79.38	491	4,582
5	72.81	457	4,296	17	94.44	527	5,582
6	68.44	458	4,097	18	59.74	444	3,450
7	52.46	301	3,213	19	90.50	623	5,079
8	70.77	484	4,809	20	93.24	596	5,735

(continued)

(continued)

Month	Distribution Cost ($000)	Sales ($000)	Orders	Month	Distribution Cost ($000)	Sales ($000)	Orders
9	82.03	517	5,237	21	69.33	463	4,269
10	74.39	503	4,732	22	53.71	389	3,708
11	70.84	535	4,413	23	89.18	547	5,387
12	54.08	353	2,921	24	66.80	415	4,161

Based on the results obtained from Excel,
(a) State the multiple regression equation.
(b) Interpret the meaning of the slopes in this problem.

12.4 Suppose a large consumer products company wants to measure the effectiveness of different types of advertising media in the promotion of its products. Specifically, two types of advertising media are to be considered: radio and television advertising and newspaper advertising (including the cost of discount coupons). A sample of 22 cities with approximately equal populations is selected for study during a test period of 1 month. Each city is allocated a specific expenditure level for both radio and television advertising as well as for newspaper advertising. The sales of the product (in thousands of dollars) and the levels of media expenditure during the test month are recorded with the following results:

STANDBY.TXT

Advertising media data.

City	Sales ($000)	Radio and Television Advertising ($000)	Newspaper Advertising ($000)	City	Sales ($000)	Radio and Television Advertising ($000)	Newspaper Advertising ($000)
1	973	0	40	12	1,577	45	45
2	1,119	0	40	13	1,044	50	0
3	875	25	25	14	914	50	0
4	625	25	25	15	1,329	55	25
5	910	30	30	16	1,330	55	25
6	971	30	30	17	1,405	60	30
7	931	35	35	18	1,436	60	30
8	1,177	35	35	19	1,521	65	35
9	882	40	25	20	1,741	65	35
10	982	40	25	21	1,866	70	40
11	1,628	45	45	22	1,717	70	40

Based on the results obtained from Excel,
(a) State the multiple regression equation.
(b) Interpret the meaning of the slopes in this problem.

12.5 The director of broadcasting operations for a television station wants to study the issue of "standby hours," hours in which unionized graphic artists at the station are paid but are not actually involved in any activity. The variables to be considered are

ADRADTV.TXT

Standby hours (Y)—the total number of standby hours per week

Total staff present (X_1)—the weekly total of people-days over a 7-day week

Remote hours (X_2)—the total number of hours worked by employees at locations away from the central plant

The results for a period of 26 weeks are as shown at the top of page 606:

Standby hours.

Week	Standby Hours	Total Staff Present	Remote Hours	Week	Standby Hours	Total Staff Present	Remote Hours
1	245	338	414	14	161	307	402
2	177	333	598	15	274	322	151
3	271	358	656	16	245	335	228
4	211	372	631	17	201	350	271
5	196	339	528	18	183	339	440
6	135	289	409	19	237	327	475
7	195	334	382	20	175	328	347
8	118	293	399	21	152	319	449
9	116	325	343	22	188	325	336
10	147	311	338	23	188	322	267
11	154	304	353	24	197	317	235
12	146	312	289	25	261	315	164
13	115	283	388	26	232	331	270

On the basis of the results obtained from Excel,
(a) State the multiple regression model.
(b) Interpret the meaning of the slopes in this problem.

12.3 PREDICTION OF THE DEPENDENT VARIABLE Y FOR GIVEN VALUES OF THE EXPLANATORY VARIABLES

Now that the multiple regression model has been fitted to these data, various procedures, analogous to those discussed for simple linear regression, can be developed. In this section we shall use the multiple regression model to predict the monthly consumption of home heating oil.

Suppose we want to predict the number of gallons of heating oil consumed in a house that has 6 inches of attic insulation during a month in which the average daily atmospheric temperature is 30°F. Using our multiple regression equation

$$\hat{Y}_i = 562.151 - 5.43658 X_{1i} - 20.0123 X_{2i}$$

with $X_{1i} = 30$ and $X_{2i} = 6$, we have

$$\hat{Y}_i = 562.151 - (5.43658)(30) - (20.0123)(6)$$

and thus

$$\hat{Y}_i = 278.9798$$

Therefore, we would estimate that an average of 278.98 gallons of heating oil would be used in houses with 6 inches of insulation when the average temperature is 30°F.

Problems for Section 12.3

Note: The problems in this section can be solved using Microsoft Excel (see Sections 11.7 and 12.14).

12.6 Referring to Problem 12.3 (distribution cost) on page 604, predict the monthly warehouse distribution costs when sales are $400,000 and the number of orders is 4,500.

WARECOST.TXT

- 12.7 Referring to Problem 12.4 (advertising media) on page 605, predict the sales for a city in which radio and television advertising is $20,000 and newspaper advertising is $20,000.

12.8 Referring to Problem 12.5 (standby hours) on page 605, predict the standby hours for a week in which the total staff present is 310 people-days and the remote hours are 400.

ADRADTV.TXT

STANDBY.TXT

12.4 MEASURING ASSOCIATION IN THE MULTIPLE REGRESSION MODEL

You may recall from Section 11.6 that once a regression model has been developed, we can compute the coefficient of determination r^2. In multiple regression, since there are at least two explanatory variables, the **coefficient of multiple determination** represents the proportion of the variation in Y that is explained by the set of explanatory variables selected. For data with two explanatory variables, the coefficient of multiple determination ($r^2_{Y.12}$) is given by

> The coefficient of multiple determination is equal to the regression sum of squares divided by the total sum of squares.
>
> $$r^2_{Y.12} = \frac{\text{SSR}}{\text{SST}} \qquad (12.3)$$

where

$$\text{SSR} = b_0 \sum_{i=1}^{n} Y_i + b_1 \sum_{i=1}^{n} X_{1i} Y_i + b_2 \sum_{i=1}^{n} X_{2i} Y_i - n \bar{Y}^2$$

$$\text{SST} = \sum_{i=1}^{n} Y_i^2 - n \bar{Y}^2$$

In the heating oil consumption example, we have already computed SSR = 228,015 and SST = 236,135 (rounded). Thus, as is displayed in the Excel output of Figure 12.2 on page 603,

$$r^2_{Y.12} = \frac{\text{SSR}}{\text{SST}} = \frac{228,015}{236,135} = .9656$$

This coefficient of multiple determination, computed as .9656, can be interpreted to mean that from the sample, 96.56% of the variation in home heating oil consumption can be explained by the variation in the average daily atmospheric temperature and the variation in the amount of attic insulation.

However, you may recall from Section 11.6 that when dealing with multiple regression models, some researchers suggest that an *adjusted* r^2 be computed to reflect both the number of explanatory variables in the model and the sample size. This is especially necessary when we are comparing two or more regression models that predict the same dependent variable but have different numbers of explanatory or predictor variables. Thus, in multiple regression, we may denote the adjusted r^2 as

$$r^2_{adj} = 1 - \left[(1 - r^2_{Y.12\ldots P}) \frac{n-1}{n-P-1} \right] \qquad (12.4)$$

where P = number of explanatory variables in the regression equation

Thus, for our monthly heating oil consumption data, since $r^2_{Y.12} = .9656$, $n = 15$, and $P = 2$,

$$r^2_{adj} = 1 - \left[(1 - r^2_{Y.12})\frac{(15-1)}{(15-2-1)}\right]$$

$$= 1 - \left[(1 - .9656)\frac{14}{12}\right]$$

$$= 1 - .04$$

$$= .96$$

Hence, 96% of the variation in monthly heating oil consumption can be explained by our multiple regression model—adjusted for number of predictors and sample size.

To study the relationship among the variables further, it is often useful to examine the correlation between each pair of variables included in the model. Such a *correlation matrix* that indicates the coefficient of correlation between each pair of variables is displayed in Table 12.2.

From Table 12.2, we observe that the correlation between the amount of heating oil consumed and temperature is −.86974, indicating a strong negative association between the variables. We can also see that the correlation between the amount of heating oil consumed and attic insulation is −.46508, indicating a moderate negative correlation between these variables. Furthermore, we note that there is virtually no correlation (.00892) between the two explanatory variables, temperature and attic insulation. Finally, the correlation coefficients along the main diagonal of the matrix (r_{YY}, r_{11}, r_{22}) are each 1.0, since there will be perfect correlation between a variable and itself.

Table 12.2 Correlation matrix for the monthly heating oil consumption example.

	Y (Heating Oil)	X_1 (Temperature)	X_2 (Attic Insulation)
Y (Heating Oil)	$r_{YY} = 1.0$	$r_{Y1} = -.86974$	$r_{Y2} = -.46508$
X_1 (Temperature)	$r_{Y1} = -.86974$	$r_{11} = 1.0$	$r_{12} = .00892$
X_2 (Attic Insulation)	$r_{Y2} = -.46508$	$r_{12} = .00892$	$r_{22} = 1.0$

Problems for Section 12.4

For Problems 12.9–12.12
    (a) Compute the coefficient of multiple determination $r^2_{Y.12}$ and interpret its meaning.
    (b) Compute the adjusted r^2.

- **12.9** Refer to Problem 12.2 (shoe durability) on page 604.
- 12.10 Refer to Problem 12.3 (distribution cost) on page 604.
- **12.11** Refer to Problem 12.4 (advertising media) on page 605.
- **12.12** Refer to Problem 12.5 (standby hours) on page 605.

12.5 RESIDUAL ANALYSIS IN MULTIPLE REGRESSION

In Section 11.10, we utilized residual analysis to evaluate whether the simple linear regression model is appropriate for the set of data being studied. When examining a multiple linear regression model with two explanatory variables, the following residual plots are of particular interest:

1. Residuals versus $\hat{Y}_i$.
2. Residuals versus X_{1i}.
3. Residuals versus X_{2i}.
4. Residuals versus time.

The first residual plot examines the pattern of residuals for the predicted values of Y. If the residuals appear to vary for different levels of the predicted Y value, it provides evidence of a possible curvilinear effect in at least one explanatory variable and/or the need to transform the dependent variable. The second and third residual plots involve the explanatory variables. Patterns in the plot of the residuals versus an explanatory variable may indicate the existence of a curvilinear effect and, therefore, lead to the possible transformation of that explanatory variable. The fourth type of plot is used to investigate patterns in the residuals when the data have been collected in time order. Associated with the residuals plot versus time, as in Section 11.11, the Durbin-Watson statistic can be computed and the existence of positive autocorrelation among the residuals can be determined.

The residual plots are available as part of the output of virtually all statistical and spreadsheet software. Figure 12.3 consists of the residual plots obtained from Excel for the monthly heating oil consumption example. We can observe from Figure 12.3 that there appears to be very little or no pattern in the relationship between the residuals and either the predicted value of Y, the value of X_1 (temperature), or the value of X_2 (attic insulation). Thus, we may conclude that the multiple linear regression model is appropriate for predicting heating oil usage.

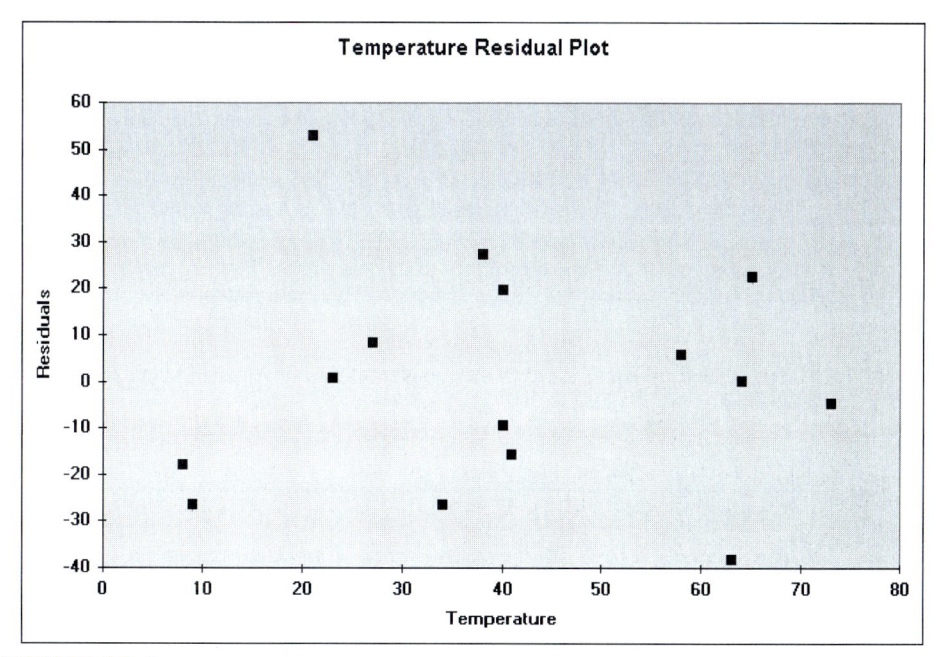

FIGURE 12.3 Residual plots for the monthly heating oil consumption model obtained from Excel. Panel A.

Insulation Residual Plot

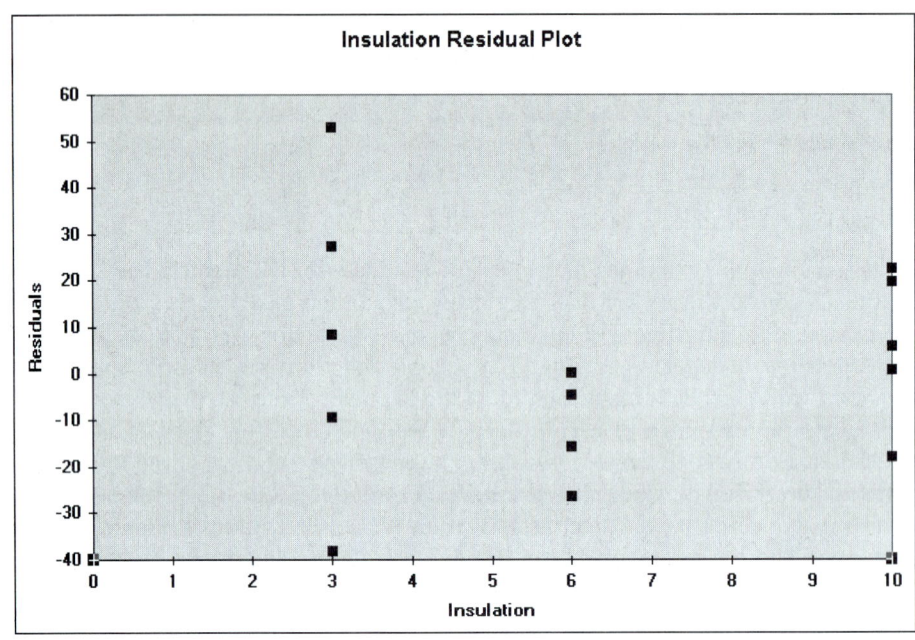

[Scatter plot of residuals vs. Insulation, ranging from about -40 to 55 on residual axis, with Insulation values at 0, 3, 6, and 10.]

FIGURE 12.3 Panel B.

Problems for Section 12.5

Note: The problems in this section can be solved using Microsoft Excel (see Sections 11.7 and 12.14).

- **12.13** (a) Referring to the distribution cost problem (pages 604, 606, and 608), perform a residual analysis on your results and determine the adequacy of the fit of the model.
  - (b) Plot the residuals against the months. Is there any evidence of a pattern in the residuals? Explain.
  - (c) Compute the Durbin-Watson statistic.
  - (d) At the .05 level of significance, is there evidence of positive autocorrelation in the residuals?

- • **12.14** Referring to the advertising media problem (pages 605, 607, and 608), perform a residual analysis on your results and determine the adequacy of the fit of the model.

- **12.15** (a) Referring to the standby hours problem (pages 605, 607, and 608), perform a residual analysis on your results and determine the adequacy of the fit of the model.
  - (b) Plot the residuals against the weeks. Is there evidence of a pattern in the residuals? Explain.
  - (c) Compute the Durbin-Watson statistic.
  - (d) At the .05 level of significance, is there evidence of positive autocorrelation in the residuals?

12.6 TESTING FOR THE SIGNIFICANCE OF THE RELATIONSHIP BETWEEN THE DEPENDENT VARIABLE AND THE EXPLANATORY VARIABLES

Now that we have used residual analysis to assure ourselves that the multiple linear regression model is appropriate, we can determine whether there is a significant relationship between the dependent variable and the set of explanatory variables. Since there is more than one explanatory variable, the null and alternative hypothesis can be set up as follows:

H_2: $\beta_1 = \beta_2 = 0$ (There is no linear relationship between the dependent variable and the explanatory variables.)

H_1: At least one $\beta_j \neq 0$ (At least one regression coefficient is not equal to 0.)

This null hypothesis may be tested with an F test, as indicated in Table 12.3. You may recall from Sections 8.5 and 8.6 that the F test is used to test the ratio of two variances. When testing for the significance of the regression coefficients, the measure of random error is called the *error variance*, so the F test is the ratio of the variance due to the regression divided by the error variance as shown in Equation (12.5):

> The F statistic is equal to the regression mean square (MSR) divided by the error mean square (MSE).
>
> $$F = \frac{MSR}{MSE} \tag{12.5}$$

where P = number of explanatory variables in the regression model

F = test statistic from an F distribution with P and $n - P - 1$ degrees of freedom.

The decision rule is

Reject H_0 at the α level of significance if $F > F_{U(P, n - P - 1)}$; otherwise do not reject H_0.

Table 12.3 ANOVA table for testing the significance of a set of regression coefficients in a multiple regression model with $P = 2$ explanatory variables.

Source	df	Sums of Squares	Mean Square (Variance)	F
Regression	P	$SSR = b_0 \sum_{i=1}^{n} Y_i + b_1 \sum_{i=1}^{n} X_{1i} Y_i + b_2 \sum_{i=1}^{n} X_{2i} Y_i - n\bar{Y}^2$	$MSR = \dfrac{SSR}{P}$	$F = \dfrac{MSR}{MSE}$
Error	$n - P - 1$	$SSE = \sum_{i=1}^{n} Y_i^2 - b_0 \sum_{i=1}^{n} Y_i - b_1 \sum_{i=1}^{n} X_{1i} Y_i - b_2 \sum_{i=1}^{n} X_{2i} Y_i$	$MSE = \dfrac{SSE}{n - P - 1}$	
Total	$n - 1$	$SST = \sum_{i=1}^{n} Y_i^2 - n\bar{Y}^2$		

The complete set of the heating oil consumption computations is shown for illustrative purposes although the completed ANOVA table is available as part of the output from Excel (see Figure 12.2). The ANOVA table is presented in Table 12.4 on page 612.

If a level of significance of .05 is chosen, we can determine from Table E.5 that the critical value on the F distribution (with 2 and 12 degrees of freedom) is 3.89, as depicted in Figure 12.4. From Equation (12.5), since $F = 168.47 > F_{U(2, 12)} = 3.89$, or since the p-value = .0001 < .05, we can reject H_0 and conclude that *at least* one of the explanatory variables (temperature and/or insulation) is related to monthly heating oil consumption.

Table 12.4 ANOVA table for testing the significance of a set of regression coefficients for the monthly heating oil consumption data.

Source	df	Sums of Squares	Mean Square (Variance)	F
Regression	2	$(562.151)(3,247.4) + (-5.43658)(98,060.1)$ $+ (-20.0123)(18,057) - 15(216.493)^2$ $= 228,014.6263$	$\dfrac{228,014.6263}{2}$ $= 114,007.31315$	$\dfrac{114,007.31315}{676.71692}$ $= 168.47$
Error	$15 - 2 - 1 = 12$	$939,175.68 - (562.151)(3,247.4)$ $- (-5.43658)(98,060.1) - (-20.0123)(18,057)$ $= 8,120.6030$	$\dfrac{8,120.6030}{12}$ $= 676.71692$	
Total	$15 - 1 = 14$	$939,175.68 - 15(216.443)^2 = 236,135.2293$		

Source: Format of Table 12.3.

FIGURE 12.4 Testing for the significance of a set of regression coefficients at the .05 level of significance with 2 and 12 degrees of freedom.

Problems for Section 12.6

- **12.16** Referring to Problem 12.2 (shoe durability) on page 604
  (a) Determine whether there is a significant relationship between long-term impact and the two explanatory variables at the .05 level of significance.
  (b) Interpret the meaning of the p-value.

WARECOST.TXT

12.17 Referring to Problem 12.3 (distribution cost) on page 604
  (a) Determine whether there is a significant relationship between distribution cost and the two explanatory variables (sales and number of orders) at the .05 level of significance.
  (b) Interpret the meaning of the p-value.

ADRADTV.TXT

- **12.18** Referring to Problem 12.4 (advertising media) on page 605
  (a) Determine whether there is a significant relationship between sales and the two explanatory variables (radio and television advertising and newspaper advertising) at the .05 level of significance.
  (b) Interpret the meaning of the p-value.

STANDBY.TXT

12.19 Referring to Problem 12.5 (standby hours) on page 606
  (a) Determine whether there is a significant relationship between standby hours and the two explanatory variables (total staff present and remote hours) at the .05 level of significance.
  (b) Interpret the meaning of the p-value.

12.7 TESTING PORTIONS OF A MULTIPLE REGRESSION MODEL

In developing a multiple regression model, the objective is to utilize only those explanatory variables that are useful in predicting the value of a dependent variable. If an explanatory variable is not helpful in making this prediction, it could be deleted from the multiple regression model, and a model with fewer explanatory variables could be used in its place.

One method for determining the contribution of an explanatory variable is called the **partial F-test criterion**. It involves determining the contribution to the regression sum of squares made by each explanatory variable after all the other explanatory variables have been included in the model. The new explanatory variable is included only if it significantly improves the model. To apply the partial F-test criterion in the home heating oil consumption example, we need to evaluate the contribution of the variable attic insulation (X_2) after average daily atmospheric temperature (X_1) has been included in the model and, conversely, we must also evaluate the contribution of the variable average daily atmospheric temperature (X_1) after attic insulation (X_2) has been included in the model.

The contribution of each explanatory variable to be included in the model can be determined by taking into account the regression sum of squares of a model that includes all explanatory variables except the one of interest, SSR (all variables except k). Thus, in general, to determine the contribution of variable k given that all other variables are already included, we would have

$$\text{SSR}(X_k | \text{all variables } except\ k) = \text{SSR}(\text{all variables } including\ k) - \text{SSR}(\text{all variables } except\ k) \quad (12.6a)$$

If, as in the monthly heating oil consumption example, there are two explanatory variables, the contribution of each can be determined from Equations (12.6b) and (12.6c).

Contribution of variable X_1 given X_2 has been included
$$\text{SSR}(X_1 | X_2) = \text{SSR}(X_1 \text{ and } X_2) - \text{SSR}(X_2) \quad (12.6b)$$

Contribution of variable X_2 given X_1 has been included
$$\text{SSR}(X_2 | X_1) = \text{SSR}(X_1 \text{ and } X_2) - \text{SSR}(X_1) \quad (12.6c)$$

The terms $\text{SSR}(X_2)$ and $\text{SSR}(X_1)$, respectively, represent the sum of squares due to regression for a model that includes *only* the explanatory variable X_2 (amount of attic insulation) and *only* the explanatory variable X_1 (average daily atmospheric temperature). Computer output obtained from Excel for these two models is presented in Figures 12.5 and 12.6.

We can observe from Figure 12.5 that
$$\text{SSR}(X_2) = 51{,}076 \text{ (rounded)}$$

and, therefore, from Equation (12.6b),
$$\text{SSR}(X_1 | X_2) = \text{SSR}(X_1 \text{ and } X_2) - \text{SSR}(X_2)$$

	A	B	C	D	E	F
1	SUMMARY OUTPUT					
2						
3	*Regression Statistics*					
4	Multiple R	0.465082527				
5	R Square	0.216301757				
6	Adjusted R Square	0.156017277				
7	Standard Error	119.3117327				
8	Observations	15				
9						
10	ANOVA					
11		df	SS	MS	F	Significance F
12	Regression	1	51076.46501	51076.46501	3.588017285	0.080660953
13	Residual	13	185058.7643	14235.28956		
14	Total	14	236135.2293			
15						
16		Coefficients	Standard Error	t Stat	P-value	
17	Intercept	345.3783784	74.69065911	4.624117426	0.000476363	
18	Insulation	-20.35027027	10.743429	-1.894206241	0.080660953	

FIGURE 12.5 Partial output of simple linear regression model for amount of heating oil consumed and amount of attic insulation (obtained from Excel).

	A	B	C	D	E	F
1	SUMMARY OUTPUT					
2						
3	*Regression Statistics*					
4	Multiple R	0.86974117				
5	R Square	0.756449704				
6	Adjusted R Square	0.737715065				
7	Standard Error	66.51246564				
8	Observations	15				
9						
10	ANOVA					
11		df	SS	MS	F	Significance F
12	Regression	1	178624.4242	178624.4242	40.37706498	2.51847E-05
13	Residual	13	57510.80511	4423.908086		
14	Total	14	236135.2293			
15						
16		Coefficients	Standard Error	t Stat	P-value	
17	Intercept	436.4382299	38.63970893	11.29507033	4.30471E-08	
18	Temperature	-5.462207697	0.859608768	-6.354295002	2.51847E-05	

FIGURE 12.6 Partial output of simple linear regression model for amount of heating oil consumed and average daily atmospheric temperature (obtained from Excel).

Table 12.5 ANOVA table dividing the regression sum of squares into components to determine the contribution of variable X_1.

Source	df	Sums of Squares	Mean Square (Variance)	F
Regression	2	228,015	114,007.5	
$\begin{Bmatrix} X_2 \\ X_1\|X_2 \end{Bmatrix}$	$\begin{Bmatrix} 1 \\ 1 \end{Bmatrix}$	$\begin{Bmatrix} 51,076 \\ 176,939 \end{Bmatrix}$	51,076 176,939	261.47
Error	12	8,120	MSE = 676.717	
Total	14	236,135		

and we have

$$SSR(X_1 \mid X_2) = 228{,}015 - 51{,}076 = 176{,}939$$

To determine whether X_1 significantly improves the model after X_2 has been included, we can now subdivide the regression sum of squares into two component parts as shown in Table 12.5.

The null and alternative hypotheses to test for the contribution of X_1 to the model would be

H_0: Variable X_1 does not significantly improve the model once variable X_2 has been included.

H_1: Variable X_1 significantly improves the model once variable X_2 has been included.

The partial F-test criterion is expressed by

$$F = \frac{SSR(X_k \mid \text{all variables } except\ k)}{MSE} \quad (12.7)$$

In Equation (12.7) F represents the F-test statistic that follows an F distribution with 1 and $n - P - 1$ degrees of freedom.

Thus, from Table 12.5, we have

$$F = \frac{176{,}939}{676.717} = 261.47$$

Since there are 1 and 12 degrees of freedom, respectively, if a level of significance of .05 is selected, we can observe from Table E.5 that the critical value is 4.75 (see Figure 12.7). Since the computed F value exceeds this critical F value (261.47 > 4.75), our decision would be to reject H_0 and conclude that the addition of variable X_1 (average daily atmospheric temperature) significantly improves a multiple regression model that already contains variable X_2 (attic insulation).

To evaluate the contribution of variable X_2 (attic insulation) to a model in which variable X_1 has been included, we need to use Equation (12.6c):

$$SSR(X_2 \mid X_1) = SSR(X_1 \text{ and } X_2) - SSR(X_1)$$

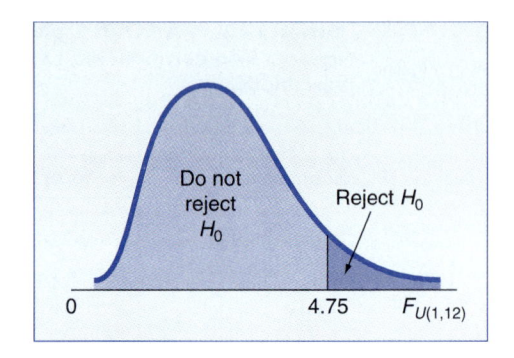

FIGURE 12.7 Testing for the contribution of a regression coefficient to a multiple regression model at the .05 level of significance with 1 and 12 degrees of freedom.

From Figures 12.2 and 12.6, we determine that
$$\text{SSR}(X_1) = 178{,}624$$

Therefore,
$$\text{SSR}(X_2 \mid X_1) = 228{,}015 - 178{,}624 = 49{,}391 \text{ (rounded)}$$

Thus, to determine whether X_2 significantly improves a model after X_1 has been included, the regression sum of squares can be subdivided into two component parts as shown in Table 12.6.

The null and alternative hypotheses to test for the contribution of X_2 to the model would be

H_0: Variable X_2 does not significantly improve the model once variable X_1 has been included.

H_1: Variable X_2 significantly improves the model once variable X_1 has been included.

Using Equation (12.7), we obtain
$$F = \frac{49{,}391}{676.717} = 72.99$$

as indicated in Table 12.6. Since there are 1 and 12 degrees of freedom, respectively, if a .05 level of significance is selected, we again observe from Figure 12.7 that the critical value of F is 4.75. Since the computed F value exceeds this critical value (72.99 > 4.75), our decision is to reject H_0 and conclude that the addition of variable X_2 (attic insulation) significantly improves the multiple regression model already containing X_1 (average daily atmospheric temperature).

Table 12.6 ANOVA table dividing the regression sum of squares into components to determine the contribution of variable X_2.

Source	df	Sums of Squares	Mean Square (Variance)	F
Regression	2	228,015	114,007.5	
$\begin{cases} X_1 \\ X_2 \mid X_1 \end{cases}$	$\begin{cases} 1 \\ 1 \end{cases}$	$\begin{cases} 178{,}624 \\ 49{,}391 \end{cases}$	178,624 49,391	72.99
Error	12	8,120	MSE = 676.717	
Total	14	236,135		

Chapter 12 Multiple Regression Models

Thus, by testing for the contribution of each explanatory variable after the other had been included in the model, we determine that each of the two explanatory variables significantly improves the model. Therefore, our multiple regression model should include both average daily atmospheric temperature X_1 and the amount of attic insulation X_2 in predicting the monthly consumption of home heating oil.

Problems for Section 12.7

Note: *Microsoft Excel should be used to solve the problems in this section (see Sections 11.7 and 12.14).*

In Problems 12.20–12.22 at the .05 level of significance, determine whether each explanatory variable makes a significant contribution to the regression model. On the basis of these results, indicate the regression model that should be utilized in the problem.

12.20 Refer to Problem 12.3 (distribution cost) on page 604.
• **12.21** Refer to Problem 12.4 (advertising media) on page 605.
12.22 Refer to Problem 12.5 (standby hours) on page 605.

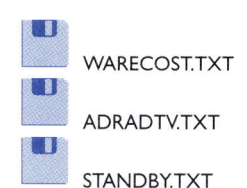

WARECOST.TXT
ADRADTV.TXT
STANDBY.TXT

12.8 INFERENCES CONCERNING THE POPULATION REGRESSION COEFFICIENTS

In Section 11.14, you may recall that tests of hypotheses were performed on the regression coefficients in a simple linear regression model to determine the significance of the relationship between X and Y. In addition, confidence intervals were used to estimate the population values of these regression coefficients. In this section, these procedures will be extended to situations involving multiple regression.

12.8.1 Tests of Hypothesis

To test the hypothesis that the population slope β_1 was 0, we used Equation (11.19):

$$t = \frac{b_1}{S_{b_1}}$$

However, this equation can be generalized for multiple regression as follows:

$$t = \frac{b_k}{S_{b_k}} \qquad (12.8)$$

where P = number of explanatory variables in the regression equation

S_{b_k} = standard error of the regression coefficient b_k

t = t statistic for a t distribution with $n - P - 1$ degrees of freedom

Since the formulas for the standard errors of the regression coefficients are unwieldy with a large numbers of variables, it is fortunate that the results are provided as part of the output obtained from computer software such as Microsoft Excel (see Figure 12.2 on page 603).

Thus, if we wish to determine whether variable X_2 (amount of attic insulation) has a significant effect on the monthly consumption of home heating oil, taking into account the average daily atmospheric temperature, the null and alternative hypotheses would be

$$H_0: \beta_2 = 0$$
$$H_1: \beta_2 \neq 0$$

From Equation (12.8), we have

$$t = \frac{b_2}{S_{b_2}}$$

and from the data of this example,

$$b_2 = -20.0123 \quad \text{and} \quad S_{b_2} = 2.3425$$

so that

$$t = \frac{-20.0123}{2.3425} = -8.5431$$

If a level of significance of .05 is selected, we check Table E.3 and find that for 12 degrees of freedom, the critical values of t are -2.1788 and $+2.1788$ (see Figure 12.8). From Figure 12.2 on page 603, we observe the p-value is .0000019073 (1.9073E-06 in scientific notation).

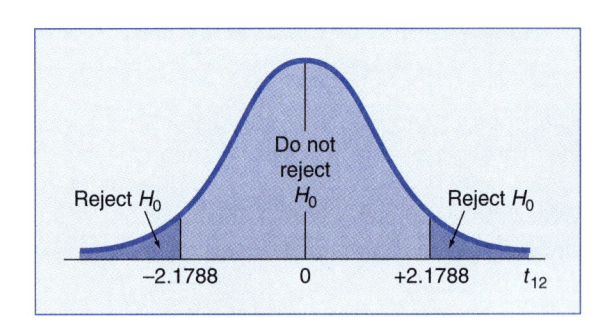

FIGURE 12.8 Testing for the significance of a regression coefficient at the .05 level of significance with 12 degrees of freedom.

Since we have $t = -8.5431 < t_{12} = -2.1788$ or the p-value of $.0000019073 < .05$, we reject H_0 and conclude that there is a significant relationship between variable X_2 (amount of attic insulation) and heating oil consumption, taking into account the average daily atmospheric temperature X_1.

In a similar manner, we can test for the significance of β_1, the slope of atmospheric temperature with the monthly consumption of heating oil. From Figure 12.2 on page 603, we find $t = -16.17 < -2.1788$ (the critical value for $\alpha = .05$) or the p-value $= .0000000016418 < .05$. Therefore, there is a significant relationship between atmospheric temperature (X_1) and heating oil consumption taking into account the attic insulation X_2.

Focusing on the interpretation of these conclusions, we note that there is a relationship between the value of the t-test statistic obtained from Equation (12.8) and the partial F-test statistic [Equation (12.7)] used to determine the contributions X_1 and X_2 to the multiple regression model. The t values were computed to be -16.17 and -8.5431, and the corresponding values of F were 261.47 and 72.99. This points up the following relationship[2] between t and F.

$$t_a^2 = F_{1,a} \tag{12.9}$$

where $\quad a =$ number of degrees of freedom

Thus, the test of significance for a particular regression coefficient is actually a test for the significance of adding a particular variable into a regression model given that the other variables have been included. Therefore, the *t* test for the regression coefficient is equivalent to testing for the contribution of each explanatory variable as discussed in Section 12.7 and the two approaches can be used interchangeably.

12.8.2 Confidence Interval Estimation

Rather than attempting to determine the significance of a regression coefficient, we may be more concerned with estimating the population value of a regression coefficient. In multiple regression analysis, a confidence interval estimate can be obtained from

$$b_k \pm t_{n-P-1} S_{b_k} \tag{12.10}$$

For example, if we wish to obtain a 95% confidence interval estimate of the population slope β_1 (the effect of average daily temperature X_1 on monthly heating oil consumption Y, holding constant the effect of attic insulation X_2), we would have, from Equation (12.10) and Figure 12.2,

$$b_1 \pm t S_{b_1}$$

Since the critical value of *t* at the 95% confidence level with 12 degrees of freedom is 2.1788 (see Table E.3), we have

$$-5.43658 \pm (2.1788)(.33622)$$

$$-5.43658 \pm .732556$$

$$-6.169136 \leq \beta_1 \leq -4.704024$$

Thus, taking into account the effect of attic insulation, we estimate that the effect of average daily atmospheric temperature is to reduce the average consumption of heating oil by between approximately 4.7 and 6.17 gallons for each 1°F increase in temperature. We have 95% confidence that this interval correctly estimates the true relationship between these variables. From a hypothesis-testing viewpoint, since this confidence interval does not include 0, the regression coefficient β_1 would have a significant effect.

Problems for Section 12.8

- **12.23** Referring to Problem 12.2 (shoe durability) on page 604
  (a) Set up a 95% confidence interval estimate of the population slope between long-term impact and forefoot impact.
  (b) At the .05 level of significance determine whether each explanatory variable makes a significant contribution to the regression model. On the basis of these results, indicate the regression model that should be used in this problem.

12.24 Referring to Problem 12.3 (distribution cost) on page 604
  (a) Set up a 95% confidence interval estimate of the population slope between distribution cost and sales.
  (b) At the .05 level of significance determine whether each explanatory variable makes a significant contribution to the regression model. On the basis of these results, indicate the regression model that should be used in this problem.
  (c) Compare the results in (b) to those of Problem 12.20 on page 617.

ADRADTV.TXT

• 12.25 Referring to Problem 12.4 (advertising media) on page 605
  (a) Set up a 95% confidence interval estimate of the population slope between sales and radio and television advertising.
  (b) At the .05 level of significance determine whether each explanatory variable makes a significant contribution to the regression model. On the basis of these results, indicate the regression model that should be used in this problem.
  (c) Compare the results in (b) to those of Problem 12.21 on page 617.

STANDBY.TXT

12.25 Referring to Problem 12.5 (standby hours) on page 605
  (a) Set up a 95% confidence interval estimate of the population slope between standby hours and total staff present.
  (b) At the .05 level of significance determine whether each explanatory variable makes a significant contribution to the regression model. On the basis of these results, indicate the regression model that should be used in this problem.
  (c) Compare the results in (b) to those of Problem 12.22 on page 617.

12.9 COEFFICIENT OF PARTIAL DETERMINATION

In Section 12.4, we discussed the coefficient of multiple determination ($r^2_{Y.12}$), which measured the proportion of the variation in Y that was explained by variation in the two explanatory variables. Now that we have examined ways in which the contribution of each explanatory variable to the multiple regression model can be evaluated, we can also compute the **coefficients of partial determination** ($r^2_{Y1.2}$ and $r^2_{Y2.1}$). The coefficients measure the proportion of the variation in the dependent variable that is explained by each explanatory variable while controlling for, or holding constant, the other explanatory variable(s). Thus, in a multiple regression model with two explanatory variables, we have

$$r^2_{Y1.2} = \frac{SSR(X_1|X_2)}{SST - SSR(X_1 \text{ and } X_2) + SSR(X_1|X_2)} \quad (12.11a)$$

and also

$$r^2_{Y2.1} = \frac{SSR(X_2|X_1)}{SST - SSR(X_1 \text{ and } X_2) + SSR(X_2|X_1)} \quad (12.11b)$$

where $SSR(X_1 | X_2)$ = sum of squares of the contribution of variable X_1 to the regression model given that variable X_2 has been included in the model

SST = total sum of squares for Y

$SSR(X_1 \text{ and } X_2)$ = regression sum of squares when variables X_1 and X_2 are both included in the multiple regression model

$SSR(X_2 | X_1)$ = sum of squares of the contribution of variable X_2 to the regression model given that variable X_1 has been included in the model

while in a multiple regression model containing several (P) explanatory variables, we have for the kth variable

$$r^2_{Yk.\,(\text{all variables } except\,k)} = \frac{\text{SSR}(X_k|\text{all variables } except\,k)}{\text{SST} - \text{SSR}(\text{all variables } including\,k) + \text{SSR}(X_k|\text{all variables } except\,k)} \quad (12.12)$$

For the monthly heating oil consumption example, we can compute

$$r^2_{Y1.2} = \frac{176{,}939}{236{,}135 - 228{,}015 + 176{,}939}$$

$$= 0.9561$$

and

$$r^2_{Y2.1} = \frac{49{,}391}{236{,}135 - 228{,}015 + 49{,}391}$$

$$= 0.8588$$

The coefficient of partial determination of variable Y with X_1 while holding X_2 constant ($r^2_{Y1.2}$) means that for a fixed (constant) amount of attic insulation, 95.61% of the variation in heating oil consumption in January can be explained by the variation in the average daily atmospheric temperature in that month. The coefficient of partial determination of variable Y with X_2 while holding X_1 constant ($r^2_{Y2.1}$) means that for a given (constant) average daily atmospheric temperature, 85.88% of the variation in heating oil consumption in January can be explained by variation in the amount of attic insulation.

Problems for Section 12.9

Note: *The problems in this section can be solved using Microsoft Excel (see Sections 11.7 and 12.14).*

12.27 Referring to Problem 12.3 (distribution cost) on page 603, compute the coefficients of partial determination $r^2_{Y1.2}$ and $r^2_{Y2.1}$ and interpret their meaning.   WARECOST.TXT

• 12.28 Referring to Problem 12.4 (advertising media) on page 604, compute the coefficients of partial determination $r^2_{Y1.2}$ and $r^2_{Y2.1}$ and interpret their meaning.   ADRADTV.TXT

12.29 Referring to Problem 12.5 (standby hours) on page 605, compute the coefficients of partial determination $r^2_{Y1.2}$ and $r^2_{Y2.1}$ and interpret their meaning.   STANDBY.TXT

12.10 THE CURVILINEAR REGRESSION MODEL

In our discussion of simple regression in Chapter 11 and multiple regression in this chapter, we have assumed that the relationship between Y and each explanatory variable is linear. However, you may recall that several different types of relationships between variables were introduced in Section 11.3. One of the more common nonlinear relationships illustrated was a curvilinear polynomial relationship between two variables (see Figure 11.3 on page 539, Panels D–F) in which Y increases (or decreases) at a *changing rate* for various values of X. This model of a polynomial relationship between X and Y can be expressed as

$$Y_i = \beta_0 + \beta_1 X_{1i} + \beta_{11} X_{1i}^2 + \epsilon_i \quad (12.13a)$$

where
$\beta_0 = Y$-intercept
$\beta_1 = $ *linear* effect on Y
$\beta_{11} = $ *curvilinear* effect on Y
$\epsilon_i = $ random error in Y for observation i

This **curvilinear regression model** is similar to the multiple regression model with two explanatory variables [see Equation (12.1a) on page 607] except that the second explanatory variable in this instance is the square of the first explanatory variable.

As in the case of multiple linear regression, the sample regression coefficients (b_0, b_1, and b_{11}) are used as estimates of the population parameters (β_0, β_1, and β_{11}). Thus, the regression equation for the curvilinear polynomial model with one explanatory variable (X_1) and a dependent variable (Y) is

$$\hat{Y}_i = b_0 + b_1 X_{1i} + b_{11} X_{1i}^2 \qquad (12.13b)$$

12.10.1 Finding the Regression Coefficients and Predicting Y

To illustrate the curvilinear regression model, let us suppose the marketing department of a large supermarket chain wants to study the price elasticity of packages of disposable razors. A sample of 15 stores with equal store traffic and product placement (i.e., at the checkout counter) is selected. Five stores are randomly assigned to each of three price levels (79, 99, and 119 cents) for the package of razors. The number of packages sold and the price at each store are presented in Table 12.7.

To help select the proper model for expressing the relationship between price and sales, a scatter diagram is plotted (Figure 12.9). An examination of Figure 12.9 indicates that the decreases in sales levels off for an increasing price. Therefore, it appears that a curvilinear model (rather than linear) may be the more appropriate choice to estimate sales based on price.

As in the case of multiple regression, the values of the three sample regression coefficients (b_0, b_1, and b_{11}) can be obtained by using computer software such as Excel (see Reference 12).

DISPRAZ.TXT

Table 12.7 Sales and price of packages of disposable razors for a sample of 15 stores.

Sales	Price (cents)	Sales	Price (cents)
142	79	115	99
151	79	126	99
163	79	77	119
168	79	86	119
176	79	95	119
91	99	100	119
100	99	106	119
107	99		

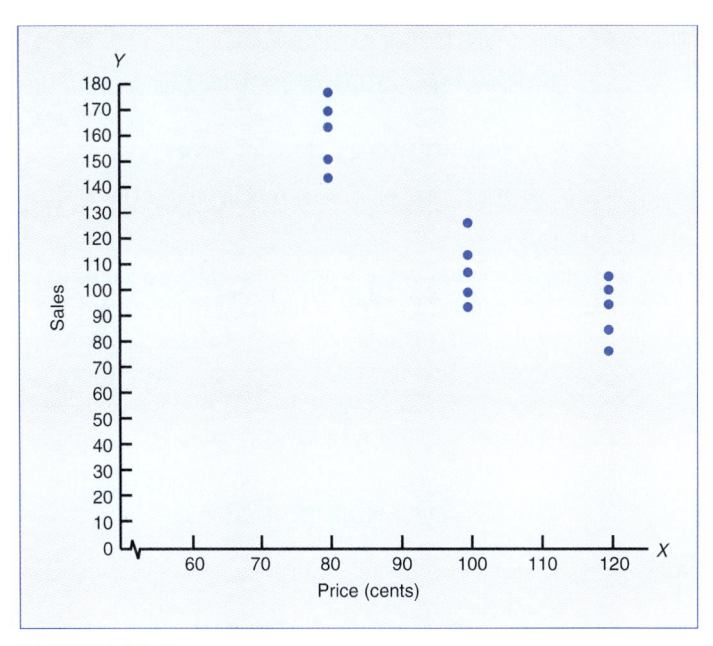

FIGURE 12.9 Scatter diagram of price (X) and sales (Y).

Figure 12.10 presents partial output from Excel for the razor sales data. Using the curvilinear model of Equation (12.13b), we observe that

$$b_0 = 729.8665 \qquad b_1 = -10.887 \qquad b_{11} = .0465$$

	A	B	C	D	E	F
1	SUMMARY OUTPUT					
2						
3	*Regression Statistics*					
4	Multiple R	0.928581176				
5	R Square	0.862263				
6	Adjusted R Square	0.839306834				
7	Standard Error	12.86986143				
8	Observations	15				
9						
10	ANOVA					
11		df	SS	MS	F	Significance F
12	Regression	2	12442.8	6221.4	37.56127994	6.82816E-06
13	Residual	12	1987.6	165.6333333		
14	Total	14	14430.4			
15						
16		Coefficients	Standard Error	t Stat	P-value	
17	Intercept	729.8665	169.2575176	4.312165924	0.001009972	
18	Price	-10.887	3.495239703	-3.114807831	0.008940633	
19	Price Squared	0.0465	0.017622784	2.638629696	0.0216284	

FIGURE 12.10 Partial output obtained for the razor sales data from Excel.

Therefore, the curvilinear model can be expressed as

$$\hat{Y}_i = 729.8665 - 10.887 X_{1i} + .0465 X^2_{1i}$$

where
$\hat{Y}_i$ = predicted average sales for store i

X_{1i} = price of disposable razors in store i

As depicted in Figure 12.11, this curvilinear regression equation is plotted on the scatter diagram to see how well the selected regression model fits the original data. From our curvilinear regression equation and Figure 12.11, the Y intercept (b_0, computed as 729.8665) has no direct interpretation for these data. To interpret the coefficients b_1 and b_{11}, we see from Figure 12.11 that the sales decrease with increasing price; nevertheless, we can also observe that these decreases in sales level off or become reduced with increasing price. This can be demonstrated by predicting average sales for packages priced at 79, 99, and 119 cents. Using our curvilinear regression equation,

$$\hat{Y}_i = 729.8665 - 10.887 X_{1i} + .0465 X^2_{1i}$$

for $X_{1i} = 79$, we have

$$\hat{Y}_i = 729.8665 - 10.887(79) + (.0465)(79)^2 = 160$$

For $X_{1i} = 99$, we have

$$\hat{Y}_i = 729.8665 - 10.887(99) + (.0465)(99)^2 = 107.8$$

For $X_{1i} = 119$, we have

$$\hat{Y}_i = 729.8665 - 10.887(119) + (.0465)(119)^2 = 92.8$$

Thus a store selling the razors for 79 cents is expected to sell 52.2 more packages than a store selling them for 99 cents, but a store selling them for 99 cents is expected to sell only 15 more packages than a store selling them for $1.19 (119 cents).

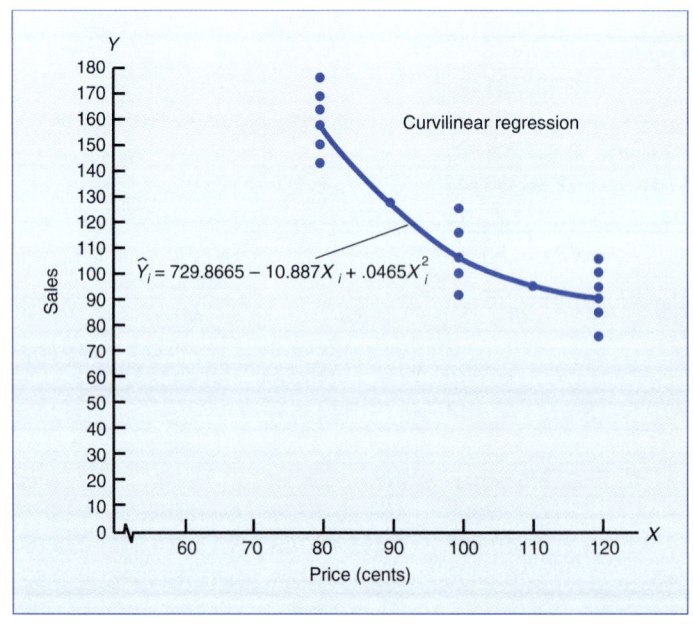

FIGURE 12.11 Scatter diagram expressing the curvilinear relationship between price and sales for the razor sales data.

12.10.2 Testing for the Significance of the Curvilinear Model

Now that the curvilinear model has been fitted to the data, we can determine whether there is a significant curvilinear relationship between sales, Y, and price, X. In a manner similar to multiple regression (see Section 12.6), the null and alternative hypotheses can be set up as follows:

H_0: $\beta_1 = \beta_{11} = 0$ (There is no relationship between X_1 and Y.)

H_1: β_1 and/or $\beta_{11} \neq 0$ (At least one regression coefficient is not equal to 0.)

The null hypothesis can be tested by using Equation (12.5):

$$F = \frac{MSR}{MSE}$$

For the razor sales data from Figure 12.10 on page 623, we have

$$F = \frac{MSR}{MSE} = \frac{6,221.4}{165.63} = 37.56$$

If a level of significance of .05 is chosen, we consult Table E.5 and find that for 2 and 12 degrees of freedom, the critical value on the F distribution is 3.89 (see Figure 12.12). Since $F = 37.56 > 3.89$, or since the p-value = .000006828 < .05, we can reject the null hypothesis (H_0) and conclude that there is a significant curvilinear relationship between sales and price of razors.

In the multiple regression model, we computed the coefficient of multiple determination $r^2_{Y.12}$ (see Section 12.4) to represent the proportion of variation in Y that is explained by variation in the explanatory variables. In curvilinear regression analysis, this coefficient can be computed from Equation (12.3):

$$r^2_{Y.12} = \frac{SSR}{SST}$$

From Figure 12.10,

$$SSR = 12,442.8 \qquad SST = 14,430.4$$

Thus,

$$r^2_{Y.12} = \frac{SSR}{SST} = \frac{12,442.8}{14,430.4} = .862$$

This coefficient of multiple determination, computed as .862, means that 86.2% of the variation in sales can be explained by the curvilinear relationship between sales and price. You

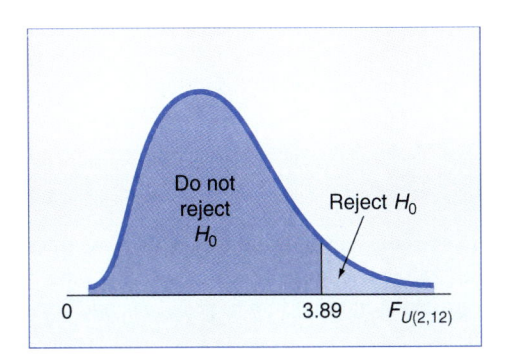

FIGURE 12.12 Testing for the existence of a curvilinear relationship at the .05 level of significance with 2 and 12 degrees of freedom.

may also recall from Section 12.4 that we computed an *adjusted* $r^2_{Y.12}$ to take into account the number of explanatory variables and the degrees of freedom. In our curvilinear regression model, $P = 2$ since we have two explanatory variables, X_1, and its square, X_1^2. Thus, using Equation (12.4) for the razor sales data, we have

$$r^2_{adj} = 1 - \left[(1 - r^2_{Y.12})\frac{(15-1)}{(15-2-1)}\right]$$

$$= 1 - \left[(1 - .862)\frac{14}{12}\right]$$

$$= 1 - .161$$

$$= .839$$

12.10.3 Testing the Curvilinear Effect

In using a regression model to examine a relationship between two variables, we would like to fit not only the most accurate model, but also the simplest model expressing that relationship. Therefore, it becomes important to examine whether there is a significant difference between the curvilinear model

$$Y_i = \beta_0 + \beta_1 X_{1i} + \beta_{11} X^2_{1i} + \epsilon_i$$

and the linear model

$$Y_i = \beta_0 + \beta_1 X_i + \epsilon_i$$

We can compare these two models by determining the regression effect of adding the curvilinear term, given that the linear term has already been included [SSR($X_1^2 | X_1$)].

You may recall that in Section 12.8.1, we used the *t* test for the regression coefficient to determine whether each particular variable made a significant contribution to the regression model. Since the standard error of each regression coefficient and its corresponding *t* statistic is available as part of the Excel output (see Figure 12.10 on page 623), we may test the significance of the contribution of the curvilinear effect with the following null and alternative hypotheses:

H_0: Including the curvilinear effect does not significantly improve the model ($\beta_{11} = 0$).

H_1: Including the curvilinear effect significantly improves the model ($\beta_{11} \neq 0$).

For our data

$$t = \frac{b_{11}}{S_{b_{11}}}$$

so that

$$t = \frac{.0465}{.01762} = 2.64$$

If a level of significance of .05 is selected, we use Table E.3 and find that with 12 degrees of freedom, the critical values are -2.1788 and $+2.1788$ (see Figure 12.13). Since $t = 2.64 > t_{12} = 2.1788$, or since the *p*-value $= .0216 < .05$, our decision would be to reject H_0 and con-

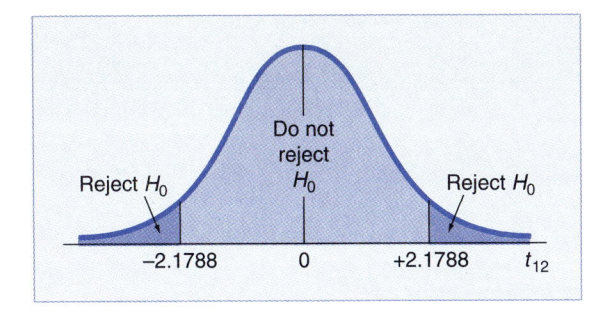

FIGURE 12.13 Testing for the contribution of the curvilinear effect to a regression model at the .05 level of significance with 12 degrees of freedom.

clude that the curvilinear model is significantly better than the linear model in representing the relationship between sales and price.

Problems for Section 12.10

Note: The problems in this section should be solved using Microsoft Excel (see Sections 11.7 and 12.14).

- 12.30 A researcher for a major oil company wishes to develop a model to predict miles per gallon based on highway speed. An experiment is designed in which a test car is driven during two trial periods at speeds ranging from 10 miles per hour to 75 miles per hour. The results are as follows:

SPEED.TXT

Observation	Miles per Gallon	Speed (miles per hour)	Observation	Miles per Gallon	Speed (miles per hour)
1	4.8	10	15	21.3	45
2	5.7	10	16	22.0	45
3	8.6	15	17	20.5	50
4	7.3	15	18	19.7	50
5	9.8	20	19	18.6	55
6	11.2	20	20	19.3	55
7	13.7	25	21	14.4	60
8	12.4	25	22	13.7	60
9	18.2	30	23	12.1	65
10	16.8	30	24	13.0	65
11	19.9	35	25	10.1	70
12	19.0	35	26	9.4	70
13	22.4	40	27	8.4	75
14	23.5	40	28	7.6	75

Assuming a curvilinear polynomial relationship between speed and mileage, based on the results obtained from Excel:
(a) Set up a scatter diagram between speed and miles per gallon.
(b) State the equation for the curvilinear model.
(c) Predict the mileage obtained when the car is driven at 55 miles per hour.
(d) Determine whether there is a significant curvilinear relationship between mileage and speed at the .05 level of significance.
(e) Interpret the meaning of the coefficient of multiple determination $r^2_{Y.12}$.
(f) Compute the adjusted r^2.
(g) Perform a residual analysis on your results and determine the adequacy of the fit of the model.

(h) At the .05 level of significance, determine whether the curvilinear model is a better fit than the linear regression model.

12.31 An industrial psychologist would like to develop a model to predict the number of typing errors based on the amount of alcoholic consumption. A random sample of 15 typists is selected with the following results:

ALCOHOL.TXT

Typist	X Alcoholic Consumption (ounces)	Y Number of Errors
1	0	2
2	0	6
3	0	3
4	1	7
5	1	5
6	1	9
7	2	12
8	2	7
9	2	9
10	3	13
11	3	18
12	3	16
13	4	24
14	4	30
15	4	22

Assuming a curvilinear relationship between alcoholic consumption and the number of errors,
(a) Set up a scatter diagram between alcoholic consumption X and number of errors Y.
(b) State the equation for the curvilinear model.
(c) Predict the number of errors made by a typist who has consumed 2.5 ounces of alcohol.
(d) Determine whether there is a significant curvilinear relationship between alcoholic consumption and the number of errors made at the .05 level of significance.
(e) Interpret the meaning of the coefficient of multiple determination $r^2_{Y.12}$.
(f) Compute the adjusted r^2.
(g) Perform a residual analysis on your results and determine the adequacy of the fit of the model.
(h) At the .05 level of significance, determine whether the curvilinear model is a better fit than the linear regression model.

12.32 Suppose an agronomist wants to design a study in which a wide range of fertilizer levels (pounds per hundred square feet) is to be used to determine whether the relationship between the yield of tomatoes and amount of fertilizer can be fit by a curvilinear model. Six rates of application are to be used: 0, 20, 40, 60, 80, and 100 pounds per hundred square feet. These rates are then randomly assigned to plots of land with the following results:

TOMYLD2.TXT

Tomato yield.

Plot	Fertilizer Application Rate (pounds per 100 square feet)	Yield (pounds)
1	0	6
2	0	9
3	20	19
4	20	24
5	40	32
6	40	38
7	60	46
8	60	50
9	80	48
10	80	54
11	100	52
12	100	58

Assuming a curvilinear relationship between the application rate and tomato yield, and based on the results obtained from Microsoft Excel
(a) Set up a scatter diagram between application rate and yield.
(b) State the regression equation for the curvilinear model.
(c) Predict the yield of tomatoes (in pounds) for a plot that has been fertilized with 70 pounds per hundred square feet of natural organic fertilizer.
(d) Determine whether there is a significant relationship between the application rate and tomato yield at the .05 level of significance.
(e) Compute the p-value in (d) and interpret its meaning.
(f) Compute the coefficient of multiple determination $r^2_{Y.12}$ and interpret its meaning.
(g) Compute the adjusted r^2.
(h) At the .05 level of significance, determine whether the curvilinear model is superior to the linear regression model.
(i) Compute the p-value in (h) and interpret its meaning.
(j) Perform a residual analysis on your results and determine the adequacy of the fit of the model.

12.33 Referring to the home heating oil consumption data of Table 12.1 on page 600
(a) Fit a multiple regression model that includes a linear relationship between heating oil consumption and temperature and a curvilinear relationship between heating oil consumption and the amount of attic insulation.
(b) Perform a residual analysis of your results and determine the adequacy of the fit of the model.
(c) At the .05 level of significance, determine whether the model that includes this curvilinear term is superior to the multiple linear regression model.

HTNGOIL.TXT

12.11 DUMMY-VARIABLE MODELS

In our discussion of multiple regression models, we have thus far assumed that each explanatory (or independent) variable is numerical. However, there are many occasions in which categorical variables need to be included in the model development process. For example, in Section 12.2, we used the atmospheric temperature and the amount of attic insulation to predict the monthly consumption of home heating oil. In addition to these numerical independent variables, we may want to include the effect of the style of the house (in terms of whether or not it is a ranch-style house) to develop a model to predict the heating oil consumption.

The use of **dummy variables** is the vehicle that permits us to consider categorical explanatory variables as part of the regression model. If a given categorical explanatory variable has two categories, then only one dummy variable will be needed to represent the two categories. The particular dummy variable (X_d) would be defined as

$$X_d = 0 \text{ if the observation is in category 1}$$

$$X_d = 1 \text{ if the observation is in category 2}$$

To illustrate the application of dummy variables in regression, let us examine a model for predicting the heating oil consumption based on atmospheric temperature (X_1), the amount of attic insulation (X_2), and whether or not the house is a ranch-style house (X_3). Thus, a dummy variable for ranch-style house (X_3) could be defined as

$$X_3 = 0 \text{ if the style is not ranch}$$

$$X_3 = 1 \text{ if the style is ranch}$$

Assuming that the slope between home heating oil consumption and atmospheric temperature (X_1) and the amount of attic insulation (X_2) is the same for both groups,[3] the regression model is

$$Y_i = \beta_0 + \beta_1 X_{1i} + \beta_2 X_{2i} + \beta_3 X_{3i} + \epsilon_i \tag{12.14}$$

where
Y_i = monthly heating oil consumption in gallons

β_0 = Y intercept

β_1 = slope of heating oil consumption with atmospheric temperature holding constant the effect of attic insulation and the house style

β_2 = slope of heating oil consumption with attic insulation holding constant the effect of atmospheric temperature and the house style

β_3 = incremental effect of the presence of a ranch-style house holding constant the effect of atmospheric temperature and attic insulation

ϵ_i = random error in Y for house i

Using the sample of 15 houses in which houses 1, 4, 6, 7, 8, 10, and 12 are ranch-style houses, the model stated in Equation (12.14) is fitted. Figure 12.14 displays partial output obtained from Excel. From this output, we note the following:

1. Holding constant the effect of attic insulation and the house style, for each additional 1°F increase in atmospheric temperature, the average oil consumption is predicted to decrease by 5.525 gallons.

2. Holding constant the effect of atmospheric temperature and the house style, for each additional 1-inch increase in attic insulation, the average oil consumption is predicted to decrease by 21.376 gallons.

3. β_3 measures the effect on oil consumption of having a ranch-style house ($X_3 = 1$) as compared with not having a ranch-style house ($X_3 = 0$). Thus, we estimate that, holding the atmospheric temperature and attic insulation constant, a ranch-style house is predicted to use 38.973 fewer gallons of heating oil per month than when the style is not ranch.

Using the results of Figure 12.14, the model for these data may be stated as

$$\hat{Y}_i = 592.5401 - 5.5251 X_{1i} - 21.3761 X_{2i} - 38.9727 X_{3i}$$

	A	B	C	D	E	F
1	SUMMARY OUTPUT					
2						
3	*Regression Statistics*					
4	Multiple R	0.994206187				
5	R Square	0.988445943				
6	Adjusted R Square	0.985294836				
7	Standard Error	15.7489393				
8	Observations	15				
9						
10	ANOVA					
11		df	SS	MS	F	Significance F
12	Regression	3	233406.9094	77802.30312	313.6821708	6.21548E-11
13	Residual	11	2728.319981	248.0290892		
14	Total	14	236135.2293			
15						
16		Coefficients	Standard Error	t Stat	P-value	
17	Intercept	592.5401168	14.33698425	41.32948091	2.02317E-13	
18	Temperature	-5.525100884	0.204431228	-27.0266971	2.07188E-11	
19	Insulation	-21.37612794	1.448019304	-14.76232249	1.34816E-08	
20	Style	-38.97266608	8.358437237	-4.662673772	0.000690709	

FIGURE 12.14 Excel output for the regression model including the dummy variable for the home heating oil consumption data.

For houses that are not ranch-style, the model reduces to

$$\hat{Y}_i = 592.5401 - 5.5251X_{1i} - 21.3761X_{2i}$$

since $X_3 = 0$. For houses that are ranch-style, the model reduces to

$$\hat{Y}_i = 553.5674 - 5.5251X_{1i} - 21.3761X_{2i}$$

since $X_3 = 1$.

Problems for Section 12.11

Note: *The problems in this section should be solved using Microsoft Excel (see Sections 11.7 and 12.14).*

12.34 Under what circumstances would we want to include a dummy variable in a regression model?

12.35 What assumption concerning the slope between the response variable Y and the explanatory variable X must be made when a dummy variable is included in a regression model?

12.36 A bank would like to develop a model to predict the total sum of money customers withdraw from automatic teller machines (ATMs) on a weekend based on the median value of homes in the vicinity of the ATM and the ATM's location (0 = not a shopping center; 1 = shopping center). A random sample of 15 ATMs is selected with the results shown at the top of page 632:

ATM2.TXT

ATM Number	Withdrawal Amount ($000)	Median Value of Homes ($000)	Location of ATM
1	12.0	225	1
2	9.9	170	0
3	9.1	153	1
4	8.2	132	0
5	12.4	237	1
6	10.4	187	1
7	12.7	245	1
8	8.0	125	1
9	11.5	215	1
10	9.7	170	0
11	11.7	223	0
12	8.6	147	0
13	10.9	197	1
14	9.4	167	0
15	11.2	210	0

On the basis of the results obtained using Microsoft Excel:
(a) State the multiple regression equation.
(b) Interpret the meaning of the slopes in this problem.
(c) Predict the average withdrawal amount for a neighborhood in which the median value of homes is $200,000 for an ATM located in a shopping center.
(d) Determine whether there is a significant relationship between the withdrawal amount and the two explanatory variables (median value of houses and the dummy variable of ATM location) at the .05 level of significance.
(e) Interpret the meaning of the coefficient of multiple determination $r^2_{Y.12}$.
(f) Compute the adjusted r^2.
(g) Perform a residual analysis of your results and determine the adequacy of the model's fit.
(h) At the .05 level of significance, determine whether each explanatory variable makes a contribution to the regression model. On the basis of these results, indicate the regression model that should be used in this problem.
(i) Set up 95% confidence interval estimates of the population slope for the relationship between the withdrawal amount and median value of homes, and for the withdrawal amount and ATM location.
(j) Compute and interpret the coefficients of partial determination.
(k) What assumption about the slope of withdrawal amount with median value of homes must be made in this problem?

PETFOOD.TXT

● 12.37 Referring to Problem 11.1 (pet food sales) on page 536, suppose that in addition to studying the effect of shelf space on the sales of pet food, the marketing manager also wants to study the effect of product placement on sales. Suppose that in stores 2, 6, 9, and 12, the pet food is placed at the front of the aisle, while in the other stores it is placed at the back of the aisle. On the basis of the results obtained from Microsoft Excel
(a) State the multiple regression equation.
(b) Interpret the meaning of the slopes in this problem.
(c) Predict the average weekly sales of pet food for a store with 8 square feet of shelf space that is situated at the back of the aisle.
(d) Determine whether there is a significant relationship between sales and the two explanatory variables (shelf space and the dummy variable aisle position) at the .05 level of significance.
(e) Interpret the meaning of the coefficient of multiple determination $r^2_{Y.12}$.
(f) Compute the adjusted r^2.
(g) Compare $r^2_{Y.12}$ with the r^2 value computed in Problem 11.16(a) on page 554 and the adjusted r^2 of (f) with the adjusted r^2 computed in Problem 11.16(b). Explain the results.

(h) At the .05 level of significance, determine whether each explanatory variable makes a contribution to the regression model. On the basis of these results, indicate the regression model that should be used in this problem.
(i) Set up 95% confidence interval estimates of the population slope for the relationship between sales and shelf space and between sales and aisle location.
(j) Compare the slope obtained in (b) with the slope for the simple linear regression model of Problem 11.6 on page 544. Explain the difference in the results.
(k) Compute the coefficients of partial determination and interpret their meaning.
(l) What assumption about the slope of shelf space with sales must be made in this problem?
(m) Perform a residual analysis on your results and determine the adequacy of the fit of the model.

12.38 Referring to Problem 11.65 (relating wins to E.R.A.) on page 596, suppose that in addition to using earned run average to predict the number of wins, Crazy Dave wants to include the league (American versus National) as an explanatory variable.
(a) State the multiple regression equation.
(b) Interpret the meaning of the slopes in this problem.
(c) Predict the average number of wins for a team with an E.R.A. of 4.00 in the American League.
(d) Determine whether there is a significant relationship between wins and the two explanatory variables (E.R.A. and league) at the .05 level of significance.
(e) Interpret the meaning of the coefficient of multiple determination $r^2_{Y.12}$.
(f) Compute the adjusted r^2.
(g) Compare $r^2_{Y.12}$ with the value computed in Problem 11.65(f). Explain the results.
(h) Perform a residual analysis on your results and determine the adequacy of the fit of the model.
(i) At the .05 level of significance, determine whether each explanatory variable makes a contribution to the regression model. On the basis of these results, indicate the regression model that should be used in this problem.
(j) Set up 95% confidence interval estimates of the population slope for the relationship between wins and E.R.A. and between wins and league.
(k) Compare the slope obtained in (b) with the slope for the simple linear regression model of Problem 11.65 on page 596. Explain the difference in the results.
(l) Compute the coefficients of partial determination and interpret their meaning.
(m) What assumption about the slope of wins with E.R.A. must be made in this problem?

12.12 OTHER TYPES OF REGRESSION MODELS

In our discussion of multiple regression models, we have thus far examined the multiple linear regression model [Equations (12.1a and b)], the curvilinear polynomial model [Equation (12.13a)], and the dummy-variable model [Equation (12.14)]. See pages 601, 621, and 630, respectively.

12.12.1 Interaction Terms in Regression Models

In the multiple linear regression model [Equation (12.1b)], we have included only terms that express a relationship between the explanatory variables and a dependent variable Y. However, in some situations the relationship between X_1 and Y changes for differing values of X_2. In such a case, an interaction term involving the product of explanatory variables may be included. With two explanatory variables, such an interaction model may be stated as

$$Y_i = \beta_0 + \beta_1 X_{1i} + \beta_2 X_{2i} + \beta_3 X_{1i} X_{2i} + \epsilon_i \qquad (12.15)$$

As an example of an interaction model, let us refer to the dummy-variable model discussed in Section 12.11 for the home heating oil consumption data. You may recall that Equation (12.14) postulated a dummy-variable model in which the slope of X_1 was constant for each category of the dummy variable (X_3). If in fact the slope of heating oil consumption with atmospheric temperature was different for houses that were ranch-style and those that were not, an interaction term consisting of the product of these two explanatory variables should be included. For this example, the model would be stated as

$$Y_i = \beta_0 + \beta_1 X_{1i} + \beta_2 X_{2i} + \beta_3 X_{3i} + \beta_4 X_{1i} X_{3i} + \epsilon_i$$

where Y_i = monthly heating oil consumption

β_0 = Y-intercept

β_1 = slope of heating oil consumption with atmospheric temperature holding constant the amount of attic insulation and house style

β_2 = slope of heating oil consumption with attic insulation holding constant the atmospheric temperature and house style

β_3 = incremental effect of having a ranch-style house holding constant the atmospheric temperature and attic insulation

β_4 = slope representing interaction of atmospheric temperature and house style

ϵ_i = random error in Y for house i

12.12.2   Using Transformations in Regression Models

In Section 11.9, we discussed the assumptions of normality, homoscedasticity, and independence of error that are involved in the regression model. In many circumstances, the effect of violations of these assumptions can be overcome by transforming the dependent variable, the explanatory variables, or both.

By using the transformed variables, we are often able to obtain a simpler model than had we maintained the original variables. By reexpressing X and/or Y, we may simplify the relationship to one that is linear in its transformation. Unfortunately, the choice of an appropriate transformation is often not an easy one to make. Among the transformations along a "ladder of powers" discussed by Tukey (see References 4 and 15) are the square-root transformation, the logarithmic transformation, and the reciprocal transformation. If a **square-root transformation** were applied to the values of each of two explanatory variables, the multiple regression model would be

$$Y_i = \beta_0 + \beta_1 \sqrt{X_{1i}} + \beta_2 \sqrt{X_{2i}} + \epsilon_i \qquad (12.16)$$

If a **logarithmic transformation** had been applied, the model would be

$$Y_i = \beta_0 + \beta_1 \ln X_{1i} + \beta_2 \ln X_{2i} + \epsilon_i \qquad (12.17)$$

If a **reciprocal transformation** were applied, the model would be

$$Y_i = \beta_0 + \beta_1 \frac{1}{X_{1i}} + \beta_2 \frac{1}{X_{2i}} + \epsilon_i \qquad (12.18)$$

In some situations, the use of a transformation can change what appears to be a nonlinear model into a linear model. For example, the multiplicative model

$$Y_i = \beta_0 X_{1i}^{\beta_1} X_{2i}^{\beta_2} \epsilon_i \qquad (12.19)$$

can be transformed (by taking natural logarithms[4] of both the dependent and explanatory variables) to the model

$$\ln Y_i = \ln \beta_0 + \beta_1 \ln X_{1i} + \beta_2 \ln X_{2i} + \ln \epsilon_i \qquad (12.20)$$

Hence, Equation (12.20) is linear in the natural logarithms. In a similar fashion, the **exponential model**

$$Y_i = e^{\beta_0 + \beta_1 X_{1i} + \beta_2 X_{2i}} \epsilon_i \qquad (12.21)$$

can also be transformed to linear form (by taking natural logarithms of both the dependent and explanatory variables). The resulting model is

$$\ln Y_i = \beta_0 + \beta_1 X_{1i} + \beta_2 X_{2i} + \ln \epsilon_i \qquad (12.22)$$

Problems for Section 12.12

Note: *The problems in this section should be solved using Microsoft Excel (see Sections 11.7 and 12.14).*

12.39 Referring to Problem 12.36 (ATM location) on page 631, suppose we wish to include a term in the multiple regression model that represents the interaction of median value of homes and ATM location. Reanalyze the data for this model. On the basis of the results obtained

(a) State the multiple regression equation.

(b) At the .05 level of significance, determine whether the inclusion of an interaction term makes a significant contribution to a model that already includes the median value of homes and the ATM location. On the basis of these results, indicate the regression model that should be used in this problem.

ATM2.TXT

12.40 Referring to the data of Problem 12.30 (gas mileage) on page 627, use a square-root transformation of the explanatory variable (speed) as in Equation (12.16) and reanalyze the data using this model. On the basis of your results

(a) State the regression equation.

(b) Predict the average mileage obtained when the car is driven at 55 miles per hour.

(c) Perform a residual analysis of your results and determine the adequacy of the fit of the model.

SPEED.TXT

(d) At the .05 level of significance, is there a significant relationship between mileage and the square root of speed?
(e) Interpret the meaning of the coefficient of determination r^2 in this problem.
(f) Compute the adjusted r^2.
(g) Compare your results with those obtained in Problem 12.30. Which model would you choose? Why?

SPEED.TXT

12.41 Referring to the data of Problem 12.30 on page 627, use a logarithmic transformation of the explanatory variable (speed) as in Equation (12.17) and reanalyze the data using this model. On the basis of your results
(a) State the regression equation.
(b) Predict the average mileage obtained when the car is driven at 55 miles per hour.
(c) Perform a residual analysis of your results and determine the adequacy of the fit of the model.
(d) At the .05 level of significance, is there a significant relationship between mileage and the natural logarithm of speed?
(e) Interpret the meaning of the coefficient of determination r^2 in this problem.
(f) Compute the adjusted r^2.
(g) Compare your results with those obtained in Problems 12.30 and 12.40. Which model would you choose? Why?

PETFOOD.TXT

• 12.42 Referring to Problem 12.37 (pet food sales) on page 632, suppose we wish to include a term in the multiple regression model that represents the interaction of shelf space and aisle location. Reanalyze the data for this model. On the basis of the results, obtained
(a) State the multiple regression equation.
(b) At the .05 level of significance, determine whether the inclusion of the interaction term makes a significant contribution to a model that already includes shelf space and aisle location. On the basis of these results, indicate the regression model that should be used in this problem.

BB95.TXT

12.43 Referring to Problem 12.38 (relating wins to E.R.A.) on page 633, suppose that Crazy Dave wishes to include a term in the multiple regression model for the interaction of E.R.A. and league. Reanalyze the data for this model. On the basis of the results obtained
(a) State the multiple regression equation.
(b) At the .05 level of significance, determine whether the inclusion of an interaction term makes a significant contribution to a model that already includes E.R.A. and league. On the basis of these results, indicate the regression model that should be used in this problem.

12.13 MULTICOLLINEARITY

One important problem in the application of multiple regression analysis involves the possible **multicollinearity** of the explanatory variables. This condition refers to situations in which some of the explanatory variables are highly correlated with each other. In such situations, collinear variables do not provide new information, and it becomes difficult to separate the effect of such variables on the dependent or response variable. In such cases, the values of the regression coefficients for the correlated variables may fluctuate drastically, depending on which variables are included in the model.

One method of measuring collinearity uses the **variance inflationary factor (VIF)** for each explanatory variable. This VIF is defined as in Equation (12.23):

$$\text{VIF}_j = \frac{1}{1 - R_j^2} \tag{12.23}$$

where R_j^2 = coefficient of multiple determination of explanatory variable X_j with all other X variables

If there are only two explanatory variables, R_j^2 is just the coefficient of determination between X_1 and X_2. If, for example, there were three explanatory variables, then R_1^2 would be the **coefficient of multiple determination** of X_1 with X_2 and X_3.

If a set of explanatory variables is uncorrelated, then VIF_j will be equal to 1. If the set is highly intercorrelated, then VIF_j might even exceed 10. Marquardt (see Reference 10) suggests that if VIF_j is greater than 10, there is too much correlation between variable X_j and the other explanatory variables. However, other researchers (see Reference 13) suggest a more conservative criterion that would employ alternatives to least squares regression if the maximum VIF_j exceeds 5.

If we reexamine our monthly heating oil consumption data, we note from Table 12.2 on page 608 that the correlation between the two explanatory variables, temperature and attic insulation, is only .00892. Therefore, since there are only two explanatory variables in the model, we may compute the VIF_j from Equation (12.23):

$$VIF_1 = VIF_2 = \frac{1}{1 - (.00892)^2}$$

$$\cong 1.00$$

Thus, we may conclude that there is no reason to suspect any multicollinearity for the heating oil consumption data.

Problems for Section 12.13

Note: *The problems in this section should be solved using Microsoft Excel (see Sections 11.7 and 12.14).*

12.44 Referring to Problem 12.3 (distribution cost) on page 604, determine the VIF for each explanatory variable in the model. Is there reason to suspect the existence of multicollinearity?

12.45 Referring to Problem 12.4 (advertising media) on page 605, determine the VIF for each explanatory variable in the model. Is there reason to suspect the existence of multicollinearity?

12.46 Referring to Problem 12.5 (standby hours) on page 605, determine the VIF for each explanatory variable in the model. Is there reason to suspect the existence of multicollinearity?

12.14 USING MICROSOFT EXCEL FOR MULTIPLE REGRESSION

In Chapter 11, we used the Data Analysis tool and its Regression option for the simple linear regression model. In this chapter we developed a variety of multiple regression models whose computations can also be done with the Regression option of the Data Analysis tool. When using the Regression option for multiple regression models, it is important to remember that the entire set of X variables must be placed in consecutive columns, since the Regression tool will allow us to specify only a single contiguous range for the X variable as noted in the next section.

12.14.1 Using the Regression Option of the Data Analysis tool for Multiple Regression

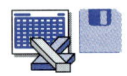

12-14-1.XLS

To illustrate the use of Microsoft Excel in multiple regression, let us return to the monthly heating oil consumption example. Open the HEATING.XLS workbook and click the Data sheet tab to make that sheet active. Select Tools | Data Analysis and select Regression from the Analysis Tools list box and click OK. In the Regression dialog box, do the following:

(a) Enter A1:A16 in the Input Y Range edit box.

(b) Enter B1:C16 in the Input X Range edit box. Note that only a single, contiguous range is allowed in this edit box. (In this example, values for temperature have been placed in column B and values for attic insulation in column C).

(c) Select the Labels check box.

(d) Select the New Worksheet Ply option button and enter Figure 12.2 as the sheet name.

(e) Select the Residuals, Standardized Residuals, and Residual Plots check boxes. Click the OK button to obtain the output similar to that displayed in Figure 12.2 on page 603.

12.14.2 Using the Correlation Option of the Data Analysis Tool for Multiple Regression

12-14-2.XLS

The Correlation option of the Data Analysis tool computes the correlation between all pairs of variables listed in the Input Range box. To obtain the correlation matrix for our Y variable and our two X variables, again open the HEATING.XLS workbook (or the workbook developed in the previous section) and click the Data Sheet tab to make that sheet the active one. Select Tools | Data Analysis and then select Correlation from the Analysis Tools list box and click OK. Enter A1:C16 in the Input Range edit box, select the Labels in First Row check box, and select the New Worksheet Ply option button, entering Corr. Matrix as the new sheet name. Click the OK button. The resulting correlation matrix (similar to Table 12.3) is displayed in Figure 12.1.Excel.

	A	B	C	D
1		Heating Oil	Temperature	Insulation
2	Heating Oil	1		
3	Temperature	-0.86974117	1	
4	Insulation	-0.465082527	0.008922039	1

FIGURE 12.1.EXCEL Correlation matrix using Microsoft Excel for the monthly heating oil consumption data.

12.14.3 Using Microsoft Excel to Obtain the Coefficients of Partial Determination

12-14-3.XLS

The coefficients of partial determination may be obtained using Equations (12.11a) and (12.11b) on page 620 along with the output of the multiple regression model produced in Section 12.14.1, the output of a regression model of insulation (X_2) with monthly heating oil

Table 12.1.Excel Design for calculating the coefficients of partial determination.

	A	B
1	Coefficients of Partial Determination	
2		
3	SSR(X1 and X2)	xxxx
4	SSR(X1)	xxxx
5	SSR(X2)	xxxx
6	SST	xxxx
7	SSR(X1 | X2)	=B3–B5
8	SSR(X2 | X1)	=B3–B4
9	RSQY1.2	=B7/(B6–B3+B7)
10	RSQY2.1	=B8/(B6–B3+B8)

consumption, and the output of a regression model of temperature (X_1) with heating oil consumption. The design for a Calculations sheet to compute the coefficients of partial determination is presented in Table 12.1.Excel.

To implement this sheet, open the workbook developed in Section 12.14.1 (or the 12-14-1.XLS workbook). Use the Data Analysis regression tool twice, following the procedure discussed in Section 11.7 and use the values in the table below. This will produce the two other regression models required by this sheet.

Regression Model	Input Y Range	Input X Range	New Worksheet Ply Name
Insulation (X_2) with heating oil	A1:A16	C1:C16	Figure 12.5
Temperature (X_1) with heating oil	A1:A16	B1:B16	Figure 12.6

(These values will produce outputs similar to Figures 12.5 and 12.6 on page 614, which can be used as alternate sources of data for this example.)

After producing the regression model outputs, insert a new worksheet and rename it Calculations. Then, copy the following:

(a) The regression sum of squares for the model that includes X_1 and X_2 from cell C12 of the Figure 12.2 sheet to cell B3.

(b) The regression sum of squares for the model that includes X_1 from cell C12 of the Figure 12.6 sheet to cell B4.

(c) The regression sum of squares for the model that includes X_2 from cell C12 of the Figure 12.5 sheet to cell B5.

(d) The total sum of squares from cell C14 of the Figure 12.2 sheet to cell B6.

Next, enter the formula =B3–B5 in cell B7 to compute the regression sum of squares for X_1 given X_2. Enter the formula =B3–B4 in cell B8 to compute the regression sum of squares of X_2 given X_1. Finally, compute the coefficients of partial determination by entering the formulas =B7/(B6–B3+B7) and =B8/(B6–B3+B8) in cells B9 and B10, respectively. Figure 12.2.Excel illustrates the resulting coefficients of partial determination.

FIGURE 12.2.EXCEL Coefficients of partial determination obtained from Microsoft Excel for the monthly heating oil consumption data.

	A	B
1	Coefficients of Partial Determination	
2		
3	SSR(X1 and X2)	228014.6
4	SSR(X1)	178624.4
5	SSR(X2)	51076.47
6	SST	236135.2
7	SSR(X1 | X2)	176938.2
8	SSR(X2 | X1)	49390.2
9	RSQY1.2	0.956119
10	RSQY2.1	0.858799

12.14.4 Using Microsoft Excel for Curvilinear Regression

12-14-4.XLS

The curvilinear regression model as expressed in Equation (12.13b) on page 622 includes both linear (X) and curvilinear (X^2) terms for each independent variable believed to have a curvilinear relationship with Y. To illustrate the development of the curvilinear model using Excel, we will refer to the razor sales example discussed in Section 12.10. Open the RAZOR.XLS workbook which contains the price and sales data of Table 12.7 on page 622. Click the Data sheet tab to make that sheet active. Insert a column between columns A and B to hold the X^2 terms (by clicking any cell in column B and then selecting Insert | Columns). The sales column becomes column C and a new, blank Column B appears. (Do *not* put the X^2 terms in a column to the right of the Sales column for reasons noted at the beginning of Section 12.14.) After entering the column heading Price Squared in cell B1, enter the formula =A2^2 in cell B2, and copy it down the column through row 16. Then use the Data Analysis Regression tool with the Input Y Range entered as C1:C16 and Input X Range entered as A1:B16. The output obtained is similar to Figure 12.10 on page 623.

12.14.5 Using Microsoft Excel for Dummy-Variable and Other Types of Regression Models

In Section 12.11, we covered the use of categorical independent variables (known as dummy variables) in regression, while in Section 12.12, we introduced a variety of other regression models. Models that include dummy variables do not require any special Excel functions. Each dummy variable needs to be coded as zero or one for each observation depending on whether the observation is in category one or category two. Interaction terms can be included in the model by defining a new variable as the product of two other variables.

Models involving transformations (see page 634) can also be developed using an appropriate Excel function. To apply the square-root transformation of the model shown in Equation (12.16) on page 634, we can use the SQRT function. To apply transformations involving the natural logarithm (ln), we can use the LN function.

12.14.6 Using Microsoft Excel to Obtain the Variance Inflationary Factor

In Section 12.13, we discussed using the variance inflationary factor (VIF) [Equation (12.23) on page 636] as a measure of the multicollinearity of an independent variable. Microsoft Excel can be used to obtain the VIF for each independent variable by using the Data Analysis Regression tool to perform a regression analysis in which the X variable of interest *is used as the dependent variable* and all the other X variables are the independent variables. For example, if we want to compute the VIF for variable X_1 when there are two independent variables, X_1 and X_2, the dependent variable would be X_1, and X_2 would be the independent variable. If we want to compute the VIF for variable X_1 when there are three independent variables, X_1, X_2, and X_3, the dependent variable would be X_1, and the independent variables would be X_2 and X_3. Then the VIF would be computed by using a formula in the format of $=1/(1-(1-r^2 \text{ value}))$ using the r^2 that can be found in cell B5 of the Regression output.

12.15 PITFALLS IN MULTIPLE REGRESSION AND ETHICAL ISSUES

12.15.1 Pitfalls in Multiple Regression

In Section 11.15, we discussed pitfalls in regression and ethical issues. Now that we have examined a variety of multiple regression models, we need to concern ourselves with some additional pitfalls related to the use of regression analysis. They include the following:

1. The need to understand that the regression coefficient for a particular independent variable is interpreted from a perspective in which the values of all other independent variables are held constant
2. The need to use residual plots for each independent variable
3. The need to evaluate interaction terms to determine whether the slope of other independent variables with the response variable is the same at each level of the dummy variable
4. The need to obtain the VIF for each independent variable before determining which independent variables should be included in the model

12.15.2 Ethical Considerations

Ethical considerations arise when a user wishing to develop forecasts manipulates the development process of the multiple regression model. The key here is intent. In addition to the situations discussed in Section 11.15.3, unethical behavior occurs when someone uses multiple regression analysis and willfully fails to (1) remove variables from consideration that exhibit a high multicollinearity with other independent variables, or (2) use methods other than least squares regression when the assumptions necessary for least squares regression have been seriously violated.

12.16 SUMMARY AND OVERVIEW

In this chapter we developed the multiple regression model, including dummy variables, transformations, and multicollinearity. To be sure you understand what we have covered, you should be able to answer the following conceptual questions:

1. How does the interpretation of the regression coefficients differ in multiple regression as compared to simple regression?
2. How does testing the significance of the entire regression model differ from testing the contribution of each independent variable in the multiple regression model?

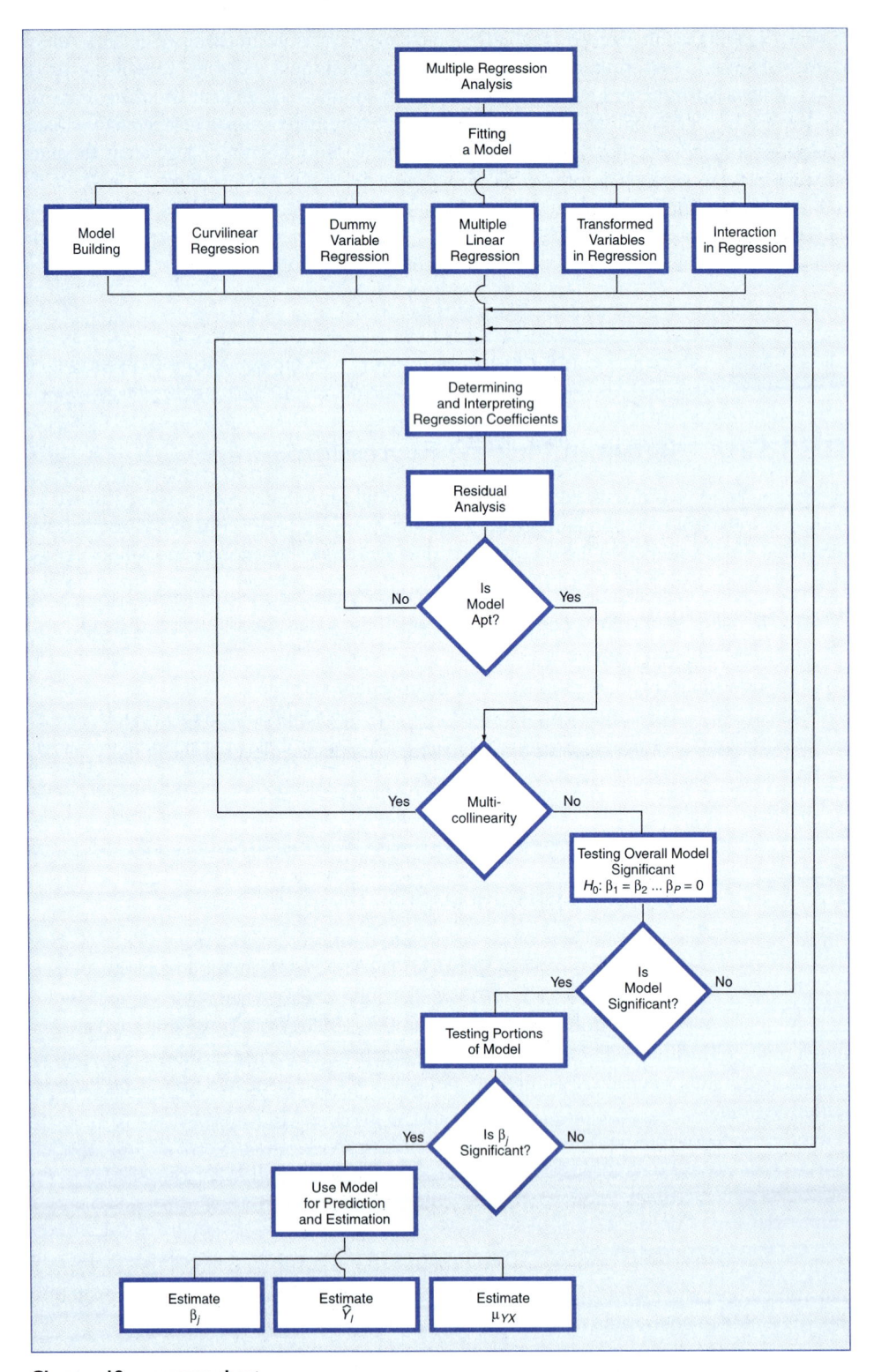

Chapter 12 summary chart.

3. How do the coefficients of partial determination differ from the coefficient of multiple determination?
4. Why and how are dummy variables used?
5. How can we evaluate whether the slope of an independent variable with the response variable is the same for each level of the dummy variable?
6. What is the purpose of using transformations in multiple regression analysis?
7. How do we evaluate whether independent variables are intercorrelated?

In Chapter 13, we will continue our discussion of forecasting by considering a variety of time-series forecasting models.

Getting It All Together

Key Terms

coefficient of multiple determination   607
coefficient of partial determination   620
curvilinear regression model   621
dummy variables   630
exponential model   635
interaction term   633
logarithmic transformation   634

multicollinearity   636
multiple regression   600
net regression coefficient   602
partial F-test criterion   613
reciprocal transformation   634
square-root transformation   634
variance inflationary factor (VIF)   636

Chapter Review Problems

Note: *Microsoft Excel can be used to solve the Chapter Review Problems.*

12.47 Crazy Dave, the well-known baseball analyst, has expanded his analysis of which variables are important in predicting a team's wins in a given season. He has collected the following data related to wins, E.R.A., runs scored, and saves for a recent season (1995):

BASEBALL.TXT

Team	Wins	E.R.A.	Runs Scored	Saves
Boston	86	4.39	791	39
Cleveland	100	3.83	840	50
Kansas City	70	4.49	629	37
Minnesota	56	5.76	703	27
Toronto	56	4.88	642	22
California	78	4.49	800	42
Seattle	78	4.52	787	39
Texas	74	4.66	691	34
Detroit	60	5.49	654	38
Chicago White Sox	68	4.85	755	36
Milwaukee	65	4.82	740	31
Oakland	67	4.93	730	34
Baltimore	71	4.31	704	29

(continued)

(continued)

Team	Wins	E.R.A.	Runs Scored	Saves
New York Yankees	79	4.56	749	35
Florida	67	4.27	673	29
Chicago Cubs	73	4.13	693	45
Cincinnati	85	4.03	747	38
San Francisco	67	4.86	652	34
Los Angeles	78	3.66	634	37
Pittsburgh	58	4.70	629	29
San Diego	70	4.13	668	35
New York Mets	69	3.88	657	36
St. Louis	62	4.09	563	38
Philadelphia	69	4.21	615	41
Atlanta	90	3.44	645	34
Montreal	66	4.11	621	42
Houston	76	4.06	747	32
Colorado	77	4.97	785	43

Evaluate the variables provided (runs, E.R.A., and saves) as possible explanatory variables to be included in the model. Be sure to include a thorough residual analysis. In addition, provide a detailed explanation of your results.

12.48 A headline on page 1 of *The New York Times* of March 4, 1990, read "Wine Equation Puts Some Noses Out of Joint." The article proceeded to explain that Professor Orley Ashenfelter, a Princeton University economist, had developed a multiple regression model to predict the quality of French Bordeaux based on the amount of winter rain, the average temperature during the growing season, and the harvest rain. The equation developed was

$$Q = -12.145 + .00117 WR + .6164 TMP - .00386 HR$$

where Q = logarithmic index of quality where 1961 equals 100

WR = winter rain (October through March) in millimeters

TMP = average temperature during the growing season (April through September), in degrees Celsius

HR = harvest rain (August to September) in millimeters

You are at a cocktail party, sipping a glass of wine, when one of your friends mentions to you that she has read the article. She asks you to explain the meaning of the coefficients in the equation and also asks you what analyses that might have been done have not been included in the article. You respond…

12.49 Referring to Problem 11.58 on page 589, the statistician for the automobile manufacturer believes that a second explanatory variable, shipping mileage, may be related to delivery time. The following shipping mileages (in hundreds of miles) for the 16-car sample is collected:

S-12-49.XLS

644 Chapter 12 Multiple Regression Models

Car	Shipping Mileage (hundreds of miles)
1	7.5
2	13.3
3	4.7
4	14.6
5	8.4
6	12.6
7	6.2
8	16.4
9	9.7
10	17.2
11	10.6
12	11.3
13	9.0
14	12.3
15	8.2
16	11.5

On the basis of the results obtained using Microsoft Excel:
(a) State the multiple regression equation.
(b) Interpret the meaning of the slopes in this problem.
(c) If a car is ordered with 10 options and has to be shipped 800 miles, what would you predict the average delivery time to be?
(d) Determine whether there is a significant relationship between delivery time and the two explanatory variables (number of options and shipping mileage) at the .05 level of significance.
(e) Indicate the *p*-values in (d) and interpret their meaning.
(f) Interpret the meaning of the coefficient of multiple determination $r^2_{Y.12}$ in this problem.
(g) Compute the adjusted r^2.
(h) At the .05 level of significance, determine whether each explanatory variable makes a significant contribution to the regression model. Based on these results, indicate the regression model that should be utilized in this problem.
(i) Indicate the *p*-values in (h) and interpret their meaning.
(j) Compute the coefficients of partial determination $r^2_{Y1.2}$ and $r^2_{Y2.1}$ and interpret their meaning.
(k) Determine the VIF for each explanatory variable in the model. Is there reason to suspect the existence of multicollinearity?
(l) Perform a residual analysis on your results and determine the adequacy of the fit of the model (relating selling price to assessed value).

12.50 Referring to Problem 11.61 on page 592, suppose that we also wish to include the time period in which the house was sold in the model. The following table represents the time (in months) in which each of the 30 houses was sold.

HOUSE1.TXT

Time period for the sample of 30 houses.

House	Time Period (months)	House	Time Period (months)
1	10	16	12
2	10	17	5
3	11	18	14
4	2	19	1
5	5	20	3
6	4	21	14
7	17	22	12
8	13	23	11
9	6	24	12
10	5	25	2
11	7	26	6
12	4	27	12
13	11	28	4
14	10	29	9
15	17	30	12

Based on the results obtained from Microsoft Excel:
(a) State the multiple regression equation.
(b) Interpret the meaning of the slopes in this equation.
(c) Predict the average selling price for a house that has an assessed value of $70,000 and was sold in time period 12.
(d) Determine whether there is a significant relationship between selling price and the two explanatory variables (assessed value and time period) at the .05 level of significance.
(e) Compute the p-value in (d) and interpret its meaning.
(f) Interpret the meaning of the coefficient of multiple determination $r^2_{Y.12}$ in this problem.
(g) Compute the adjusted r^2.
(h) At the .05 level of significance, determine whether each explanatory variable makes a significant contribution to the regression model. Based on these results, indicate the regression model that should be used in this problem.
(i) Compute the p-values in (h) and interpret their meaning.
(j) Set up a 95% confidence interval estimate of the true population slope between selling price and assessed value. How does the interpretation of the slope here differ from Problem 11.61(k)?
(k) Compute the coefficients of partial determination $r^2_{Y1.2}$ and $r^2_{Y2.1}$ and interpret their meaning.
(l) Determine the VIF for each explanatory variable in the model. Is there reason to suspect the existence of multicollinearity?
(m) Perform a residual analysis on your results and determine the adequacy of the fit of the model.

BET.TXT

12.51 Referring to the data of Problem 11.59 (race track betting) on page 590, suppose we wish to fit a curvilinear model to predict the amount bet based on attendance. Based on the results obtained from Microsoft Excel:
(a) State the regression equation.
(b) Predict the average amount bet for a day in which attendance is 30,000.
(c) Determine whether there is a significant relationship between attendance and amount bet at the .05 level of significance.
(d) Compute the p-value in (c) and interpret its meaning.
(e) Compute the coefficient of multiple determination $r^2_{Y.12}$ and interpret its meaning.

(f) Compute the adjusted r^2.
(g) At the .05 level of significance, determine whether the curvilinear model is superior to the linear regression model.
(h) Compute the *p*-value in (g) and interpret its meaning.
(i) Perform a residual analysis on your results and determine the adequacy of the fit of the model.

12.52 Referring to Problem 11.61 on page 592, suppose that in addition to using assessed value to predict sales price, we also want to use information concerning whether the house is brand new. Houses 1, 2, 10, 14, 18, 20, 22, 24, 25, 26, 28, and 30 are brand new. On the basis of the results obtained from Microsoft Excel:

HOUSE1.TXT

(a) State the multiple regression equation.
(b) Interpret the meaning of the slopes in this problem.
(c) Predict the average selling price for a brand new house with an assessed value of $75,000.
(d) Determine whether there is a significant relationship between selling price and the two explanatory variables (assessed value and whether the house is brand new) at the .05 level of significance.
(e) Compute the *p*-value in (d) and interpret its meaning.
(f) Interpret the meaning of the coefficient of multiple determination $r^2_{Y.12}$.
(g) Compute the adjusted r^2.
(h) At the .05 level of significance, determine whether each explanatory variable makes a significant contribution to the regression model. Based on these results, indicate the regression model that should be used in this problem.
(i) Compute the *p*-values in (h) and interpret their meaning.
(j) Set up a 95% confidence interval estimate of the population slope between selling price and assessed value.
(k) Compute the coefficients of partial determination $r^2_{Y1.2}$ and $r^2_{Y2.1}$ and interpret their meaning.
(l) Determine the VIF for each explanatory variable in the model. Is there reason to suspect the existence of multicollinearity?
(m) What assumption about the slope of selling price and assessed value must be made in this problem?
(n) At the .05 level of significance, determine whether the inclusion of an interaction term makes a significant contribution to the model that already contains assessed value and whether the house is brand new. On the basis of these results, indicate the regression model that should be used in this problem.
(o) Compute the *p*-value in (n) and interpret its meaning.
(p) Perform a residual analysis on your results and determine the adequacy of the fit of the model.

Case Study F—The Mountain States Potato Company

The Mountain States Potato Company is a potato-processing firm in eastern Idaho. A by-product of the process, called a filter cake, is sold to area feedlots as cattle feed. (Recently, the feedlot owners complained that the cattle are not gaining weight; they believe the problem may be the filter cake purchased from the Mountain States Potato Company.)

Initially, all that is known of the filter cake system is that historical records show that the solids have been running in the neighborhood of 11.5% in years past. Presently, the solids are running in the 8–9% range. Several additions have been made to the plant in recent years that had significantly increased the water and solids volume and the clarifier temperature. What is actually affecting the solids is a mystery, but since the plant needs to get rid of its solid waste, something has to be done quickly. The only practical solution is to determine some way to get the solids content back up to the previous levels. Individuals involved in the process are asked to identify variables that might be manipulated that could in turn affect the solids content. This review turns up two variables that would affect solids. The variables are

SOLIDS   Percent solids in filter cake.

PH   Acidity. This indicates bacterial action in the clarifier. As bacterial action progresses, organic acids are produced that can be measured using pH. This is controlled by the downtime of the system.

LOWER   Pressure of vacuum line below fluid line on rotating drum.

Data obtained by monitoring the process several times daily for 20 days are reported below. Develop a regression model to predict the percent of solids. Write an executive summary of your findings to the president of the Mountain Potato Company.

POTATO.TXT

Obs.	SOLIDS	PH	LOWER	Obs.	SOLIDS	PH	LOWER
1	9.7	3.7	13	28	15.5	4.3	13
2	9.4	3.8	17	29	13.1	4.0	17
3	10.5	3.8	14	30	11.0	4.0	14
4	10.9	3.9	14	31	12.5	4.2	15
5	11.6	4.3	17	32	11.7	4.2	14
6	10.9	4.2	16	33	11.9	4.4	15
7	11.0	4.3	16	34	11.7	3.4	8
8	10.7	3.9	15	35	17.8	4.3	12
9	11.8	3.6	8	36	11.8	4.5	14
10	9.7	4.0	18	37	10.0	3.7	12
11	11.6	4.0	12	38	10.3	3.7	15
12	10.9	3.9	15	39	9.8	3.8	14
13	10.0	3.8	17	40	10.0	3.7	13
14	10.3	3.8	13	41	10.6	4.1	14
15	10.1	3.6	17	42	11.2	3.9	13
16	9.9	3.8	17	43	10.9	3.7	13
17	9.5	3.5	17	44	11.0	4.1	13
18	10.5	3.8	15	45	11.0	4.1	14
19	10.8	3.9	15	46	11.7	4.5	14
20	10.4	3.9	14	47	11.8	4.4	13
21	10.9	4.0	15	48	12.0	4.2	13
22	11.2	4.4	17	49	11.8	4.6	14
23	9.5	3.8	17	50	11.1	4.0	14
24	10.7	3.9	15	51	11.6	3.9	14
25	10.1	3.8	15	52	11.0	4.0	14
26	10.5	3.8	17	53	11.2	3.9	15
27	10.9	4.0	15	54	11.0	4.2	14

Source: Midwest Society for Case Research, 1994.

Endnotes

1. Regression models that involve categorical dependent variables are called logistic regression models (see References 4 and 9).
2. The relationship between t and F indicated in Equation (12.9) holds when t is a two-tailed test.
3. If the two groups have different slopes, an *interaction* term needs to be included in the model (see Section 12.12 and Reference 4).
4. The natural logarithm, usually abbreviated as ln, is the logarithm to the base e, the mathematical constant equal to 2.71828.

References

1. Andrews, D. F., and D. Pregibon, "Finding the Outliers That Matter," *Journal of the Royal Statistical Society,* Ser. B., 1978, Vol. 40, pp. 85–93.
2. Atkinson, A. C., "Robust and Diagnostic Regression Analysis," *Communications in Statistics*, 1982, Vol. 11, pp. 2559–2572.
3. Belsley, D. A., E. Kuh, and R. Welsch, *Regression Diagnostics: Identifying Influential Data and Sources of Collinearity* (New York: John Wiley, 1980).
4. Berenson, M. L., and D. M. Levine, *Basic Business Statistics: Concepts and Applications,* 6th ed. (Englewood Cliffs, NJ: Prentice-Hall, 1996).
5. Cook, R. D., and S. Weisberg, *Residuals and Influence in Regression* (New York: Chapman and Hall, 1982).
6. Dillon, W. R., and M. Goldstein, *Multivariate Analysis: Methods and Applications*, 2d ed. (New York: John Wiley, 1988).
7. Hoaglin, D. C., and R. Welsch, "The Hat Matrix in Regression and ANOVA," *The American Statistician*, 1978, Vol. 32, pp. 17–22.
8. Hocking, R. R., "Developments in Linear Regression Methodology: 1959–1982," *Technometrics*, 1983, Vol. 25, pp. 219–250.
9. Hosmer, D., and S. Lemeshow, *Applied Logistic Regression* (New York: John Wiley, 1989).
10. Marquardt, D. W., "You Should Standardize the Predictor Variables in Your Regression Models," discussion of "A Critique of Some Ridge Regression Methods," by G. Smith and F. Campbell, *Journal of the American Statistical Association*, 1980, Vol. 75, pp. 87–91.
11. Marquardt, D. W., and R. D. Snee, "Ridge Regression in Practice," *The American Statistician*, 1975, Vol. 29, pp. 3–19.
12. *Microsoft Excel Version 7* (Redmond, WA: Microsoft Corp., 1996).
13. Snee, R. D., "Some Aspects of Nonorthogonal Data Analysis, Part I. Developing Prediction Equations," *Journal of Quality Technology*, 1973, Vol. 5, pp. 67–79.
14. Tukey, J. W., "Data Analysis, Computation and Mathematics," *Quarterly Journal of Applied Mathematics*, 1972, Vol. 30, pp. 51–65.
15. Tukey, J. W., *Exploratory Data Analysis* (Reading, MA: Addison-Wesley, 1977).
16. Velleman, P. F., and R. Welsch, "Efficient Computing of Regression Diagnostics," *The American Statistician*, 1981, Vol. 35, pp. 234–242.

chapter 13

Time-Series Forecasting for Annual Data

CHAPTER OBJECTIVE To introduce a variety of time-series models for forecasting annual data.

13.1 INTRODUCTION

In the preceding two chapters we discussed the topic of regression analysis as a tool for model building and prediction. In this chapter we shall develop other forecasting methods useful for data collection on an annual basis.

13.2 THE IMPORTANCE OF BUSINESS FORECASTING

13.2.1 Introduction to Forecasting

Since economic and business conditions vary over time, business leaders must find ways to keep abreast of the effects that such changes will have on their operations. One technique that business leaders may use as an aid in planning for future operational needs is **forecasting**. Although numerous forecasting methods have been devised, they all have one common goal—to make predictions of future events so that these projections can then be incorporated into the decision-making process. As examples, the government must be able to forecast such things as unemployment, inflation, industrial production, and expected revenues from personal and corporate income taxes in order to formulate its policies, and the marketing department of a large retailing corporation must be able to forecast product demand, sales revenues, consumer preferences, inventory, and so on, in order to make timely decisions regarding its strategic planning.

13.2.2 Types of Forecasting Methods

There are basically two approaches to forecasting: *qualitative* and *quantitative*. Qualitative forecasting methods are especially important when historical data are unavailable, as would be the case, for example, if the marketing department wants to predict the sales of a new product. Qualitative forecasting methods are considered to be highly subjective and judgmental. These include the *factor listing method, expert opinion*, and the *Delphi technique* (see Reference 4). On the other hand, quantitative forecasting methods make use of historical data. The goal is to study past happenings to better understand the underlying structure of the data and thereby provide the means necessary for predicting future occurrences.

Quantitative forecasting methods can be subdivided into two types: *time series* and *causal*. Causal forecasting methods involve the determination of factors that relate to the variable to be predicted. These include multiple regression analysis with lagged variables, econometric modeling, leading indicator analysis, and diffusion indexes and other economic barometers (see References 5 and 7). On the other hand, time-series forecasting methods involve the projection of future values of a variable based entirely on the past and present observations of that variable. It is these methods that we shall be concerned with here.

13.2.3 Introduction to Time-Series Analysis

A **time series** is a set of numerical data that is obtained at regular periods over time.

For example, the *daily* closing prices of a particular stock on the New York Stock Exchange constitutes a time series. Other examples of economic or business time series are the *monthly* publication of the Consumer Price Index (CPI); the *quarterly* statements of gross national product (GNP); as well as the *annually* recorded total sales revenues of a particular firm. Time series, however, are not restricted to economic or business data. As an example, the Dean of Students at your college may wish to study trends in enrollment during the past decade.

13.2.4 Objectives of Time-Series Analysis

The basic assumption underlying time-series analysis is that the factors that have influenced patterns of activity in the past and present will continue to do so in more or less the same manner in the future. Thus, the major goals of time-series analysis are to identify and isolate these influencing factors for predictive (forecasting) purposes as well as for managerial planning and control.

13.3 COMPONENT FACTORS OF THE CLASSICAL MULTIPLICATIVE TIME-SERIES MODEL

13.3.1 Introduction

To achieve these goals, many mathematical models have been devised for exploring the fluctuations among the component factors of a time series. Perhaps the most fundamental is the **classical multiplicative model** for data recorded annually, quarterly, or monthly. It is this model that will be considered in this text.

To demonstrate the classical multiplicative time-series model, Figure 13.1 presents the net sales for Eastman Kodak Company from 1970 to 1992. If we may characterize these time-series data, it is clear that net sales has shown a tendency to increase over this 23-year period. This overall long-term tendency or impression (of upward or downward movements) is known as a **trend**.

However, trend is not the only component factor influencing either these particular data or other annual time series. Two other factors, the *cyclical* component and the *irregular* component, are also present in the data. The **cyclical component** depicts the up-and-down swings or movements through the series. Cyclical movements vary in length, usually lasting from 2 to 10 years; differ in intensity or amplitude; and are often correlated with a business cycle. In some years the values will be higher than what would be predicted by a simple trend line (i.e., they are at or near the *peak* of a cycle), whereas in other years the values will be lower than

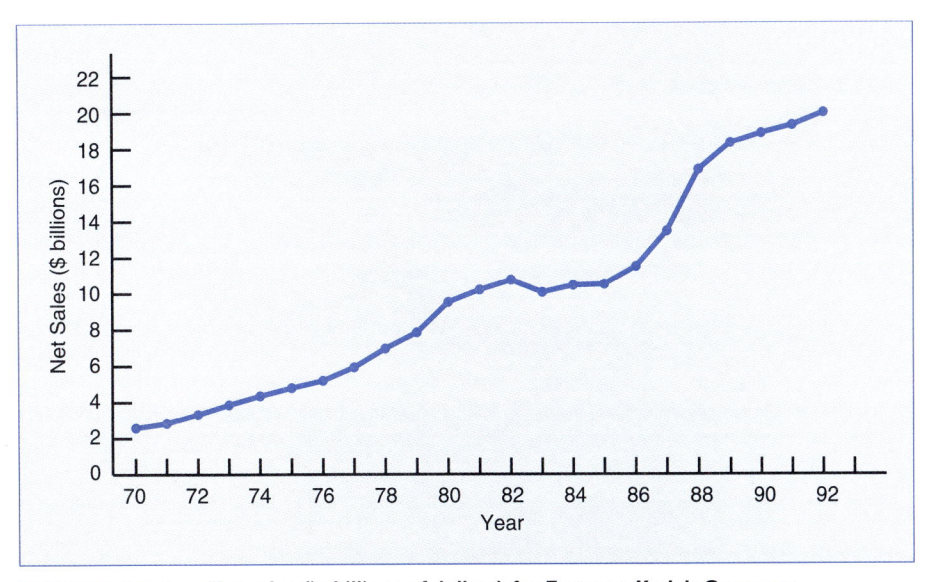

FIGURE 13.1 Net sales (in billions of dollars) for Eastman Kodak Company (1970–1992). Source: Moody's Handbook of Common Stocks, 1980, 1989, 1993.

what would be predicted by a trend line (i.e., they are at or near the bottom or *trough* of a cycle). Any observed data that do not follow the trend curve modified by the cyclical component are indicative of the **irregular or random component**. When data are recorded monthly or quarterly, in addition to the trend, cyclical, and irregular components, an additional component called the *seasonal factor* could be considered (see References 1, 2, and 6).

13.3.2 The Classical Multiplicative Time-Series Model

We have thus far mentioned that there are three or four component factors, respectively, which influence an economic or business time series. These are summarized in Table 13.1. The classical multiplicative time-series model states that any observed value in a time series is the *product* of these influencing factors; that is, when the data are obtained annually, an observation Y_i recorded in the year i may be expressed as

$$Y_i = T_i \cdot C_i \cdot I_i \tag{13.1}$$

where in the year i, T_i = value of the trend component

C_i = value of the cyclical component

I_i = value of the irregular component

On the other hand, when the data are obtained either quarterly or monthly, an observation Y_i recorded in time period i may be given as

$$Y_i = T_i \cdot S_i \cdot C_i \cdot I_i \tag{13.2}$$

where, in the time period i, T_i, C_i, and I_i are the values of the trend, cyclical, and irregular components, respectively, and S_i is the value of the seasonal component.

The first step in a time-series analysis is to plot the data and observe their tendencies over time. We must first determine whether there appears to be a long-term upward or down-

Table 13.1 Factors influencing time-series data.

Component	Classification of Component	Definition	Reason for Influence	Duration
Trend	Systematic	Overall or persistent, long-term upward or downward pattern of movement	Changes in technology, population, wealth, value	Several years
Seasonal	Systematic	Fairly regular periodic fluctuations that occur within each 12-month period year after year	Weather conditions, social customs, religious customs	Within 12 months (or monthly or quarterly data)
Cyclical	Systematic	Repeating up-and-down swings or movements through four phases: from peak (prosperity) to contraction (recession) to trough (depression) to expansion (recovery or growth)	Interactions of numerous combinations of factors influencing the economy	Usually 2–10 years with differing intensity for a complete cycle
Irregular	Unsystematic	The erratic or "residual" fluctuations in a time series that exist after taking into account the systematic effects—trend, seasonal, and cyclical	Random variations in data or due to unforeseen events such as strikes, hurricanes, floods, political assassinations, etc.	Short duration and nonrepeating

ward movement in the series (i.e., a trend) or whether the series seems to oscillate about a horizontal line over time. If the latter is the case (i.e., there is no long-term upward or downward trend), then the method of moving averages or the method of exponential smoothing may be employed to smooth the series and provide us with an overall long-term impression (see Section 13.4). On the other hand, if a trend is actually present, a variety of time-series forecasting methods can be considered (see Sections 13.5 and 13.6) when dealing with annual data.

13.4 SMOOTHING THE ANNUAL TIME SERIES: MOVING AVERAGES AND EXPONENTIAL SMOOTHING

Table 13.2 presents the annual worldwide factory sales (in millions of units) of cars, trucks, and buses manufactured by General Motors Corporation (GM) over the 23-year period from 1970 to 1992, and Figure 13.2 is a time-series plot of these data. When we examine annual data such as these, our visual impression of the overall long-term tendencies or trend movements in the series is obscured by the amount of variation from year to year. It then becomes difficult to judge whether any long-term upward or downward trend effect really exists in the series.

In situations such as these, the method of *moving averages* or the method of *exponential smoothing* may be used to smooth a series and thereby provide us with an overall impression of the pattern of movement in the data over time.

Table 13.2 GM factory sales (in millions of units*) (1970–1992).

Year	Factory Sales	Year	Factory Sales
1970	5.3	1982	6.2
1971	7.8	1983	7.8
1972	7.8	1984	8.3
1973	8.7	1985	9.3
1974	6.7	1986	8.6
1975	6.6	1987	7.8
1976	8.6	1988	8.1
1977	9.1	1989	7.9
1978	9.5	1990	7.5
1979	9.0	1991	7.0
1980	7.1	1992	7.2
1981	6.8		

Source: *Moody's Handbook of Common Stocks*, 1980, 1989, 1993. *From all sources including passenger cars, trucks and buses, and overseas plants.

13.4.1 Moving Averages

The method of moving averages for smoothing a time series is highly subjective and dependent on the length of the period selected for constructing the averages. To eliminate the cyclical fluctuations, the period chosen should be an integer value that corresponds to (or is a multiple of) the estimated average length of a cycle in the series.

But what are moving averages and how are they computed?

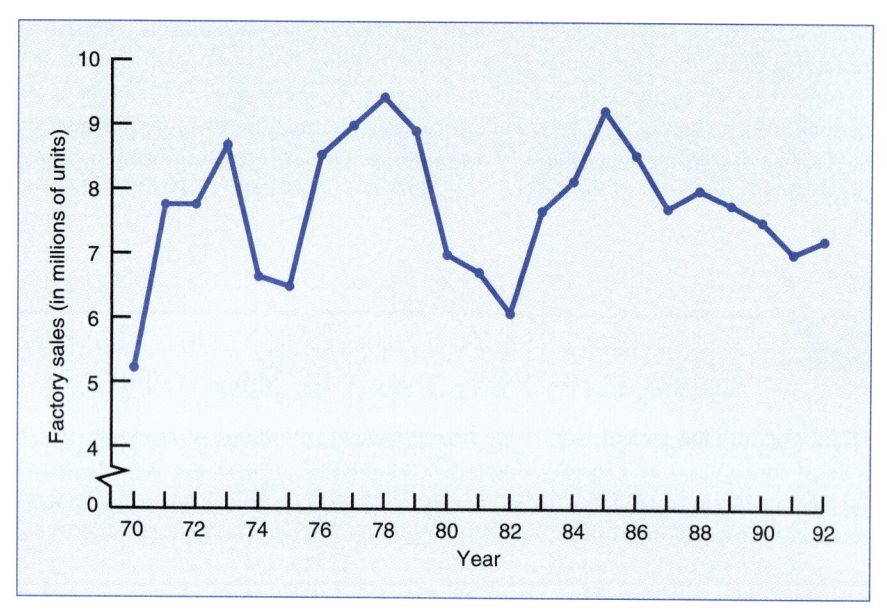

FIGURE 13.2 **GM factory sales (in millions of units). (1970–1992).** *Source:* Data are taken from Table 13.2.

Moving averages for a chosen period of length L consist of a series of arithmetic means computed over time such that each mean is calculated for a sequence of observed values having that particular length L.

For example, 5-year moving averages consist of a series of means obtained over time by averaging out consecutive sequences containing five observed values. In general, for any series composed of n years, a moving average of length L [given by the symbol $MA_i(L)$] can be computed at year i as follows:

$$MA_i(L) = \frac{1}{L} \sum_{t=(1-L)/2}^{(L-1)/2} Y_{i+t} \qquad (13.3)$$

where $\qquad L =$ an *odd* number of years

$$i = \left(\frac{L-1}{2}\right) + 1, \left(\frac{L-1}{2}\right) + 2, \ldots, n - \left(\frac{L-1}{2}\right)$$

To illustrate the use of Equation (13.3), suppose we desire to compute 5-year moving averages from a series containing $n = 11$ years. Since $L = 5$, then $i = 3, 4, 5, 6, 7, 8, 9$. Therefore, we have

$$MA_3(5) = (1/5)(Y_1 + Y_2 + Y_3 + Y_4 + Y_5)$$
$$MA_4(5) = (1/5)(Y_2 + Y_3 + Y_4 + Y_5 + Y_6)$$
$$MA_5(5) = (1/5)(Y_3 + Y_4 + Y_5 + Y_6 + Y_7)$$
$$MA_6(5) = (1/5)(Y_4 + Y_5 + Y_6 + Y_7 + Y_8)$$
$$MA_7(5) = (1/5)(Y_5 + Y_6 + Y_7 + Y_8 + Y_9)$$

$$MA_8(5) = (1/5)(Y_6 + Y_7 + Y_8 + Y_9 + Y_{10})$$
$$MA_9(5) = (1/5)(Y_7 + Y_8 + Y_9 + Y_{10} + Y_{11})$$

We note that when the chosen period of length L is an odd number, moving average $MA_i(L)$ at year i is centered on i, the middle year in the consecutive sequence of L yearly values used to compute it. Thus, with $L = 5$, $MA_3(5)$ is centered on the third year, $MA_4(5)$ is centered on the fourth year, ..., and $MA_9(5)$ is centered on the ninth year. We also note that no moving averages can be obtained for the first $(L-1)/2$ years or the last $(L-1)/2$ years of the series. Thus, for a 5-year moving average, we cannot make computations for the first 2 years or the last 2 years of the series.

Let us now take another look at GM factory sales data for the 23-year period from 1970 to 1992. Table 13.3 presents the annual data along with the computations for the 3- and 7-year moving averages. Both of these constructed series are plotted in Figure 13.3 with the original data.

In practice, to compute 3-year moving averages, we first obtain a series of 3-year moving totals as indicated in column (3) of Table 13.3 and then divide each of these totals by 3. The results are given in column (4). For example, since our observed time series was first recorded in 1970, the first 3-year moving total consists of the sum of the first three annually recorded values—5.3, 7.8, and 7.8. This moving total, 20.9, is then centered so that the

Table 13.3 The 3- and 7-year moving averages of GM factory sales (1970–1992).

(1) Year	(2) Factory Sales (in millions of units)	(3) 3-Year Moving Total	(4) 3-Year Moving Average	(5) 7-Year Moving Total	(6) 7-Year Moving Average
1970	5.3	—	—	—	—
1971	7.8	20.9	6.97	—	—
1972	7.8	24.3	8.10	—	—
1973	8.7	23.2	7.73	51.5	7.36
1974	6.7	22.0	7.33	55.3	7.90
1975	6.6	21.9	7.30	57.0	8.14
1976	8.6	24.3	8.10	58.2	8.31
1977	9.1	27.2	9.07	56.6	8.09
1978	9.5	27.6	9.20	56.7	8.10
1979	9.0	25.6	8.53	56.3	8.04
1980	7.1	22.9	7.63	55.5	7.93
1981	6.8	20.1	6.70	54.7	7.81
1982	6.2	20.8	6.93	54.5	7.78
1983	7.8	22.3	7.43	54.1	7.73
1984	8.3	25.4	8.47	54.8	7.83
1985	9.3	26.2	8.73	56.1	8.01
1986	8.6	25.7	8.50	57.8	8.26
1987	7.8	24.5	8.17	57.5	8.21
1988	8.1	23.8	7.93	56.2	8.03
1989	7.9	23.5	7.83	54.1	7.73
1990	7.5	22.4	7.47	—	—
1991	7.0	21.7	7.23	—	—
1992	7.2	—	—	—	—

Source: Data are taken from Table 13.2.

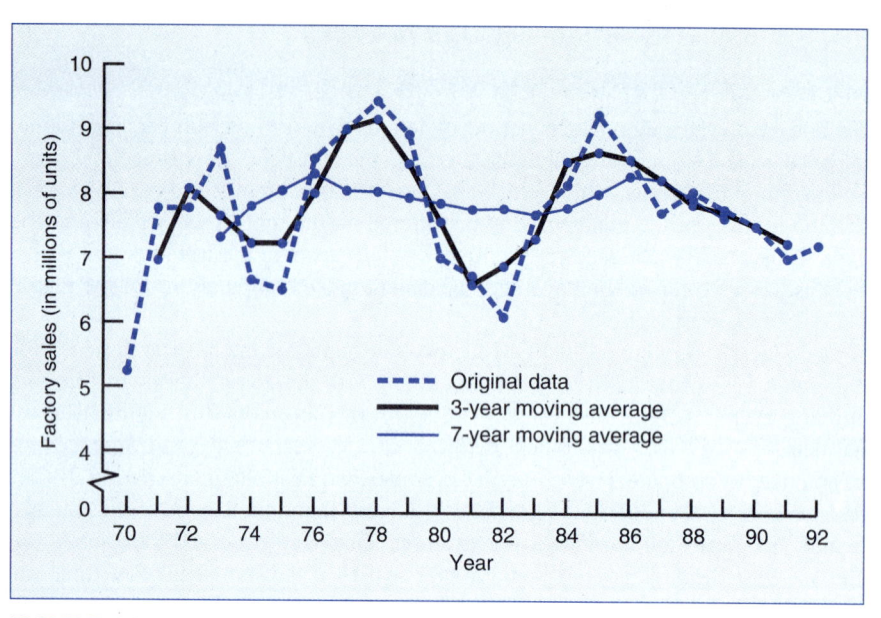

FIGURE 13.3 Plotting the 3- and 7-year moving averages for the GM factory sales data. *Source:* Data are taken from Table 13.3.

recording is made against the year 1971. To obtain the moving total for the year 1972—which consists of the observed annual sales data for the years 1971, 1972, and 1973—we add the next observed value in the time series (year 1973) to the previous moving total and then subtract the first (oldest) value in the series. This process continues so that the 3-year moving total for any particular year i in the series represents the sum of the observed value for the year i along with the observed values for the year preceding it and the year following it. However, with 7-year moving totals, the result computed and recorded for the year i consists of the observed value in the time series for year i plus the three observed values that precede it and the three observed values that follow it. To "move" the 7-year total from one year to the next, we add on to the previous total the next observed value in the time series and remove the oldest value that had appeared in the previous total. This process continues through the series. The 7-year moving averages are then obtained by dividing the series of moving totals by 7.

We note from columns (3) and (4) of Table 13.3 that, in obtaining the 3-year moving averages, no result can be computed for the first or last observed value in the time series. We also see from columns (5) and (6) that, when we compute 7-year moving averages, there are no results for the first three observed values or the last three values. This occurs because the first 7-year moving total for the data at hand consists of factory sales during the years 1970–1976, which is centered at 1973, and the last moving total consists of factory sales recorded in 1986 through 1992, which is centered at 1989.

From Figure 13.3, we can see that the 7-year moving averages smooth the series a great deal more than do the 3-year moving averages, since the period is of longer duration. Unfortunately, however, as we previously noted, the longer the period, the fewer the number of moving average values that can be computed and plotted. Therefore, selecting moving averages with periods of length greater than 7 years is usually undesirable since too many computed data points would be missing at the beginning and end of the series, making it more difficult to obtain an overall impression of the entire series.

13.4.2 Exponential Smoothing

Exponential smoothing is another technique that may be used to smooth a time series and thereby provide us with an impression as to the overall long-term movements in the data. In addition, the method of exponential smoothing can be utilized for obtaining short-term (one period into the future) forecasts for time series for which it is questionable as to what type of long-term trend effect, if any, is present in the data. In this respect, the technique possesses a distinct advantage over the method of moving averages.

The method of exponential smoothing derives its name from the fact that it provides us with an *exponentially weighted* moving average through the time series; that is, throughout the series each smoothing calculation or forecast is dependent on all previously observed values. This is another advantage over the method of moving averages, which does not take into account all the observed values in this manner. With exponential smoothing, the weights assigned to the observed values decrease over time so that when a calculation is made, the most recently observed value receives the highest weight, the previously observed value receives the second highest weight, and so on, with the initially observed value receiving the lowest weight. Although the magnitude of computations involved may seem formidable, exponential smoothing as well as moving average methods are available among the procedures provided in Excel (see Section 13.4.3).

If we focus on the smoothing aspects of the technique (rather than the forecasting aspects), the formulas developed for exponentially smoothing a series in any time period i are based only on three terms—the presently observed value in the time series Y_i, the previously computed exponentially smoothed value E_{i-1}, and some subjectively assigned weight or smoothing coefficient W. Thus, to smooth a series at any time period i, we have the following expression:[1]

$$E_i = WY_i + (1 - W)E_{i-1} \quad (13.4)$$

where
E_i = value of the exponentially smoothed series being computed in time period i

E_{i-1} = value of the exponentially smoothed series already computed in time period $i - 1$

Y_i = observed value of the time series in period i

W = subjectively assigned weight or smoothing coefficient (where $0 < W < 1$)

The choice of the smoothing coefficient or weight that we should assign to our time series is quite important since it will affect our results. Unfortunately, this selection is rather subjective. However, in regard to smoothing ability, we may observe from Figures 13.3 and 13.4 (pages 658 and 661) that a series of L-term moving averages is related to an exponentially smoothed series having weight W as follows:

$$W = \frac{2}{L + 1} \quad (13.5)$$

or

$$L = \frac{2}{W} - 1 \qquad (13.6)$$

From Equations (13.5) and (13.6), we note that with respect to smoothing ability, similarities are found between the 3-year series of moving averages (Figure 13.3) and the exponentially smoothed series with weight $W = .50$ (see Figure 13.4). In addition, we see that the series of 7-year moving averages (Figure 13.3) corresponds to the exponentially smoothed series with weight $W = .25$ (see Figure 13.4). By examining how our two smoothing series (one with $W = .25$ and the other with $W = .50$) fit the observed data in Figure 13.4, we can see that the choice of a particular smoothing coefficient W is dependent on the purpose of the user. If we desire only to smooth a series by eliminating unwanted cyclical and irregular variations, we should select a small value for W (closer to 0). On the other hand, if our goal is forecasting, we should choose a large value for W (closer to 1). In the former case, the overall long-term tendencies of the series will be apparent; in the latter case, future short-term directions may be more adequately predicted.

● **Smoothing** Table 13.4 presents the exponentially smoothed values (using smoothing coefficients of $W = .50$ and $W = .25$) for annual factory sales at GM over the 23-year period 1970–1992. As previously indicated, the two smoothed series are plotted in Figure 13.4 along with the original time-series data.

To demonstrate the computations for the exponentially smoothed values as shown in Table 13.4, let us consider the series with a smoothing coefficient of $W = .25$. As a starting point, we use the initial observed value $Y_{1970} = 5.3$ as our first smoothed value ($E_{1970} = 5.3$). Now using the observed value of the time series for the year 1971 ($Y_{1971} = 7.8$), we smooth the series for the year 1971 by computing

$$E_{1971} = WY_{1971} + (1 - W)E_{1970}$$
$$= (.25)(7.8) + (.75)(5.3) = 5.93 \text{ million}$$

To smooth the series for the year 1972, we have

$$E_{1972} = WY_{1972} + (1 - W)E_{1971}$$
$$= (.25)(7.8) + (.75)(5.93) = 6.39 \text{ million}$$

To smooth the series for the year 1973, we have

$$E_{1973} = WY_{1973} + (1 - W)E_{1972}$$
$$= (.25)(8.7) + (.75)(6.39) = 6.97 \text{ million}$$

This process continues until exponentially smoothed values have been obtained for all 23 years in the series as shown in Table 13.4 and Figure 13.4.

● **Forecasting** To use the exponentially weighted moving average for purposes of forecasting rather than for smoothing, we take the smoothed value in our current period of time (say time period i) as our projected estimate of the observed value of the time series in the following time period, $i + 1$—that is,

$$\hat{Y}_{i+1} = E_i \qquad (13.7)$$

Table 13.4 Exponentially smoothed series of GM factory sales (1970–1992).

Year	Factory Sales (in millions of units)	W = .50	W = .25
1970	5.3	5.30	5.30
1971	7.8	6.55	5.93
1972	7.8	7.18	6.39
1973	8.7	7.94	6.97
1974	6.7	7.32	6.90
1975	6.6	6.96	6.82
1976	8.6	7.78	7.27
1977	9.1	8.44	7.72
1978	9.5	8.97	8.17
1979	9.0	8.98	8.37
1980	7.1	8.04	8.06
1981	6.8	7.42	7.74
1982	6.2	6.81	7.36
1983	7.8	7.31	7.47
1984	8.3	7.80	7.68
1985	9.3	8.55	8.08
1986	8.6	8.58	8.21
1987	7.8	8.19	8.11
1988	8.1	8.14	8.11
1989	7.9	8.02	8.05
1990	7.5	7.76	7.92
1991	7.0	7.38	7.69
1992	7.2	7.29	7.57

Source: Data are taken from Table 13.2.

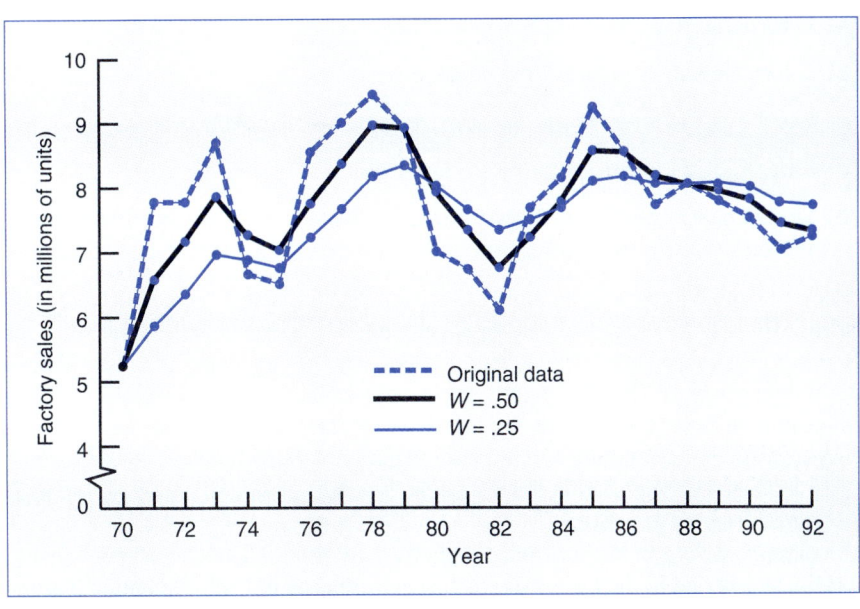

FIGURE 13.4 Plotting the exponentially smoothed series (W = .50 and W = .25) for the GM factory sales data. *Source:* Data are taken from Table 13.4.

For example, to forecast the number of units sold from all General Motors Corporation plants during the year 1993, we would use the smoothed value for the year 1992 as its estimate. From Table 13.4, for a smoothing coefficient of $W = .50$, that projection is 7.29 million units.

Once the observed data for the year 1993 become available, we can use Equation (13.4) to make a forecast for the year 1994 by obtaining the smoothed value for 1993 as follows:

$$E_{1993} = WY_{1993} + (1 - W)E_{1992}$$

Current smoothed value = (W)(current observed value)
+ (1 − W)(previous smoothed value)

or, in terms of forecasting

$$\hat{Y}_{1994} = WY_{1993} + (1 - W)\hat{Y}_{1993}$$

New forecast = (W)(current observed value)
+ (1 − W)(current forecast)

13.4.3 Using Microsoft Excel for Moving Averages and Exponential Smoothing

In this section we have used moving averages and exponentially smoothing techniques to smooth a set of time-series data. Since the computations involved in these procedures are repetitive and tedious, it is fortunate that Microsoft Excel has Moving Average and Exponential Smoothing options as part of the Data Analysis tool. The use of these options will be illustrated by using the GM factory sales data presented in Table 13.2 on page 655.

● **Moving Averages** To illustrate the computation of moving averages, open the FACTORY.XLS workbook and click on the Data sheet tab. Note that the year and factory sales data from Table 13.2 have been entered into columns A and B. Select Tools | Data Analysis, then select Moving Average from the Analysis Tools list box and click OK.

13–4.3.XLS

In the Moving Average dialog box, do the following:

(a) Enter B1:B24 in the Input Range edit box.

(b) Select the Labels in First Row check box.

(c) Enter 3 in the Interval edit box to specify a 3-year moving average.

(d) Enter C2:C24 in the Output Range edit box.

(e) Do *not* check the Chart Output box, since we will use the Chart Wizard to plot the output from the moving averages and exponential smoothing procedures.

The dialog box should now appear similar to the one illustrated in Figure 13.1.Excel. Click the OK button. Microsoft Excel generates 3-year moving averages and places them in column C.

Repeat the same procedure for 7-year moving averages changing the Interval value to 7 and the output range to D2:D24. Note that in both cases, Microsoft Excel has improperly placed the moving averages at the end of the series of years and not in the center of the series. (The 3-year averages were placed in the row of the third year—and not in the second—and the 7-year averages were placed in the row of the seventh year—and not in the fourth.) These errors can be corrected as follows:

For 3-year moving averages: Select cell C2 (the first cell of the output range). Then select Edit | Delete and in the Delete dialog box select the Shift Cells Up option button. Click OK. Go to cell C24 (the last cell of the output range) and enter #N/A.

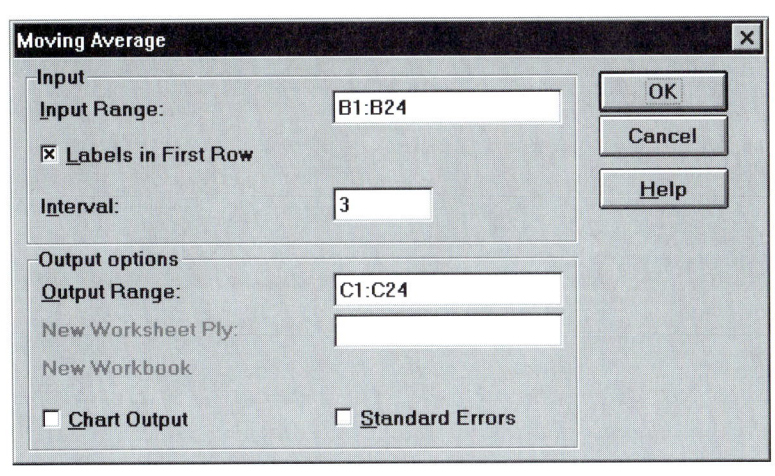

FIGURE 13.1.EXCEL Moving average dialog box for the GM factory sales data.

For the 7-year moving averages: Select the cell range D2:D4 (the first three cells of the output range) and use the same delete shift cells up command. Then enter #N/A in the range D22:D24 (the last three cells of the output range). The Data sheet should now appear similar to the first four columns of the sheet illustrated in Figure 13.3.Excel.

- **Exponential Smoothing** To illustrate the computation of exponentially smoothed values, with the Data sheet still active, select Tools | Data Analysis and then select Exponential Smoothing from the Analysis Tools list box and click OK. In the Exponential Smoothing dialog box, do the following:

  (a) Enter B1:B24 in the Input Range edit box.

  (b) Select the Labels check box.

  (c) To obtain exponentially smoothed values for W = .25, enter .75 in the Damping factor edit box since the Damping factor is defined as the complement of the smoothing coefficient (1 − W).

  (d) Enter E2:E24 in the Output Range edit box.

  (e) Do *not* check the Chart Output box, since we will use the Chart Wizard to plot the output from the moving averages and exponential smoothing procedures. The dialog box should now appear similar to the one illustrated in Figure 13.2.Excel. Click the OK button.

Microsoft Excel generates the exponentially smoothed values for W = .25 and places them in column E.

Repeat the same procedure for W = .50, changing the Damping factor to .50 (i.e., 1 − .50) and the output range to F2:F24. Note that in both cases Microsoft Excel has provided exponentially smoothed forecasts one period into the future. To adjust these columns so that they mirror those presented in Table 13.4 on page 661, shift the cells up one row for both columns (as was done for the 3-year moving averages) and then copy the formulas in cells E23 and F23 to cells E24 and F24, respectively.

The moving averages and exponentially smoothed values obtained are illustrated in Figure 13.3.Excel on page 664.

Now that we have obtained the moving averages and exponentially smoothed values, we can use the Chart Wizard (see Section 2.7) to obtain the time-series plot similar to the ones

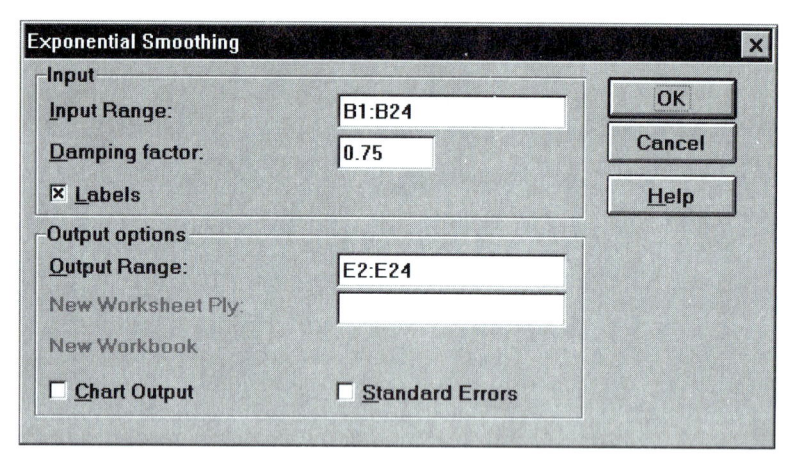

FIGURE 13.2. EXCEL Exponential Smoothing dialog box for the GM factory sales data.

	A	B	C	D	E	F
1	Year	Sales	MA 3yr	MA 7yr	ESV (W=.25)	ESV (W=.50)
2	1970	5.3	#N/A	#N/A	5.3	5.3
3	1971	7.8	6.966667	#N/A	5.925	6.55
4	1972	7.8	8.1	#N/A	6.39375	7.175
5	1973	8.7	7.733333	7.357143	6.9703125	7.9375
6	1974	6.7	7.333333	7.9	6.902734375	7.31875
7	1975	6.6	7.3	8.142857	6.827050781	6.959375
8	1976	8.6	8.1	8.314286	7.270288086	7.7796875
9	1977	9.1	9.066667	8.085714	7.727716064	8.43984375
10	1978	9.5	9.2	8.1	8.170787048	8.969921875
11	1979	9.0	8.533333	8.042857	8.378090286	8.984960938
12	1980	7.1	7.633333	7.928571	8.058567715	8.042480469
13	1981	6.8	6.7	7.814286	7.743925786	7.421240234
14	1982	6.2	6.933333	7.785714	7.35794434	6.810620117
15	1983	7.8	7.433333	7.728571	7.468458255	7.305310059
16	1984	8.3	8.466667	7.828571	7.676343691	7.802655029
17	1985	9.3	8.733333	8.014286	8.082257768	8.551327515
18	1986	8.6	8.566667	8.257143	8.211693326	8.575663757
19	1987	7.8	8.166667	8.214286	8.108769995	8.187831879
20	1988	8.1	7.933333	8.028571	8.106577496	8.143915939
21	1989	7.9	7.833333	7.728571	8.054933122	8.02195797
22	1990	7.5	7.466667	#N/A	7.916199841	7.760978985
23	1991	7.0	7.233333	#N/A	7.687149881	7.380489492
24	1992	7.2	#N/A	#N/A	7.565362411	7.290244746

FIGURE 13.3. EXCEL Moving averages and exponentially smoothed values obtained from Microsoft Excel for the GM factory sales data.

shown in Figures 13.3 and 13.4 (pages 658 and 661). Again with the Data sheet still active, select Insert | Chart | As New Sheet. The entries and choices in the five dialog boxes for the Chart Wizard would be

Dialog box 1: Enter the range A1:F24 for the six variables—and click the Next button.

Dialog box 2: Select the *XY* Scatter chart and click the Next button.

Dialog box 3: Select format 2, which shows points plotted as parts of a connecting line and click the Next button.

Dialog box 4: Select the Columns option button and enter 1 in the First Columns edit box and 1 in the First Rows edit box. Click the Next button.

Dialog box 5: Select the Yes option button and enter the Chart title Factory Sales, Years in the Category (*X*) edit box, and Sales in the Value (*Y*) edit box. Click the Finish button.

Rename the sheet Time Series. Figure 13.4.Excel presents the resulting time-series chart. Note that a key on the right side of the chart indicates which line represents the 3-year moving average, (MA 3 yr); the 7-year moving average, (MA 7 yr); and the two exponentially smoothed sets of values, ESV($W = .25$) and ESV($W = .50$).

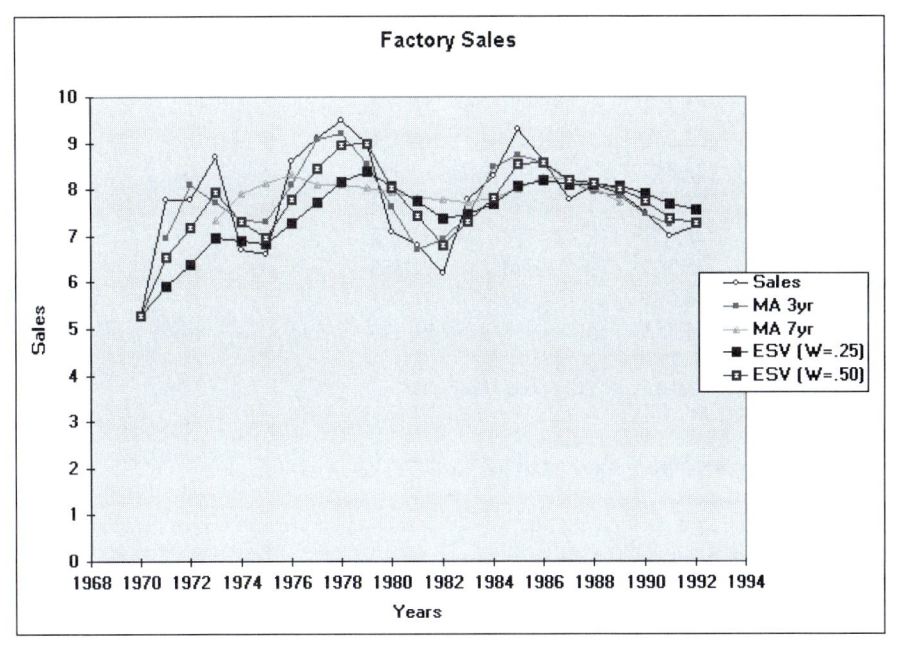

FIGURE 13.4.EXCEL Time-series charts for moving averages and exponential smoothing obtained from Microsoft Excel for the GM factory sales data.

▲ WHAT IF EXAMPLE

The workbook developed in this section allows us to explore the effect of changes in the time period of the moving average and the weight for the exponentially smoothed values. In Figure 13.3.Excel, we observe the differences in the results obtained between the 3- and 7-year moving averages, and the exponentially smoothed values based on $W = .25$ and $W = .50$. Other time periods and weighting coefficients could be selected, and the Data Analysis tool could be used to obtain new moving averages and exponentially smoothed values.

Problems for Section 13.4

Note: *Microsoft Excel should be used to solve the problems in this section.*

MEDINCUS.TXT

13.1 The following data represent the median income of families in the United States (in 1990 dollars) for all races, for whites, and for blacks, for the 14-year period from 1977 to 1990.

Median family income in the United States (1977–1990).

Year	All Races	Whites	Blacks
1977	34,528	36,104	20,625
1978	35,361	36,821	21,808
1979	35,262	36,796	20,836
1980	33,346	34,743	20,103
1981	32,190	33,814	19,074
1982	31,738	33,322	18,417
1983	32,378	33,905	19,108
1984	33,251	34,827	19,411
1985	33,689	35,410	20,390
1986	35,129	36,740	20,993
1987	35,632	37,260	21,177
1988	35,565	37,470	21,355
1989	36,062	37,919	21,301
1990	35,353	36,915	21,423

Source: U.S. Department of Commerce, Bureau of the Census, Table B-28.

For each of the three sets of data (all races, whites, and blacks):
(a) Plot the data on a chart.
(b) Fit a 3-year moving average to your data and plot the results on your chart.
(c) Using a smoothing coefficient of .50, exponentially smooth the series and plot your results on your chart.
(d) What is your exponentially smoothed forecast for the trend in 1991?
(e) Do (c) using a smoothing constant of .25.
(f) Using the results of (e), what is your exponentially smoothed forecast for the trend in 1991?
(g) Compare the results of (d) and (f).
(h) **ACTION** Go to your library and record the actual 1991 value from the table available from the U.S. Department of Commerce. Compare your results to the forecast you made in (d). Discuss.
(i) **ACTION** Write a letter to one of your United States Senators explaining the trend in median family income for each of the two groups and all races combined for the period from 1977 to 1990.

13.2 The following data represent the annual earnings per share of TRW Inc. over the 23-year period 1970–1992:

Earnings per share at TRW Inc. (1970–1992).

Year	Earnings per Share	Year	Earnings per Share
1970	2.39	1982	5.49
1971	1.85	1983	5.53
1972	2.22	1984	6.66
1973	2.95	1985	3.79
1974	2.76	1986	7.25
1975	3.08	1987	4.01
1976	4.02	1988	4.23
1977	4.77	1989	4.31
1978	5.42	1990	3.39
1979	5.86	1991	2.30
1980	6.15	1992	3.09
1981	6.60		

Source: Moody's Handbook of Common Stocks, 1980, 1989, 1993.

(a) Plot the data on a chart.
(b) Fit a 3-year moving average to the data and plot the results on your chart.
(c) Using a smoothing coefficient of $W = .50$, exponentially smooth the series and plot the results on your chart.
(d) What is your exponentially smoothed forecast for the trend in 1993?
(e) Do (c) using a smoothing coefficient of $W = .25$.
(f) Using the results of (e), what is your exponentially smoothed forecast for 1993?
(g) Compare the results of (d) and (f).

● 13.3 The following data represent the annual number of employees (in thousands) in an oil supply company for the years 1976–1995:

Number of employees (in thousands).

Year	Number	Year	Number	Year	Number
1976	1.45	1984	2.06	1992	1.88
1977	1.55	1985	1.80	1993	2.00
1978	1.61	1986	1.73	1994	2.08
1979	1.60	1987	1.77	1995	1.88
1980	1.74	1988	1.90		
1981	1.92	1989	1.82		
1982	1.95	1990	1.65		
1983	2.04	1991	1.73		

(a) Plot the data on a chart.
(b) Fit a 3-year moving average to the data and plot the results on your chart.
(c) Using a smoothing coefficient of $W = .50$, exponentially smooth the series and plot the results on your chart.
(d) What is your exponentially smoothed forecast for the trend in 1996?
(e) Do (c) using a smoothing coefficient of $W = .25$.
(f) Using the results of (e), what is your exponentially smoothed forecast for the trend in 1996?
(g) Compare the results of (d) and (f).

13.4 The following data represent the annual sales dollars (in millions) for a food-processing company for the years 1970–1995.

Annual sales dollars (millions).

Year	Sales Dollars	Year	Sales Dollars	Year	Sales Dollars
1970	41.6	1979	53.2	1988	36.4
1971	48.0	1980	53.3	1989	38.4
1972	51.7	1981	51.6	1990	42.6
1973	55.9	1982	49.0	1991	34.8
1974	51.8	1983	38.6	1992	28.4
1975	57.0	1984	37.3	1993	23.9
1976	64.4	1985	43.8	1994	27.8
1977	60.8	1986	41.7	1995	42.1
1978	56.3	1987	38.3		

(a) Plot the data on a chart.
(b) Fit a 7-year moving average to the data and plot the results on your chart.
(c) Using a smoothing coefficient of $W = .25$, exponentially smooth the series and plot the results on your chart.
(d) What is your exponentially smoothed forecast for the trend in 1996?
(e) Do (c) using a smoothing coefficient of $W = .50$.
(f) Using the results of (e), what is your exponentially smoothed forecast for the trend in 1996?
(g) Compare the results of (d) and (f).

13.5 TIME-SERIES ANALYSIS OF ANNUAL DATA: LEAST SQUARES TREND FITTING AND FORECASTING

The component factor of a time series most often studied is trend. Primarily, we study trend for predictive purposes; that is, we may wish to study trend directly as an aid in making intermediate and long-range forecasting projections. Secondly, we may wish to study trend in order to isolate and then eliminate its influencing effects on the time-series model as a guide to short-run (1 year or less) forecasting of general business cycle conditions. As depicted in Figure 13.1 on page 653, to obtain some visual impression of the overall long-term movements in a time series, we construct a chart in which the observed data (dependent variable) are plotted on the vertical axis. If it appears that a straight-line trend could be adequately fitted to the data, the two most widely used methods of trend fitting are the method of least squares (see Section 11.4) and the method of *double exponential smoothing* (References 2 and 4). If the time-series data indicate some long-run downward or upward curvilinear movement, the two most widely used trend-fitting methods are the methods of least squares (see Section 12.10) and *triple exponential smoothing* (References 2 and 4). In this section we shall focus on least squares methods for fitting linear and curvilinear trends as guides to forecasting. In Section 13.6, other, more elaborate, forecasting approaches will be described.

13.5.1 The Linear Model

You may recall from Section 11.4 that the least squares method permits us to fit a straight line **(linear trend model)** of the form

$$\hat{Y}_i = b_0 + b_1 X_i \qquad (13.8)$$

such that the values we calculate for the two coefficients—the intercept b_0 and the slope b_1—result in the sum of squared differences between each observed value Y_i and each predicted value $\hat{Y}_i$ along the trend line being minimized. To obtain such a line, recall that in linear regression analysis we computed the slope from b_1 and b_0 [see Equations (11.3) and (11.4) on page 541]. Once this is accomplished and the line $\hat{Y}_i = b_0 + b_1 X_i$ is obtained, we can substitute values for X into Equation (13.8) to predict various values for Y.

When using the method of least squares for fitting trends in time series, our interpretation of the results can be simplified if we code the X values. Thus, the first observation in our time series is selected as the origin and assigned a code value of $X = 0$. All successive observations are then assigned consecutively increasing integer codes: 1, 2, 3, ..., so that the nth and last observation in the series has $n - 1$. For example, for time-series data recorded annually over 23 years, the first year will be assigned a coded value of 0, the second year will be coded as 1, the third year will be coded as 2, ..., and the final (23rd) year will be coded as 22.

The annual time series presented in Table 13.5 and plotted in Figure 13.1 on page 653 represents the net sales (in billions of dollars) for the Eastman Kodak Company over the 23-year period 1970–1992. Coding the consecutive X values 0–22 and then using Microsoft Excel (see Figure 13.5 on page 670), we determine that

$$\hat{Y}_i = 1.2011 + 0.8003 X_i$$

where the origin is 1970 and X units = 1 year.

The Y intercept $b_0 = 1.2011$ is the fitted trend value reflecting the net sales (in billions of dollars) at Eastman Kodak during the origin or base year, 1970. The slope $b_1 = 0.8003$ indicates that net sales are increasing at a rate of 0.8003 billion dollars per year.

To project the trend in the net sales to the year 1993, we substitute $X = 23$, the code for 1993, into the equation and our forecast is

$$\hat{Y}_{24} = 1.2011 + (0.8003)(23) = 19.6 \text{ billions of dollars}$$

Table 13.5 Net sales (in billions of dollars) for Eastman Kodak Company (1970–1992).

Year	Net Sales (billions of dollars)	Year	Net Sales (billions of dollars)
1970	2.8	1982	10.8
1971	3.0	1983	10.2
1972	3.5	1984	10.6
1973	4.0	1985	10.6
1974	4.6	1986	11.5
1975	5.0	1987	13.3
1976	5.4	1988	17.0
1977	6.0	1989	18.4
1978	7.0	1990	18.9
1979	8.0	1991	19.4
1980	9.7	1992	20.1
1981	10.3		

Source: Moody's Handbook of Common Stocks, 1980, 1989, 1993.

	A	B	C	D	E	F	G
1	SUMMARY OUTPUT						
2							
3	*Regression Statistics*						
4	Multiple R	0.97113048					
5	R Square	0.94309442					
6	Adjusted R Square	0.94038463					
7	Standard Error	1.36468165					
8	Observations	23					
9							
10	ANOVA						
11		*df*	*SS*	*MS*	*F*	*Significance F*	
12	Regression	1	648.160089	648.16009	348.03232	1.50099E-14	
13	Residual	21	39.1094763	1.862356			
14	Total	22	687.269565				
15							
16		*Coefficients*	*Std. Error*	*t Stat*	*P-value*	*Lower 95%*	*Upper 95%*
17	Intercept	1.20108696	0.55103993	2.1796732	0.0408085	0.055136495	2.347037418
18	Coded Year	0.80029644	0.0428984	18.655624	1.501E-14	0.71108432	0.889508565

FIGURE 13.5 Partial Excel output for fitting linear regression model to forecast annual net sales at Eastman Kodak Company.

The fitted trend line projected to 1993 is plotted in Figure 13.6 along with the original time series. A careful examination of Figure 13.6 reveals that a marked increase has occurred in the more recent years of the series. Perhaps, then, a curvilinear trend model would better fit the series? Two such models—a *quadratic* trend model and an *exponential* trend model—are presented in Sections 13.5.2 and 13.5.3, respectively.

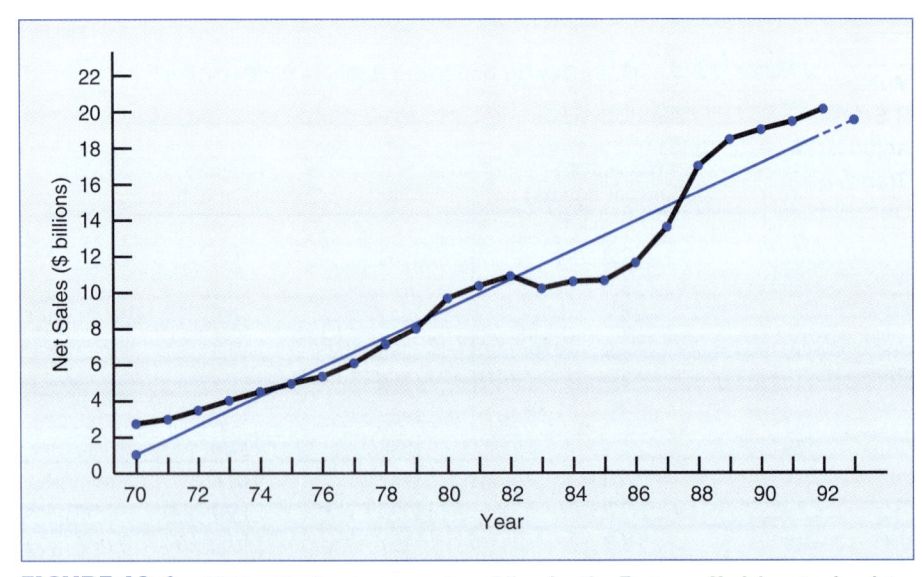

FIGURE 13.6 Fitting the least squares trend line for the Eastman Kodak net sales data.

670 Chapter 13 Time-Series Forecasting for Annual Data

13.5.2 The Quadratic Model

A **quadratic trend model** or *second-degree polynomial* is the simplest of the curvilinear models. Using the least squares method of Section 12.10, we may fit a quadratic trend equation of the form

$$\hat{Y}_i = b_0 + b_1 X_i + b_{11} X_i^2 \qquad (13.9)$$

where
b_0 = estimated Y intercept
b_1 = estimated *linear* effect on Y
b_{11} = estimated *curvilinear* effect on Y

Once again, we may use Excel to perform the computations necessary to obtain the least squares fit. Figure 13.7 provides Excel output for the quadratic model representing annual net sales at Eastman Kodak. From this, we determine that

$$\hat{Y}_i = 2.9217 + 0.3087 X_i + 0.0223 X_i^2$$

where the origin is 1970 and X units = 1 year.

To use the quadratic trend equation for forecasting purposes, we substitute the appropriate coded X values into this equation. For example, to predict the trend in net sales for the year 1993 (i.e., X = 23), we have

$$\hat{Y}_{24} = 2.9217 + 0.3087(23) + 0.0223(23^2)$$
$$= 21.82 \text{ billions of dollars}$$

	A	B	C	D	E	F	G
1	SUMMARY OUTPUT						
2							
3	Regression Statistics						
4	Multiple R	0.98429139					
5	R Square	0.96882953					
6	Adjusted R Square	0.96571249					
7	Standard Error	1.034952					
8	Observations	23					
9							
10	ANOVA						
11		df	SS	MS	F	Significance F	
12	Regression	2	665.847053	332.92353	310.8165	8.65831E-16	
13	Residual	20	21.4225127	1.0711256			
14	Total	22	687.269565				
15							
16		Coefficients	Std. Error	t Stat	P-value	Lower 95%	Upper 95%
17	Intercept	2.92173913	0.59492619	4.9110952	8.434E-05	1.68074543	4.162732831
18	Coded Year	0.30868154	0.12527938	2.4639452	0.0229285	0.047353443	0.570009629
19	Year Squared	0.02234613	0.00549915	4.0635577	0.0006064	0.010875102	0.033817162

FIGURE 13.7 Partial Excel output for fitting a quadratic regression model to forecast annual net sales at Eastman Kodak.

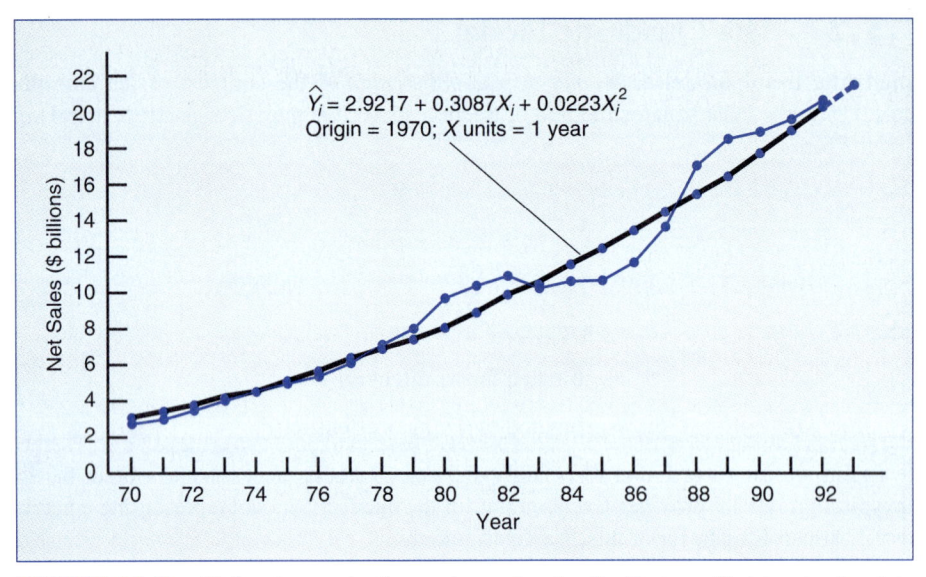

FIGURE 13.8 Fitting the quadratic trend equation for the Eastman Kodak net sales data.

The fitted quadratic trend equation projected to 1993 is plotted in Figure 13.8 together with the original time series.

13.5.3 The Exponential Model

When a series appears to be *increasing at an increasing rate* such that the *percent difference* from observation to observation is *constant*, we may use an **exponential trend model**; its equation takes the form

$$\hat{Y}_i = b_0 b_1^{X_i} \qquad (13.10)$$

where
b_0 = estimated Y intercept

$(b_1 - 1) \times 100\%$ = estimated annual *compound growth rate* (in percent)

If we take the logarithm (base 10) of both sides of Equation (13.10), we have

$$\log \hat{Y}_i = \log b_0 + X_i \log b_1 \qquad (13.11)$$

Since Equation (13.11) is linear in form, we may use the method of least squares by working with the log Y_i values instead of the Y_i values and obtain the slope (log b_1) and Y intercept (log b_0). Once again, we may access Excel to accomplish the necessary calculations.

Figure 13.9 represents Excel output for an exponential model of annual net sales at Eastman Kodak. From this we determine that

$$\log \hat{Y}_i = 0.49949 + 0.0389 X_i$$

where the origin is 1970 and X units = 1 year.

	A	B	C	D	E	F	G
1	SUMMARY OUTPUT						
2							
3	*Regression Statistics*						
4	Multiple R	0.98487486					
5	R Square	0.96997848					
6	Adjusted R Square	0.96854889					
7	Standard Error	0.04750978					
8	Observations	23					
9							
10	ANOVA						
11		df	SS	MS	F	Significance F	
12	Regression	1	1.53149236	1.5314924	678.49828	1.7978E-17	
13	Residual	21	0.04740077	0.0022572			
14	Total	22	1.57889313				
15							
16		Coefficients	Std. Error	t Stat	P-value	Lower 95%	Upper 95%
17	Intercept	0.4994909	0.01918381	26.037113	1.813E-17	0.459595986	0.539385814
18	Coded Year	0.03890157	0.00149346	26.047999	1.798E-17	0.035795758	0.042007388

FIGURE 13.9 Partial Excel output for fitting an exponential regression model to forecast annual net sales at Eastman Kodak Company.

The values for b_0 and b_1 may be obtained by taking the antilog of the regression coefficients in this equation:

$$b_0 = \text{antilog } 0.49949 = 3.155$$

$$b_1 = \text{antilog } 0.0389 = 1.0937$$

Thus, the fitted exponential trend equation can be expressed as

$$\hat{Y}_i = (3.155)(1.0937)^{X_i}$$

where the origin is 1970 and X units = 1 year.

The Y-intercept $b_0 = 3.155$ is the fitted trend value representing net sales in the base year 1970. The value $(b_1 - 1) \times 100\% = 9.37\%$ is the annual compound growth rate in net sales at Eastman Kodak.

For forecasting purposes, we may substitute the appropriate coded X values into either of the two equations. For example, to predict the trend in net sales for the year 1993 (i.e., X = 23), we have

$$\log \hat{Y}_{24} = 0.49949 + (0.0389)(23) = 1.3937$$

$$\hat{Y}_{24} = \text{antilog } 1.3937 = 24.76 \text{ billion dollars}$$

or

$$\hat{Y}_{24} = (3.155)(1.0937)^{23} = 24.76 \text{ billion dollars}$$

The fitted exponential trend equation projected to 1993 is plotted in Figure 13.10 together with the original time series.

We have now seen the annual net sales data at Eastman Kodak fitted by three different models: linear, quadratic, and exponential. In Section 13.7, we will compare the results of

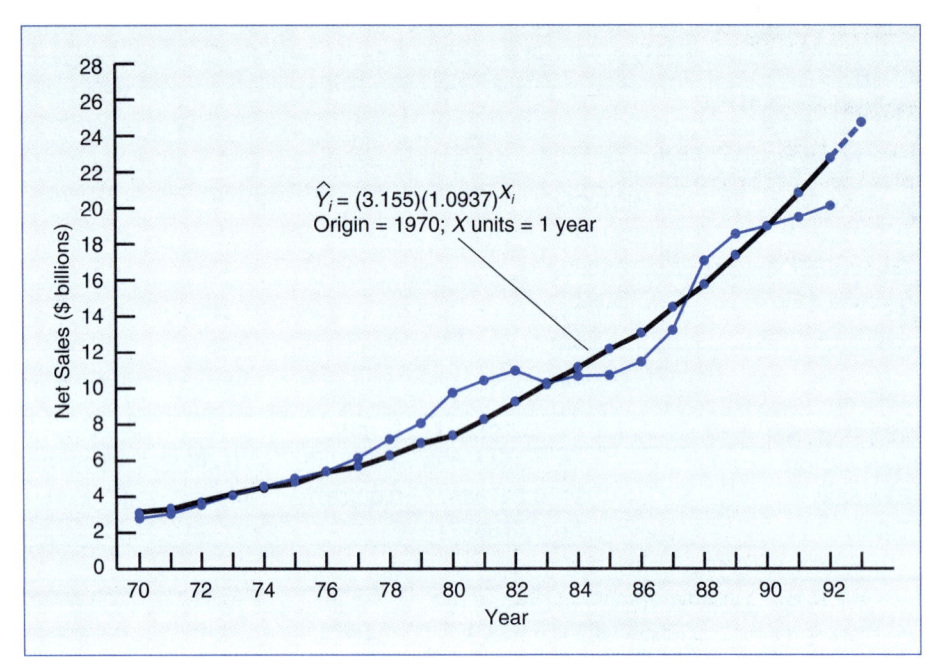

FIGURE 13.10 Fitting the exponential trend equation for the Eastman Kodak net sales data.

these and other forecasting models to determine, a posteriori, the best fit. In Problems 13.26 and 13.27 on pages 697 and 698, you will have the opportunity to use a priori methods to determine appropriate models.

13.5.4 Using Microsoft Excel for Least Squares Trend Fitting

In Chapters 11 and 12, we used the Data Analysis tool and its Regression option for the simple linear regression model and for a variety of multiple regression models. In this section we developed time-series models assuming either a linear, quadratic, or exponential trend. The computations for these models can also be done with the Regression option of the Data Analysis tool.

To illustrate the use of Excel for these three time-series models, open the NET-SALES.XLS workbook that contains the Eastman Kodak net sales data presented in Table 13.5 on page 665. You may recall that in Section 13.5.1, we suggested simplifying the interpretation of the results by coding the years (the X values) consecutively, starting with the first year of 1970 equal to 0 and continuing until 1992 is coded as 22. Click on the Data sheet tab and note that coded year values have been included on the sheet in column B, between the original year and net sales data. With this additional column, we are ready to use the Data Analysis tool to develop the linear trend, quadratic trend, and exponential trend models.

13-5-4.XLS

● **Linear Trend Model** Use the Data Analysis Regression tool as discussed in Section 11.7 to generate the linear trend model. Enter the following information in the Regression dialog box:

(a) Enter C1:C24 as the Input Y Range.

(b) Enter B1:B24 as the Input X Range.

(c) Select the Labels check box.

(d) Enter Figure 13.5 as the name of the New Worksheet Ply.

(e) Select the Residuals and Residual Plots check boxes.

The output produced will be similar to Figure 13.5 on page 670.

● **Quadratic Trend Model** The use of the Regression option of the Data Analysis tool is similar to the curvilinear linear regression model discussed in Section 12.14.4. The curvilinear regression model as expressed in Equation (13.9) includes both linear (X) and curvilinear (X^2) terms for each independent variable. Thus, we need to create a new variable that consists of coded years squared. As mentioned in Section 12.14, since all the X variables in a model need to be located in adjacent columns, we place the coded years squared variable in column C and move the sales variable to column D. We do this by selecting any cell in column C and then selecting Insert | Columns. The sales column becomes column D and a new, blank column C appears. After entering the column heading Year Squared in cell C1, enter the formula =B2^2 in cell C2, and copy it down the column through cell C24.

With the Data sheet still active, again use the Data Analysis | Regression tool. For this model, enter D1:D24 as the Input (Y) Range, B1:C24 as the Input (X) Range, and Figure 13.7 as the name of the New Worksheet Ply. As before, also select the Labels, Residuals, and Residual Plots check boxes.

The output produced will be similar to Figure 13.7 on page 671.

● **Exponential Trend Model** For the exponential trend model, the use of the Regression option of the Data Analysis tool is similar to the linear trend model except that the logarithm (to the base 10) of sales (the Y variable) is used instead of sales. Create this log (sales) variable in column E by first entering the column heading LogSales in cell E1 and then entering the formula =LOG10(D2) in cell E2 and copying it down the column through cell E24.

With the Data sheet still active, again use the Data Analysis Regression tool. For this model enter E1:E24 as the Input (Y) Range, B1:B24 as the Input (X) Range, and Figure 13.9 as the name of the New Worksheet Ply. As before also select the Labels, Residuals, and Residual Plots check boxes.

The output produced will be similar to Figure 13.9 on page 673.

Problems for Section 13.5

Note: The problems in this section should be solved using Microsoft Excel.

● 13.5 The following data represent the annual net sales (in billions of dollars) of Upjohn Co. over the 23-year period from 1970 to 1992:

UPJOHN.TXT

Net sales at Upjohn Co. (1970–1992).

Year	Sales	Year	Sales	Year	Sales	Year	Sales
1970	0.4	1976	1.0	1982	1.8	1988	2.7
1971	0.4	1977	1.1	1983	1.7	1989	2.9
1972	0.5	1978	1.3	1984	1.9	1990	3.0
1973	0.7	1979	1.5	1985	2.0	1991	3.4
1974	0.8	1980	1.8	1986	2.3	1992	3.6
1975	0.9	1981	1.9	1987	2.5		

Source: Moody's Handbook of Common Stocks, 1980, 1989, 1993.

(a) Plot the data on a chart.
(b) Fit a least squares linear trend line to the data and plot the line on your chart.
(c) What are your trend forecasts for the years 1993, 1994, 1995, and 1996?

13.6 The following data represent the annual net operating revenues (in billions of dollars) of Coca-Cola Co. over the 23-year period from 1970 to 1992:

Operating revenues at Coca-Cola Co. (1970–1992).

Year	Revenues	Year	Revenues
1970	1.6	1982	5.9
1971	1.7	1983	6.6
1972	1.9	1984	7.2
1973	2.1	1985	7.9
1974	2.5	1986	7.0
1975	2.9	1987	7.7
1976	3.1	1988	8.3
1977	3.6	1989	9.0
1978	4.3	1990	10.2
1979	4.5	1991	11.6
1980	5.3	1992	13.0
1981	5.5		

Source: Moody's Handbook of Common Stocks, 1980, 1989, 1993.

(a) Plot the data on a chart.
(b) Fit a least squares linear trend line to the data and plot the line on your chart.
(c) What are your trend forecasts for the years 1993, 1994, 1995, and 1996?

13.7 The following data represent the annual net sales (in billions of dollars) of the Georgia-Pacific Corp. over the 23-year period from 1970 to 1992:

Net sales at Georgia-Pacific Corp. (1970–1992).

Year	Sales	Year	Sales	Year	Sales	Year	Sales
1970	1.1	1976	3.0	1982	5.4	1988	9.5
1971	1.3	1977	3.7	1983	6.5	1989	10.1
1972	1.8	1978	4.4	1984	6.7	1990	12.7
1973	2.2	1979	5.2	1985	6.7	1991	11.5
1974	2.4	1980	5.0	1986	7.2	1992	11.8
1975	2.4	1981	5.4	1987	8.6		

Source: Moody's Handbook of Common Stocks, 1980, 1989, 1993.

(a) Plot the data on a chart.
(b) Fit a least squares linear trend line to the data and plot the line on your chart.
(c) What are your trend forecasts for the years 1993, 1994, 1995, and 1996?

13.8 The following data represent average SAT scores (verbal and math) for males and females, along with the average total SAT score for the 20-year period from 1972 to 1991:

SAT scores (1972–1991).

	Male		Female		
Year	Verbal	Math	Verbal	Math	Total
1972	454	505	452	461	937
1973	446	502	443	460	926
1974	447	501	442	459	924
1975	437	495	431	449	906
1976	433	497	430	446	903
1977	431	497	427	445	899
1978	433	494	425	444	897
1979	431	493	423	443	894
1980	428	491	420	443	890
1981	430	492	418	443	890
1982	431	493	421	443	893
1983	430	493	420	445	893
1984	433	495	420	449	897
1985	437	499	425	452	906
1986	437	501	426	451	906
1987	435	500	425	453	906
1988	435	498	422	455	904
1989	434	500	421	454	903
1990	429	499	419	455	900
1991	426	497	418	453	896

SAT.TXT

Source: New York Times, August 27, 1991, p. A20 and August 28, 1990.

For each of the five variables (average SAT score in verbal and math for males and females, along with the average total score):
(a) Plot the data on a chart.
(b) Fit a quadratic trend equation to the data.
(c) What are your trend forecasts for the years 1992, 1993, and 1994?

13.9 The following data represent the receipts and expenditures for state and local governments for the 22-year period from 1970 to 1991.

STATE.TXT

State and local government expenditures (1970–1991).

Year	Receipts	Expenditures	Surplus or Deficit
1970	129.0	127.2	1.80
1971	145.3	142.8	2.50
1972	169.7	156.3	13.40
1973	185.3	171.9	13.40
1974	200.6	193.5	7.10
1975	225.6	221.0	4.60
1976	253.9	239.3	14.60
1977	281.9	256.3	25.60
1978	309.3	278.2	31.10
1979	330.6	305.4	25.20
1980	361.4	336.6	24.80

(continued)

(continued)

Year	Receipts	Expenditures	Surplus or Deficit
1981	390.8	362.3	28.50
1982	409.0	382.1	26.90
1983	443.4	403.2	40.20
1984	492.2	434.1	58.10
1985	528.7	472.6	56.10
1986	571.2	517.0	54.20
1987	594.3	554.2	40.10
1988	631.3	593.0	38.30
1989	677.0	635.9	41.10
1990	724.5	698.8	25.70
1991	770.6	741.1	29.50

Source: U.S. Department of Commerce, Bureau of Economic Analysis, Table B-77.

For each of the three variables (receipts, expenditures, and surplus or deficit):
(a) Plot the data on a chart.
(b) Fit a linear trend equation to the data.
(c) Fit a quadratic trend equation to the data.
(d) Using the models fit in (b) and (c), make annual forecasts for 1992, 1993, and 1994.
(e) **ACTION** Go to your library and record the actual 1992, 1993, and 1994 values from the table available from the Department of Commerce. Compare your results to those in (d). Discuss.

13.10 The following data represent the annual net sales (in billions of dollars) of Black & Decker Corp. over the 23-year period from 1970 to 1992:

Net sales of Black & Decker Corp. (1970–1992).

Year	Sales	Year	Sales
1970	0.3	1982	1.2
1971	0.3	1983	1.2
1972	0.3	1984	1.5
1973	0.4	1985	1.7
1974	0.6	1986	1.8
1975	0.7	1987	1.9
1976	0.7	1988	2.3
1977	0.8	1989	3.2
1978	1.0	1990	4.8
1979	1.2	1991	4.7
1980	1.2	1992	4.8
1981	1.2		

Source: Moody's Handbook of Common Stocks, 1980, 1989, 1993.

(a) Plot the data on a chart.
(b) Fit an exponential trend equation to the data and plot the curve on your chart.
(c) What are your trend forecasts for the years 1993, 1994, 1995, and 1996?

13.11 The following data represent the number of employees (in thousands) on nonagricultural payrolls for the 42-year period from 1950 to 1991. The data are divided into goods-producing industries, nongovernment service producing industries, federal government, and state and local government.

EMPLOYEE.TXT

Total employment on nonagricultural payrolls (1950–1991).

Year	Goods Producing	Nongovernment Services	Federal Government	State and Local Government
1950	45,197	20,665	1,928	4,098
1951	47,819	21,471	2,302	4,087
1952	48,793	21,987	2,420	4,188
1953	50,202	22,483	2,305	4,340
1954	48,990	22,488	2,188	4,563
1955	50,641	23,214	2,187	4,727
1956	52,369	23,988	2,209	5,069
1957	52,853	24,273	2,217	5,399
1958	51,324	23,972	2,191	5,648
1959	53,268	24,774	2,233	5,850
1960	54,189	25,402	2,270	6,083
1961	53,999	25,548	2,279	6,315
1962	55,549	26,208	2,340	6,550
1963	56,653	26,787	2,358	6,868
1964	58,283	27,682	2,348	7,248
1965	60,765	28,765	2,378	7,696
1966	63,901	29,959	2,564	8,220
1967	65,803	31,104	2,719	8,672
1968	67,897	32,321	2,737	9,102
1969	70,384	33,828	2,758	9,437
1970	70,880	34,748	2,731	9,823
1971	71,214	35,397	2,696	10,185
1972	73,675	36,674	2,684	10,649
1973	76,790	38,166	2,663	11,068
1974	78,265	39,301	2,724	11,446
1975	76,945	39,660	2,748	11,937
1976	79,382	41,159	2,733	12,138
1977	82,471	42,999	2,727	12,399
1978	86,697	45,441	2,753	12,919
1979	89,823	47,416	2,773	13,174
1980	90,406	48,507	2,886	13,375
1981	91,156	49,628	2,772	13,259
1982	89,566	49,916	2,739	13,098
1983	90,200	50,996	2,774	13,096
1984	94,496	53,746	2,807	13,216
1985	97,519	56,266	2,875	13,519
1986	99,525	58,274	2,899	13,794
1987	102,200	60,482	2,943	14,067
1988	105,536	62,977	2,971	14,415
1989	108,329	65,228	2,988	14,791
1990	109,971	66,692	3,085	15,237
1991	108,975	66,720	2,965	15,469

Source: U.S. Department of Labor, Bureau of Labor Statistics, Table B-41.

For each of the four variables (employees in goods-producing industries, nongovernment service producing industries, federal government, and state and local government):
(a) Plot the data on a chart.
(b) Fit a linear trend equation to the data.
(c) Fit an exponential trend equation to the data.
(d) Using the models fit in (b) and (c), make annual forecasts for 1992, 1993, and 1994.
(e) **ACTION** Go to your library and record the actual 1992, 1993, and 1994 values from the table available from the Department of Labor. Compare your results to those in (d). Discuss.

13.6 Autoregressive Modeling for Trend Fitting and Forecasting

13.6.1 Model Development

Another useful approach to forecasting with annual time-series data is based on **autoregressive modeling**.[2] Frequently, we find that the values of a series of data at particular points in time are highly correlated with the values that precede and succeed them. A first-order autocorrelation refers to the magnitude of the association between consecutive values in a time series. A second-order autocorrelation refers to the magnitude of the relationship between values two periods apart. Thus, a pth-order autocorrelation refers to the size of the correlation between values in a time series that are p periods apart. To obtain a better historical fit of our data and, at the same time, make useful forecasts of their future behavior, we may take advantage of the potential autocorrelation features inherent in such data by considering autoregressive modeling methods.

A set of autoregressive models are expressed by Equations (13.12), (13.13), and (13.14).

First-Order autoregressive model:
$$Y_i = A_0 + A_1 Y_{i-1} + \delta_i \tag{13.12}$$

Second-Order autoregressive model:
$$Y_i = A_0 + A_1 Y_{i-1} + A_2 Y_{i-2} + \delta_i \tag{13.13}$$

pth-Order autoregressive model:
$$Y_i = A_0 + A_1 Y_{i-1} + A_2 Y_{i-2} + \cdots + A_p Y_{i-p} + \delta_i \tag{13.14}$$

where
Y_i = the observed value of the series at time i
Y_{i-1} = the observed value of the series at time $i - 1$
Y_{i-2} = the observed value of the series at time $i - 2$
Y_{i-p} = the observed value of the series at time $i - p$
A_0 = fixed parameter to be estimated from least squares regression analysis
$A_1, A_2, \ldots, A_p$ = autoregressive parameters to be estimated from least squares regression analysis
δ_i = a nonautocorrelated random (error) component (with 0 mean and constant variance)

We note that the first-order autoregressive model [Equation (13.12)] is similar in form to the simple linear regression model [Equation (11.1) on page 540] and the pth-order autoregressive model [Equation (13.14)] is similar in form to the multiple linear regression model [Equation (12.1a) on page 601]. In the regression models, the regression parameters were given by the symbols $\beta_0, \beta_1, \ldots, \beta_p$, with corresponding statistics denoted by $b_0, b_1, \ldots, b_p$. In the autoregressive models, the analogous parameters are given by the symbols $A_0, A_1, \ldots, A_p$ with corresponding estimates denoted by $a_0, a_1, \ldots, a_p$.

A first-order autoregressive model [Equation (13.12)] is concerned with only the correlation between consecutive values in a series. A second-order autoregressive model [Equation (13.13)] considers the effects of the relationship between consecutive values in a series as well as the correlation between values two periods apart. A pth-order autoregressive model [Equation (13.14)] deals with the effects of relationships between consecutive values, values two periods apart, and so on—up to values p periods apart. The selection of an appropriate autoregressive model, then, is not easy task. We must weigh the advantages due to parsimony (see Section 13.7.3) with the concern of failing to take into account important autocorrelation behavior inherent in the data. On the other hand, we must be equally concerned with selecting a high-order model requiring the estimation of numerous, unnecessary parameters—especially if n, the number of observations in the series, is not too large. The reason for this is that p out of n data values will be lost in obtaining an estimate of A_p when comparing each data value Y_i with its "fairly near neighbor" Y_{i-p}, which is p periods apart (i.e., the comparisons are Y_{1+p} versus Y_1, Y_{2+p} versus Y_2, …, and Y_n versus Y_{n-p}). To illustrate this, suppose we have the following series of $n = 7$ consecutive values:

$$31, \quad 34, \quad 37, \quad 35, \quad 36, \quad 43, \quad 40$$

Comparison schema for first- and second-order autoregressive models are established in the following table:

i	First-Order Autoregressive Model (Y_i versus Y_{i-1})	Second-Order Autoregressive Model (Y_i versus Y_{i-1} and Y_i versus Y_{i-2})
1	31 ↔ …	31 ↔ … and 31 ↔ …
2	34 ↔ 31	34 ↔ 31 and 34 ↔ …
3	37 ↔ 34	37 ↔ 34 and 37 ↔ 31
4	35 ↔ 37	35 ↔ 37 and 35 ↔ 34
5	36 ↔ 35	36 ↔ 35 and 36 ↔ 37
6	43 ↔ 36	43 ↔ 36 and 43 ↔ 35
7	40 ↔ 43	40 ↔ 43 and 40 ↔ 36
	(1 comparison is lost for regression analysis)	(2 comparisons are lost for regression analysis)

Once a model is selected and least squares regression methods are used to obtain estimates of the parameters, the next step is to determine the appropriateness of this model. Either we can select a given pth-order autoregressive model based on previous experiences with similar data, or else, as a starting point, a model with several parameters and then eliminate the parameters that do not contribute significantly. In this latter approach, the following test for the significance of the highest-order autoregressive parameter in the fitted model may be used:

$$H_0: A_p = 0 \quad \text{(The highest-order parameter is 0.)}$$

against the two-sided alternative

$$H_1: A_p \neq 0 \quad \text{(The parameter } A_p \text{ is significantly meaningful.)}$$

The test statistic, obtainable from Excel (which provides estimates of regression coefficients and standard errors), is approximated by

$$Z \cong \frac{a_p}{S_{a_p}} \qquad (13.15)$$

where a_p = the estimate of the highest-order parameter A_p in the autoregressive model

S_{a_p} = the standard deviation of a_p

Using an α level of significance, the decision rule is to reject H_0 if $Z > +Z_{\alpha/2}$ (the upper-tail critical value from a standardized normal distribution) or if $Z < -Z_{\alpha/2}$ (the lower-tail critical value from a standardized normal distribution), and not to reject H_0 if $-Z_{\alpha/2} \leq Z \leq +Z_{\alpha/2}$.

If the null hypothesis that $A_p = 0$ is not rejected, we may conclude that the selected model contains too many estimated parameters. The highest-order term would then be discarded, and an autoregressive model of order $p - 1$ would be obtained through least squares regression. A test of the hypothesis that the new highest-order term is 0 would then be repeated.

This testing and modeling procedure continues until we reject H_0. When this occurs, we know that our highest-order parameter is significant, and we are ready to use the particular model for forecasting purposes.

The fitted pth-order autoregressive model has the following form:

$$\hat{Y}_i = a_0 + a_1 Y_{i-1} + a_2 Y_{i-2} + \cdots + a_p Y_{i-p} \qquad (13.16)$$

where
$\hat{Y}_i$ = the fitted value of the series at time i

Y_{i-1} = the observed value of the series at time $i - 1$

Y_{i-2} = the observed value of the series at time $i - 2$

Y_{i-p} = the observed value of the series at time $i - p$

$a_0, a_1, a_2, \ldots, a_p$ = regression estimates of the parameters $A_0, A_1, A_2, \ldots, A_p$

To forecast j years into the future from the current nth time period, we have

$$\hat{Y}_{n+j} = a_0 + a_1 \hat{Y}_{n+j-1} + a_2 \hat{Y}_{n+j-2} + \cdots + a_p \hat{Y}_{n+j-p} \qquad (13.17)$$

where $a_0, a_1, a_2, \ldots, a_p$ = regression estimates of the parameters $A_0, A_1, A_2, \ldots, A_p$

j = number of years into the future

$\hat{Y}_{n+j-p}$ = forecast of Y_{n+j-p} from the current time period for $j - p > 0$

$\hat{Y}_{n+j-p}$ = is observed value Y_{n+j-p} for $j - p \leq 0$

Thus, to make forecasts j years into the future from, say, a p = third-order autoregressive model, we need only the most recent $p = 3$ observed data values Y_n, Y_{n-1}, and Y_{n-2} and the estimates of the parameters A_0, A_1, A_2, and A_3 from a multiple regression program. To forecast 1 year ahead, Equation (13.17) becomes

$$\hat{Y}_{n+1} = a_0 + a_1 Y_n + a_2 Y_{n-1} + a_3 Y_{n-2}$$

To forecast 2 years ahead, Equation (13.17) becomes

$$\hat{Y}_{n+2} = a_0 + a_1 \hat{Y}_{n+1} + a_2 Y_n + a_3 Y_{n-1}$$

To forecast 3 years ahead, Equation (13.17) becomes

$$\hat{Y}_{n+3} = a_0 + a_1 \hat{Y}_{n+2} + a_2 \hat{Y}_{n+1} + a_3 Y_n$$

To forecast 4 years ahead, Equation (13.17) becomes

$$\hat{Y}_{n+4} = a_0 + a_1 \hat{Y}_{n+3} + a_2 \hat{Y}_{n+2} + a_3 \hat{Y}_{n+1}$$

and so on.

To demonstrate the autoregressive modeling technique, let us return once more to the Eastman Kodak net sales data presented in Table 13.5 (on page 669) and plotted in Figure 13.1 (on page 653). Table 13.6 displays the setup for first-, second-, and third-order autoregressive models. All the columns in this table are needed for fitting third-order autoregressive models. The last column would be omitted when fitting second-order autoregressive models, and the last two columns would be eliminated when fitting first-order autoregressive models.

Table 13.6 Developing first-, second-, and third-order autoregressive models on net sales for Eastman Kodak Company (1970–1992).

Year	i	"Dependent" Variable Y_i	Y_{i-1}	Y_{i-2}	Y_{i-3}
1970	1	2.8	*	*	*
1971	2	3.0	2.8	*	*
1972	3	3.5	3.0	2.8	*
1973	4	4.0	3.5	3.0	2.8
1974	5	4.6	4.0	3.5	3.0
1975	6	5.0	4.6	4.0	3.5
1976	7	5.4	5.0	4.6	4.0
1977	8	6.0	5.4	5.0	4.6
1978	9	7.0	6.0	5.4	5.0
1979	10	8.0	7.0	6.0	5.4
1980	11	9.7	8.0	7.0	6.0
1981	12	10.3	9.7	8.0	7.0
1982	13	10.8	10.3	9.7	8.0
1983	14	10.2	10.8	10.3	9.7
1984	15	10.6	10.2	10.8	10.3
1985	16	10.6	10.6	10.2	10.8
1986	17	11.5	10.6	10.6	10.2
1987	18	13.3	11.5	10.6	10.6
1988	19	17.0	13.3	11.5	10.6
1989	20	18.4	17.0	13.3	11.5
1990	21	18.9	18.4	17.0	13.3
1991	22	19.4	18.9	18.4	17.0
1992	23	20.1	19.4	18.9	18.4

Thus, we note that either $p = 1$, 2, or 3 observations out of $n = 23$ are lost in the comparisons needed for developing these three autoregressive models.

Using Excel (Reference 8), the following third-order autoregressive model is fitted to the Eastman Kodak net sales data (see Figure 13.11):

$$\hat{Y}_i = 0.446 + 1.534 Y_{i-1} - 0.739 Y_{i-2} + 0.218 Y_{i-3}$$

where the origin is 1973 and Y units = 1 year.

Next, we might consider testing for the significance of the highest-order parameter. On the other hand, if our experience with similar data permit us to hypothesize that a third-order autoregressive model is appropriate for this time series, our fitted model can be used directly for forecasting purposes without the need for testing for parameter significance. Therefore, to demonstrate the forecasting procedure for our third-order autoregressive model, we use the estimates

$$a_0 = 0.446 \qquad a_1 = 1.534 \qquad a_2 = -0.739 \qquad a_3 = 0.218$$

as well as the three most current data values

$$Y_{21} = 18.9 \qquad Y_{22} = 19.4 \qquad Y_{23} = 20.1$$

Our forecasts of net sales at Eastman Kodak for the years 1993–1996 are obtained from Equation (13.17) as follows:

$$\hat{Y}_{n+j} = 0.446 + 1.534 \hat{Y}_{n+j-1} - 0.739 \hat{Y}_{n+j-2} + 0.218 \hat{Y}_{n+j-3}$$

1993 (1 year ahead): $\hat{Y}_{24} = 0.446 + (1.534)(20.1) - (0.739)(19.4) + (0.218)(18.9)$
$\qquad\qquad\qquad\qquad\quad = 21.0$ billions of dollars

SUMMARY OUTPUT

Regression Statistics

Multiple R	0.98952169
R Square	0.97915317
Adjusted R Square	0.97524439
Standard Error	0.82513136
Observations	20

ANOVA

	df	SS	MS	F	Significance F
Regression	3	511.654532	170.55151	250.50096	1.17979E-13
Residual	16	10.8934681	0.6808418		
Total	19	522.548			

	Coefficients	Std. Error	t Stat	P-value	Lower 95%	Upper 95%
Intercept	0.44591849	0.43007893	1.0368294	0.3152321	-0.465807903	1.357644874
X Variable 1	1.53433287	0.24499378	6.2627422	1.133E-05	1.014969382	2.053696366
X Variable 2	-0.7393539	0.42263675	-1.749384	0.0993772	-1.635303566	0.156595804
X Variable 3	0.21794709	0.2688183	0.8107599	0.4293991	-0.351922105	0.787816295

FIGURE 13.11 Partial Excel output for the third-order autoregressive model for the Eastman Kodak net sales data.

1994 (2 years ahead): $\hat{Y}_{25} = 0.446 + (1.534)(21.0) - (0.739)(20.1) + (0.218)(19.4)$
= 21.9 billions of dollars

1995 (3 years ahead): $\hat{Y}_{26} = 0.446 + (1.534)(21.9) - (0.739)(21.0) + (0.218)(20.1)$
= 22.9 billions of dollars

1996 (4 years ahead): $\hat{Y}_{27} = 0.446 + (1.534)(22.9) - (0.739)(21.9) + (0.218)(21.0)$
= 23.8 billions of dollars

Prior to forecasting, however, most researchers would have preferred to test for the significance of the parameters of a fitted model. Using the output from Excel (Figure 13.11), the highest-order parameter estimate a_3 for the fitted third-order autoregressive model is 0.218 (rounded) with a standard deviation S_{a_3} of 0.269 (rounded).

To test

$$H_0: A_3 = 0$$

against

$$H_1: A_3 \neq 0$$

we have, from Equation (13.15),

$$Z \cong \frac{a_3}{S_{a_3}} = \frac{0.218}{0.269} = 0.81$$

Using a .05 level of significance, the two-tailed test has critical Z values of ±1.96. Since Z = +0.81 < +1.96, the upper-tail critical value under the standardized normal distribution (Table E.2), we may not reject H_0, and we would conclude that the third-order parameter of the autoregressive model is not significantly important and can be deleted.

Using Excel once again, a second-order autoregressive model is obtained, and partial output is displayed in Figure 13.12.

The second-order autoregressive model is

$$\hat{Y}_i = 0.473 + 1.453 Y_{i-1} - 0.455 Y_{i-2}$$

where the origin is 1972 and Y units = 1 year.

From the Excel output, the highest-order parameter estimate a_2 is –0.455 (rounded) with a standard deviation $S_{a_2} = 0.225$ (rounded).

To test

$$H_0: A_2 = 0$$

against

$$H_1: A_2 \neq 0$$

we have, from Equation (13.15),

$$Z \cong \frac{a_2}{S_{a_2}} = \frac{-0.455}{0.225} = -2.02$$

Testing again at the .05 level of significance, since Z = –2.02 < –1.96, we may reject H_0, and we would conclude that the second-order parameter of the autoregressive model is significantly important and it should be included in the model.

Our model-building approach has led to the selection of the second-order autoregressive model as the most appropriate for the given data. Using the estimates $a_0 = 0.473$, $a_1 = 1.453$, and $a_2 = -0.455$, as well as the two most recent data values $Y_{22} = 19.4$ and $Y_{23} = 20.1$, our

	A	B	C	D	E	F	G
1	SUMMARY OUTPUT						
2							
3	*Regression Statistics*						
4	Multiple R	0.99011492					
5	R Square	0.98032756					
6	Adjusted R Square	0.97814174					
7	Standard Error	0.79389844					
8	Observations	21					
9							
10	ANOVA						
11		df	SS	MS	F	Significance F	
12	Regression	2	565.34744	282.673718	448.492859	4.4129E-16	
13	Residual	18	11.344945	0.63027473			
14	Total	20	576.69238				
15							
16		Coefficients	Std. Error	t Stat	P-value	Lower 95%	Upper 95%
17	Intercept	0.47278714	0.3825234	1.23596923	0.23235716	-0.330865306	1.2764396
18	X Variable 1	1.45297485	0.2131838	6.81559598	2.2162E-06	1.005091894	1.9008578
19	X Variable 2	-0.4548002	0.2248946	-2.0222815	0.05826536	-0.927286609	0.0176862

FIGURE 13.12 Partial Excel output for the second-order autoregressive model for the Eastman Kodak net sales data.

forecasts of net sales at Eastman Kodak for the years 1993–1996 are obtained from Equation (13.19) as follows:

$$\hat{Y}_{n+j} = 0.473 + 1.453\hat{Y}_{n+j-1} - 0.455\hat{Y}_{n+j-2}$$

1993 (1 year ahead): $\hat{Y}_{24} = 0.473 + (1.453)(20.1) - 0.455(19.4)$
= 20.8 billions of dollars

1994 (2 years ahead): $\hat{Y}_{25} = 0.473 + (1.453)(20.8) - (0.455)(20.1)$
= 21.4 billions of dollars

1995 (3 years ahead): $\hat{Y}_{26} = 0.473 + (1.453)(21.4) - (0.455)(20.8)$
= 22.1 billions of dollars

1996 (4 years ahead): $\hat{Y}_{27} = 0.473 + (1.453)(22.1) - (0.455)(21.4)$
= 22.7 billions of dollars

The data and the forecasts are plotted in Figure 13.13.

13.6.2 Using Microsoft Excel for Autoregressive Modeling

In this section we discussed the use of autoregressive models for trend fitting and forecasting. These models involve independent variables that represent the dependent variable that is lagged by one, two, three, or more time periods. In order to use the Data Analysis Regression option, we need to create these lagged variables. To accomplish this, open the NETSALES.XLS workbook, click on the Data sheet tab to make that sheet active, and do the following:

13–6–2.XLS

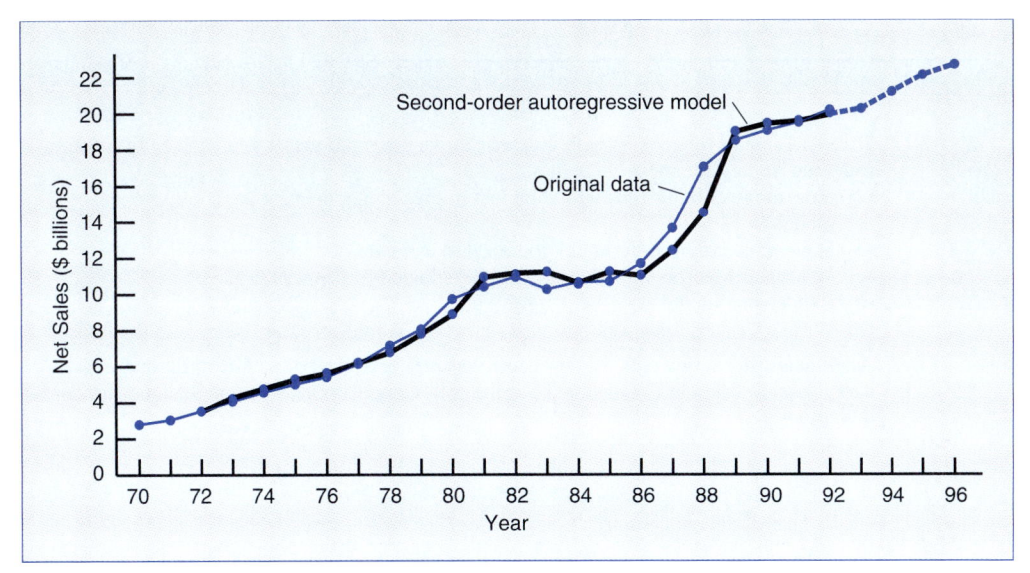

FIGURE 13.13 Using a second-order autoregressive model for the Eastman Kodak net sales.

❶ To create an X variable that is lagged from the sales (Y) variable by one time period, enter the formula =C2 in cell D3 and copy this formula down the column through row 24.

❷ To create an X variable that is lagged from the sales (Y) variable by two time periods, enter the formula =C2 in cell E4 and copy this formula down the column through row 24.

❸ To create an X variable that is lagged from the sales (Y) variable by three time periods, enter the formula =C2 in cell F5 and copy this formula down the column through row 24.

> Entering #N/A in cells D2, E2, E3, F2, F3, and F4 that were left empty in the above procedure, while not necessary, will prevent their use in the regression procedures below should an incorrect range that includes any of these cells be inadvertently entered in the Regression dialog box.

With the Data sheet still active, use the Data Analysis | Regression command to obtain the third-order autoregressive model. Enter C5:C24 in the Input Y Range edit box, enter D5:F24 in the Input (X) range edit box, and Figure 13.11 as the name for the New Worksheet Ply. (The ranges begin in row 5 because we need to exclude the first three observations.)

Select the Residuals and Residual Plots boxes, but this time do *not* select the Labels check box. Click the OK button to obtain the output similar to Figure 13.11 on page 684.

If it is necessary to obtain the second-order model because the third-order autoregressive term was not significant. repeat the procedure with the Data sheet active using C4:C24 as the Input Y range value and D4:E24 as the Input X range value and supply a unique name for the New Worksheet Ply. (Remember *not* to select the Labels check box and to begin the ranges with row 4 in order to exclude the first two observations.) The output obtained will be similar to Figure 13.12 on page 686.

13.6 Autoregressive Modeling for Trend Fitting and Forecasting

Problems for Section 13.6

Note: *The problems in this section should be solved using Microsoft Excel.*

- **13.12** Given an annual time series with 40 consecutive observations, if you were to fit a fifth-order autoregressive model:
  - (a) How many comparisons would be lost in the development of the autoregressive model?
  - (b) How many parameters would you need to estimate?
  - (c) Which of the original 40 values would you need for forecasting?
  - (d) Express the model.
  - (e) Write a general equation to indicate how you would forecast j years into the future.

- **13.13** An annual time series with 17 consecutive values is obtained. A third-order autoregressive model is fitted to the data and has the following estimated parameters and standard deviations:

$$a_0 = 4.50 \quad a_1 = 1.80 \quad a_2 = .70 \quad a_3 = .20$$

$$S_{a_1} = .50 \quad S_{a_2} = .30 \quad S_{a_3} = .10$$

  At the .05 level of significance, test the appropriateness of the fitted model.

- **13.14** Refer to Problem 13.13. The three most recent observations are

$$Y_{15} = 23 \quad Y_{16} = 28 \quad Y_{17} = 34$$

  - (a) Forecast the series for the next year and the following year.
  - (b) Suppose, when testing for the appropriateness of the fitted model in Problem 13.13, the standard deviations were

$$S_{a_1} = .45 \quad S_{a_2} = .35 \quad S_{a_3} = .15$$

  (1) What would you conclude?
  (2) Discuss how you would proceed if forecasting were still your main objective.

For problems 13.15–13.19:
  - (a) Fit a third-order autoregressive model and test for the significance of the third-order autoregressive parameter. (Use $\alpha = .05$.)
  - (b) If necessary, fit a second-order autoregressive model and test for the significance of the second-order autoregressive parameter. (Use $\alpha = .05$.)
  - (c) If necessary, fit a first-order autoregressive model and test for the significance of the first-order autoregressive parameter. (Use $\alpha = .05$.)
  - (d) If appropriate, provide annual forecasts from 1993 through 1996.

- **13.15** Refer to Problem 13.2 (earnings per share at TRW Inc.) on page 667.
- **13.16** Refer to Problem 13.5 (net sales at Upjohn Co.) on page 675.
- **13.17** Refer to Problem 13.6 (net operating revenues of Coca-Cola Co.) on page 676.
- **13.18** Refer to Problem 13.7 (net sales at Georgia-Pacific Corp.) on page 676.
- **13.19** Refer to Problem 13.10 (net sales at Black & Decker Corp.) on page 678.

13.7 CHOOSING AN APPROPRIATE FORECASTING MODEL

In Sections 13.5 and 13.6, six alternative time-series forecasting methods were developed. In Section 13.5, we studied three commonly used models based on the method of least squares—the linear model, the quadratic model, and the exponential model. In Section 13.6, we covered three autoregressive models—the first-, second-, and third-order models.

A major question must be answered at this time: Is there a *best* model? That is, among such models as these, which *one* should we select if we are interested in time-series forecasting? Three approaches are offered as guidelines for model selection:

1. Perform a residual analysis.
2. Measure the magnitude of the residual error.
3. Use the principle of parsimony.

These are the most widely used methods for determining the adequacy of a particular forecasting model; they are based on a judgment of how well the model has fit a given set of time-series data. These methods assume that future movements in the series can be projected by a study of past behavior patterns.

13.7.1 Residual Analysis

You may recall from our study of regression analysis in Sections 11.10 and 12.5 that differences between the observed and fitted data are known as *residuals*. Thus, for the *i*th year in an annual time series of *n* years, the residual is defined as

$$e_i = (Y_i - \hat{Y}_i) \tag{13.18}$$

where
Y_i = observed value in year *i*
$\hat{Y}_i$ = fitted value in year *i*

Once a particular model has been fitted to a given time series, we may plot the residuals over the *n* time periods. As depicted in Figure 13.14(a), if the particular model fits adequately, the residuals represent the irregular component of the time series, and they should therefore be randomly distributed throughout the series. On the other hand, as illustrated in the three remaining

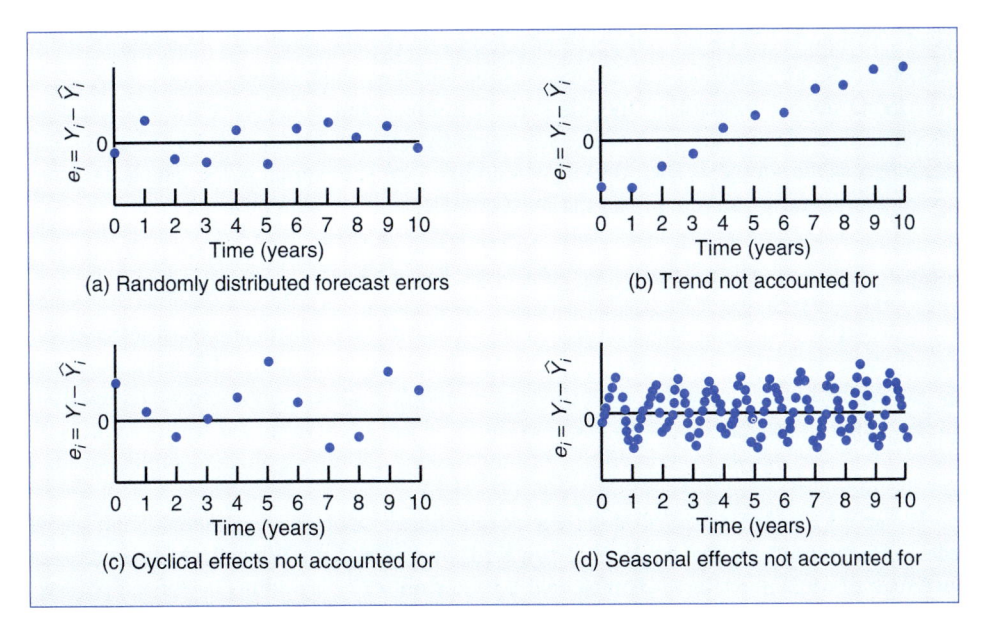

FIGURE 13.14   A residual analysis for studying error patterns.

panels of Figure 13.14, if the particular model does not fit adequately, the residuals may demonstrate some systematic pattern such as a failure to account for trend (b), a failure to account for cyclical variation (c), or, with monthly data, a failure to account for seasonal variation (d).

13.7.2 Measuring the Magnitude of the Residual Error

If, after performing a residual analysis, we still believe that two or more models appear to fit the data adequately, then a second method used for model selection is based on some measure of the magnitude of the residual error. Numerous measures have been proposed (see Reference 2), and unfortunately, there is no consensus among researchers as to which particular measure is best for determining the most appropriate forecasting model.

A measure that most researchers seem to prefer for assessing the appropriateness of various forecasting models is the **mean absolute deviation (MAD)**:

$$\text{MAD} = \frac{\sum_{i=1}^{n} |Y_i - \hat{Y}_i|}{n} \tag{13.19}$$

For a particular model, the MAD is a measure of the average of the absolute discrepancies between the actual and fitted values in a given time series. If a model were to fit the past time-series data *perfectly*, the MAD would be zero, while if a model were to fit the past time-series data *poorly*, the MAD would be large. Hence, when comparing the merits of two or more forecasting models, the one with the *minimum* MAD can be selected as most appropriate on the basis of past fits of the given time series.

13.7.3 Principle of Parsimony

If, after performing a residual analysis and comparing the obtained MAD measures, we still believe that two or more models appear to adequately fit the data, then a third method for model selection based on the **principle of parsimony** can be used. This principle states that we should select the *simplest* model that gets the job done adequately.

Among the six forecasting models studied in this chapter, the least squares linear and quadratic models and the first-order autoregressive model would be regarded by most researchers as the simplest. Their ranking would likely be in the order given. The least squares exponential model would qualify as the most complex of the techniques presented.

13.7.4 A Comparison of Four Forecasting Methods

To illustrate the model selection process, we again consider the Eastman Kodak net sales data. We will compare four of the forecasting methods described in Sections 13.5 and 13.6: the linear model, the quadratic model, the exponential model, and the second-order autoregressive model. (There is no need to study the third-order autoregressive model for this time series since as you may recall from Section 13.6, this model did not significantly improve the fit over the simpler second-order autoregressive model.)

Figure 13.15 displays the residual plots for the linear, quadratic, exponential, and second-order autoregressive models on the Eastman Kodak net sales data. When drawing conclusions from such residual plots, caution must be used since only 23 data points have been observed.

From Figure 13.15(a), (b), and (c), we note that the cyclical effects are unaccounted for in each of the least squares models. However, the residual plots for the quadratic and expo-

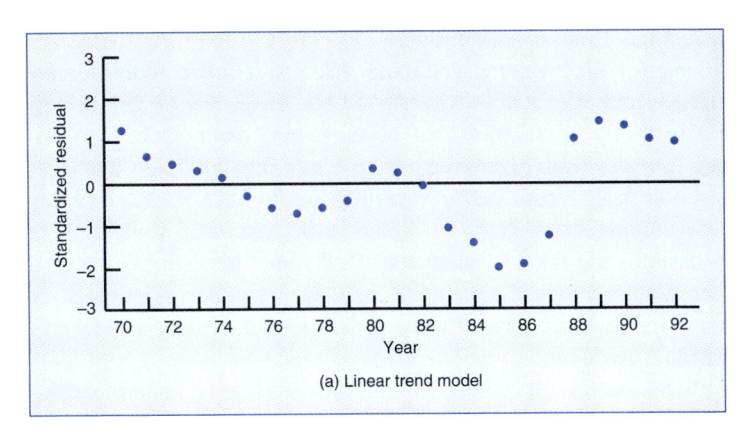

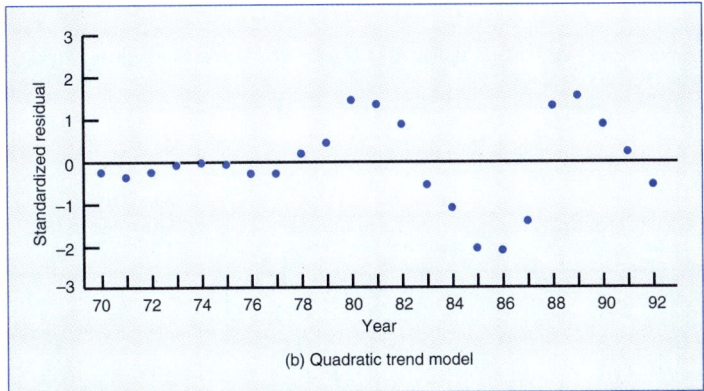

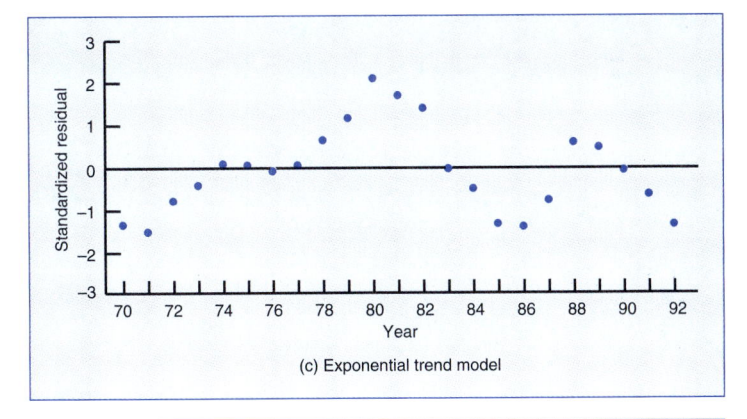

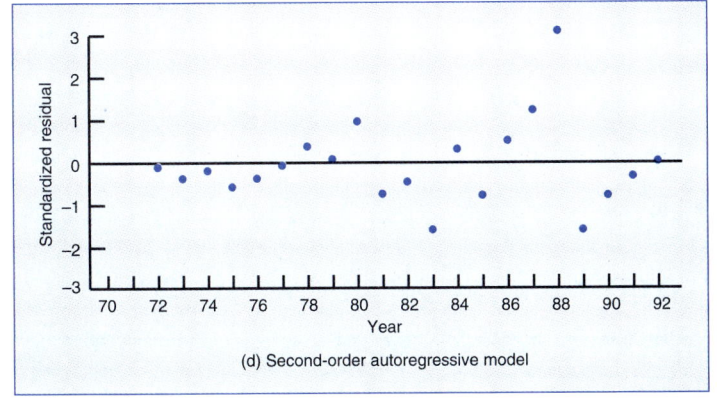

FIGURE 13.15 **Residual plots for four forecasting methods.** *Source:* Data are taken from Table 13.7 on page 692.

13.7 Choosing an Appropriate Forecasting Model **691**

nential models seem to suggest that these models provide a better fit to the series than does the linear model because (b) and (c) display more randomness (less systematic pattern) in the residuals over the first 8 years of the series. On the other hand, the increasing (wider) amplitude observed in the later years of all residual plots may be suggesting that none of the models examined here performs outstandingly with respect to capturing the large net sales movements that have occurred in these recent years. Nevertheless, from (d), we can see that the second-order autoregressive model exhibits the least amount of systematic structure.

To summarize, on the basis of the residual analyses of all four forecasting models, it would appear that the second-order autoregressive model may be most appropriate and the linear model the least appropriate. To verify this, let us compare the four models with respect to the magnitude of their residual errors.

Table 13.7 provides the actual values (Y_i), along with the fitted values ($\hat{Y}_i$) and the residuals (e_i) for each of the four models. In addition, the MAD for each model is displayed.

Table 13.7 Comparison of four forecasting methods using the mean absolute deviation (MAD) for the Eastman Kodak net sales data.

		\multicolumn{8}{c|}{Forecasting Method}							
	Net Sales	\multicolumn{2}{c	}{Linear}	\multicolumn{2}{c	}{Quadratic}	\multicolumn{2}{c	}{Exponential}	\multicolumn{2}{c	}{Second-Order Autoregressive}
Year	Y_i	$\hat{Y}_i$	e_i	$\hat{Y}_i$	e_i	$\hat{Y}_i$	e_i	$\hat{Y}_i$	e_i
1970	2.8	1.20	1.60	2.92	−0.12	3.16	−0.36	—	—
1971	3.0	2.00	1.00	3.25	−0.25	3.45	−0.45	—	—
1972	3.5	2.80	0.70	3.63	−0.13	3.78	−0.28	3.56	−0.06
1973	4.0	3.60	0.40	4.05	0.05	4.13	−0.13	4.19	−0.19
1974	4.6	4.40	0.20	4.51	0.09	4.52	0.08	4.69	−0.09
1975	5.0	5.20	−0.20	5.02	0.02	4.94	0.06	5.34	−0.34
1976	5.4	6.00	−0.60	5.58	−0.18	5.41	0.01	5.65	−0.25
1977	6.0	6.80	−0.80	6.18	−0.18	5.91	0.09	6.04	0.04
1978	7.0	7.60	−0.60	6.82	0.18	6.47	0.53	6.73	0.27
1979	8.0	8.40	−0.40	7.51	0.49	7.07	0.93	7.91	0.09
1980	9.7	9.20	0.50	8.24	1.46	7.74	1.96	8.91	0.78
1981	10.3	10.00	0.30	9.02	1.28	8.46	1.84	10.93	−0.63
1982	10.8	10.80	0.00	9.84	0.96	9.25	1.55	11.03	−0.27
1983	10.2	11.60	−1.40	10.71	−0.51	10.12	0.08	11.48	−1.28
1984	10.6	12.41	−1.81	11.62	−1.02	11.07	0.47	10.38	0.22
1985	10.6	13.21	−2.61	12.58	−1.98	12.11	−1.51	11.24	−0.64
1986	11.5	14.01	−2.51	13.58	−2.08	13.24	1.74	11.05	0.47
1987	13.3	14.81	−1.51	14.63	−1.33	14.48	−1.18	12.36	0.94
1988	17.0	15.61	1.39	15.72	1.28	15.84	1.16	14.57	2.43
1989	18.4	16.41	1.99	16.85	1.55	17.32	1.08	19.12	−0.72
1990	18.9	17.21	1.69	18.03	0.87	18.95	0.05	19.48	−0.58
1991	19.4	18.01	1.39	19.26	0.14	20.72	−1.32	19.57	−0.17
1992	20.1	18.81	1.29	20.53	−0.43	22.66	−2.56	20.06	0.04
Absolute Sum			24.9		16.7		19.5		10.42
MAD		\multicolumn{2}{c	}{$\dfrac{24.9}{23} = 1.08$}	\multicolumn{2}{c	}{$\dfrac{16.7}{23} = 0.720$}	\multicolumn{2}{c	}{$\dfrac{19.41}{23} = 0.844$}	\multicolumn{2}{c	}{$\dfrac{10.42}{21} = 0.496$}

A comparison of the MAD for each of the models clearly indicates that the simplest model, the linear model, is, for this time series, the one with the poorest fit. We can also see that the other two least squares models (quadratic and exponential) do not show sufficient improvement over the linear model. As anticipated from the residual analysis (Figure 13.15 on page 691), the model with the smallest MAD is the second-order autoregressive model. On this basis, the second-order autoregressive model is the one we should select.

13.7.5 Model Selection: A Warning

Once a particular forecasting model is selected, it becomes imperative that we appropriately monitor it. After all, the objective in selecting the model is to be able to project or forecast future movements in a set of time-series data. Unfortunately, such forecasting models are generally poor at detecting changes in the underlying structure of the time series. It is important then that such projections be examined together with those obtained through other types of forecasting methods (such as the use of leading indicators). As soon as a *new* data value (Y_t) is observed in time period t, it must be compared with its projection ($\hat{Y}_t$). If the difference is too large, the forecasting model should be revised. Such *adaptive-control procedures* are described in Reference 2.

13.7.6 Using Microsoft Excel to Obtain the Mean Absolute Deviation

13-7-6.XLS

In this section we used Equation (13.19) on page 690 to compute the mean absolute deviation (MAD) for each of the forecasting models we studied. Although the Data Analysis Regression option does not compute the MAD, as long as the residuals have been computed, we can use an Excel formula instead. The Excel formula =ABS(*residual*) can be used for all the models in this chapter except for the exponential trend model. Enter this formula in column D of the sheet that contains the regression output, in the row corresponding to the first observation and then copy it through the last row of data. Once this has been done, the MAD can be computed by entering the formula =AVERAGE(*range of absolute residuals*) in a row following the last row of the range containing the absolute value of the residuals. To illustrate this computation for the linear trend model of Figure 13.5 and Section 13.5.4, open the workbook you developed in Section 13.5.4 (or 13-5-4.XLS) and click on the Figure 13.5 sheet tab to make that sheet active. Scroll down to the Residual Output portion of the sheet (row 22). Then do the following:

❶ Enter the heading Abs.Residuals in cell D24.

❷ Enter the formula =ABS(C25) in cell D25 and copy this formula down the column to row 47.

❸ Enter the label MAD in cell C48.

❹ In cell D48, enter the formula =AVERAGE(D25:D47) to compute the value of the mean absolute deviation 1.082445437.

The computation of the MAD for the exponential trend model is more complicated because both the predicted sales and the residuals are expressed not in sales, but as the logarithm (to the base 10) of the sales. Thus, we first need to compute the predicted sales, equal to antilogarithm of the log of predicted sales, and subtract this value from the sales to obtain the residual. Modifications to the Figure 13.9 sheet (containing the regression output) to accomplish these computations is presented in Table 13.2.Excel.

Table 13.2.Excel Modifications for obtaining the mean absolute deviation for the exponential trend.

Model

	D	E	F
24	Predicted Sales	Sales	Abs.Residuals
25	=POWER(10,B25)	=Data!D2	=ABS(E25–D25)
26	=POWER(10,B26)	=Data!D3	=ABS(E26–D26)
27	.	.	.
47	=POWER(10,B47)	=Data!D24	=ABS(E47–D47)
48		MAD:	= AVERAGE(F25:F47)

To obtain the antilog of the log of the predicted value, we can use the Excel POWER function whose format is

$$\text{POWER}(number, power)$$

where   *number* = the number you want to raise to a specified power

power = the power to which the number is to be raised

For example, =POWER(10,2) would equal 100 because the antilog of 2 base 10 is 10^2 which equals 100. Thus, to obtain the MAD, we would do the following:

❶ To obtain the predicted sales (the antilog of the log of the predicted sales) in cell D25, enter the formula =POWER(10,B25) and copy this formula down through cell D47.

❷ Copy the actual sales from the Data sheet into column E by entering the formula =Data!D2 in cell E25 and copying this formula down through cell E47.

❸ Obtain the absolute value of the residuals by entering the formula =ABS(E25–D25) in cell F25 and copying this formula down through cell F47.

❹ Obtain the MAD in cell F48 (after entering the label MAD in cell E48) by entering the formula =AVERAGE(F25:F47).

The computed value of the mean absolute deviation, equal to 0.843966753, is now displayed in cell F48.

Problems for Section 13.7

Note: The problems in this section should be solved using Microsoft Excel.

For Problems 13.20–13.24:
(a) Perform a residual analysis for each fitted model.
(b) Compute the MAD for each fitted model.
(c) On the basis of (a), (b), and parsimony, which model would you select for purposes of forecasting? Discuss.

13.20 Refer to Problems 13.2 and 13.15 (earnings per share at TRW Inc.) on pages 667 and 688.

• 13.21 Refer to Problems 13.5 and 13.16 (net sales at Upjohn Co.) on pages 675 and 688.

13.22 Refer to Problems 13.6 and 13.17 (net operating revenues at Coca-Cola Co.) on pages 676 and 688.

13.23 Refer to Problems 13.7 and 13.18 (net sales at Georgia-Pacific Corp.) on pages 676 and 688.

• 13.24 Refer to Problems 13.10 and 13.19 (net sales at Black & Decker Corp.) on pages 678 and 688.

13.8 PITFALLS CONCERNING TIME-SERIES ANALYSIS

The value of such forecasting methodology as time-series analysis, which utilizes past and present information as a guide to the future, was recognized and most eloquently expressed more than two centuries ago by the American statesman Patrick Henry, who said

> I have but one lamp by which my feet are guided, and that is the lamp of experience. I know no way of judging the future but by the past. [*Speech at Virginia Convention (Richmond) March 23, 1775*]

If it were true (as time-series analysis assumes) that the factors that have affected particular patterns of economic activity in the past and present will continue to do so in a similar manner in the future, time-series analysis, by itself, would certainly be a most appropriate and effective forecasting tool as well as an aid in the management control of present activities. However, critics of classical time-series methods have argued that these techniques are overly naïve and mechanical; that is, a mathematical model based on the past should not be utilized for mechanically extrapolating trends into the future without considering personal judgments, business experiences, or changing technologies, habits, and needs. (See Problem 13.25 on page 696.) Thus, in recent years, econometricians have been concerned with including such factors in developing highly sophisticated computerized models of economic activity for forecasting purposes. Such forecasting methods, however, are beyond the scope of this text (See References 1–5 and 7).

Nevertheless, as we have seen from the preceding sections of this chapter, time-series methods provide useful guides for projecting future trends (on a long- and short-term basis). If used properly—in conjunction with other forecasting methods as well as business judgment and experience—time-series methods will continue to be an excellent managerial tool for decision making.

13.9 SUMMARY AND OVERVIEW

As can be seen in the chapter summary chart, we have developed numerous approaches for time-series forecasting, including moving averages, exponential smoothing, the linear, quadratic, and exponential trend models, and the autoregressive model. To be sure you understand what we have covered, you should be able to answer the following conceptual questions:

1. How do the time-series forecasting models developed in this chapter differ from the regression and multiple regression models considered in Chapters 11 and 12?
2. What is the difference between moving averages and exponential smoothing?
3. Under what circumstances would the exponential trend model be most likely to be appropriate?
4. How do autoregressive modeling approaches differ from the other approaches to forecasting?
5. What are the alternative approaches to choosing an appropriate forecasting model?

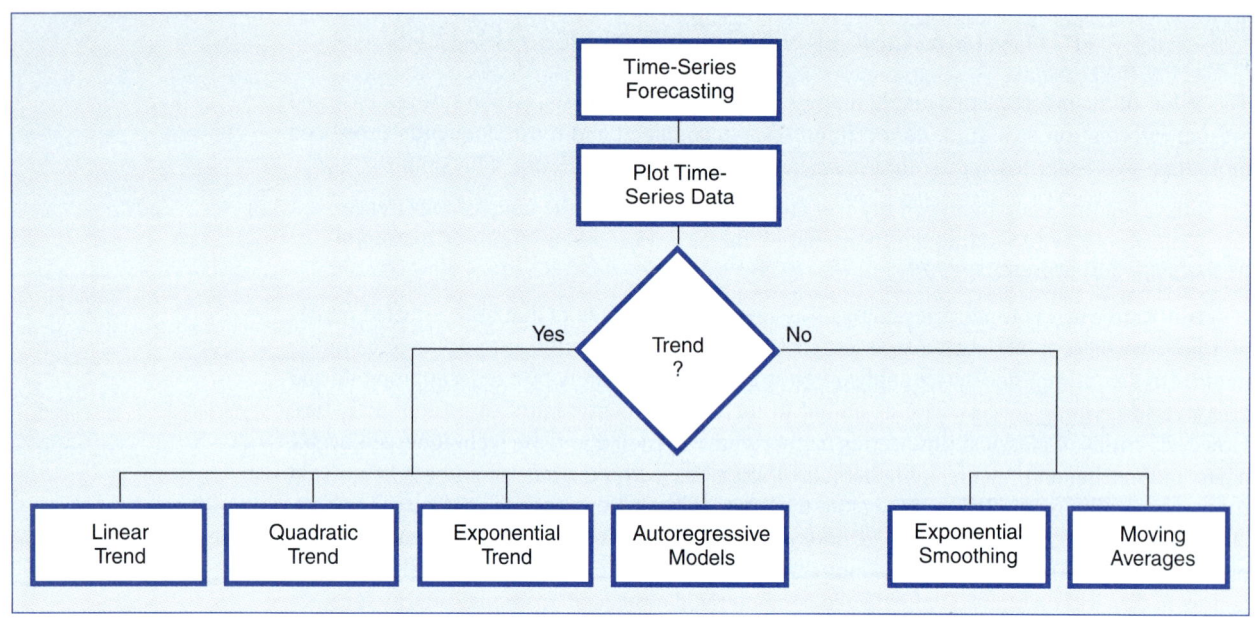

Chapter 13 summary chart.

Getting It All Together

Key Terms

autoregressive modeling 680
classical multiplicative model 653
cyclical component 653
exponential smoothing 659
exponential trend model 672
irregular or random component 654
linear trend model 668

mean absolute deviation (MAD) 690
moving averages 656
principle of parsimony 690
quadratic trend model 671
time series 652
trend 653

Chapter Review Problems

Note: *The Chapter Review Problems should be solved using Microsoft Excel.*

13.25 The following data represent the annual incidence rates (per 100,000 persons) of reported acute poliomyelitis over 5-year periods from 1915 to 1955.

Incidence rates of reported acute poliomyelitis.

Year	1915	1920	1925	1930	1935	1940	1945	1950	1955
Rate	3.1	2.2	5.3	7.5	8.5	7.4	10.3	22.1	17.6

Source: Data are taken from B. Wattenberg, ed., *The Statistical History of the United States: From Colonial Times to the Present* (Series B303), (New York: Basic Books, 1976).

(a) Plot the data on a chart.

(b) Fit a least squares linear trend line to the data and plot the line on your chart.
(c) What are your trend forecasts for the years 1960, 1965, and 1970?
(d) **ACTION** Go to your library and, using the above reference, look up the reported incidence rates of acute poliomyelitis for the years 1960, 1965, and 1970. Record your results.
(e) Why are the mechanical trend extrapolations from your least squares model not useful? Discuss.

13.26 If a linear trend model were to perfectly fit a time series, then the *first differences* would be constant. That is, the differences between consecutive observations in the series would be the same throughout:

$$Y_2 - Y_1 = Y_3 - Y_2 = \cdots = Y_{i+1} - Y_i = \cdots = Y_n - Y_{n-1}$$

If a quadratic trend model were to perfectly fit a time series, then the *second differences* would be constant. That is,

$$[(Y_3 - Y_2) - (Y_2 - Y_1)] = [(Y_4 - Y_3) - (Y_3 - Y_2)]$$
$$= \cdots = [(Y_{i+2} - Y_{i+1}) - (Y_{i+1} - Y_i)]$$
$$= \cdots = [(Y_n - Y_{n-1}) - (Y_{n-1} - Y_{n-2})]$$

If an exponential trend model were to perfectly fit a time series, then the *percentage differences* between consecutive observations would be constant. That is,

$$\left(\frac{Y_2 - Y_1}{Y_1}\right) \times 100\% = \left(\frac{Y_3 - Y_2}{Y_2}\right) \times 100\%$$
$$= \left(\frac{Y_{i+1} - Y_i}{Y_i}\right) \times 100\%$$
$$= \cdots$$
$$= \left(\frac{Y_n - Y_{n-1}}{Y_{n-1}}\right) \times 100\%$$

Although we should not expect a perfectly fitting model for any particular set of time-series data, we nevertheless can evaluate the first differences, second differences, and percentage differences for a given series as a guide for determining an appropriate model to choose.

For each of the time-series data sets presented in the following table:
(a) Determine the most appropriate model to fit.
(b) Develop this trend equation.
(c) Forecast the trend value for the year 1999.

	Year									
	1985	1986	1987	1988	1989	1990	1991	1992	1993	1994
Time series I	10.0	15.1	24.0	36.7	53.8	74.8	100.0	129.2	162.4	199.0
Time series II	30.0	33.1	36.4	39.9	43.9	48.2	53.2	58.2	64.5	70.7
Time series III	60.0	67.9	76.1	84.0	92.2	100.0	108.0	115.8	124.1	132.0

13.27 A time-series plot often aids the forecaster in determining an appropriate model to use. For each of the time-series data sets presented as shown at the top of page 698:

(a) Plot the observed data (Y) over time (X) and plot the logarithm of the observed data (log Y) over time (X) to determine whether a linear trend model or an exponential trend model is more appropriate. (*Hint:* Recall from Section 13.5.3 that if the plot of log Y versus X appears to be linear, an exponential trend model provides an appropriate fit.
(b) Develop this trend equation.
(c) Forecast the trend value for the year 1999.

	Year									
	1985	1986	1987	1988	1989	1990	1991	1992	1993	1994
Time series I	100.0	115.2	130.1	144.9	160.0	175.0	189.8	204.9	219.8	235.0
Time series II	100.0	115.2	131.7	150.8	174.1	200.0	230.8	266.1	305.5	351.8

AGREV.TXT

13.28 The following data represent the annual gross revenues (in millions of dollars) obtained by a utility company for the years 1981–1994:

Annual gross revenues.

Year	1981	1982	1983	1984	1985	1986	1987	1988	1989	1990	1991	1992	1993	1994
Gross revenues	13.0	14.1	15.7	17.0	18.4	20.9	23.5	26.2	29.0	32.8	36.5	41.0	45.4	50.8

(a) Compare the first differences, second differences, and percent differences (see Problem 13.26) to determine the most appropriate model to fit.
(b) Develop this trend equation.
(c) What has been the annual growth in gross revenues over the 14 years?
(d) Forecast the trend value for the year 1999.

ADTIME.TXT

13.29 The following data represent the annual revenues (in millions of dollars) of an advertising agency for the years 1975–1994:

Annual revenues (millions of dollars).

Year	Revenues	Year	Revenues	Year	Revenues
1975	51.0	1982	93.0	1989	100.9
1976	54.1	1983	102.8	1990	110.9
1977	56.4	1984	98.0	1991	133.3
1978	58.1	1985	83.6	1992	192.8
1979	69.5	1986	81.0	1993	234.0
1980	79.2	1987	87.0	1994	238.9
1981	89.2	1988	102.9		

(a) Plot the data over time and plot the logarithm of the data over time to determine whether a linear trend model or an exponential trend model is more appropriate (see Problem 13.27).
(b) Develop this trend equation.
(c) What has been the annual growth in advertising revenues over the 20 years?
(d) Forecast the trend value for the year 1998.

For Problems 13.30–13.33:
- (a) Plot the data on a chart.
- (b) Fit a least squares linear trend line to the data.
- (c) Fit a quadratic trend equation to the data.
- (d) Fit an exponential trend equation to the data.
- (e) Fit a third-order autoregressive model and test for the significance of the third-order autoregressive parameter. (Use $\alpha = .05$.)
- (f) If necessary, fit a second-order autoregressive model and test for the significance of the second-order autoregressive parameter. (Use $\alpha = .05$.)
- (g) If necessary, fit a first-order autoregressive model and test for the significance of the first-order autoregressive parameter. (Use $\alpha = .05$.)
- (h) Perform a residual analysis for each of the fitted models in (b)–(e) and the most appropriate autoregressive model in (f)–(g).
- (i) Compute the MAD for each corresponding model in (h).
- (j) On the basis of (h), (i), and parsimony, which model would you select for purposes of forecasting? Discuss.
- (k) Using the selected model in (j), make an annual forecast from 1993 through 1996.

13.30 The following data represent the annual total sales (in billions of dollars) of International Business Machines Corp. (IBM) over the 23-year period from 1970 to 1992:

Total sales at IBM (1970–1992).

Year	Sales	Year	Sales	Year	Sales
1970	7.5	1978	21.1	1986	52.2
1971	8.3	1979	22.9	1987	55.3
1972	9.5	1980	26.2	1988	59.7
1973	11.0	1981	29.1	1989	62.7
1974	12.7	1982	34.4	1990	43.9
1975	14.4	1983	40.2	1991	37.0
1976	16.3	1984	45.9	1992	33.8
1977	18.1	1985	50.1		

Source: Moody's Handbook of Common Stocks, 1980, 1989, 1993.

13.31 The following data represent the annual total revenues (in billions of dollars) of McDonald's Corp. over the 22-year period from 1971 to 1992:

Total revenues at McDonald's Corp. (1971–1992).

Year	Revenues	Year	Revenues	Year	Revenues
1971	0.3	1979	1.9	1987	4.9
1972	0.4	1980	2.2	1988	5.6
1973	0.6	1981	2.5	1989	6.1
1974	0.7	1982	2.8	1990	5.0
1975	1.0	1983	3.1	1991	4.9
1976	1.2	1984	3.4	1992	5.1
1977	1.4	1985	3.8		
1978	1.7	1986	4.2		

Source: Moody's Handbook of Common Stocks, 1980, 1989, 1993.

SEARS.TXT

13.32 The following data represent the annual total revenues (in billions of dollars) of Sears, Roebuck & Co. over the 23-year period from 1970 to 1992:

Total revenues at Sears, Roebuck & Co. (1970–1992).

Year	Revenues	Year	Revenues	Year	Revenues
1970	8.9	1978	22.9	1986	42.3
1971	9.3	1979	24.5	1987	45.9
1972	10.0	1980	25.2	1988	50.3
1973	11.0	1981	27.4	1989	53.8
1974	12.3	1982	30.0	1990	50.3
1975	13.1	1983	35.9	1991	50.9
1976	17.7	1984	38.8	1992	52.3
1977	19.6	1985	40.7		

Source: Moody's Handbook of Common Stocks, 1980, 1989, 1993.

XEROX.TXT

13.33 The following data represent the annual net sales (in billions of dollars) of Xerox Corporation over the 23-year period from 1970 to 1992:

Annual net sales at Xerox Corp. (1970–1992).

Year	Revenues	Year	Revenues	Year	Revenues
1970	1.7	1978	5.9	1986	13.3
1971	2.0	1979	6.9	1987	15.1
1972	2.4	1980	8.0	1988	16.4
1973	3.0	1981	8.5	1989	17.6
1974	3.5	1982	8.5	1990	13.6
1975	4.1	1983	8.3	1991	13.8
1976	4.4	1984	8.6	1992	14.7
1977	5.1	1985	9.0		

Case Study G—Currency Trading

As a member of a financial firm hired by a group of investors to undertake trading in various currencies, you have been assigned the task of studying the long-term trends in the exchange rates of the Canadian dollar, the French franc, the German mark, the Japanese yen, and the English pound. All currencies are expressed in units per U.S. dollar except the English pound, which is expressed in cents per pound. The data in the table on page 701 have been collected for the 26-year period from 1967 to 1992.

Develop forecasting models for the exchange rate of each of these five currencies based on these data. Be sure to indicate which forecasting model you have chosen for each currency and what limitations there are to the model. Provide forecasts for the years 1995, 1996, and 1997 for each currency. Write an executive summary for a presentation that is scheduled to be made to the investor's group next week.

Exchange Rates of Five Currencies in Terms of the U.S. Dollar.

CURRENCY.TXT

Year	Canadian Dollar	French Franc	German Mark	Japanese Yen	English Pound
1967	1.0789	4.9206	3.9865	362.13	275.04
1968	1.0776	4.9529	3.9920	360.55	239.35
1969	1.0769	5.1999	3.9251	358.36	239.01
1970	1.0444	5.5288	3.6465	358.16	239.15
1971	1.0099	5.5100	3.4830	347.79	244.42
1972	0.9907	5.0444	3.1886	303.13	250.34
1973	1.0002	4.4535	2.6715	271.31	245.25
1974	0.9780	4.8107	2.5868	291.84	234.03
1975	1.0175	4.2877	2.4614	296.78	222.17
1976	0.9863	4.7825	2.5185	296.45	180.48
1977	1.0633	4.9161	2.3236	268.62	174.49
1978	1.1405	4.5091	2.0097	210.39	191.84
1979	1.1713	4.2567	1.8343	219.02	212.24
1980	1.1693	4.2251	1.8175	226.63	232.46
1981	1.1990	5.4397	2.2632	220.63	202.43
1982	1.2344	6.5794	2.4281	249.06	174.80
1983	1.2325	7.6204	2.5539	237.55	151.59
1984	1.2952	8.7356	2.8455	237.46	133.68
1985	1.3659	8.9800	2.9420	238.47	129.74
1986	1.3896	6.9257	2.1705	168.35	146.77
1987	1.3259	6.0122	1.7981	144.60	163.98
1988	1.2306	5.9595	1.7570	128.17	178.13
1989	1.1842	6.3802	1.8808	138.07	163.82
1990	1.1668	5.4467	1.6166	145.00	178.41
1991	1.1460	5.6468	1.6610	134.59	176.74
1992	1.2085	5.2935	1.5618	126.78	176.63

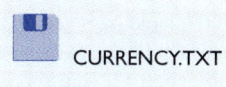

Source: Board of Governors of the Federal Reserve System, Table B-107.

Endnotes

1. That all i observed values in the time series are included in the computation of the exponentially smoothed value in time period i can be seen by noting that the present smoothed value is calculated using the smoothed value of the previous period, and that value, in turn, was calculated using the smoothed value from its previous period, and so on. Algebraically, this can be stated as follows:

In time period 1,
$$E_1 = Y_1$$

In time period 2,
$$E_2 = WY_2 + (1 - W)E_1 = WY_2 + (1 - W)Y_1$$

In time period 3,
$$E_3 = WY_3 + (1 - W)E_2 = WY_3 + (1 - W)[WY_2 + (1 - W)Y_1]$$
$$= WY_3 + W(1 - W)Y_2 + (1 - W)^2 Y_1$$

In general, in time period i,
$$E_i = WY_i + (1 - W)E_{i-1}$$
$$= WY_i + W(1 - W)Y_{i-1} + W(1 - W)^2 Y_{i-2}$$
$$+ \cdots + (1 - W)^{(i-1)} Y_1$$

Thus, we see that over time, as the integer value i gets large, the weights assigned to the earlier (older) values in the time series may become negligible.

2. It should be noted that the exponential smoothing model of Section 13.4.2 and the autoregressive models of Section 13.6 are special cases of *autoregressive integrated moving average (ARIMA)* models developed by Box and Jenkins (References 3 and 6). The Box-Jenkins approach, however, is beyond the scope of this text.

References

1. Berenson, M. L., and D. M. Levine, *Basic Business Statistics: Concepts and Applications*, 6th ed. (Englewood Cliffs, NJ: Prentice-Hall, 1996).
2. Bowerman, B. L., and R. T. O'Connell, *Forecasting and Time-Series*, 3d ed. (North Scituate, MA: Duxbury Press, 1993).
3. Box, G. E. P., and G. M. Jenkins, *Time Series Analysis: Forecasting and Control*, 2d ed. (San Francisco, CA: Holden-Day, 1977).
4. Brown, R. G., *Smoothing, Forecasting, and Prediction* (Englewood Cliffs, NJ: Prentice-Hall, 1963).
5. Chambers, J. C., S. K. Mullick, and D. D. Smith, "How to Choose the Right Forecasting Technique," *Harvard Business Review*, Vol. 49, No. 4, July–August 1971, pp. 45–74.
6. Frees, E. W. *Data Analysis Using Regression Models: The Business Perspective* (Englewood Cliffs, NJ: Prentice-Hall, 1996).
7. Mahmoud, E., "Accuracy in Forecasting: A Survey," *Journal of Forecasting*, Vol. 3, 1984, pp. 139–159.
8. *Microsoft Excel Version 7* (Redmond, WA: Microsoft Corp., 1996).
9. Wilson, J. H., and B. Keating, *Business Forecasting* (Homewood, IL: Richard D. Irwin, 1990).

Answers to Selected Problems (•)

Chapter 1

1.6   (a) discrete numerical, ratio
      (b) categorical, nominal
      (c) discrete numerical, ratio
      (d) continuous numerical, ratio
      (e) categorical, nominal
      (f) continuous numerical, ratio
      (g) categorical, nominal
      (h) discrete numerical, ratio
      (i) continuous numerical, ratio
      (j) categorical, nominal
      (k) categorical, nominal

1.20 (a) continuous numerical, ratio
      (b) discrete numerical, ratio
      (c) categorical, nominal
      (d) continuous numerical, ratio
      (e) categorical, nominal
      (f) discrete numerical, ratio
      (g) categorical, nominal

1.21 (a) categorical, nominal
      (b) categorical, nominal
      (c) continuous numerical, ratio
      (d) continuous numerical, ratio
      (e) categorical, nominal
      (f) categorical, nominal
      (g) discrete numerical, ratio
      (h) discrete numerical, ratio
      (i) continuous numerical, ratio
      (j) continuous numerical, ratio
      (k) categorical, nominal
      (l) discrete numerical, ratio
      (m) continuous numerical, ratio
      (n) continuous numerical, ratio
      (o) categorical, nominal
      (p) continuous numerical, ratio

1.24 $N = 93$ $n = 15$; sample *without* replacement
Row 29: 12 47 83 76 22 65 93 10 61 36 89 58 86 92 71

Chapter 2

2.1 (a)
```
 9 | 147
10 | 02238
11 | 135566777
12 | 223489
13 | 02
```
(c) Stem-and-leaf display shows how data distribute and cluster.

2.4 (b) Stem-and-leaf display: Book values
```
0L | 4
0H | 5566667777788888888999999
1L | 000000001122334
1H | 555668
2L | 3
```

2.10 (b) Frequencies are 4 7 9 13 9 5 3

2.11 Classes are 0 < 5, 5 < 10, etc. Frequencies are 1 27 15 6 1

2.15 The percentages are 8 14 18 26 18 10 6

2.16 The percentages are 2 54 30 12 2

2.26 (a) The cumulative frequencies are 0 4 11 20 33 42 47 50

(b) The cumulative percentages are 0 8 22 40 66 84 94 100

2.27 (a) The cumulative frequencies are 0 1 28 43 49 50

(b) The cumulative percentages are 0 2 56 86 98 100

2.43 (a)

	Education Level			
Financial Condition	H.S. Degree or Lower	Some College	Degree or Higher	Totals
Worse off now than before	60.7	30.0	18.1	43.4
No difference	24.2	45.6	19.5	27.2
Better off now than before	15.1	24.4	62.4	29.4
Totals	100.0	100.0	100.0	100.0

(b) More people felt that they were worse off than better off.

2.47 (a) The classes are: 0 < 10; 10 < 20; etc.
The frequencies from the ASE are 16 6 1 1 1 0 0 0 0
The frequencies from the NYSE are 13 8 15 7 2 3 1 0 1

Chapter 3

3.1 (a)

	Set 1	Set 2
mean	4	14
median	3	13
mode	2	12
midrange	6	16
midhinge	3.5	13.5

(b) The observations in Set 1 are each 10 units less than the observations in Set 2.

3.8 (a) Revised stem-and-leaf display

```
 2 | 8
 3 | 458
 4 | 1
 5 | 01157
 6 | 28
 7 | 16
 8 | 5
 9 |
10 |
11 | 9
12 |
13 | 12
14 | 19
15 | 9
```

(b) mean = 7.78; median = 6.20; mode = 5.10; midrange = 9.35; midhinge = 8.53

(c) The midrange and midhinge are largest.

3.12 (a)

	Set 1	Set 2
range	8	8
IQR	3	3
variance	8.33	8.33
S	2.89	2.89
CV	72.2%	20.6%

3.16 (a) range = 13.10; interquartile range = 7.95; variance = 17.95; $S = 4.24$; CV = 54.5%

(b) The majority of the data fall within ±4.24 of the mean.

3.19 (a) and (b)   For each data set, the data are positive or right-skewed since mean > median.

3.21 The data are positive or right-skewed since mean > median.

3.25 (a) Five-number summary: 2.80 4.55 6.20 12.50 15.90

(b) and (c)   The data are right-skewed.

3.28 (a) mean = 6.0; median = 6.5; mode = 8.0; midrange = 6.0; midhinge = 5.5

(b) range = 10.0; interquartile range = 5.0; variance = 9.40; $\sigma_x = 3.07$; $CV_{pop} = 51.1\%$

(c) No. Data are approximately symmetrical.

3.33 (a) mean = $41.78; median = $42.00; midrange = $58.50; midhinge = $39.88; range = $97.00; IQR = $29.75; $S = 21.30; CV = 51.0%. Five-number summary: $10.00 $25.00 $42.00 $54.75 $107.00

Chapter 4

4.2 (a) Having a bank credit card, since there is only one criterion satisfied.

(b) Having a bank credit card *and* having a travel and entertainment credit card, since two criteria are involved.

(c) Not having a bank credit card is the complement of having a bank credit card, since it involves all events other than having a bank credit card.

(d) It satisfies two criteria, having a bank credit card and having a travel and entertainment credit card.

4.4 (b) Enjoying shopping for clothing is a simple event since it satisfies one criterion.
(c) A male who enjoys shopping for clothing is a joint event since it satisfies two criteria.
(d) Not enjoying shopping for clothing is the complement.

4.7 (a) $P(B) = 120/200$
(b) $P(B') = 80/200$
(c) $P(T) = 75/200$
(d) $P(T) = 125/200$

4.9 (a) 240/500
(b) 360/500
(c) 260/500
(d) 140/500

4.12 (a) $P(B$ and $T) = 60/200$
(b) $P(B'$ and $T) = 15/200$
(c) $P(B'$ and $T') = 65/200$

4.14 (a) 224/500
(b) 104/500
(c) 136/500

4.19 (a) 135/200
(b) 140/200
(c) 200/200 = 1.0

4.21 (a) 396/500
(b) 276/500
(c) 500/500 = 1.0

4.24 (a) 60/120
(b) 60/125
(c) Since $P(T \mid B) = 60/120 \neq P(T) = 75/200$; not statistically independent.

4.26 (a) 36/260
(b) 136/360
(c) $P(\text{Enjoys} \mid \text{female}) = 224/260 \neq P(\text{Enjoys}) = 360/500$; not statistically independent.

4.29 $(120/200)(75/200) \neq 60/200$; not statistically independent.

4.31 $(360/500)(240/500) \neq (136/500)$; not statistically independent.

4.37 (a) A: 1.00; B: 3.00
(b) A: 1.22; B: 1.22
(c) A: Right-skewed; B: Left-skewed

4.41 (a) 7
(b) $\sigma^2 = 5.8333$
$\sigma = 2.42$
(d) −.056
(e) Lose 5.6¢ per bet.
(f) Gain 5.6¢ per bet.

4.43 (b) $E(500) = \$500$; $E(1,000) = \$800$; $E(2,000) = \$600$
Purchase 1,000 pounds.

(c) $E(500) = \$750$; $E(1,000) = \$1,250$; $E(2,000) = \$1,250$
Purchase 1,000 or 2,000 pounds.

(d) $E(500) = \$500$; $E(1,000) = \$600$; $E(2,000) = 0$
Purchase 1,000 pounds.

4.59 (a) .0778

(b) .6826

(c) $P(X = 0) = .0102$ $P(X = 1) = .0768$ $P(X = 2) = .2304$.

(d) Distribution is slightly left skewed.

(e) $P(X = 5) = .32768$ $P(X \geq 3) = .94208$ $P(X = 0) = .00032$ $P(X = 1) = .0064$ $P(X = 2) = .0512$

(f) $P(X = 5) = .59049$ $P(X \geq 3) = .99144$ $P(X = 0) = .00001$ $P(X = 1) = .00045$ $P(X = 2) = .0081$

4.62 (a) .2851

(b) .1606

(c) .7149

(d) .2945

(e) $P(X < 5) = .4405$ $P(X = 5) = .1755$ $P(X \geq 5) = .5595$ $P(X = 4 \text{ or } 5) = .3509$

4.67 (a) (1) 80/200

(2) 55/200

(3) 125/200

(b) 55/80

(c) $P(< 5 \mid \text{College grad}) = 55/80 \neq P(< 5) = 130/200$; not statistically independent).

4.73 (a) (1) .6496

(2) .1503

(3) .1493

(b) 7; $P(X = 7) = .2668$

(c) 1.449

(d) $P(X \geq 7) = .8791$ $P(X < 6) = .0328$ $P(X = 9 \text{ or } 10) = .3758$ $\mu = 8$ $P(X = 8) = .3020$ $\sigma = 1.2649$

4.74 (a) .8171

(b) .1667

(c) .0162

(d) $P(X = 0) = .9044$ $P(X = 1) = .0914$ $P(X \geq 2) = .0043$

Chapter 5

5.3 (a) (1) .3599   (2) .6401
(3) .0832   (4) .9168
(5) .8599   (6) .5832
(7) .4431   (8) .5569

(b) (1) .1401   (2) .4168
(3) .3918   (4) .8349
(5) .1151

(c) 0

(d) −1.00

(e) +1.00

5.6 (a) .4082
- (b) .0669
- (c) 25.08%
- (d) 749.2 trucks
- (e) $Z = -0.84$ so that $X = 39.92$ thousand miles.
- (f) .4453, .0603, .1809, 719.1, 41.58

5.10 (a) .1587
- (b) .0466
- (c) .7865
- (d) 46.4 hours
- (e) 40.0 hours
- (f) 6.7 hours
- (g) .3085, .0968, .4796, 52.81, 40, 13.59

5.14 Area under normal curve covered: .1429 .2857 .4286 .5714 .7143 .8571
Standardized normal quantile value: −1.07 −0.57 −0.18 +0.18 +0.57 +1.07

5.15 (a) $\bar{X} = \$106.80$
$S = \$38.16$
- (b) Five-number summary: $40 $80 $100 $135 $200
- (c) Slightly right skewed

5.28 $\mu = 1.30$; $\sigma = 0.04$
- (a) .1915
- (b) .1747
- (c) 1.2664 to 1.3336
- (d) (1) $\mu_{\bar{X}} = 1.30$; $\sigma_{\bar{X}} = 0.01$
  - (2) normal
  - (3) .4772
  - (4) .15735
  - (5) 1.2916 to 1.3084
- (e) and (f) Since samples of size 16 are being taken, rather than individual values (samples of $n = 1$), because $\sigma_{\bar{X}} = \sigma/\sqrt{n}$, more values lie closer to the mean with the increased sample size and fewer values lie far away from the mean.
- (g) They are equally likely to occur (probability = .1587) since, as n increases, more sample means will be closer to the population mean.

5.31 (a) .2486
- (b) .0918
- (c) .1293 and .2514
- (d) A percent defective above 10.5% is more likely to occur since it is only 0.33 standard deviation above the population value of 10%.

5.36 .14833

5.38 .2549 and .0823

5.43 Height: $P(X > 67) = .2119$
Weight: $P(X > 135) = .1587$
Weight is slightly the more unusual characteristic

Chapter 6

6.5 (a) $.9877 \leq \mu \leq 1.0023$

(b) Since the value of 1.0 is included in the interval, there is no reason to believe that the average is below 1.0.

(c) No, since σ is known and $n = 50$, from the central limit theorem we may assume that $\bar{X}$ is normally distributed.

(d) An individual value of .98 is only .75 standard deviation below the sample mean of .995. The confidence interval represents the estimate of the average of a sample of 50, not an individual value.

(e) $.9895 \leq \mu \leq 1.0005$
Same answer in (b)

6.9 $\$1,067.40 \leq \mu \leq \$1,332.60$

6.13 (a) $87.769 \leq \mu \leq 109.964$

(b) That the waiting time is normally distributed

(c) $83.42 \leq \mu \leq 140.98$. The confidence interval is much wider.

6.20 $.2246 \leq p \leq .3754$

6.22 $.342 \leq p \leq .478$

6.29 $n = 97$

6.31 (a) $n = 167$

(b) $n = 97$

6.37 $n = 323$

6.38 $n = 271$

6.42 (a) $322.62 \leq \mu \leq 377.38$

(b) $n = 92$

(c) $322.97 < \mu < 377.03$, $n = 88$

6.44 (a) $.2285 \leq p \leq .3715$

(b) $n = 214$

(c) $.2265 \leq p \leq .3735$, $n = 239$

6.54 (a) $14.085 \leq \mu \leq 16.515$

(b) $.530 \leq p \leq .820$

(c) $n = 25$

(d) $n = 784$

Chapter 7

7.12 (a) $-1.96 < Z = -0.80 < +1.96$ or p-value $= .423 > .05$. Don't reject H_0. There is no evidence that the average amount dispensed is different from 8 ounces.

(b) $Z = -2.40 < -1.96$ or p-value $= .0162 < .05$. Reject H_0.

(c) $Z = -2.26 < -1.96$ or p-value $= .0237 < .05$. Reject H_0.

7.16 p-value $= .423$

7.21 (a) $H_0: \mu \geq 2.8$; $H_1: \mu < 2.8$

(b) $Z = -1.75 < -1.645$. Reject H_0. The average is significantly less than 2.8, and we can therefore conclude that there is evidence that the process is not working properly.

7.24 p-value $= .0401$

7.27 (a) $t = 3.30 > t_{35} = 2.0301$. Reject H_0. There is evidence that the EER is different from 9.0.

(b) The data are approximately normally distributed.

(c) p-value is .0022.

(d) $t = 2.275 > 2.0301$

7.29 (a) $t = 3.552 > t_{99} = 1.9842$. Reject H_0. There is evidence that the average balance is different from $75.

(b) p-value = .0006

(c) $t = 2.254 > 1.9842$
p-value = .0264 < .05

(d) $-1.9842 < t = 1.61 < 1.9842$ or
p-value = .11 > .05. Don't Reject H_0.

7.38 (a) $Z = -1.60 > -2.33$. Don't reject H_0. There is no evidence that the proportion is less than .25.

(b) p-value = .0545.

(c) $Z = -2.92 < -2.33$ or p-value = .0035 < .01. Reject H_0.

7.39 (a) $Z = +2.93 > +1.645$. Reject H_0. There is evidence that the proportion is different from .30.

(b) p-value = .0034.

(c) $-1.645 < Z = 1.543 < 1.645$ or p-value = .1228 > .10. Don't reject H_0.

Chapter 8

8.1 (a) $Z = +0.39 < +1.96$. Don't reject H_0. There is no evidence of a difference in the average life of bulbs produced by the two machines.

(b) p-value = .6966

8.3 (a) $t = +1.91 > +1.645$. Reject H_0. There is evidence of a difference in output between the day and evening shift.

(b) p-value = .0576

8.8 (a) $t = -2.19 < -2.0106$. Reject H_0. There is evidence of a difference in the average talking time prior to recharge. The newly developed battery lasts longer.

(b) Normality in each population and equality of variances.

(c) p-value = .0334

8.10 No. Since $78 < T_1 = 84 < 132$, don't reject H_0.

8.14 (a) Yes. Let the nickel-cadmium sample be group 1. Thus,
$T_1 = 502.5$; since $Z = -2.62 < -1.96$, reject H_0.

(b) Equal variability in the two populations.

(c) The results are all similar.

8.16 (a) $F_{L(9,9)} = 0.248 < F = 0.811 < F_{U(9,9)} = 4.03$. Don't reject H_0.

(b) One is the reciprocal of the other if the sample sizes are equal.

(c) Reject H_0 if $F > F_{U(9,9)} = 3.18$.

(d) Reject H_0 if $F < F_{L(9,9)} = 0.314$.

8.21 (a) No. We don't reject H_0 since $F_{L(24,24)} = 0.441 < F = 0.867 < F_{U(24,24)} = 2.27$. Note that the p-value = .7295

(b) Depending on the assumption of underlying normality in the populations, either the pooled-variance t test or the Wilcoxon rank sum test is more appropriate.

8.25 (d) $F = 4.22 > F_{U(3,28)} = 2.95$. Reject H_0. There is evidence of a difference.

(e) Program A is superior to B and C.

8.26 $F = 10.30 > F_{U(4,17)} = 2.96$. Reject H_0. Alloy 2 is weakest.

8.31 $H = 0.635 < \chi^2 = 9.210$. Don't reject H_0.

8.34 $H = 9.51 > \chi^2 = 7.815$. Reject H_0.

8.37 (a) [$104.64 ≤ μ ≤ $115.36]

  (b) $t = -10.0 < -2.7969$. Reject H_0. The average monthly balance of Plan A accounts is not equal to $105.

  (c) $F_{L(24,49)} = 0.37 < F = 1.125 < F_{U(24,49)} = 2.40$. Don't reject H_0. There is no evidence of a difference in the variances between Plan A and Plan B.

  (d) $t = -9.903 < -2.6449$. Reject H_0. There is evidence of a difference in the average monthly balance between Plan A and Plan B. Plan A's balance is lower.

  (e) The respective p-values are .0000, .708, .0000.

Chapter 9

9.2 (a) $Z = +2.37 > +1.96$. Reject H_0. There is evidence of a difference in the proportion of women in the two groups who eat dinner in a restaurant during the work week.

  (b) p-value = .0178.

9.4 (a) $Z = +2.58 > +1.645$. Reject H_0. There is evidence that a high temperature setting is preferred.

  (b) p-value = .005.

9.8 $\chi^2 = 5.617 > X^2 = 3.841$. Reject H_0.

9.13 (a) $\chi^2 = 1.125 < \chi_2^2 = 9.210$. Don't reject H_0. There is no evidence of a difference in attitude toward the trimester among the various groups.

9.17 (a) $\chi^2 = 11.2949 > \chi_2^2 = 9.210$. Reject H_0. There is evidence of a difference.

9.21 (a) $\chi^2 = 22.780 > \chi_8^2 = 15.507$. Reject H_0. There is a relationship between type of area of residence and manufacturer preference in an automobile purchase.

  (b) p-value = .0037.

  (c) $\chi^2 = 34.58$ and p-value = .0003

9.23 (a) $X^2 = 9.82 < X_4^2 = 13.277$. Don't reject H_0. There is no evidence of a relationship between commuting time and stress.

  (b) p-value = .0436.

9.25 (a) $Z = -5.00 < -2.33$. Reject H_0. There is evidence that the proportion of male medical doctors having heart attacks is lower for those who had the aspirin than for those who did not have the aspirin.

  (b) p-value = .00000029. There is very little likelihood that these results could have occurred if the aspirin did not reduce the incidence of heart attacks.

  (c) The χ^2 test is not appropriate because we have a directional alternative hypothesis.

  (d) $Z = -2.848 < -2.33$, p-value = .0022

Chapter 10

10.6 (a) $\bar{p} = .1145$; LCL = .0522; UCL = .1768. The proportion of late arrivals on day 13 is substantially out of control. Possible special causes of this value should be investigated. In addition, the next four highest points all occur on Friday.

  (b) $\bar{X} = 26.9$; UCL = 41.54; and LCL = 12.26.

  (c) The results are exactly the same. The p chart expresses the results in terms of the proportion, and the np chart expresses the results in terms of the number of successes.

  (d) The snowstorm would explain why the proportion of late arrivals was so high on day 13.

  (e) p chart: $\bar{p} = .1102$; LCL = .0489, UCL = .1715; np chart: $\bar{X} = 25.9$; LCL = 11.498, UCL = 40.302, the point is still above the UCL.

10.10 (a) $\bar{p} = .01288$; UCL = .01753; LCL = .00823. Although none of the points are outside the control limits, there is evidence of a pattern over time, since the last eight points are all above the mean and most of the earlier points are below the mean. Thus, the special causes that might be contributing to this pattern should be investigated before any change in the system of operation is contemplated.

(b) Once special causes have been eliminated and the process is stable, process flow and fishbone diagrams of the process should be developed to increase process knowledge. Then Deming's 14 points can be applied to improve the system.

10.17 (a) $\bar{R} = 271.57$; UCL = 574.20. LCL does not exist. There are no points outside the control limits and no evidence of a pattern in the range chart. $\bar{\bar{X}} = 198.67$; UCL = 355.31; and LCL = 42.03. There are no points outside the control limits and no evidence of a pattern in the $\bar{X}$ chart.

10.18 (a) $\bar{R} = 4.8325$, $\bar{\bar{X}} = 3.549$, For R chart $LCL = 1.0786$, and $UCL = 8.5864$. For $\bar{X}$ chart: $LCL = 2.06$, and $UCL = 5.038$. The process appears to be in control since there are no points outside the lower and upper control limits and there is no pattern in the results over time.

(b) Since the process is in control, it is up to management to reduce the common-cause variation by application of the 14 points of the Deming theory of management by process. In addition, process knowledge would be enhanced by the development of process flow and fishbone diagrams.

Chapter 11

11.6 (a) $b_0 = 1.45$; $b_1 = .074$

(b) For each increase of 1 foot of shelf space, sales will increase by $7.40 per week.

(c) $\hat{Y}_i = 2.042$ or $204.20

(d) $b_0 = 1.5$; $b_1 = .068$; $\hat{Y}_i = 2.044$ or $204.40

11.8 (a) $b_0 = 12.6786$; $b_1 = 1.9607$

(b) The Y intercept b_0 (equal to 12.6786) represents the portion of the worker-hours that is not affected by variation in lot size. The slope b_1 (equal to 1.96) means that for each increase in lot size of one unit, worker-hours are predicted to increase by 1.96.

(c) $\hat{Y}_i = 100.91$

(d) Lot size varied from 20 to 80 so that predicting for a lot size of 100 would be extrapolating beyond the range of the X variable.

(e) $b_0 = 13.1607$; $b_1 = 1.9125$; $\hat{Y}_i = 99.22$

11.11. $S_{YX} = 0.308$

11.13 $S_{YX} = 4.71$

11.16 (a) $r^2 = .684$; 68.4% of the variation in sales can be explained by variation in shelf space.

(b) $r^2 = .652$

11.18 (a) $r^2 = .9878$. 98.78% of the variation in worker-hours can be explained by variation in lot size.

(b) $r^2_{adj} = .987$

11.24 $r = +.827$

11.26 $r = +.9939$

11.31 Based on a residual analysis, the model appears to be adequate.

11.33 Based on a residual analysis, the model appears to be adequate.

11.37 The data have been collected for a single time period from a set of stores. There is no sequential nature about the set of stores. Therefore, it is not necessary to compute the Durbin-Watson statistic.

11.41 $1.835 \leq \mu_{YX} \leq 2.249$

11.43 $98.597 \leq \mu_{YX} \leq 103.223$

11.46 (a) $1.447 \leq Y_I \leq 2.637$

11.48 (b) This is an estimate for an individual response rather than an average predicted value.

11.48 (a) $92.20 \leq Y_I \leq 109.62$

(b) This is an estimate for an individual response rather than an average predicted value.

11.51 $t = 4.653 > t_{10} = 1.8125$. Reject H_0. There is evidence of a linear relationship.

11.53 $t = 31.15 > t_{12} = 1.7823$. Reject H_0. There is evidence of a significant relationship between lot size and worker-hours.

11.58 (b) $b_0 = 21.9256; b_1 = +2.0687$

(c) If the cars have no options, delivery time averages approximately 22 days; for each option ordered delivery time increases by 2.0687 days.

(d) 55.0248 days

(e) $S_{YX} = 3.0448$

(f) $r^2 = .9575$; 95.75% of the variation in delivery time can be explained by variation in the number of options ordered.

(g) $r = +.9785$

(h) $r_{adj} = .955$

(i) $53.1115 \leq \mu_{YX} \leq 56.9381$

(j) $48.22 \leq Y_I \leq 61.83$

(k) $t = 17.769 > t_{14} = 2.1448$. Reject H_0. There is evidence of a linear relationship.

(l) $+1.8187 \leq \beta_1 \leq +2.3187$

(m) Based on a residual analysis, the model appears to be adequate.

11.63 (a) $b_0 = +0.30; b_1 = +.00487$

(b) The slope b_1 can be interpreted to mean that for each increase of one point in GMAT score, GPI is predicted to increase by .00487 point (or for each increase of 100 points in GMAT score, GPI is predicted to increase by 0.487 point). The Y-intercept b_0 represents the portion of the GPI that varies with factors other than GMAT score.

(c) $\hat{Y} = 3.222$

(d) $S_{YX} = .158$

(e) $r^2 = .793$; 79.3% of the variation in GPI can be explained by variation in GMAT score.

(f) $r = +.891$

(g) $r^2_{adj} = .781$

(h) $t = 8.31 > t_{18} = 2.1009$. Reject H_0. There is evidence of a linear relationship between GMAT score and GPI.

(i) $3.143 \leq \mu_{YX} \leq 3.301$

(j) $2.881 \leq Y_I \leq 3.563$

(k) $+.00364 \leq \beta_1 \leq +.00610$

(l) There is widespread scatter in the residual plot, but there is no pattern in the relationship between the residuals and X_i.

(m) $b_0 = +.2582, b_1 = .00494, Y_i = 3.2209, S_{YX} = .147, r^2 = .82, r^2_{adj} = .81, t = 9.059 > 2.1009, 3.147 \leq \mu_{YX} \leq 3.295, 2.903 \leq Y_I \leq 3.538, .00379 \leq \beta_1 \leq .00608$

Chapter 12

12.2 (a) $\hat{Y}_i = -.02686 + .79116 X_{1i} + .60484 X_{2i}$

(b) For a given midsole impact, each increase of one unit in forefoot impact-absorbing capability results in an increase in the long-term ability to absorb shock by .79116 units. For a given forefoot impact-absorbing capability, each increase in one unit in midsole impact results in an increase in the long-term ability to absorb shock by .60484 unit.

12.4 (a) $\hat{Y}_i = 156.4 + 13.081 X_{1i} + 16.795_{2i}$ where X_1 = radio and television advertising in thousands of dollars and X_2 = newspaper advertising in thousands of dollars.

(b) Holding the amount of newspaper advertising constant, for each increase of $1,000 in radio and television advertising, sales are predicted to increase by $13,081. Holding the amount of radio and television advertising constant, for each increase of $1,000 in newspaper advertising, sales are predicted to increase by $16,795.

12.7 $\hat{Y}_i = \$753.95$

12.9 (a) $r^2_{Y.12} = .9421$; 94.21% of the variation in the long-term ability to absorb shock can be explained by variation in forefoot impact-absorbing capability and variation in midsole impact.

(b) $r^2_{adj} = .9263$

12.11 (a) $r^2_{Y.12} = .809$; 80.9% of the variation in sales can be explained by variation in radio and television advertising and in newspaper advertising.

(b) $r^2_{adj} = .789$

12.12 (a) $r^2_{Y.12} = .490$; 49.0% of the variation in standby hours can be explained by variation in the total staff and remote hours.

(b) $r^2_{adj} = .446$

12.14 There appears to be a curvilinear relationship in the plot of the residuals against both radio and television advertising and against newspaper advertising. Thus, curvilinear terms for each of these explanatory variables should be considered for inclusion in the model.

12.16 $F = 97.69 > F_{U(2,12)} = 3.89$. Reject H_0. At least one of the independent variables is related to the dependent variable Y.

12.18 $F = 40.16 > F_{U(2,19)} = 3.52$. Reject H_0. There is a significant relationship between sales and radio and television advertising and newspaper advertising.

12.21 $t = 7.43 > t_{19} = 2.093$ and $t = 5.67 > t_{19} = 2.093$. Each explanatory variable makes a significant contribution and should be included in the model.

12.23 (a) $.654 \le \beta_1 \le .928$

(b) $t = 12.57 > 2.1788$ and $t = 8.43 > 2.1788$. Both variables make a significant contribution to the model.

12.25 (a) $9.399 \le \beta_1 \le 16.763$

(b) $t = 7.43 > 2.093$ and $t = 5.67 > 2.093$. Both variables make a significant contribution to the model.

12.28 $r^2_{Y1.2} = .7442$. For a given amount of newspaper advertising, 74.42% of the variation in sales can be explained by variation in radio and television advertising. $r^2_{Y2.1} = .6283$. For a given amount of radio and television advertising, 62.83% of the variation in sales can be explained by variation in newspaper advertising.

12.30 (a) $\hat{Y}_i = -.0145$

(c) $\hat{Y}_i = 18.52$

(d) $F = 141.46 > F_{U(2,25)} = 3.39$. Reject H_0. There is evidence of a curvilinear relationship between speed and miles per gallon.

(e) $r^2_{Y.12} = .9188$; 91.88% of the variation in miles per gallon can be explained by the curvilinear relationship with speed.

(f) $r^2_{adj} = .912$

(g) The model indicates only positive residuals for intermediate values of $(X_i - \bar{X})$.

(h) $t = -16.63 < -2.0595$. The curvilinear effect makes a significant contribution to the model.

12.37 (a) $\hat{Y}_i = 1.30 + .074X_{1i} + .45X_{2i}$, where X_1 = shelf space, X_2 = 0 for back of aisle and 1 for front of aisle.

(b) Holding constant the effect of aisle location, for each additional foot of shelf space, predicted sales increase by .074 hundred dollars ($7.40). For a given amount of shelf space, a front-of-aisle location increases average sales by .45 hundred dollars ($45).

(c) $\hat{Y}_i = 1.892$ or $189.20

(d) $F = 28.562 > F_{U(2,9)} = 4.26$. Reject H_0. There is evidence of a relationship between sales and the two independent variables.

(e) $r^2_{Y.12} = .864$; 86.4% of the variation in sales can be explained by variation in shelf space and variation in aisle location.

(f) $r^2_{adj} = .834$

(g) $r^2_{Y.12} = .864$ while $r^2 = .684$; $r^2_{adj} = .834$ as compared to .652. The inclusion of the aisle-location variable has resulted in an increase in the r^2.

(h) $t = 6.72 > t_9 = 2.2622$ and $t = 3.45 > t_9 = 2.2622$. Therefore, each explanatory variable makes a significant contribution and should be included in the model.

(i) $.049 \le \beta_1 \le .099$
$.155 \le \beta_2 \le .745$

(j) The slope $b_1 = .074$ is unchanged in this case because shelf space and aisle location are not correlated.

(k) $r^2_{Y1.2} = .834$. Holding constant the effect of aisle location, 83.4% of the variation in sales can be explained by variation in shelf space; $r^2_{Y2.1} = .569$; for a given amount of shelf space 56.9% of the variation in sales can be explained by variation in aisle location.

(l) That the slope of shelf space and sales is the same regardless of whether the aisle location is front or back.

(m) Based on a residual analysis, the model appears adequate.

12.42 (a) $\hat{Y}_i = 1.20 + .082X_{1i} + .75X_{2i} - .024X_{1i}X_{2i}$, where X_1 = shelf space, $X_2 = 0$ for back of aisle, and 1 for front of aisle.

(b) $t = -1.03 > t_8 = -2.306$. Don't reject H_0. No evidence that the interaction term makes a contribution to the model. Therefore, we should use the $\hat{Y}_i = b_0 + b_1 X_{1i} + b_2 X_{2i}$ model.

12.49 (a) $\hat{Y}_i = 16.19567 + 2.03779 X_{1i} + 0.56262 X_{2i}$

(b) For a given shipping mileage, each additional option ordered increases delivery time by 2.03779 days. For a given number of options, each 100-mile increase in shipping mileage increases delivery time by 0.56262 day.

(c) 41.07 days

(d) $F = 270.58 > F_{U(2,13)} = 3.81$, so we may reject H_0. There is a significant relationship between delivery time and number of options ordered and shipping mileage.

(e) $r^2_{Y.12} = .9765$; 97.65% of the variation in delivery time can be explained by variation in the number of options ordered and the shipping mileage.

(f) $r^2_{adj} = .973$

(g) $F = 509.16 > F_{U(1,13)} = 4.67$ and $F = 10.53 > F_{U(1,13)} = 4.67$. Each independent variable makes a significant contribution and should be included in the model.

(h) $.18794 \le \beta_2 \le .93730$

(j) $r^2_{Y1.2} = .9751$. For a given shipping mileage, 97.51% of the variation in delivery time can be explained by variation in the number of options.
$r^2_{Y2.1} = .4474$. For a given number of options, 44.74% of the variation in delivery time can be explained by variation in shipping mileage.

(k) $VIF_1 = 1.0$; $VIF_2 = 1.0$. There is no reason to suspect the existence of multicollinearity.

(l) On the basis of a residual analysis, the model appears adequate.

Chapter 13

13.3 (b), (c), and (e)

Pd.	Year	Y_i	3-Year Moving Total	3-Year Moving Avg.	(W = .50) ϵ_i	(W = .25)
1	1976	1.45	**	**	1.45	1.45
2	1977	1.55	4.61	1.54	1.50	1.48
3	1978	1.61	4.76	1.59	1.55	1.51
4	1979	1.60	4.95	1.65	1.58	1.53
5	1980	1.74	5.26	1.75	1.66	1.58
6	1981	1.92	5.61	1.87	1.79	1.67
7	1982	1.95	5.91	1.97	1.87	1.74
8	1983	2.04	6.05	2.02	1.95	1.81
9	1984	2.06	5.90	1.97	2.01	1.88
10	1985	1.80	5.59	1.86	1.90	1.86
11	1986	1.73	5.30	1.77	1.82	1.82
12	1987	1.77	5.40	1.80	1.79	1.81
13	1988	1.90	5.49	1.83	1.85	1.83
14	1989	1.82	5.37	1.79	1.83	1.83
15	1990	1.65	5.20	1.73	1.74	1.79
16	1991	1.73	5.26	1.75	1.74	1.77
17	1992	1.88	5.61	1.87	1.81	1.80
18	1993	2.00	5.96	1.99	1.90	1.85
19	1994	2.08	5.96	1.99	1.99	1.91
20	1995	1.88	**	**	1.94	1.90

(d) $\hat{Y}_{1996} = \epsilon_{1995} = 1.94$

(f) $\hat{Y}_{1996} = 1.90$

13.5 (b) $\hat{Y}_i = 0.216 + 0.139 X_i$, where origin = 1970 and X units = 1 year.

(c) 1993: 3.413
1994: 3.552
1995: 3.691
1996: 3.830

13.10 (b) $\log \hat{Y}_i = -.51371 + .0532 X_i$ or $\hat{Y}_i = (.3062)(1.1303)^{Xi}$ where origin = 1970 and X units = 1 year.

(c) 1993: 5.12
1994: 5.79
1995: 6.55
1996: 7.40

13.12 (a) 5

(b) 6

(c) The most recent five observed values—Y_{36}, Y_{37}, Y_{38}, Y_{39}, and Y_{40}.

(d) $\hat{Y}_i = a_0 + a_1 Y_{i-1} + a_2 Y_{i-2} + \cdots + a_5 Y_{i-5}$

(e) $\hat{Y}_{n+j} = a_0 + a_1 \hat{Y}_{n+j-1} + a_2 \hat{Y}_{n+j-2} + \cdots + a_5 \hat{Y}_{n+j-5}$.

13.16 First-order model is chosen.

13.19 Third-order model is chosen.

13.21 $\text{MAD}^A_1 = 0.11$

13.24 $\text{MAD}_{ET} = 0.25$; $\text{MAD}^A_3 = 0.21$

APPENDIX A

Review of Arithmetic and Algebra

A.1 RULES FOR ARITHMETIC OPERATIONS

The following is a summary of various rules for arithmetic operations with each rule illustrated by a numerical example.

Rule	Example
1. $a + b = c$ and $b + a = c$	$2 + 1 = 3$ and $1 + 2 = 3$
2. $a + (b + c) = (a + b) + c$	$5 + (7 + 4) = (5 + 7) + 4 = 16$
3. $a - b = c$ but $b - a \neq c$	$9 - 7 = 2$ but $7 - 9 = -2$
4. $a \times b = b \times a$	$7 \times 6 = 6 \times 7 = 42$
5. $a \times (b + c) = (a \times b) + (a \times c)$	$2 \times (3 + 5) = (2 \times 3) + (2 \times 5) = 16$
6. $a \div b \neq b \div a$	$12 \div 3 \neq 3 \div 12$
7. $\dfrac{a + b}{c} = \dfrac{a}{c} + \dfrac{b}{c}$	$\dfrac{7 + 3}{2} = \dfrac{7}{2} + \dfrac{3}{2} = 5$
8. $\dfrac{a}{b + c} \neq \dfrac{a}{b} + \dfrac{a}{c}$	$\dfrac{3}{4 + 5} \neq \dfrac{3}{4} + \dfrac{3}{5}$
9. $\dfrac{1}{a} + \dfrac{1}{b} = \dfrac{b + a}{ab}$	$\dfrac{1}{3} + \dfrac{1}{5} = \dfrac{5 + 3}{(3)(5)} = \dfrac{8}{15}$
10. $\dfrac{a}{b} \times \dfrac{c}{d} = \dfrac{a \times c}{b \times d}$	$\dfrac{2}{3} \times \dfrac{6}{7} = \dfrac{2 \times 6}{3 \times 7} = \dfrac{12}{21}$
11. $\dfrac{a}{b} \div \dfrac{c}{d} = \dfrac{a \times d}{b \times c}$	$\dfrac{5}{8} \div \dfrac{3}{7} = \dfrac{5 \times 7}{8 \times 3} = \dfrac{35}{24}$

A.2 RULES FOR ALGEBRA: EXPONENTS AND SQUARE ROOTS

The following is a summary of various rules for arithmetic operations with each rule illustrated by a numerical example:

Rule	Example
1. $X^a \cdot X^b = X^{a+b}$	$4^2 \cdot 4^3 = 4^5$
2. $(X^a)^b = X^{ab}$	$(2^2)^3 = 2^6$
3. $\dfrac{X^a}{X^b} = X^{a-b}$	$\dfrac{3^5}{3^3} = 3^2$
4. $\dfrac{X^a}{X^a} = X^0 = 1$	$\dfrac{3^4}{3^4} = 3^0 = 1$
5. $\sqrt{XY} = \sqrt{X}\sqrt{Y}$	$\sqrt{(25)(4)} = \sqrt{25}\sqrt{4} = 10$
6. $\sqrt{\dfrac{X}{Y}} = \dfrac{\sqrt{X}}{\sqrt{Y}}$	$\sqrt{\dfrac{16}{100}} = \dfrac{\sqrt{16}}{\sqrt{100}} = .40$

APPENDIX B

Summation Notation

Since the operation of addition occurs so frequently in statistics, the special symbol Σ (sigma) is used to denote "taking the sum of." Suppose, for example, that we have a set of n values for some variable X. The expression $\sum_{i=1}^{n} X_i$ means that these n values are to be added together. Thus,

$$\sum_{i=1}^{n} X_i = X_1 + X_2 + X_3 + \cdots X_n$$

The use of the summation notation can be illustrated by the following problem. Suppose we have five observations of a variable X: $X_1 = 2$, $X_2 = 0$, $X_3 = -1$, $X_4 = 5$, and $X_5 = 7$. Thus,

$$\sum_{i=1}^{5} X_i = X_1 + X_2 + X_3 + X_4 + X_5 = 2 + 0 + (-1) + 5 + 7 = 13$$

In statistics we are also frequently involved with summing the squared values of a variable. Thus,

$$\sum_{i=1}^{n} X_i^2 = X_1^2 + X_2^2 + X_3^2 + \cdots + X_n^2$$

and, in our example, we have

$$\sum_{i=1}^{5} X_i^2 = X_1^2 + X_2^2 + X_3^2 + X_4^2 + X_5^2$$
$$= 2^2 + 0^2 + (-1)^2 + 5^2 + 7^2$$
$$= 4 + 0 + 1 + 25 + 49$$
$$= 79$$

We should realize here that $\sum_{i=1}^{n} X_i^2$, the summation of the squares, is not the same as $\left(\sum_{i=1}^{n} X_i\right)^2$, the square of the sum, that is,

$$\sum_{i=1}^{n} X_i^2 \neq \left(\sum_{i=1}^{n} X_i\right)^2$$

In our example the summation of squares is equal to 79. This is not equal to the square of the sum, which is $13^2 = 169$.

Another frequently used operation involves the summation of the product. That is, suppose we have two variables, X and Y, each having n observations. Then

$$\sum_{i=1}^{n} X_i Y_i = X_1 Y_1 + X_2 Y_2 + X_3 Y_3 + \cdots + X_n Y_n$$

Continuing with our previous example, suppose there is also a second variable Y whose five values are $Y_1 = 1$, $Y_2 = 3$, $Y_3 = -2$, $Y_4 = 4$, and $Y_5 = 3$. Then

$$\sum_{i=1}^{5} X_i Y_i = X_1 Y_1 + X_2 Y_2 + X_3 Y_3 + X_4 Y_4 + X_5 Y_5$$
$$= (2)(1) + (0)(3) + (-1)(-2) + (5)(4) + (7)(3)$$
$$= 2 + 0 + 2 + 20 + 21$$
$$= 45$$

In computing $\sum_{i=1}^{n} X_i Y_i$ we must realize that the first value of X is multiplied by the first value of Y, the second value of X is multiplied by the second value of Y, and so on. These cross products are then summed to obtain the desired result. However, we should note here that the summation of cross products is not equal to the product of the individual sums, that is,

$$\sum_{i=1}^{n} X_i Y_i \neq \left(\sum_{i=1}^{n} X_i\right)\left(\sum_{i=1}^{n} Y_i\right)$$

In our example, $\sum_{i=1}^{5} X_i = 13$ and $\sum_{i=1}^{5} Y_i = 1 + 3 + (-2) + 4 + 3 = 9$ so that $\left(\sum_{i=1}^{5} X_i\right)\left(\sum_{i=1}^{5} Y_i\right) = (13)(9) = 117$. This is not the same as $\sum_{i=1}^{n} X_i Y_i$, which equals 45.

Before studying the four basic rules of performing operations with summation notation, it is helpful to present the values for each of the five observations of X and Y in a tabular format.

Observation	X_i	Y_i
1	2	1
2	0	3
3	-1	-2
4	5	4
5	7	3
	$\sum_{i=1}^{5} X_i = 13$	$\sum_{i=1}^{5} Y_i = 9$

Rule 1: The summation of the values of two variables is equal to the sum of the values of each summed variable.

$$\sum_{i=1}^{n} (X_i + Y_i) = \sum_{i=1}^{n} X_i + \sum_{i=1}^{n} Y_i$$

Thus, in our example,

$$\sum_{i=1}^{5} (X_i Y_i) = (2 + 1) + (0 + 3) + (-1 + (-2)) + (5 + 4) + (7 + 3)$$

$$= 3 + 3 + (-3) + 9 + 10$$

$$= 22 = \sum_{i=1}^{5} X_i + \sum_{i=1}^{5} Y_i = 13 + 9 = 22$$

Rule 2: The summation of a difference between the values of two variables is equal to the difference between the summed values of the variables.

$$\sum_{i=1}^{n}(X_i - Y_i) = \sum_{i=1}^{n} X_i - \sum_{i=1}^{n} Y_i$$

Thus, in our example,

$$\sum_{i=1}^{5}(X_i - Y_i) = (2-1) + (0-3) + (-1-(-2)) + (5-4) + (7-3)$$

$$= 1 + (-3) + 1 + 1 + 4$$

$$= 4 = \sum_{i=1}^{5} X_i - \sum_{i=1}^{5} Y_i = 13 - 9 = 4$$

Rule 3: The summation of a constant times a variable is equal to that constant times the summation of the values of the variable.

$$\sum_{i=1}^{n} cX_i = c\sum_{i=1}^{n} X_i$$

where c is a constant.

Thus, in our example, if $c = 2$,

$$\sum_{i=1}^{5} cX_i = \sum_{i=1}^{5} 2X_i = (2)(2) + (2)(0) + (2)(-1) + (2)(5) + (2)(7)$$

$$= 4 + 0 + (-2) + 10 + 14$$

$$= 26 = 2\sum_{i=1}^{5} X = (2)(13) = 26$$

Rule 4: A constant summed n times will be equal to n times the value of the constant.

$$\sum_{i=1}^{n} c = nc$$

where c is a constant. Thus, if the constant $c = 2$ is summed five times, we would have

$$\sum_{i=1}^{5} c = 2 + 2 + 2 + 2 + 2$$

$$= 10 = (5)(2) = 10$$

To illustrate the use of these summation rules, we may demonstrate one of the mathematical properties pertaining to the average or arithmetic mean (see Section 3.4.1), that is,

$$\sum_{i=1}^{n} \left(X_i - \overline{X} \right) = 0$$

This property states that the summation of the differences between each observation and the arithmetic mean is zero. This can be proven mathematically in the following manner.

1. From Equation (3.1),

$$\overline{X} = \frac{\sum_{i=1}^{n} X_i}{n}$$

Thus, using summation rule 2, we have

$$\sum_{i=1}^{n} \left(X_i - \overline{X} \right) = \sum_{i=1}^{n} X_i - \sum_{i=1}^{n} \overline{X}$$

2. Since, for any fixed set of data, $\overline{X}$ can be considered a constant, from summation rule 4, we have

$$\sum_{i=1}^{n} \overline{X} = n\overline{X}$$

Therefore,

$$\sum_{i=1}^{n} \left(X_i - \overline{X} \right) = \sum_{i=1}^{n} X_i - n\overline{X}$$

3. However, from Equation (3.1), since

$$\overline{X} = \frac{\sum_{i=1}^{n} X_i}{n} \quad \text{then} \quad n\overline{X} = \sum_{i=1}^{n} X_i$$

Therefore,

$$\sum_{i=1}^{n} \left(X_i - \overline{X} \right) = \sum_{i=1}^{n} X_i - \sum_{i=1}^{n} X_i$$

Thus we have shown that

$$\sum_{i=1}^{n} \left(X_i - \overline{X} \right) = 0$$

Problem

Suppose that there are six observations for the variables X and Y such that $X_1 = 2$, $X_2 = 1$, $X_3 = 5$, $X_4 = -3$, $X_5 = 1$, $X_6 = -2$, and $Y_1 = 4$, $Y_2 = 0$, $Y_3 = -1$, $Y_4 = 2$, $Y_5 = 7$, and $Y_6 = -3$. Compute each of the following:

(a) $\sum_{i=1}^{6} X_i$

(b) $\sum_{i=1}^{6} Y_i$

(c) $\sum_{i=1}^{6} X_i^2$

(d) $\sum_{i=1}^{6} Y_i^2$

(e) $\sum_{i=1}^{6} X_i Y_i$

(f) $\sum_{i=1}^{6} (X_i + Y_i)$

(g) $\sum_{i=1}^{6} (X_i - Y_i)$

(h) $\sum_{i=1}^{6} (X_i - 3Y_i + 2X_i^2)$

(i) $\sum_{i=1}^{6} (cX_i)$, where $c = -1$

(j) $\sum_{i=1}^{6} (X_i - 3Y_i + c)$, where $c = +3$

References

1. Bashaw, W. L., *Mathematics for Statistics* (New York: Wiley, 1969).
2. Lanzer, P., *Video Review of Arithmetic* (Roslyn Heights, NY: Video Aided Instruction, 1990).
3. Levine, D., *Video Review of Statistics* (Roslyn Heights, NY: Video Aided Instruction, 1989).
4. Shane, H., *Video Review of Elementary Algebra* (Roslyn Heights, NY: Video Aided Instruction, 1990).

APPENDIX C

Statistical Symbols and Greek Alphabet

C.1 STATISTICAL SYMBOLS

+	add	×	multiply
−	subtract	÷	divide
=	equals	≠	not equal
≅	approximately equal to		
>	greater than	<	less than
≥ or ⩾	greater than or equal to	≤ or ⩽	less than or equal to

C.2 GREEK ALPHABET

Greek Letter	Greek Name	English Equivalent	Greek Letter	Greek Name	English Equivalent
A α	Alpha	a	N ν	Nu	n
B β	Beta	b	Ξ ξ	Xi	x
Γ γ	Gamma	g	O o	Omicron	ŏ
Δ δ	Delta	d	Π π	Pi	p
E ε	Epsilon	ĕ	P ρ	Rho	r
Z ζ	Zeta	z	Σ σ	Sigma	s
H η	Eta	ē	T τ	Tau	t
Θ θ	Theta	th	Y υ	Upsilon	u
I ι	Iota	i	Φ φ	Phi	ph
K κ	Kappa	k	X χ	Chi	ch
Λ λ	Lambda	l	Ψ ψ	Psi	ps
M μ	Mu	m	Ω ω	Omega	ō

APPENDIX

D

Special Data Sets (for Team Projects)

D.1 SPECIAL DATA SET 1

This is a file containing data for the states of Texas, North Carolina, and Pennsylvania regarding out-of-state tuition payments. There are 60 schools in Texas, 45 in North Carolina, and 90 in Pennsylvania. To use the file, note the following codes for the data:

- Out-of-state tuition charges (in $000)
- Type of institution: 1 = private; 2 = public
- Location of institution: 1 = rural; 2 = suburb; 3 = urban
- Academic calendar: 1 = qtr.; 2 = sem.; 3 = tri.; 4 = 414; 5 = other
- Classification of institution: 1 = NLA; 2 = NU; 3 = RLA; 4 = RU; 5 = SS
- State: 1 = Texas; 2 = North Carolina; 3 = Pennsylvania

Moreover, for academic calendar: Qtr. = quarter, Sem. = semester; Tri. = Trimester; 414 = 4–1–4. For institutional classification: NLA = national liberal arts school; NU = national university; RLA = regional liberal arts school; RU = regional university; SS = school with special focus.

NCC&U.TXT
PENNC&U.TXT
TEXASC&U.TXT

NC-1.XLS
NC-2.XLS
PA-POP-1.XLS
PA-SAM-1.XLS
PA-SAM-2.XLS
TEXAS-1.XLS
TEXAS-2.XLS
TEXAS-3.XLS

School	Tuition (in $000)	Type	Setting	Calendar	Class
Texas					
Abilene Christian Univ.	7.2	Private	Suburb	Sem.	RU
Angelo State Univ.	4.9	Public	Urban	Sem.	RU
Austin College	10.7	Private	Urban	414	NLA
Baylor Univ.	10.4	Private	Urban	Sem.	NU
Concordia Lutheran College	6.4	Private	Urban	Sem.	RLA
Dallas Baptist Univ.	4.8	Private	Urban	414	RLA
East Texas Baptist Univ.	4.7	Private	Urban	414	RLA
East Texas State Univ.	4.6	Private	Urban	Qtr.	RU
Hardin Simmons Univ.	6.0	Private	Urban	Sem.	RU
Houston Baptist Univ.	5.4	Private	Urban	Qtr.	RU
Howard Payne Univ.	4.8	Private	Rural	Sem.	RLA
Huston-Tillotson College	4.7	Private	Urban	Sem.	RLA
Incarnate Word College	8.3	Private	Urban	Sem.	RLA
Jarvis Christian College	3.8	Private	Rural	Sem.	RLA
Lamar Univ.	4.8	Public	Urban	Sem.	RU
LeTourneau Univ.	8.3	Private	Urban	Sem.	RLA

(continued)

School	Tuition (in $000)	Type	Setting	Calendar	Class
Texas (Continued)					
Lubbock Christian Univ.	6.4	Private	Urban	Sem.	RLA
McMurry Univ.	6.6	Private	Urban	414	RLA
Midwestern State Univ.	4.5	Public	Urban	Sem.	RU
Our Lady of the Lake	8.0	Private	Urban	Sem.	RU
Paul Quinn College	3.6	Private	Urban	Sem.	RLA
Prairie View A&M Univ.	2.4	Public	Rural	Sem.	RU
Rice Univ.	8.5	Private	Urban	Sem.	NU
St. Edward's Univ.	8.8	Private	Urban	Sem.	RU
St. Mary's Univ.	7.7	Private	Urban	Sem.	RU
Sam Houston State Univ.	4.9	Public	Rural	Sem.	RU
Schreiner College	8.6	Private	Rural	414	RLA
Southern Methodist Univ.	12.0	Private	Suburb	Sem.	NU
Southwest Texas State Univ.	4.9	Public	Urban	Sem.	RU
Southwestern Adventist	7.0	Private	Rural	Sem.	RLA
Southwestern Univ.	11.0	Private	Suburb	Sem.	RLA
Stephen F. Austin State U.	4.9	Public	Rural	Sem.	RU
Sul Ross State Univ.	3.9	Public	Rural	Sem.	RU
Tarleton State Univ.	4.9	Public	Rural	Sem.	RU
Texas A&I Univ.	4.4	Public	Rural	Sem.	RU
Texas A&M Univ.	4.9	Public	Urban	Sem.	NU
Texas A&M at Galveston	4.9	Public	Urban	Sem.	RU
Texas Christian Univ.	8.0	Private	Urban	Sem.	NC
Texas College	3.6	Private	Urban	Sem.	RLA
Texas Lutheran College	7.4	Private	Urban	Sem.	RLA
Texas Southern Univ.	7.9	Public	Urban	Sem.	RU
Texas Tech Univ.	4.9	Public	Urban	Sem.	RU
Texas Wesleyan Univ.	5.8	Private	Urban	Sem.	RU
Texas Woman's Univ.	3.9	Public	Urban	Sem.	RU
Trinity Univ.	11.6	Private	Urban	Sem.	RU
U. of Dallas	10.3	Private	Suburb	Sem.	NLA
U. of Houston	3.4	Public	Urban	Sem.	NU
U. of Houston-Downtown	3.9	Public	Urban	Sem.	RU
U. of Mary Hardin-Baylor	5.0	Private	Suburb	Sem.	RLA
U. of North Texas	3.9	Public	Urban	Sem.	RU
U. of St. Thomas	8.0	Private	Urban	Sem.	RU
U. of Texas at Arlington	3.5	Public	Suburb	Sem.	NU
U. of Texas at Austin	4.9	Public	Urban	Sem.	NU
U. of Texas at Dallas	5.8	Public	Suburb	Sem.	RU
U. of Texas at El Paso	4.1	Public	Urban	Sem.	RU
U. of Texas-Pan American	3.5	Public	Urban	Sem.	RU
U. of Texas, San Antonio	3.9	Public	Urban	Sem.	RU
Wayland Baptist Univ.	4.8	Private	Urban	414	RU
West Texas State Univ.	5.9	Public	Rural	Sem.	RU
Wiley College	3.6	Private	Urban	Sem.	RLA

(continued)

School	Tuition (in $000)	Type	Setting	Calendar	Class
North Carolina					
Appalachian State Univ.	6.5	Public	Rural	Sem.	RU
Barber Scotia College	4.0	Private	Urban	Sem.	RLA
Barton College	7.1	Private	Urban	Sem.	RLA
Belmont Abbey College	8.3	Private	Suburb	Sem.	RLA
Bennett College	5.4	Private	Urban	Sem.	RLA
Campbell Univ.	7.6	Private	Rural	Sem.	RU
Catawba College	9.0	Private	Suburb	Sem.	RLA
Davidson College	15.7	Private	Suburb	Sem.	NLA
Duke Univ.	16.7	Private	Urban	Sem.	NU
East Carolina Univ.	6.4	Public	Urban	Sem.	RU
Elizabeth City State Univ.	5.0	Public	Rural	Sem.	RU
Elon College	8.5	Private	Suburb	414	RU
Fayetteville State Univ.	5.7	Public	Urban	Sem.	RU
Gardner Webb Univ.	7.7	Private	Rural	Sem.	RU
Greensboro College	7.2	Private	Urban	Sem.	RLA
Guilford College	12.4	Private	Suburb	Sem.	NLA
High Point Univ.	7.1	Private	Urban	Sem.	RU
Johnson C. Smith Univ.	5.5	Private	Urban	Sem.	RLA
Lenoir-Rhyne College	9.7	Private	Suburb	Sem.	RLA
Livingston College	4.4	Private	Urban	Sem.	RLA
Mars Hill College	7.0	Private	Rural	Sem.	RLA
Meredith College	6.3	Private	Urban	Sem.	RU
Methodist College	8.3	Private	Urban	Sem.	RLA
N.C. A&T Univ.	6.9	Public	Urban	Sem.	RU
N.C. Central Univ.	5.7	Public	Urban	Sem.	RU
N.C. School of the Arts	7.6	Public	Urban	Tri.	SS
N.C. State Univ.	7.9	Public	Urban	Sem.	NU
N.C. Wesleyan College	7.9	Private	Suburb	Sem.	RLA
Pembrooke State Univ.	6.0	Public	Rural	Sem.	RU
Pfeiffer College	8.2	Private	Rural	Sem.	RLA
Queens College	10.4	Private	Suburb	Sem.	RLA
St. Andrews Presbyterian Coll.	9.9	Private	Rural	414	RLA
St. Augustine's College	3.9	Private	Urban	Sem.	RU
Salem College	9.8	Private	Urban	414	RU
Shaw Univ.	8.2	Private	Urban	Sem.	RU
U. of N.C. at Asheville	5.6	Public	Urban	Sem.	RU
U. of N.C. at Chapel Hill	7.9	Public	Urban	Sem.	NU
U. of N.C. at Charlotte	6.4	Public	Urban	Sem.	RU
U. of N.C. at Greensboro	7.4	Public	Urban	Sem.	NU
U. of N.C. at Wilmington	7.0	Public	Suburb	Sem.	RU
Wake Forest Univ.	13.0	Private	Suburb	Sem.	RU
Warren Wilson College	8.7	Private	Rural	Sem.	RLA
Western Carolina Univ.	6.4	Public	Rural	Sem.	RU
Wingate College	6.7	Private	Suburb	Sem.	RU
Winston–Salem State Univ.	7.4	Private	Urban	Sem.	RU

(continued)

School	Tuition (in $000)	Type	Setting	Calendar	Class
Pennsylvania					
Albright College	14.9	Private	Suburb	414	NLA
Allegheny College	16.4	Private	Rural	Sem.	NLA
Al'town College St. Francis	9.3	Private	Rural	Sem.	RLA
Alvernia College	8.4	Private	Urban	Sem.	RLA
Beaver College	12.3	Private	Suburb	Sem.	RU
Bloomsburg Univ. of Pa.	4.9	Public	Rural	Sem.	RU
Bryn Mawr College	17.1	Private	Suburb	Sem.	NLA
Bucknell Univ.	17.7	Private	Rural	Sem.	NLA
Cabrini College	9.7	Private	Suburb	Sem.	RLA
California Univ. of Pa.	6.3	Public	Rural	Sem.	RU
Carlow College	9.4	Private	Urban	Sem.	RLA
Carnegie Mellon Univ.	17.0	Private	Urban	Sem.	NU
Cedar Crest College	13.7	Private	Suburb	Sem.	RLA
Chatham College	12.6	Private	Urban	414	NLA
Chestnut Hill College	9.5	Private	Suburb	Sem.	NLA
Cheyney Univ.	2.7	Public	Rural	Sem.	RU
Clarion Univ. of Pa.	4.4	Public	Rural	Sem.	RU
College Misericordia	10.0	Private	Suburb	Sem.	RLA
Delaware Valley College	11.6	Private	Suburb	Sem.	RU
Dickinson College	17.7	Private	Urban	Sem.	NLA
Drexel Univ.	11.7	Private	Urban	Qtr.	NU
Duquesne Univ.	10.6	Private	Urban	Sem.	NU
East Stroudsburg Univ.	4.9	Public	Rural	Sem.	RU
Eastern College	10.6	Private	Suburb	Sem.	RLA
Edinboro Univ.	6.1	Public	Rural	Sem.	RU
Elizabethtown College	13.2	Private	Suburb	Sem.	RU
Franklin & Marshall College	22.3	Private	Urban	Sem.	NLA
Gannon Univ.	9.1	Private	Urban	Sem.	RU
Geneva College	8.9	Private	Suburb	Sem.	RLA
Gettysburg College	18.9	Private	Rural	Sem.	NLA
Grove City College	5.0	Private	Rural	Sem.	RU
Gwynedd Mercy College	10.2	Private	Suburb	Sem.	RU
Hahnemann Univ.	8.3	Private	Urban	Sem.	NU
Haverford College	17.9	Private	Suburb	Sem.	NLA
Holy Family College	8.3	Private	Urban	Sem.	RLA
Immaculata College	9.4	Private	Suburb	Sem.	RLA
Indiana Univ. of Pa.	4.9	Public	Rural	Sem.	RU
Juniata College	14.2	Private	Rural	Sem.	NLA
King's College	10.2	Private	Urban	Sem.	RU
Kutztown Univ.	8.3	Public	Rural	Sem.	RU
LaRoche College	8.4	Private	Suburb	Sem.	RU
LaSalle Univ.	11.5	Private	Urban	Sem.	RU
Lafayette College	17.9	Private	Urban	Sem.	NLA
Lebanon Valley College	13.3	Private	Rural	Sem.	NLA
Lehigh Univ.	17.8	Private	Urban	Sem.	NU
Lincoln Univ.	4.0	Public	Rural	Tri.	RLA
Lock Haven Univ. of Pa.	6.1	Public	Rural	Sem.	RU

(continued)

School	Tuition (in $000)	Type	Setting	Calendar	Class
Pennsylvania (Continued)					
Lycoming College	13.0	Private	Rural	414	RLA
Mansfield Univ.	5.5	Public	Rural	Sem.	RU
Marywood College	9.6	Private	Suburb	Sem.	RU
Mercyhurst College	9.3	Private	Suburb	Other	RU
Messiah College	9.7	Private	Suburb	414	RU
Millersville Univ. of Pa.	7.7	Public	Suburb	414	RU
Moore College Art and Design	13.5	Private	Urban	Sem.	SS
Moravian College	14.3	Private	Urban	Sem.	RU
Muhlenberg College	16.4	Private	Suburb	Sem.	NLA
Neumann College	10.0	Private	Suburb	Sem.	NLA
Penn State U. at Erie	10.1	Public	Urban	Sem.	RU
Penn State U. at College Park	9.6	Public	Urban	Sem.	NU
Philadelphia College Tex. & Sci.	11.7	Private	Suburb	Sem.	RU
Point Park College	9.3	Private	Urban	Sem.	RU
Robert Morris College	6.0	Private	Urban	Sem.	SS
Rosemont College	10.7	Private	Suburb	Sem.	RLA
St. Francis College	10.3	Private	Rural	Sem.	RU
St. Joseph's Univ.	11.9	Private	Urban	Sem.	RU
St. Vincent College	10.2	Private	Rural	Sem.	RLA
Seton Hill College	10.2	Private	Suburb	Sem.	RLA
Shippensburg Univ.	6.1	Public	Rural	Sem.	RU
Slippery Rock Univ.	7.7	Public	Rural	Sem.	RU
Susquehanna Univ.	15.6	Private	Rural	Sem.	RU
Swarthmore College	18.3	Private	Suburb	Sem.	NLA
Temple Univ.	9.1	Public	Urban	Sem.	NU
Thiel College	10.4	Private	Rural	Sem.	RLA
Univ. of the Arts	11.2	Private	Urban	Sem.	SS
Univ. of Pennsylvania	16.1	Private	Urban	Sem.	NU
Univ. of Pittsburgh	10.3	Public	Urban	Sem.	NU
U. of Pittsburgh at Bradford	9.7	Public	Rural	Sem.	RLA
U. of Pittsburgh at Greensburg	10.3	Public	Suburb	Sem.	RLA
U. of Pittsburgh at Johnstown	9.7	Public	Suburb	Sem.	RU
U. of Scranton	10.7	Private	Urban	414	RU
Ursinus College	14.1	Private	Suburb	Sem.	NLA
Villanova Univ.	15.2	Private	Suburb	Sem.	RU
Washington and Jefferson College	15.4	Private	Suburb	414	NLA
Waynesburg College	8.4	Private	Rural	Sem.	RU
West Chester Univ.	6.1	Public	Suburb	Sem.	RU
Westminster College	11.4	Private	Rural	414	RLA
Widener Univ.	11.7	Private	Suburb	Sem.	RU
Wilkes Univ.	9.5	Private	Suburb	Sem.	RU
Wilson College	11.4	Private	Rural	Tri.	RLA
York College of Pa.	4.8	Private	Suburb	Sem.	RU

Source: "America's Best Colleges, 1994 College Guide," *U.S. News & World Report,* extracted from College Counsel 1993 of Natick, Mass. Reprinted by special permission, *U.S. News & World Report,* © 1993 by *U.S. News & World Report* and by College Counsel.

D.2 SPECIAL DATA SET 2

CEREAL.TXT

This is a file containing data for a sample of $n = 84$ ready-to-eat cereals. To use the file, note the following:

- Type product (H for high fiber, M for moderate fiber, L for low fiber)
- Cost per serving (in cents)
- Weight per serving (in ounces)
- Calories per serving
- Sugar per serving (in grams)

Ready-to-Eat Cereal	Type Product	Cost	Weight	Calories	Sugar
All-Bran with Extra Fiber	H	38	2.0	100	0
Fiber One	H	34	2.0	120	0
Bran Buds	H	21	1.5	110	12
100% Bran	H	23	1.5	110	9
All-Bran Original	H	23	1.5	110	8
100% Organic Raisin Bran Flakes	H	51	2.0	180	17
Uncle Sam	H	28	2.0	220	0
Bran Flakes	H	23	1.5	140	8
Bran Flakes	H	21	1.5	140	8
Crunchy Corn Bran	H	28	1.5	140	9
Fiberwise	H	43	1.5	140	8
Raisin Bran	H	30	2.0	180	21
Multi Bran Chex	H	25	1.5	140	9
Shredded Wheat 'N Bran	H	28	1.5	140	0
Fruit & Fibre Peaches, Raisins Almonds & Oat Clusters	H	38	2.0	180	12
Fruitful Bran	H	43	2.0	180	18
Raisin Bran	H	29	1.75	160	16
Shredded Wheat	H	29	1.67	160	0
Cracklin' Oat Bran	H	44	2.0	220	14
Skinner's Raisin Bran	H	27	2.0	220	12
Frosted Wheat Squares	H	40	2.0	200	12
Grape-Nuts	H	29	2.0	220	6
Shredded Wheat Spoon Size	M	28	1.5	140	0
Common Sense Oat Bran	M	27	1.33	130	8
Frosted Mini-Wheats	M	28	1.25	130	8
Grape-Nuts Flakes	M	19	1.25	110	6
Whole Grain Total	M	23	1.0	100	3
Whole Grain Wheat Chex	M	24	1.5	150	5
Whole Grain Wheaties	M	16	1.0	100	3
Total Raisin Bran	M	30	1.5	140	14
Raisin Nut Bran	M	37	2.0	220	16
Raisin Squares	M	38	2.0	180	12
Oatios with Extra Oat Bran	M	27	1.0	110	2
Nutri-Grain Almond Raisin	M	47	2.0	210	11
Crispy Wheats 'N Raisins	M	24	1.33	130	13
Life	M	26	1.5	150	9
Multi Grain Cheerios	M	24	1.0	100	6

(continued)

Ready-to-Eat Cereal	Type Product	Cost	Weight	Calories	Sugar
Oat Squares	M	34	2.0	200	12
Mueslix Crispy Blend	M	50	2.25	240	19
Cheerios	M	15	.8	90	0
Cinnamon Oat Squares	M	35	2.0	220	14
Clusters	M	44	2.0	220	25
100% Natural Whole Grain with Raisins (Low Fat)	M	39	2.0	220	14
Honey Bunches of Oats with Almonds	M	29	1.5	180	9
Low-Fat Granola with Raisins	M	36	1.67	180	14
Basic 4	M	38	1.75	170	11
Just Right with Fruit & Nuts	M	42	1.75	190	12
Apple Cinnamon Cheerios	M	25	1.33	150	13
Honey Nut Cheerios	M	26	1.33	150	13
Oatmeal Raisin Crisp	M	47	2.5	260	20
Nut & Honey Crunch	M	29	1.5	170	12
Puffed Rice	L	13	.5	50	0
Puffed Wheat	L	15	.5	50	0
Kix	L	16	.67	70	2
Honey-Comb	L	17	.75	80	8
Corn Flakes	L	10	1.0	100	2
Product 19	L	23	1.0	100	3
Rice Chex	L	17	1.0	100	2
Apple Jacks	L	23	1.0	110	14
Cocoa Puffs	L	21	1.0	110	13
Cookie-Crisp Chocolate Chip	L	25	1.0	110	13
Corn Chex	L	19	1.0	110	3
Corn Pops	L	22	1.0	110	12
Crispix	L	21	1.0	110	3
Froot Loops	L	21	1.0	110	13
Lucky Charms	L	22	1.0	110	12
Marshmallow Alpha-Bits	L	23	1.0	110	14
Rice Krispies	L	18	1.0	110	3
Special K	L	22	1.0	110	3
Teenage Mutant Ninja Turtles	L	23	1.0	110	11
Total Corn Flakes	L	28	1.0	110	3
Trix	L	24	1.0	110	12
Fruity Pebbles	L	24	1.25	130	14
Wheaties Honey Gold	L	20	1.33	130	13
Cocoa Krispies	L	28	1.33	150	15
Cocoa Pebbles	L	28	1.33	150	17
Frosted Flakes	L	20	1.33	150	15
Golden Grahams	L	27	1.33	150	12
Kenmei Rice Bran	L	22	1.33	150	5
Smacks	L	24	1.33	150	20
Triples	L	22	1.33	150	4
Cap'n Crunch	L	22	1.33	160	16
Cap'n Crunch's Crunch Berries	L	23	1.33	160	16
Cinnamon Toast Crunch	L	26	1.33	160	12

Source: "Cereals: Which Belong in Your Bowl?" Copyright 1992 by Consumers Union of United States, Inc., Yonkers, NY 10703. Adapted by permission from *Consumer Reports,* November 1992, pp. 693–695. Although this data set originally appeared in *Consumer Reports*, the selective adaptation and resulting conclusions presented are those of the authors and are not sanctioned or endorsed in any way by Consumers Union, the publisher of *Consumer Reports.*

D.3 SPECIAL DATA SET 3

FRAGRAN.TXT

This is a file containing data for a sample of $n = 83$ fragrances. To use the file, note the following:

- Gender (W = women and M = men with respective coded values of 1 and 2)
- Type of fragrance (P = perfume, C = cologne, O = other; the respective coded values are 1, 2, 3)
- Cost in dollars per ounce
- Intensity or strength of the fragrance (VS = very strong, S = strong, Me = medium, and Mi = mild; the respective coded values are 1, 2, 3, 4)

Fragrance	Gender	Type	Cost	Intensity
Gio	W	P	300	S
Cabochard	W	P	175	S
"Delicious"	W	P	190	S
Chanel No. 5	W	C	19	S
Obsession	W	P	180	S
Sublime	W	P	250	VS
Vivid	W	P	230	S
Chanel No. 5	W	O	32	S
Dune	W	P	185	S
Ninja	W	C	8	S
Safari	W	C	19	Mi
Soft Musk	W	C	7	Me
360	W	P	200	S
Tresor	W	P	220	Mi
Venzia	W	P	320	VS
Coco	W	P	215	S
Opium	W	O	30	VS
Oscar de la Renta	W	P	200	Mi
Volupte	W	P	220	Me
Wild Heart	W	C	9	S
"An Impression of Chanel # 5"	W	C	4	Me
Chanel No. 5	W	P	215	S
Chloe	W	P	170	Mi
Mesmerize	W	C	9	Me
Obsession	W	C	21	Mi
Aliage Sport Fragrance	W	C	12	Me
Chanel No. 5	W	O	19	Mi
Passion	W	P	185	S
Charlie	W	C	8	S
Realm Women	W	O	29	VS
Shalimar	W	P	205	VS
White Diamonds	W	P	225	VS
Wind Song	W	C	10	Me
Incognito	W	C	13	S
L'Air du Temps	W	P	260	Me
Ma Griffe	W	P	152	VS
Navy	W	C	11	S

(continued)

Fragrance	Gender	Type	Cost	Intensity
White Linen	W	P	170	Mi
Samsara	W	P	210	Me
Tuscany per Donna	W	P	190	VS
Versus	W	O	22	S
Halston	W	P	272	S
Red	W	C	23	S
Catalyst	W	P	250	S
Escape	W	C	23	S
Liz Claiborne	W	P	135	VS
Chantilly	W	C	2	S
Amarige	W	P	280	S
Feminite du Bois	W	P	240	S
Opium	W	P	205	Me
Giorgio	W	C	23	VS
Primo	W	C	7	S
Anais Anais	W	P	210	Me
Chloe Narcisse	W	P	225	VS
Donna Karan New York	W	P	350	VS
Tribu	W	P	220	VS
Our Version of White Diamonds	W	C	8	VS
Calyx	W	C	26	Me
Jean Naté	W	C	7	S
White Diamonds	W	O	23	VS
Emeraude	W	C	6	Mi
Poison	W	C	19	VS
Joy	W	P	300	VS
Angelfire	W	C	9	Mi
Caliente	W	C	11	Mi
Dewberry	W	O	19	S
Drakkar Noir	M	O	15	VS
Lancer	M	C	4	S
Eternity for Men	M	C	12	VS
Realm for Men	M	C	29	S
Preferred Stock	M	C	9	S
Gravity	M	C	10	S
Obsession for Men	M	C	10	S
Escape for Men	M	O	21	S
Old Spice	M	C	2	S
Tribute	M	C	7	S
Egoiste	M	C	16	S
Stetson	M	C	7	Me
English Leather	M	C	3	Me
Safari for Men	M	O	14	VS
Aramis	M	C	9	S
Brut	M	C	4	VS
Polo	M	C	10	VS

Source: "How to Buy a Fragrance." Copyright 1993 by Consumers Union of United States, Inc., Yonkers, NY 10703. Adapted by permission from *Consumer Reports,* December 1993, pp. 772–773. Although this data set originally appeared in *Consumer Reports,* the selective adaptation and resulting conclusions presented are those of the authors and are not sanctioned or endorsed in any way by Consumers Union, the publisher of *Consumer Reports.*

APPENDIX E

Tables

	Table	Page
E.1	Table of Random Numbers	E–2
E.2	The Standardized Normal Distribution	E–4
E.3	Critical Values of t	E–5
E.4	Critical Values of χ^2	E–7
E.5	Critical Values of F	E–8
E.6	Lower and Upper Critical Values T_1 of Wilcoxon Rank Sum Test	E–12
E.7	Critical Values of the Studentized Range Q	E–13
E.8	Control Chart Factors	E–15
E.9	Critical Values d_L and d_U of the Durbin-Watson Statistic D	E–16

TABLE E.1 Table of Random Numbers

Row	00000 12345	00001 67890	11111 12345	11112 67890	22222 12345	22223 67890	33333 12345	33334 67890
01	49280	88924	35779	00283	81163	07275	89863	02348
02	61870	41657	07468	08612	98083	97349	20775	45091
03	43898	65923	25078	86129	78496	97653	91550	08078
04	62993	93912	30454	84598	56095	20664	12872	64647
05	33850	58555	51438	85507	71865	79488	76783	31708
06	97340	03364	88472	04334	63919	36394	11095	92470
07	70543	29776	10087	10072	55980	64688	68239	20461
08	89382	93809	00796	95945	34101	81277	66090	88872
09	37818	72142	67140	50785	22380	16703	53362	44940
10	60430	22834	14130	96593	23298	56203	92671	15925
11	82975	66158	84731	19436	55790	69229	28661	13675
12	39087	71938	40355	54324	08401	26299	49420	59208
13	55700	24586	93247	32596	11865	63397	44251	43189
14	14756	23997	78643	75912	83832	32768	18928	57070
15	32166	53251	70654	92827	63491	04233	33825	69662
16	23236	73751	31888	81718	06546	83246	47651	04877
17	45794	26926	15130	82455	78305	55058	52551	47182
18	09893	20505	14225	68514	46427	56788	96297	78822
19	54382	74598	91499	14523	68479	27686	46162	83554
20	94750	89923	37089	20048	80336	94598	26940	36858
21	70297	34135	53140	33340	42050	82341	44104	82949
22	85157	47954	32979	26575	57600	40881	12250	73742
23	11100	02340	12860	74697	96644	89439	28707	25815
24	36871	50775	30592	57143	17381	68856	25853	35041
25	23913	48357	63308	16090	51690	54607	72407	55538
26	79348	36085	27973	65157	07456	22255	25626	57054
27	92074	54641	53673	54421	18130	60103	69593	49464
28	06873	21440	75593	41373	49502	17972	82578	16364
29	12478	37622	99659	31065	83613	69889	58869	29571
30	57175	55564	65411	42547	70457	03426	72937	83792
31	91616	11075	80103	07831	59309	13276	26710	73000
32	78025	73539	14621	39044	47450	03197	12787	47709
33	27587	67228	80145	10175	12822	86687	65530	49325
34	16690	20427	04251	64477	73709	73945	92396	68263
35	70183	58065	65489	31833	82093	16747	10386	59293
36	90730	35385	15679	99742	50866	78028	75573	67257
37	10934	93242	13431	24590	02770	48582	00906	58595
38	82462	30166	79613	47416	13389	80268	05085	96666
39	27463	10433	07606	16285	93699	60912	94532	95632
40	02979	52997	09079	92709	90110	47506	53693	49892
41	46888	69929	75233	52507	32097	37594	10067	67327
42	53638	83161	08289	12639	08141	12640	28437	09268
43	82433	61427	17239	89160	19666	08814	37841	12847
44	35766	31672	50082	22795	66948	65581	84393	15890
45	10853	42581	08792	13257	61973	24450	52351	16602
46	20341	27398	72906	63955	17276	10646	74692	48438
47	54458	90542	77563	51839	52901	53355	83281	19177
48	26337	66530	16687	35179	46560	00123	44546	79896
49	34314	23729	85264	05575	96855	23820	11091	79821
50	28603	10708	68933	34189	92166	15181	66628	58599

(continued)

	Column								
Row	00000 12345	00001 67890	11111 12345	11112 67890	22222 12345	22223 67890	33333 12345	33334 67890
51	66194	28926	99547	16625	45515	67953	12108	57846
52	78240	43195	24837	32511	70880	22070	52622	61881
53	00833	88000	67299	68215	11274	55624	32991	17436
54	12111	86683	61270	58036	64192	90611	15145	01748
55	47189	99951	05755	03834	43782	90599	40282	51417
56	76396	72486	62423	27618	84184	78922	73561	52818
57	46409	17469	32483	09083	76175	19985	26309	91536
58	74626	22111	87286	46772	42243	68046	44250	42439
59	34450	81974	93723	49023	58432	67083	36876	93391
60	36327	72135	33005	28701	34710	49359	50693	89311
61	74185	77536	84825	09934	99103	09325	67389	45869
62	12296	41623	62873	37943	25584	09609	63360	47270
63	90822	60280	88925	99610	42772	60561	76873	04117
64	72121	79152	96591	90305	10189	79778	68016	13747
65	95268	41377	25684	08151	61816	58555	54305	86189
66	92603	09091	75884	93424	72586	88903	30061	14457
67	18813	90291	05275	01223	79607	95426	34900	09778
68	38840	26903	28624	67157	51986	42865	14508	49315
69	05959	33836	53758	16562	41081	38012	41230	20528
70	85141	21155	99212	32685	51403	31926	69813	58781
71	75047	59643	31074	38172	03718	32119	69506	67143
72	30752	95260	68032	62871	58781	34143	68790	69766
73	22986	82575	42187	62295	84295	30634	66562	31442
74	99439	86692	90348	66036	48399	73451	26698	39437
75	20389	93029	11881	71685	65452	89047	63669	02656
76	39249	05173	68256	36539	20250	68686	05947	09335
77	96777	33605	29481	20063	09398	01843	35139	61344
78	04860	32918	10798	50492	52655	33359	94713	28393
79	41613	42375	00403	03656	77580	87772	86877	57085
80	17930	00794	53836	53692	67135	98102	61912	11246
81	24649	31845	25736	75231	83808	98917	93829	99430
82	79899	34061	54308	59358	56462	58166	97302	86828
83	76801	49594	81002	30397	52728	15101	72070	33706
84	36239	63636	38140	65731	39788	06872	38971	53363
85	07392	64449	17886	63632	53995	17574	22247	62607
86	67133	04181	33874	98835	67453	59734	76381	63455
87	77759	31504	32832	70861	15152	29733	75371	39174
88	85992	72268	42920	20810	29361	51423	90306	73574
89	79553	75952	54116	65553	47139	60579	09165	85490
90	41101	17336	48951	53674	17880	45260	08575	49321
91	36191	17095	32123	91576	84221	78902	82010	30847
92	62329	63898	23268	74283	26091	68409	69704	82267
93	14751	13151	93115	01437	56945	89661	67680	79790
94	48462	59278	44185	29616	76537	19589	83139	28454
95	29435	88105	59651	44391	74588	55114	80834	85686
96	28340	29285	12965	14821	80425	16602	44653	70467
97	02167	58940	27149	80242	10587	79786	34959	75339
98	17864	00991	39557	54981	23588	81914	37609	13128
99	79675	80605	60059	35862	00254	36546	21545	78179
00	72335	82037	92003	34100	29879	46613	89720	13274

Source: Partially extracted from The Rand Corporation, *A Million Random Digits with 100,000 Normal Deviates* (Glencoe, IL: The Free Press, 1955).

TABLE E.2 The Standardized Normal Distribution

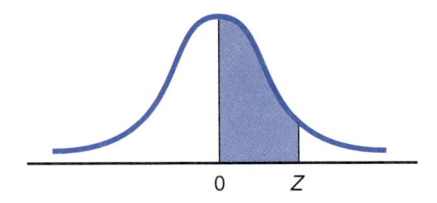

Entry represents area under the standard normal distribution from the mean to Z

Z	.00	.01	.02	.03	.04	.05	.06	.07	.08	.09
0.0	.0000	.0040	.0080	.0120	.0160	.0199	.0239	.0279	.0319	.0359
0.1	.0398	.0438	.0478	.0517	.0557	.0596	.0636	.0675	.0714	.0753
0.2	.0793	.0832	.0871	.0910	.0948	.0987	.1026	.1064	.1103	.1141
0.3	.1179	.1217	.1255	.1293	.1331	.1368	.1406	.1443	.1480	.1517
0.4	.1554	.1591	.1628	.1664	.1700	.1736	.1772	.1808	.1844	.1879
0.5	.1915	.1950	.1985	.2019	.2054	.2088	.2123	.2157	.2190	.2224
0.6	.2257	.2291	.2324	.2357	.2389	.2422	.2454	.2486	.2518	.2549
0.7	.2580	.2612	.2642	.2673	.2704	.2734	.2764	.2794	.2823	.2852
0.8	.2881	.2910	.2939	.2967	.2995	.3023	.3051	.3078	.3106	.3133
0.9	.3159	.3186	.3212	.3238	.3264	.3289	.3315	.3340	.3365	.3389
1.0	.3413	.3438	.3461	.3485	.3508	.3531	.3554	.3577	.3599	.3621
1.1	.3643	.3665	.3686	.3708	.3729	.3749	.3770	.3790	.3810	.3830
1.2	.3849	.3869	.3888	.3907	.3925	.3944	.3962	.3980	.3997	.4015
1.3	.4032	.4049	.4066	.4082	.4099	.4115	.4131	.4147	.4162	.4177
1.4	.4192	.4207	.4222	.4236	.4251	.4265	.4279	.4292	.4306	.4319
1.5	.4332	.4345	.4357	.4370	.4382	.4394	.4406	.4418	.4429	.4441
1.6	.4452	.4463	.4474	.4484	.4495	.4505	.4515	.4525	.4535	.4545
1.7	.4554	.4564	.4573	.4582	.4591	.4599	.4608	.4616	.4625	.4633
1.8	.4641	.4649	.4656	.4664	.4671	.4678	.4686	.4693	.4699	.4706
1.9	.4713	.4719	.4726	.4732	.4738	.4744	.4750	.4756	.4761	.4767
2.0	.4772	.4778	.4783	.4788	.4793	.4798	.4803	.4808	.4812	.4817
2.1	.4821	.4826	.4830	.4834	.4838	.4842	.4846	.4850	.4854	.4857
2.2	.4861	.4864	.4868	.4871	.4875	.4878	.4881	.4884	.4887	.4890
2.3	.4893	.4896	.4898	.4901	.4904	.4906	.4909	.4911	.4913	.4916
2.4	.4918	.4920	.4922	.4925	.4927	.4929	.4931	.4932	.4934	.4936
2.5	.4938	.4940	.4941	.4943	.4945	.4946	.4948	.4949	.4951	.4952
2.6	.4953	.4955	.4956	.4957	.4959	.4960	.4961	.4962	.4963	.4964
2.7	.4965	.4966	.4967	.4968	.4969	.4970	.4971	.4972	.4973	.4974
2.8	.4974	.4975	.4976	.4977	.4977	.4978	.4979	.4979	.4980	.4981
2.9	.4981	.4982	.4982	.4983	.4984	.4984	.4985	.4985	.4986	.4986
3.0	.49865	.49869	.49874	.49878	.49882	.49886	.49889	.49893	.49897	.49900
3.1	.49903	.49906	.49910	.49913	.49916	.49918	.49921	.49924	.49926	.49929
3.2	.49931	.49934	.49936	.49938	.49940	.49942	.49944	.49946	.49948	.49950
3.3	.49952	.49953	.49955	.49957	.49958	.49960	.49961	.49962	.49964	.49965
3.4	.49966	.49968	.49969	.49970	.49971	.49972	.49973	.49974	.49975	.49976
3.5	.49977	.49978	.49978	.49979	.49980	.49981	.49981	.49982	.49983	.49983
3.6	.49984	.49985	.49985	.49986	.49986	.49987	.49987	.49988	.49988	.49989
3.7	.49989	.49990	.49990	.49990	.49991	.49991	.49992	.49992	.49992	.49992
3.8	.49993	.49993	.49993	.49994	.49994	.49994	.49994	.49995	.49995	.49995
3.9	.49995	.49995	.49996	.49996	.49996	.49996	.49996	.49996	.49997	.49997

TABLE E.3 Critical Values of t

For a particular number of degrees of freedom, entry represents the critical value of t corresponding to a specified upper tail area (α).

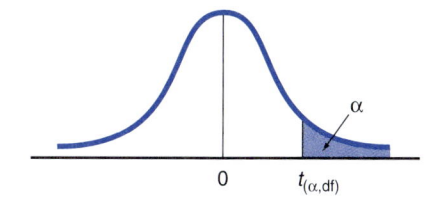

Degrees of Freedom	.25	.10	.05	.025	.01	.005
1	1.0000	3.0777	6.3138	12.7062	31.8207	63.6574
2	0.8165	1.8856	2.9200	4.3027	6.9646	9.9248
3	0.7649	1.6377	2.3534	3.1824	4.5407	5.8409
4	0.7407	1.5332	2.1318	2.7764	3.7469	4.6041
5	0.7267	1.4759	2.0150	2.5706	3.3649	4.0322
6	0.7176	1.4398	1.9432	2.4469	3.1427	3.7074
7	0.7111	1.4149	1.8946	2.3646	2.9980	3.4995
8	0.7064	1.3968	1.8595	2.3060	2.8965	3.3554
9	0.7027	1.3830	1.8331	2.2622	2.8214	3.2498
10	0.6998	1.3722	1.8125	2.2281	2.7638	3.1693
11	0.6974	1.3634	1.7959	2.2010	2.7181	3.1058
12	0.6955	1.3562	1.7823	2.1788	2.6810	3.0545
13	0.6938	1.3502	1.7709	2.1604	2.6503	3.0123
14	0.6924	1.3450	1.7613	2.1448	2.6245	2.9768
15	0.6912	1.3406	1.7531	2.1315	2.6025	2.9467
16	0.6901	1.3368	1.7459	2.1199	2.5835	2.9208
17	0.6892	1.3334	1.7396	2.1098	2.5669	2.8982
18	0.6884	1.3304	1.7341	2.1009	2.5524	2.8784
19	0.6876	1.3277	1.7291	2.0930	2.5395	2.8609
20	0.6870	1.3253	1.7247	2.0860	2.5280	2.8453
21	0.6864	1.3232	1.7207	2.0796	2.5177	2.8314
22	0.6858	1.3212	1.7171	2.0739	2.5083	2.8188
23	0.6853	1.3195	1.7139	2.0687	2.4999	2.8073
24	0.6848	1.3178	1.7109	2.0639	2.4922	2.7969
25	0.6844	1.3163	1.7081	2.0595	2.4851	2.7874
26	0.6840	1.3150	1.7056	2.0555	2.4786	2.7787
27	0.6837	1.3137	1.7033	2.0518	2.4727	2.7707
28	0.6834	1.3125	1.7011	2.0484	2.4671	2.7633
29	0.6830	1.3114	1.6991	2.0452	2.4620	2.7564
30	0.6828	1.3104	1.6973	2.0423	2.4573	2.7500
31	0.6825	1.3095	1.6955	2.0395	2.4528	2.7440
32	0.6822	1.3086	1.6939	2.0369	2.4487	2.7385
33	0.6820	1.3077	1.6924	2.0345	2.4448	2.7333
34	0.6818	1.3070	1.6909	2.0322	2.4411	2.7284
35	0.6816	1.3062	1.6896	2.0301	2.4377	2.7238
36	0.6814	1.3055	1.6883	2.0281	2.4345	2.7195
37	0.6812	1.3049	1.6871	2.0262	2.4314	2.7154
38	0.6810	1.3042	1.6860	2.0244	2.4286	2.7116
39	0.6808	1.3036	1.6849	2.0227	2.4258	2.7079
40	0.6807	1.3031	1.6839	2.0211	2.4233	2.7045
41	0.6805	1.3025	1.6829	2.0195	2.4208	2.7012
42	0.6804	1.3020	1.6820	2.0181	2.4185	2.6981
43	0.6802	1.3016	1.6811	2.0167	2.4163	2.6951
44	0.6801	1.3011	1.6802	2.0154	2.4141	2.6923
45	0.6800	1.3006	1.6794	2.0141	2.4121	2.6896
46	0.6799	1.3002	1.6787	2.0129	2.4102	2.6870
47	0.6797	1.2998	1.6779	2.0117	2.4083	2.6846
48	0.6796	1.2994	1.6772	2.0106	2.4066	2.6822
49	0.6795	1.2991	1.6766	2.0096	2.4049	2.6800
50	0.6794	1.2987	1.6759	2.0086	2.4033	2.6778

(continued)

Appendix E Tables

Degrees of Freedom	Upper Tail Areas (α)					
	.25	.10	.05	.025	.01	.005
51	0.6793	1.2984	1.6753	2.0076	2.4017	2.6757
52	0.6792	1.2980	1.6747	2.0066	2.4002	2.6737
53	0.6791	1.2977	1.6741	2.0057	2.3988	2.6718
54	0.6791	1.2974	1.6736	2.0049	2.3974	2.6700
55	0.6790	1.2971	1.6730	2.0040	2.3961	2.6682
56	0.6789	1.2969	1.6725	2.0032	2.3948	2.6665
57	0.6788	1.2966	1.6720	2.0025	2.3936	2.6649
58	0.6787	1.2963	1.6716	2.0017	2.3924	2.6633
59	0.6787	1.2961	1.6711	2.0010	2.3912	2.6618
60	0.6786	1.2958	1.6706	2.0003	2.3901	2.6603
61	0.6785	1.2956	1.6702	1.9996	2.3890	2.6589
62	0.6785	1.2954	1.6698	1.9990	2.3880	2.6575
63	0.6784	1.2591	1.6694	1.9983	2.3870	2.6561
64	0.6783	1.2949	1.6690	1.9977	2.3860	2.6549
65	0.6783	1.2947	1.6686	1.9971	2.3851	2.6536
66	0.6782	1.2945	1.6683	1.9966	2.3842	2.6524
67	0.6782	1.2943	1.6679	1.9960	2.3833	2.6512
68	0.6781	1.2941	1.6676	1.9955	2.3824	2.6501
69	0.6781	1.2939	1.6672	1.9949	2.3816	2.6490
70	0.6780	1.2938	1.6669	1.9944	2.3808	2.6479
71	0.6780	1.2936	1.6666	1.9939	2.3800	2.6469
72	0.6779	1.2934	1.6663	1.9935	2.3793	2.6459
73	0.6779	1.2933	1.6660	1.9930	2.3785	2.6449
74	0.6778	1.4931	1.6657	1.9925	2.3778	2.6439
75	0.6778	1.2929	1.6654	1.9921	2.3771	2.6430
76	0.6777	1.2928	1.6652	1.9917	2.3764	2.6421
77	0.6777	1.2926	1.6649	1.9913	2.3758	2.6412
78	0.6776	1.2925	1.6646	1.9908	2.3751	2.6403
79	0.6776	1.2924	1.6644	1.9905	2.3745	2.6395
80	0.6776	1.2922	1.6641	1.9901	2.3739	2.6387
81	0.6775	1.2921	1.6639	1.9897	2.3733	2.6379
82	0.6775	1.2920	1.6636	1.9893	2.3727	2.6371
83	0.6775	1.2918	1.6634	1.9890	2.3721	2.6364
84	0.6774	1.2917	1.6632	1.9886	2.3716	2.6356
85	0.6774	1.2916	1.6630	1.9883	2.3710	2.6349
86	0.6774	1.2915	1.6628	1.9879	2.3705	2.6342
87	0.6773	1.2914	1.6626	1.9876	2.3700	2.6335
88	0.6773	1.2912	1.6624	1.9873	2.3695	2.6329
89	0.6773	1.2911	1.6622	1.9870	2.3690	2.6322
90	0.6772	1.2910	1.6620	1.9867	2.3685	2.6316
91	0.6772	1.2909	1.6618	1.9864	2.3680	2.6309
92	0.6772	1.2908	1.6616	1.9861	2.3676	2.6303
93	0.6771	1.2907	1.6614	1.9858	2.3671	2.6297
94	0.6771	1.2906	1.6612	1.9855	2.3667	2.6291
95	0.6771	1.2905	1.6611	1.9853	2.3662	2.6286
96	0.6771	1.2904	1.6609	1.9850	2.3658	2.6280
97	0.6770	1.2903	1.6607	1.9847	2.3654	2.6275
98	0.6770	1.2902	1.6606	1.9845	2.3650	2.6269
99	0.6770	1.2902	1.6604	1.9842	2.3646	2.6264
100	0.6770	1.2901	1.6602	1.9840	2.3642	2.6259
110	0.6767	1.2893	1.6588	1.9818	2.3607	2.6213
120	0.6765	1.2886	1.6577	1.9799	2.3578	2.6174
∞	0.6745	1.2816	1.6449	1.9600	2.3263	2.5758

TABLE E.4 Critical Values of χ^2

For a particular number of degrees of freedom, entry represents the critical value of χ^2 corresponding to a specified upper tail area, (α).

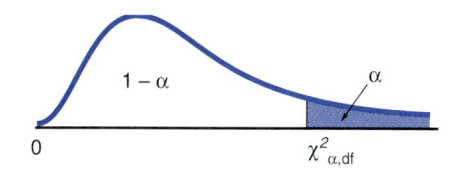

Upper Tail Areas α

Degrees of Freedom	.995	.99	.975	.95	.90	.75	.25	.10	.05	.025	.01	.005
1			0.001	0.004	0.016	0.102	1.323	2.706	3.841	5.024	6.635	7.879
2	0.010	0.020	0.051	0.103	0.211	0.575	2.773	4.605	5.991	7.378	9.210	10.597
3	0.072	0.115	0.216	0.352	0.584	1.213	4.108	6.251	7.815	9.348	11.345	12.838
4	0.207	0.297	0.484	0.711	1.064	1.923	5.385	7.779	9.488	11.143	13.277	14.860
5	0.412	0.554	0.831	1.145	1.610	2.675	6.626	9.236	11.071	12.833	15.086	16.750
6	0.676	0.872	1.237	1.635	2.204	3.455	7.841	10.645	12.592	14.449	16.812	18.548
7	0.989	1.239	1.690	2.167	2.833	4.255	9.037	12.017	14.067	16.013	18.475	20.278
8	1.344	1.646	2.180	2.733	3.490	5.071	10.219	13.362	15.507	17.535	20.090	21.955
9	1.735	2.088	2.700	3.325	4.168	5.899	11.389	14.684	16.919	19.023	21.666	23.589
10	2.156	2.558	3.247	3.940	4.865	6.737	12.549	15.987	18.307	20.483	23.209	25.188
11	2.603	3.053	3.816	4.575	5.578	7.584	13.701	17.275	19.675	21.920	24.725	26.757
12	3.074	3.571	4.404	5.226	6.304	8.438	14.845	18.549	21.026	23.337	26.217	28.299
13	3.565	4.107	5.009	5.892	7.042	9.299	15.984	19.812	22.362	24.736	27.688	29.819
14	4.075	4.660	5.629	6.571	7.790	10.165	17.117	21.064	23.685	26.119	29.141	31.319
15	4.601	5.229	6.262	7.261	8.547	11.037	18.245	22.307	24.996	27.488	30.578	32.801
16	5.142	5.812	6.908	7.962	9.312	11.912	19.369	23.542	26.296	28.845	32.000	34.267
17	5.697	6.408	7.564	8.672	10.085	12.792	20.489	24.769	27.587	30.191	33.409	35.718
18	6.265	7.015	8.231	9.390	10.865	13.675	21.605	25.989	28.869	31.526	34.805	37.156
19	6.844	7.633	8.907	10.117	11.651	14.562	22.718	27.204	30.144	32.852	36.191	38.582
20	7.434	8.260	9.591	10.851	12.443	15.452	23.828	28.412	31.410	34.170	37.566	39.997
21	8.034	8.897	10.283	11.591	13.240	16.344	24.935	29.615	32.671	35.479	38.932	41.401
22	8.643	9.542	10.982	12.338	14.042	17.240	26.039	30.813	33.924	36.781	40.289	42.796
23	9.260	10.196	11.689	13.091	14.848	18.137	27.141	32.007	35.172	38.076	41.638	44.181
24	9.886	10.856	12.401	13.848	15.659	19.037	28.241	33.196	36.415	39.364	42.980	45.559
25	10.520	11.524	13.120	14.611	16.473	19.939	29.339	34.382	37.652	40.646	44.314	46.928
26	11.160	12.198	13.844	15.379	17.292	20.843	30.435	35.563	38.885	41.923	45.642	48.290
27	11.808	12.879	14.573	16.151	18.114	21.749	31.528	36.741	40.113	43.194	46.963	49.645
28	12.461	13.565	15.308	16.928	18.939	22.657	32.620	37.916	41.337	44.461	48.278	50.993
29	13.121	14.257	16.047	17.708	19.768	23.567	33.711	39.087	42.557	45.722	49.588	52.336
30	13.787	14.954	16.791	18.493	20.599	24.478	34.800	40.256	43.773	46.979	50.892	53.672

For larger values of degrees of freedom (df) the expression $Z = \sqrt{2\chi^2} - \sqrt{2(df) - 1}$ may be used and the resulting upper tail area can be obtained from the table of the standardized normal distribution (Table E.2).

TABLE E.5 Critical Values of F

For a particular combination of numerator and denominator degrees of freedom, entry represents the critical values of F corresponding to a specified upper tail area (α).

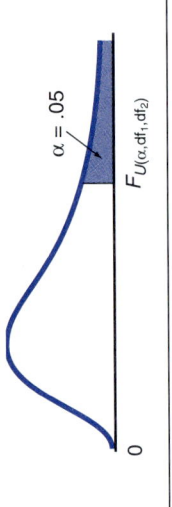

$\alpha = .05$

$F_{U(\alpha, df_1, df_2)}$

Numerator df_1

Denominator df_2	1	2	3	4	5	6	7	8	9	10	12	15	20	24	30	40	60	120	∞
1	161.4	199.5	215.7	224.6	230.2	234.0	236.8	238.9	240.5	241.9	243.9	245.9	248.0	249.1	250.1	251.1	252.2	253.3	254.3
2	18.51	19.00	19.16	19.25	19.30	19.33	19.35	19.37	19.38	19.40	19.41	19.43	19.45	19.45	19.46	19.47	19.48	19.49	19.50
3	10.13	9.55	9.28	9.12	9.01	8.94	8.89	8.85	8.81	8.79	8.74	8.70	8.66	8.64	8.62	8.59	8.57	8.55	8.53
4	7.71	6.94	6.59	6.39	6.26	6.16	6.09	6.04	6.00	5.96	5.91	5.86	5.80	5.77	5.75	5.72	5.69	5.66	5.63
5	6.61	5.79	5.41	5.19	5.05	4.95	4.88	4.82	4.77	4.74	4.68	4.62	4.56	4.53	4.50	4.46	4.43	4.40	4.36
6	5.99	5.14	4.76	4.53	4.39	4.28	4.21	4.15	4.10	4.06	4.00	3.94	3.87	3.84	3.81	3.77	3.74	3.70	3.67
7	5.59	4.74	4.35	4.12	3.97	3.87	3.79	3.73	3.68	3.64	3.57	3.51	3.44	3.41	3.38	3.34	3.30	3.27	3.23
8	5.32	4.46	4.07	3.84	3.69	3.58	3.50	3.44	3.39	3.35	3.28	3.22	3.15	3.12	3.08	3.04	3.01	2.97	2.93
9	5.12	4.26	3.86	3.63	3.48	3.37	3.29	3.23	3.18	3.14	3.07	3.01	2.94	2.90	2.86	2.83	2.79	2.75	2.71
10	4.96	4.10	3.71	3.48	3.33	3.22	3.14	3.07	3.02	2.98	2.91	2.85	2.77	2.74	2.70	2.66	2.62	2.58	2.54
11	4.84	3.98	3.59	3.36	3.20	3.09	3.01	2.95	2.90	2.85	2.79	2.72	2.65	2.61	2.57	2.53	2.49	2.45	2.40
12	4.75	3.89	3.49	3.26	3.11	3.00	2.91	2.85	2.80	2.75	2.69	2.62	2.54	2.51	2.47	2.43	2.38	2.34	2.30
13	4.67	3.81	3.41	3.18	3.03	2.92	2.83	2.77	2.71	2.67	2.60	2.53	2.46	2.42	2.38	2.34	2.30	2.25	2.21
14	4.60	3.74	3.34	3.11	2.96	2.85	2.76	2.70	2.65	2.60	2.53	2.46	2.39	2.35	2.31	2.27	2.22	2.18	2.13
15	4.54	3.68	3.29	3.06	2.90	2.79	2.71	2.64	2.59	2.54	2.48	2.40	2.33	2.29	2.25	2.20	2.16	2.11	2.07
16	4.49	3.63	3.24	3.01	2.85	2.74	2.66	2.59	2.54	2.49	2.42	2.35	2.28	2.24	2.19	2.15	2.11	2.06	2.01
17	4.45	3.59	3.20	2.96	2.81	2.70	2.61	2.55	2.49	2.45	2.38	2.31	2.23	2.19	2.15	2.10	2.06	2.01	1.96
18	4.41	3.55	3.16	2.93	2.77	2.66	2.58	2.51	2.46	2.41	2.34	2.27	2.19	2.15	2.11	2.06	2.02	1.97	1.92
19	4.38	3.52	3.13	2.90	2.74	2.63	2.54	2.48	2.42	2.38	2.31	2.23	2.16	2.11	2.07	2.03	1.98	1.93	1.88
20	4.35	3.49	3.10	2.87	2.71	2.60	2.51	2.45	2.39	2.35	2.28	2.20	2.12	2.08	2.04	1.99	1.95	1.90	1.84
21	4.32	3.47	3.07	2.84	2.68	2.57	2.49	2.42	2.37	2.32	2.25	2.18	2.10	2.05	2.01	1.96	1.92	1.87	1.81
22	4.30	3.44	3.05	2.82	2.66	2.55	2.46	2.40	2.34	2.30	2.23	2.15	2.07	2.03	1.98	1.94	1.89	1.84	1.78
23	4.28	3.42	3.03	2.80	2.64	2.53	2.44	2.37	2.32	2.27	2.20	2.13	2.05	2.01	1.96	1.91	1.86	1.81	1.76
24	4.26	3.40	3.01	2.78	2.62	2.51	2.42	2.36	2.30	2.25	2.18	2.11	2.03	1.98	1.94	1.89	1.84	1.79	1.73
25	4.24	3.39	2.99	2.76	2.60	2.49	2.40	2.34	2.28	2.24	2.16	2.09	2.01	1.96	1.92	1.87	1.82	1.77	1.71
26	4.23	3.37	2.98	2.74	2.59	2.47	2.39	2.32	2.27	2.22	2.15	2.07	1.99	1.95	1.90	1.85	1.80	1.75	1.69
27	4.21	3.35	2.96	2.73	2.57	2.46	2.37	2.31	2.25	2.20	2.13	2.06	1.97	1.93	1.88	1.84	1.79	1.73	1.67
28	4.20	3.34	2.95	2.71	2.56	2.45	2.36	2.29	2.24	2.19	2.12	2.04	1.96	1.91	1.87	1.82	1.77	1.71	1.65
29	4.18	3.33	2.93	2.70	2.55	2.43	2.35	2.28	2.22	2.18	2.10	2.03	1.94	1.90	1.85	1.81	1.75	1.70	1.64
30	4.17	3.32	2.92	2.69	2.53	2.42	2.33	2.27	2.21	2.16	2.09	2.01	1.93	1.89	1.84	1.79	1.74	1.68	1.62
40	4.08	3.23	2.84	2.61	2.45	2.34	2.25	2.18	2.12	2.08	2.00	1.92	1.84	1.79	1.74	1.69	1.64	1.58	1.51
60	4.00	3.15	2.76	2.53	2.37	2.25	2.17	2.10	2.04	1.99	1.92	1.84	1.75	1.70	1.65	1.59	1.53	1.47	1.39
120	3.92	3.07	2.68	2.45	2.29	2.17	2.09	2.02	1.96	1.91	1.83	1.75	1.66	1.61	1.55	1.50	1.43	1.35	1.25
∞	3.84	3.00	2.60	2.37	2.21	2.10	2.01	1.94	1.88	1.83	1.75	1.67	1.57	1.52	1.46	1.39	1.32	1.22	1.00

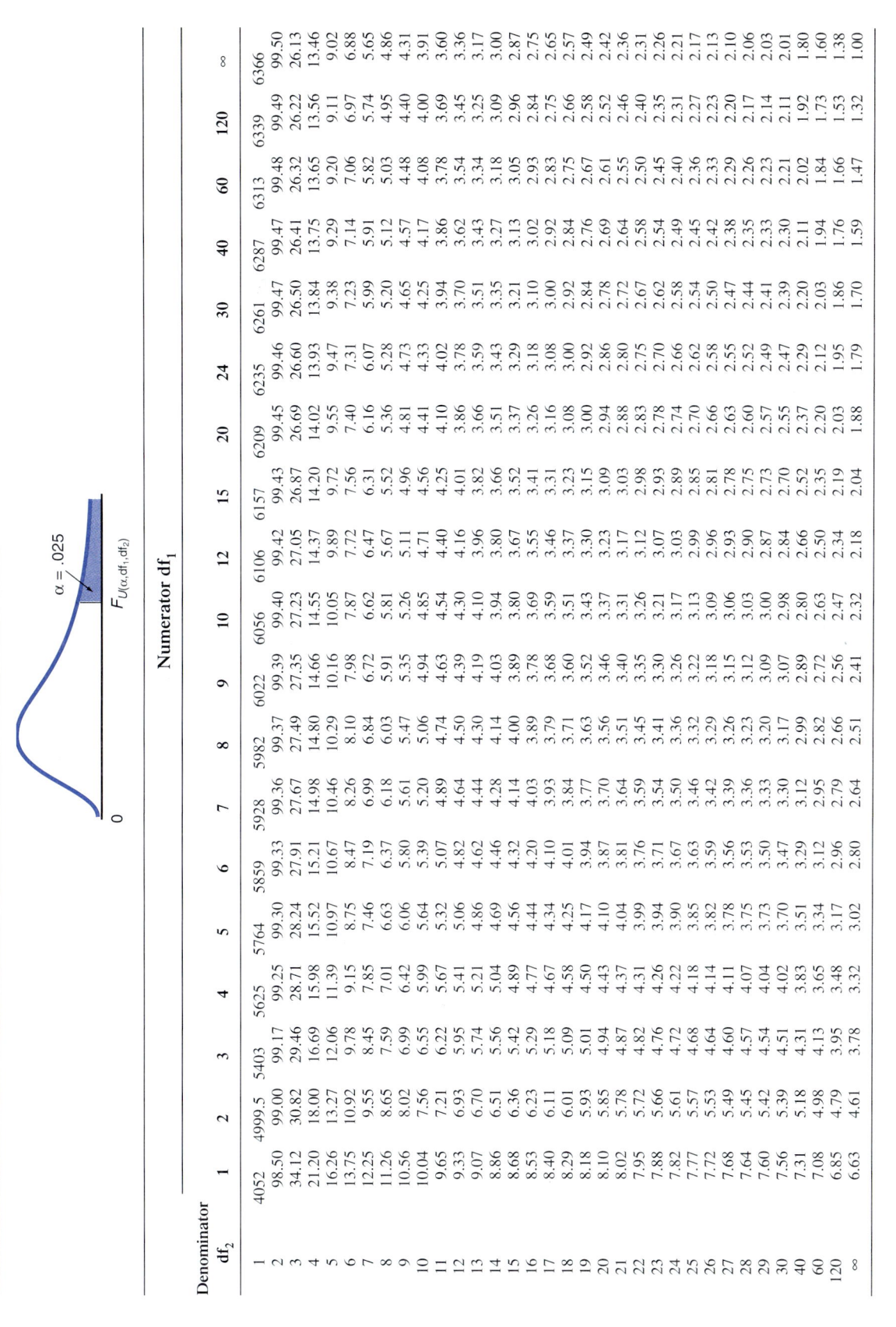

TABLE E.5 (Continued)

$\alpha = .025$

$F_{U(\alpha, df_1, df_2)}$

Denominator df_2	\multicolumn{20}{c}{Numerator df_1}
	1	2	3	4	5	6	7	8	9	10	12	15	20	24	30	40	60	120	∞
1	4052	4999.5	5403	5625	5764	5859	5928	5982	6022	6056	6106	6157	6209	6235	6261	6287	6313	6339	6366
2	98.50	99.00	99.17	99.25	99.30	99.33	99.36	99.37	99.39	99.40	99.42	99.43	99.45	99.46	99.47	99.47	99.48	99.49	99.50
3	34.12	30.82	29.46	28.71	28.24	27.91	27.67	27.49	27.35	27.23	27.05	26.87	26.69	26.60	26.50	26.41	26.32	26.22	26.13
4	21.20	18.00	16.69	15.98	15.52	15.21	14.98	14.80	14.66	14.55	14.37	14.20	14.02	13.93	13.84	13.75	13.65	13.56	13.46
5	16.26	13.27	12.06	11.39	10.97	10.67	10.46	10.29	10.16	10.05	9.89	9.72	9.55	9.47	9.38	9.29	9.20	9.11	9.02
6	13.75	10.92	9.78	9.15	8.75	8.47	8.26	8.10	7.98	7.87	7.72	7.56	7.40	7.31	7.23	7.14	7.06	6.97	6.88
7	12.25	9.55	8.45	7.85	7.46	7.19	6.99	6.84	6.72	6.62	6.47	6.31	6.16	6.07	5.99	5.91	5.82	5.74	5.65
8	11.26	8.65	7.59	7.01	6.63	6.37	6.18	6.03	5.91	5.81	5.67	5.52	5.36	5.28	5.20	5.12	5.03	4.95	4.86
9	10.56	8.02	6.99	6.42	6.06	5.80	5.61	5.47	5.35	5.26	5.11	4.96	4.81	4.73	4.65	4.57	4.48	4.40	4.31
10	10.04	7.56	6.55	5.99	5.64	5.39	5.20	5.06	4.94	4.85	4.71	4.56	4.41	4.33	4.25	4.17	4.08	4.00	3.91
11	9.65	7.21	6.22	5.67	5.32	5.07	4.89	4.74	4.63	4.54	4.40	4.25	4.10	4.02	3.94	3.86	3.78	3.69	3.60
12	9.33	6.93	5.95	5.41	5.06	4.82	4.64	4.50	4.39	4.30	4.16	4.01	3.86	3.78	3.70	3.62	3.54	3.45	3.36
13	9.07	6.70	5.74	5.21	4.86	4.62	4.44	4.30	4.19	4.10	3.96	3.82	3.66	3.59	3.51	3.43	3.34	3.25	3.17
14	8.86	6.51	5.56	5.04	4.69	4.46	4.28	4.14	4.03	3.94	3.80	3.66	3.51	3.43	3.35	3.27	3.18	3.09	3.00
15	8.68	6.36	5.42	4.89	4.56	4.32	4.14	4.00	3.89	3.80	3.67	3.52	3.37	3.29	3.21	3.13	3.05	2.96	2.87
16	8.53	6.23	5.29	4.77	4.44	4.20	4.03	3.89	3.78	3.69	3.55	3.41	3.26	3.18	3.10	3.02	2.93	2.84	2.75
17	8.40	6.11	5.18	4.67	4.34	4.10	3.93	3.79	3.68	3.59	3.46	3.31	3.16	3.08	3.00	2.92	2.83	2.75	2.65
18	8.29	6.01	5.09	4.58	4.25	4.01	3.84	3.71	3.60	3.51	3.37	3.23	3.08	3.00	2.92	2.84	2.75	2.66	2.57
19	8.18	5.93	5.01	4.50	4.17	3.94	3.77	3.63	3.52	3.43	3.30	3.15	3.00	2.92	2.84	2.76	2.67	2.58	2.49
20	8.10	5.85	4.94	4.43	4.10	3.87	3.70	3.56	3.46	3.37	3.23	3.09	2.94	2.86	2.78	2.69	2.61	2.52	2.42
21	8.02	5.78	4.87	4.37	4.04	3.81	3.64	3.51	3.40	3.31	3.17	3.03	2.88	2.80	2.72	2.64	2.55	2.46	2.36
22	7.95	5.72	4.82	4.31	3.99	3.76	3.59	3.45	3.35	3.26	3.12	2.98	2.83	2.75	2.67	2.58	2.50	2.40	2.31
23	7.88	5.66	4.76	4.26	3.94	3.71	3.54	3.41	3.30	3.21	3.07	2.93	2.78	2.70	2.62	2.54	2.45	2.35	2.26
24	7.82	5.61	4.72	4.22	3.90	3.67	3.50	3.36	3.26	3.17	3.03	2.89	2.74	2.66	2.58	2.49	2.40	2.31	2.21
25	7.77	5.57	4.68	4.18	3.85	3.63	3.46	3.32	3.22	3.13	2.99	2.85	2.70	2.62	2.54	2.45	2.36	2.27	2.17
26	7.72	5.53	4.64	4.14	3.82	3.59	3.42	3.29	3.18	3.09	2.96	2.81	2.66	2.58	2.50	2.42	2.33	2.23	2.13
27	7.68	5.49	4.60	4.11	3.78	3.56	3.39	3.26	3.15	3.06	2.93	2.78	2.63	2.55	2.47	2.38	2.29	2.20	2.10
28	7.64	5.45	4.57	4.07	3.75	3.53	3.36	3.23	3.12	3.03	2.90	2.75	2.60	2.52	2.44	2.35	2.26	2.17	2.06
29	7.60	5.42	4.54	4.04	3.73	3.50	3.33	3.20	3.09	3.00	2.87	2.73	2.57	2.49	2.41	2.33	2.23	2.14	2.03
30	7.56	5.39	4.51	4.02	3.70	3.47	3.30	3.17	3.07	2.98	2.84	2.70	2.55	2.47	2.39	2.30	2.21	2.11	2.01
40	7.31	5.18	4.31	3.83	3.51	3.29	3.12	2.99	2.89	2.80	2.66	2.52	2.37	2.29	2.20	2.11	2.02	1.92	1.80
60	7.08	4.98	4.13	3.65	3.34	3.12	2.95	2.82	2.72	2.63	2.50	2.35	2.20	2.12	2.03	1.94	1.84	1.73	1.60
120	6.85	4.79	3.95	3.48	3.17	2.96	2.79	2.66	2.56	2.47	2.34	2.19	2.03	1.95	1.86	1.76	1.66	1.53	1.38
∞	6.63	4.61	3.78	3.32	3.02	2.80	2.64	2.51	2.41	2.32	2.18	2.04	1.88	1.79	1.70	1.59	1.47	1.32	1.00

TABLE E.5 (Continued)

$\alpha = .01$

$F_{U(\alpha, df_1, df_2)}$

Numerator df_1

Denominator df_2	1	2	3	4	5	6	7	8	9	10	12	15	20	24	30	40	60	120	∞
1	4052	4999.5	5403	5625	5764	5859	5928	5982	6022	6056	6106	6157	6209	6235	6261	6287	6313	6339	6366
2	98.50	99.00	99.17	99.25	99.30	99.33	99.36	99.37	99.39	99.40	99.42	99.43	99.45	99.46	99.47	99.47	99.48	99.49	99.50
3	34.12	30.82	29.46	28.71	28.24	27.91	27.67	27.49	27.35	27.23	27.05	26.87	26.69	26.60	26.50	26.41	26.32	26.22	26.13
4	21.20	18.00	16.69	15.98	15.52	15.21	14.98	14.80	14.66	14.55	14.37	14.20	14.02	13.93	13.84	13.75	13.65	13.56	13.46
5	16.26	13.27	12.06	11.39	10.97	10.67	10.46	10.29	10.16	10.05	9.89	9.72	9.55	9.47	9.38	9.29	9.20	9.11	9.02
6	13.75	10.92	9.78	9.15	8.75	8.47	8.26	8.10	7.98	7.87	7.72	7.56	7.40	7.31	7.23	7.14	7.06	6.97	6.88
7	12.25	9.55	8.45	7.85	7.46	7.19	6.99	6.84	6.72	6.62	6.47	6.31	6.16	6.07	5.99	5.91	5.82	5.74	5.65
8	11.26	8.65	7.59	7.01	6.63	6.37	6.18	6.03	5.91	5.81	5.67	5.52	5.36	5.28	5.20	5.12	5.03	4.95	4.86
9	10.56	8.02	6.99	6.42	6.06	5.80	5.61	5.47	5.35	5.26	5.11	4.96	4.81	4.73	4.65	4.57	4.48	4.40	4.31
10	10.04	7.56	6.55	5.99	5.64	5.39	5.20	5.06	4.94	4.85	4.71	4.56	4.41	4.33	4.25	4.17	4.08	4.00	3.91
11	9.65	7.21	6.22	5.67	5.32	5.07	4.89	4.74	4.63	4.54	4.40	4.25	4.10	4.02	3.94	3.86	3.78	3.69	3.60
12	9.33	6.93	5.95	5.41	5.06	4.82	4.64	4.50	4.39	4.30	4.16	4.01	3.86	3.78	3.70	3.62	3.54	3.45	3.36
13	9.07	6.70	5.74	5.21	4.86	4.62	4.44	4.30	4.19	4.10	3.96	3.82	3.66	3.59	3.51	3.43	3.34	3.25	3.17
14	8.86	6.51	5.56	5.04	4.69	4.46	4.28	4.14	4.03	3.94	3.80	3.66	3.51	3.43	3.35	3.27	3.18	3.09	3.00
15	8.68	6.36	5.42	4.89	4.56	4.32	4.14	4.00	3.89	3.80	3.67	3.52	3.37	3.29	3.21	3.13	3.05	2.96	2.87
16	8.53	6.23	5.29	4.77	4.44	4.20	4.03	3.89	3.78	3.69	3.55	3.41	3.26	3.18	3.10	3.02	2.93	2.84	2.75
17	8.40	6.11	5.18	4.67	4.34	4.10	3.93	3.79	3.68	3.59	3.46	3.31	3.16	3.08	3.00	2.92	2.83	2.75	2.65
18	8.29	6.01	5.09	4.58	4.25	4.01	3.84	3.71	3.60	3.51	3.37	3.23	3.08	3.00	2.92	2.84	2.75	2.66	2.57
19	8.18	5.93	5.01	4.50	4.17	3.94	3.77	3.63	3.52	3.43	3.30	3.15	3.00	2.92	2.84	2.76	2.67	2.58	2.49
20	8.10	5.85	4.94	4.43	4.10	3.87	3.70	3.56	3.46	3.37	3.23	3.09	2.94	2.86	2.78	2.69	2.61	2.52	2.42
21	8.02	5.78	4.87	4.37	4.04	3.81	3.64	3.51	3.40	3.31	3.17	3.03	2.88	2.80	2.72	2.64	2.55	2.46	2.36
22	7.95	5.72	4.82	4.31	3.99	3.76	3.59	3.45	3.35	3.26	3.12	2.98	2.83	2.75	2.67	2.58	2.50	2.40	2.31
23	7.88	5.66	4.76	4.26	3.94	3.71	3.54	3.41	3.30	3.21	3.07	2.93	2.78	2.70	2.62	2.54	2.45	2.35	2.26
24	7.82	5.61	4.72	4.22	3.90	3.67	3.50	3.36	3.26	3.17	3.03	2.89	2.74	2.66	2.58	2.49	2.40	2.31	2.21
25	7.77	5.57	4.68	4.18	3.85	3.63	3.46	3.32	3.22	3.13	2.99	2.85	2.70	2.62	2.54	2.45	2.36	2.27	2.17
26	7.72	5.53	4.64	4.14	3.82	3.59	3.42	3.29	3.18	3.09	2.96	2.81	2.66	2.58	2.50	2.42	2.33	2.23	2.13
27	7.68	5.49	4.60	4.11	3.78	3.56	3.39	3.26	3.15	3.06	2.93	2.78	2.63	2.55	2.47	2.38	2.29	2.20	2.10
28	7.64	5.45	4.57	4.07	3.75	3.53	3.36	3.23	3.12	3.03	2.90	2.75	2.60	2.52	2.44	2.35	2.26	2.17	2.06
29	7.60	5.42	4.54	4.04	3.73	3.50	3.33	3.20	3.09	3.00	2.87	2.73	2.57	2.49	2.41	2.33	2.23	2.14	2.03
30	7.56	5.39	4.51	4.02	3.70	3.47	3.30	3.17	3.07	2.98	2.84	2.70	2.55	2.47	2.39	2.30	2.21	2.11	2.01
40	7.31	5.18	4.31	3.83	3.51	3.29	3.12	2.99	2.89	2.80	2.66	2.52	2.37	2.29	2.20	2.11	2.02	1.92	1.80
60	7.08	4.98	4.13	3.65	3.34	3.12	2.95	2.82	2.72	2.63	2.50	2.35	2.20	2.12	2.03	1.94	1.84	1.73	1.60
120	6.85	4.79	3.95	3.48	3.17	2.96	2.79	2.66	2.56	2.47	2.34	2.19	2.03	1.95	1.86	1.76	1.66	1.53	1.38
∞	6.63	4.61	3.78	3.32	3.02	2.80	2.64	2.51	2.41	2.32	2.18	2.04	1.88	1.79	1.70	1.59	1.47	1.32	1.00

TABLE E.5 (Continued)

$F_{U(\alpha, df_1, df_2)}$, $\alpha = .005$

Denominator df_2	Numerator df_1																
	1	2	3	4	5	6	7	8	9	10	12	15	20	24	30	40	60	120	∞
1	16211	20000	21615	22500	23056	23437	23715	23925	24091	24224	24426	24630	24836	24940	25044	25148	25253	25359	25465
2	198.5	199.0	199.2	199.2	199.3	199.3	199.4	199.4	199.4	199.4	199.4	199.4	199.4	199.5	199.5	199.5	199.5	199.5	199.5
3	55.55	49.80	47.47	46.19	45.39	44.84	44.43	44.13	43.88	43.69	43.39	43.08	42.78	42.62	42.47	42.31	42.15	41.99	41.83
4	31.33	26.28	24.26	23.15	22.46	21.97	21.62	21.35	21.14	20.97	20.70	20.44	20.17	20.03	19.89	19.75	19.61	19.47	19.32
5	22.78	18.31	16.53	15.56	14.94	14.51	14.20	13.96	13.77	13.62	13.38	13.15	12.90	12.78	12.66	12.53	12.40	12.27	12.14
6	18.63	14.54	12.92	12.03	11.46	11.07	10.79	10.57	10.39	10.25	10.03	9.81	9.59	9.47	9.36	9.24	9.12	9.00	8.88
7	16.24	12.40	10.88	10.05	9.52	9.16	8.89	8.68	8.51	8.38	8.18	7.97	7.75	7.65	7.53	7.42	7.31	7.19	7.08
8	14.69	11.04	9.60	8.81	8.30	7.95	7.69	7.50	7.34	7.21	7.01	6.81	6.61	6.50	6.40	6.29	6.18	6.06	5.95
9	13.61	10.11	8.72	7.96	7.47	7.13	6.88	6.69	6.54	6.42	6.23	6.03	5.83	5.73	5.62	5.52	5.41	5.30	5.19
10	12.83	9.43	8.08	7.34	6.87	6.54	6.30	6.12	5.97	5.85	5.66	5.47	5.27	5.17	5.07	4.97	4.86	4.75	4.64
11	12.23	8.91	7.60	6.88	6.42	6.10	5.86	5.68	5.54	5.42	5.24	5.05	4.86	4.76	4.65	4.55	4.44	4.34	4.23
12	11.75	8.51	7.23	6.52	6.07	5.76	5.52	5.35	5.20	5.09	4.91	4.72	4.53	4.43	4.33	4.23	4.12	4.01	3.90
13	11.37	8.19	6.93	6.23	5.79	5.48	5.25	5.08	4.94	4.82	4.64	4.46	4.27	4.17	4.07	3.97	3.87	3.76	3.65
14	11.06	7.92	6.68	6.00	5.56	5.26	5.03	4.86	4.72	4.60	4.43	4.25	4.06	3.96	3.86	3.76	3.66	3.55	3.44
15	10.80	7.70	6.48	5.80	5.37	5.07	4.85	4.67	4.54	4.42	4.25	4.07	3.88	3.79	3.69	3.58	3.48	3.37	3.26
16	10.58	7.51	6.30	5.64	5.21	4.91	4.69	4.52	4.38	4.27	4.10	3.92	3.73	3.64	3.54	3.44	3.33	3.22	3.11
17	10.38	7.35	6.16	5.50	5.07	4.78	4.56	4.39	4.25	4.14	3.97	3.79	3.61	3.51	3.41	3.31	3.21	3.10	2.98
18	10.22	7.21	6.03	5.37	4.96	4.66	4.44	4.28	4.14	4.03	3.86	3.68	3.50	3.40	3.30	3.20	3.10	2.99	2.87
19	10.07	7.09	5.92	5.27	4.85	4.56	4.34	4.18	4.04	3.93	3.76	3.59	3.40	3.31	3.21	3.11	3.00	2.89	2.78
20	9.94	6.99	5.82	5.17	4.76	4.47	4.26	4.09	3.96	3.85	3.68	3.50	3.32	3.22	3.12	3.02	2.92	2.81	2.69
21	9.83	6.89	5.73	5.09	4.68	4.39	4.18	4.01	3.88	3.77	3.60	3.43	3.24	3.15	3.05	2.95	2.84	2.73	2.61
22	9.73	6.81	5.65	5.02	4.61	4.32	4.11	3.94	3.81	3.70	3.54	3.36	3.18	3.08	2.98	2.88	2.77	2.65	2.55
23	9.63	6.73	5.58	4.95	4.54	4.26	4.05	3.88	3.75	3.64	3.47	3.30	3.12	3.02	2.92	2.82	2.71	2.60	2.48
24	9.55	6.66	5.52	4.89	4.49	4.20	3.99	3.83	3.69	3.59	3.42	3.25	3.06	2.97	2.87	2.77	2.66	2.55	2.43
25	9.48	6.60	5.46	4.84	4.43	4.15	3.94	3.78	3.64	3.54	3.37	3.20	3.01	2.92	2.82	2.72	2.61	2.50	2.38
26	9.41	6.54	5.41	4.79	4.38	4.10	3.89	3.73	3.60	3.49	3.33	3.15	2.97	2.87	2.77	2.67	2.56	2.45	2.33
27	9.34	6.49	5.36	4.74	4.34	4.06	3.85	3.69	3.56	3.45	3.28	3.11	2.93	2.83	2.73	2.63	2.52	2.41	2.29
28	9.28	6.44	5.32	4.70	4.30	4.02	3.81	3.65	3.52	3.41	3.25	3.07	2.89	2.79	2.69	2.59	2.48	2.37	2.25
29	9.23	6.40	5.28	4.66	4.26	3.98	3.77	3.61	3.48	3.38	3.21	3.04	2.86	2.76	2.66	2.56	2.45	2.33	2.21
30	9.18	6.35	5.24	4.62	4.23	3.95	3.74	3.58	3.45	3.34	3.18	3.01	2.82	2.73	2.63	2.52	2.42	2.30	2.18
40	8.83	6.07	4.98	4.37	3.99	3.71	3.51	3.35	3.22	3.12	2.95	2.78	2.60	2.50	2.40	2.30	2.18	2.06	1.93
60	8.49	5.79	4.73	4.14	3.76	3.49	3.29	3.13	3.01	2.90	2.74	2.57	2.39	2.29	2.19	2.08	1.96	1.83	1.69
120	8.18	5.54	4.50	3.92	3.55	3.28	3.09	2.93	2.81	2.71	2.54	2.37	2.19	2.09	1.98	1.87	1.75	1.61	1.43
∞	7.88	5.30	4.28	3.72	3.35	3.09	2.90	2.74	2.62	2.52	2.36	2.19	2.00	1.90	1.79	1.67	1.53	1.36	1.00

Source: Reprinted from E. S. Pearson and H. O. Hartley, eds., *Biometrika Tables for Statisticians*, 3rd ed., 1966, by permission of the *Biometrika* Trustees.

TABLE E.6 Lower and Upper Critical Values T_1 of Wilcoxon Rank Sum Test

	α		n_1						
n_2	One-Tailed	Two-Tailed	4	5	6	7	8	9	10
4	.05	.10	11,25						
	.025	.05	10,26						
	.01	.02	—,—						
	.005	.01	—,—						
5	.05	.10	12,28	19,36					
	.025	.05	11,29	17,38					
	.01	.02	10,30	16,39					
	.005	.01	—,—	15,40					
6	.05	.10	13,31	20,40	28,50				
	.025	.05	12,32	18,42	26,52				
	.01	.02	11,33	17,43	24,54				
	.005	.01	10,34	16,44	23,55				
7	.05	.10	14,34	21,44	29,55	39,66			
	.025	.05	13,35	20,45	27,57	36,69			
	.01	.02	11,37	18,47	25,59	34,71			
	.005	.01	10,38	16,49	24,60	32,73			
8	.05	.10	15,37	23,47	31,59	41,71	51,85		
	.025	.05	14,38	21,49	29,61	38,74	49,87		
	.01	.02	12,40	19,51	27,63	35,77	45,91		
	.005	.01	11,41	17,53	25,65	34,78	43,93		
9	.05	.10	16,40	24,51	33,63	43,76	54,90	66,105	
	.025	.05	14,42	22,53	31,65	40,79	51,93	62,109	
	.01	.02	13,43	20,55	28,68	37,82	47,97	59,112	
	.005	.01	11,45	18,57	26,70	35,84	45,99	56,115	
10	.05	.10	17,43	26,54	35,67	45,81	56,96	69,111	82,128
	.025	.05	15,45	23,57	32,70	42,84	53,99	65,115	78,132
	.01	.02	13,47	21,59	29,73	39,87	49,103	61,119	74,136
	.005	.01	12,48	19,61	27,75	37,89	47,105	58,122	71,139

Source: Adapted from Table 1 of F. Wilcoxon and R. A. Wilcox, *Some Rapid Approximate Statistical Procedures* (Pearl River, N.Y., Lederle Laboratories, 1964), with permission of the American Cyanamid Company.

TABLE E.7 Critical Values[a] of the Studentized Range Q

Upper 5% points ($\alpha = .05$)

v \ η	2	3	4	5	6	7	8	9	10	11	12	13	14	15	16	17	18	19	20
1	18.0	27.0	32.8	37.1	40.4	43.1	45.4	47.4	49.1	50.6	52.0	53.2	54.3	55.4	56.3	57.2	58.0	58.8	59.6
2	6.09	8.3	9.8	10.9	11.7	12.4	13.0	13.5	14.0	14.4	14.7	15.1	15.4	15.7	15.9	16.1	16.4	16.6	16.8
3	4.50	5.91	6.82	7.50	8.04	8.48	8.85	9.18	9.46	9.72	9.95	10.15	10.35	10.52	10.69	10.84	10.98	11.11	11.24
4	3.93	5.04	5.76	6.29	6.71	7.05	7.35	7.60	7.83	8.03	8.21	8.37	8.52	8.66	8.79	8.91	9.03	9.13	9.23
5	3.64	4.60	5.22	5.67	6.03	6.33	6.58	6.80	6.99	7.17	7.32	7.47	7.60	7.72	7.83	7.93	8.03	8.12	8.21
6	3.46	4.34	4.90	5.31	5.63	5.89	6.12	6.32	6.49	6.65	6.79	6.92	7.03	7.14	7.24	7.34	7.43	7.51	7.59
7	3.34	4.16	4.68	5.06	5.36	5.61	5.82	6.00	6.16	6.30	6.43	6.55	6.66	6.76	6.85	6.94	7.02	7.09	7.17
8	3.26	4.04	4.53	4.89	5.17	5.40	5.60	5.77	5.92	6.05	6.18	6.29	6.39	6.48	6.57	6.65	6.73	6.80	6.87
9	3.20	3.95	4.42	4.76	5.02	5.24	5.43	5.60	5.74	5.87	5.98	6.09	6.19	6.28	6.36	6.44	6.51	6.58	6.64
10	3.15	3.88	4.33	4.65	4.91	5.12	5.30	5.46	5.60	5.72	5.83	5.93	6.03	6.11	6.20	6.27	6.34	6.40	6.47
11	3.11	3.82	4.26	4.57	4.82	5.03	5.20	5.35	5.49	5.61	5.71	5.81	5.90	5.99	6.06	6.14	6.20	6.26	6.33
12	3.08	3.77	4.20	4.51	4.75	4.95	5.12	5.27	5.40	5.51	5.62	5.71	5.80	5.88	5.95	6.03	6.09	6.15	6.21
13	3.06	3.73	4.15	4.45	4.69	4.88	5.05	5.19	5.32	5.43	5.53	5.63	5.71	5.79	5.86	5.93	6.00	6.05	6.11
14	3.03	3.70	4.11	4.41	4.64	4.83	4.99	5.13	5.25	5.36	5.46	5.55	5.64	5.72	5.79	5.85	5.92	5.97	6.03
15	3.01	3.67	4.08	4.37	4.60	4.78	4.94	5.08	5.20	5.31	5.40	5.49	5.58	5.65	5.72	5.79	5.85	5.90	5.96
16	3.00	3.65	4.05	4.33	4.56	4.74	4.90	5.03	5.15	5.26	5.35	5.44	5.52	5.59	5.66	5.72	5.79	5.84	5.90
17	2.98	3.63	4.02	4.30	4.52	4.71	4.86	4.99	5.11	5.21	5.31	5.39	5.47	5.55	5.61	5.68	5.74	5.79	5.84
18	2.97	3.61	4.00	4.28	4.49	4.67	4.82	4.96	5.07	5.17	5.27	5.35	5.43	5.50	5.57	5.63	5.69	5.74	5.79
19	2.96	3.59	3.98	4.25	4.47	4.65	4.79	4.92	5.04	5.14	5.23	5.32	5.39	5.46	5.53	5.59	5.65	5.70	5.75
20	2.95	3.58	3.96	4.23	4.45	4.62	4.77	4.90	5.01	5.11	5.20	5.28	5.36	5.43	5.49	5.55	5.61	5.66	5.71
24	2.92	3.53	3.90	4.17	4.37	4.54	4.68	4.81	4.92	5.01	5.10	5.18	5.25	5.32	5.38	5.44	5.50	5.54	5.59
30	2.89	3.49	3.84	4.10	4.30	4.46	4.60	4.72	4.83	4.92	5.00	5.08	5.15	5.21	5.27	5.33	5.38	5.43	5.48
40	2.86	3.44	3.79	4.04	4.23	4.39	4.52	4.63	4.74	4.82	4.91	4.98	5.05	5.11	5.16	5.22	5.27	5.31	5.36
60	2.83	3.40	3.74	3.98	4.16	4.31	4.44	4.55	4.65	4.73	4.81	4.88	4.94	5.00	5.06	5.11	5.16	5.20	5.24
120	2.80	3.36	3.69	3.92	4.10	4.24	4.36	4.48	4.56	4.64	4.72	4.78	4.84	4.90	4.95	5.00	5.05	5.09	5.13
∞	2.77	3.31	3.63	3.86	4.03	4.17	4.29	4.39	4.47	4.55	4.62	4.68	4.74	4.80	4.85	4.89	4.93	4.97	5.01

TABLE E.7 (Continued)

Upper 1% points ($\alpha = .01$)

ν \ η	2	3	4	5	6	7	8	9	10	11	12	13	14	15	16	17	18	19	20
1	90.0	135	164	186	202	216	227	237	246	253	260	266	272	277	282	286	290	294	298
2	14.0	19.0	22.3	24.7	26.6	28.2	29.5	30.7	31.7	32.6	33.4	34.1	34.8	35.4	36.0	36.5	37.0	37.5	37.9
3	8.26	10.6	12.2	13.3	14.2	15.0	15.6	16.2	16.7	17.1	17.5	17.9	18.2	18.5	18.8	19.1	19.3	19.5	19.8
4	6.51	8.12	9.17	9.96	10.6	11.1	11.5	11.9	12.3	12.6	12.8	13.1	13.3	13.5	13.7	13.9	14.1	14.2	14.4
5	5.70	6.97	7.80	8.42	8.91	9.32	9.67	9.97	10.24	10.48	10.70	10.89	11.08	11.24	11.40	11.55	11.68	11.81	11.93
6	5.24	6.33	7.03	7.56	7.97	8.32	8.61	8.87	9.10	9.30	9.49	9.65	9.81	9.95	10.08	10.21	10.32	10.43	10.54
7	4.95	5.92	6.54	7.01	7.37	7.68	7.94	8.17	8.37	8.55	8.71	8.86	9.00	9.12	9.24	9.35	9.46	9.55	9.65
8	4.74	5.63	6.20	6.63	6.96	7.24	7.47	7.68	7.87	8.03	8.18	8.31	8.44	8.55	8.66	8.76	8.85	8.94	9.03
9	4.60	5.43	5.96	6.35	6.66	6.91	7.13	7.32	7.49	7.65	7.78	7.91	8.03	8.13	8.23	8.32	8.41	8.49	8.57
10	4.48	5.27	5.77	6.14	6.43	6.67	6.87	7.05	7.21	7.36	7.48	7.60	7.71	7.81	7.91	7.99	8.07	8.15	8.22
11	4.39	5.14	5.62	5.97	6.25	6.48	6.67	6.84	6.99	7.13	7.25	7.36	7.46	7.56	7.65	7.73	7.81	7.88	7.95
12	4.32	5.04	5.50	5.84	6.10	6.32	6.51	6.67	6.81	6.94	7.06	7.17	7.26	7.36	7.44	7.52	7.59	7.66	7.73
13	4.26	4.96	5.40	5.73	5.98	6.19	6.37	6.53	6.67	6.79	6.90	7.01	7.10	7.19	7.27	7.34	7.42	7.48	7.55
14	4.21	4.89	5.32	5.63	5.88	6.08	6.26	6.41	6.54	6.66	6.77	6.87	6.96	7.05	7.12	7.20	7.27	7.33	7.39
15	4.17	4.83	5.25	5.56	5.80	5.99	6.16	6.31	6.44	6.55	6.66	6.76	6.84	6.93	7.00	7.07	7.14	7.20	7.26
16	4.13	4.78	5.19	5.49	5.72	5.92	6.08	6.22	6.35	6.46	6.56	6.66	6.74	6.82	6.90	6.97	7.03	7.09	7.15
17	4.10	4.74	5.14	5.43	5.66	5.85	6.01	6.15	6.27	6.38	6.48	6.57	6.66	6.73	6.80	6.87	6.94	7.00	7.05
18	4.07	4.70	5.09	5.38	5.60	5.79	5.94	6.08	6.20	6.31	6.41	6.50	6.58	6.65	6.72	6.79	6.85	6.91	6.96
19	4.05	4.67	5.05	5.33	5.55	5.73	5.89	6.02	6.14	6.25	6.34	6.43	6.51	6.58	6.65	6.72	6.78	6.84	6.89
20	4.02	4.64	5.02	5.29	5.51	5.69	5.84	5.97	6.09	6.19	6.29	6.37	6.45	6.52	6.59	6.65	6.71	6.76	6.82
24	3.96	4.54	4.91	5.17	5.37	5.54	5.69	5.81	5.92	6.02	6.11	6.19	6.26	6.33	6.39	6.45	6.51	6.56	6.61
30	3.89	4.45	4.80	5.05	5.24	5.40	5.54	5.65	5.76	5.85	5.93	6.01	6.08	6.14	6.20	6.26	6.31	6.36	6.41
40	3.82	4.37	4.70	4.93	5.11	5.27	5.39	5.50	5.60	5.69	5.77	5.84	5.90	5.96	6.02	6.07	6.12	6.17	6.21
60	3.76	4.28	4.60	4.82	4.99	5.13	5.25	5.36	5.45	5.53	5.60	5.67	5.73	5.79	5.84	5.89	5.93	5.98	6.02
120	3.70	4.20	4.50	4.71	4.87	5.01	5.12	5.21	5.30	5.38	5.44	5.51	5.56	5.61	5.66	5.71	5.75	5.79	5.83
∞	3.64	4.12	4.40	4.60	4.76	4.88	4.99	5.08	5.16	5.23	5.29	5.35	5.40	5.45	5.49	5.54	5.57	5.61	5.65

*Range/$S_Y \sim Q_{1-\alpha, \eta, \nu}$; η is the size of the sample from which the range is obtained, and ν is the number of degrees of freedom of S_Y.

Source: Reprinted from E. S. Pearson and H. O Hartley, eds., Table 29 of Biometrika Tables for Statisticians, Vol. 1, 3rd ed., 1966, by permission of the Biometrika Trustees, London.

TABLE E.8 Control Chart Factors

Number of Observations in Sample	d_2	d_3	D_3	D_4	A_2
2	1.128	0.853	0	3.267	1.880
3	1.693	0.888	0	2.575	1.023
4	2.059	0.880	0	2.282	0.729
5	2.326	0.864	0	2.114	0.577
6	2.534	0.848	0	2.004	0.483
7	2.704	0.833	0.076	1.924	0.419
8	2.847	0.820	0.136	1.864	0.373
9	2.970	0.808	0.184	1.816	0.337
10	3.078	0.797	0.223	1.777	0.308
11	3.173	0.787	0.256	1.744	0.285
12	3.258	0.778	0.283	1.717	0.266
13	3.336	0.770	0.307	1.693	0.249
14	3.407	0.763	0.328	1.672	0.235
15	3.472	0.756	0.347	1.653	0.223
16	3.532	0.750	0.363	1.637	0.212
17	3.588	0.744	0.378	1.622	0.203
18	3.640	0.739	0.391	1.609	0.194
19	3.689	0.733	0.404	1.596	0.187
20	3.735	0.729	0.415	1.585	0.180
21	3.778	0.724	0.425	1.575	0.173
22	3.819	0.720	0.435	1.565	0.167
23	3.858	0.716	0.443	1.557	0.162
24	3.895	0.712	0.452	1.548	0.157
25	3.931	0.708	0.459	1.541	0.153

Source: Reprinted from ASTM-STP 15D by kind permission of the American Society for Testing and Materials.

TABLE E.9 Critical Values d_L and d_U of the Durbin-Watson Statistic D (Critical Values Are One-Sided)[a].

	$\alpha = .05$										$\alpha = .01$									
	$P=1$		$P=2$		$P=3$		$P=4$		$P=5$		$P=1$		$P=2$		$P=3$		$P=4$		$P=5$	
n	d_L	d_U	d_L	d_U	d_L	d_U	d_L	d_U	d_L	d_U	d_L	d_U	d_L	d_U	d_L	d_U	d_L	d_U	d_L	d_U
15	1.08	1.36	.95	1.54	.82	1.75	.69	1.97	.56	2.21	.81	1.07	.70	1.25	.59	1.46	.49	1.70	.39	1.96
16	1.10	1.37	.98	1.54	.86	1.73	.74	1.93	.62	2.15	.84	1.09	.74	1.25	.63	1.44	.53	1.66	.44	1.90
17	1.13	1.38	1.02	1.54	.90	1.71	.78	1.90	.67	2.10	.87	1.10	.77	1.25	.67	1.43	.57	1.63	.48	1.85
18	1.16	1.39	1.05	1.53	.93	1.69	.82	1.87	.71	2.06	.90	1.12	.80	1.26	.71	1.42	.61	1.60	.52	1.80
19	1.18	1.40	1.08	1.53	.97	1.68	.86	1.85	.75	2.02	.93	1.13	.83	1.26	.74	1.41	.65	1.58	.56	1.77
20	1.20	1.41	1.10	1.54	1.00	1.68	.90	1.83	.79	1.99	.95	1.15	.86	1.27	.77	1.41	.68	1.57	.60	1.74
21	1.22	1.42	1.13	1.54	1.03	1.67	.93	1.81	.83	1.96	.97	1.16	.89	1.27	.80	1.41	.72	1.55	.63	1.71
22	1.24	1.43	1.15	1.54	1.05	1.66	.96	1.80	.86	1.94	1.00	1.17	.91	1.28	.83	1.40	.75	1.54	.66	1.69
23	1.26	1.44	1.17	1.54	1.08	1.66	.99	1.79	.90	1.92	1.02	1.19	.94	1.29	.86	1.40	.77	1.53	.70	1.67
24	1.27	1.45	1.19	1.55	1.10	1.66	1.01	1.78	.93	1.90	1.04	1.20	.96	1.30	.88	1.41	.80	1.53	.72	1.66
25	1.29	1.45	1.21	1.55	1.12	1.66	1.04	1.77	.95	1.89	1.05	1.21	.98	1.30	.90	1.41	.83	1.52	.75	1.65
26	1.30	1.46	1.22	1.55	1.14	1.65	1.06	1.76	.98	1.88	1.07	1.22	1.00	1.31	.93	1.41	.85	1.52	.78	1.64
27	1.32	1.47	1.24	1.56	1.16	1.65	1.08	1.76	1.01	1.86	1.09	1.23	1.02	1.32	.95	1.41	.88	1.51	.81	1.63
28	1.33	1.48	1.26	1.56	1.18	1.65	1.10	1.75	1.03	1.85	1.10	1.24	1.04	1.32	.97	1.41	.90	1.51	.83	1.62
29	1.34	1.48	1.27	1.56	1.20	1.65	1.12	1.74	1.05	1.84	1.12	1.25	1.05	1.33	.99	1.42	.92	1.51	.85	1.61
30	1.35	1.49	1.28	1.57	1.21	1.65	1.14	1.74	1.07	1.83	1.13	1.26	1.07	1.34	1.01	1.42	.94	1.51	.88	1.61
31	1.36	1.50	1.30	1.57	1.23	1.65	1.16	1.74	1.09	1.83	1.15	1.27	1.08	1.34	1.02	1.42	.96	1.51	.90	1.60
32	1.37	1.50	1.31	1.57	1.24	1.65	1.18	1.73	1.11	1.82	1.16	1.28	1.10	1.35	1.04	1.43	.98	1.51	.92	1.60
33	1.38	1.51	1.32	1.58	1.26	1.65	1.19	1.73	1.13	1.81	1.17	1.29	1.11	1.36	1.05	1.43	1.00	1.51	.94	1.59
34	1.39	1.51	1.33	1.58	1.27	1.65	1.21	1.73	1.15	1.81	1.18	1.30	1.13	1.36	1.07	1.43	1.01	1.51	.95	1.59
35	1.40	1.52	1.34	1.58	1.28	1.65	1.22	1.73	1.16	1.80	1.19	1.31	1.14	1.37	1.08	1.44	1.03	1.51	.97	1.59
36	1.41	1.52	1.35	1.59	1.29	1.65	1.24	1.73	1.18	1.80	1.21	1.32	1.15	1.38	1.10	1.44	1.04	1.51	.99	1.59
37	1.42	1.53	1.36	1.59	1.31	1.66	1.25	1.72	1.19	1.80	1.22	1.32	1.16	1.38	1.11	1.45	1.06	1.51	1.00	1.59
38	1.43	1.54	1.37	1.59	1.32	1.66	1.26	1.72	1.21	1.79	1.23	1.33	1.18	1.39	1.12	1.45	1.07	1.52	1.02	1.58
39	1.43	1.54	1.38	1.60	1.33	1.66	1.27	1.72	1.22	1.79	1.24	1.34	1.19	1.39	1.14	1.45	1.09	1.52	1.03	1.58
40	1.44	1.54	1.39	1.60	1.34	1.66	1.29	1.72	1.23	1.79	1.25	1.34	1.20	1.40	1.15	1.46	1.10	1.52	1.05	1.58
45	1.48	1.57	1.43	1.62	1.38	1.67	1.34	1.72	1.29	1.78	1.29	1.38	1.24	1.42	1.20	1.48	1.16	1.53	1.11	1.58
50	1.50	1.59	1.46	1.63	1.42	1.67	1.38	1.72	1.34	1.77	1.32	1.40	1.28	1.45	1.24	1.49	1.20	1.54	1.16	1.59
55	1.53	1.60	1.49	1.64	1.45	1.68	1.41	1.72	1.38	1.77	1.36	1.43	1.32	1.47	1.28	1.51	1.25	1.55	1.21	1.59
60	1.55	1.62	1.51	1.65	1.48	1.69	1.44	1.73	1.41	1.77	1.38	1.45	1.35	1.48	1.32	1.52	1.28	1.56	1.25	1.60
65	1.57	1.63	1.54	1.66	1.50	1.70	1.47	1.73	1.44	1.77	1.41	1.47	1.38	1.50	1.35	1.53	1.31	1.57	1.28	1.61
70	1.58	1.64	1.55	1.67	1.52	1.70	1.49	1.74	1.46	1.77	1.43	1.49	1.40	1.52	1.37	1.55	1.34	1.58	1.31	1.61
75	1.60	1.65	1.57	1.68	1.54	1.71	1.51	1.74	1.49	1.77	1.45	1.50	1.42	1.53	1.39	1.56	1.37	1.59	1.34	1.62
80	1.61	1.66	1.59	1.69	1.56	1.72	1.53	1.74	1.51	1.77	1.47	1.52	1.44	1.54	1.42	1.57	1.39	1.60	1.36	1.62
85	1.62	1.67	1.60	1.70	1.57	1.72	1.55	1.75	1.52	1.77	1.48	1.53	1.46	1.55	1.43	1.58	1.41	1.60	1.39	1.63
90	1.63	1.68	1.61	1.70	1.59	1.73	1.57	1.75	1.54	1.78	1.50	1.54	1.47	1.56	1.45	1.59	1.43	1.61	1.41	1.64
95	1.64	1.69	1.62	1.71	1.60	1.73	1.58	1.75	1.56	1.78	1.51	1.55	1.49	1.57	1.47	1.60	1.45	1.62	1.42	1.64
100	1.65	1.69	1.63	1.72	1.61	1.74	1.59	1.76	1.57	1.78	1.52	1.56	1.50	1.58	1.48	1.60	1.46	1.63	1.44	1.65

[a] n = number of observations; P = number of independent variables.

Source: This table is reproduced from *Biometrika*, Vol. 41 (1951), pp. 173 and 175, with the permission of the *Biometrika* Trustees.

APPENDIX

F

Documentation for Diskette Files

F.1 DISKETTE OVERVIEW

The 3.5-inch High Density, MS-DOS diskette that accompanies this text contains all of the data and Excel workbook files used in the text, stored in a special compressed format. Data file and the Excel workbook icons (see below) in the margin of the text name the appropriate diskette file for a particular topic or end of section problem. Windows users can use the install program on the diskette to decompress and copy all of these files to a hard disk that has at least 3 MB free disk space. (A special version of the diskette that can install all the files on a Macintosh system is available from your local Prentice-Hall sales representative.)

Data file   Excel workbook file

The diskette also includes the Sampl1-4 folder (directory) that contains uncompressed duplicates of the Excel workbook files for chapters 1S through 4. These files are available for use by either Windows or Macintosh users without running the install program by following the instructions given at the end of this appendix. (Full instructions for installing the compressed files on a Windows system are also given there.)

F.2 FILE CONTENTS

The diskette that accompanies this text contains two types of files. Files with the MS-DOS extension .TXT are text files (sometimes known as ASCII files) that contain raw data sets. These files are designed to be used with the selected text problems that contain the diskette icon in the margin. Files with the MS-DOS extension .XLS are Microsoft Excel workbook files. Workbook files exist for every Excel section in the text as well as for selected text problems that contain the Excel solution icon.

All files are compatible with the three versions of Microsoft Excel featured in this text: version 5 for the Macintosh, version 5 (5.0 or 5.0c) for Windows 3.1, and version 7 for Windows 95. The .XLS files are **not** compatible with version 4 or earlier of Microsoft Excel. (Read the Preface for reasons to avoid using version 4.)

Presented first is an alphabetic listing of the raw data text files, followed by a listing of the Excel workbook files by category. Version 7 users may wish to read the note that follows this listing that explains a special feature of the version 7 File Open dialog box that can be used in conjunction with the workbook files on the diskette.

Raw Data Text Files

Name	Description
ABSSPC.TXT	Number of students absent and total number of students registered over 36 days.
ADAMHA.TXT	Amount of funding (in millions of dollars) for 21 schools.
ADRADTV.TXT	Sales (in thousands of dollars), radio/TV ads (in thousands of dollars), and newspaper ads (in thousands of dollars) for 22 cities.
ADTIME.TXT	Data showing year and annual revenues (in millions of dollars) over the 20-year period 1975 through 1994.
AGREV.TXT	Data showing year and annual gross revenues (in millions of dollars) over the 14-year period 1981 through 1994.
ALCOHOL.TXT	Alcohol consumption (in ounces) and number of errors made by 15 typists.
AMPHRS.TXT	Capacity (in ampere-hours) for 20 batteries.
ATM1.TXT	Cash withdrawals by 25 customers at a local bank.
ATM2.TXT	Amount withdrawn (in thousands of dollars), median value of homes (in thousands of dollars), and location of ATM (Not a shopping center=0, Shopping center=1) for 15 ATM locations.
AUTOBAT.TXT	Lengths of life (in months) for 28 automobile batteries.
AUTOREP.TXT	Day, average service time, and range in service time (over a 20-day period).
BANKTIME.TXT	Waiting times of four bank customers per day for 20 days.
BASEBALL.TXT	Wins, E.R.A, runs scored, league (American=0, National=1) for 28 teams.
BATFAIL.TXT	Times to failure (in hours) for groups of five batteries exposed to four pressure levels (low=1, normal=2, high=3, very high=4).
BB95.TXT	Wins, E.R.A., league (American=0, National=1) for 28 teams.
BDECKER.TXT	Data showing year and net sales (in billions of dollars) at Black & Decker Corp. over the 23-year period 1970 through 1992.
BET.TXT	Attendance (in thousands) and betting volume (in millions of dollars) for 15 days.
BLACKJ.TXT	Profits (in dollars) from five sessions in each of four strategies (dealer's=1, five count=2, basic ten count=3, advanced ten count=4).
BROKER.TXT	Performance scores of three customer representatives based on gender (males=1, females=2) and four background (professionals=1, business school graduates=2, salesmen=3, brokers=4) combinations.
BULBLIFE.TXT	30 days of data on the mean life, range, and subgroup number for subgroups of five bulbs.
BULBS.TXT	Length of life of 40 light bulbs from Manufacturer A (=1) and 40 light bulbs from Manufacturer B (=2).
C&U.TXT	Tuition charges along with 4 categorical variables: type (private=1, public=2), setting (rural=1, suburb=2, urban=3), calendar (quarter=1, semester=2, trimester=3, 414=4), and classification (NLA=1, NU=2, RLA=3, RU=4, SS=5) for 195 schools in 3 states (Texas=1, North Carolina=2, Pennsylvania=3).
CANCER.TXT	Cancer incidence rate in all 50 states.
CEREAL.TXT	Data on 84 ready-to-eat cereals with 6 variables: type product (high fiber=1, moderate fiber=2, low fiber=3), cost (in cents per serving), weight (in ounces per serving), calories per serving, and sugar (in grams per serving). This is Special Data Set 2.
CITY.TXT	Prices (in dollars) of hairdressers and shirts in 9 cities.
COKE.TXT	Data showing year and operating revenues (in billions of dollars) at Coca-Cola Co. over the 23-year period 1970 through 1992.

(continued)

Name	Description
COLASPC.TXT	Day, total number of cans filled, and number of unacceptable cans (over a 22-day period).
COMPFAIL.TXT	Failure time ranks for samples of eight stereo components under three different temperature levels (150=1, 200=2, 250=3).
COOKIES.TXT	Calories in 10 types of chocolate chip cookies.
CORDPHON.TXT	Prices of 32 corded telephone models from Problem 4.76.
CURRENCY.TXT	Data showing year and average annual exchange rates (against the U.S. dollar) for the Canadian dollar, French franc, German mark, Japanese yen, and English pound.
DELIVERY.TXT	Delivery time (in days), number of options ordered, and shipping mileage (in hundreds of miles) for 16 cars.
DENTAL.TXT	Annual family dental expenses for 10 employees.
DISCOUNT.TXT	The amount of discount taken from 13 invoices.
DISPRAZ.TXT	Price (in cents) and number of packages sold at 15 stores.
DRESSES.TXT	The historical cost and audited value for 15 dresses.
EER.TXT	Average energy efficiency ratings (EER) for 36 air conditioning units.
EMPLOYEE.TXT	Data showing year, number of employees (in thousands) in goods-producing industries, in non-government service producing industries, in federal government, and in state and local government over the 42-year period 1950 through 1991.
EMPSAT.TXT	Responses to 11 questions by 122 employees in Kalosha Industries Employee Satisfaction Survey.
ERRORSPC.TXT	Number of nonconforming items and number of accounts processed over 39 days.
FLASHBAT.TXT	Time to failure (in hours) for 13 flashlight batteries.
FOODTIME.TXT	Data showing year and annual sales (in millions of dollars) over the 26-year period 1968 through 1993.
FOULSPC.TXT	Number of foul shots made and number taken over 40 days.
FRAGRANC.TXT	Data on 83 fragrances with 4 variables: gender (women=1, men=2), type of product (perfume=1, cologne=2, "other"=3), cost (in dollars per ounce), intensity (very strong=1, strong=2, medium=3, mild=4). This is Special Data Set 3.
FUNRAISE.TXT	Amounts pledged (in thousands of dollars) in a fund raising campaign by 9 alumni.
GAPAC.TXT	Data showing year and net sales (in billions of dollars) at Georgia-Pacific Corp. over the 23-year period 1970 through 1992.
GAS1.TXT	Amount of gasoline filled (in gallons) in 24 automobiles.
GASERVE.TXT	Length of time (in days) to establish gasoline heating service in 15 houses.
GAUGE.TXT	Prices of 30 pencil tire pressure gauges.
GMAT.TXT	GMAT scores for 20 applicants to an MBA program.
GPIGMAT.TXT	GMAT scores and GPI for 20 students.
GROCERY.TXT	Grocery bills for 28 customers in a supermarket.
HMO.TXT	Waiting times (in minutes) for 25 patients in an HMO.
HOSPADM.TXT	Day, number of admissions, average processing time (in hours), range of processing times, and proportion of laboratory rework (over a 30-day period).
HOUSE1.TXT	Selling price (in thousands of dollars), assessed value (in thousands of dollars), type (new=0, old=1), and time period of sale for 30 houses.
HOUSE2.TXT	Assessed value (in thousands of dollars) and heating area (in thousands of square feet) for 15 houses.

(continued)

Name	Description
HTNGOIL.TXT	Temperature (in degrees Fahrenheit), attic insulation (in inches), and monthly consumption of heating oil (in gallons).
IBM.TXT	Data showing year and annual total sales (in billions of dollars) at IBM over the 23-year period 1970 through 1992.
ICECREAM.TXT	Daily temperature (in degrees Fahrenheit) and sales (in thousands of dollars) for 21 days.
INDPSYCH.TXT	Reaction time rankings of assembly-line workers for three groups of 9, 8, and 8 using different assembly methods.
INSPGM.TXT	Test scores of eight employees in each of four groups (A=1, B=2, C=3, D=4).
INTRANK.TXT	Preference ranks for 10 MBA students (=0) and 12 MPH students (=1).
LIMO.TXT	Distance (in miles) and time (in minutes) for 12 trips.
LOCATE.TXT	Sales volume (in thousands of dollars) at two stores based on three aisle locations (front=1, middle=2, rear=3) and three shelf heights (top=1, middle=2, bottom=3).
MAILORD.TXT	Monthly billing records of 20 unpaid accounts.
MAILSPC.TXT	Day, total number of packages, and late packages (over a 20-day period).
MCDONALD.TXT	Data showing year and annual total revenues (in billions of dollars) at McDonald's Corp. over the 22-year period 1971 through 1992.
MEDICARE.TXT	The difference in Medicare reimbursement for 15 office visits.
MEDINCUS.TXT	Data showing year, median income for all races, median income for whites, and median income for blacks over the 14-year period 1977 through 1990.
MKTEST.TXT	Test scores for 19 students on a marketing exam.
NCC&U.TXT	Tuition charges at 45 North Carolina schools along with 4 categorical variables: type (private=1, public=2), setting (rural=1, suburb=2, urban=3), calendar (quarter=1, semester=2, trimester=3, 414=4), and focus (NLA=1, NU=2, RLA=3, RU=4, SS=5).
NCC&UT.TXT	Tuition charges (only) at 45 North Carolina schools.
NEWSCIRC.TXT	Sunday and Daily circulation (in thousands) for 34 newspapers.
NICKBAT.TXT	Talking times (in minutes) prior to recharging for 25 cadmium batteries (=0) and 25 metal hydride batteries (=1).
NYBANKS.TXT	Annual yields (in percent) for 10 commercial banks (=0) and 10 savings banks (=1).
NYSEAX.TXT	Stock prices for 25 AE (=1) and 50 NYSE (=2) companies.
OILSUPP.TXT	Data showing year and number of employees (in thousands) over the 20-year period 1976 through 1995.
OILUSE.TXT	Data on annual amount of heating oil consumed (in gallons) in a sample of 35 single family houses.
PACKAGE.TXT	Number of customers and weekly sales (in thousands of dollars) at 20 stores.
PEN.TXT	Product ratings of three ball-point pens under gender (males=1, females=2) and five advertisement (A=1, B=2, C=3, D=4, E=5) combinations from Case Study C.
PENEAD.TXT	Product ratings of ball-point pens by 6 adults (=0) and 8 HS students (=1) from Case Study C.
PENNC&U.TXT	Tuition charges at 90 Pennsylvania schools along with 4 categorical variables: type (private=1, public=2), setting (rural=1, suburb=2, urban=3), calendar (quarter=1, semester=2, trimester=3, 414=4), and focus (NLA=1, NU=2, RLA=3, RU=4, SS=5).
PETFOOD.TXT	Data on shelf space (in feet), weekly sales (in hundreds of dollars), and aisle location (back=0, front=1).

(continued)

Name	Description
PHLEVEL.TXT	Day, average pH level, and range in pH level (over 21 days).
PLASTIC.TXT	Hardness measurements (in Brinell units) for 50 plastic blocks.
POTATO.TXT	Percent solids content in filter cake, acidity (in pH) and lower pressure for 54 measurements.
PREP.TXT	Tuition charges at 15 NE (=0) and 15 MW (=1) prep schools.
RAISINS.TXT.	Weights (in ounces) of 30 consecutively filled packages of raisins.
RESID1.TXT	Data on 10 consecutive time periods and the corresponding residuals.
RESID2.TXT	Data on 15 consecutive time periods and the corresponding residuals.
RRSPC.TXT	Number of late arrivals and total arrivals over 20 days.
SALARY2.TXT	Weekly salary (in dollars), longevity (in months), and age (in years) for 16 employees.
SALESVOL.TXT	Sales volume (in thousands of dollars) for 12 sales persons paid on an hourly rate (=0) and 12 paid on a commission basis (=1).
SAT.TXT	Data showing year, average verbal score and average math score for males and females, and average total score on the SAT, over the 20-year period 1972 through 1991.
SCALES.TXT	Prices of 39 bathroom scales.
SEARS.TXT	Data showing year and annual total revenues (in billions of dollars) at Sears, Roebuck & Co. over the 23-year period 1970 through 1992.
SHAMPOO.TXT	Cost per ounce (in cents) for 31 normal (=1) and 29 fine (=2) hair products.
SITE.TXT	Store number, square footage, and sales (in thousands of dollars) in 14 stores.
SOUP.TXT	Data on 47 brands of canned soup with 7 variables: product (chicken noodle=1, vegetable=2, tomato=3), type (CC=1, CR=2, DC=3, DI=4) cost (in cents per 8-oz serving), calories (per 8-oz serving), fat (in grams per 8-oz serving), calories as a percentage of fat (per 8-oz serving) from Case Study B.
SPEED.TXT	Miles per gallon, speed (in miles per hour), and speed difference for 28 road tests.
STANDBY.TXT	Standby hours, staff, remote hours, Dubner hours, and labor hours for 26 weeks.
STATE.TXT	Data showing year, receipts, expenditures, and surplus (all in billions of dollars) for state and local governments over the 22-year period 1970 through 1991.
STOCK1.TXT	Book values of 50 stocks.
STUDIO.TXT	Monthly rents for 10 studio apartments in Manhattan (=0) and 10 studio apartments in Brooklyn Heights (=1).
SUPERMKT.TXT	Prices (in dollars) of colas and of pain relievers in 9 cities.
TAPES.TXT	Prices of 17 Type IV (metal) audiotapes.
TAX.TXT	Quarterly sales tax receipts (in thousands of dollars) for all 50 business establishments.
TELESPC.TXT	Number of orders and number of corrections over 30 days.
TESTRANK.TXT	Rank scores for 10 students taught by a "traditional" method (T=0) and 10 students taught by an "experimental" method (E=1).
TEXASC&U.TXT	Tuition charges at 60 Texas schools along with 4 categorical variables: type (private=1, public=2), setting (rural=1, suburb=2, urban=3), calendar (quarter=1, semester=2, trimester=3, 414=4), and focus (NLA=1, NU=2, RLA=3, RU=4, SS=5).
TOMYIELD.TXT	Yield (in pounds) and amount of fertilizer (in pounds per 100 square feet) on 10 plots of land.
TOMYLD2.TXT	Amount of fertilizer (in pounds per 100 square feet) and yield (in pounds) for 12 plots of land.

(continued)

Name	Description
TOOTHPST.TXT	Cost and cleaning test score for 39 brands of tubed toothpaste.
TRAIN1.TXT	Time (in minutes) early or late with respect to scheduled arrival for 10 trains.
TRAIN2.TXT	Time (in minutes) early or late with respect to scheduled arrival for 10 LIRR trains (=0) and 12 NJT trains (=1).
TRAINING.TXT	Assembly times (in seconds) for 16 employees receiving computer-assisted, individual-based training (=0) and 16 employees receiving team-based training (=1).
TRW.TXT	Data showing year and earnings per share at TRW Inc. over the 23-year period 1970 through 1992.
UPJOHN.TXT	Data showing year and net sales (in billions of dollars) at Upjohn Co. over the 23-year period 1970 through 1992.
UTILITY.TXT	Utility charges for 50 three-bedroom apartments.
VB.TXT	Time (in minutes) for 9 students to write and run a visual basic program.
WARECOST.TXT	Distribution cost (in thousands of dollars), sales (in thousands of dollars), and number of orders for 24 months.
WAREHSE.TXT	Number of units handled per day over 30 days by each of five employees.
WATER.TXT	Annual water consumption (in thousands of gallons) for 15 families.
WEATHER.TXT	Temperatures recorded (in degrees Fahrenheit) for 22 cities, over 2 days.
WORKATT.TXT	Attitude scale scores for three trainees based on gender (males=1, females=2) and three room color (green=1, blue=2, red=3) combinations.
WORKHRS.TXT	Lot sizes and worker-hours for 14 production runs.
XEROX.TXT	Data showing year and annual net sales (in billions of dollars) at Xerox Corp. over the 23-year period 1970 through 1992.

F.3 MICROSOFT EXCEL WORKBOOK FILES

Four types of Microsoft Excel workbook files are included on the diskette that accompanies this text. Every Excel workbook file included contains an Overview sheet that serves as the introduction and table of contents for the workbook.

1. Microsoft Excel files containing the completed workbooks discussed and implemented in the Excel sections of this text. The MS-DOS primary name of these files indicates the section associated with the file. For example, 2-7-2.XLS is the workbook implemented in Section 2.7.2. (There are two files for Section 3.4.6.).

1S.XLS	4-11-4.XLS	8-4-4.XLS	11-12-2.XLS
2-7-2.XLS	4-13-3.XLS	8-5-5.XLS	11-13-2.XLS
2-7-3.XLS	4-14-4.XLS	8-6.XLS	12-14-1.XLS
2-7-4.XLS	5-5.XLS	8-7-4.XLS	12-14-2.XLS
2-12-2.XLS	5-6-5.XLS	9-2-4.XLS	12-14-3.XLS
2-12-3.XLS	5-7-2.XLS	9-6.XLS	12-14-4.XLS
3-4-6POP.XLS	5-11-3.XLS	10-6-4.XLS	13-4-3.XLS
3-4-6SAM.XLS	6-10.XLS	10-8-4.XLS	13-5-4.XLS
3-5-5.XLS	7-11-1.XLS	11-6-3.XLS	13-6-2.XLS
3-7.XLS	7-11-2.XLS	11-7.XLS	13-7-6.XLS
3-8-3.XLS	7-11-3.XLS	11-8-2.XLS	
4-9.XLS	8-3-5.XLS	11-11-4.XLS	

2. Microsoft Excel workbook files that are the starting points for the implementation of other workbooks in selected Excel sections of this text.

FACTORY.XLS	NETSALES.XLS	SALES.XLS
FILL-1.XLS	PA-POP-1.XLS	TEXAS-1.XLS
FILL-2.XLS	PA-SAM-1.XLS	TEXAS-2.XLS
HEATING.XLS	PA-SAM-2.XLS	TEXAS-3.XLS
LUGGAGE.XLS	PACKAGE.XLS	YIELD-1.XLS
NC-1.XLS	PEANUT.XLS	YIELD-2.XLS
NC-2.XLS	RAZORS.XLS	

3. Microsoft Excel workbook files that contain solutions to selected text problems. The MS-DOS primary name of these files indicates the problem associated with the file. For example, S-2-13.XLS is the workbook implemented as a solution to Problem 2.13. (S-SITE.XLS contains the solutions to the series of site selection problems in Chapter 11.)

S-2-6.XLS	S-3-14.XLS	S-9-3.XLS
S-2-13.XLS	S-4-44.XLS	S-9-9.XLS
S-2-18.XLS	S-4-51.XLS	S-10-22.XLS
S-2-23.XLS	S-5-41.XLS	S-11-64.XLS
S-2-29.XLS	S-6-58.XLS	S-SITE.XLS
S-2-53.XLS	S-7-42.XLS	S-12-49.XLS
S-3-3.XLS	S-8-41.XLS	S-13-31.XLS

4. Microsoft Excel workbooks that contain Visual Basic for Application modules that serve special purposes:

MOUSING.XLS	Allows users to practice their mousing skills.
STEMLEAF.XLS	Aids in the creation of a stem-and-leaf plot.

Special Note to Version 7 Users

The File Open dialog box in Microsoft Excel version 7 contains a Properties button that, when selected, can display summary information about a file. Every workbook file on the diskette that accompanies this text has been coded so that the statistical topic and the appropriate chapter or section associated with the workbook file will be displayed when this button is selected. Figure F.1.Excel illustrates the Properties feature in use. (To enable your workbooks

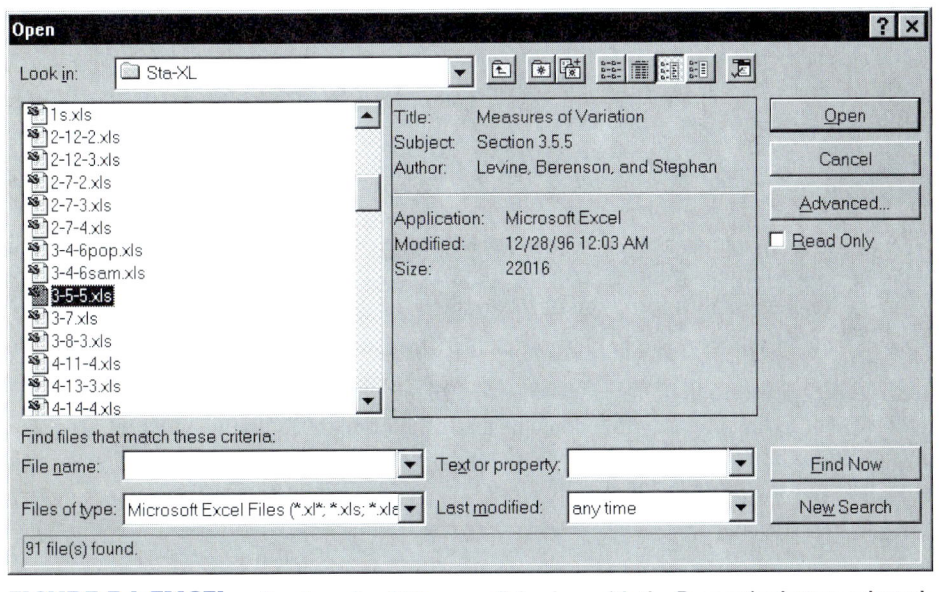

FIGURE F.1.EXCEL Excel version 7 file open dialog box with the Properties button selected.

to display similar information, select File | Properties and click the Summary tab in the Properties dialog box that appears. Version 5 users who anticipate upgrading to version 7 can select File | Summary Info to enter the same type of information.)

F.4 DISKETTE INSTALLATION INSTRUCTIONS

Making a Backup Copy

Before installing any files from this diskette, a backup copy of the diskette or selected files should be created. The following table gives abbreviated copying instructions for the three environments under which Microsoft Excel runs. (Ask your instructor or see your system documentation for further details.)

Environment	Instructions
Windows	Create a backup copy of the diskette.
Windows 95	Insert diskette. Select the My Computer icon. Select the icon representing the appropriate diskette drive. Select the command File	Copy Disk.
Windows 3.1	Insert diskette. Start the File Manager program. Select Disk	Copy Disk.
Macintosh	Create back up copies of the Sampl1-4 folder files. Create a hard disk folder to receive files. Start Apple File Exchange. Insert diskette. Select folder to receive files and select the Sampl1-4 files. Click the Translate button. Continue until all files have been copied. Quit Apple File Exchange. Copy contents of hard disk folder with diskette files to a blank diskette.

Using the Windows Install Program

Windows users can use the install program on the diskette that accompanies this text to decompress and copy all of the data and Excel workbook files used in this text to a hard disk that has at least 3 MB free disk space. Using a backup copy of the diskette, do the following:

1. Insert the diskette in the first (A:) diskette drive.
2. Select File | Run from the Windows Program Manager menu (or press the Start button and select Run in Windows 95).
3. In the Run dialog box, type A:INSTALL and click the OK button. An "Installing files" message appears. Click OK to continue.
4. Accept or change the installation directory into which all files will be copied. (The default directory is C:\PRENHALL\STA-XL). Click the Unzip button to install. A thermometer gauge will inform you of the progress of the installation.
5. When the installation is complete, click the Close button.

After the files are installed, you may want to change the default file location in Microsoft Excel to point to the directory that contains the installed files. To do this in Excel, first select Tools | Options and then, in the Options dialog box, select the General tab. Change the value of the "Default file location" to the directory created by the install process.

Index

A

A_2 factor, 518
α (alpha), probability of Type I error, 343
α (level of significance), 343
a posteriori analysis, 424
Addition rule, 178
Adjusted r^2, 549–550, 607–608
Algebraic notation, 121, A2
Alternative hypothesis, 341
Among-group variation (SSA), 417
Analysis of variance (ANOVA)
    completely randomized design model (*see* one-way ANOVA)
    ANOVA table, 419
    degrees of freedom, 417
    F table, E-8–E-11
    measures of variation
        SSA, 417
        SST, 416
        SSW, 417
    one-way F test, 418
    Tukey-Kramer procedure, 424–425
Analytical studies, 6
Answers to Selected Problems, 703–716
A priori classical probability, 172
Arithmetic mean (*see* Mean)
Assignable causes of variation (*see* Special causes)
Assumptions
    of classical parametric procedures, 381
    of nonparametric procedures, 381
    of regression analysis, 561–562
    of the analysis of variance, 425–426
    of the χ^2 test for independence, 473
    of the χ^2 test of equality of proportions, 464
    of the confidence interval estimate, 297
    of the F test for equality of variances, 408–409
    of the Kruskal-Wallis rank test, 434
    of the one-way F test, 425–426
    of the t test for the difference in two means, 387
    of the t test for the mean (σ unknown), 357–358
    of the Wilcoxon rank sum test, 395
    of the Z test for the mean (σ known), 345
    of the Z test for the proportion, 361
Attribute charts, 498
Auditing, 314
Autocorrelation, 567
    Durbin-Watson test, 570–571
Autoregressive modeling, 680–686
Average, 121

B

Bar chart, 85–86, 100–102
Behrens-Fisher problem, 387
β (beta), probability of Type II error, 343
β_1 (beta), regression coefficients
    confidence interval for, 581, 619
    test for, 580–581, 617–618
Binomial distribution, 197
    characteristics, 201–203
    development, 199–201
    mathematical expression for, 200
    mean, 202
    standard deviation, 203
Bootstrapping estimation, 301–302
Box-and-whisker plot, 149–150
Brainstorming, 490
Business cycle, phases, 653

C

Categorical data, 9–11
Categorical response, 10
Categorical variable, 9
Cause and effect diagram (*see* Fishbone diagram)
Cell, 174
Census, 9
Central limit theorem, 268
Central tendency, (*see* Measures of)
Certain event, 172
"Chart junk," 104–105
Charts
    for categorical data
        bar chart, 85–86, 100–102
        Pareto diagram, 88–89, 104
        pie chart, 86–87, 102–103
    for numerical data
        box-and-whisker plot, 149–150
        histogram, 67–68
        ogive (cumulative polygon), 71–73
        polygon, 68–73
    scatter diagram, 534–535
    for studying a process
        control charts, 488, 498–508, 515–523
        Fishbone diagram, 489–490
        process flow diagram, 491–493
χ^2 test
    for independence, 469–473
    for the differences among c proportions, 463–466
    for the differences between two proportions, 456–462
Chi-square (χ^2)
    contingency table, 458
    degrees of freedom, 458
    distribution, 458–459
    relationship to standardized normal distribution, 462
    table of, E-7
Choosing the forecasting model
    forecasting error (MAD), 690, 692–694
    principle of parsimony, 690
    residual analysis, 689
Class boundaries, 62
Classes
    number of groupings, 61
    subjectivity in selection, 63
Class groupings (categories), 61
Classical multiplicative time-series model, 653–654
Classical procedures (*see* Parametric methods)
Class intervals, 61–62
Class midpoint, 63
Coefficient of correlation, 557–559
    computational formula, 558
    test of significance for, 582–583
Coefficient of determination, 549
    multiple determination, 607
    partial determination, 620–621
Coefficient of variation, 141–142
Collectively exhaustive, 177
Combinations, 199
Common causes of variation, 488
Complement, 173
Completely randomized design model [*see* Analysis of variance (ANOVA)]
Components of time series, 653–654
Compound growth rate, 672
Conditional probability, 181

Confidence band effect, 575
Confidence coefficient $(1 - \alpha)$, 294, 343
Confidence interval, 294
Confidence interval estimate
    for a finite population, 311–312
    of the mean (μ known), 292–296
    of the mean (μ unknown), 297–300
    of the mean response in regression, 574–576
    of the population slope, 581–582, 619
    of the population total, 314–315
    total difference, 315–316
    of the proportion, 304–305
Connection between confidence intervals and hypothesis testing, 350–351
Consistency, 262–263
Contingency tables, 92–93, 174, 457
Continuous data, 10–12
Continuous numerical data, 10–12
Continuous numerical variable, 10
Continuous probability density functions, 226
Contribution of curvilinear effect, 626–627
Contribution of independent variables
    by comparing regression models, 613–614
    test for, 615
Control chart, 487
    for the proportion and number of nonconforming items, 498–504
    for the mean, 517–519
    for the range, 515–517
    table of factors, E-15
    theory of, 487–489
Correlation analysis, 534
Correlation matrix, 607
Critical range, 424
Critical region, 342
Critical value, 294, 341–343
Cross-classification tables (*see* Contingency tables)
Cumulative distribution, 70
Cumulative percentage distribution, 70–71
Cumulative percentage polygon, 71–73
Curvilinear effect, 622
Curvilinear regression model, 621–626
Cyclical component, 653–654

D

d_2 factor, 515
d_3 factor, 515

D_3 factor, 516
D_4 factor, 516
Data, 8
Data presentation (*see* Charts; Tables)
Data snooping, 373
Data sources
    conducting a survey, 9
    designing an experiment, 9
    published material, 8–9
Data types
    categorical data, 9–11
    numerical data, 11–12
Deciles, 169
Degrees of freedom, 299
Delphi technique, 652
Deming's 14 points, 495–498
Deming, W. Edwards, 495
Dependent variable (*see* Response variable)
Descriptive statistics, 5
Descriptive summary measures, 121
Difference estimation, 315–316
Directional tests (*see* One-tailed tests)
Discrete data, 9–10
Discrete distributions, 187
Discrete numerical data, 9–10
Dispersion, (*see* Measures of dispersion)
Dot scale, 122
Dummy variables, 630
Dummy variable models, 630–631
Durbin-Watson test, 570–571
    table for, E-16

E

Efficiency, 262
Empirical classical probability, 172
Empirical rule, 158
Enumerative studies, 6
Estimators, properties of, 261–262
Ethical issues, 20–21
Event, 172–174
Expected frequency (f_e), 458
Expected monetary value, 190–191
Expected value of a random variable, 188–189
Explained variation, 547–548
Explanatory variable, 534
Exploratory data analysis, 148
Exponential distribution, 258–259
Exponential smoothing, 659–662
Exponential trend model, 672–674
Extrapolation, 543–544
Extreme values (*see* Outliers)

F

Factor, 414
Factorial, 200
F distribution, 407
Finite population correction factor, 283–284, 311
First quartile, 127
Fishbone (Ishikawa) diagram, 489–490
Fisher, R. A., 405
Fisher's exact test, 461
Five-number summary, 148
Forecasting, 652
Frame [*see* Population frame (listing)]
Frequency distributions, 61
    establishing the boundaries of the class, 62–63
    obtaining the class intervals, 61
    selecting the number of classes, 61
    subjectivity in selecting classes, 63
Frequency histogram (*see* Histogram)
F tables, E-8–E-11
    obtaining a lower critical value, 407
F test
    for the contribution of a predictor variable, 615–616
    for the differences among c means, 418
    for the equality of two variances, 405–409
    for the significance of a multiple regression model, 611–612

G

Gaussian distribution (*see* Normal distribution)
General multiplication rule, 183
GIGO, 9
Gosset, W. S., 297
Grand mean (*see* Overall mean)
Greek alphabet, C-1

H

Histogram, 67–68
Homogeneity of proportions, 457
Homogeneity of variance (*see* Homoscedasticity)
Homoscedasticity, 426, 562
Hypergeometric distribution, 207–208
Hypothesis-testing methodology
    procedure, 340–345
    p-value approach, 348–350
    steps, 347–348

I

Independence, 182
Independence of error, 425–426, 562
Independent events, 182
Independent variable (*see* Explanatory variable)
Inferential statistics, 5
Interaction terms, 633–634
Intercept (*see* Y intercept)
Interpolation, 543
Interquartile range, 136
Interrelationship between t and F, 618
interrelationship between Z and χ^2, 462
Interval estimates, 292
Interval scaling, 11–12
Inverse normal scores transformation, 249–251
Irregular component, 654
Ishikawa, Kaoru, 489

J

Joint event, 173
Joint probability, 177
Juran, Joseph, 487

K

Kruskal-Wallis rank test for c independent samples, 433–437

L

Law of large numbers, 263
Least-squares method, 540–541
Least squares trend fitting, 668
    for exponential models, 672–674
    for linear models, 668–670
    for quadratic models, 671–672
Left skew, 145
Less-than curve (*see* Ogive)
Level of confidence, 294
Level of measurement, 10–12
Level of significance, 343
Levels (*see* Groups)
Linear relationship, 538
Linear (trend) model, 668–670
Location, 121
Logarithmic transformation, 634
Lower control limit, 488

M

MAD (mean absolute deviation), 690–693
Malcolm Baldrige Award, 486
Management
    by control, 486
    by directing, 486
    by doing, 486
    by process, 487
Management planning tools
    Fishbone (Ishikawa) diagram, 489–490
    Process flow diagram, 491–494
Marginal probability, 176
Mathematical expectation, 188–189
Mean, 121
Mean square
    MSA (among), 417
    MSE, 611
    MSR, 611
    MST, 418
    MSW (within), 418
Measurement error, 20
Measurement scales
    interval and ratio, 11–12
    nominal and ordinal, 10
Measures of central tendency, 121
    arithmetic mean, 121–122
    median, 124
    midhinge, 127
    midrange, 126
    mode, 125–126
Measures of dispersion, 135
    coefficient of variation, 141–142
    interquartile range (midspread), 136–137
    range, 135–136
    standard deviation, 137–141
    variance, 137–141
Median, 124
Midhinge, 127
Midrange, 126
Midspread, 136
Mode, 125–126
Model, 196
Moving averages, 655–658
Multicollinearity, 636–637
Multiple comparisons
    Tukey-Kramer procedure, 424–425
Multiple linear regression model, 600–602
Multiplication rule, 183–185
Multiplication rule for independent events, 184
Mutually exclusive, 177

N

Negative binomial distribution, 208–209
Net regression coefficient, 602
Nominal scaling, 10–11
Noncentral tendency (or location), 127
Nonparametric methods 380–381
Nonparametric test procedures
    Kruskal-Wallis rank test for c independent samples, 433–436
    Wilcoxon rank sum test, 395–400
Nonprobability sample, 14
Nonresponse error, 19
Normal distribution
    applications, 233–242
    evaluating the properties, 247–248
    goodness-of-fit, 247
    mathematical expression for, 228
    parameters of, 228
    properties of, 227
    table of, E-4
    using the table of, 231–233
Normal equations, 541
Normal probability density function, 227
Normal probability plot, 248–257
np chart, 502–503
Null event, 172
Null hypothesis, 340–341
Null set, 180
Numerical data, 9–10
Numerical random variable, 9–10
Numerical response, 9
Numerical variables, 10

O

Observed frequency (f_0), 458
Ogive (Cumulative polygon), 71–73
One-factor analysis of variance, 414
One-sample tests, 345–363
One-tailed tests, 351–353
One-way ANOVA F test, 414–424
Operational definition, 12
Ordered array, 52–53
Ordinal scaling, 10
Outliers, 120
Overall mean, 416

P

p chart, 498–502
Parameter, 5
Parametric (or classical) methods, 380
Pareto diagram, 88–89, 104
Parsimony, principle of, 690
Partial F test criterion, 613–617
p value, 348–350
Percentage distribution, 65
Percentage polygon, 68–69
Percentiles, 169
Pie chart, 86–87, 102–103
Point estimates, 292

Poisson distribution, 209–211
Polls, 331–332
Polygons, 68–73
Polynomial, 621
Pooled estimate ($\bar{p}$) of common population proportion, 451
Pooled variance, 383
Pooled-variance t test for differences in two means, 382–387
Population, 5
  mean of, 153
  standard deviation of, 153
  total of, 314
  variance of, 153
Population frame (listing), 15
Positioning point formulas (for Q_1, Median, Q_3), 124, 127
Power of a statistical test $(1 - \beta)$, 344
Prediction, 543–544
Prediction interval for an individual response, 579
Predictor variable (see Explanatory variable)
Primary source, 8
Probability, 172
Probability density function, 226
Probability distribution
  for a discrete random variable, 187
Probability of a Type I error (α), 343
Probability of a Type II error (β), 343
Probability sample, 14
Process, 489
Process flow diagram, 491–493
Proportion of successes, 304, 499
Published sources of data, 8–9
p values, 348–349, 354

Q

Q_1: first quartile, 127
Q_3: third quartile, 127
Quadratic trend model, 671–672
Quantiles, 248–249
Quartiles, 127–128

R

R chart, 515–517
Randomization, 372
Randomness, 425–426
Random numbers, table of, E-2–E-3
Random variable, categorical, 9
Random variable, numerical, 10
  continuous, 10
  discrete, 10
Range, 135–136
Ranks, 395

Ratio scaling, 11–12
Raw form, 52,120
Reciprocal transformation, 634–635
Rectangular-shaped distribution, 270
Red bead experiment, 512–514
Region of nonrejection, 342
Region of rejection, 342
Regression analysis, 534
Regression coefficients, 540–541
Regression sum of squares
  measures of variation
    SSE, 548
    SSR, 548
    SST, 548
Rejection region, 342
Relative frequency distribution, 65
Resampling distribution, 302
Residual analysis, 562–566, 608–609, 689–690
Residuals, 563
Response variable, 534
Revised stem-and-leaf display, 54
Right-skew, 145
Robust, 381
Rule of combinations, 199
Rules for algebra, A-2
Rules for arithmetic operations, A-2

S

Sample, 5
Sample mean ($\bar{X}$), 121–122
Sample proportion (p_s), 360
Samples
  nonprobability, 14
  probability, 14
    simple random, 14–17
Sample selection, 16–17
Sample size
  determination,
    for a finite population, 312–313
    for the mean, 306–307
    for a proportion, 308–310
Sample space, 172
Sample standard deviation (S), 140
Sample variance (S^2), 140
Sampling distribution,
  of the mean,
    for nonnormal populations, 268–272
    for normal populations, 264–268
    properties, 261–263
  of the proportion, 273–275
Sampling error, 19–20, 306
Sampling *with* and *without* replacement, 15

Satterthwaite, F.E., 387
Scatter diagram, 534–535
Scientific notation, 148
Seasonal component, 654
Secondary source, 8
Shape, 144–145
Shewhart, W. A., 486
Shewhart cycle, 496
Simple event, 172
Simple linear regression model, 541–543
Simple probability, 176
Simple random sample, 14
Skewness
  left-skewed (negative), 145
  right-skewed (positive), 145
  zero-skewed (symmetry), 145
Slope, 538
Special causes of variation, 487–488
Standard deviation
  definitional formula, 137–138
  "calculator" formula, 140–141
  of a discrete random variable, 190
Standard error
  of the estimate, 545
  of the mean, 263–264
  of the proportion, 274
  of the regression coefficient, 581
Standard normal quantiles, 249–250
Standardized normal (Z) distribution, 229
Standardized residuals, 563
Standardizing the normal distribution, 229
Statistic, 5
Statistical independence, 182
Statistical inference, 260
Statistical thinking, 4,
Stem-and-leaf display, 53–55
Student's t distribution (*see t* distribution)
Studentized range, 424
Studentized residuals, 563
Subjective probability, 172
Sum of Squares Among (SSA), 417
Sum of Squares due to regression (SSR), 547–548
Sum of Squares Error (SSE), 548
Sum of Squares Total (SST), 416, 547–548
Sum of Squares Within (SSW), 417
Summary table, 85
Summation notation, B-1–B-5
Symmetry, 144

T

t distribution
  normal approximation, 298
  properties of, 297
  t table, E-5–E-6
Table of random numbers, 16–17
Tables
  for categorical data
    contingency table, 92–93, 174
    summary table, 84–85
  for numerical data
    cumulative distribution, 70–71
    frequency distribution, 60–63
    percentage distribution, 65–67
    relative frequency distribution, 65–67
Target population (*see* Population frame)
Taylor, Frederick W., 487
Test of hypothesis
  for independence in the $r \times c$ table, 469–473
  for portions of the multiple regression model, 613–616
  for positive autocorrelation, 570–571
  for the correlation coefficient, 580–581
  for the curvilinear effect, 626–627
  for the curvilinear regression model, 625
  for the differences among *c* proportions, 463–466
  for the differences among *c* means, 414–424
  for the differences among *c* medians, 433–436
  for the differences between two means, 382–387
  for the differences between two medians, 395–400
  for the differences between two proportions, 450–453, 456–462
  for the equality of two variances, 405–409
  for the existence of correlation, 582–583
  for the highest order autoregressive parameter, 681–682
  for the population mean
    σ known, 345–346
    σ unknown, 354–358
  for the population slope, 580–581
  for the population proportion, 360–362
  for a regression coefficient in multiple regression, 617–619
  for the significance of a regression model, 580–581, 611
  p-value approach, 348–350
  using the χ^2 test, 463–473
  using the normal approximation, 450–453
Test of significance (*see* Test of hypothesis)
Test statistic, 347–348
Theoretical frequency [*see* Expected frequency (f_e)]
Third quartile, 127
Tied observations, 395
Time series, 652
Total quality management (TQM), 4
Total variation (SST), 416
TQM (*see* Total quality management)
Transformation formula, 229
Transformations for regression models
  exponential, 635
  logarithmic, 634
  reciprocal, 635
  square-root, 634
Trend component, 654
Trend effect, 653
t test
  for the correlation coefficient, 582–583
  for the difference between two means, 382–387
  for the mean, 354–358
  for the slope, 580–581, 617
Tukey-Kramer procedure, 424–425
  table for, E-13–E-14
Two-tailed tests, 346
Type I error, 343
Type II error, 343

U

Unbiasedness, 261
Unexplained variation, 547–548
Uniform (rectangular-shaped) distribution, 196
Universe (*see* Population)
Upper control limit, 488

V

Variables, 9
Variance, 137–138
  among groups (*see* Mean square)
  of a discrete random variable, 189–190
  within groups (*see* Mean square)
Variance inflationary factor (VIF), 636–637
Variation (*see* Measures of dispersion)
VIF (*see* Variance inflationary factor)

W

Width of class interval, 61
Wilcoxon rank sum test, 395–400
  table of, E-12
Within group variation (SSW), 417

X

$\bar{X}$ chart, 517–519

Y

Y intercept, 539

Z

Z test
  for differences in two proportions, 450–453
  for the mean, 345–346
  for the proportion, 360–363

Index I-5

Index for Microsoft® Excel

A

Absolute address, 76
Active cell, 38
Advanced data filter, 401
Auditing, 327–331
Autocorrelation, 572–573
Average function, 128

B

Bars, 32
Bin range, 77
Binomdist (binomial) function, 203
Buttons, 32
Box-and-whisker plot, 150–152

C

Calculations sheet, 34
Cancel button, 31
Chart Wizard used for
    Bar chart, 100–102
    Box-and-whisker plot, 150–152
    Control charts, 506–508, 521–523
    Exponential smoothing, 664–665
    Frequency polygon, 80–83
    Moving averages, 664–665
    Pareto diagram, 104
    Pie chart, 102–103
    Trend, 550–553
Cell address, 37
Cell highlight, 38
Cell range, 40
Cells, 37
Check box, 31
Chidist function, 439
Chiinv function, 439, 477
Chitest function, 476
Click, 29
Confidence functions, 319
Confidence intervals, 317–322, 576–578
Control charts
    p and np charts, 504–508
    R chart, 519–523
    X chart, 519–523
Copying, 40–41
Correcting errors, 38
Correl function, 560
Count function, 129

D

Data analysis installation, 34
Data analysis tool used for

Cumulative frequency distributions, 78–80
Descriptive statistics, 146–148
Exponential smoothing, 663–664
F test for the difference between two variances, 409–410
Frequency distributions, 77–79
Generate random numbers, 278–283
Histogram, 78–80
Moving averages, 662
One-way ANOVA, 427
Regression, 554–556
Selecting a simple random sample, 277–278
Time series analysis, 674–675, 686–687
two-sample t test assuming equal variances, 388
Data sheet, 34
    design, 35–36
    enhancing appearance, 41–42
    entering values, 37
Data | Sort command, 55–56
Dialog boxes, 30
    Advanced filter, 402
    Data analysis, 78
    Exponential smoothing, 664
    File | open, 30
    File | page setup, 44
    File | print, 31
    File | Save As, 39
    Format | 1 Column group, 79
    Histogram, 78
    Moving averages, 662–663
    Regression, 555
    Scenario manager, 46
    Sort, 56
    Trendline options tab, 551
    Trendline type tab, 551
Diskette,
    File contents, F-1–F-6
    installation, F-8
    overview, F-1
    workbook files, F-6–F-7
Double-click, 29
Drag, 29
Drop-down list box, 31
Durbin-Watson statistic, 572–573

E

Edit box, 30
Entering values, 37

Expected values, 192–193
Expondist (exponential distribution) function, 259
Exponential smoothing, 663–665

F

Fdist function, 412
File | open dialog box, 30
File | page setup dialog box, 44
File | print dialog box, 31
File | Save As dialog box, 39
Finv function, 411
Formulas, 39–40
Frequency function, 74–75
Functions, 42
    Average, 128
    Binomdist, 203
    Chidist, 439
    Chiinv, 439, 477
    Chitest, 476
    Confidence, 319
    Correl, 560
    Count, 129
    Expondist, 259
    Fdist, 412
    Finv, 411
    Frequency, 74–75
    Hypgeomdist, 211
    If, 130, 365
    If and, 402
    If or, 412
    Max, 43, 129
    Median, 128
    Min, 43, 129
    Mode, 128
    Negbinomdist, 213
    Normdist , 244
    Normsdist, 244
    Norminv, 245
    Normsinv, 244
    Poisson, 214
    Power, 694
    Rand, 276
    Randbetween, 276
    Roundup, 323
    Small, 129
    Stdev function, 142
    Stdevp function, 158
    Standardize, 244
    Sum, 42
    Sumsq function, 572
    Sumxmy2 function, 572

Tdist, 368
Tinv, 320
Trend, 550–553
Var function, 142
Varp function, 158
Vlookup, 520

G

Gap width edit box, 79
Generating random numbers, 278–283

H

Header/footer tab choices, 44
Hypgeomdist (hypergeometric) function, 211–212

I

Icons, 29
If function, 130
Increase decimal button, 76

M

margins tab choices, 44
Max function, 43, 129
mean absolute deviation, 693–694
Median function, 128
Menu bar, 32
Min function, 43, 129
Mode function, 128
Mouse, 29
Mouse pointer, 29
Moving averages, 662–663
Microsoft Office, 3
Multiple regression, 637–641

N

Negbinomdist (negative binomial) function, 213–214
Normal probabilities, 244–246
Normal probability plot, 254–257
Normdist function, 244
Normsdist function, 244
Norminv function, 245
Normsinv function, 244

O

OK button, 31
One-way ANOVA, 426–430
Operators, 40
Option buttons, 31
Orientation options buttons, 44
Overview sheet, 34

P

Page tab choices, 44
Pivot tableWizard used for
    One-Way summary tables, 95–98
    Contingency tables, 98–99
    Statistics classified by groups, 131–132
Poisson function, 214–216
Power function, 694
Probabilities, 186
Pull-down menu, 32
Prediction interval, 579–580
Print-outs
    Customizing, 44

R

Rand function, 276
Randbetween function, 276
Referring to other sheets, 43
Regression, 550–556
Renaming sheets, 35
Residual analysis, 567
Right-click, 29
Roundup function, 323

S

sample size determination, 322–325
Scaling options buttons, 44
Scenario manager, 45–47, 76
Scroll bar, 31
Scrollable list box, 31
Select, 29
Selecting a simple random sample, 276–277
Setting up the application window, 33–34
Sheet tab, 35
Sheet tab choices, 44
Small function, 129
Sorting data, 55–56
Sort dialog box, 56
Spreadsheet, 28
Spinner buttons, 31
Stdev (standard deviation) function, 142
Stacked data, 388–389, 410, 427–428
Standardize function, 244
Sum function, 42
Sumsq function, 572
Sumxmy2 function, 572

T

Tdist function, 368
Tests of hypothesis
    Chi-square tests, 475–477
    for a proportion, 370–371
    for the difference between two variances, 409–413
    Kruskal-Wallis test, 437–439
    One sample Z test for the mean (σ known), 364–366
    One sample t test for the mean (σ unknown), 366–369
    One-way ANOVA, 426–430
    Pooled variance t test, 387–391
    Wilcoxon rank sum test, 400–404
    Z test for the difference between two proportions, 453–454
Text files, 47
Text Import Wizard, 47–49
Time series analysis, 674–675
Tinv function, 320
Title bars, 32
Toolbars, 33
Trend function, 550–553
Trendline feature, 553

V

Var (variance) function, 142
Vlookup function, 520

W

Windows, 29
Windowing environment, 29
Workbook, 28
Worksheet, 28

License Agreement and Limited Warranty

READ THE FOLLOWING TERMS AND CONDITIONS CAREFULLY BEFORE OPENING THIS DATA DISK PACKAGE. THIS IS AN AGREEMENT BETWEEN YOU AND PRENTICE-HALL, INC. (THE "COMPANY"). BY OPENING THIS SEALED PACKAGE, YOU ARE AGREEING TO BE BOUND BY THESE TERMS AND CONDITIONS. IF YOU DO NOT AGREE WITH THESE TERMS AND CONDITIONS, DO NOT OPEN THE DISK PACKAGE. PROMPTLY RETURN THE DISK PACKAGE TO THE COMPANY.

1. GRANT OF LICENSE: In consideration of your purchase of textbooks and/or other materials published by the Company, and your agreement to abide by the terms and conditions of this Agreement, the Company grants to you a nonexclusive right to use and display the copy of the enclosed data disk (hereinafter "the SOFTWARE"). The Company reserves all rights not expressly granted to you under this Agreement. This license is not a sale of the original SOFTWARE or any copy to you.

2. USE RESTRICTIONS: You may not sell or license copies of the SOFTWARE or the Documentation to others or transfer or distribute it, except to instructors and students in your school who are users of the Company textbook that accompanies this SOFTWARE.

3. LIMITED WARRANTY AND DISCLAIMER OF WARRANTY: The Company makes no warranties about the SOFTWARE, which is provided "AS-IS." IF THE DISK IS DEFECTIVE IN MATERIALS OR WORKMANSHIP, YOUR ONLY REMEDY IS TO RETURN IT TO THE COMPANY, WHICH SHALL REPLACE IT UNLESS THE COMPANY DETERMINES IN GOOD FAITH THAT THE DISK HAS BEEN MISUSED OR IMPROPERLY INSTALLED, REPAIRED, ALTERED OR DAMAGED. THE COMPANY DISCLAIMS ALL WARRANTIES, EXPRESS OR IMPLIED, INCLUDING WITHOUT LIMITATION, THE IMPLIED WARRANTIES OF MERCHANTABILITY AND FITNESS FOR A PARTICULAR PURPOSE. THE COMPANY DOES NOT WARRANT, GUARANTEE OR MAKE ANY REPRESENTATION REGARDING THE USE OR THE RESULTS OF THE USE OF THE SOFTWARE. IN NO EVENT, SHALL THE COMPANY OR ITS EMPLOYEES, AGENTS, SUPPLIERS OR CONTRACTORS BE LIABLE FOR ANY INCIDENTAL, INDIRECT, SPECIAL OR CONSEQUENTIAL DAMAGES ARISING OUT OF OR IN CONNECTION WITH THE LICENSE GRANTED UNDER THIS AGREEMENT INCLUDING, WITHOUT LIMITATION, LOSS OF USE, LOSS OF DATA, LOSS OF INCOME OR PROFIT, OR OTHER LOSSES SUSTAINED AS A RESULT OF INJURY TO ANY PERSON, OR LOSS OF OR DAMAGE TO PROPERTY, OR CLAIMS OF THIRD PARTIES, EVEN IF THE COMPANY OR AN AUTHORIZED REPRESENTATIVE OF THE COMPANY HAS BEEN ADVISED OF THE POSSIBILITY OF SUCH DAMAGES.

SOME JURISDICTIONS DO NOT ALLOW THE LIMITATION OF IMPLIED WARRANTIES OR LIABILITY FOR INCIDENTAL, INDIRECT, SPECIAL OR CONSEQUENTIAL DAMAGES, SO THE ABOVE LIMITATIONS MAY NOT ALWAYS APPLY. THE WARRANTIES IN THIS AGREEMENT GIVE YOU SPECIFIC LEGAL RIGHTS AND YOU MAY ALSO HAVE OTHER RIGHTS WHICH VARY IN ACCORDANCE WITH LOCAL LAW.

To the Student:

Every NEW copy of this text includes a disk containing all the Microsoft® Excel workbooks and data files used in the text. This disk is essential, and we highly recommend that you purchase a NEW text that includes the disk.